ollins

White Rose Maths

Key Stage 3
Student Book 1

Ian Davies and Caroline Hamilton

William Collins' dream of knowledge for all began with the publication of his first book in 1819. A self-educated mill worker, he not only enriched millions of lives, but also founded a flourishing publishing house. Today, staying true to this spirit, Collins books are packed with inspiration, innovation and practical expertise.
They place you at the centre of a world of possibility and give you exactly what you need to explore it.

Collins. Freedom to teach.

Published by Collins
An imprint of HarperCollins*Publishers*
The News Building
1 London Bridge Street
London
SE1 9GF

HarperCollins Publishers
Macken House
39/40 Mayor Street Upper
Dublin 1, D01 C9W8
Ireland

Browse the complete Collins catalogue at www.collins.co.uk

10 9 8

ISBN: 978-0-00-840088-0

British Library Cataloguing-in-Publication Data
A catalogue record for this publication is available from the British Library.

Authors and series editors: Ian Davies and Caroline Hamilton
Publisher: Katie Sergeant
Product manager: Jennifer Hall
Editors: Karl Warsi, Julie Bond, Phil Gallagher, Tim Jackson
Proofreader: Tim Jackson
Answer checkers: Laurice Suess and Jess White
Project manager: Karen Williams
Cover designer: Kneath Associates Ltd
Internal designer and illustrator: Ken Vail Graphic Design Ltd
Typesetter: Ken Vail Graphic Design Ltd
Production controller: Katharine Willard
Printed and bound by Grafica Veneta in Italy

Text acknowledgements
The publishers gratefully acknowledge the permission granted to reproduce the copyright material in this book. Every effort has been made to trace copyright holders and to obtain their permission for the use of copyright material. The publishers will gladly receive any information enabling them to rectify any error or omission at the first opportunity. p361 Walpole, S.C., Prieto-Merino, D., Edwards, P. *et al*. The weight of nations: an estimation of adult human biomass. *BMC Public Health* **12**, 439 (2012). https://doi.org/10.1186/1471-2458-12-439

Photo acknowledgements
The publishers wish to thank the following for permission to reproduce photographs. Every effort has been made to trace copyright holders and to obtain their permission for the use of copyright materials. The publishers will gladly receive any information enabling them to rectify any error or omission at the first opportunity. p151 WAYHOME studio/Shutterstock; p360 maglyvi/Shutterstock; p361 VectorMine/Shutterstock.

This book is produced from independently certified FSC™ paper to ensure responsible forest management.

For more information visit: www.harpercollins.co.uk/green

Contents

Contents

Introduction

How to use this book

Welcome to the **Collins White Rose Maths Key Stage 3** course. We hope you enjoy your learning journey. Here is a short guide to how to get the most out of this book.

Ian Davies and Caroline Hamilton, authors and series editors

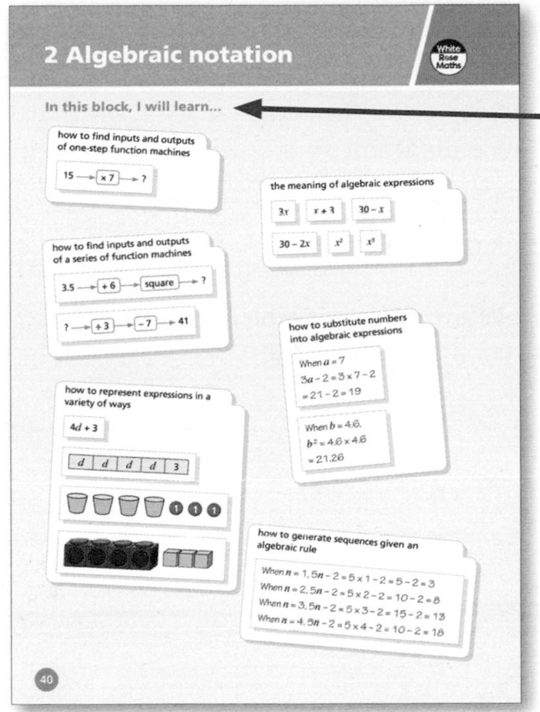

2 Algebraic notation

In this block, I will learn...

how to find inputs and outputs of one-step function machines

15 → ×7 → ?

the meaning of algebraic expressions

$3x$ $x+3$ $30-x$ $30-2x$ x^2 x^3

how to find inputs and outputs of a series of function machines

3.5 → +6 → square → ?

? → +3 → −7 → 41

how to substitute numbers into algebraic expressions

When $a = 7$
$3a - 2 = 3 \times 7 - 2$
$= 21 - 2 = 19$

When $b = 4.6$
$b^2 = 4.6 \times 4.6$
$= 21.26$

how to represent expressions in a variety of ways

$4d + 3$

d	d	d	d	3

how to generate sequences given an algebraic rule

When $n = 1$, $5n - 2 = 5 \times 1 - 2 = 5 - 2 = 3$
When $n = 2$, $5n - 2 = 5 \times 2 - 2 = 10 - 2 = 8$
When $n = 3$, $5n - 2 = 5 \times 3 - 2 = 15 - 2 = 13$
When $n = 4$, $5n - 2 = 5 \times 4 - 2 = 10 - 2 = 18$

40

Block overviews Each block of related chapters starts with a visual introduction to the key concepts and learning you will encounter.

Small steps The learning for each chapter is broken down into small steps to ensure progression and understanding. The Ⓗ symbol against a chapter or small step indicates more challenging content.

Key words Important terms are defined at the start of the chapter, and are highlighted the first time that they appear in the text. Definitions of all key terms are provided in the glossary at the back of the book.

Are you ready? Remind yourself of the maths you already know with these questions, before you move on to the new content of the chapter.

Models and representations Familiarise yourself with the key visual representations that you will use through the chapter.

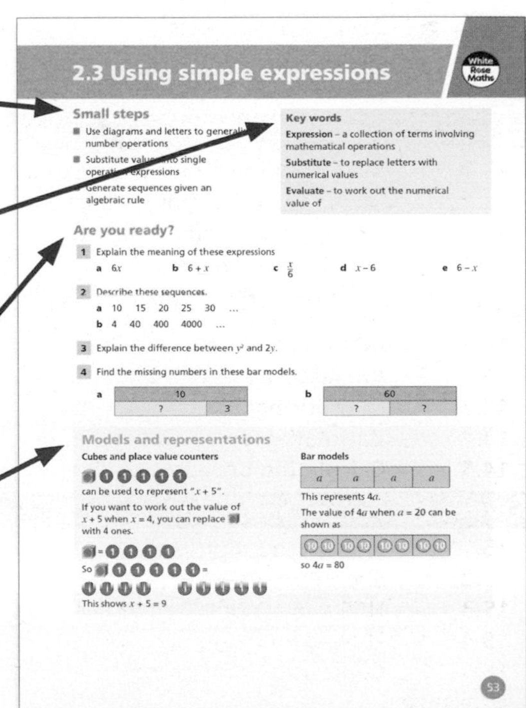

2.3 Using simple expressions

Small steps

- Use diagrams and letters to generali[illegible] number operations
- Substitute valu[illegible] single operat[illegible] expressions
- Generate sequences given an algebraic rule

Key words

Expression – a collection of terms involving mathematical operations

Substitute – to replace letters with numerical values

Evaluate – to work out the numerical value of

Are you ready?

1 Explain the meaning of these expressions

a $6x$ b $6 + x$ c $\frac{x}{6}$ d $x - 6$ e $6 - x$

2 Describe these sequences.

a 10 15 20 25 30 ...
b 4 40 400 4000 ...

3 Explain the difference between y^2 and $2y$.

4 Find the missing numbers in these bar models.

a

10	
?	3

b

60	
?	?

Models and representations

Cubes and place value counters

can be used to represent "$x + 5$".

If you want to work out the value of $x + 5$ when $x = 4$, you can replace [cube] with 4 ones.

So

This shows $x + 5 = 9$

Bar models

a	a	a	a

This represents $4a$.

The value of $4a$ when $a = 20$ can be shown as

so $4a = 80$

53

Introduction

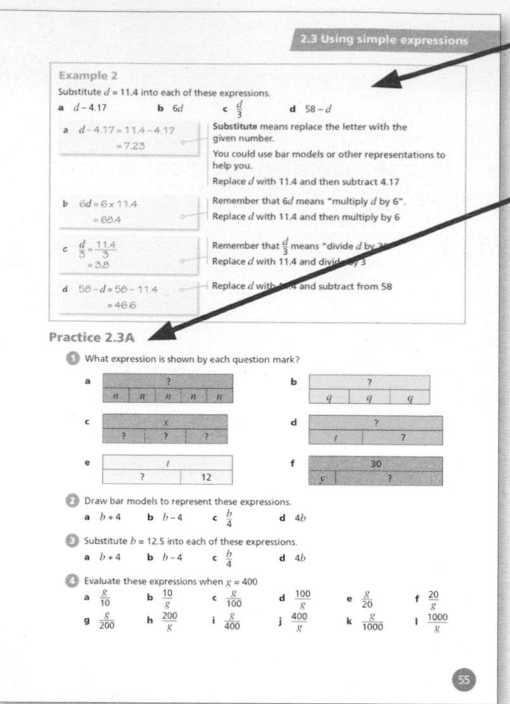

Example 2

Substitute $d = 11.4$ into each of these expressions.

a $d - 4.17$ **b** $6d$ **c** $\frac{d}{3}$ **d** $58 - d$

a $d - 4.17 = 11.4 - 4.17$
$= 7.23$

Substitute means replace the letter with the given number.
You could use bar models or other representations to help you.
Replace d with 11.4 and then subtract 4.17

b $6d = 6 \times 11.4$
$= 68.4$

Remember that $6d$ means "multiply d by 6".
Replace d with 11.4 and then multiply by 6

c $\frac{d}{3} = \frac{11.4}{3}$
$= 3.8$

Remember that $\frac{d}{3}$ means "divide d by 3"
Replace d with 11.4 and divide by 3

d $58 - d = 58 - 11.4$
$= 46.6$

Replace d with 11.4 and subtract from 58

Practice 2.3A

1. What expression is shown by each question mark?

   a [bar: ? ; n n n n n] **b** [bar: ? ; q q q]
   c [bar: x ; ? ? ?] **d** [bar: ? ; t 7]
   e [bar: t ; ? 12] **f** [bar: 30 ; y ?]

2. Draw bar models to represent these expressions.
   a $b + 4$ **b** $b - 4$ **c** $\frac{b}{4}$ **d** $4b$

3. Substitute $b = 12.5$ into each of these expressions.
   a $b + 4$ **b** $b - 4$ **c** $\frac{b}{4}$ **d** $4b$

4. Evaluate these expressions when $g = 400$
   a $\frac{g}{10}$ **b** $\frac{10}{g}$ **c** $\frac{g}{100}$ **d** $\frac{100}{g}$ **e** $\frac{g}{20}$ **f** $\frac{20}{g}$
   g $\frac{g}{200}$ **h** $\frac{200}{g}$ **i** $\frac{g}{400}$ **j** $\frac{400}{g}$ **k** $\frac{g}{1000}$ **l** $\frac{1000}{g}$

55

Worked examples Learn how to approach different types of questions with worked examples that clearly walk you through the process of answering, using lots of visual representations.

Practice Put what you've just learnt into practice. Icons suggest tools or skills to help you approach a question:

- – use manipulatives such as multi-link cubes or Cuisenaire rods
- – draw a bar model
- – draw a diagram
- – discuss with a partner
- – think deeply

There is a **What do you think?** section at the end of every practice exercise.

Consolidate Reinforce what you've learnt in the chapter with additional practice questions.

Stretch Take the learning further and challenge yourself to apply it in new ways.

Reflect Look back over what you've learned to make sure you understand and remember the key points.

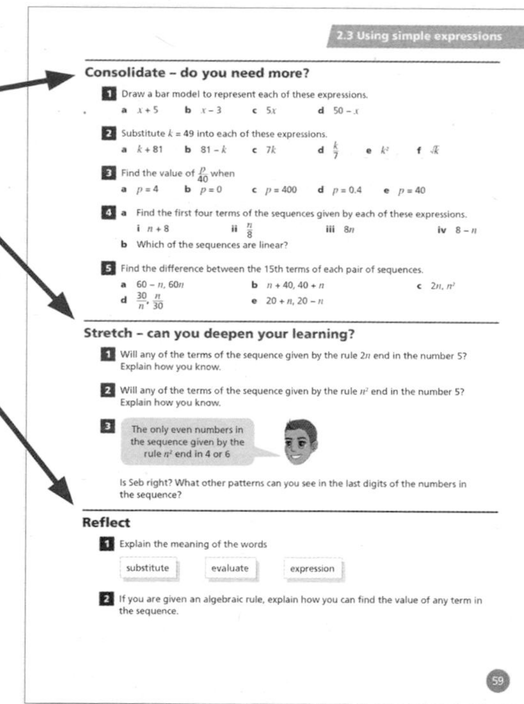

Consolidate – do you need more?

1. Draw a bar model to represent each of these expressions.
   a $x + 5$ **b** $x - 3$ **c** $5x$ **d** $50 - x$
2. Substitute $k = 49$ into each of these expressions.
   a $k + 81$ **b** $81 - k$ **c** $7k$ **d** $\frac{k}{7}$ **e** k^2 **f** $\sqrt{k}$
3. Find the value of $\frac{p}{40}$ when
   a $p = 4$ **b** $p = 0$ **c** $p = 400$ **d** $p = 0.4$ **e** $p = 40$
4. **a** Find the first four terms of the sequences given by each of these expressions.
   i $n + 8$ **ii** $\frac{n}{8}$ **iii** $8n$ **iv** $8 - n$
   b Which of the sequences are linear?
5. Find the difference between the 15th terms of each pair of sequences.
   a $60 - n$, $60n$ **b** $n + 40$, $40 + n$ **c** $2n$, n^2
   d $\frac{30}{n}$, $\frac{n}{30}$ **e** $20 + n$, $20 - n$

Stretch – can you deepen your learning?

1. Will any of the terms of the sequence given by the rule $2n$ end in the number 5? Explain how you know.
2. Will any of the terms of the sequence given by the rule n^2 end in the number 5? Explain how you know.
3. The only even numbers in the sequence given by the rule n^2 end in 4 or 6

   Is Seb right? What other patterns can you see in the last digits of the numbers in the sequence?

Reflect

1. Explain the meaning of the words
   substitute evaluate expression
2. If you are given an algebraic rule, explain how you can find the value of any term in the sequence.

59

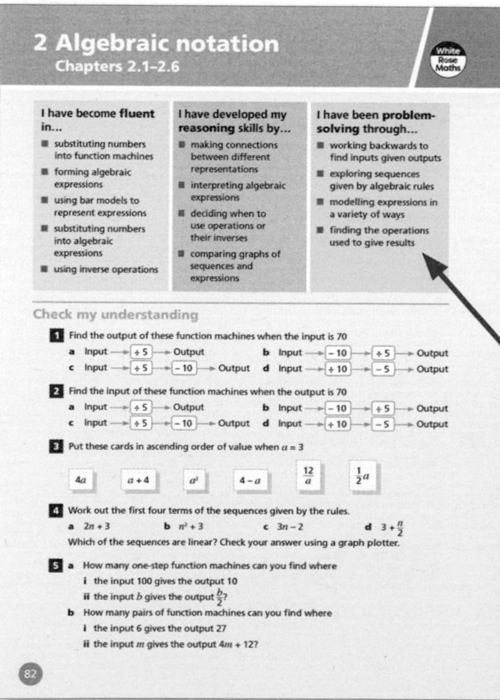
2 Algebraic notation
Chapters 2.1–2.6

White Rose Maths

I have become fluent in...
- substituting numbers into function machines
- forming algebraic expressions
- using bar models to represent expressions
- substituting numbers into algebraic expressions
- using inverse operations

I have developed my reasoning skills by...
- making connections between different representations
- interpreting algebraic expressions
- deciding when to use operations or their inverses
- comparing graphs of sequences and expressions

I have been problem-solving through...
- working backwards to find inputs given outputs
- exploring sequences given by algebraic rules
- modelling expressions in a variety of ways
- finding the operations used to give results

Check my understanding

1. Find the output of these function machines when the input is 70
   a Input → $\div 5$ → Output **b** Input → -10 → $\div 5$ → Output
   c Input → $\div 5$ → -10 → Output **d** Input → $\div 10$ → -5 → Output
2. Find the input of these function machines when the output is 70
   a Input → $\div 5$ → Output **b** Input → -10 → $\div 5$ → Output
   c Input → $\div 5$ → -10 → Output **d** Input → $\div 10$ → -5 → Output
3. Put these cards in ascending order of value when $a = 3$
   $4a$ $a + 4$ a^2 $4 - a$ $\frac{12}{a}$ $\frac{1}{2}a$
4. Work out the first four terms of the sequences given by the rules.
   a $2n + 3$ **b** $n^2 + 3$ **c** $3n - 2$ **d** $3 + \frac{n}{2}$
   Which of the sequences are linear? Check your answer using a graph plotter.
5. **a** How many one-step function machines can you find where
   i the input 100 gives the output 10
   ii the input b gives the output $\frac{b}{2}$?
   b How many pairs of function machines can you find where
   i the input 6 gives the output 27
   ii the input m gives the output $4m + 12$?

82

Block summary Bring together the learning at the end of the learning block with fluency, reasoning and problem solving statements and review questions.

Answers Check your work using the answers provided at the back of the book.

1 Sequences

White Rose Maths

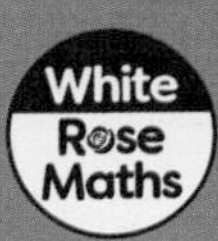

In this block, I will learn...

how to continue sequences given in a variety of forms

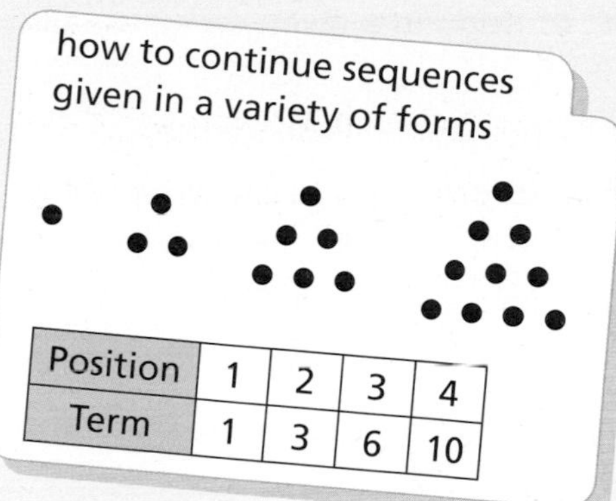

Position	1	2	3	4
Term	1	3	6	10

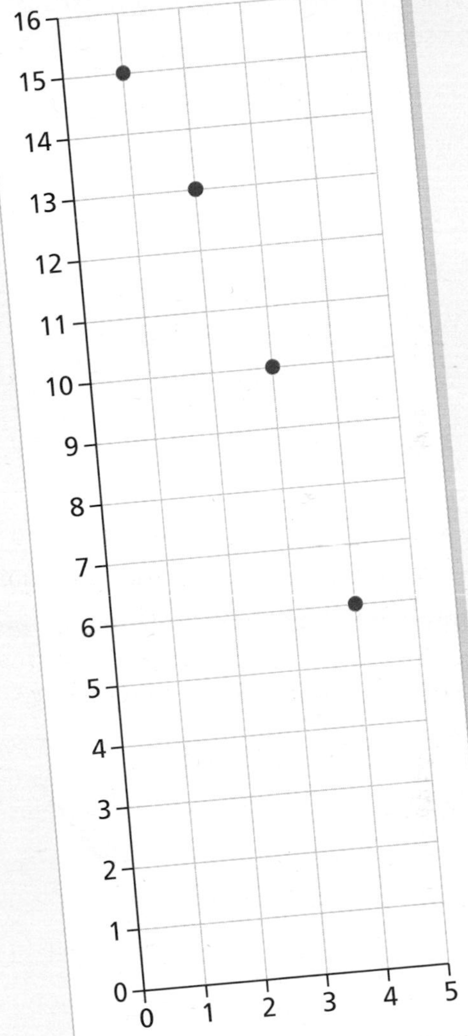

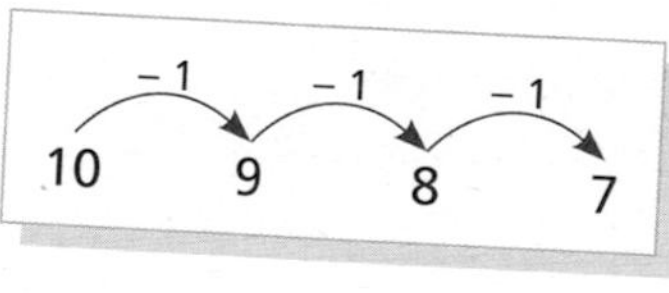

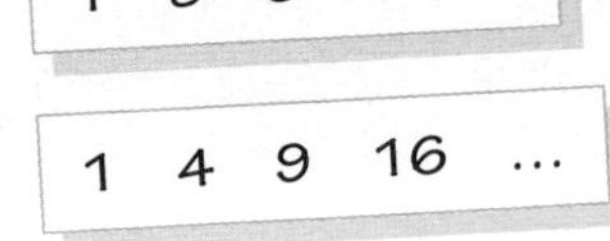

how to draw graphs of sequences

how to generate sequences given rules

■ First term: 5
Term-to-term rule:
double the previous term

how to describe sequences

how to identify linear and non-linear sequences

1 3 5 7 ...

1 4 9 16 ...

how to find missing terms in sequences (H)

6 □ 9

6 □ □ 9

6 □ □ □ 9

1.1 Continuing sequences: shapes

Small steps

- Describe and continue a sequence given diagrammatically
- Predict and check the next term(s) of a sequence

Key words

Describe – say what you see, or what is happening

Sequence – a list of items in a given order, usually following a rule

Term – one of the members of a sequence

Predict – use given information to say what will come next

Check – find out if you are correct

Are you ready?

1 What comes next in these repeating patterns?

a Red Yellow Red Yellow Red Yellow …

b ABBABBABBAB …

c 1 0 1 0 1 0 1 0 …

d ABACADAE …

2 Four children take turns playing a game. Continue the sequence for the next three turns.

Huda Ali Charlie Rob Huda Ali Charlie Rob Huda Ali

3 What would come next in each list?

a 1 22 333 4444 55555 …

b 1 10 100 1000 10000 …

4 What would be the next three items in this list? 1st 2nd 3rd 4th ….

Models and representations

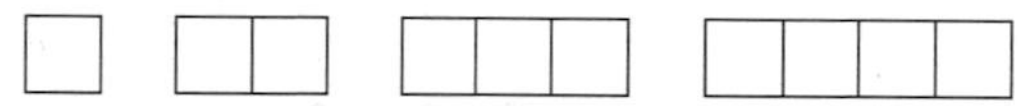

You can model a **sequence** of shapes by drawing each **term** or by using solid objects such as cubes.

Here are some other sequences.

- Get out of bed
- Shower
- Get dressed
- Have breakfast
- Clean teeth
- Check bag
- Leave for school

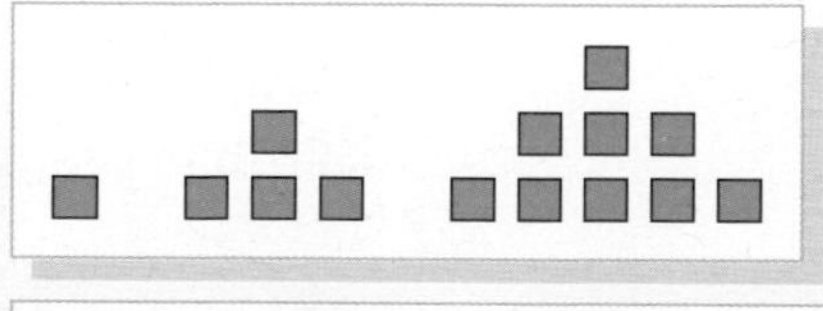

10, 20, 30, 40, …

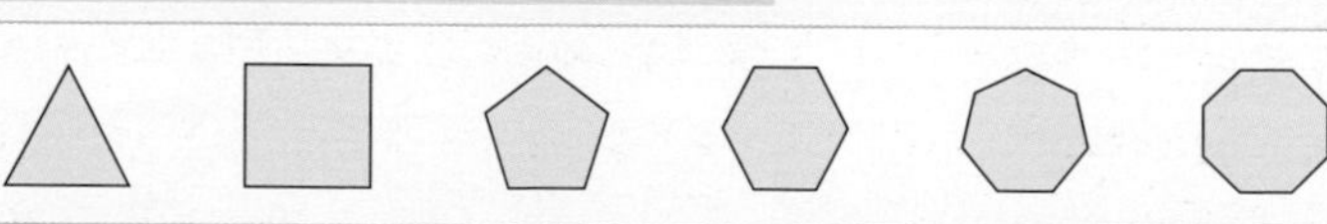

1, −1, 1, −1, 1, −1 …

1, 3, 5, 7, …

You can **describe** sequences by saying what you see.

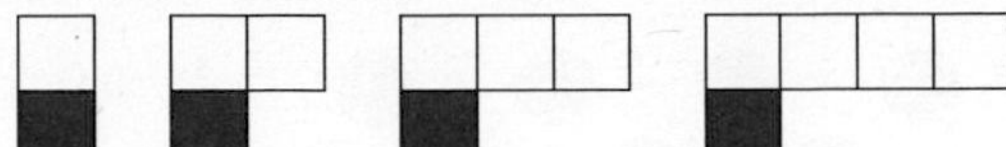

- The first term in this sequence has one white square above one red square.
- The second term has two white squares above one red square.
- The third term has three white squares above one red square.

You can see that:

- There is always one red square.
- There is one white square in the first term, and every other term has one more white square than the previous term.

You can **predict** that the fifth term will have one red square with five white squares.

You can **check** your answer by drawing the fifth term and seeing if it fits the pattern.

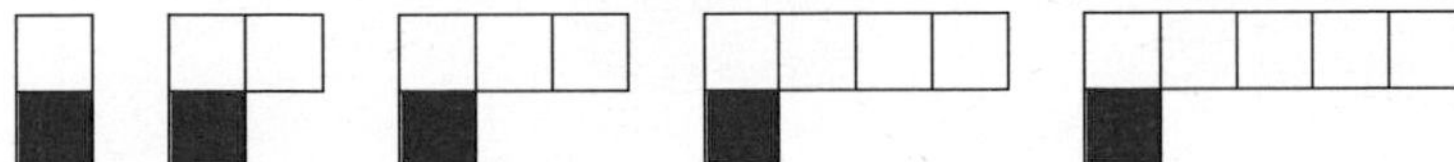

Yes, it does!

Practice 1.1A

1 **a** Draw this sequence and describe it.

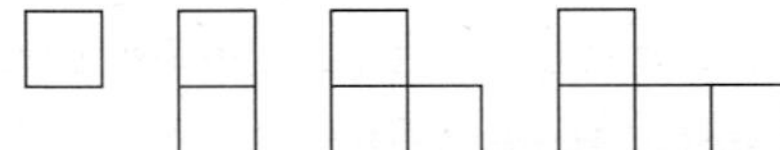

b Draw the next two terms in the sequence. Discuss with a partner how you know what they will be.

2 **a** Describe what is happening in this sequence.

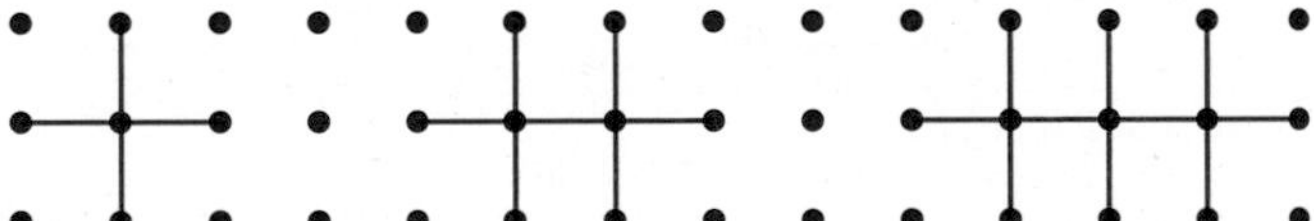

b Draw the next two terms.

3 **a** Describe this sequence.

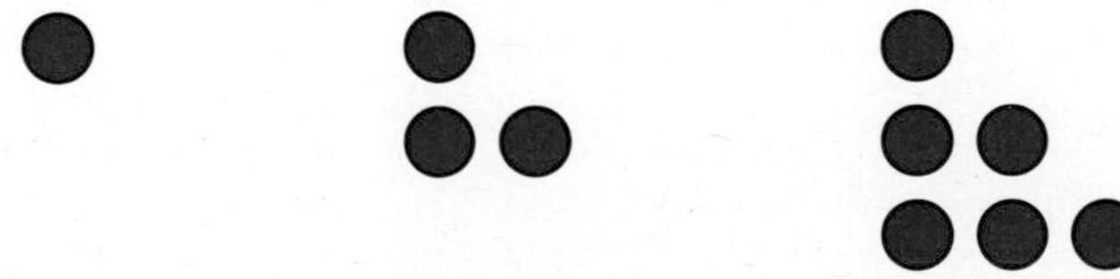

b Filipo predicts there will be 14 circles in the fifth term of the sequence.

Draw the fourth and fifth terms of the sequence and check whether Filipo is correct.

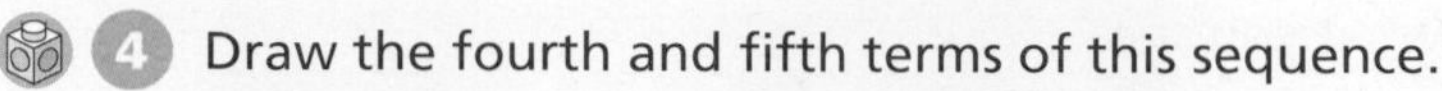

4 Draw the fourth and fifth terms of this sequence.

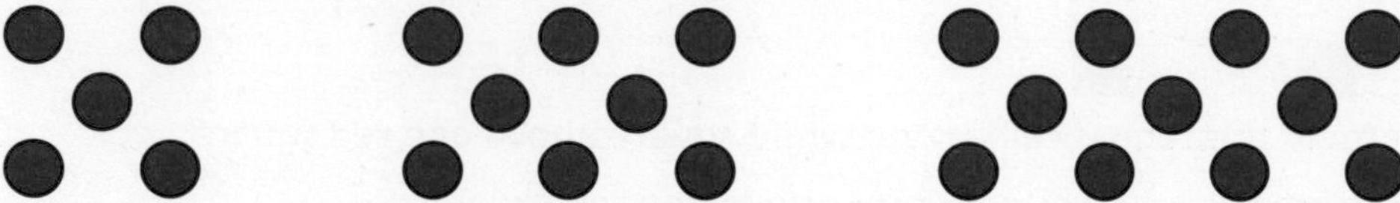

What do you think?

Faith and Benji have been asked to continue the sequence.

Here is Faith's answer:

Here is Benji's answer:

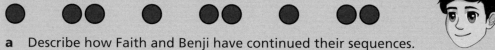

a Describe how Faith and Benji have continued their sequences.

b How many more ways can you find to continue the sequence? Fully describe each of your rules.

c Discuss with a partner – how many terms do you need to know before there is only one way to continue a sequence?

Example 1

Here is a sequence of patterns.

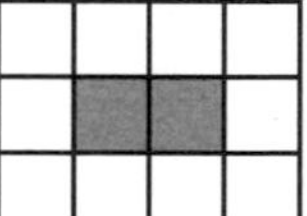

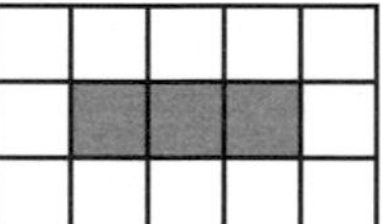

a Describe in words how:

i the number of orange squares changes

ii the number of white squares changes

iii the total number of squares changes.

b Describe what the fifth term of the sequence will look like. Check your answer by drawing.

a i The number of orange squares increases by 1 every time.

ii The number of white squares increases by 2 every time.

iii The total number of squares increases by 3 every time.

Count the different-coloured squares in each pattern and look for a rule.

b The fourth term will have 4 orange squares and 14 white squares – 18 squares in total.

So the fifth term will have 5 orange squares and 16 white squares – 21 squares in total.

Continue the rule with the fourth and fifth terms.

The drawing shows my prediction is correct.
Check by drawing.

Practice 1.1B

1 Here is a sequence of matchstick patterns.

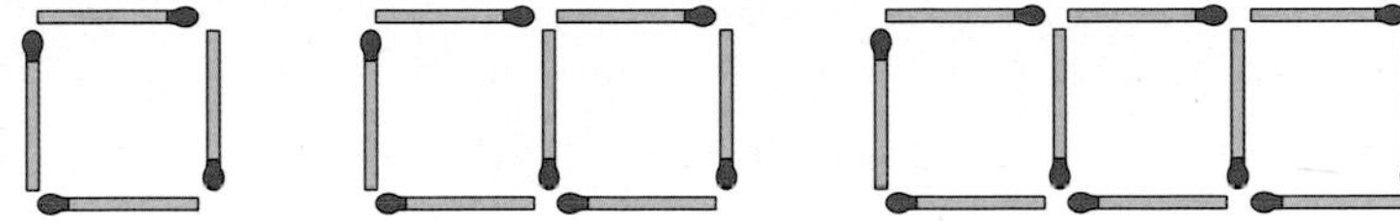

a How many squares are there in each term of the sequence?

b How many matchsticks are there in each term of the sequence?

c Predict the number of matchsticks in the next three terms, and draw them to check.

2 Here is another sequence of matchstick patterns.

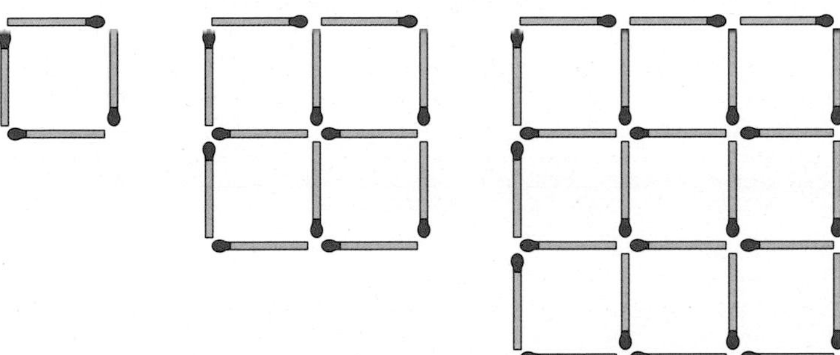

a How many 1 by 1 squares are there in each term of the sequence?

b How many matchsticks are there in each term of the sequence?

c Predict the number of matchsticks in the next three terms, and draw them to check.

3 Here is a way of arranging chairs at rectangular tables.

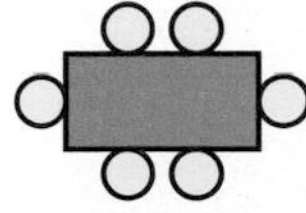
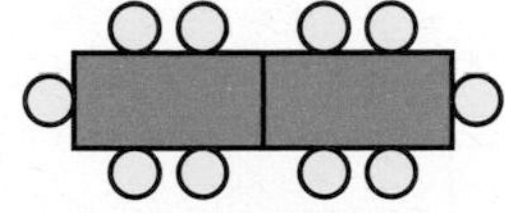
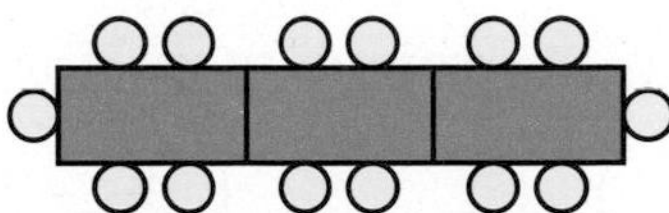

a Describe how the numbers of tables, the number of chairs and the total number of pieces of furniture change.

b Draw the next three terms in the sequence.

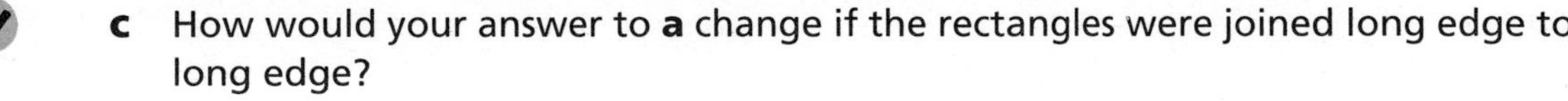

c How would your answer to **a** change if the rectangles were joined long edge to long edge?

4 Look at these patterns

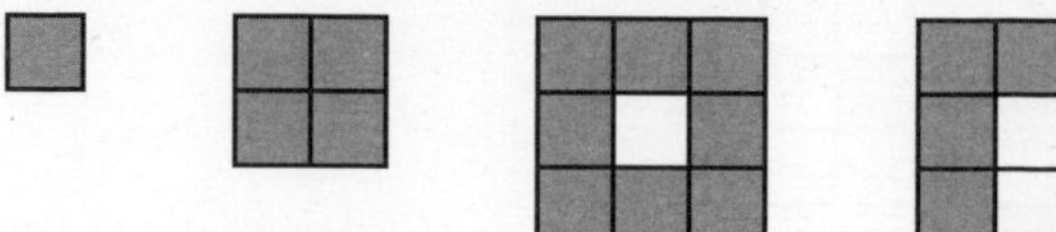

a Draw the next three terms in the sequence.

b Describe how the number of grey squares and the number of white squares change.

What do you think?

1 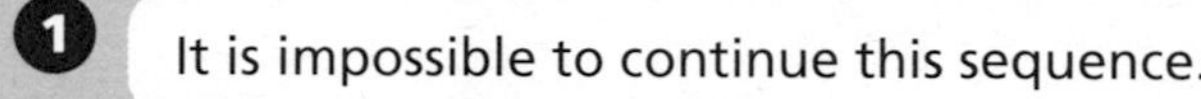
It is impossible to continue this sequence.

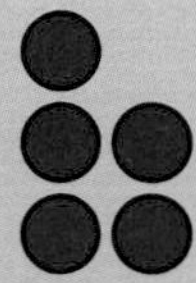

1st term 2nd term 3rd term

Do you agree with Marta?

2 The third pattern of a sequence is shown.

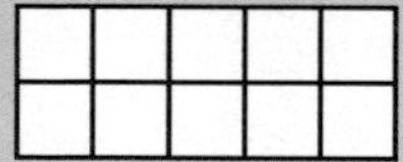

What might the first and second terms be? How many different possibilities can you find?

Consolidate – do you need more?

For each sequence

a predict the number of squares/counters in the next two terms

b draw the patterns to see if you were right.

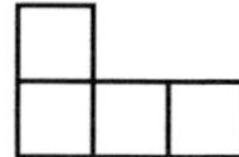

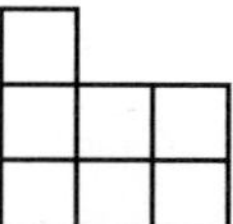

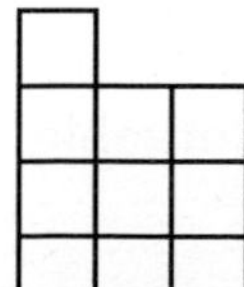

2

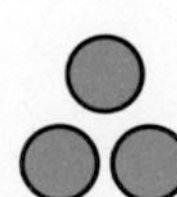

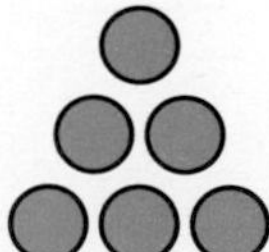

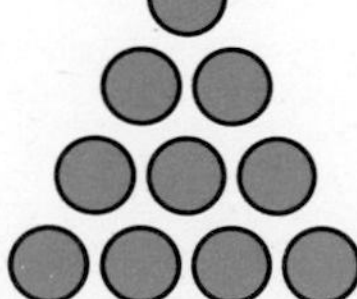

 **3**

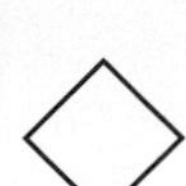 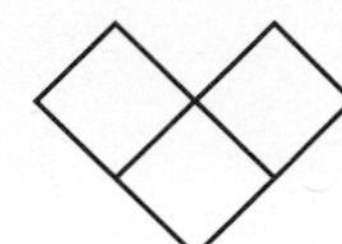 

Stretch – can you deepen your learning?

1 What's the same and what's different about these two sequences?

A

 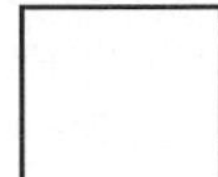 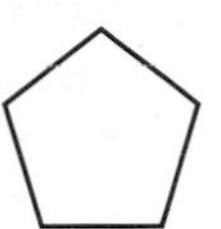 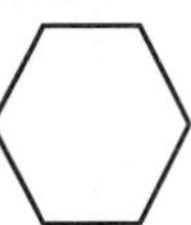 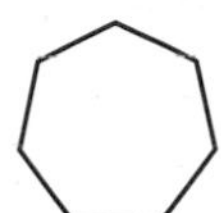 

B

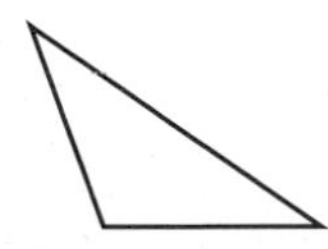   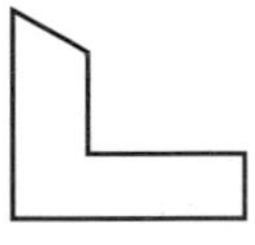  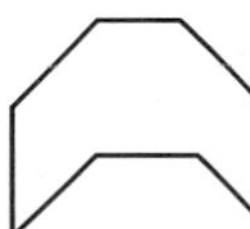

2 Describe how this sequence changes.

 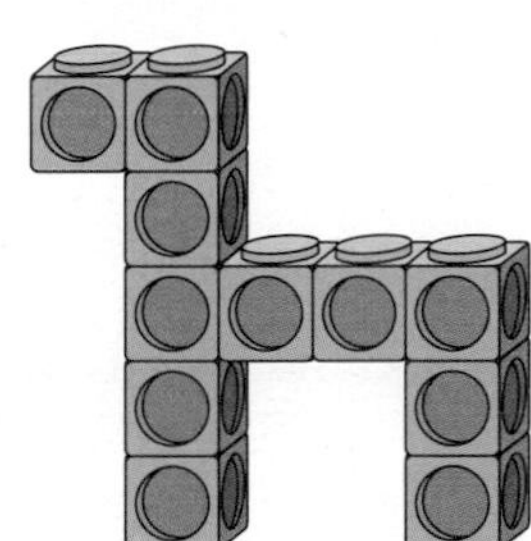 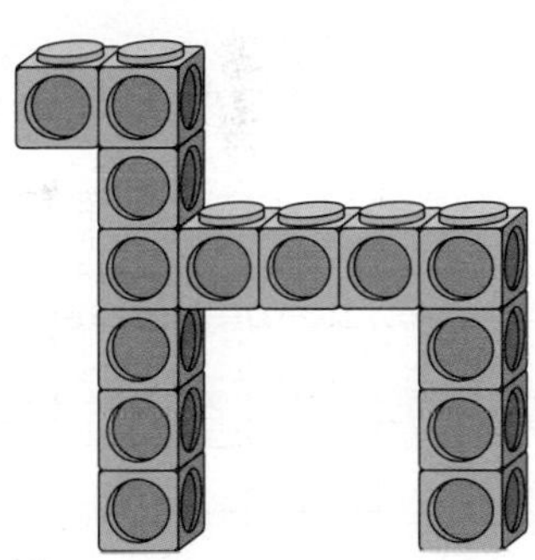

Reflect

Design your own sequence of shapes. Label your design to show the meaning of the words below.

Sequence   Term   Prediction   Check

1.2 Continuing linear sequences

Small steps

- Predict and check the next term(s) of a sequence
- Continue numerical linear sequences
- Explain the term-to-term rule of numerical sequences in words

Key words

Difference – the gap between numbers in a sequence

Constant – not changing

Successive – coming after another term in a sequence

Linear sequence – a sequence whose terms are increasing or decreasing by a constant difference

Non-linear sequence – a sequence whose terms are not increasing or decreasing by a constant difference

Increasing (or ascending) sequence – a sequence where every term is greater than the previous term

Decreasing (or descending) sequence – a sequence where every term is smaller than the previous term

Term-to-term rule – a rule that describes how you get from one term of a sequence to the next

Are you ready?

1 Find the difference between each pair of numbers.

a 10 and 15 **b** 10 and 4 **c** 17.3 and 18.7 **d** 102.3 and 98.7

2 Describe what is happening in each of these sequences.

a 10 20 30 40 50 …

b 10 20 40 80 160 …

3 The number of squares in each term of this sequence increases by two each time.

 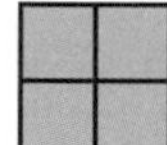 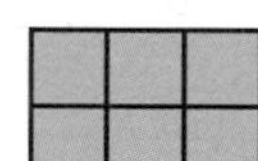

How many squares will there be in

a the fourth term of the sequence

b the fifth term of the sequence?

4 The number of triangles in each term of this sequence decreases by three each time.

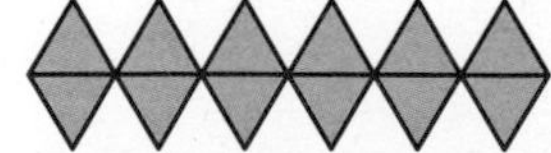 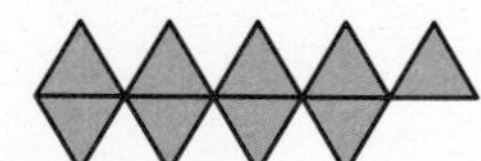 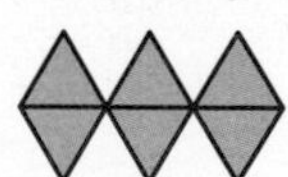

How many triangles will there be in

a the fourth term of the sequence

b the fifth term of the sequence?

Models and representations

5   7   9   11   13   ...

You can model a sequence of numbers by using cubes, counters or drawings.

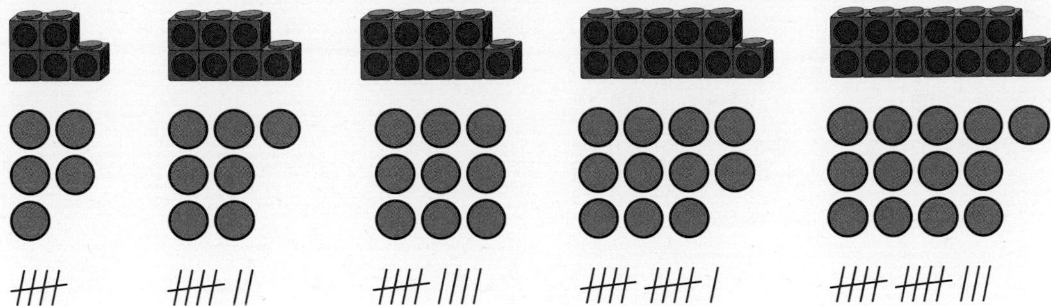

Example 1

What's the same and what's different about these three sequences?

Sequence A	10	12	14	16	...
Sequence B	10	12	16	22	...
Sequence C	10	9	8	7	...

Some things that are the same

- All the sequences start with 10.
- All the sequences contain only whole numbers.

Some things that are different

- Sequences A and B are increasing, but sequence C is decreasing.
- Sequence A goes up 2 every time, but sequence B goes up 2 then 4, then 6 and so on, whilst sequence C goes down 1 every time.

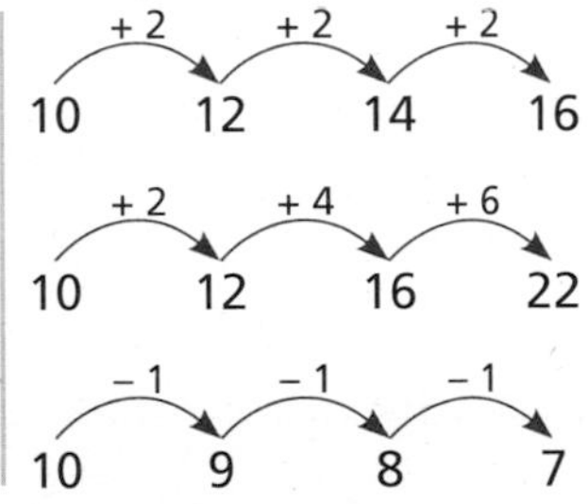

For sequence A, the "**term-to-term rule**" is "Add 2 every time".

For sequence C, the "term-to-term rule" is "Subtract 1 every time".

Sequences like A and C that have a **constant difference** between each **successive** pair of terms are called **linear** – you will see why in *1.4 Representing sequences*. If a sequence is not linear, you call it **non-linear** – which makes a lot of sense!

Example 2

Which of these sequences are linear?

a 30 26 22 18 ...

b 3.2 4.3 5.4 6.5 ...

c

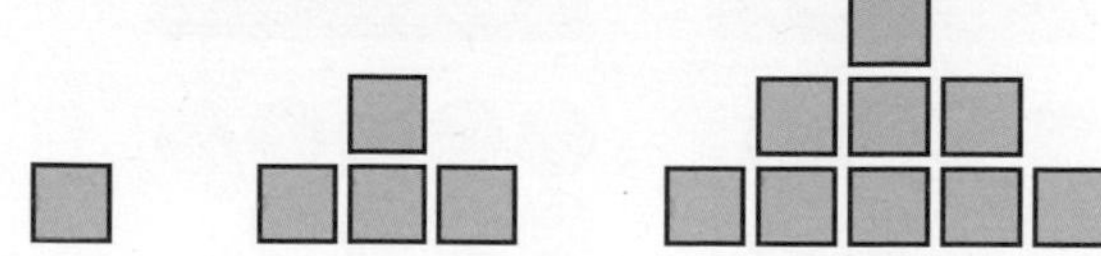

d 100 117 134 151 ...

a 30 26 22 18 ...

The terms of the sequence decrease by 4 every time, so the sequence is linear.

30 (−4) 26 (−4) 22 (−4) 18 (−4) ...

b 3.2 4.3 5.4 6.5 ...

The terms of the sequence increase by 1.1 every time, so the sequence is linear.

3.2 (+1.1) 4.3 (+1.1) 5.4 (+1.1) 6.5 (+1.1) ...

c The difference between the terms of the sequence changes so the sequence is not linear.

+3, +5

d 100 117 134 151 ...

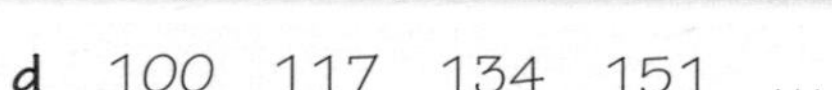

The terms of the sequence increase by 17 every time, so the sequence is linear.

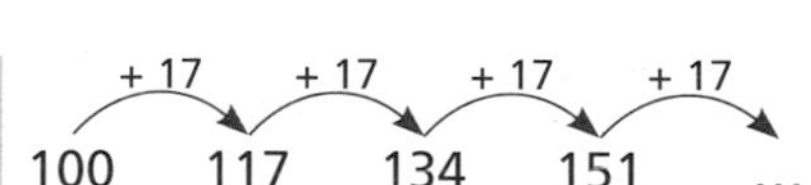

Example 3

The first two terms of a **linear sequence** are 9.3 and 8.75. Find the fifth term of the sequence.

The difference between the first two terms of the sequence is 9.3 − 8.75 = 0.55

So the third term is 8.75 − 0.55 = 8.2

So the fourth term is 8.2 − 0.55 = 7.65

So the third term is 7.65 − 0.55 = 7.1

You are told that the sequence is linear so you know that the difference between successive terms will always be the same.

As the sequence is decreasing, you subtract 0.55 each time to find the next terms.

Practice 1.2A

A calculator is particularly useful for this exercise.

1. Describe the differences between successive terms in each sequence.

   a 40 47 54 61 68 …

   b 3.86 4.27 4.68 5.09 5.50 …

   c 1000 984 968 952 936 …

   d 8.5 8.16 7.81 7.45 7.08 …

   e 2 1002 2002 3002 4002 …

   f 980 979 975 966 950 …

   g $\frac{11}{100}$ $\frac{19}{100}$ $\frac{27}{100}$ $\frac{35}{100}$ $\frac{43}{100}$ ….

   Which of the sequences in **a** to **g** are linear?

2. All these sequences are linear. For each sequence

   i state the term-to-term rule **ii** work out the next term.

   a 94 108 …

   b 78 53 …

   c 1.48 3.97 …

   d 50 29.2 …

   e 100 1000 …

3. **a** How many linear sequences can you find that start with 64, 68 …?

   b How many linear sequences can you find that start with 64 and have a constant difference of 4?

4. You are given this information about a sequence

   - The sequence is linear.
   - The first term is 18.7
   - It is an **increasing sequence**.
   - The difference between the terms is 2.06

   Work out the second, third and fourth terms of the sequence.

5. Repeat question 4 with the sequence as a **decreasing sequence**.

6. State whether the total number of squares in each of these patterns form a linear sequence. Explain your answers.

   a 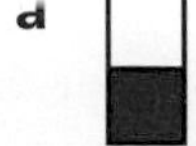 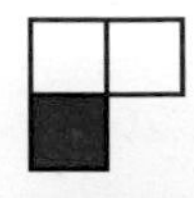 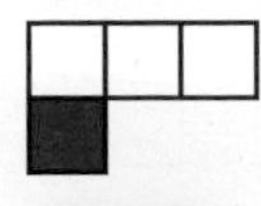 

   b   

c

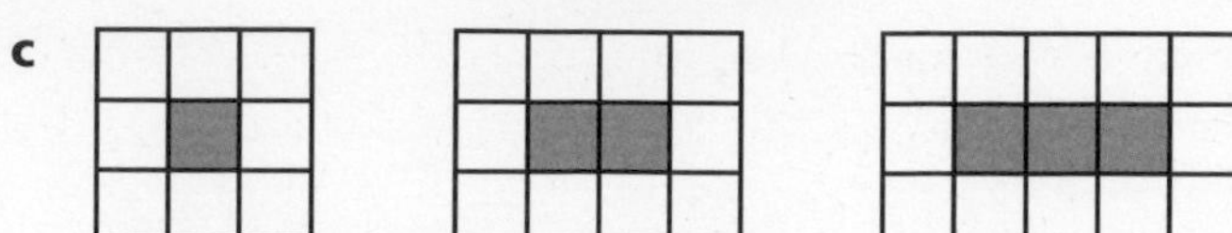

7 A sequence starts at 15. The term-to-term rule is "add on 32 to the previous term". Work out the fourth term of the sequence.

What do you think?

Ali says that if you know the first two terms of a linear sequence, then you can work out any of the other terms. Is he correct?

Given a linear sequence you can also work out the value of terms *before* the ones that you know.

Example 4

a The second and third terms of a linear sequence are 39 and 71. Work out the first term and the fourth term of the sequence.

b The fourth and fifth terms of a linear sequence are 9.7 and 12.2. Work out the first term of the sequence.

a The difference between the terms is $71 - 39 = 32$

The first term will be $39 - 32 = 7$

As the sequence is increasing, the first term will be 32 *less* than the second term.

The fourth term will be $71 + 32 = 103$

The fourth term will be 32 *more* than the third term.

b The difference between the terms is $12.2 - 9.7 = 2.5$

term	1st	2nd	3rd	4th	5th
	2.2	4.7	7.2	9.7	12.2

−2.5 −2.5 −2.5 −2.5

Subtracting 2.5 each time you can find the first term.

Practice 1.2B

1 Work out the missing terms in each of these linear sequences.

a ☐ 17 30 ☐ …

b ☐ 9.2 8.1 ☐ …

c ☐ ☐ 150 209 …

d ☐ $6\frac{3}{7}$ $6\frac{5}{7}$ ☐ …

2 All the sequences in this question are linear. They all have a constant difference of 48, and they are all increasing sequences. Work out the values of the missing terms.

a 200 ☐ ☐ ☐

b ☐ 200 ☐ ☐

c ☐ ☐ 200 ☐

d ☐ ☐ ☐ 200

3. All the sequences in this question are linear. They all have a constant difference of 48, and they are all decreasing sequences. Work out the values of the missing terms.

a 200 ☐ ☐ ☐ **b** ☐ 200 ☐ ☐

c ☐ ☐ 200 ☐ **d** ☐ ☐ ☐ 200

4. The fourth term of a linear sequence is 8000, and the fifth term is 6850. Work out the first term of the sequence.

5. The fourth term of a linear sequence is 17, and the fifth term is 19.8. Work out the difference between the first and the sixth terms of the sequence.

What do you think?

1. A linear sequence has constant difference of 30. The fourth term of the sequence is 200. Jackson thinks he can work out the first term, but Flo says there is not enough information. Who do you agree with? Explain your answer.

2. Explain why this sequence is not linear.

   5 10 5 10 5 10 …

Consolidate – do you need more?

1. For each sequence of patterns, state whether it represents a linear sequence.

   a

   b

   c

2. Say whether each of these sequences is linear or not. How do you know?

   a 50 40 30 20 … **b** 1.8 1.9 2.1 2.4 …

   c 7.62 7.6 7.58 7.56 … **d** 85 95 115 125

3. Each of the sequences is linear. Work out the next term

   a 72 99 ☐ **b** 99 72 ☐ **c** 7.2 9.9 ☐

   d 9.9 7.2 ☐ **e** 720 990 ☐ **f** 990 720 ☐

Stretch – can you deepen your learning?

1 Create linear sequences of integers such that

a the final digit of every term is a 2

b the final digits alternate between 8 and one other digit.

2 The first two terms of a sequence are 85 and 103

Which of these calculations will tell you the value of the hundredth term of the sequence?

Add 18 × 100 to the first term	Multiply the second term by 50	Add 18 × 98 to the second term

Reflect

Describe in your own words how you can tell whether a sequence of numbers is linear or not.

1.3 Continuing non-linear sequences

Small steps

- Predict and check the next term(s) of a sequence
- Continue numerical non-linear sequences
- Explain the term-to-term rule of numerical sequences in words

Key words

Geometric sequence – A sequence is geometric if the value of each successive term is found by multiplying or dividing the previous term by the same number

Fibonacci sequence – The next term in a Fibonacci sequence is found by adding the previous two terms together

Are you ready?

1 State whether each sequence is linear or not. How do you know?

a 8 10 13 17 …

b 80 77 74 71 …

2 Write the first four terms of the linear sequence that starts

5 10 …

3 Write the first four terms of the linear sequence that starts

19 17 …

4 Make up three different linear sequences whose second term is 12

Models and representations

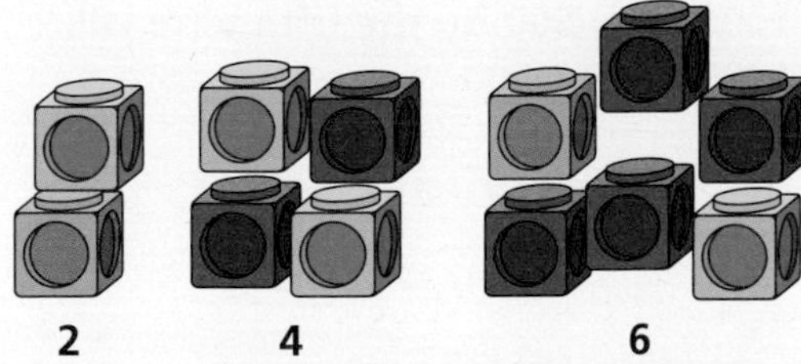

You can model a sequence by using cubes, counters or drawings to represent the numbers.

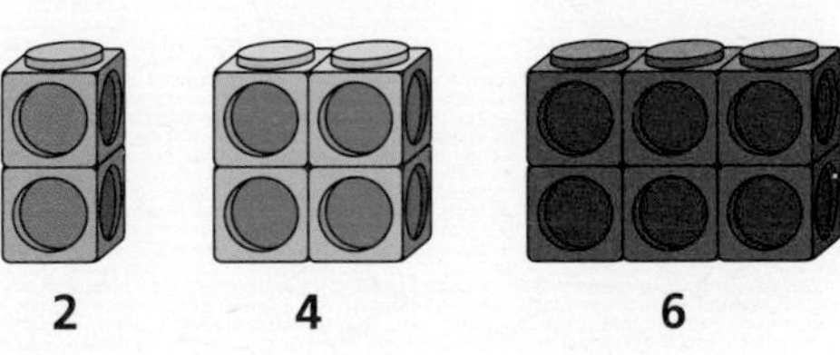

If you organise the cubes, it might help you to see the difference between terms.

1.3 Continuing non-linear sequences

Example 1

1 These sequences are all non-linear. For each sequence

- describe the term-to-term rule
- predict the next term.

Sequence A 20 30 50 80 …

Sequence B 8000 4000 2000 1000 …

Sequence C 1.7 1.8 1.7 1.8 …

Sequence D 1 3 9 27 …

Sequence A

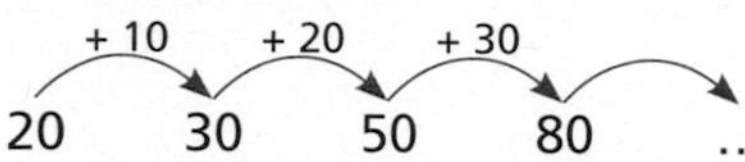

The difference between terms is increasing by 10 each time.

The difference between the fourth and fifth terms will be 30 + 10 = 40

So the fifth term will be 80 + 40 = 120

This is the term-to-term rule.

Sequence B

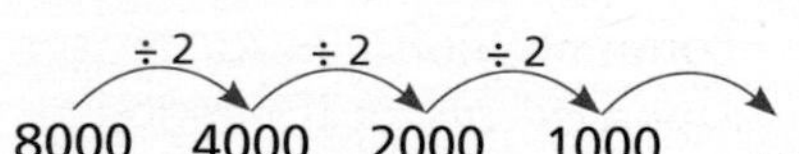

Each term is half of the previous term.

So the fifth term will be 1000 ÷ 2 = 500

In sequence B, the common divisor is 2.

Alternatively, you could look at the difference between terms.

– 4000 – 2000 – 1000

8000 4000 2000 1000 …

This is halving each time, so the difference between the fourth and fifth terms will be 1000 ÷ 2 = 500. So the fifth term will be 1000 – 500 = 500

Sequence C

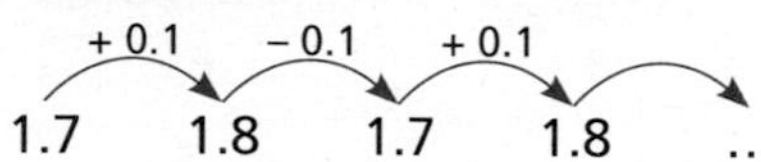

The sequence increases by 0.1, then decreases by 0.1, then increases by 0.1

Next it will decrease by 0.1 again, so the fifth term is 1.8 – 0.1 = 1.7

You could say that the sequence is alternating.

Sequence D

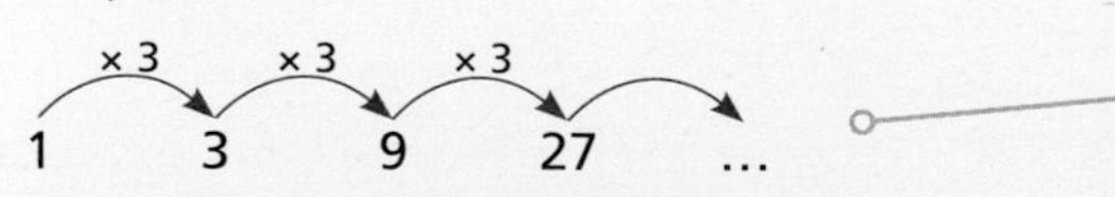

Each term is three times the size of the previous term.

So the fifth term will be 27 × 3 = 81

In sequence D, the common multiplier is 3

Sequences B and D are geometric – the next term is found by multiplying or dividing the previous term by the same number.

The next term in a **Fibonacci sequence** is found by adding the previous two terms together.

Example 2

Find the next three terms of these Fibonacci sequences.

a 1 1 2 3 5 ...

b 5 7 12 19 ...

a The next term is 3 + 5 = 8

Then 5 + 8 = 13

Then 8 + 13 = 21

The next three terms are 8, 13 and 21

b The next term is 12 + 19 = 31

Then 19 + 31 = 50

Then 31 + 50 = 81

The next three terms are 31, 50 and 81

1 1 2 3 5

1 + 1 = 2

1 + 2 = 3

2 + 3 = 5

Each term is the sum of the previous two terms.

Practice 1.3A

If a sequence is non-linear, look for patterns in the differences between terms, or look for a common multiplier or divisor. Use a calculator if needed.

1. Find the first four terms of the sequences given by these rules.

   a First term = 10 Term-to-term rule: double the previous term.

   b First term = 10 Term-to-term rule: halve the previous term.

   c First term = 8 Term-to-term rule: add 1, then add 2, then add 3 and so on.

   d First term = 67 Term-to-term rule: add 11, then subtract 11, then add 11, then subtract 11 and so on.

2. State whether each sequence is linear or non-linear.

   a 5 10 15 20 ...

   b 5 10 20 40 ...

   c 5 4 3 2 ...

   d 5 4 2.5 0.5 ...

   e 5 4 3.2 2.56 ...

   f 5 1 5 1 ...

   g 5 5 5 5

3. The first two terms of a sequence are 5 and 20.

   Find the next three terms if the sequence is

   a linear **b** geometric **c** Fibonacci.

4 Describe the rule for each sequence, and calculate the next term.

a 1 5 25 125 625 …

b 6400 3200 1600 800 400 …

c 100 81 64 49 36 …

d 1 8 27 64 125 …

e 1 $\frac{1}{10}$ $\frac{1}{100}$ $\frac{1}{1000}$ $\frac{1}{10\,000}$ …

f 7 10 17 27 44 …

5 The first two terms of a sequence are 100 and 50. Find some ways of continuing the sequence. Compare your answers with a partner.

What do you think?

1 A sequence starts 3 6 12 24 …

The rule is you add the term onto itself each time. 3 + 3 = 6, 6 + 6 = 12 and so on.

The rule is you double the previous term. 3 × 2 = 6, 6 × 2 = 12 and so on.

Both Marta and Zach are correct. Whose strategy do you prefer?

2 A sequence starts 2 12 …

How many different rules can you find for generating more terms of the sequence? Which of your sequences will produce a number greater than 100 most quickly?

Example 3

The second and third terms of a **geometric sequence** are 40 and 80.

Work out the first term and the fourth term of the sequence.

80 ÷ 40 = 2

So the term-to-term rule is "multiply the previous term by 2". — First find the term-to-term rule.

The first term will be 40 ÷ 2 = 20

The fourth term will be 80 × 2 = 160

Example 4

A sequence starts with 100.

The term-to-term rule is "multiply the previous term by 4, then subtract 50".

Which term in the sequence is the first to be greater than 10 000?

Second term = 100 × 4 − 50 = 350

Third term = 350 × 4 − 50 = 1350

Fourth term = 1350 × 4 − 50 = 5350

Fifth term = 5350 × 4 − 50 = 21 350

The fifth term is the first term to be greater than 10 000

You can just work out each term, starting with the first term which you are given.

Practice 1.3B

1. Work out the missing terms in these geometric sequences.

   a ☐ 60 180 ☐ …
   b ☐ 30 15 ☐ …
   c ☐ ☐ 8 80 …
   d ☐ 30 45 ☐ …

2. The term-to-term rule of a sequence is "multiply the previous term by 2, then subtract 1".

   a Find the first five terms of the sequence if the first term is

   i 3 **ii** 8 **iii** 8.5 **iv** 1

   b What do you notice about the terms of the sequence in **a** parts **i** and **ii**?

3. Repeat question 2 using the term-to-term rule "multiply the previous term by 2, then add 1".

   What's the same and what's different about your answers?

4. The second and third terms of a sequence are 80 and 16. Find the first term if the sequence is

   a linear

   b geometric.

5. The fourth term of a geometric sequence is 8000, and the fifth term of the sequence is 1000

   Work out the difference between the first and the sixth terms of the sequence.

6. **a** Write down the next five terms of the Fibonacci sequences that start

   i 1 1 … **ii** 1 2 … **iii** 2 2 …

   b What's the same and what's different about the sequences you found in **a**?

What do you think?

1. A geometric sequence starts with 3. The term-to-term rule is "multiply the previous term by 5". The sequence starts 3 15 75 375

   What do you notice about the last digits of the numbers in the sequence?

   Investigate what happens with other starting numbers and geometric rules.

2. How many terms of a sequence do you need to know to decide whether it is linear or non-linear?

Consolidate – do you need more?

1 For each sequence of patterns, state whether it represents a linear or a non-linear sequence.

a

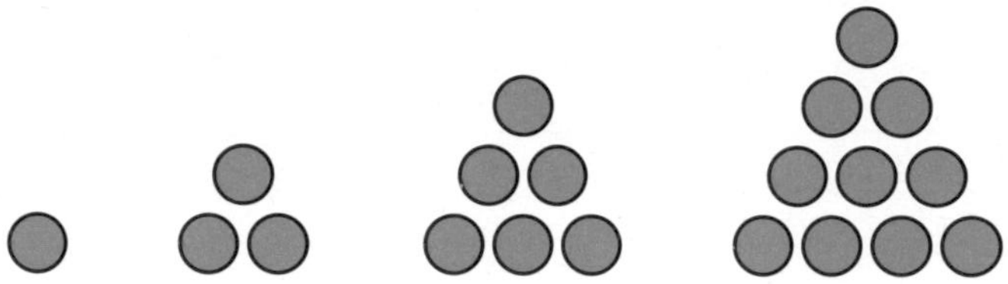

b

c

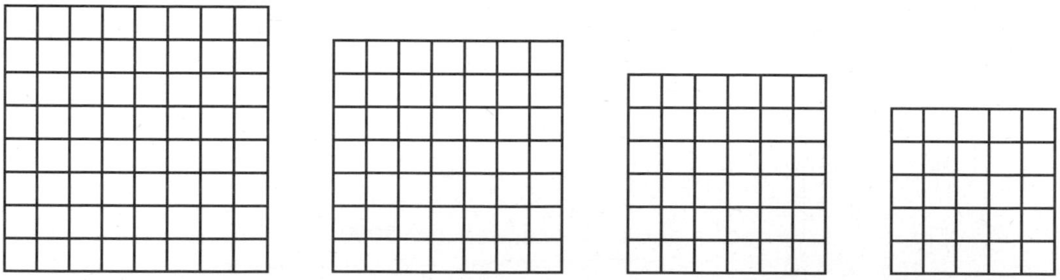

d

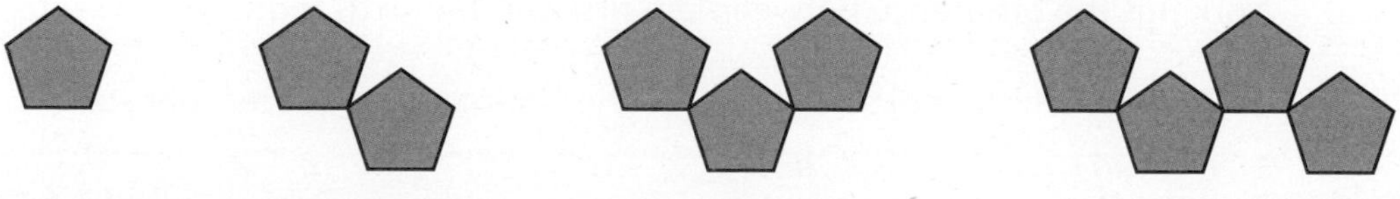

2 Which of these sequences are linear? How do you know?

a 100 200 300 400 500 …

b 1000 2000 3000 4000 5000 …

c 100 1000 10 000 100 000 1 000 000 …

d 10 10 20 20 30 30 ...

e 10 20 30 50 80 ...

f 10 20 40 70 110 ...

g 1000 900 800 700 600 ...

h 100 90 81 72.9 65.61 ...

3 All these sequences are geometric. For each sequence

i work out the term-to-term rule

ii work out the next term.

a 60 180 540 1620 ...

b 80 8 0.8 0.08 ...

c 200 1000 5000 25 000 ...

d 65 536 16 384 4096 1024 ...

e 248 832 20 736 1728 144 ....

f 0.006 0.6 60 6000 ....

4 The first and second terms of a sequence are 30 and 45. Find the next term if the sequence is

a linear

b geometric

c Fibonacci.

5 Find the first term of a sequence if the second and third terms are 30 and 45, and the sequence is

a linear

b geometric

c Fibonacci.

Stretch – can you deepen your learning?

A sequence starts with 56. The next term is found by halving the previous term and adding 3. Use a calculator to work out several more terms in the sequence. What do you notice?

Investigate using the same rule but with different starting numbers. You can use a spreadsheet to help you.

What happens if instead of adding 3, you add a different number? Can you predict what will happen to the sequences?

Reflect

Describe in your own words the differences between linear, geometric and Fibonacci sequences.

1.4 Representing sequences

Small steps

- Recognise the difference between linear and non-linear sequences
- Represent sequences in **tabular** and graphical forms
- Explain the term-to-term rule of numerical sequences in words

Key words

Tabular – organised into a table

Graph – a diagram showing how values change

Axis – a line on a graph that you can read values from

Are you ready?

1 Here is a sequence.

5 8 11 14 …

a What is the third term of the sequence?

b What would the fifth term of the sequence be?

c In what position in the sequence is the number 8?

2 Which of these two sequences is linear and which is non-linear?

Sequence A 1 11 21 31 41 …

Sequence B 1 10 100 1000 10 000 …

3 The first term of a sequence is 12. The term-to-term rule is "add 5 every time". Work out the second, third and fourth terms of the sequence.

4 Make up a descending linear sequence whose first term is 12

Models and representations

Sequences can be represented in lots of different ways.

You can use a list. 2 5 8 11 …

You can use objects.

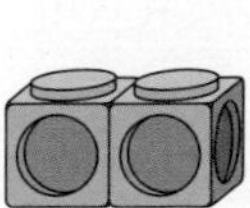

You can use a table.

Position	1	2	3	4
Term	2	5	8	11

You can use a **graph**.

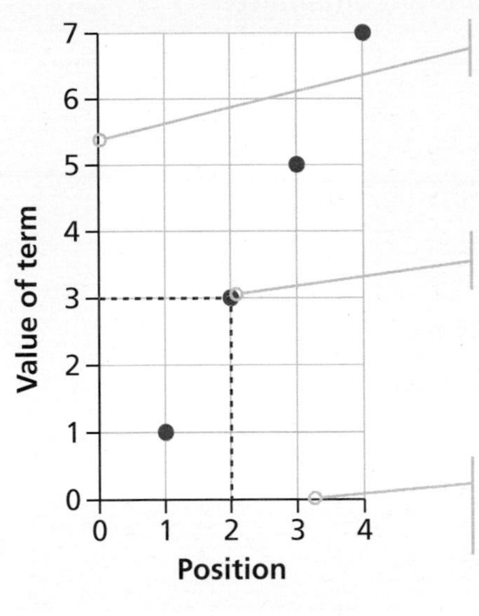

You can read the value of a term from the vertical **axis**.

This point shows the second term in the sequence is 3.

The horizontal axis shows the position of the term in the sequence.

You can use graphing software to plot the points of sequences.

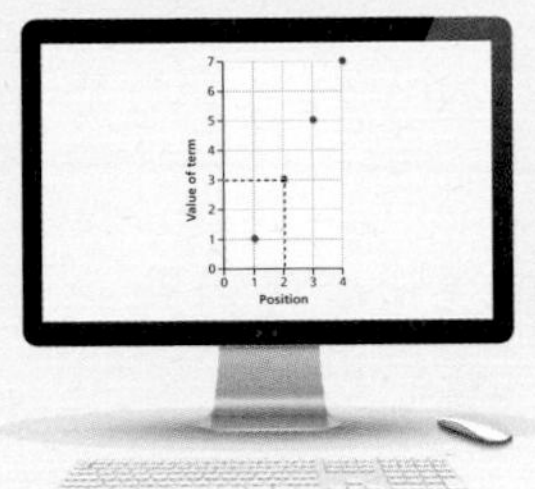

Notice in the graph above, the points lie in the direction of a straight line. Sequences with a constant difference are called linear because they always form straight lines on a graph.

Example 1

A sequence is given in the table.

Position	1	2	3	4
Term	1	3	5	7

a Draw a graph of the sequence.

b Work out the sixth term of the sequence.

a

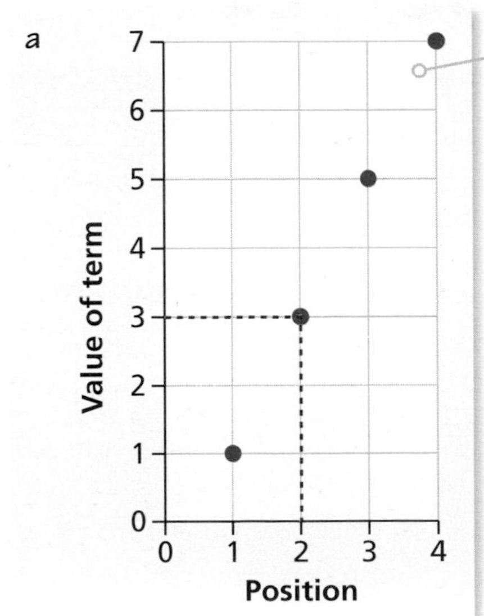

From the table you can see the coordinates to draw your graph.

The first term is 1 so the first point to plot is (1, 1).

The second term is 3 so you need to plot (2, 3).

In the same way you can also plot the points (3, 5) and (4, 7).

Looking at the table and coordinates, you can see that the horizontal axis needs to go as far as 4 and the vertical axis as far as 7

You can now draw the graph.

b The sequence increases by a constant difference of 2

The fifth term will be 7 + 2 = 9, so the sixth term will be 9 + 2 = 11

First work out the constant difference.

Practice 1.4A

1 a Draw a graph to represent the sequence shown in each table.

A

Position	1	2	3	4
Term	0	2	4	6

B

Position	1	2	3	4
Term	3	4	5	6

C

Position	1	2	3	4
Term	10	8	6	4

b For each sequence in part **a**

i state the term-to-term rule

ii work out the sixth term.

2 a Represent each sequence in a table, then draw a graph.

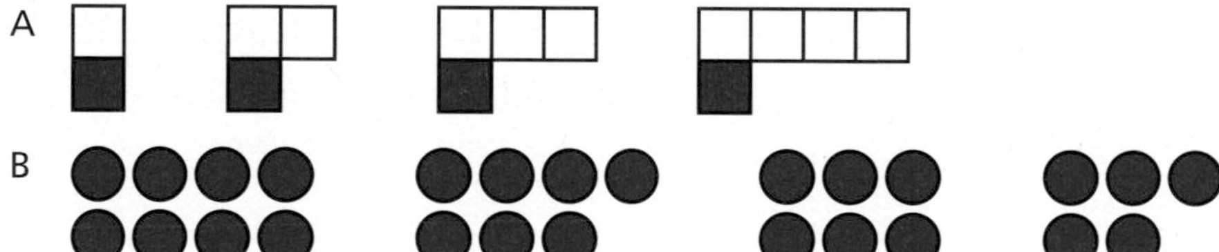

b For each sequence in part **a**

i state the term-to-term rule

ii draw the pattern representing the sixth term.

3

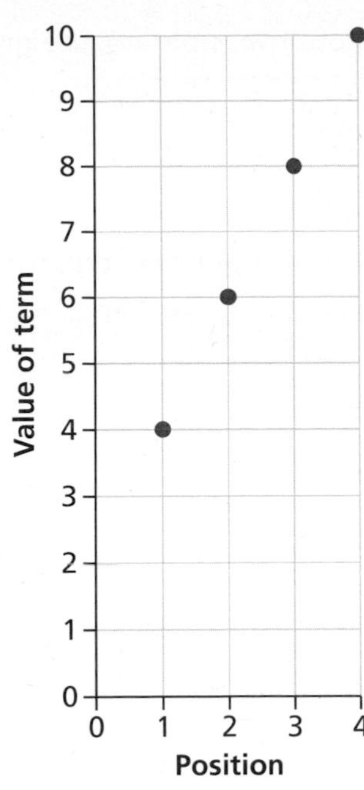

a Copy and complete the table for the graph.

Position	1	2	3	4
Term				

b Describe the sequence in words.

c Work out the seventh term of the sequence.

d Will the sequence have any odd numbers? Explain your answer.

4

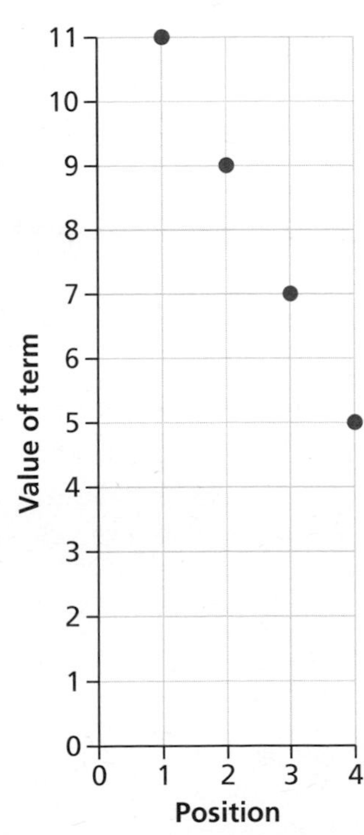

a Copy and complete the table for the graph.

Position	1	2	3	4
Term				

b Describe the sequence in words.

c Work out the sixth term of the sequence.

d Will the sequence have any even numbers? Explain your answer.

5 Ed is talking about a sequence that he has created.

a Explain why it is impossible to draw Ed's sequence from his description.

My sequence goes up in threes.

b Given that the first term of Ed's sequence is 5, copy and complete the table.

Position	1	2	3	4
Term				

6

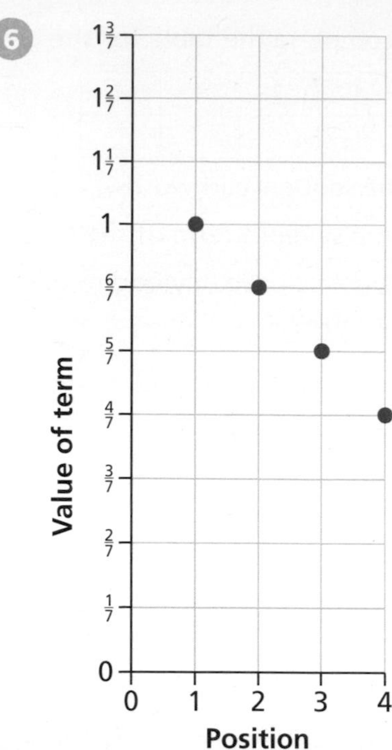

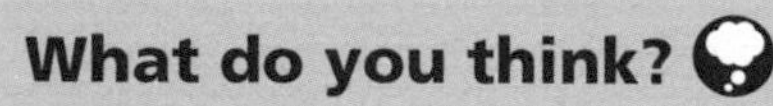

a Copy and complete the table for the graph.

Position	1	2	3	4
Term				

b Describe the sequence in words.

c Work out the fifth term of the sequence.

d Which term of the sequence will be 0?

What do you think?

1

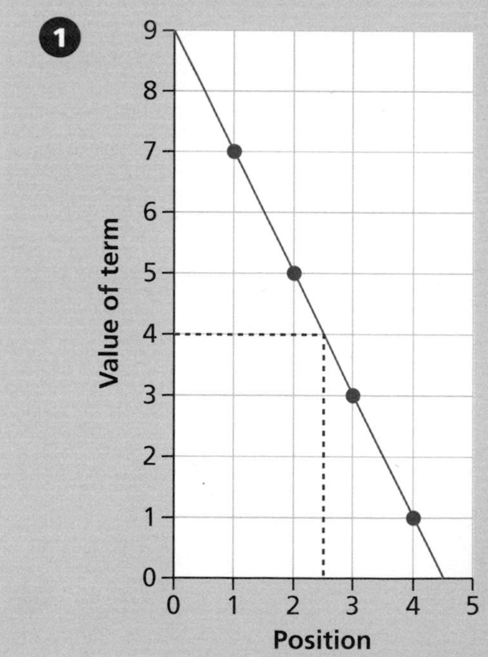

I can use the graph to work out which term in the sequence has value 4

a Explain why Faith is wrong.

b Even though the points lie in the direction of a straight line, does it make sense to join them? Give a reason for your answer.

2 Make up three different sequences that start 1, 3 …

Each of your sequences should have at least four terms.

Draw the graphs for your three sequences.

What's the same and what's different about your graphs?

You can also plot graphs of non-linear sequences.

Example 2

a Draw a graph of the sequence represented by the patterns of dots.

b How can you tell from the graph that the sequence is non-linear?

a

Position	1	2	3	4	5
Term	1	2	4	8	16

First find the values of the terms and put them in a table.

The first term has one dot, the second term has two dots, the third term has four dots and so on.

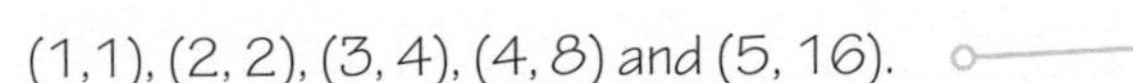

(1,1), (2, 2), (3, 4), (4, 8) and (5, 16).

Now list the coordinates of the points on the graph.

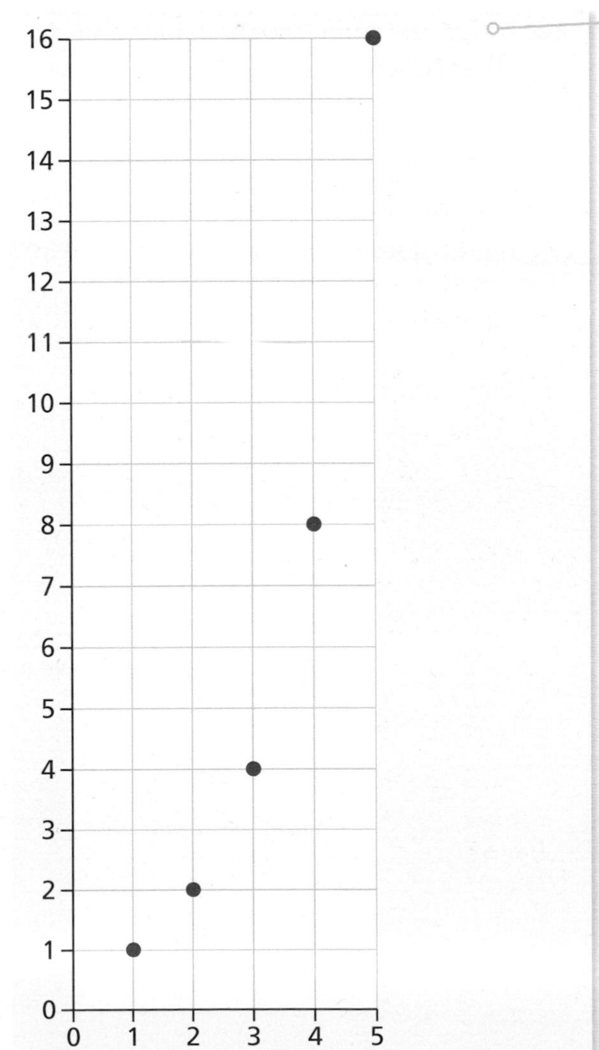

Now draw the graph.

b The points on the graph do not form a straight line. The sequence is non-linear.

Keep your explanations clear and concise.

Practice 1.4B

1 a Draw the next two terms in the sequence.

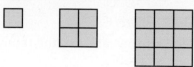

b Copy and complete the table

Position	1	2	3	4	5
Term	1	4			

c Draw a graph representing the first five terms of the sequence.

2

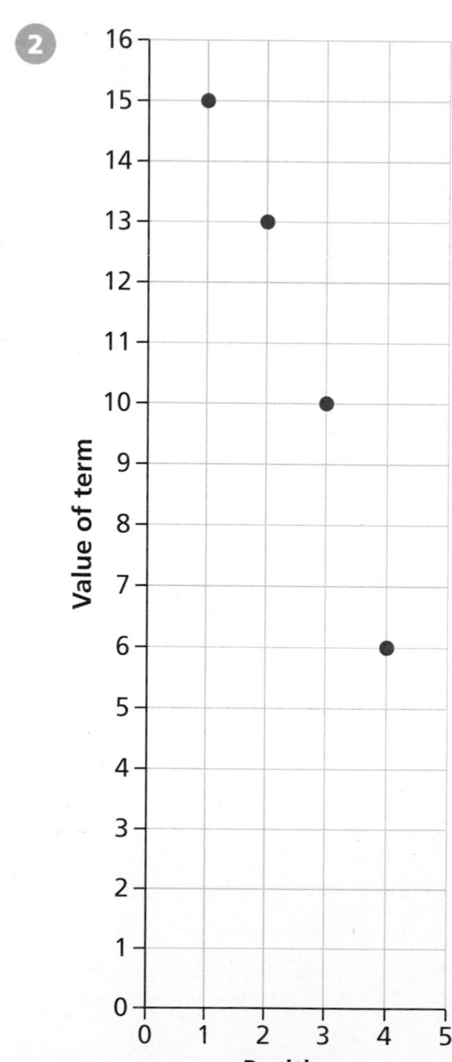

a Copy and complete the table to show the sequence.

Position	1	2	3	4
Term	15			

b Describe how the terms in the sequence change.

c Work out the fifth term of the sequence.

3    

a Copy and complete the table to show the sequence.

Position	1	2	3	4
Term	1			

b Describe how the terms in the sequence change.

c Work out the fifth term of the sequence.

4 Decide whether or not the points on the graph representing each sequence will lie in a straight line.

a 100 200 300 400 500 …

b 1000 2000 3000 4000 5000 …

c 100 1000 10 000 100 000 1 000 000 …

d 10 10 20 20 30 30 …

e 10 20 30 50 80 …

f 10 20 40 70 110 …

g 1000 900 800 700 600 …

h 100 90 81 72.9 65.61 …

5 a Work out the next four terms of the Fibonacci sequences that begin:

i 1 1 …

ii 1 2 …

iii 1 3 …

b Plot graphs of the sequences in part **a**.

For question **5b**, make sure you work out how far the axes need to extend before you draw your graph.

What do you think?

1 Write the first five terms of the sequence that starts 1, 2, … if the sequence is

a linear b geometric c Fibonacci

2 Draw the graphs of the sequences in question 1 and compare them. What's the same and what's different?

Consolidate – do you need more?

1 **a** Draw a graph to represent the sequence shown in each table.

A

Position	1	2	3	4
Term	10	9	8	7

B

Position	1	2	3	4
Term	10	11	12	13

C

Position	1	2	3	4
Term	10	9	7	4

b For each sequence

i state the term-to-term rule

ii work out the fifth term.

2

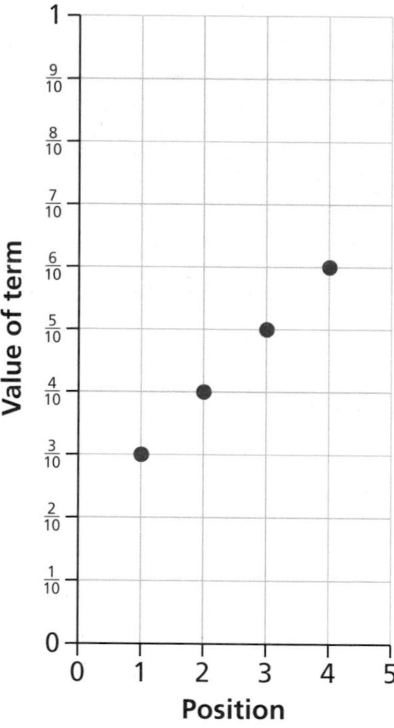

a Copy and complete the table for the graph.

Position	1	2	3	4
Term				

b Describe the sequence.

c Work out the fifth and sixth terms of the sequence.

d Which term of the sequence will be one whole?

3 Draw a graph to represent the first five terms of this linear sequence.

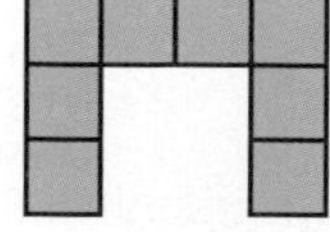

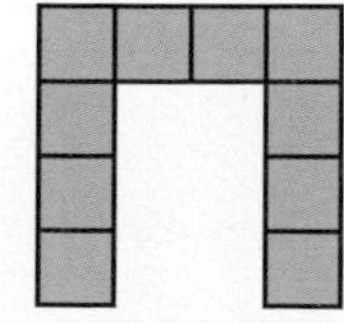

4 Decide whether the points on the graphs of each of the sequences will lie in a straight line or not.

In question 4, you do not need to draw the graphs.

a 2 6 10 14 …

b 14 10 6 2 …

c 2 4 6 8 …

d 2 4 8 16 …

e 8 6 4 2 …

f 16 8 4 2 …

g $\frac{1}{5}$ $\frac{2}{5}$ $\frac{3}{5}$ $\frac{4}{5}$ …

h $\frac{1}{5}$ $\frac{2}{5}$ $\frac{4}{5}$ $\frac{8}{5}$ …

Stretch – can you deepen your learning?

The first 4 terms of a sequence are shown on a graph.

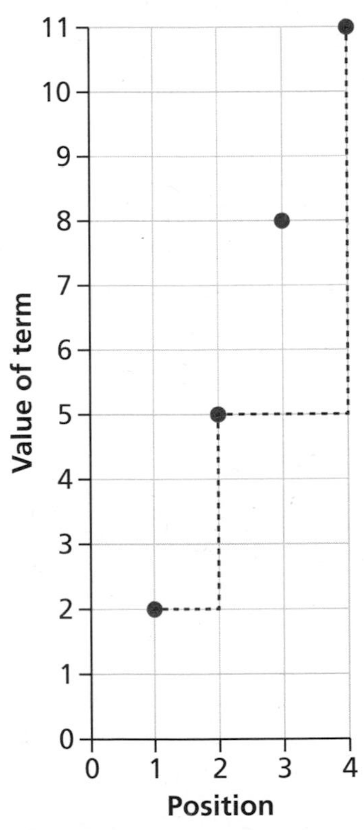

When you go one square along the horizontal axis on the graph, you go three squares up the vertical axis, and when you go two squares along the horizontal axis on the graph you go six squares up the vertical axis. That's because the sequence goes up in threes, and $2 \times 3 = 6$

Faith

a What will happen if you go three squares along the horizontal axis?

b Investigate what happens with graphs of other linear sequences. What about non-linear sequences?

Reflect

How can you tell from looking at a sequence whether the points will lie in a straight line or not when plotted on a graph?

Ⓗ 1.5 Finding missing terms in sequences

Small steps

- Find missing numbers within sequences Ⓗ
- Explain the term-to-term rule of numerical sequences in words
- Recognise the difference between linear and non-linear sequences

Key words

Trial and improvement – A method of finding a solution to a mathematical problem where you make a guess (a trial), see if it works in the problem, and then refine it to get closer to the actual answer (improvement)

Are you ready?

1 5 and 10 are the second and third terms of a linear sequence.

- **a** What is the fourth term of the sequence?
- **b** What is the first term of the sequence?
- **c** What's the same and what's different about how you worked your answers out?

2 5 and 10 are the second and third terms of a geometric sequence.

- **a** What is the fourth term of the sequence?
- **b** What is the first term of the sequence?
- **c** What's the same and what's different about how you worked your answers out?

3 5 and 10 are the second and third terms of a Fibonacci sequence.

- **a** What is the fourth term of the sequence?
- **b** What is the first term of the sequence?
- **c** What's the same and what's different about how you worked your answers out?

4 Here are the first five terms of a sequence.

7 12 17 22 27

- **a** How many terms are there between the first and fifth terms?
- **b** How many differences are there between the first and fifth terms?

Models and representations

You can use bar models to work out differences.

<table>
<tr><td>20</td><td>?</td></tr>
<tr><td colspan="2">30</td></tr>
</table>

The difference between the bars is 10, so ? = 10

<table>
<tr><td>20</td><td>?</td><td>?</td></tr>
<tr><td colspan="3">30</td></tr>
</table>

The difference between the bars is still 10, but there are two equal missing parts.
Each missing part is 10 ÷ 2 = 5

Example 1

Find the missing terms in these linear sequences.

a 1 13 ☐ **b** 1 ☐ 13 **c** 1 ☐ ☐ 13

a 1 13 ☐

$13 - 1 = 12$

$13 + 12 = 25$

a You can work out the difference between the first two terms.

As the sequence is linear, the third term is found by adding 12 to the second term.

b 1 ☐ 13

1	?	?
13		

b This time there are two differences between 1 and 13. As the sequence is linear, these differences must be equal. You can show this as a bar model.

The bar model does not have to be to scale – it just helps us to see what calculation we need to do.

$13 - 1 = 12$

$12 \div 2 = 6$

$1 + 6 = 7$

$13 - 6 = 7$

Missing term is 7

The two differences must add up to $13 - 1 = 12$

As they are equal, they must be $12 \div 2 = 6$

So you can find the missing second term by adding six to the first; $1 + 6 = 7$

(Note you could also do $13 - 6 = 7$)

Check: 1, 7, 13 is linear with a constant difference of 6

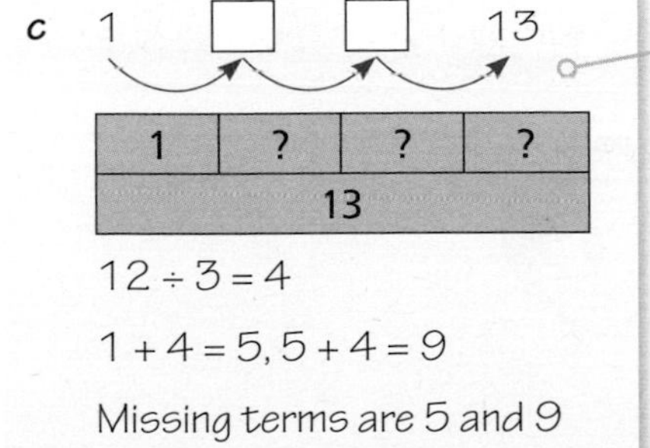

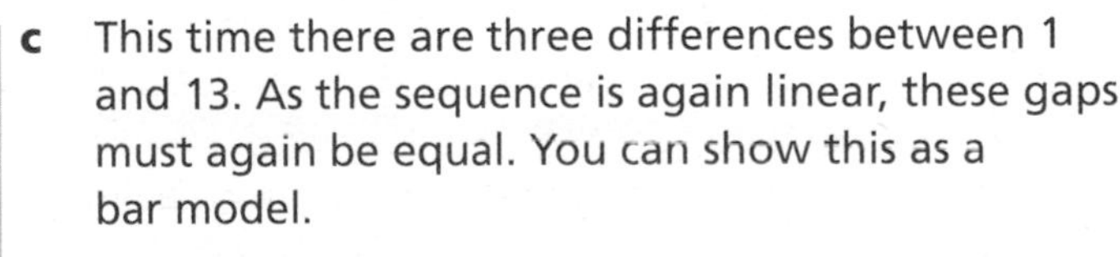

c This time there are three differences between 1 and 13. As the sequence is again linear, these gaps must again be equal. You can show this as a bar model.

The three differences must add up to $13 - 1 = 12$

As they are equal, they must be $12 \div 3 = 4$

So you can find the missing terms
$1 + 4 = 5$, $5 + 4 = 9$

Check: 1, 5, 9, 13 is linear with a constant difference of 4

Example 2

The sixth term of a linear sequence is 25 and the first term is 10. What is the third term?

1st	2nd	3rd	4th	5th	6th
10	☐	☐	☐	☐	25

$25 - 10 = 15$

$15 \div 5 = 3$

So the sequence is 10, 13, 16, 19, 22, 25, …
and the third term is 16

Between the first and sixth terms there are five differences.

$25 - 10 = 15$, so 5 differences altogether add to 15

So each difference is $15 \div 5 = 3$

Practice 1.5A

1. Find the missing terms in each linear sequence.

   a 10 40 ☐

   b 10 ☐ 40

   c 10 ☐ ☐ 40

   d 10 ☐ ☐ ☐ 40

   e 10 ☐ ☐ ☐ ☐ 40

   f 10 ☐ ☐ ☐ ☐ ☐ 40

   g 40 ☐ ☐ ☐ ☐ ☐ 10

   h 40 ☐ ☐ 10

2. A linear sequence starts 4 10 16 …

The sequence goes up in sixes, so the tenth term will be 4 + 10 × 6 = 4 + 60 = 64

   a Explain why Flo is wrong.

   b Without working out the terms in between, find the correct value of the tenth term of the sequence.

3. A linear sequence starts 20 24 28 …

The second term is 24 so the 20th term will be 10 × 24 = 240

   a Explain why Ed is wrong.

   b Without working out the terms in between, find the correct value of the twentieth term of the sequence.

4. Work out

   a The tenth term of the linear sequence that starts 7 17 …

   b The hundredth term of the linear sequence that starts 7 17 …

   c The thousandth term of the linear sequence that starts 7 17 …

   d The thousandth term of the linear sequence that starts 7 18 …

   e The thousandth term of the linear sequence that starts 7 19 …

5. The fifth term of a linear sequence is 60 and the second term is 24. Work out the seventh term of the sequence.

6. The fifth term of a linear sequence is 24 and the second term is 60. Work out the seventh term of the sequence.

What do you think?

1 A sequence starts 0.4 0.8 1.2 ...

a Which is the first term of the sequence that will be an integer?

b How often will integers occur after that?

c Will there be any odd integers in the sequence? Explain your answer.

2 How many different linear sequences can you create whose first term is 100 and whose terms include 20?

3 How many different linear sequences can you create whose first term is 100 and whose terms include both 90 and 80?

You can also work out missing terms in non-linear sequences.

Example 3

A geometric sequence starts 4 □ 36

Find the fourth term of the sequence

You can see the sequence is increasing so you know each term is found by multiplying by a number greater than 1.

Try × 2: $4 \times 2 = 8, 8 \times 2 = 16 \neq 36$

Try × 3: $4 \times 3 = 12, 12 \times 3 = 36$

So the term-to-term rule is "Multiply the previous term by 3".

The fourth term must be $36 \times 3 = 108$

This method is called **trial and improvement**. Trial and improvement is sometimes useful in cases like Example 3. It is not always recommended; for example, if the multiplier is not an easy integer.

Example 4

A Fibonacci sequence starts 2 □ 10 □

Find the missing terms in the sequence.

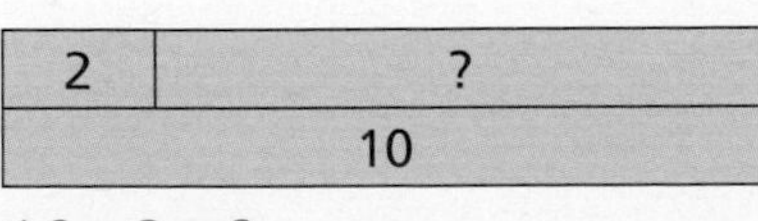

In a Fibonacci sequence you add the two previous terms to make the next term.

The missing second term must be 10 – 2 = 8

$10 - 2 = 8$

So the sequence starts 2 8 10 □

The fourth term must be $8 + 10 = 18$

Practice 1.5B

1. Find the missing terms in each geometric sequence.

   a 10 40 ☐

   b 10 ☐ 40

   c 40 10 ☐

   d 40 ☐ 10

   e 3 ☐ 12

   f 3 ☐ 75

   g 3 ☐ ☐ 192

   h 1000 ☐ 10

   i 100 ☐ 4

   j 800 ☐ ☐ 12.5

2. Find the missing terms in these geometric sequences.

   a 1 ☐ 64

   b 1 ☐ ☐ 64

   c 1 ☐ ☐ ☐ ☐ ☐ 64

3. Find the missing terms in these Fibonacci sequences.

   a 1 ☐ 2 ☐

   b ☐ ☐ 7 11

   c 1 ☐ 3 ☐

   d 1 ☐ 8 ☐

   e ☐ ☐ 1 2

   f ☐ ☐ ☐ 14 22 ☐

4. The first term of a sequence is 10 and the third term is 90. Find the second term if the sequence is

   a linear b geometric c Fibonacci.

5. The second and third terms of a sequence are 12 and 18. Find the first term if the sequence is

   a linear b geometric c Fibonacci.

What do you think?

1. Find the missing terms in these Fibonacci sequences.

   a 1 ☐ ☐ 11

   b 8 ☐ ☐ 12

   c 3 ☐ ☐ ☐ 27

2. How many sequences can you find with first term 2 and fourth term 16? Can you find a linear, geometric and Fibonacci example?

Consolidate – do you need more?

1 The second and third terms of some linear sequences are shown.

For each sequence, find the first term and the fourth term.

a ☐ 10 15 ☐

b ☐ 10 5 ☐

c ☐ 1 0.5 ☐

2 The second and fourth terms of some linear sequences are shown.

For each sequence, find the first term and the third term.

a ☐ 10 ☐ 16

b ☐ 12 ☐ 16

c ☐ 14 ☐ 16

d ☐ 15 ☐ 16

e ☐ 10 ☐ 18

f ☐ 11 ☐ 18

g ☐ 11 ☐ 19

3 A linear sequence starts 5 10 …

a What is the 10th term of the sequence?

b What is the 20th term of the sequence?

c What is 47th term of the sequence?

4 Find the missing terms in these linear sequences.

a 6 30 ☐

b 6 ☐ 30

c 6 ☐ ☐ 30

d 6 ☐ ☐ ☐ 30

e 6 ☐ ☐ ☐ ☐ ☐ 30

f 6 ☐ ☐ ☐ ☐ ☐ ☐ ☐ 30

5 Find the missing terms in these geometric sequences.

a 1 ☐ 16

b 1 ☐ ☐ ☐ 16

c 1 ☐ 81

d 1 ☐ ☐ ☐ 81

e 6 ☐ 96

f 8 ☐ ☐ 125

g 8 ☐ ☐ 27

6 The first and third terms of a sequence are 2 and 10

a Find three different rules that might describe the sequence.

b State the fourth term for each of your rules.

Stretch – can you deepen your learning?

1 Find the missing terms in these sequences.

> You will need to recall your Key Stage 2 study of fractions here – equivalence, adding, subtracting and multiplying. You will study these in detail later in this book.

a $\frac{1}{10}$ ☐ ☐ $\frac{2}{5}$

b ☐ $\frac{1}{10}$ ☐ $\frac{1}{40}$

c 0.1 ☐ ☐ 0.0001

d 1.8 ☐ ☐ ☐ 0.8 ☐

2 If you are given the first and third terms of a sequence, is it always possible to find the second term so that the sequence is

a linear

b geometric

c Fibonacci?

Reflect

1 Describe how you go about finding the missing terms in a linear sequence.

2 What's the same and what's different when finding the missing terms in a geometric or Fibonacci sequence?

1 Sequences
Chapters 1.1–1.5

I have become fluent in...

- finding the next term in sequences
- using a rule to form numerical sequences
- using tables and graphs to represent sequences
- using a calculator to work out terms in a sequence.

I have developed my reasoning skills by...

- making connections between information given in a wide variety of forms
- making and testing ideas about patterns and relationships
- identifying whether a sequence is linear or not
- working out the term-to-term rules for sequences.

I have been problem-solving through...

- spotting and describing patterns
- finding more than one way to continue a sequence
- representing sequences in a variety of forms
- working out missing terms in sequences. (H)

Check my understanding

1 **a** Draw the next pattern in this sequence.

b Predict how many circles there will be in the sixth term of the sequence. Draw the sixth term and check if you were right.

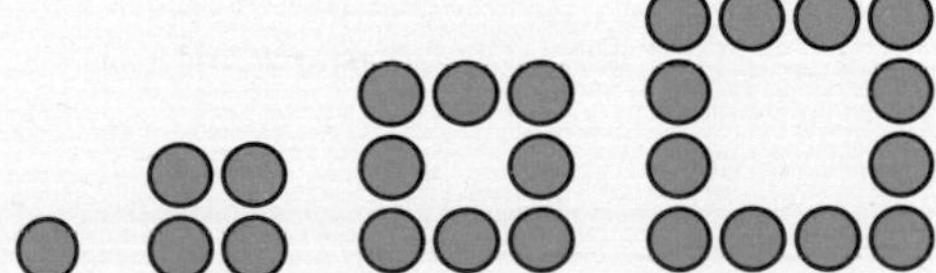

2 Find the first four terms of these sequences, where the first term is 5

a Add 7 to the previous term.

b Multiply the previous term by 6.

c Subtract 2 from the previous term and multiply the result by 3.

3 **a** Describe the sequences

i 80, 70, 60, 50, ... **ii** 80, 40, 20, 10, ... **iii** 80, 85, 95, 110, ...

b Are the sequences linear or non-linear? How do you know?

4 The first two terms of a sequence are 4 and 8. Find the next two terms if the sequence is

a linear **b** geometric **c** Fibonacci.

5 Draw a graph of the first five terms of the sequence shown in the table.

Position	1	2	3	4	5
Term	15	14	12	9	5

6 **a** Find the missing terms in these sequences, explaining your reasoning. (H)

i 6 ☐ 12 **ii** 6 ☐ ☐ 12

iii 6 ☐ ☐ ☐ 12 **iv** 12 ☐ ☐ ☐ 6

b Is it possible to have more than one answer? Why or why not?

2 Algebraic notation

In this block, I will learn...

how to find inputs and outputs of one-step function machines

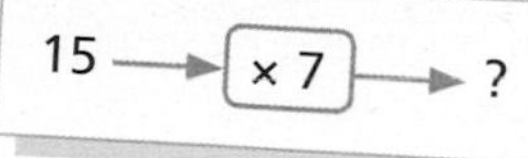

the meaning of algebraic expressions

$3x$ $x + 3$ $30 - x$

$30 - 2x$ x^2 x^3

how to find inputs and outputs of a series of function machines

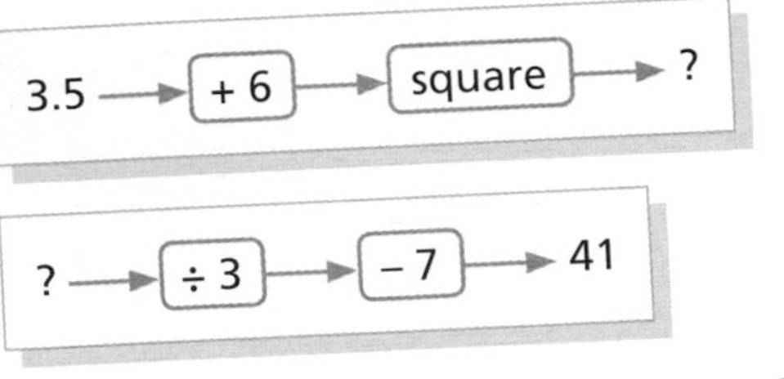

how to substitute numbers into algebraic expressions

When $a = 7$
$3a - 2 = 3 \times 7 - 2$
$= 21 - 2 = 19$

When $b = 4.6$,
$b^2 = 4.6 \times 4.6$
$= 21.26$

how to represent expressions in a variety of ways

$4d + 3$

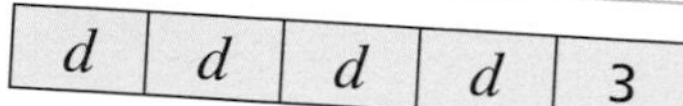

how to generate sequences given an algebraic rule

When $n = 1$, $5n - 2 = 5 \times 1 - 2 = 5 - 2 = 3$
When $n = 2$, $5n - 2 = 5 \times 2 - 2 = 10 - 2 = 8$
When $n = 3$, $5n - 2 = 5 \times 3 - 2 = 15 - 2 = 13$
When $n = 4$, $5n - 2 = 5 \times 4 - 2 = 10 - 2 = 18$

2.1 Inputting to one-step function machines

Small steps

- Given a numerical input, find the output of a one-step function machine
- Use diagrams and letters to generalise number operations
- Use diagrams and letters with one-step function machines

Key words

Function – a relationship with an input and an output

Variable – a numerical quantity that might change, often denoted by a letter, for example x or t

Coefficient – a number in front of a variable, for example for $4x$ the coefficient of x is 4

Term – a single number or variable or a number and variable combined by multiplication or division

Are you ready?

1 Which of these are mathematical operations?

$+$ $\times$ x $-$ 7 $\div$

2 Which of these calculations have the same answer?

$5 + 5 + 5$ 3×5 5×3

3 Write a shorter calculation that has the same value as the ones shown.

a $6 + 6 + 6 + 6$ **b** $7 + 7 + 7 + 7 + 7 + 7 + 7 + 7$ **c** $5 + 5 + 5 + 5 + 5 + 5 + 5$

4 **a** What does "7 squared" mean?

b How do you square a number on your calculator?

c Work out the squares of **i** 10 **ii** 20 **iii** 5.6 **iv** 203

Models and representations

Function machines

You can use "function machine" diagrams to model **functions**.

Input ⟶ [+ 7] ⟶ Output

If the input is 3, then the machine gives the answer 10.

You can use letters, counters, cubes or bars to represent **terms**: these all show $3t$

Letters

$t + t + t$

Counters

(t) (t) (t)

Cubes

Bar models

t	t	t

$3t$ means "3 lots of t" and can be calculated by multiplying the value of t by 3. You write $3t$ as shorthand for $3 \times t$. The **coefficient** of t is 3.

Example 1

Input → [+ 6] → Output

Find the output of the function machine if the input is

a 10 **b** ▭ **c** m

a $10 + 6 = 16$ — The machine adds 6 to 10

b ▭ $+ 6$ — The machine adds 6 to ▭.
As the value of ▭ can vary, you cannot work out a numerical answer like you could in part **a**, so the answer is just ▭ + 6

c $m + 6$ — The machine adds 6 to m.
As the value of m can vary, you cannot work out a numerical answer like you could in part **a**, so the answer is just $m + 6$

Unlike with multiplication, do not leave out the sign when representing additions to and subtractions from a **variable**. So write "6 more than m" as $m + 6$ and "3 less than t" as $t - 3$

Example 2

Input → [× 5] → Output

Find the output of the function machine if the input is

a 10 **b** ▭ **c** p

a $10 \times 5 = 50$ — The machine multiplies 10 by 5

b ▭▭▭▭▭ — The machine multiplies ▭ by 5 so you now have ▭▭▭▭▭.

c $p \times 5 = 5p$ — The machine multiplies p by 5. Notice you write $5p$, not $p5$

Practice 2.1A

1. Find the output when you input 50 into each of these machines.
   - **a** Input → [+ 20] → Output
   - **b** Input → [− 20] → Output
   - **c** Input → [× 5] → Output
   - **d** Input → [÷ 20] → Output
   - **e** Input → [+ 0.8] → Output
   - **f** Input → [− 0.8] → Output
   - **g** Input → [× 0.8] → Output
   - **h** Input → [÷ 0.8] → Output

2. Find the outputs when you input 0, 1, 2, 3, 4 and 5 into each of these machines.
   - **a** Input → [+ 5] → Output
   - **b** Input → [− 5] → Output
   - **c** Input → [× 5] → Output
   - **d** Input → [Subtract from 5] → Output

3 Find the outputs when you input 17.4, 103 and 358 into each of these machines.

a Input → × 4.5 → Output **b** Input → + 11.9 → Output

c Input → – 23.7 → Output **d** Input → + 8500 → Output

4 Find the output when you input [cubes] into each of these machines.

a Input → × 2 → Output **b** Input → + 2 → Output

c Input → ÷ 2 → Output

5 Find the output when you input x into each of these machines.

a Input → + 4 → Output **b** Input → – 4 → Output

c Input → × 4 → Output

d Input → Subtract from 4 → Output

6 Write each of these using algebraic notation.

a five times y **b** five more than y **c** five less than y

What do you think?

1 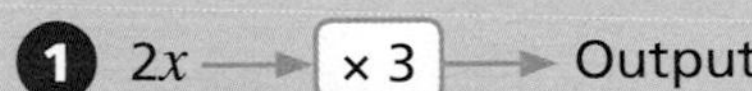
2x
× 3
Output

Flo
If the input is $2x$ then the output is $32x$ as you do not write × signs in algebra.

Seb
I don't agree. If you have $2x$ three times, then the answer should be $6x$

a Who do you agree with?

Find the output when the input is

b $5p$ **c** $10t$ **d** $\frac{1}{3}h$

2 Find two different function machines so that the output for 20 is greater than the output for 200. Investigate for other numbers and rules.

3 Rob says that the output of the function machine "+ 7" will always be greater than the input. Do you agree with Rob? Explain why or why not.

You know from Key Stage 2 that fractions are a result of division. When you do divisions with letters, you use fraction notation to show your answers.

If you represent x as [bar]

then $\frac{x}{2}$ would be [half bar]

and $\frac{x}{3}$ would be [third bar]

You can also multiply two variables together.

Notation. You write $a \div 4$ as $\frac{a}{4}$. In the same way, you write $3 \div b$ as $\frac{3}{b}$

Notation. In the same way as you write $5 \times d$ as $5d$, you write $c \times d$ as cd.

2.1 Inputting to one-step function machines

When you multiply a number by itself, it is called "squaring", so 6 × 6 is 6^2 ("six squared"). You can also square letters.

Notation. You do not write $a \times a$ as aa. Instead, write it as a^2, which you say as "a squared".

Example 3

Input → $\times t$ → Output

Find the output of the function machine if the input is

a 10 **b** m **c** t

a $10 \times t = 10t$ — The machine multiplies 10 by t.

b $m \times t = mt$ — The machine multiplies m by t.

c $t \times t = t^2$ — The machine multiplies t by t.

Example 4

Input → $\div 6$ → Output

Find the output of the function machine if the input is

a 30 **b** 10.38 **c** k **d** $5m$

a $30 \div 6 = 5$ — The machine divides 30 by 6

b $10.38 \div 6 = 1.73$ — The machine divides 10.38 by 6

c $k \div 6 = \frac{k}{6}$ — The machine divides k by 6, so the output is $k \div 6$, which you write as $\frac{k}{6}$

d $5m \div 6 = \frac{5m}{6}$ — The machine divides $5m$ by 6

Practice 2.1B

1 **a** Find the output when you input 60 into each of these machines.

i Input → $\times 2$ → Output **ii** Input → $\times b$ → Output

iii Input → $\div 2$ → Output **iv** Input → $\div b$ → Output

v Input → $+ b$ → Output **vi** Input → $- b$ → Output

b Find the output when you input x into each of these machines.

i Input → $\times 2$ → Output **ii** Input → $\times b$ → Output

iii Input → $\div 2$ → Output **iv** Input → $\div b$ → Output

v Input → $+ b$ → Output **vi** Input → $- b$ → Output

c Find the output when you input b into each of these machines.

i Input → $\times 2$ → Output **ii** Input → $\times b$ → Output

iii Input → $\div 2$ → Output **iv** Input → $\div b$ → Output

v Input → $+ b$ → Output **vi** Input → $- b$ → Output

2 Find the outputs when you input 0, 6, 47 and t into each of these machines.

a Input → $+ 2$ → Output
b Input → $- 3$ → Output
c Input → $\times 5$ → Output
d Input → square → Output
e Input → $\times m$ → Output
f Input → $\div 4$ → Output
g Input → $\div p$ → Output

3 a Find the outputs when you input x into each of these machines.

i Input → $+ 5$ → Output
ii Input → $\times 3$ → Output
iii Input → $\times x$ → Output
iv Input → $\times y$ → Output
v Input → $\div 2$ → Output
vi Input → $\div x$ → Output

b Find the outputs when you input $4x$ into the same machines.

4 Filipo says $\frac{1}{2}x$ and $\frac{x}{2}$ are the same, but Junaid says they are different.
Who do you agree with? Why?

What do you think?

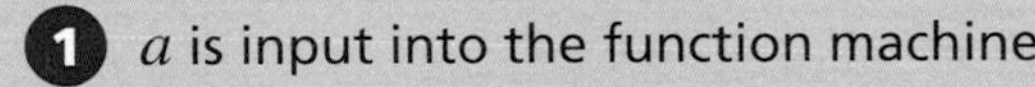

1 a is input into the function machine

a → $\times b$ → Output

Rhys thinks that the output should be ab.

Huda thinks that the output should be ba.

Samira thinks that it does not matter.

a Who do you agree with?

b What if the machine had been $\div b$?

Consolidate – do you need more?

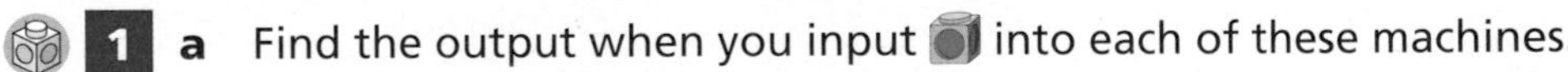

1 a Find the output when you input [cube] into each of these machines.

i Input → $\times 2$ → Output
ii Input → + [cube] → Output
iii Input → $+ 2$ → Output
iv Input → − [cube] → Output

b Find the output when you input [cube][cube] into each of these machines.

i Input → $\times 2$ → Output
ii Input → $\div 2$ → Output
iii Input → $+ 2$ → Output
iv Input → − [cube] → Output

c Find the output when you input $2a$ into each of these machines.

i Input → $\times 2$ → Output
ii Input → $\div 2$ → Output
iii Input → $+ 2$ → Output
iv Input → $- a$ → Output

d Find the output when you input 108 into each of these machines.

i Input → $\times 2$ → Output
ii Input → $\div 2$ → Output
iii Input → $+ 2$ → Output
iv Input → $- a$ → Output

e Find the output when you input 0.16 into each of these machines.

i Input → [× 2] → Output

ii Input → [÷ 2] → Output

iii Input → [+ 2] → Output

iv Input → [$-a$] → Output

2 Write these terms without mathematical operations.

a $3 \times a$ **b** $b \times 7$ **c** $c \div 5$ **d** $4 \div d$

e $e + e + e + e + e$ **f** $f \times f$ **g** $g \times h$

3 Explain the meaning of these terms. An example is given for you.

Example: $3m$ means the product of 3 and m

a $4p$ **b** $\frac{f}{5}$ **c** h^2 **d** $t - 5$

e $5 + d$ **f** $k + 4$ **g** $6 - n$

4 Write down the coefficient of x for each of these terms.

a $6x$ **b** $8x$ **c** $12x$ **d** $\frac{1}{2}x$

5 Find the outputs when you input 100, 0.06, 32 and p into each of these machines.

a Input → [+ 30] → Output

b Input → [− 30] → Output

c Input → [× 2] → Output

d Input → [square] → Output

e Input → [$\times m$] → Output

f Input → [÷ 4] → Output

g Input → [$\div p$] → Output

Stretch – can you deepen your learning?

1 Find inputs for which both machines in each pair give

i the same output **ii** different outputs

a Input → [− 10] → Output Input → [Subtract from 10] → Output

b Input → [× 2] → Output Input → [square] → Output

2 **a** What do you think it means to cube a number?

b Find the outputs of the function machine for the inputs 1, 0.2, 10 and x.

Input → [cube] → Output

Reflect

1 Explain the meaning of the terms on these cards.

$5x$ $\frac{x}{5}$ $x + 5$ $x - 5$ x^2 $\frac{5}{x}$

2 What's the same and what's different about how you write algebraic terms involving addition, subtraction, multiplication and division?

2.2 Working backwards with one-step function machines

Small steps

- Use inverse operations to find the input given the output
- Use diagrams and letters to generalise number operations
- Use diagrams and letters with one-step function machines

Key word

Inverse – the opposite of a mathematical operation; it reverses the process

Are you ready?

1 **a** If you add 10 to a number, what do you need to do to your answer to get back to the original number?

b If you subtract 10 from a number, what do you need to do to your answer to get back to the original number?

c If you multiply a number by 10, what do you need to do to your answer to get back to the original number?

d If you divide a number by 10, what do you need to do to your answer to get back to the original number?

2 Write these terms without mathematical operations.

a $8 \times a$ **b** $b \times 5$ **c** $5 \div c$ **d** $d \div 5$ **e** $e + e + e$ **f** $f \times f$ **g** $h \times g$

3 Write these using algebraic notation.

a Double x **b** x divided by 3 **c** 4 more than x **d** 5 less than x

Models and representations

Function machines

Input → [× 2] → Output

If the output is 12, then the input was 6 6 ← [÷ 2] ← 12

You can find the *input* when you know the *output* of a function machine

Example 1

Input → [+ 2] → Output

Find the input of the function machine if the output is

a 10 **b** [bar] (1) (1) (1) (1) (1) **c** $m + 5$ **d** h

a $10 - 2 = 8$

The machine added 2 to the input, so the input must be 2 less than the output.

You can check your answer by using 8 as the input.

b [ ] (1) (1) (1) — If the output is [ ] and five ones, the input must have two less, which is [ ] and three ones.

c $m+5-2=m+3$ — The input must have been 2 less, which is m plus 3 ones.

d $h-2$ — 2 less than h.

In the above example, you used the fact that the **inverse** of "add 2" is "subtract 2".

Example 2

The output of each of these function machines is 10. Find the input for each machine.

a Input → [– 20] → 10 **b** Input → [× 20] → 10

c Input → [÷ 20] → 10 **d** Input → [$-p$] → 10

a $10+20=30$ — The inverse of "subtract 20" is "add 20".

b $10\div 20=0.5$ — The inverse of "multiply by 20" is "divide by 20".

c $10\times 20=200$ — The inverse of "divide by 20" is "multiply by 20".

d $10+p$ — The inverse of "subtract p" is "add p".

The inverse of addition is subtraction.
The inverse of subtraction is addition.
The inverse of multiplication is division.
The inverse of division is multiplication.

Practice 2.2A

1 Write down the inverse of these operations.

a + 5 **b** – 3 **c** × 8 **d** ÷ 6

2 The output for each function machine is 50. Work out the inputs.

a ? → [+ 5] → 50 **b** ? → [– 5] → 50

c ? → [× 5] → 50 **d** ? → [÷ 5] → 50

3 The output for each function machine is 60. Work out the inputs.

a ? → [× 1.25] → 60 **b** ? → [+ 1.25] → 60

c ? → [÷ 1.25] → 60 **d** ? → [– 1.25] → 60

4 Find the input for each of the function machines.

a ? → + 4.5 → 166.5
b ? → − 3.72 → 12.67
c ? → + 12.89 → 13.108
d ? → ÷ 9.2 → 1.37
e ? → ÷ 74 → 74
f ? → + 74 → 74
g ? → × 74 → 74
h ? → − 74 → 74
i ? → − 74 → 0
j ? → × 74 → 0

5 Find the missing numbers.

a ? → − 4.35 → 16.2
b 16.2 → − 4.35 → ?
c ? → + 503 → 8101
d 8101 → − 503 → ?
e 2200 → ÷ 11 → ?
f ? → × 11 → 22 000
g ? → × 0.3 → 600
h ? → ÷ 0.3 → 600
i 0 → + 0.3 → ?
j ? → × 0.3 → 0

6 The output for each function machine is t. Work out the inputs, giving your answer in correct algebraic notation.

a ? → − 7 → t
b ? → × 2 → t
c ? → ÷ 7 → t
d ? → + 7 → t

7 The output for each function machine is $12t$. Work out each input, giving your answer in correct algebraic notation.

a ? → × 2 → $12t$
b ? → ÷ 2 → $12t$
c ? → − $2t$ → $12t$
d ? → + $2t$ → $12t$

What do you think?

1 The output for each function machine is $c + 4$. Work out the inputs.

a ? → + 2 → $c + 4$
b ? → − 2 → $c + 4$
c ? → × 2 → $c + 4$
d ? → ÷ 2 → $c + 4$

2 The output for each function machine is $c - 4$. Work out the inputs.

a ? → + 2 → $c - 4$
b ? → − 2 → $c - 4$
c ? → × 2 → $c - 4$
d ? → ÷ 2 → $c - 4$

3 The output for each function machine is $2c + 4$. Work out the inputs.

a ? → + 2 → $2c + 4$
b ? → − 2 → $2c + 4$
c ? → × 2 → $2c + 4$
d ? → ÷ 2 → $2c + 4$

The inverse of squaring is "square rooting" the square root of a number is the number that needs to be squared (multiplied by itself) to give the number.

Notation.

The symbol for square root is $\sqrt{\ }$.

$\sqrt{16}$ means "the square root of 16, which is 4."

Example 3

Input → square → Output

Find the input of the function machine if the output is

a 100 **b** 36 **c** t^2 **d** 2.25

a $\sqrt{100} = 10$ — The square root of 100 is the number that when squared is 100. $10 \times 10 = 100$, so the input must be 10

b $\sqrt{36} = 6$ — $6 \times 6 = 36$, so the input is 6

c $\sqrt{t^2} = t$ — $t \times t = t^2$, so the input is t.

d $\sqrt{2.25} = 1.5$ — If you do not recognise the square root of a number, you can find it using a calculator.

Some calculators give the answer $\frac{3}{2}$ for $\sqrt{2.25}$, which is just another way of writing 1.5. Investigate how your calculator works, and how to change fractional answers to decimal answers.

Positive numbers can be found by squaring two different numbers. You will look at this later in the book when you explore directed numbers.

Practice 2.2B

1 Input → square → Output

Find the input of this function machine if the output is

a 25 **b** 81 **c** 144 **d** 900

e 396 900 **f** 0.25 **g** $\frac{1}{9}$

2 $4^2 = 16$ and $5^2 = 25$

So none of the numbers between 16 and 25 have a square root.

Do you agree with Ed? Explain your answer.

3 Find the square roots of these numbers. Give your answers to the nearest whole number.

a 1000 **b** 850 **c** Five thousand and five

4 Find the missing numbers for this function machine.

a

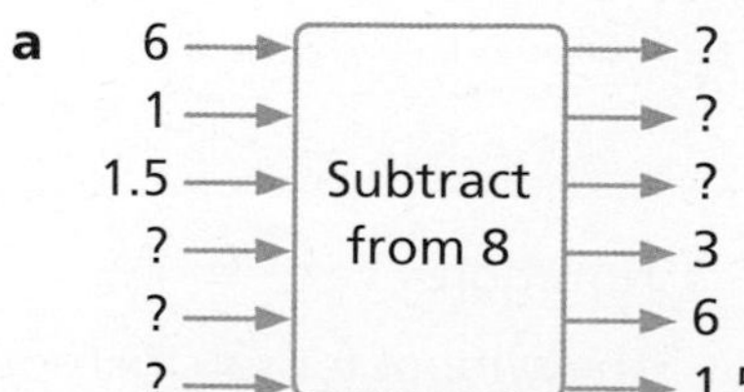

b Investigate for other inputs and outputs. What do you think the inverse of "Subtract from 8" might be?

5 The output for each function machine is 64. Work out the inputs.

a ? → [× 16] → 64

b ? → [÷ 2] → 64

c ? → [square] → 64

d ? → [− 404] → 64

e ? → [+ 11.8] → 64

What do you think?

1 There are two numbers whose square roots are the same as the numbers themselves. Find the numbers.

2 **a** What is the inverse of square root?

b What is the inverse of "cube"?

c Find the missing numbers.

i ? → [square] → 16

ii ? → [$\sqrt{\ }$] → 16

iii ? → [cube] → 64

iv ? → [$\sqrt[3]{\ }$] → 64

v ? → [square] → 1 000 000

vi ? → [cube] → 1 000 000

Consolidate – do you need more?

1 Write down the inverse of these operations.

a × 4 **b** − 9 **c** ÷ 0.5 **d** + 3.5 **e** square

2 Find the inputs for each of these function machines.

a ? → [× 18] → 666

b ? → [+ 492] → 723

c ? → [− 5.24] → 8.73

d ? → [÷ 2.15] → 6.8

3 The output for each function machine is 1000. Work out the inputs.

a ? → [+ 1000] → 1000

b ? → [− 1000] → 1000

c ? → [× 1000] → 1000

d ? → [÷ 1000] → 1000

4 The output for each function machine is 81. Work out the inputs.

a ? → [square] → 81

b ? → [× 3] → 81

c ? → [− 81] → 81

d ? → [÷ 9] → 81

e ? → [+ 36] → 81

f ? → [$\sqrt{\ }$] → 81

5 Find the missing values.

a ? → [× 4] → $4b$

b x → [+ 3] → [+ 6] → ?

c ? → [× d] → $5d$

d g → [+ 10] → [− 5] → ?

e ? → [+ 3] → $g + 10$

f $5b$ → [÷ 5] → ?

g ? → [square] → k^2

h ? → [$\sqrt{\ }$] → 4

i ? → [÷ 2] → $12f$

j ? → [Subtract from 40] → 10

k ? → [square] → 256

l 10 → [Subtract from 40] → ?

Stretch – can you deepen your learning?

1

The function machine "+ 0" always has the same output as input.

Abdullah is right. Can you find any other function machines that have the same property?

2 Investigate finding the inverse of functions on your calculator.

Input → [cube] → Output

Reflect

1 What does the word "inverse" mean?

2 How would you explain to a younger student how to find the input of a function machine given the output?

3 Make a list of all the operations that you know the inverse of, and give their inverses.

2.3 Using simple expressions

Small steps

- Use diagrams and letters to generalise number operations
- Substitute values into single operation expressions
- Generate sequences given an algebraic rule

Key words

Expression – a collection of terms involving mathematical operations

Substitute – to replace letters with numerical values

Evaluate – to work out the numerical value of

Are you ready?

1 Explain the meaning of these expressions

a $6x$ **b** $6 + x$ **c** $\frac{x}{6}$ **d** $x - 6$ **e** $6 - x$

2 Describe these sequences.

a 10 15 20 25 30 …

b 4 40 400 4000 …

3 Explain the difference between y^2 and $2y$.

4 Find the missing numbers in these bar models.

a

10	
?	3

b

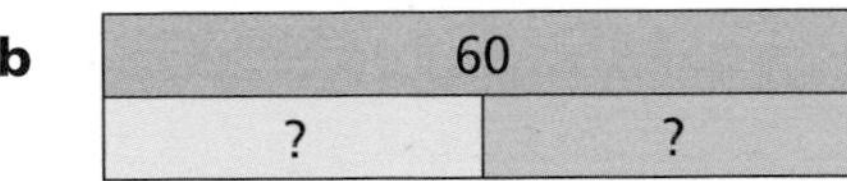

Models and representations

Cubes and place value counters

can be used to represent "$x + 5$".

If you want to work out the value of $x + 5$ when $x = 4$, you can replace the cube with 4 ones.

= 1 1 1 1

So the cube 1 1 1 1 1 =

1 1 1 1 1 1 1 1 1

This shows $x + 5 = 9$

Bar models

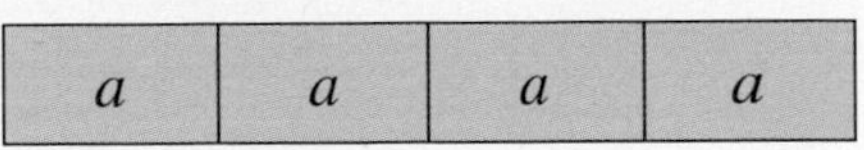

This represents $4a$.

The value of $4a$ when $a = 20$ can be shown as

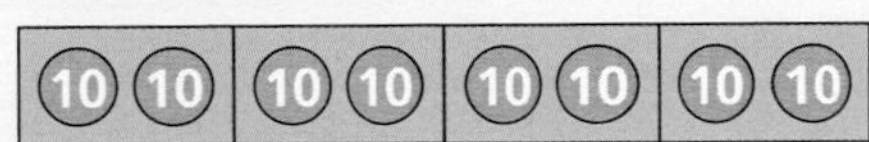

so $4a = 80$

Example 1

Work out the value of each of these **expressions** when $n = 40$.

a $3n$ **b** $n + 90$ **c** $\frac{n}{2}$ **d** $n - 27$

a $3 \times 40 = 120$

$3n$ means "multiply the value of n by 3". You can think of this in several ways.

n	n	n
40	40	40

$40 + 40 + 40 = 3 \times 40 = 120$

n	n	n

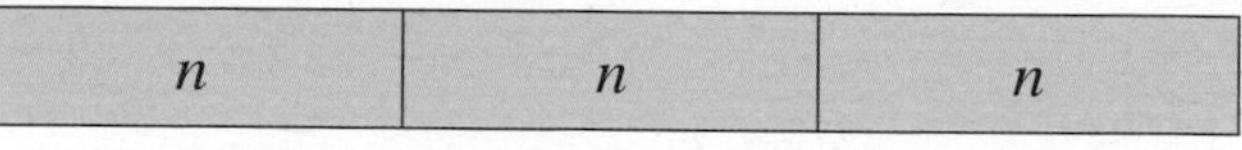

$40 + 40 + 40 = 3 \times 40 = 120$

b $40 + 90 = 130$

$n + 90$ means add 90 to the value of n.

n	90
40	90

$40 + 90 = 130$

10 10 10 10 : 10 10 10 10 10 10 10 10 10

$90 + 40 = 130$

c $40 \div 2 = 20$

$\frac{n}{2}$ means "divide the value of n by 2".

n	
$\frac{n}{2}$	$\frac{n}{2}$

40	
20	20

$40 \div 2 = 20$

d $40 - 27 = 13$

$n - 27$ means subtract 27 from the value of n.

Start with n

n

Subtract 27

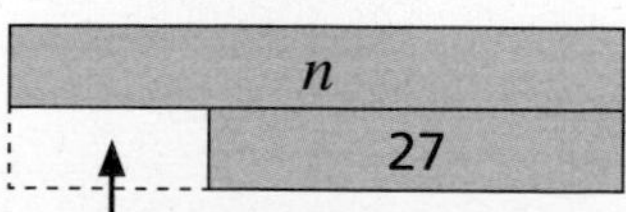

This section represents $n - 27$

When $n = 40$

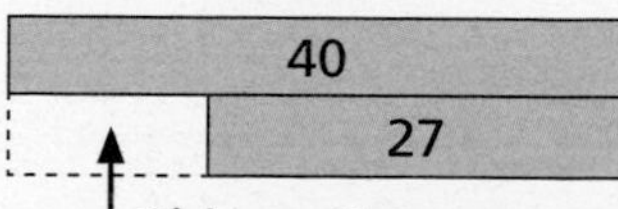

This section represents $40 - 27 = 13$

Do you think it matters whether the bar models are to scale?

Example 2

Substitute $d = 11.4$ into each of these expressions.

a $d - 4.17$ **b** $6d$ **c** $\frac{d}{3}$ **d** $58 - d$

a $d - 4.17 = 11.4 - 4.17$
$= 7.23$

Substitute means replace the letter with the given number.

You could use bar models or other representations to help you.

Replace d with 11.4 and then subtract 4.17

b $6d = 6 \times 11.4$
$= 68.4$

Remember that $6d$ means "multiply d by 6".

Replace d with 11.4 and then multiply by 6

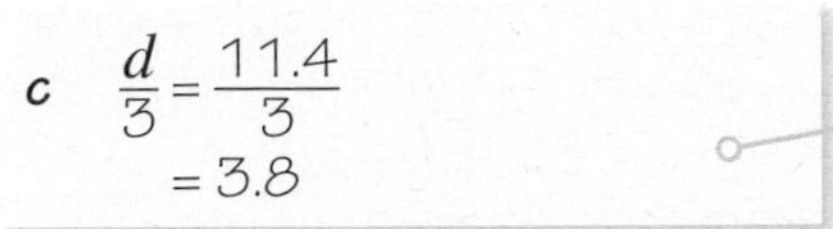

c $\frac{d}{3} = \frac{11.4}{3}$
$= 3.8$

Remember that $\frac{d}{3}$ means "divide d by 3".

Replace d with 11.4 and divide by 3

d $58 - d = 58 - 11.4$
$= 46.6$

Replace d with 11.4 and subtract from 58

Practice 2.3A

1 What expression is shown by each question mark?

a

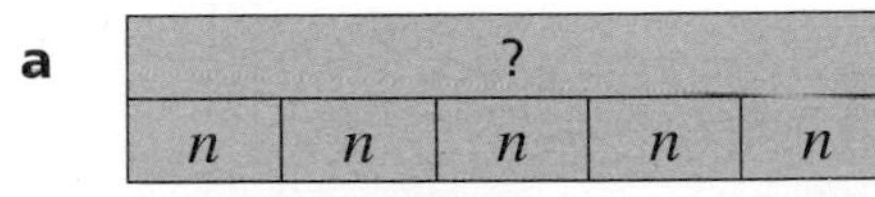

b

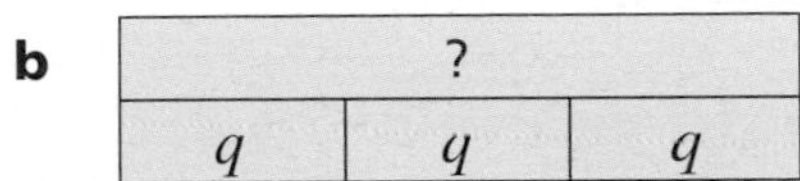

c

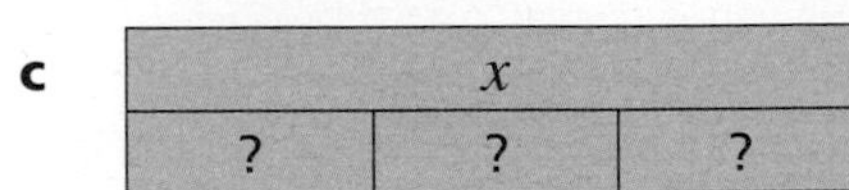

d

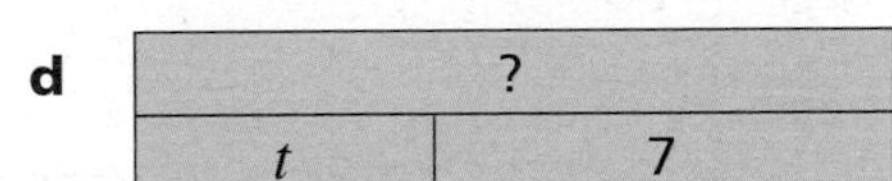

e

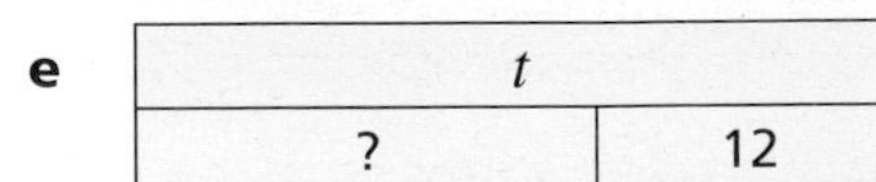

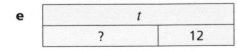

f

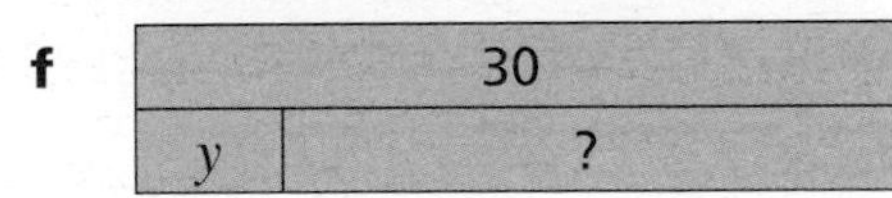

2 Draw bar models to represent these expressions.

a $b + 4$ **b** $b - 4$ **c** $\frac{b}{4}$ **d** $4b$

3 Substitute $b = 12.5$ into each of these expressions.

a $b + 4$ **b** $b - 4$ **c** $\frac{b}{4}$ **d** $4b$

4 Evaluate these expressions when $g = 400$

a $\frac{g}{10}$ **b** $\frac{10}{g}$ **c** $\frac{g}{100}$ **d** $\frac{100}{g}$ **e** $\frac{g}{20}$ **f** $\frac{20}{g}$

g $\frac{g}{200}$ **h** $\frac{200}{g}$ **i** $\frac{g}{400}$ **j** $\frac{400}{g}$ **k** $\frac{g}{1000}$ **l** $\frac{1000}{g}$

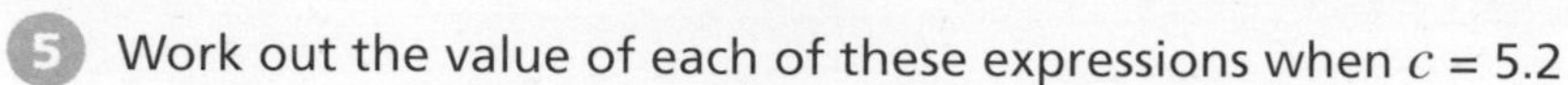

5 Work out the value of each of these expressions when $c = 5.2$

a $c + 5.2$ **b** $\frac{c}{5.2}$ **c** $c - 5.2$ **d** $5.2c$

e $5.2 - c$ **f** $\frac{5.2}{c}$ **g** $5.2 + c$

6 **a** Match the expressions that are equal when $x = 10$

$10x$ | $\frac{x}{10}$ | $x + 10$ | $10 - x$ | $\frac{10}{x}$ | $10 + x$ | $x - 10$

b Will any of the pairs of expressions in part **a** be equal for all values of x?

7 **a** What does x^2 mean?

b Work out x^2 for each of these values of x.

i $x = 8$ **ii** $x = 80$ **iii** $x = 0.8$ **iv** $x = 800$

8 **a** What does $\sqrt{t}$ mean?

b Work out $\sqrt{t}$ for each of these values of t.

i $t = 324$ **ii** $t = 3.24$ **iii** $t = 0.004$ **iv** $t = 400$

9 Are the expressions on each pair of cards always equal, sometimes equal or never equal in value?

a $p + 7$ | $7 + p$ **b** $5p$ | $\frac{p}{5}$ **c** $p - 2$ | $2 - p$

What do you think?

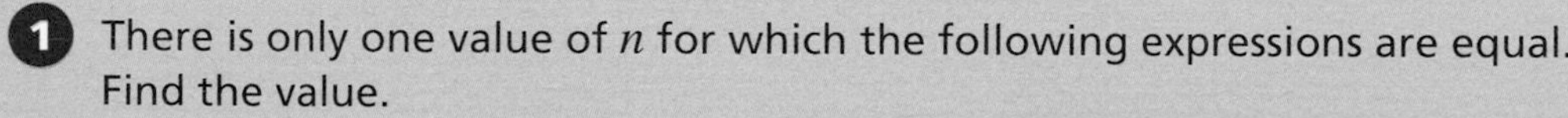

1 There is only one value of n for which the following expressions are equal. Find the value.

$4n$ | $\frac{n}{4}$

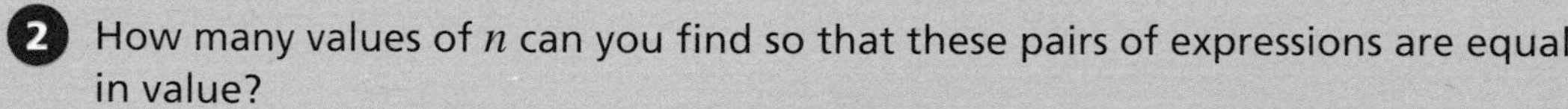

2 How many values of n can you find so that these pairs of expressions are equal in value?

a n | n^2 **b** $2n$ | n^2 **c** $n + 3$ | $2n$ **d** $n + 3$ | $3n$

3

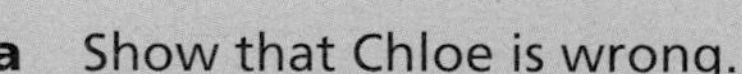

a Show that Chloe is wrong.

b Find a value for p for which $\frac{p}{5}$ and $\frac{5}{p}$ are equal.

If you substitute the whole numbers in order into the same expression, you get a sequence.

Remember that a sequence is a list of items in a given order, usually following a rule.

Example 3

a Substitute $n = 1$, $n = 2$, $n = 3$, $n = 4$ and $n = 5$ into the expression $4n$ and describe the sequence formed by your answers.

b What is the tenth term of the sequence?

a When $n = 1, 4n = 4 \times 1 = 4$

When $n = 2, 4n = 4 \times 2 = 8$

When $n = 3, 4n = 4 \times 3 = 12$

When $n = 4, 4n = 4 \times 4 = 16$

When $n = 5, 4n = 4 \times 4 = 20$

The sequence formed is 4, 8, 12, 16, 20 – the first five multiples of 4.

If you carried on substituting whole number values of n, you would get more multiples of 4

The difference between the terms in the sequence is 4, and the sequence is linear and ascending.

b $4 \times 10 = 40$

For the 10th term in the sequence, n has the value 10

Can you remember what "difference" means? Can you remember what a linear sequence is?

Example 4

a Substitute $n = 1$, $n = 2$, $n = 3$, $n = 4$ and $n = 5$ into the expression $4 - n$ and describe the sequence formed by your answers.

b What is the tenth term of the sequence?

a When $n = 1, 4 - n = 4 - 1 = 3$

When $n = 2, 4 - n = 4 - 2 = 2$

When $n = 3, 4 - n = 4 - 3 = 1$

When $n = 4, 4 - n = 4 - 4 = 0$

When $n = 5, 4 - n = 4 - 5 = -1$

It is OK to get a negative answer. We will look at negative numbers in Block 9.

The sequence formed is 3, 2, 1, 0, –1

The difference between the terms in the sequence is 1 and the sequence is linear and descending.

b $4 - 10 = -6$

For the 10th term in the sequence, n has the value 10

n is often used in expressions to represent sequences.

Practice 2.3B

1 **a** Substitute $n = 1$, $n = 2$, $n = 3$, $n = 4$ and $n = 5$ into each of these expressions.

i $n + 4$ **ii** $n + 0.4$ **iii** $40 + n$ **iv** $40 - n$

b Describe the sequences formed by your answers.

What's the same and what's different?

c Find the 10th term and the 100th term of each sequence in part **a**.

2 **a** Substitute $n = 1$, $n = 2$, $n = 3$, $n = 4$ and $n = 5$ into each of these expressions.

i $6n$ **ii** $n + 6$ **iii** $\frac{6}{n}$ **iv** $\frac{n}{6}$

b Describe the sequences formed by your answers.

What's the same and what's different?

c Find the 10th term and the 100th term of each sequence in part **a**.

3 **a** Substitute $n = 1$, $n = 2$, $n = 3$, $n = 4$ and $n = 5$ into the expression n^2.

b Describe the sequence formed by your answers.

c Find the 10th term and the 100th term of the sequence in part **a**.

d Now substitute $n = 1$, $n = 2$, $n = 3$, $n = 4$ and $n = 5$ into the expression n^3.

e Describe the sequence formed by your answers.

f Find the 10th term and the 100th term of the sequence in part **d**.

g What's the same and what's different about the sequences formed by substituting into n^2 and n^3?

4 **a** Find the 10th term of the sequences given by these expressions.

i $n + 10$ **ii** $n - 10$ **iii** $10n$ **iv** $\frac{n}{10}$ **v** $\frac{10}{n}$ **vi** $10 + n$

b Predict whether the 100th term will be greater than or smaller than the 10th term for each sequence.

c Now work out the 100th term of each sequence in part **a** to see if you were right.

5 **a** Substitute $n = 1$, $n = 2$, up to $n = 10$ into the expression 2^n.

2^n means 2 multiplied by itself n times, for example $2^4 = 2 \times 2 \times 2 \times 2$

b What do you notice about the last digits of the numbers in this sequence?

c Can you predict the last digit of the 20th term of the sequence?

d Can you predict the last digit of the 20th term of the sequence given by 10^n?

What do you think?

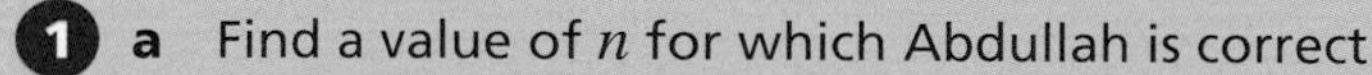

1 **a** Find a value of n for which Abdullah is correct.

b Find a value of n for which Abdullah is incorrect.

c Which sequence is greater more often?

The terms of the sequence 3^n are greater than those of the sequence n^3

2 Compare the sequences given by 2^n and n^2.

Consolidate – do you need more?

1 Draw a bar model to represent each of these expressions.

a $x + 5$ **b** $x - 3$ **c** $5x$ **d** $50 - x$

2 Substitute $k = 49$ into each of these expressions.

a $k + 81$ **b** $81 - k$ **c** $7k$ **d** $\frac{k}{7}$ **e** k^2 **f** $\sqrt{k}$

3 Find the value of $\frac{p}{40}$ when

a $p = 4$ **b** $p = 0$ **c** $p = 400$ **d** $p = 0.4$ **e** $p = 40$

4 **a** Find the first four terms of the sequences given by each of these expressions.

i $n + 8$ **ii** $\frac{n}{8}$ **iii** $8n$ **iv** $8 - n$

b Which of the sequences are linear?

5 Find the difference between the 15th terms of each pair of sequences.

a $60 - n$, $60n$ **b** $n + 40$, $40 + n$ **c** $2n$, n^2

d $\frac{30}{n}$, $\frac{n}{30}$ **e** $20 + n$, $20 - n$

Stretch – can you deepen your learning?

1 Will any of the terms of the sequence given by the rule $2n$ end in the number 5? Explain how you know.

2 Will any of the terms of the sequence given by the rule n^2 end in the number 5? Explain how you know.

3

The only even numbers in the sequence given by the rule n^2 end in 4 or 6

Is Seb right? What other patterns can you see in the last digits of the numbers in the sequence?

Reflect

1 Explain the meaning of the words

substitute | evaluate | expression

2 If you are given an algebraic rule, explain how you can find the value of any term in the sequence.

2.4 Inputs and outputs with two-step function machines

Small steps

- Find numerical inputs and outputs for a series of two function machines
- Use diagrams and letters with a series of two function machines
- Use diagrams and letters to generalise number operations

Key word

Two-step – when a calculation involves two processes rather than one

Are you ready?

1 Find the output of the function machine if the input is

a 5 **b** 100 **c** ▭ **d** a

Input → [+ 6] → Output

2 Find the input of the function machine if the output is

a 6 **b** ▭ **c** 2 **d** b

Input → [− 2] → Output

3 **a** What is the inverse of "× 4"?

b What is the inverse of "÷ 3"?

4 Here is a number puzzle.

"Think of a number. Double it. Add on 11".

a What is the answer to the puzzle if the number you think of is 10?

b What is the answer to the puzzle if the number you think of is 100?

c What is the answer to the puzzle if the number you think of is 0?

5 Write these expressions without mathematical operations.

a $a + a + a + a$ **b** $b \div 2$ **c** $c \times 4$ **d** $20 \div d$ **e** $e \times e$

6 Write an expression for

a two more than x **b** two more than $2x$ **c** three less than $5x$

Models and representations

Function machines	Two-step function machines
Input → [× 5] → Output	Input → [× 2] → [+ 3] → Output
Objects	**Letters**
[algebra tiles] (1) (1) (1)	If $t = 5$, $4t$ represents $4 \times 5 = 20$

Example 1

Input → ×2 → +3 → Output

Find the output of the function machine if the input is

a 7 **b** [object] **c** m **d** [object] 1 **e** $m + 1$

a 7 → ×2 → 14
14 → +3 → 17

Work out the answer by going from left to right applying each function machine in turn.

The output of the first machine is 14. This becomes the input of the second machine.

You could show this on one line:

7 → ×2 → 14 → +3 → 17

b

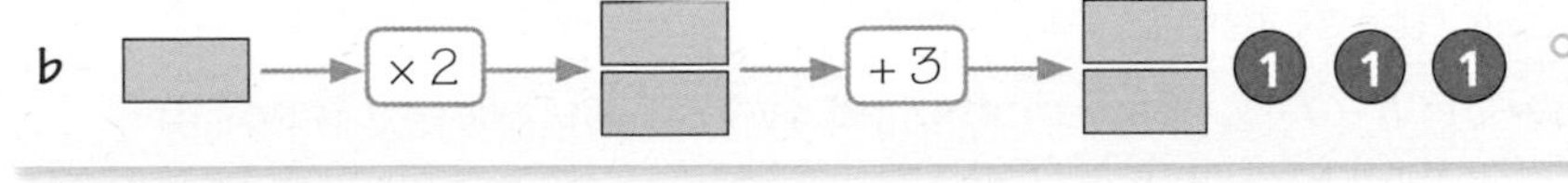

It works the same way with objects.

c m → ×2 → $2m$ → +3 → $2m + 3$

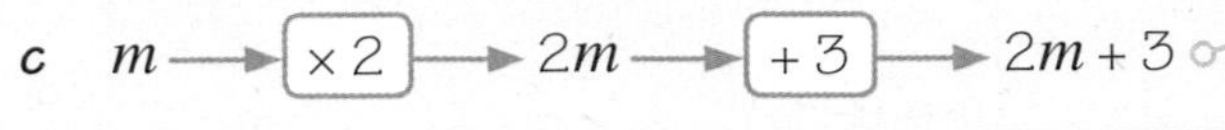

You need to remember the correct notation when using letters.

Remember that you write $m \times 2$ as $2m$ not m^2.

d

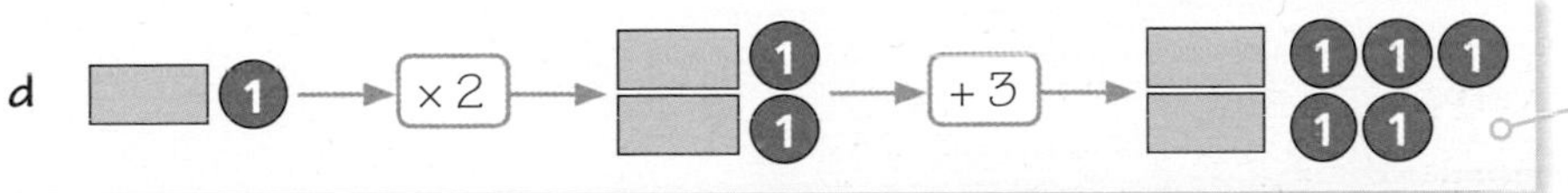

Remember you can also use objects and ones.

e $m + 1$ → ×2 → $2m + 2$ → +3 → $2m + 5$

Notice how similar working with $m + 1$ is to the result in part **d**.

Example 2

Find the ouput for each of these pairs of function machines if the input is 24.

a Input → ×2 → +2 → Output

b Input → +2 → ×2 → Output

c Input → ÷2 → +2 → Output

d Input → +2 → ÷2 → Output

e Input → ×2 → $+a$ → Output

f Input → $+a$ → ×2 → Output

g Input → $+a$ → ÷2 → Output

a 24 → ×2 → 48 → +2 → 50

Work through each machine in turn.

b 24 → +2 → 26 → ×2 → 52

c 24 → ÷2 → 12 → +2 → 14

d 24 → +2 → 26 → ÷2 → 13

e 24 → [× 2] → 48 → [+ a] → $48 + a$

f 24 → [+ a] → $24 + a$ → [× 2] → $2(24 + a)$ or $48 + 2a$

There are two possible answers. For the second machine, you could multiply each term in the expression $24 + a$ by 2 to get $48 + 2a$, or you could write $2(24 + a)$ which means "2 times the value of $24 + a$" in the same way that $2m$ means "2 times m".

When multiplying an expression by a number, you put the expression in brackets and put the number first, for example

"$3 + a$" multiplied by 4 is written as $4(3 + a)$

"$b - 7$" multiplied by 3 is written as $3(b - 7)$.

This shows that the whole of "$b - 7$" is being multiplied by 3 so you know it is not the same as $3b - 7$ (3 times b then take away 7).

You will learn much more about brackets in Book 2.

g 24 → [+ a] → $24 + a$ → [÷ 2] → $\frac{24 + a}{2}$

Notice that the answer to part **g** is $\frac{24 + a}{2}$ and not $24 + \frac{a}{2}$.
This shows that you are dividing the answer of the full expression by 2.

Practice 2.4A

1 Find the output when you input 50 into each of these pairs of function machines.

a Input → [+ 10] → [÷ 2] → Output

b Input → [÷ 2] → [+ 10] → Output

c Input → [+ 2] → [÷ 10] → Output

d Input → [÷ 10] → [+ 2] → Output

2 **a** Find the outputs when you input 0, 1, 2, 3, 4 and 5 into this pair of function machines.

Input → [+ 5] → [× 3] → Output

b What type of sequence do your six answers form?

3 I think of a number. I multiply my number by 4 and then subtract 2.

a Draw a pair of function machines to show the number puzzle.

b Find the output of the pair of machines if the input is

i 12 **ii** 12.8 **iii** 120 **iv** m

4 I think of a number. I subtract 2 from my number and then multiply the result by 4.

a Draw a pair of function machines to show the number puzzle.

b Find the output of the pair of machines if the input is

i 12 **ii** 12.8 **iii** 120 **iv** m

Hint – you could use brackets in your answer to part **iv**

5 I think of a number. I divide my number by 4 and then subtract 2.

a Draw a pair of function machines to show the number puzzle.

b Find the output of the pair of machines if the input is

i 12 **ii** 12.8 **iii** 120 **iv** m

6 I think of a number. I subtract 2 from my number and then divide by 4.

a Draw a pair of function machines to show the number puzzle.

b Find the output of the pair of machines if the input is

i 12 **ii** 12.8 **iii** 120 **iv** m

What do you think?

1 Darius thinks that if you input the same number into these pairs of function machines you will always get the same output.

Input → × 3 → + 2 → Output   Input → × 2 → + 3 → Output

a Find some inputs to show that Darius is wrong.

b Find one input that does give the same answer for both pairs of machines.

2 Faith thinks that if you input the same number into these pairs of function machines you will never get the same output.

Input → × 5 → + 1 → Output   Input → + 1 → × 5 → Output

a Choose some numbers to input into the pairs of function machines to demonstrate that Faith is correct.

b Input [cube] into the pairs of function machines to prove that Faith is correct.

3 Which of these pairs of function machines could be replaced by a single machine? State the replacement single machine in each case.

a Input → × 3 → × 2 → Output   **b** Input → + 3 → + 2 → Output

c Input → × 3 → + 2 → Output   **d** Input → + 3 → × 2 → Output

e Input → × 3 → ÷ 2 → Output   **f** Input → ÷ 3 → ÷ 2 → Output

2.4 Inputs and outputs with two-step function machines

Remember you can find the input of a one-step function machine when given the output by using inverse functions. You can do the same with two (or more) function machines.

Example 3

I think of a number, double it and then add 7. My answer is 23. What number was I thinking of?

$23 - 7 = 16$ — The last step before the answer 23 is "add 7". The inverse of "add 7" is "subtract 7".

$16 \div 2 = 8$ — Before that, I had doubled my number. The inverse of "double" (or × 2) is "divide by 2".

You can check by carrying out the steps in the puzzle:

8 doubled is 16; 16 add 7 is 23, so it is correct.

This strategy is often called "working backwards" as you reverse the order of the steps in the original calculation in order to find the starting number.

"Working forwards" 8 → × 2 → 16 → + 7 → 23

"Working backwards" 23 → − 7 → 16 → ÷ 2 → 8

Example 4

The output of each of these pairs of function machines is 10. Find the input in each case.

a Input → [+ 4] → [÷ 2] → Output

b Input → [× 2] → [+ 3] → Output

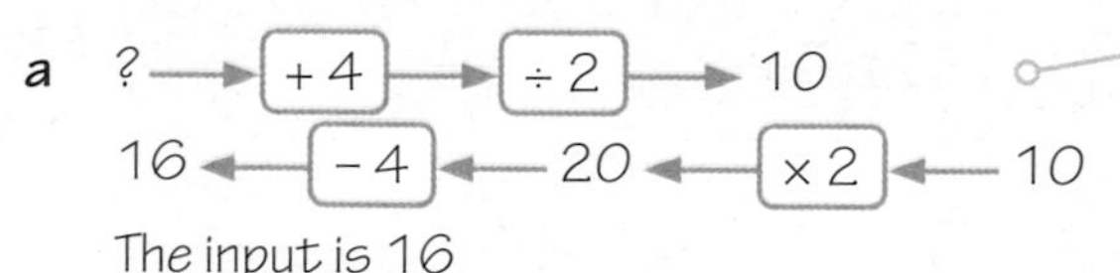

The inverse of "+ 4" is "− 4" and the inverse of "÷ 2" is "× 2".

Working backwards gives you the inverse function machines.

b ? → [× 2] → [+ 3] → 10

3.5 ← [÷ 2] ← 7 ← [− 3] ← 10

The input is 3.5

The inverse of "× 2" is "÷ 2" and the inverse of "+ 3" is "− 3".

Working backwards gives you the inverse function machines.

You can make your answer the input to the original function machine to check that it gives the correct output.

$16 + 4 = 20$, $20 \div 2 = 10$. Correct.

$3.5 \times 2 = 7$, $7 + 3 = 10$. Correct.

Practice 2.4B

1. I think of a number, add on 5 and double the result.
   - **a** Show this using a series of function machines.
   - **b** If my answer is 30, what number did I start with?
   - **c** If my answer is 110, what number did I start with?
   - **d** If my answer is 11, what number did I start with?

2. Input → [× 2] → [+ 6] → Output

   Find the input to this pair of function machines if the output is

   a 30 **b** 60 **c** 144 **d** $4x + 6$ **e** $2x + 6$ **f** 6

3. The output of all these pairs of function machines is 40. Find the inputs.
   - **a** Input → [+ 10] → [÷ 2] → Output
   - **b** Input → [÷ 2] → [+ 10] → Output
   - **c** Input → [+ 2] → [÷ 10] → Output
   - **d** Input → [÷ 10] → [+ 2] → Output

4. The output of each pair of function machines is [ten-rod] ① ① ① ① ① ①

   Find the inputs.
   - **a** Input → [× 3] → [+ 6] → Output
   - **b** Input → [÷ 10] → [+ 2] → Output

5. The output of this pair of function machines is [ten-rod] ① ① ① ① ① ①

   Find the input.

   Input → [− 1] → [× 3] → Output

6. Find the missing values or expressions.
   - **a** □ → [× 2] → □ → [+ □] → $2a + 5$
   - **b** □ → [× 2] → □ → [+ 7] → $4a + 7$
   - **c** □ → [× a] → □ → [+ 4] → $a^2 + 4$

7. The output of all these pairs of function machines is $12a + 16$. Find the inputs.
   - **a** Input → [× 2] → [+ 8] → $12a + 16$
   - **b** Input → [+ 8] → [× 2] → $12a + 16$
   - **c** Input → [× 4] → [+ 12] → $12a + 16$
   - **d** Input → [× 12] → [− 8] → $12a + 16$

What do you think?

1 Investigate these pairs of function machines.

Input → ÷ 2 → × 2 → Output   Input → + 7 → − 7 → Output

a What do you notice about the inputs and the outputs? Will this always be the case?

b Design a sequence of three function machines that will give the same result.

Consolidate – do you need more?

1 I think of a number, subtract 10 and triple the result.

a Show this using a series of function machines.

b If the input is 30, what is the output?

c If the output is 30, what is the input?

2 I think of a number, triple it and subtract 10.

a Show this using a series of function machines.

b If the input is 20, what is the output?

c If the output is 20, what is the input?

3 Find the output when you input a into each of these pairs of function machines.

a Input → + 5 → × 2 → Output

b Input → × 2 → + 5 → Output

c Input → + 2 → × 5 → Output

d Input → × 5 → + 2 → Output

e Input → + 5 → ÷ 2 → Output

f Input → ÷ 2 → + 5 → Output

g Input → + 2 → ÷ 5 → Output

h Input → ÷ 5 → + 2 → Output

4 Find the output when you input b into each of these pairs of function machines.

a Input → × 2 → × 3 → Output

b Input → × 3 → × 2 → Output

c Input → + 2 → + 3 → Output

d Input → + 3 → + 2 → Output

What do you notice?

5 **a** Find the outputs when you input 24, 60 and 480 into this pair of function machines.

Input → ÷ 3 → ÷ 2 → Output

b What is the output if you input x into the pair of machines?

Stretch – can you deepen your learning?

1 Investigate these pairs of function machines

Input → square → × 2 → Output

Input → × 2 → square → Output

a Is there an input for which both pairs of machines give the same output?

b Is there a connection between the outputs that the pairs of machines give for a given input?

2 Repeat question **1** using these pairs of function machines.

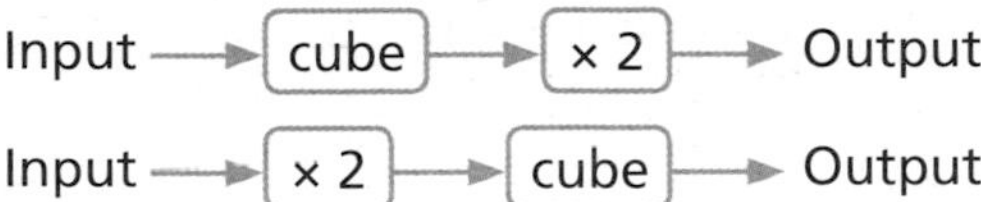

Reflect

1 Why does order matter when inputting into a series of function machines?

2 When we know the output and we are finding the input for a series of function machines, why do we "work backwards"?

2.5 Using multi-step expressions

Small steps

- Substitute values into two-step expressions
- Represent one- and two-step functions graphically
- Generate sequences given an algebraic rule

Key word

Expression – a collection of terms involving mathematical operations

Are you ready?

1 Find the values of the expressions when $x = 3$

a $3x$ **b** $3 + x$ **c** $\frac{x}{3}$ **d** $x - 3$

e $3 - x$ **f** $\frac{3}{x}$ **g** x^2

2 Find the 20th term of the sequences given by each of these rules.

a $2n$ **b** $n + 4$ **c** $\frac{200}{n}$ **d** $\frac{n}{20}$ **e** $200 - n$

3 Work out the output of this function machine if the input is 5

Input → × 10 → Output

4 What expressions are represented by the question marks? Choose your answers from the list.

$2x - 5$ $5 - 2x$ $2x + 5$

a

?		
x	x	5

b

x	x
?	5

c

5		
?	x	x

Models and representations

These **cubes** and **place value counters** can be used to represent "$2x + 7$", where each cube represents x.

When $x = 4$

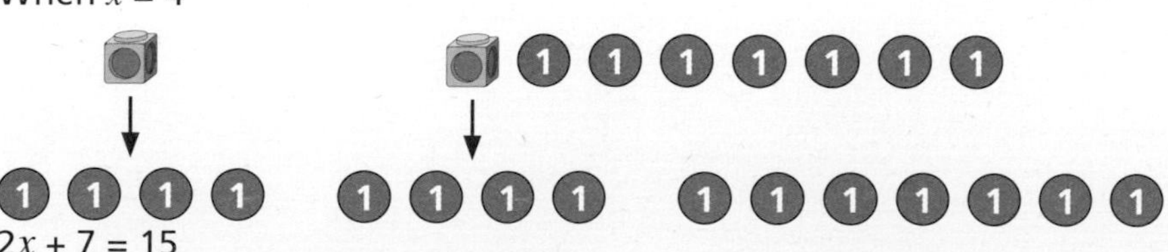

$2x + 7 = 15$

This **bar model** represents $3a + 2$

When $a = 11$,

$3a + 2 = 35$

Example 1

Work out the value of these expressions when $a = 80$

a $2a + 9$ **b** $3a - 9$ **c** $\frac{a}{2} + 17$ **d** $500 - 3a$

a $2a + 9$
$= 2 \times 80 + 9$
$= 160 + 9$
$= 169$

Replace a with 80

Remember that $2a$ stands for "a multiplied by 2".

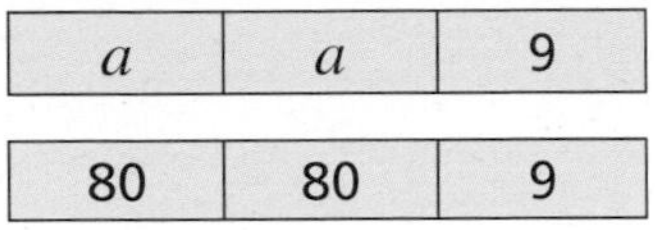

a	a	9

80	80	9

b $3a - 9$
$= 3 \times 80 - 9$
$= 240 - 9$
$= 231$

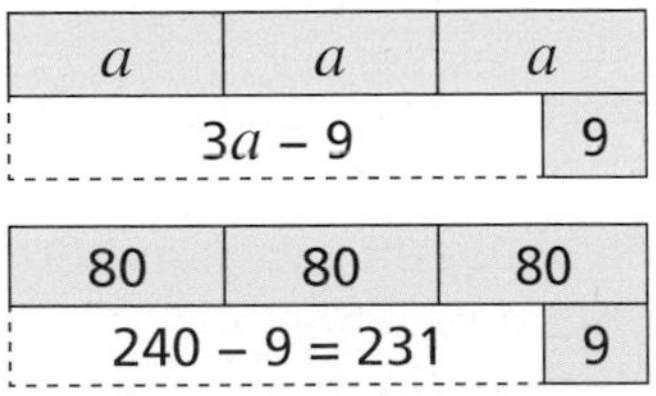

a	a	a
$3a - 9$		9

80	80	80
240 – 9 = 231		9

c $\frac{a}{2} + 17$
$= \frac{80}{2} + 17$
$= 40 + 17$
$= 57$

Remember that $\frac{a}{2}$ means "a divided by 2" so $\frac{80}{2}$ means "80 divided by 2".

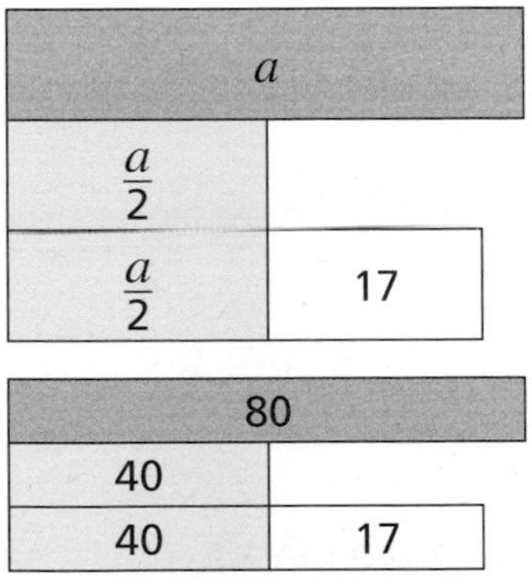

a	
$\frac{a}{2}$	
$\frac{a}{2}$	17

80	
40	
40	17

d $500 - 3a$
$= 500 - 3 \times 80$
$= 500 - 240$
$= 260$

Start with 500

500

Subtract $3a$

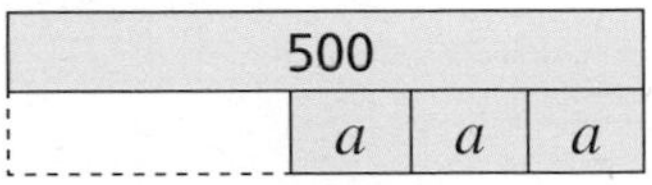

500			
	a	a	a

When $a = 80$

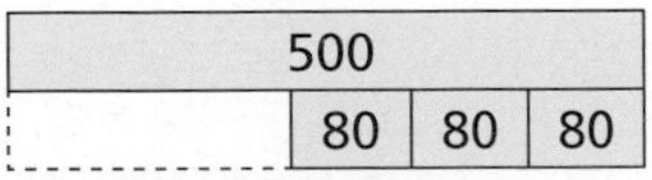

500			
	80	80	80

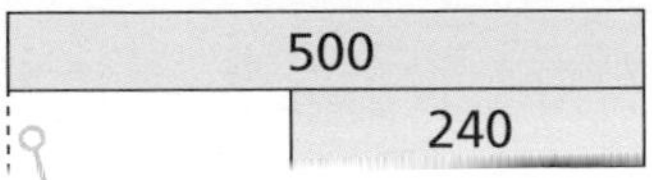

500	
	240

The blank section represents 500 – 240 = 260

Example 2

Find the difference between $a^2 + 4$ and $2a - 4$ when $a = 3.5$

$a^2 + 4$
$= 3.5 \times 3.5 + 4$
$= 12.25 + 4$
$= 16.25$

First, work out the values of the two expressions when $a = 3.5$

Remember that a^2 means $a \times a$.

$2a - 4$
$= 2 \times 3.5 - 4$
$= 7 - 4$
$= 3$

$16.25 - 3 = 13.25$

The difference between two numbers is found by subtracting the smaller from the larger.

Number	
Number	Difference

Example 3

Evaluate $5p^3$ when $p = 3$

When $p = 3, p^3 = 3 \times 3 \times 3 = 27$
So $5p^3 = 5 \times 27 = 135$

$5p^3$ means $5 \times p^3$

Practice 2.5A

1. What expression is shown by each question mark?

a

a	a	a	a	5
?				

b

12	a	a
?		

c

a	a	a
?		2

d

a	a	a	a
?			6

e

10		
?	a	a

2. Draw bar models to represent each of these expressions.

a $3b + 4$ **b** $3b - 4$ **c** $\frac{b}{4} + 7$ **d** $10 + 4b$ **e** $10 - 4b$

3. Substitute $b = 8.17$ into each of these expressions.

a $3b + 4$ **b** $3b - 4$ **c** $\frac{b}{4} + 7$ **d** $10 + 4b$ **e** $10 - 4b$

4 Evaluate these expressions when $g = 80$

a $\frac{g}{10} + 5$ **b** $\frac{g}{10} - 5$ **c** $10g + 5$ **d** $5 + \frac{g}{10}$ **e** $10g - 5$

f $10 - \frac{g}{20}$ **g** $10 - \frac{20}{g}$ **h** $10 + \frac{40}{g}$ **i** $\frac{10}{g} + 40$ **j** $g^2 - 1400$

5 Work out the value of each of these expressions when $c = 5.2$

a $c + 5.2$ **b** $2c + 5.2$ **c** $c - 5.2$ **d** $2c - 5.2$ **e** $5.2 + 3c$

f $20 - c$ **g** $20 - 2c$ **h** $c^2 - 20$ **i** $20 + c^2$ **j** $20 + \frac{c}{2}$

6 Sven and Ali are working out the value of $3a^2$ when $a = 10$

Sven

$a^2 = 10 \times 10 = 100$
So $3a^2 = 3 \times 100 = 300$

Ali

$3a = 3 \times 10 = 30$
So $3a^2 = 30 \times 30 = 900$

a Who do you agree with?

b Still using $a = 10$, find the difference between the values of $2a^3$ and $5a^2$.

7 Find the first five terms of each of the sequences given by these rules.

a $4n + 1$ **b** $4n + 3$ **c** $4n - 3$ **d** $6n - 1$ **e** $6n + 2$

f $6n - 4$ **g** $\frac{n}{2} + 4$ **h** $\frac{n}{2} - 3$ **i** $5 - \frac{n}{2}$

8 **a** Use a graph plotter to draw graphs of the sequences in question **7**. How many of them are linear?

b What's the same and what's different about the sequences in question **7**?

What do you think?

1 Darius says that the sequences given by these rules are different, but Charlie says they are the same.

$3n + 2$ $2n + 3$

Who do you agree with? Explain your answer.

2 For each pair of rules, state whether the sequences are the same or different.

a $3n + 4$ $4 + 3n$

b $2n + 1$ $n^2 + 1$

c $n + 3$ $2n$

d $n + 3$ $3n$

3 Compare the graphs of the sequences given by the rules in question **1** and question **2**.

You can also work out expressions involving brackets or more than one letter.

Example 4

Given that $x = 5$ and $y = 8$, find the values of

a $x + y$ **b** xy **c** $\frac{x}{y}$ **d** $\frac{y}{x}$ **e** $9(y - x)$

a $x + y$
$= 5 + 8$
$= 13$

Replace both letters with the values given.

b xy
$= 5 \times 8$
$= 40$

Remember that xy means $x \times y$.

c $\frac{x}{y}$
$= \frac{5}{8}$
$= 0.625$

Divide x by y.

d $\frac{y}{x}$
$= \frac{8}{5}$
$= 1.6$

Divide y by x.

e $9(y - x)$
$= 9 \times (8 - 5)$
$= 9 \times 3$
$= 27$

Just like $9a$ means $9 \times a$, $9(y - x)$ means $9 \times (y - x)$
So you work out $y - x$ and multiply the result by 9

You will look at calculations with brackets in more detail later in the year when you study the order of operations.

Practice 2.5B

1 Substitute $a = 25$ and $b = 40$ to find the values of

a $a + b$ **b** $b - a$ **c** ab **d** $\frac{b}{a}$ **e** $\frac{a}{b}$
f $\frac{2a}{b}$ **g** $\frac{2b}{a}$

2 Given that $x = 6.5$ and $y = 11.3$, evaluate

a $3x - y$ **b** $3y - x$ **c** $3(y - x)$ **d** $3(x - y)$ **e** $x^2 - y$
f $y^2 - x$ **g** $xy - 3$ **h** $yx - 3$ **i** $x^2 + 2y$ **j** $x^2 - 2y$

3 Find the value of each of these expressions when $c = 100$

a $\frac{c}{5} + 1$ **b** $\frac{c + 1}{5}$ **c** $\frac{c}{4} + 1$ **d** $\frac{c + 1}{4}$ **e** $\frac{c}{10} + 1$ **f** $\frac{c + 1}{10}$
g $\frac{c + 10}{5}$ **h** $\frac{c}{5} + 10$ **i** $\frac{c^2 + 10}{5}$ **j** $\frac{c^2}{5} + 10$ **k** $\frac{c^2 - 10}{5}$ **l** $\frac{c^2}{5} - 10$
m $\sqrt{c}$ **n** $10\sqrt{c}$ **o** $\frac{\sqrt{c}}{5}$ **p** $\sqrt{c} + 5$ **q** $5 + \sqrt{c}$ **r** $1 + 5\sqrt{c}$

4 **a** Find the difference between $4a + 3$ and $4(a + 3)$ when $a = 2$

b Find the difference between $4a + 3$ and $4(a + 3)$ when $a = 10$

c Find the difference between $4a + 3$ and $4(a + 3)$ when $a = 100$

d Find the difference between $4a + 3$ and $4(a + 3)$ when $a = 0.2$

e What do you notice?

f Write down the difference between $4a + 3$ and $4(a + 3)$ when $a = 786.174$

When a question says "Write down" it means you should not need to do any calculations to work out the answer, you should be able to write it down from previous knowledge.

5 **a** Which of these expressions has the greatest value and which has the least value when $m = 4$ and $n = 0.2$?

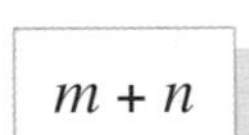

$\frac{m}{n}$ $\frac{n}{m}$ $10m - n$ $\frac{m^2}{n}$ $\frac{m}{n^2}$

b State whether the same expressions are greatest and least when

i $m = 0.4$ and $n = 0.2$ **ii** $m = 40$ and $n = 0.2$

iii $m = 4$ and $n = 0.02$

What do you think?

1 Which will be greater $3(p + 2)$ or $3p + 2$? Will this be true for all values of p? Draw bar models to help explain why.

2 $a = 60$ and $b = 10$

Kate thinks that $\frac{a}{2b} = 3$ but Mario thinks that $\frac{a}{2b} = 300$

Who do you agree with? What mistake do you think the other person made?

3 Work out $3k^2$ and $(3k)^2$ when

a $k = 10$ **b** $k = 0.1$ **c** $k = 100$

d Find a value of k for which $3k^2$ and $(3k)^2$ will be equal.

Consolidate – do you need more?

1 **a** Substitute $n = 1$ to $n = 5$ to find the first five terms of each of these sequences given by these rules.

i $3n + 7$ **ii** $7n + 3$ **iii** $3 + 7n$ **iv** $n^2 + 7$

v $n^2 - 1$ **vi** $\frac{n}{2} + 1$ **vii** $\frac{n}{2} - 1$ **viii** $10 - \frac{n}{2}$

ix $100 - n^2$

b Which of the sequences are not linear?

c What is the same about the rules of the sequences that are not linear?

2 Work out the value of each of these expressions when $a = 5$

a $10a + 3$ **b** $\frac{10}{a} + 3$ **c** $10a - 3$ **d** $\frac{a}{10} + 3$

e $10(a + 3)$ **f** $\frac{10}{a + 3}$ **g** $10(a - 3)$ **h** $3 + \frac{a}{10}$

i $\frac{10 - a}{3}$ **j** $10 - \frac{3}{a}$ **k** $3 + \frac{10}{a}$ **l** $3 - \frac{10}{a}$

3 Evaluate these expressions when $p = 12$ and $q = 0.3$

a $p + 2q$ **b** $q + 2p$ **c** $2(p + q)$ **d** $2(p - q)$

e $p(q + 2)$ **f** $q(p - 2)$ **g** $\frac{p}{q} + 2$ **h** $\frac{p}{q} - 2$

i $\frac{p}{q + 2}$ **j** $\frac{p + 2}{q}$ **k** $\frac{p + q}{2}$ **l** $\frac{q}{p} + 2$

m $2 - \frac{q}{p}$ **n** $\frac{p - 2}{q}$ **o** $p - 2q$ **p** $pq + 2$

q $pq - 2$ **r** $2 + \frac{p}{q}$ **s** $\frac{p^2}{q}$ **t** $\frac{q^2}{p}$

4 If $x = 3$ and $y = 0$ find the value of

a $x + y$ **b** $x + 2y$ **c** $x - y$ **d** $x - 2y$

e xy **f** $\frac{y}{x}$ **g** $\frac{x}{2 + y}$ **h** $\frac{x}{2 - y}$

i $x^2 + y$ **j** $(x + y)^2$ **k** $x + y^2$ **l** $x(y + 2)$

5 If $m = 9$ and $n = 16$ find the values of

a $\sqrt{m} + \sqrt{n}$ **b** $\sqrt{m + n}$ **c** $m\sqrt{n}$ **d** $n\sqrt{m}$ **e** $\sqrt{mn}$

Stretch – can you deepen your learning?

1 a Work out the value of each of these expressions when $a = 9$ and $b = 3$

i ab **ii** a^b **iii** b^a **iv** $b^{\sqrt{a}}$ **v** $\sqrt{a^b}$

b Which of the expressions in part a would be equal in value if $a = 4$ and $b = 2$?

2 Work out the value of each of these expressions when $x = 16$ and $y = 100$

a $x(x + y)$ **b** $y(y - x)$ **c** $3x(x + y)$ **d** $x(3x + y)$

e $\sqrt{x}(\sqrt{x} + y)$ **f** $\sqrt{x}(y - \sqrt{x})$ **g** $\sqrt{x}(\sqrt{y} - \sqrt{x})$ **h** $\sqrt{xy}(x - y)$

i $\sqrt{4x}(\sqrt{x} + 4y)$ **j** $\frac{x + y}{\sqrt{x}}$ **k** $\frac{\sqrt{x} + y}{y}$ **l** $\frac{\sqrt{x} + y}{\sqrt{y}}$

3 Look back at question **2**. Make up other complex expressions involving x, y, $\sqrt{x}$ and $\sqrt{y}$ that have integer solutions. Challenge a partner to evaluate your expressions correctly.

Reflect

Explain the steps you would take to work out the value of each of these expressions.

$a + 2b$	$2a + b$	$2(a + b)$	$a + b^2$	$(a + b)^2$
$a + \sqrt{b}$	$\sqrt{a + b}$	$\sqrt{a + 2b}$	$\sqrt{a} + 2b$	$2\sqrt{a} + b$

2.6 Decoding expressions

Small steps

- Find the function machine given a simple expression
- Find the function machines given a two-step expression

Key words

Expression – a collection of terms involving mathematical operations

Are you ready?

1 Find the values of these expressions when $a = 5$

a $a - 5$ **b** $5 - a$ **c** $\frac{a}{5}$ **d** $\frac{5}{a}$

e a^2 **f** $a + 5$ **g** $5 + a$

2 Work out the output of the function machines if the input is

a 10 **b** n

i Input → × 5 → Output

ii Input → ÷ 5 → Output

iii Input → + 5 → Output

iv Input → – 5 → Output

3 Work out the output of the pairs of function machines if the input is

a 4 **b** m

i Input → × 3 → + 2 → Output

ii Input → ÷ 2 → + 3 → Output

iii Input → + 3 → × 2 → Output

iv Input → – 3 → ÷ 2 → Output

4 Explain why the expressions $3m + 2$ and $3(m + 2)$ are different.

Models and representations

Function machines

If you know the input and output, then you can work out the operation represented by the function machine.

Bar models

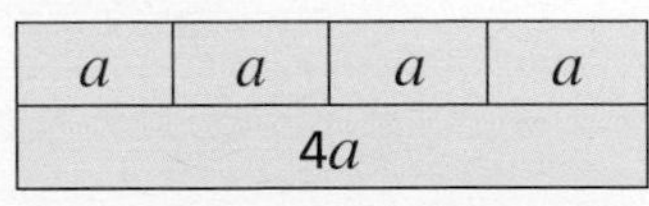

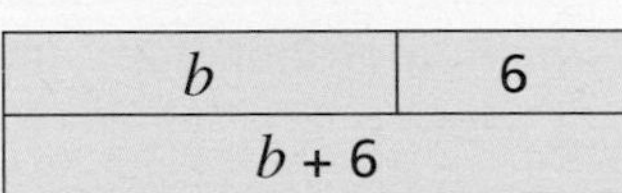

These are useful for showing relationships.

2.6 Decoding expressions

Example 1

What operation might this function machine be representing?

5 → [?] → 10

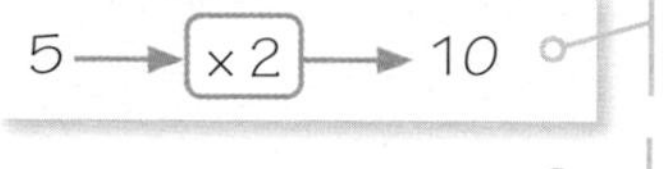

There is more than one possible answer.

As 10 is greater than 5 the function machine might have been doing an addition.

5	?
10	

5 + 5 = 10 so one possible solution would be that the function machine is "+ 5".

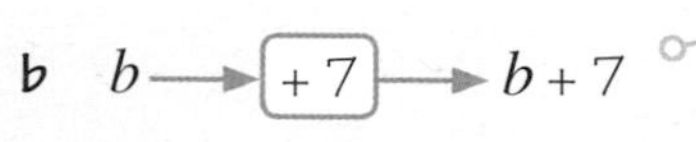

You could also interpret this as 5 × 2 (five times 2) or 2 × 5 (two lots of 5), so another possible function machine is "× 2".

There are other possible solutions involving negative numbers and fractions.

Example 2

What operations might these function machines be representing?

a a → [?] → $3a$

b b → [?] → $b + 7$

c c → [?] → $\frac{c}{4}$

d $d + 2$ → [?] → $d + 6$

a a → [× 3] → $3a$

$3a$ means $3 \times a$ (or $a \times 3$) so the function machine could be "× 3".

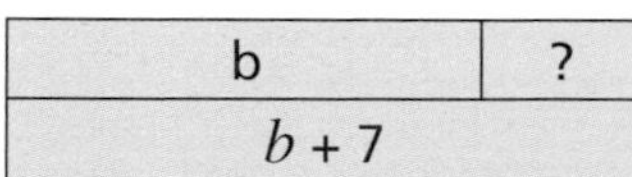

Another possibility would be "+ $2a$" (because $a + 2a$ altogether is $3a$). You will look at this type of operation in the next section "Equality and equivalence".

b b → [+ 7] → $b + 7$

To get from b to $b + 7$, the machine must have added 7. So the operation could be "+ 7".

b	?
$b + 7$	

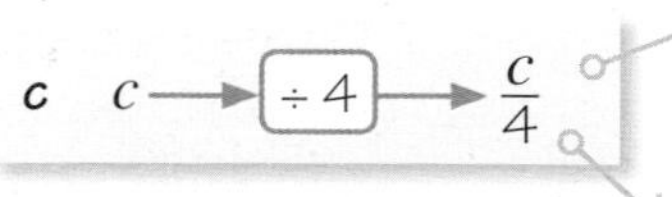

Remember that $\frac{c}{4}$ means $c \div 4$, so the function machine could be "÷ 4".

Another possibility would be "$-\frac{3}{4}c$".

d $d+2 \rightarrow \boxed{+4} \rightarrow d+6$

From $d + 2$ to $d + 6$ is an increase.

$d + 2$	?
$d + 6$	

You can split the numbers added to d into ones.

d	1	1	?			
d	1	1	1	1	1	1

You need four more ones to get from $d + 2$ to $d + 6$
So the function machine is "+ 4".

Practice 2.6A

1 What operation might each of these function machines be representing? Find one possibility for each.

a $7 \rightarrow \boxed{?} \rightarrow 10$
b $7 \rightarrow \boxed{?} \rightarrow 70$
c $7 \rightarrow \boxed{?} \rightarrow 1$
d $7 \rightarrow \boxed{?} \rightarrow 21$
e $7 \rightarrow \boxed{?} \rightarrow 0.7$
f $7 \rightarrow \boxed{?} \rightarrow 7000$
g $7 \rightarrow \boxed{?} \rightarrow 7$

2 What operation might each of these function machines be representing? Find one possibility for each.

a $t \rightarrow \boxed{?} \rightarrow 2t$
b $t \rightarrow \boxed{?} \rightarrow t - 4$
c $t \rightarrow \boxed{?} \rightarrow \frac{t}{5}$
d $t \rightarrow \boxed{?} \rightarrow t + 3$
e $t \rightarrow \boxed{?} \rightarrow 3 + t$
f $t \rightarrow \boxed{?} \rightarrow t^2$
g $t \rightarrow \boxed{?} \rightarrow \sqrt{t}$
h $t \rightarrow \boxed{?} \rightarrow at$

3 All the functions in this question are multiplications or divisions. Use your calculator and your knowledge of inverse operations to find each function machine.

a $4 \rightarrow \boxed{?} \rightarrow 0.16$
b $4 \rightarrow \boxed{?} \rightarrow 32\,000$
c $4 \rightarrow \boxed{?} \rightarrow 0.004$
d $4 \rightarrow \boxed{?} \rightarrow 2.8$
e $4 \rightarrow \boxed{?} \rightarrow 0.012$
f $4 \rightarrow \boxed{?} \rightarrow 4004$
g $4 \rightarrow \boxed{?} \rightarrow 40\,040$

4 What operation might these function machines be representing? Find one possibility for each.

a $2a \rightarrow \boxed{?} \rightarrow 20a$
b $a + 6 \rightarrow \boxed{?} \rightarrow a + 10$
c $3a \rightarrow \boxed{?} \rightarrow a$
d $10a \rightarrow \boxed{?} \rightarrow \frac{10a}{3}$
e $a - 4 \rightarrow \boxed{?} \rightarrow a$
f $a - 4 \rightarrow \boxed{?} \rightarrow a + 4$
g $\frac{a}{2} \rightarrow \boxed{?} \rightarrow a$

5 Find two possible operations for this function machine.

$1 \rightarrow \boxed{?} \rightarrow 0$

What do you think?

1 Lydia says that there are three possible operations for this function machine.

2 → [?] → 4

Can you find all three?

2 **a** How many function machines can you find that have input a and output a?

a → [?] → a

b How many function machines can you find that have input 0 and output 0?

0 → [?] → 0

c How many function machines can you find that have input 1 and output 1?

1 → [?] → 1

In the next example, you will find the operations for a pair of function machines.

Example 3

Find the missing operations for these function machines if

a they are both additions

b the first machine is a multiplication.

5 → [?] → [?] → 12

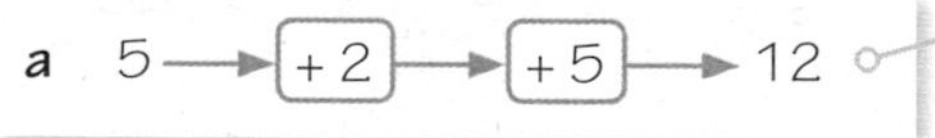

a 5 → [+ 2] → [+ 5] → 12

Using the fact 12 – 5 = 7, you know that the two machines must combine together to give "+ 7". There are an infinite number of possibilities.

5 → [+ 5] → [+ 2] → 12

5 → [+ 0] → [+ 7] → 12

5 → [+ 3.8] → [+ 3.2] → 12

b 5 → [× 1] → 5 → [+ 7] → 12

Again, there are an infinite number of possibilities!

Try the first machine as "× 1".

So the second machine has input 5 and output 12: this could be "+ 7".

5 → [× 2] → 10 → [+ 2] → 12

Try the first machine as "× 2".

So the second machine has input 10 and output 12: this could be "+ 2".

5 → [× 3] → 15 → [– 3] → 12

Try the first machine as "× 3".

So the second machine has input 15 and output 12: this could be "– 3".

And so on!

Example 4

Find the missing operations represented by these function machines.

$a \rightarrow$ [?] $\rightarrow$ [?] $\rightarrow 3a + 12$

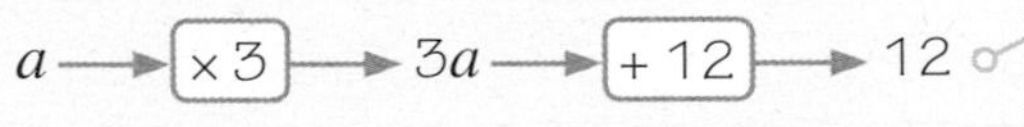

The final output includes $3a$, so try "× 3" as the first machine.

So the second machine has input $3a$ and output $3a + 12$. This could be "+ 12". There are many other possibilities.

Practice 2.6B

1 In this question, the first machine in each pair is always a multiplication. Find some possible operations.

a 3 → [× ?] → [?] → 10

b 3 → [× ?] → [?] → 20

c 3 → [× ?] → [?] → 3

d 3 → [× ?] → [?] → 2

e 3 → [× ?] → [?] → 0

f 3 → [× ?] → [?] → 100

g 3 → [× ?] → [?] → 3001

2 Now find possible operations if the second machine is a multiplication.

a 3 → [?] → [× ?] → 10

b 3 → [?] → [× ?] → 20

c 3 → [?] → [× ?] → 3

d 3 → [?] → [× ?] → 2

e 3 → [?] → [× ?] → 0

f 3 → [?] → [× ?] → 100

g 3 → [?] → [× ?] → 3001

3 Bobbie says that the machines in the diagram below should be "× 3" and "+ 9".

Ali says that the machines should be "+ 3" and "× 3".

$n \rightarrow$ [?] $\rightarrow$ [?] $\rightarrow 3n + 9$

Who do you agree with? Use symbols or diagrams to justify your answer.

4 Find some possible machines with the given inputs and outputs.

a $a \rightarrow$ [?] $\rightarrow$ [?] $\rightarrow 2a + 10$

b $a \rightarrow$ [?] $\rightarrow$ [?] $\rightarrow 10 + 2a$

c $a \rightarrow$ [?] $\rightarrow$ [?] $\rightarrow 2a + 8$

d $a \rightarrow$ [?] $\rightarrow$ [?] $\rightarrow \frac{a}{2} + 4$

e $a \rightarrow$ [?] $\rightarrow$ [?] $\rightarrow 6a$

f $a \rightarrow$ [?] $\rightarrow$ [?] $\rightarrow \frac{a + 1}{2}$

g $a \rightarrow$ [?] $\rightarrow$ [?] $\rightarrow a^2 - 7$

h $a \rightarrow$ [?] $\rightarrow$ [?] $\rightarrow ab + c$

5 Huda inputs 2 into this pair of function machines. Her output is 11

Lydia inputs 5 into the same pair of function machines. Her output is 20

Input → [× ?] → [?] → Output

What might the operations be?

6 Find some possible machines with these given inputs and outputs.

a $x \rightarrow$ [÷ ?] $\rightarrow$ [?] $\rightarrow x$

b $x \rightarrow$ [?] $\rightarrow$ [?] $\rightarrow 2x^2$

c $x \rightarrow$ [?] $\rightarrow$ [?] $\rightarrow 4x^2$

d $x \rightarrow$ [× ?] $\rightarrow$ [?] $\rightarrow 6x$

e $x \rightarrow$ [+ ?] $\rightarrow$ [?] $\rightarrow 3x + 3$

f $x \rightarrow$ [?] $\rightarrow$ [?] $\rightarrow x + \frac{3}{4}$

g $x \rightarrow$ [÷ ?] $\rightarrow$ [?] $\rightarrow \frac{x + 3}{4}$

What do you think?

1 One of the operations for each of these pairs of function machines is "square". Find possible operations for each input and output.

a $3 \rightarrow$ [?] $\rightarrow$ [?] $\rightarrow 36$

b $1 \rightarrow$ [?] $\rightarrow$ [?] $\rightarrow 25$

c $0 \rightarrow$ [?] $\rightarrow$ [?] $\rightarrow 9$

d $2 \rightarrow$ [?] $\rightarrow$ [?] $\rightarrow 100$

2 Find *two* possible sets of operations for each pair of function machines.

a $a \rightarrow$ [?] $\rightarrow$ [?] $\rightarrow 2a + 4$

b $b \rightarrow$ [?] $\rightarrow$ [?] $\rightarrow \frac{b}{2} + 6$

c $2c \rightarrow$ [?] $\rightarrow$ [?] $\rightarrow 4c + 12$

d $d \rightarrow$ [?] $\rightarrow$ [?] $\rightarrow 4d^2$

Consolidate – do you need more?

1 What operations might be represented by these function machines? Find one possibility for each.

a $6 \rightarrow$ [?] $\rightarrow 60$

b $6 \rightarrow$ [?] $\rightarrow 36$

c $6 \rightarrow$ [?] $\rightarrow 0$

d $6 \rightarrow$ [?] $\rightarrow 0.6$

e $6 \rightarrow$ [?] $\rightarrow 150$

f $6 \rightarrow$ [?] $\rightarrow 6a$

g $6 \rightarrow$ [?] $\rightarrow \frac{6}{a}$

2 What operations might be represented by these function machines? Find one possibility for each.

a $t \rightarrow$ [?] $\rightarrow tv$

b $t \rightarrow$ [?] $\rightarrow t - 4$

c $t \rightarrow$ [?] $\rightarrow 9t$

d $t \rightarrow$ [?] $\rightarrow 1$

e $t \rightarrow$ [?] $\rightarrow t^2$

3 What operations might be represented by these function machines? Find one possibility for each.

a $2g \rightarrow$ [?] $\rightarrow 10g$

b $\frac{g}{4} \rightarrow$ [?] $\rightarrow g$

c $g^2 \rightarrow$ [?] $\rightarrow g$

d $g + 7 \rightarrow$ [?] $\rightarrow g + 2$

e $g + 7 \rightarrow$ [?] $\rightarrow g - 2$

4 Find some possible function machines with each of these given inputs and outputs.

a $m \rightarrow$? $\rightarrow$? $\rightarrow m - 7$

b $m \rightarrow$? $\rightarrow$? $\rightarrow 2m - 1$

c $m \rightarrow$? $\rightarrow$? $\rightarrow 6 + m^2$

d $m \rightarrow$? $\rightarrow$? $\rightarrow \frac{6 + m}{2}$

e $m \rightarrow$? $\rightarrow$? $\rightarrow 3\sqrt{m}$

f $m \rightarrow$? $\rightarrow$? $\rightarrow 25m^2$

Stretch – can you deepen your learning?

1 Find some possible function machines with each of these given inputs and outputs.

a $a \rightarrow$? $\rightarrow$? $\rightarrow 10 - 2a$

b $b \rightarrow$? $\rightarrow$? $\rightarrow \frac{b^2}{4}$

c $c \rightarrow$? $\rightarrow$? $\rightarrow d$

2 **a** How many one-step function machines can you find so that

i the output is always 0

ii the output is always 1

iii the output is always the same as the input?

b How many pairs of function machines can you find so that

i the output is always 0

ii the output is always 1

iii the output is always the same as the input?

Reflect

Explain how you can find the operations or function machines to make each of these expressions.

$3a + 2$	$3(a + 2)$	$2 + 3a$	$3(a + 2)$	$a^2 + 3$

What's the same and what's different?

2 Algebraic notation
Chapters 2.1–2.6

I have become fluent in…

- substituting numbers into function machines
- forming algebraic expressions
- using bar models to represent expressions
- substituting numbers into algebraic expressions
- using inverse operations

I have developed my reasoning skills by…

- making connections between different representations
- interpreting algebraic expressions
- deciding when to use operations or their inverses
- comparing graphs of sequences and expressions

I have been problem-solving through…

- working backwards to find inputs given outputs
- exploring sequences given by algebraic rules
- modelling expressions in a variety of ways
- finding the operations used to give results

Check my understanding

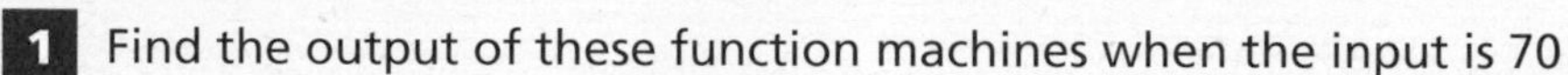

1 Find the output of these function machines when the input is 70

a Input → $\div 5$ → Output
b Input → -10 → $\div 5$ → Output
c Input → $\div 5$ → -10 → Output
d Input → $\div 10$ → -5 → Output

2 Find the input of these function machines when the output is 70

a Input → $\div 5$ → Output
b Input → -10 → $\div 5$ → Output
c Input → $\div 5$ → -10 → Output
d Input → $\div 10$ → -5 → Output

3 Put these cards in ascending order of value when $a = 3$

$4a$ | $a + 4$ | a^2 | $4 - a$ | $\frac{12}{a}$ | $\frac{1}{2}a$

4 Work out the first four terms of the sequences given by the rules.

a $2n + 3$ **b** $n^2 + 3$ **c** $3n - 2$ **d** $3 + \frac{n}{2}$

Which of the sequences are linear? Check your answer using a graph plotter.

5 **a** How many one-step function machines can you find where

i the input 100 gives the output 10

ii the input b gives the output $\frac{b}{2}$?

b How many pairs of function machines can you find where

i the input 6 gives the output 27

ii the input m gives the output $4m + 12$?

3 Equality and equivalence

In this block, I will learn...

how to find the fact family of a mathematical statement

$28 + 68 = 96$
$68 + 28 = 96$
$96 - 28 = 68$
$96 - 68 = 28$

$1.8 \times 7 = 12.6$
$7 \times 1.8 = 12.6$
$12.6 \div 7 = 1.8$
$12.6 \div 1.8 = 7$

how to solve one-step equations

$\frac{a}{4.8} = 3.2$
$a = 3.2 \times 4.8$
$a = 15.36$

the difference between like and unlike terms

x^2 $2x$ $-2x$

how to simplify expressions

$8p + 4q - 3p$
$\equiv 8p - 3p + 4q$
$\equiv 5p + 4q$

when two expressions are equivalent

$6m + 4m \equiv 5m \times 2$

$p + p \equiv 2p$

the difference between equality and equivalence

$5m = 20$

$4 \times 5m \equiv 20m$

3.1 Using the equals sign

Small steps

- Understand the meaning of equality
- Understand and use fact families, numerically and algebraically

Key words

Equal – having the same value. We use the sign = between numbers and calculations that are equal in value, and the sign ≠ when they are not

Equation – a statement showing that two things are equal

Fact family – a list of related facts from one calculation

Commutative – when an operation can be in any order

Are you ready?

1 Sort these calculations into groups of equal value.

6 + 3	12 – 5	2 × 5	5 + 4	5 + 5
10 – 1	8 + 1	6 + 4	18 ÷ 2	2 + 5

2 Which of these expressions are equal when the value of a is 5?

$2a$ $\quad$ $15 - a$ $\quad$ $\frac{100}{a}$ $\quad$ $a^2 - 10$ $\quad$ $4a - 10$

3 Put the correct symbol, = or ≠, between these calculations.

a 3 + 4 ◯ 4 + 3 $\qquad$ **b** 7 – 2 ◯ 8 – 3

c 9 + 7 ◯ 10 + 8 $\qquad$ **d** 6 – 4 ◯ 7 – 3

4 Are the statements true or false?

a 8 + 6 = 6 + 8 $\qquad$ **b** 8 – 6 = 6 – 8 $\qquad$ **c** 8 × 6 = 6 × 8 $\qquad$ **d** 8 ÷ 6 = 6 ÷ 8

Models and representations

Bar models are really useful to show equality. If two amounts are **equal** in value, then you can represent this by bars of equal length.

Consider 6 + 2 and 2 + 6

6	2
2	6

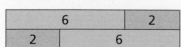

The bar model shows visually that the two number facts have equal value.

You can illustrate the result shown above for any pair of numbers.

First use a bar to represent the number a ... ... and another bar to represent the number b

Now compare $a + b$ with $b + a$

a	b

b	a

The bars are equal in length, so $a + b = b + a$

It does not matter what order you add numbers, the total is the same.

The same is true for multiplication:

$a \times b = b \times a$, which you can write as $ab = ba$

You say that addition and multiplication are **commutative**.

You will look at the commutative property of addition and multiplication in more detail later when you study calculations.

Example 1

Which of these statements are true?

a $13 + 7 = 4 \times 5$

b $19 + 5 = 18 + 6$

c $364 + 99 = 364 + 100 - 1$

d $37 \times 9 = 370 - 37$

a $13 + 7 = 20$
$4 \times 5 = 20$
As both calculations have the answer 20, the statement $13 + 7 = 4 \times 5$ is true.

You can use your calculator or knowledge of number facts here.

b $19 + 5 = 24$, and $18 + 6 = 24$
So the statement $19 + 5 = 18 + 6$ is true.

You can also see this is true from the structure of the **equation** – here is a bar model to show why.

The left-hand side of the equation is 19 + 5

19	5

19 = 18 + 1 so

18	1	5

1 + 5 = 6 so

18	6

The total is the same. So 19 + 5 = 18 + 6 is true.

c $364 + 99 = 364 + 100 - 1$
The statement is true.

Again, instead of doing the calculations, just use the fact that 99 = 100 – 1 to see that the statement is true. You are adding an equal amount to 364 on both sides of the equation, so the equation must be true.

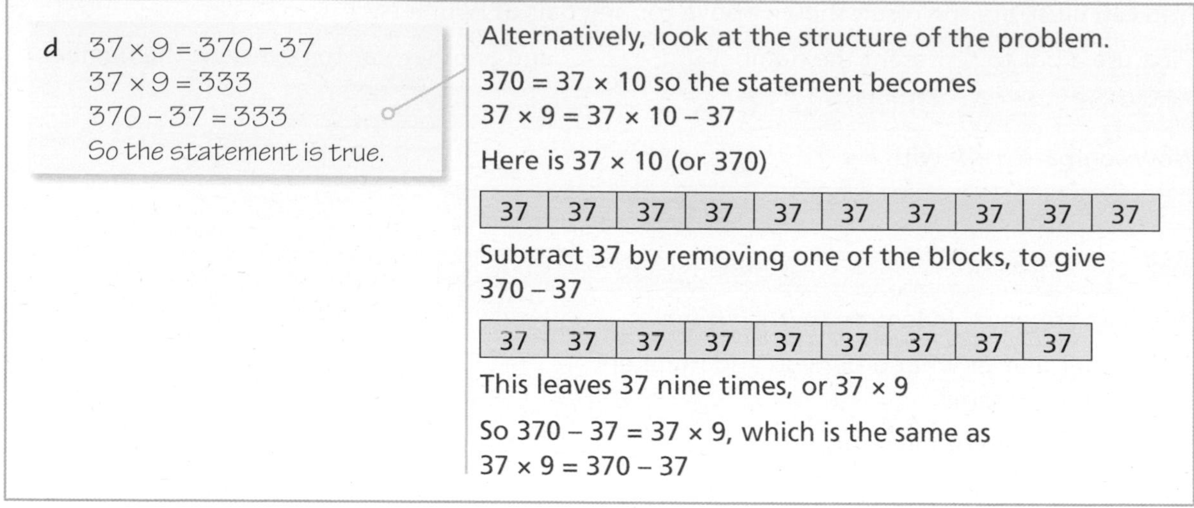

d $37 \times 9 = 370 - 37$

$37 \times 9 = 333$

$370 - 37 = 333$

So the statement is true.

Alternatively, look at the structure of the problem.

$370 = 37 \times 10$ so the statement becomes

$37 \times 9 = 37 \times 10 - 37$

Here is 37×10 (or 370)

37	37	37	37	37	37	37	37	37	37

Subtract 37 by removing one of the blocks, to give $370 - 37$

37	37	37	37	37	37	37	37	37

This leaves 37 nine times, or 37×9

So $370 - 37 = 37 \times 9$, which is the same as $37 \times 9 = 370 - 37$

If two things are equal it does not matter what order you write the equation, they have the same value. So if $a = b$, then $b = a$.

If $370 - 37 = 37 \times 9$ then $37 \times 9 = 370 - 37$ as well.

Example 2

What equations with addition does this number wall show?

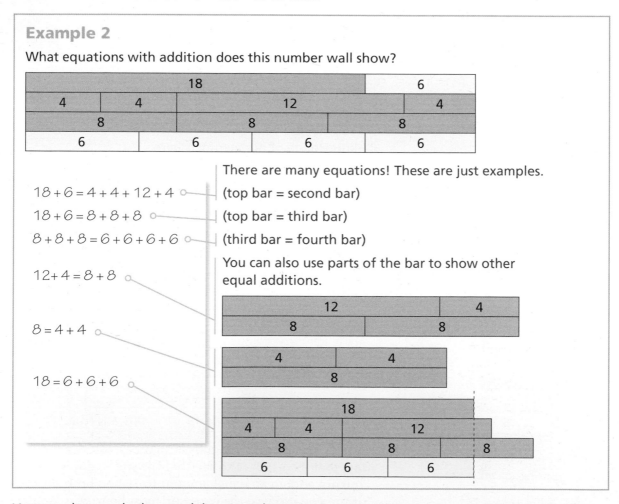

18 | 6

4 | 4 | 12 | 4

8 | 8 | 8

6 | 6 | 6 | 6

There are many equations! These are just examples.

$18 + 6 = 4 + 4 + 12 + 4$ (top bar = second bar)

$18 + 6 = 8 + 8 + 8$ (top bar = third bar)

$8 + 8 + 8 = 6 + 6 + 6 + 6$ (third bar = fourth bar)

You can also use parts of the bar to show other equal additions.

$12 + 4 = 8 + 8$

12 | 4

8 | 8

$8 = 4 + 4$

4 | 4

8

$18 = 6 + 6 + 6$

18

4 | 4 | 12

8 | 8 | 8

6 | 6 | 6

You can also use the bar model to see subtractions – you will have a look at this after the first practice section.

Example 3

Find the missing numbers.

a $10 + 9 = ? + 14$

b $? \div 4 = 20 \times 3$

a $10 + 9 = ? + 14$

left-hand side : $10 + 9 = 19$

The right-hand side must be equal to 19 as well.

So $14 + ? = 19$

$? = 5$

You might just "spot" or know the answer, ? = 5

Or you can ask "What number adds to 14 to give 19?".

Or you can use a function machine

? → + 14 → 19

You know the inverse of "+ 14" is "– 14", so ? = 19 – 14 = 5

b $? \div 4 = 20 \times 3$

So $? \div 4 = 60$

So $? = 60 \times 4 = 240$

This time the calculation on the right-hand side is complete 20 × 3 = 60

The left-hand side must equal 60 as well.

Again there are lots of ways of thinking about this, such as the inverse of ÷ 4 is × 4,

so ? = 240

In this section, you can sometimes "spot" the missing numbers quite easily.

Practice 3.1A

1 Are these statements true or false?

a $17 + 8 = 5 \times 5$ **b** $30 \times 7 = 280 - 60$ **c** $50^2 = 10\,000 \div 4$

d $10.3 \times 2.6 = 2.06 \times 13$ **e** $500 - 127 = 265 + 108$ **f** $300 \div 0.4 = 250 \times 3$

2 By looking at the structure of the calculations, decide which of the following are true.

a $87 + 9 = 88 + 10$ **b** $345 + 29 = 344 + 30$ **c** $612 - 9 = 613 - 10$

d $504 - 99 = 503 - 100$ **e** $20 \times 10 = 2 \times 10 \times 10$ **f** $600 \div 2 = 300 \div 4$

g $99 \times 60 = 6000 - 60$ **h** $38 \times 101 = 3800 - 38$ **i** $700 - 7 = 99 \times 7$

3 Use the number wall to write down some addition equations.

8	8	8	8				
12	12	8					
6	6	6	6	6	2		
4	4	4	4	4	4	4	4
9	9	4	4	4	2		

4 Amina says "116 + 93 = 115 + 94". Without doing the calculations, explain why Amina is right.

5 Match the pairs of calculations with same the overall result.

Add 100 and subtract 1	Subtract 99
Add 1 and subtract 100	Subtract 11
Subtract 1 and subtract 10	Add 99
Subtract 10 and add 1	Subtract 9
Subtract 1 and add 10	Add 9

6 Find the missing numbers.

a $37 + 84 = 36 + \square$ **b** $101 - 17 = 100 - \square$

c $10 \times 3 = 29 + \square$ **d** $712 - 99 = \square - 100$

e $228 - \square = 230 - 105$ **f** $80 \times 5 = \square \times 10$

What do you think?

1 Amina says that if $a = b$ then $a + c = b + c$

Investigate Amina's claim with some numbers for a, b and c. Try to illustrate the result using a bar model.

2 Rob says that if $a = b$ then $a - c = b - c$

Investigate Rob's claim with some numbers for a, b and c. Try to illustrate the result using a bar model.

3 Sven says that if $a + b = a + c$ then b and c are equal.

Investigate Sven's claim with some numbers for a, b and c. Try to illustrate the result using a bar model.

Now you are going to look at fact families – lists of facts that use the same numbers.

You have already looked at the additions from a bar model like this one:

8	5
13	

It shows both 8 + 5 = 13 and 5 + 8 = 13

Remember that addition is commutative so you can write the numbers either way around.

You can also see subtractions in the bar model.

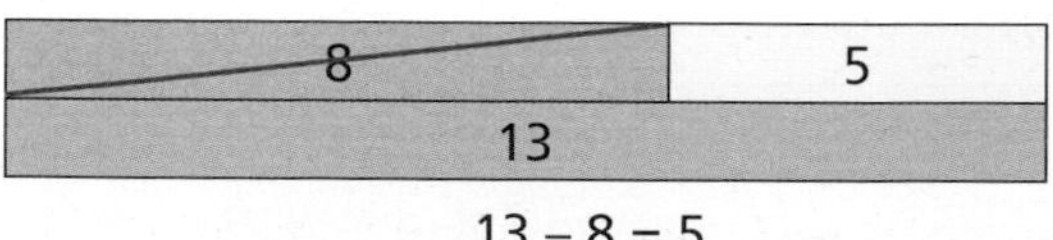

13 – 8 = 5

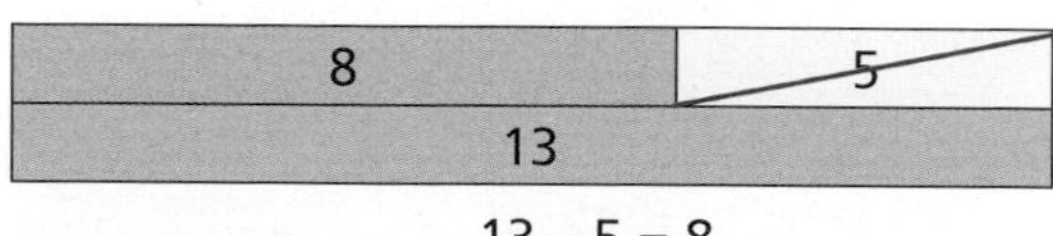

13 – 5 = 8

This completes the addition/subtraction **fact family** for 5, 8 and 13

5 + 8 = 13	8 + 5 = 13
13 – 8 = 5	13 – 5 = 8

You could write these as 13 = 5 + 8, 8 = 13 – 5 and so on, but these are the same facts with the equals sign in a different place, not 'new' facts.

You can also make fact families for multiplication and division.

10				
5		5		
2	2	2	2	2

This shows the multiplication and division fact family for 2, 5 and 10

$2 \times 5 = 10$	$5 \times 2 = 10$
$10 \div 2 = 5$	$10 \div 5 = 2$

You will explore fact families again when you study calculations later in the year.

Example 4

Write the fact family for these bar models.

a

7	21
28	

b

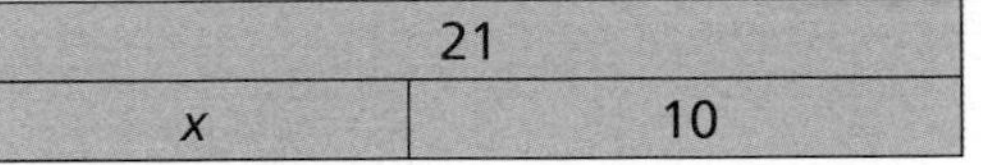

c

24			
6	6	6	6

a Working left to right, $7 + 21 = 28$

Also $21 + 7 = 28$

$28 - 7 = 21$

Likewise, $28 - 21 = 7$

7	21
28	

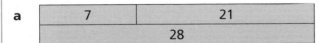

Addition is commutative.

If you subtract the 7 from the whole 28 the remaining part is 21

b Working left to right, $x + 10 = 21$

Also $10 + x = 21$

$21 - x = 10$

Likewise, $21 - 10 = x$

21	
x	10

Addition is commutative

If you subtract the x from the whole 21 the remaining part is 10

This fact gives an easy calculation from which you can work out $x = 11$. Now check this value in the other fact equations.

c

$6 \times 4 = 24$

$4 \times 6 = 24$

$24 \div 4 = 6$

$24 \div 6 = 4$

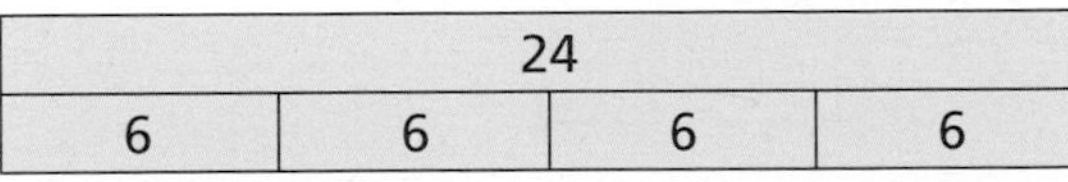

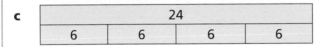

"6 four times"

"4 lots of 6"

As multiplication is commutative, both facts are true.

"What is each part if you share 24 into 4 equal parts?"

"How many groups of 6 are there in 24?"

Practice 3.1B

1 Write the addition/subtraction fact families for these bar models.

a

7	10
17	

b

23	59
82	

c

7.4	
4.6	3.8

d

x	53
81	

e

2.6	y
11.3	

f

9.4	a
b	

g

p	q
r	

2 For each statement, write the other three facts that complete the fact family.

a $72 + 96 = 168$ **b** $17.3 - 8.8 = 8.5$

c $c + 104 = 172$ **d** $130 - d = 51$

3 Write the multiplication/division fact families for these bar models.

a

15		
5	5	5

b

30				
x	x	x	x	x

c

7.1	7.1	7.1
21.3		

d

280		
y	y	y

e

a	a
18.62	

f

1.7			
p	p	p	p

4 For each statement, write the other three facts that complete the fact family.

a $72 \times 5 = 360$ **b** $80 \div 4 = 20$

c $5 \times c = 108$ **d** $x \div 7 = 103.4$

e $4e = 13.2$ **f** $\frac{12}{f} = 3$

What do you think?

1 For each statement, write the other three facts that complete the fact family.

a $\frac{1}{2} + \frac{1}{3} = \frac{5}{6}$ **b** $0.5 - \frac{1}{5} = 0.3$

c $\frac{3}{4} \times 12 = 9$ **d** $x \div \frac{1}{3} = 2$

2 Can you find a fact family with squares and square roots?

Consolidate – do you need more?

1 Are these statements true or false?

a $6 \times 7 = 21 \times 2$ **b** $500 - 180 = 80 \times 4$ **c** $800 \div 2 = 20^2$

d $17 \times 5 = 34 \times 10$ **e** $12^2 = 36 \times 4$ **f** $1.8 \times 2.3 = 18 \times 23 \times 10$

2 By looking at the structure of the calculations, state whether the following are true or false.

a $203 + 99 = 202 + 100$ **b** $100 \times 20 = 50 \times 40$ **c** $356 - 99 = 355 - 100$

d $164 + 99 = 154 + 109$ **e** $600 \div 4 = 600 \div 2 \div 2$ **f** $56 \times 99 = 56 \times 90 + 56 \times 9$

g $99 \times 12 = 120 - 12$ **h** $102 \times 45 = 4500 + 90$ **i** $201 - 40 = 202 - 39$

3 Copy and complete the facts shown by the bar model.

42					
a	a	a	a	a	a

42 = □ × □ 42 = □ × □

42 ÷ □ = □ 42 ÷ □ = □

4 Draw a bar model to show $86 - 37 = 49$

List other facts that the bar model shows.

5 Write the fact families for the calculations.

a $7.8 + 6.7 = 14.5$ **b** $400 - 128 = 272$ **c** $6 \times 9.3 = 55.8$

d $836 \div 11 = 76$ **e** $a + 9.7 = 23.4$ **f** $305 - b = 127$

g $c \times 15 = 57$ **h** $98 \div d = 140$

Stretch – can you deepen your learning?

1 Junaid says that if $a = b$ then $ka = kb$. Is Junaid correct?

2 Without working out the calculations, find the missing numbers and operations. Explain your thinking.

a $68 \div 5 = 68 \times 2 \div \square$

b $307 \times \square = 3070 \div 2$

c $a \div 4 = a \div 2 \bigcirc \square$

d $b \bigcirc \square = b \div 2 \times 100$

Reflect

1 Explain the meaning of the word 'equation' in your own words.

2 Explain how you can use one fact to find the rest of a fact family.

3.2 One-step linear equations

Small steps

- Solve one-step linear equations involving +/– using inverse operations
- Solve one-step linear equations involving ×/÷ using inverse operations
- Understand and use fact families, numerically and algebraically

Key words

Linear equation – an equation with a simple unknown like a, b, or x

Solve – find a value that makes an equation true

Solution – a value you can substitute in place of the unknown in an equation to make it true

Are you ready?

1 For each pair of calculations, state whether they are equal in value or not.

a 17 + 28 and 28 + 17 **b** 10 – 3 and 3 – 10

c 6 × 10 and 10 × 6 **d** 5 ÷ 2 and 2 ÷ 5

2 Write down the inverse of each operation.

a – 12 **b** + 12 **c** ÷ 12 **d** × 12

3 Find the value of each expression when $y = 30$

a $90 - y$ **b** $90 + y$ **c** $90y$ **d** $\frac{90}{y}$ **e** $\frac{y}{90}$

4 Use the fact that 186 + 473 = 659 to find

a 473 + 186 **b** 659 – 473 **c** 659 – 186

5 Use the fact that 5893 ÷ 71 = 83 to find

a 71 × 83 **b** 5893 ÷ 83 **c** 83 × 71

Models and representations

A **linear equation** is usually represented using letters and symbols.

$x + 7.49 = 12.103$

You can use many other ways to show the equation.

A bar model

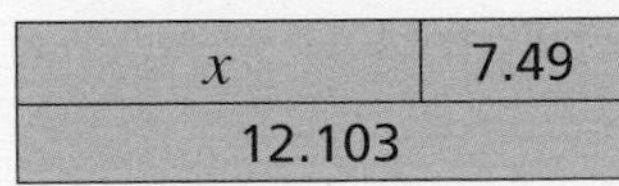

A part-whole model

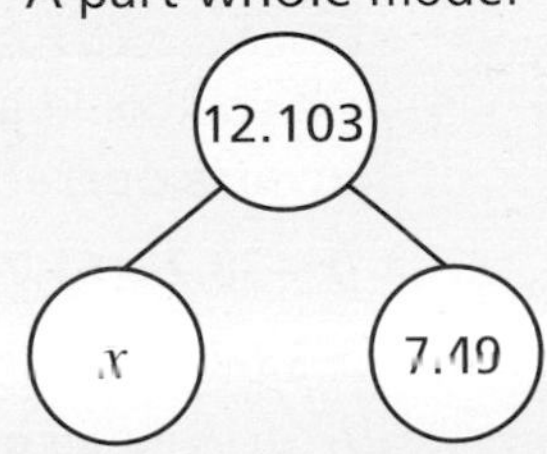

A balance

Or a function machine

$x \rightarrow$ + 7.49 $\rightarrow$ 12.103

We call these equations one-step as we only need to perform one calculation to get the answer.

Equations with integer values can also be represented using objects like cubes and counters.

could represent $3 + a = 8$

Example 1

Solve the equations.

a $a + 47 = 903$ **b** $61.2 = b - 38.7$ **c** $8 = \frac{c}{19}$ **d** $45d = 3294$

a $a + 47 = 903$

$a = 903 - 47 = 856$

You could use a function machine

$a \rightarrow$ + 47 $\rightarrow$ 903

The inverse of "+ 47" is "– 47"

Or you could use a bar model

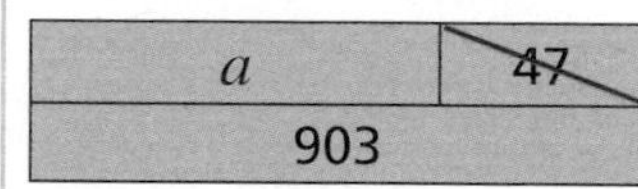

Or you could use a balance

The scale will stay balanced if you subtract 47 from each side.

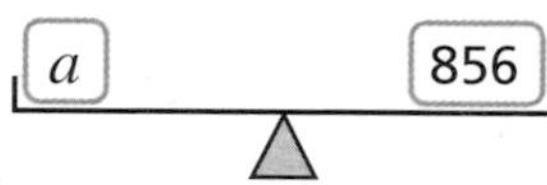

These all show $a = 903 - 47 = 856$

b $61.2 = b - 38.7$

$b = 61.2 + 38.7 = 99.9$

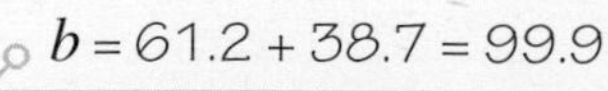

This also links to the 'balance' method of "doing the same operation to both sides of the equation".

There are many ways of solving this equation.
Try a bar model

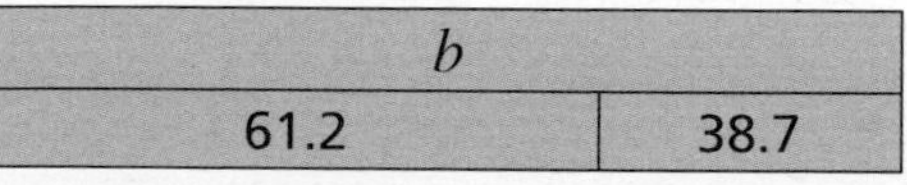

Or you could just use abstract symbols and inverse operations

$b - 38.7 = 61.2$

$+ 38.7 \quad + 38.7$

$b \quad = 99.9$

c $8 = \frac{c}{19}$

This is the same as $\frac{c}{19} = 8$

You could use a function machine

$c \rightarrow$ [÷ 19] $\rightarrow 8$

The inverse of "÷ 19" is "× 19".

Or use abstract symbols and inverse operations

$\frac{c}{19} = 8$

× 19 × 19

$c = 152$

$c = 8 \times 19 = 152$

Notice a bar model would have been awkward here because of the 19 parts, but bar models could be used for smaller divisions.

d $45d = 3294$

Remember $45d$ means $45 \times d$

So as a function machine

$d \rightarrow$ [× 45] $\rightarrow 3294$

The inverse of "× 45" is "÷ 45".

Alternatively, using symbols and inverse operations

$45d = 3294$

÷ 45 ÷ 45

$d = 73.2$

$d = 3294 \div 45 = 73.2$

See the link to the balance method.

You can check the answer to an equation by substituting back into the original equation.

For example, for the equation $a + 47 = 903$ you found $a = 856$

Check: 856 + 47 = 903, which is correct.

Check the answers to parts **b, c** and **d** are also correct.

Practice 3.2A

1 For each diagram, write down

i an equation **ii** the fact family **iii** the value of the letter

a

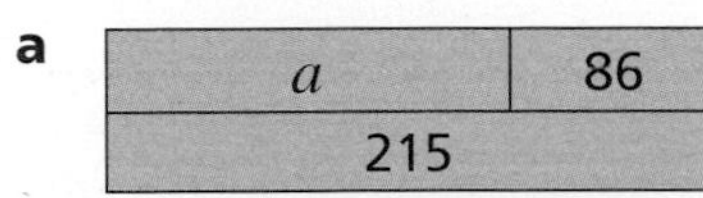

b

390				
b	b	b	b	b

c

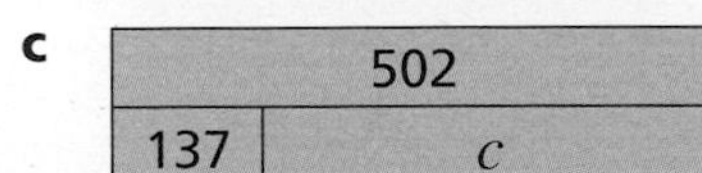

d

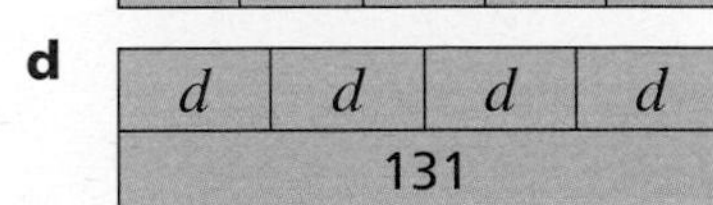

e

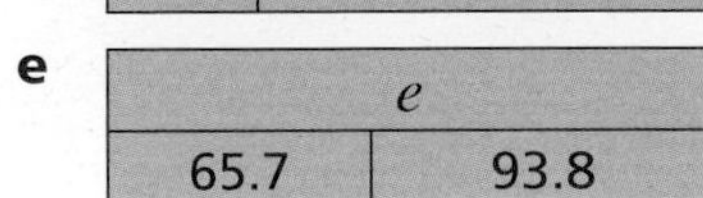

f

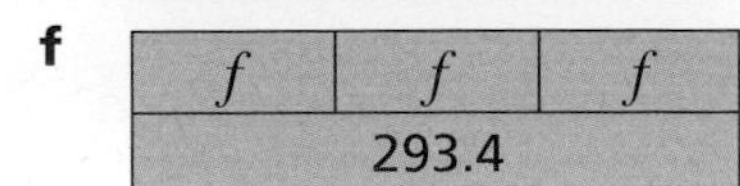

g

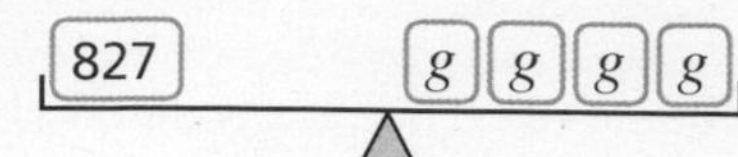

h

i

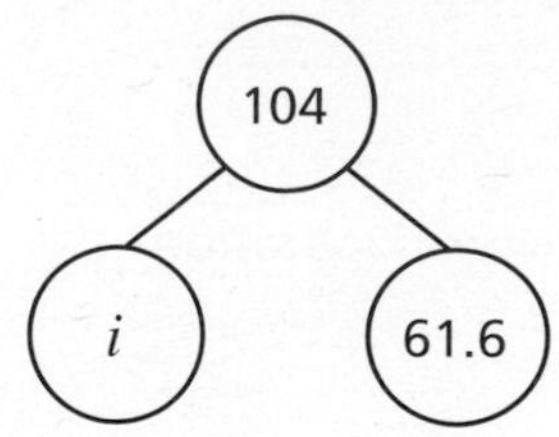

j

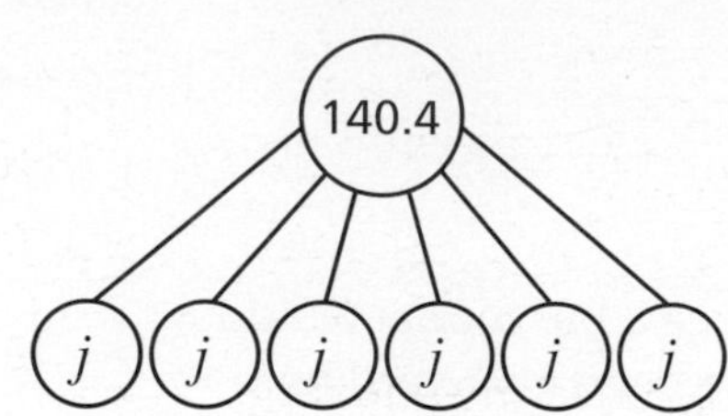

2 Draw a diagram to show each of these equations.

a $a + 12 = 40$ **b** $3b = 108$ **c** $c - 12.8 = 4.2$ **d** $\frac{d}{2} = 17.8$

Solve the equations.

3 Solve each equation.

a $a + 81 = 103$ **b** $506 = b + 92$ **c** $311 + c = 508$

d $2.7 = 1.3 + d$ **e** $6783 + e = 9265$ **f** $0.614 = 0.37 + f$

4 Solve

a $a - 19 = 40$ **b** $80 = b - 31.5$ **c** $c - 11.3 = 0.82$

5 Find the values of the letters.

a $8a = 92$ **b** $107 = 10b$ **c** $110.4 = 6c$ **d** $18d = 22.14$

6 Solve

a $\frac{a}{3} = 3$ **b** $17 = \frac{b}{5}$ **c** $206 = \frac{c}{3.5}$ **d** $\frac{d}{38} = 2.4$

7 Solve the equations.

a $45 = 300a$ **b** $3.9 = b - 11.8$

c $\frac{c}{6} = 108$ **d** $9.3 = 11.5 + d$

e $e - 9 = 0$ **f** $\frac{f}{3.5} = 8.4$

g $687 + g = 807$ **h** $0.3h = 513$

What do you think?

1 Kate says this equation is impossible because the answer is the same as the number you started with.

$a + 19.3 = 19.3$

What do you think?

2 Amina says this equation is also impossible because the answer is the same as the number you started with.

$483b = 483$

What do you think?

3 What might the values of a and b be if $a + b = 17$?

4 Solve the equations.

a $12a = 0$ **b** $48b = 0$ **c** $0 = 34c$ **d** $0 = \frac{d}{17}$ **e** $\frac{e}{23} = 0$

5 What might the values of f and g be if $fg = 0$?

Now you're going to solve equations that look a little more complicated.

Example 2

Solve the equations.

a $18.7 - a = 9.56$ **b** $\frac{58.5}{b} = 39$

a $18.7 - a = 9.56$

This is more complicated to do in a 'balance' method so you could draw a bar model.

Start with 18.7

18.7

Subtract a

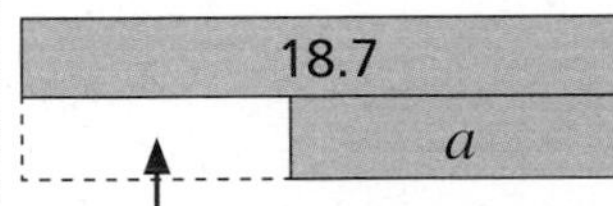

This section represents $18.7 - a$

But you know $18.7 - a = 9.56$, so you can complete the bar model

18.7	
9.56	a

$a + 9.56 = 18.7$

$9.56 + a = 18.7$

$18.7 - 9.56 = a$

$18.7 - 9.56 = 9.14$

So $a = 9.14$

Now write down the fact family for $18.7 - a = 9.56$

The last fact gives you a calculation from which you can find a.

Check: $18.7 - 9.14 = 9.56$

Check in the original equation – correct!

b $\frac{58.5}{b} = 39$

$b \times 39 = 58.5$,

$39 \times b = 58.5$

and $58.5 \div 39 = b$

So $b = 58.5 \div 39 = 1.5$

Check: $\frac{58.5}{b} = \frac{58.5}{1.5} = 39$

You do not have to draw the bar model if you remember how to write the fact family. In this case, the large numbers make a bar model difficult.

$\frac{58.5}{b}$ means "$58.5 \div b = 39$"

which leads to the fact family for the equation.

Correct!

Practice 3.2B

1 Solve the equations.

a $37.2 - a = 25.8$ **b** $6.82 + b = 25.8$ **c** $c + 19.7 = 28.3$

d $d - 26 = 58$ **e** $73 - e = 27.1$ **f** $f - 6.8 = 13.7$

2 Solve the equations.

a $\frac{a}{29} = 24$ **b** $\frac{336}{b} = 48$ **c** $24c = 408$ **d** $37 = \frac{d}{24}$

e $28 = \frac{336}{e}$ **f** $\frac{150}{f} = 20$ **g** $18 = \frac{g}{18}$ **h** $\frac{87}{h} = 87$

3 Find the values of the letters.

a $36 = 78 - a$ **b** $72 - b = 41$ **c** $\frac{12}{c} = 40$ **d** $31.25 = \frac{25}{d}$

e $\frac{e}{17} = 12$ **f** $38 = f - 29$ **g** $\frac{100}{g} = 1.25$ **h** $37 = 25h$

4 Form and solve equations to find the numbers in each of these puzzles.

a I think of a number and add 23. The answer is 81

b I think of a number and subtract 23. The answer is 81

c I think of a number and subtract it from 23. The answer is 12

d I think of a number and divide it by 23. The answer is 12

e I think of a number and multiply it by 23. The answer is 483

f I think of a number and divide 23 by my number. The answer is 4.6

What do you think?

1 Abdullah is trying to solve each of the following. Explain why this is not possible in each case.

a $5 + 7 = 12$ **b** $\frac{b}{3}$ **c** $12c$ **d** $d + 9$ **e** $f = g$

2 Explain the difference between solving the equations $x - 103 = 24$ and $103 - x = 24$

3 Explain the difference between solving the equations $\frac{g}{12} = 8$ and $\frac{12}{g} = 8$

Consolidate – do you need more?

1 **a** Draw a bar model to show each of the equations.

i $a + 34 = 51$ **ii** $b - 34 = 51$

iii $3c = 255$ **iv** $\frac{d}{3} = 5.7$

b Solve your equations in part **a** to find the values of the letters.

2 Write and solve the equations represented by the diagrams.

a

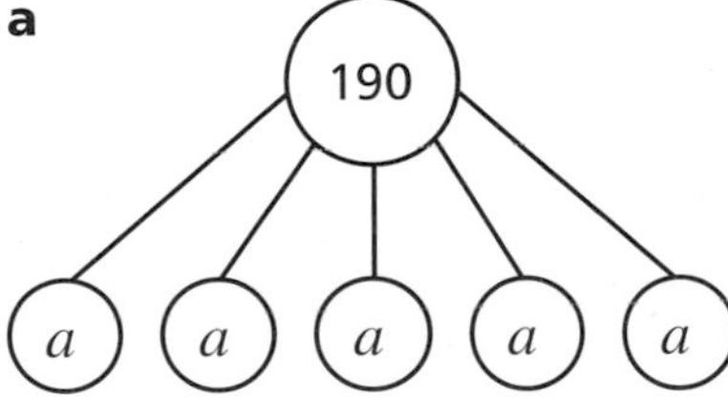

b

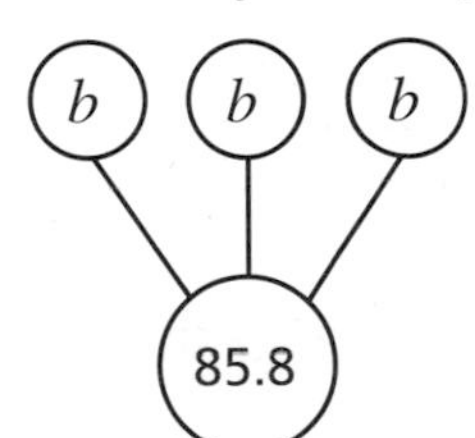

c

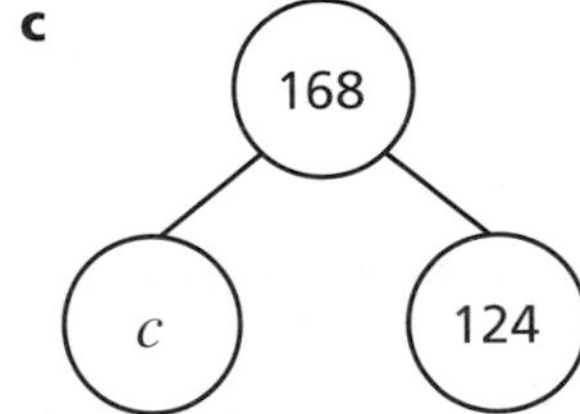

d

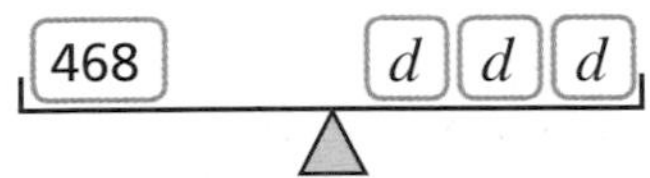

e

e	47
513	

f

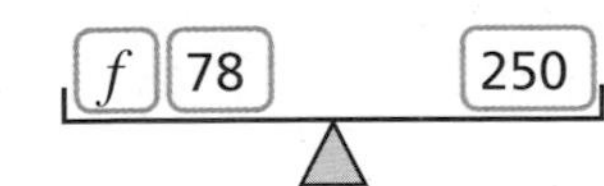

g

g			
$\frac{g}{4}$	$\frac{g}{4}$	$\frac{g}{4}$	$\frac{g}{4}$
12.8			

3 Find the value of x in each equation.

a $\frac{x}{8} = 18$ **b** $5x = 420$ **c** $87 + x = 400$

d $87 - x = 12$ **e** $x - 12 = 87$ **f** $42 = \frac{x}{12}$

g $12 = \frac{42}{x}$ **h** $103 = x + 94$ **i** $72 + x = 101$

j $87 = 6x$ **k** $52 = \frac{13}{x}$ **l** $4.2 = x - 19.7$

4 Solve the equations.

a $3a = 984$ **b** $3 + b = 984$ **c** $c - 3 = 984$ **d** $984 = \frac{d}{3}$

e $984 - e = 3$ **f** $f - 984 = 3$ **g** $3 = \frac{984}{g}$

Stretch – can you deepen your learning?

1 Form and solve equations for each of these situations.

a The area of a rectangle is 153 cm^2. The length is 17 cm. Find the width.

b The area of a square is 529 mm^2. Find the length.

2 Mario says the solution to $\frac{x}{3} = 6$ is $x = 2$. What mistake do you think Mario has made?

3 The solution to a one-step equation is $x = 17$. How many different types of equation can you find? For example, $x + \square = \square$

Reflect

Explain how to solve a one-step equation. How do you know which operation to use?

3.3 Like and unlike terms

Small steps

- Understand the meaning of like terms and unlike terms
- Substitute values into single operation expressions
- Understand the meaning of equivalence

Key words

Like terms – terms whose variables are the same, for example $7x$ and $12x$

Unlike terms – terms whose variables are not exactly the same, for example $7x$ and 12 or $5a$ and $5a^2$

Equivalent – two expressions are equivalent if they always have exactly the same value, for example $x + x$ is equivalent to $2x$

Are you ready?

1 What's the same and what's different about these three expressions?

$5a$ $\qquad \frac{5}{a}$ $\qquad \frac{a}{5}$

2 Are these two expressions the same or different? Why?

b^2 $\qquad 2b$

3 Work out the value of each expression when $c = 5$

a $c + 10$ **b** $10 + c$ **c** $10c$ **d** $\frac{c}{10}$ **e** $\frac{10}{c}$

Models and representations

You can use letters to represent numbers: $a = 7$, $b = 4$

You can also use objects or bars to represent numbers. For example, you could use  to represent the number a and ● to represent the number b.

Algebra tiles are commonly used as well, for example

▢ represents 1 ▯ represents x and  represents x^2

Example 1

Is each pair of terms like or unlike?

a $5a$, $12a$ **b** $3b$, $3c$ **c** $2c$, c^2

d $5d$, $-3d$ **e** 8, 163 **f** $7f$, $\frac{f}{7}$

a $5a$ and $12a$ are **like terms**. — Both terms involve the letter a and only the letter a.

b $3b$ and $3c$ are **unlike terms**. — Although both terms include the number 3, one has the letter b and the other has the letter c.

c $2c$ and c^2 are unlike terms. — Although both terms include the letter c, one has c and the other has c^2.

d $5d$ and $-3d$ are like terms.

Although one term is positive and the other is negative, they are both multiples of d.

e 8 and 163 are like terms.

Although there are no letters, they are both of the same form as they are just numbers.

f $7f$ and $\frac{f}{7}$ are like terms.

You can think of $\frac{f}{7}$ as $\frac{1}{7}f$ so both terms relate to f.

Example 2

a Substitute $x = 1$ into each expression. Which are equal in value?

i $2x$ **ii** $x + x$ **iii** x^2 **iv** $\sqrt{x}$

v $x + 1$ **vi** $1 + x$ **vii** $x - 1$ **viii** $1 - x$

b Substitute $x = 4$ into the expressions in **a**. Which are equal in value now?

c Will any of the expressions always be equal in value?

a When $x = 1$,

i $2x = 2 \times 1 = 2$ ii $x + x = 1 + 1 = 2$

iii $x^2 = 1 \times 1 = 1$ iv $\sqrt{x} = \sqrt{1} = 1$

v $x + 1 = 1 + 1 = 2$ vi $1 + x = 1 + 1 = 2$

vii $x - 1 = 1 - 1 = 0$ viii $1 - x = 1 - 1 = 0$

Substitute $x = 1$ into each expression.

When $x = 1$, $2x$, $x + x$, $x + 1$ and $1 + x$ are all equal

x^2 and $\sqrt{x}$ are also equal

$x - 1$ and $1 - x$ are also equal

b When $x = 4$,

$2x = 2 \times 4 = 8$ $x + x = 4 + 4 = 8$

$x^2 = 4 \times 4 = 16$ $\sqrt{x} = \sqrt{4} = 2$

$x + 1 = 4 + 1 = 5$ $1 + x = 1 + 4 = 5$

$x - 1 = 4 - 1 = 3$ $1 - x = 1 - 4 = -3$

We will be looking at negative numbers later in the year.

When $x = 4$, $2x$ and $x + x$ are equal

$x + 1$ and $1 + x$ are also equal

c $2x$ and $x + x$ are equivalent. They will always have the same value, whatever the value of x.

$2x$ and $x + x$ are always equal in value (**equivalent**), as multiplying a number by 2 and adding a number to itself are the same thing.

x	x
$2x$	

$x + 1$ and $1 + x$ are always equivalent too, because addition is commutative.

It does not matter which order you add the same two numbers, you will always get the same answer.

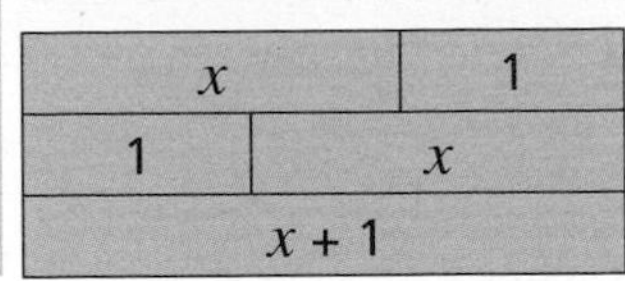

Practice 3.3A

1. Which of these terms are mathematically like $5a$?

   $3a$ 5 $-5a$ $-3a$ a^5

2. In each set of three, which term is unlike the other two?

   a $2x$ x^2 $-2x$ **b** $3b$ b 3 **c** $6ab$ $7ab$ $\frac{a}{b}$

   d -3 -10 $-3t$ **e** $\frac{1}{2}y$ $\frac{1}{2}$ y

3. Write five terms that are like

   a $6d$ **b** $-2v^2$ **c** ab

4. Sort these terms into groups of like terms.

   4 $4b$ $4a$ $4ab$ $4a^2$ $-4a^2$ -4 $40b$ $-40ab$ $\frac{1}{4}a$

   $-4a$ 41 $-14b$ $4b^2$ $-4ab$ $40ab$ $40a$ $-40b$ $-4b^2$ $\frac{1}{2}b$

5. Here are six expressions.

   $2y$ $y + 2$ y^2 $0.5y$ $2 + y$ $\frac{1}{2}y$

   a Substitute $y = 2$ into each of the expressions. Which expressions are equal in value?

   b Substitute $y = 10$ into each of the expressions. Which expressions are equal in value?

   c Which of the expressions will always be equal in value?

6. Match any pairs of equivalent expressions. There may be some expressions with more than one match or that do not have a match.

   $5x$ $5 + x$ $\frac{x}{5}$

   $5 - x$ $\frac{1}{5}x$ $x + 5$

   $\frac{5}{x}$ $x - 5$ $2x + 3x$ $x + x + x + x + x$

   If you are not sure, test with some different values of x.

What do you think?

1. Write three different expressions that are equivalent to $c + c + c$.

2. Explain why x^3 and $3x$ are unlike terms. You could use algebra tiles or other objects to help you.

Consolidate – do you need more?

1 Write five terms that are like

a n **b** $-2p$ **c** $5a^3$

2 Which of these terms are mathematically like $6b$?

$\frac{1}{2}b$ b^6 6 $-6b$ $-3b$ $\frac{b}{6}$

3 Decide whether each pair is like or unlike.

a $5a$ $5b$ **b** $3b$ $4b$ **c** 11 2 **d** $-7d$ $21d$

e e^2 f^2 **f** 27 $27f$ **g** $8xy$ $9xy$ **h** $3z^2$ $3z^{-3}$

4 Here are seven expressions.

$3m$ $3 - m$ m^3 $m + m + m$ $m - 3$ m^2 $6 + m$

a Substitute $m = 3$ into each of the expressions. Which expressions are equal in value?

b Substitute $m = 1$ into each of the expressions. Which expressions are equal in value?

c Which of the expressions will always be equal in value?

5 Match any pairs of equivalent expressions. There may be some expressions with more than one match or that do not have a match.

$2t$ t^2 $0.5t$ $3t - t$ $2 + t$

$t \times t$ $t + t$ $\frac{t}{2}$ $t + 2$ t

Stretch – can you deepen your learning?

1 Are ab and ba like or unlike terms? Explain your thinking.

2 **a** Zach is wrong. Find a value of t that shows this.

b There is one value of t for which $2t - t$ is equal to 2 Find this value.

$2t - t$ is equivalent to 2 because you have taken the t from the $2t$ which leaves 2

Reflect

What is the difference between "equal to" and "equivalent to"?

3.4 Using the equivalence sign

Small steps

- Understand the meaning of equivalence
- Understand the meaning of like and unlike terms
- Simplify algebraic expressions by collecting like terms, using the ≡ symbol

Key words

Is equivalent to – is shown as ≡ (the same way as = means "is equal to")

Collect like terms – put like terms in an expression together as a single term

Simplify – rewrite in a simpler form, for example rewrite $8 \times h$ as $8h$

Are you ready?

1 Write three terms that are mathematically like $7a$.

2 Are these pairs of terms like or unlike?

a a^2 $2a^2$ **b** $7b$ $-4b$

c $11ab$ $11bc$ **d** 8 6

3 In each set, which term is unlike the other two?

a $3x$ $3y$ $-3x$ **b** $2b$ b^2 $10b$ **c** $9ab$ $6ab$ $3bc$

4 Write down the answers to these calculations.

a 2 + 6 + 1 **b** 2 + 6 – 1 **c** 6 + 2 + 1 **d** 6 – 2 – 1

Models and representations

You have seen that you can use pictures or objects to represent letters in algebraic expressions.

If [cube] represents the number a and [counter] represents the number b

then:

[cube] [counter] [counter] means $a + 2b$

  means $3a + 4b$

You can also use algebra tiles.

[small square] represents 1 [rectangle] represents x and [large square] represents x^2

So [3 rectangles] [2 small squares] represents $3x + 2$ and [large square] [2 rectangles] represents $x^2 + 2x$

In the last section you explored the idea of expressions being equivalent.

3.4 Using the equivalence sign

The sign ≡ means "**is equivalent to**". For example,

$x + x \equiv 2x$ $\quad$ $x \times 4 \equiv 4x$ $\quad$ $a + b \equiv b + a$

The expressions on either side of the sign are always equal, no matter what the value of the variable(s).

Example 1

Simplify each expression by writing it in correct algebraic notation.

a $a \times 4$ $\quad$ **b** $b + b + b + b + b$ $\quad$ **c** $c \times c$

d $d \times d \times d$ $\quad$ **e** $q \times p$

a $\quad a \times 4 \equiv 4a$

You write this without the multiplication sign and with the number before the letter.

b $\quad b + b + b + b + b \equiv 5b$

Altogether you are adding five lots of b so the total is the same as $5 \times b$ or $b \times 5$

c $\quad c \times c \equiv c^2$

Remember this is called "c squared".

d $\quad d \times d \times d \equiv d^3$

$d \times d \times d$ is called "d cubed".

Think back to working out the area of squares and volumes of cubes at Key Stage 2. Is there a connection?

e $\quad q \times p \equiv pq$

As with a number multiplied by a letter, you omit the multiplication sign. Also, you usually write letters in a term in alphabetical order.

Example 2

Simplify each expression by writing as a single term.

a $p + 4p + 2p$ $\quad$ **b** $6t - 2t$ $\quad$ **c** $2x^2 + 3x^2$

a $\quad p + 4p + 2p \equiv 7p$

Use ● to stand for p.

So you have

Altogether you have

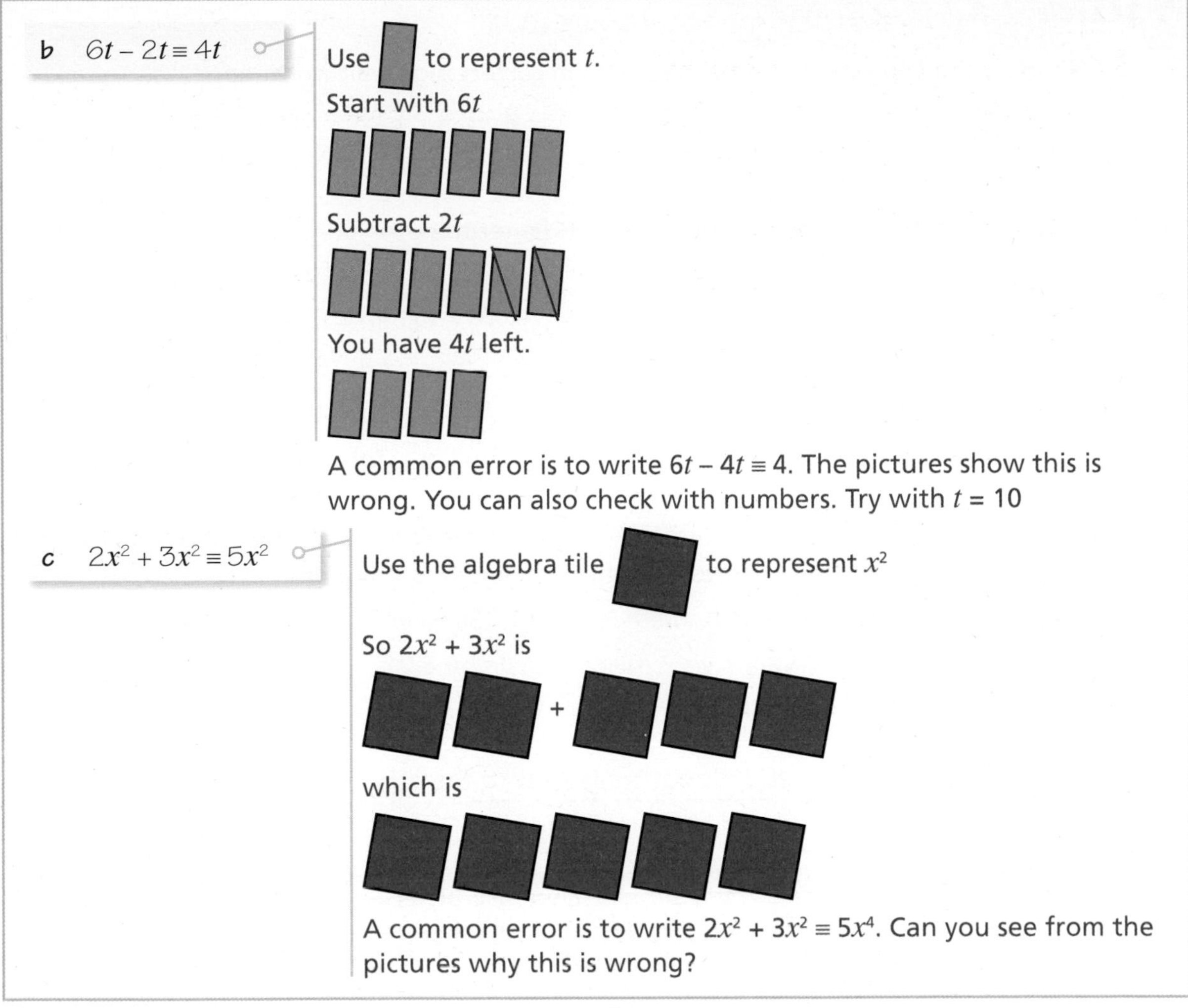

b $6t - 2t \equiv 4t$

Use [tile] to represent t.

Start with $6t$

Subtract $2t$

You have $4t$ left.

A common error is to write $6t - 4t \equiv 4$. The pictures show this is wrong. You can also check with numbers. Try with $t = 10$

c $2x^2 + 3x^2 \equiv 5x^2$

Use the algebra tile [tile] to represent x^2

So $2x^2 + 3x^2$ is

\+

which is

A common error is to write $2x^2 + 3x^2 \equiv 5x^4$. Can you see from the pictures why this is wrong?

Notice in all three cases, when you are adding or subtracting like terms, you only need to add or subtract the coefficients of the terms.

Example 3

Are the following statements true or false?

a $t + 3t \equiv 4t$ **b** $5 \times t \times 2 \equiv 10t$ **c** $t^2 - t \equiv 2$

a $t + 3t \equiv 4t$ is true

You can prove this using manipulatives or pictures and verify by checking with numbers.

b $5 \times t \times 2 \equiv 10t$ is true

Work from left to right.

$5 \times t \equiv 5t$

$5t \times 2 \equiv 10t$

c $t^2 - t \equiv 2$ is false

t^2 and t are unlike terms, so you can't subtract directly.

You could also check with a value, say $t = 10$

$10^2 - 10 = 100 - 10 = 90$, not 2

So the statement is false.

Practice 3.4A

1. Simplify the expressions by writing them in correct algebraic notation.

   a $7 \times t$ **b** $6 \times x$ **c** $h \times 9$ **d** $y + y + y + y$ **e** $n + n + n$

   f $a \times a$ **g** $a \times b$ **h** $g \times 3$ **i** $g \times f$ **j** $g \times g$

2. Simplify the expressions by writing as a single term.

   a **i** $5a + 3a$ **ii** $5a - 3a$

   iii $ab + ab$ **iv** $3a - a$

   v $a + 3a$ **vi** $4a + 6a + 2a$

   vii $4a + 6a - 2a$ **viii** $6a - 4a + 2a$

   ix $6a - 4a - 2a$ **x** $7ab + 2ab - 4ab$

   b **i** $5 \times 3a$ **ii** $3 \times 5a$ **iii** $a \times a$ **iv** $2 \times a \times a$ **v** $2 \times ab$

3. Write down any of these expressions that are equivalent to $4m$.

$4 + m$	$m \times 4$	$3m + m$	$\frac{4}{m}$
$2m \times 2$	$6m - 2$	$2m + 2$	$6m - 2m$
$2 \times 2 \times m$	$2 \times m \times m$	$2m \times 2m$	$8m - 6m + 2m$

4. In each set, which expression is **not** equivalent to the other two?

a	$3a$	$a \times a \times a$	$a + a + a$
b	$6b - 3b$	$2b + 1$	$b \times 3$
c	$c \times 2$	$c \times c$	$2 \times c$
d	$d + d$	$8d \div 4$	$9d - 6d$

5. Are the statements true or false? Explain your answers.

   a $3t - t \equiv t$ **b** $4 \times y \equiv y + y + y + y$ **c** $5n - n \equiv 5$

   d $4x + 3x \equiv 7x$ **e** $2 \times 5p \equiv 6p + 4p$ **f** $7p + 3p \equiv 10p^2$

   g $6ab - b \equiv 6a$

6. Jakub has done his homework. All of his simplifications are wrong.

   Write down the correct simplifications.

   a $3a + 7a \equiv 10a^2$ b $8b - 4b \equiv 2b$ c $6c - 2c \equiv 4$

   d $2d \times 3d \equiv 6d$ e $3x + x \equiv 3x^2$

What do you think? 

1 

I think $2a - a - a$ simplifies to $0a$

I think $2a - a - a$ simplifies to 0

Who do you agree with, Jackson or Flo? Explain your reasoning.

2 Find ten different expressions that simplify to $6p$.

3 Is it possible to simplify $2ab + 3ba$? Explain how you can or why you cannot.

You can also simplify expressions that involve more than one set of like terms.

Example 4

Simplify these expressions.

a $3a + 4b + 2c + 5a$ **b** $x^2 + 5x - 2x + 3x^2$ **c** $a + b + 2b - a$

a $3a + 4b + 2c + 5a$

$\equiv 3a + 5a + 4b + 2c$

$\equiv 8a + 4b + 2c$

Use ● to represent a, ▯ to represent b and ▶ to represent c

Then:

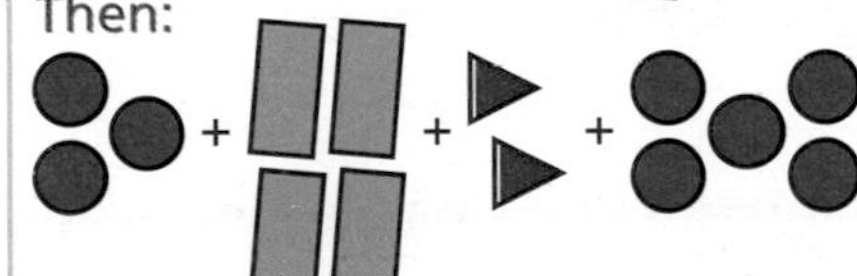

You can rearrange the objects:

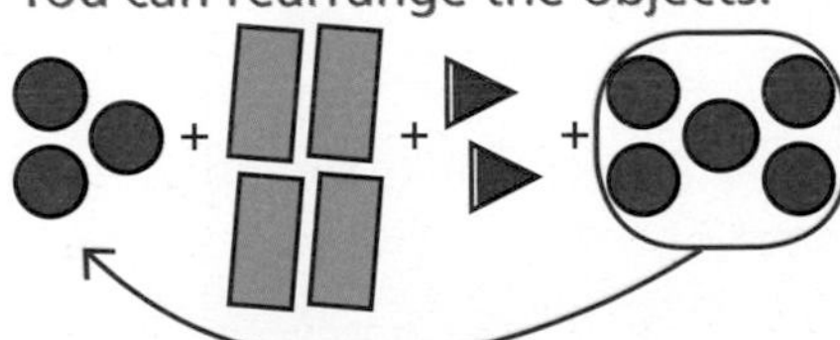

Just like collecting the shapes together you can collect the like terms together.

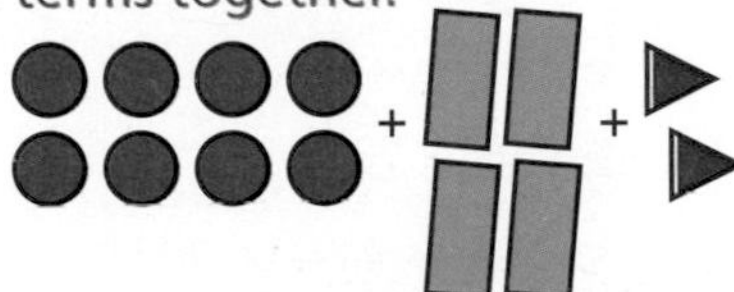

b $x^2 + 5x - 2x + 3x^2$

$\equiv x^2 + 3x^2 + 5x - 2x$

$\equiv 4x^2 + 3x$

Collect the like terms together.

Notice the signs of the terms do not change when you reorder. The first expression has x^2, $+ 5x$, $- 2x$, and $+ 3x^2$ and so does the reordered expression.

You can tackle this using pictures: 

c $a + b + 2b - a$

$\equiv a - a + b + 2b$

$\equiv 3b$

Collect the like terms together.

There are no a terms left so you don't say "$0a+3b$" or "$3b+0$". You only include the letters you are still working with.

It is possible to have answers like $-3a$, but you will not look at examples like this until you have covered negative numbers later in the book.

Practice 3.4B

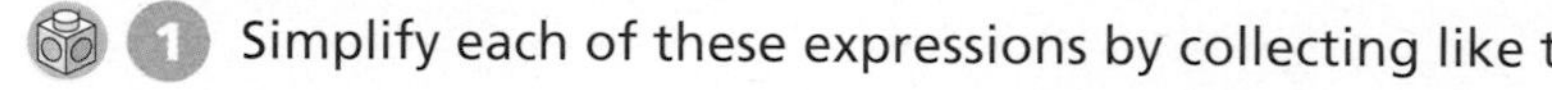

1 Simplify each of these expressions by collecting like terms.

a $5a + 4b + 3a$ **b** $5a + 4b - 3a$ **c** $5a + 4b + 3b$

d $5a + 4b - 3b$ **e** $5a + 4b + 3a + 2b$ **f** $5a + 4b - 3a + 2b$

g $5a + 4b - 3a - 2b$ **h** $5a + 4b + 3b + 2a$ **i** $5a + 4b - 3b + 2a$

j $5a + 4b - 3b - 2a$ **k** $5b + 4a + 3b + 2a$ **l** $5b + 4a + 3b - 2a$

m $5b - 4a + 3b + 2a$ **n** $5b + 4a + 3b + 2$ **o** $5b + 4a - 3b + 2$

p $5b + 4 + 3b + 2$ **q** $5b + 4 - 3b + 2$ **r** $5b + 4 - 3b - 2$

s $5a + 4 + 3b + 2$ **t** $5a + 4 + 3b - 2$

2 (Rectangle with length $3p + 2q$ and width $p + q$)

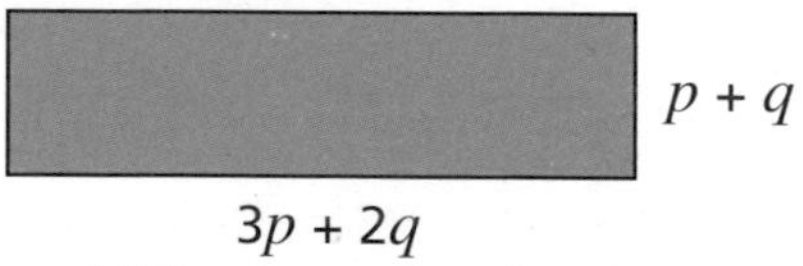

a Write an expression in terms of p and q for the perimeter of the rectangle.

b Write an expression in terms of p and q for the difference between the length and width of the rectangle.

3 Rhys is simplifying $3x + 2x^2 + 4x$. He thinks the answer is $9x^4$. Explain why Rhys is wrong, and give the correct answer.

4 Write the following expressions in their simplest form.

a $2 + 3x + 4$ **b** $2x + 3x + 4$ **c** $2x + 3x + 4x$

d $2 + 3x + 4x$ **e** $x + x^2 + 2x$ **f** $2x + x^2 + 2x$

g $2x + 2x^2 + 2x$ **h** $2x + 2x^2 + 2x + x^2$ **i** $2x + 2x^2 + 2x + 2$

j $2x + 2x^2 + 2x + 2x^2$ **k** $3x^2 + 4x - x^2 + 4x$ **l** $3x^2 + 4x - 2x^2 - 4x$

m $3x^2 + 4x + x^2 - 2x$ **n** $3x^2 + 4x - 3x^2 + 2x$ **o** $3x^2 + 4 - x^2 + 4$

p $3x^2 - 4 - x^2 + 4$

5 Are the statements true or false?

a $3a + 2b \equiv 5ab$ **b** $3a + 2b \equiv 6ab$ **c** $3x^2 - x \equiv 3x$

d $3x^2 - 3 \equiv x^2$ **e** $3x - x \equiv 3$ **f** $3x - x \equiv 2x$

g $5ab - a \equiv 5b$ **h** $2ab - ab \equiv ab$ **i** $5ab + a \equiv 5a^2b$

j $5ab - 3ba \equiv 2ab$

What do you think?

1 Simplify the expressions.

a $18m \div 3 - 10m \div 2$

b $\frac{n}{5} \times 20 - n \times 3$

c $3a \times a + a \times 2 \times a$

d $8 \times a \times b - 3 \times b \times a - a \times 2 \times b$

2 Find ten different expressions that simplify to $3a + 4b$.

Consolidate – do you need more?

1 Simplify the expressions by writing them in correct algebraic notation.

a $a \times 4$

b $b + b + b + b$

c $9 \times c$

d $d \times d$

e $e + e + e$

f $f \times 7$

g $g \times h$

h $j \times j \times j$

i $6 \times mn$

2 Simplify each expression by writing it as a single term.

a **i** $6a + 4a$

ii $6a - 4a$

iii $6ab - 2ab$

iv $6a - a$

v $a + 6a$

vi $4ab + 6ab + 2ab$

vii $4ab + 6ab - 2ab$

b **i** $6 \times 4a$

ii $6a \times 4$

iii $a \times b$

iv $2 \times a \times b$

v $ab \times 2$

3 Which of the expressions below are equivalent to $2n$?

$3n - n$	$n \times 2$	$n + n$	$2 \times n$
$4n - 2$	$n \times n$	$4n \div 2$	$2 + n$

4

Jackson: $3x - 2x \equiv 1x$

Emily: $3x - 2x \equiv 1$

Chloe: $3x - 2x \equiv x$

Who do you agree with? Why?

5 Simplify these expressions by collecting like terms.

a $2x + 2x + 2y$
b $x + y + x$
c $x + y - x$
d $x + 2y + x - y$
e $3x + 4y + 2x + y$
f $3x + 4y - 2x + y$
g $3x + 4y - 2x - y$
h $3x + 4y - 2x - 2y$
i $3xy + 4y + 2xy + y$
j $3x + 4xy - 2x + xy$
k $3xy + 4xy - 2xy + x$
l $3xy + 4y - 2xy - 2y$

6 Write down any of the expressions that cannot be simplified.

$3x^2 + 3x$ $\quad$ $3x^2 + 3x^2$ $\quad$ $3x + 3x^2$ $\quad$ $3x^2 + 6x^3 - 3x - x^2$

7 **a** The sides of a square are $3a$ cm long. What is the perimeter of the square?

b The sides of an equilateral triangle are $(2x + 3y)$ cm long. What is the perimeter of the equilateral triangle?

c The perimeter of a regular hexagon is $24n$ cm. How long is each side of the regular hexagon?

8 All these simplifications are wrong. Find the correct simplifications.

a $2a + 2a \equiv 4a^2$ $\quad$ b $6b + 2c + 3b + 5c \equiv 16bc$ $\quad$ c $5d^2 + d + d \equiv 6d^2$

d $3ab + 4ab + 2b + 3a \equiv 7ab + 5ba$ $\quad$ e $8x^2 + 4x - 4x^2 - 2x^2 \equiv 12x^2 - 2x$

Stretch – can you deepen your learning?

1 Although x^2 and $2x$ are not equivalent, there are two values of x for which the expressions are equal. Find the two values.

Generalise to find values for which x^2 and kx are equal.

2 Work out the two expressions below for several values of y.

$2y + 10$ $\quad$ $2(y + 5)$

Do you think the expressions are equivalent? Explain your answer.

Investigate further, for example with $3(x + 4)$.

You will learn about expressions with brackets in Book 2.

Reflect

1 Describe how you simplify an expression involving all like terms.

2 Describe how you simplify an expression involving like and unlike terms.

3 Equality and equivalence

Chapters 3.1–3.4

White Rose Maths

I have become **fluent** in...	I have developed my **reasoning skills by...**	I have been **problem-solving through...**
■ Solving one-step equations ■ Writing expressions in the simplest form ■ Simplifying expressions by collecting like terms ■ Recognising equivalent expressions	■ Using one fact to find a fact family ■ Identifying how to solve equations given in different forms ■ Determining whether terms are like or unlike ■ Knowing when to use = and when to use ≡	■ Forming and solving equations from real-life problems ■ Finding unknown numbers in problems presented in many forms ■ Finding expressions that are equivalent to a given expression

Check my understanding

1 Given that 357 + 574 = 931, write down which of these statements must also be true.

574 + 357 = 931   931 = 357 + 574   931 – 357 = 574   931 – 574 = 357

2 Write the fact families for these statements.

a $19 \times 72 = 1368$ **b** $2128 \div 38 = 56$

c $93 + y = x$ **d** $48 - b = c$

3 a Write down the equation that each diagram shows.

i

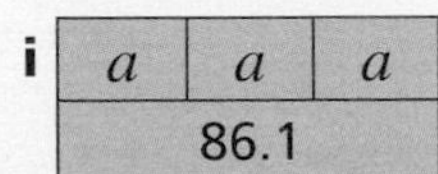

ii

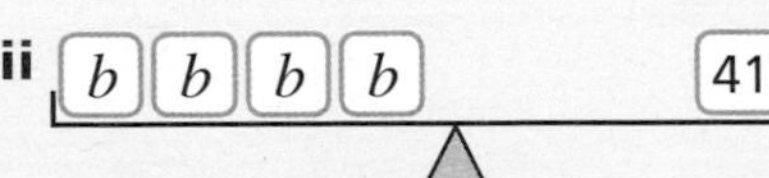

iii

84.2

c   47.7

b Solve each of your equations to find the value of the letter.

4 a Form an equation for each of these number puzzles.

i I think of a number and divide it by 12. The answer is 3.6

ii When I divide 360 by my number the answer is 12

iii I think of a number and subtract it from 36. The answer is 19

b Solve the equations formed in part **a**.

5 Identify the unlike term in each set

a $9a, 5d, 5a$ **b** $3b, b^3, -3b$ **c** $\frac{1}{2}c^2, \frac{c}{2}, 2c^2$

6 Write down any expressions shown that are equivalent to $2m$.

$4m \div 2$ $4m - 2$ $m \times 2$ $22m - 2$ $5m - 3m$ $m + m$ $22m \div 2$ $\frac{m}{2} \times 2$ $m \times m$

7 Simplify these expressions by collecting like terms.

a $3a + 4b + 2a$ **b** $3a + 4b - 2a$ **c** $3a + 4a + a^2$

d $4a + 3a^2 - 2a + a^2$ **e** $3ab + 2ba + 5ab - ba$

4 Place value and ordering

In this block, I will learn...

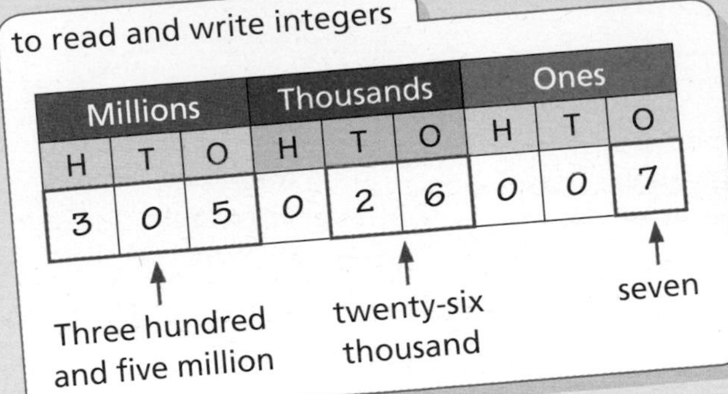

Millions			Thousands			Ones		
H	T	O	H	T	O	H	T	O
3	0	5	0	2	6	0	0	7

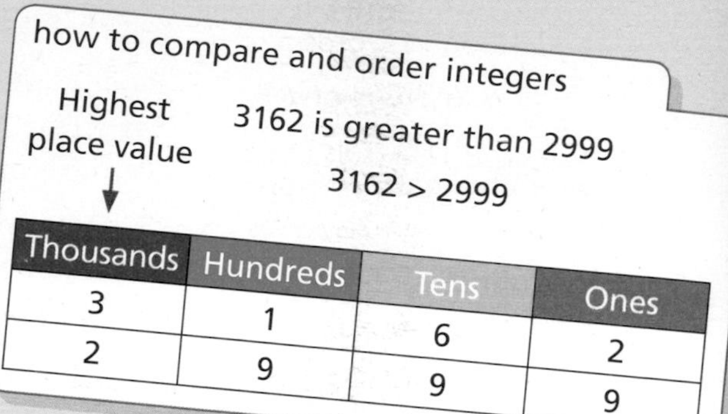

Thousands	Hundreds	Tens	Ones
3	1	6	2
2	9	9	9

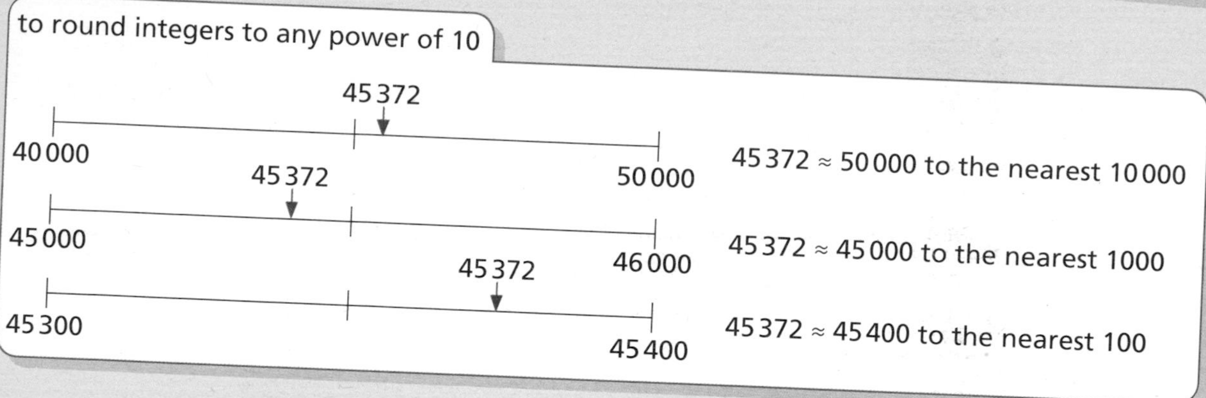

to round integers to any power of 10
45 372
40 000
50 000
45 372 ≈ 50 000 to the nearest 10 000
45 372
45 000
46 000
45 372 ≈ 45 000 to the nearest 1000
45 372
45 300
45 400
45 372 ≈ 45 400 to the nearest 100

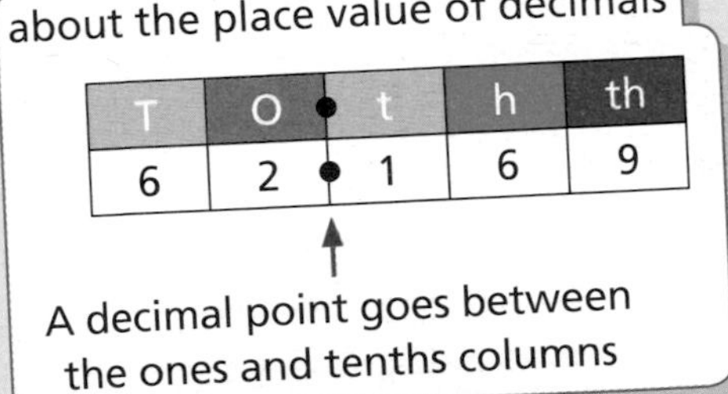

T	O	t	h	th
6	2 •	1	6	9

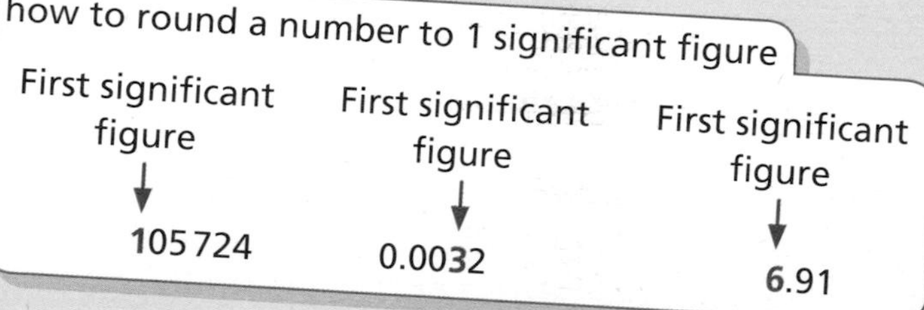

to write ordinary numbers in standard index form and vice versa (H)

Ordinary number 300 000
Standard index form 3×10^5

Ordinary number 0.00003
Standard index form 3×10^{-5}

4.1 Reading and writing integers

Small steps

- Recognise the place value of any number in an integer up to one billion
- Understand and write integers up to one billion in words and figures
- Position integers on a number line
- Work out intervals on a number line

Key words

Integer – a whole number

Digits – the numerals used to form a number

Represent – draw or show

Partition – break a number into its parts

Estimate – give an approximate answer

Are you ready?

1 Write down all the integers from the list below.

7   3.2   30 638 $2\frac{1}{2}$ 11 million

2 **a** What is the number shown on the place value grid?

Thousands			Ones		
H	T	O	H	T	O
		●		●●●	●●●●●

b Write the number in words and in figures.

3 What does the 6 represent in each number?

a 620   **b** 706   **c** 12 962   **d** 106 000

4 Copy and complete with numbers that make these number sentences true.

a 2736 = 2000 + 700 + 30 + ______

b 3095 = 3000 + ______ + ______

c 10 000 + 6000 + 500 + 8 = __________

5 What is the size of the interval on each number line?

a

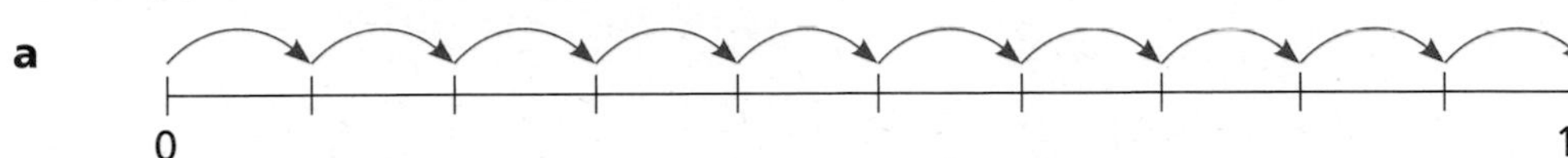

b

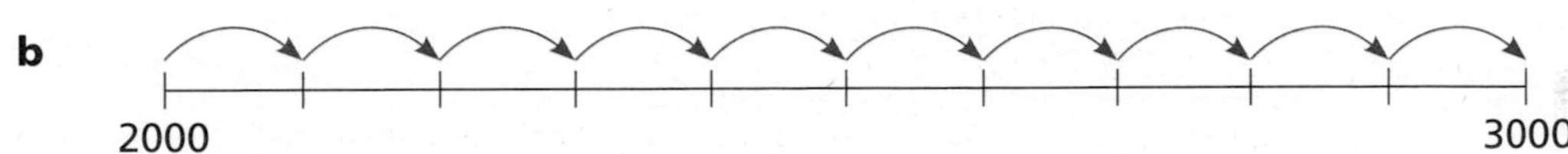

c

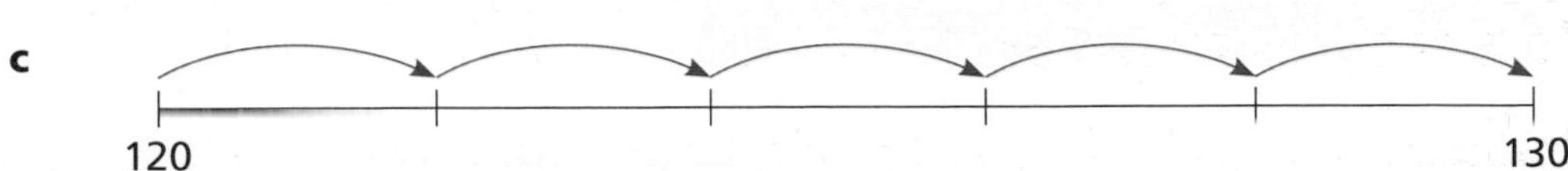

Models and representations

Thousands			Ones		
H	T	O	H	T	O
●●		●●●●●●●	●●●	●●●●●	●●

Place value grids are a useful way to **represent** numbers. They can also help you to read and write numbers.

Number lines can help you see the position of a number.

+ 100 + 100 + 100 + 100 + 100 + 100 + 100 + 100 + 100 + 100

2000 — 3000

You can use a place value grid and counters to represent numbers up to a billion.

The number of counters in each column tells you the value of that column. If there are no counters in a column it represents zero. Zero in a number acts as a place holder.

You can also use a place value grid to help you write a number in words.

Example 1

What is the value of the 7 in the number 207 352?

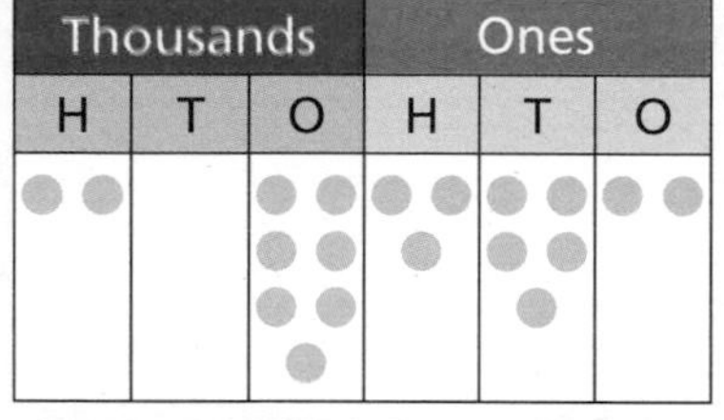

Thousands			Ones		
H	T	O	H	T	O
●●		●●●●●●●	●●●	●●●●●	●●

Represent the number in a place value grid.

7 is in the thousands column so this represents 7000 or 7 thousands.

Practice 4.1A

1 Write down the numbers that are represented in these place value grids.

a

Thousands			Ones		
H	T	O	H	T	O
		●●●●●	●●●●	●●●●●	●●●

b

Thousands			Ones		
H	T	O	H	T	O
	●	●●●●			●●●●●●●

c

Millions			Thousands			Ones		
H	T	O	H	T	O	H	T	O
		●	●●			●●●	●	●●●●

2 Represent each of these numbers in a place value grid.

Millions			Thousands			Ones		
H	T	O	H	T	O	H	T	O

a 612 **b** 2005 **c** 15 340 **d** 300 000

3 Write down the value of the 2 in each of these numbers.

a 725 **b** 3725 **c** 57 725 **d** 108 725

What do you notice?

4 **a** Write down a 4-digit number with 7 in the thousands column.

b Write down a 5-digit number with 7 in the thousands column.

c Is it possible to write down a 3-digit number with 7 in the thousands column?

5 Mario uses a place value grid to help him write down a number.

Thousands			Ones		
H	T	O	H	T	O
	●●			●	●●●●●●●●

Mario says he has made the number 218.

a Describe the mistake that Mario has made.

b What number has Mario made?

6 Write down the digit that is in the hundreds column in each number.

a 2932 **b** 205 306 **c** 1 406 211 **d** 12 000 000

7 Write down the number that has been partitioned each time.

a 500 000 + 60 000 + 3000 + 200 + 90 + 1

b 200 000 + 80 000 + 400 + 20 + 6

c 3 000 000 + 100 000 + 7000 + 90 + 2

d 3 000 000 + 100 000 + 7000 + 92

8 Partition each of these numbers.

a 23 452 **b** 692 007 **c** 4 103 560 **d** 15 300 000

Compare your answers to another student's. Did you do the same partition?

What do you think?

Emily is describing a number.

My number is between 200 000 and 300 000
- It is even.
- Each of the **digits** is a different number.
- It has 6 thousands.
- The digit in the hundreds column is 3 more than the digit in the ten thousands column.

What could Emily's number be?

Example 2

Write the number 12 603 054 in words.

Millions			Thousands			Ones		
H	T	O	H	T	O	H	T	O
	1	2	6	0	3	0	5	4

Put the number into a place value grid.

There are 12 millions, 603 thousands and 54 ones.

The number is twelve million, six hundred and three thousand and fifty-four.

Practice 4.1B

1. Write each of these numbers in words.

a

Thousands			Ones		
H	T	O	H	T	O
3	5	0	7	2	5

b

Thousands			Ones		
H	T	O	H	T	O
5	0	0	3	0	0

c

Millions			Thousands			Ones		
H	T	O	H	T	O	H	T	O
	2	6	9	1	1	1	0	3

d

Millions			Thousands			Ones		
H	T	O	H	T	O	H	T	O
1	3	6	2	4	8	3	3	3

2. Write each of these numbers in words.

a 751 b 751 000 c 751 000 000

What do you notice?

3 Write each of these numbers in words.

a 203 **b** 4203 **c** 14 203 **d** 314 203

What do you notice?

4 Write each of these numbers in words.

a 3502 **b** 73 112 **c** 903 700 **d** 17 420 030

5 Write the number or numbers in each statement in figures.

a Thirty-six thousand Lego® bricks are made every minute.

b Last year a hundred and fifty-seven million people visited a Disney® theme park.

c The distance to the Moon from Earth is three hundred and eighty-four thousand, four hundred kilometres.

d The population of China in 2017 was one billion, three hundred and eighty-six thousand living across nine point six million square kilometres of land.

What do you think? 

Bobbie is at a football game.

An announcement is made over the loud speaker but she only hears part of it.

Bobbie hears this: *"Today we have a record attendance! In total *** thousand, one hundred and thirty-six people are in the stadium."*

What could the attendance be? Write down one possibility.

What do the rest of the class think?

What is the same about all the numbers? What is different?

Example 3

Write down the number that each arrow is pointing to.

The number line goes from 20 000 to 30 000, which is an increase of 10 000

There are 10 equal sections

$10000 \div 10 = 1000$

The number line goes up in thousands.

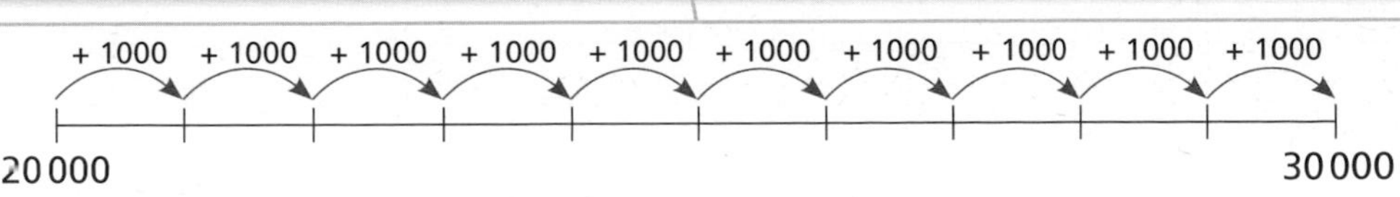

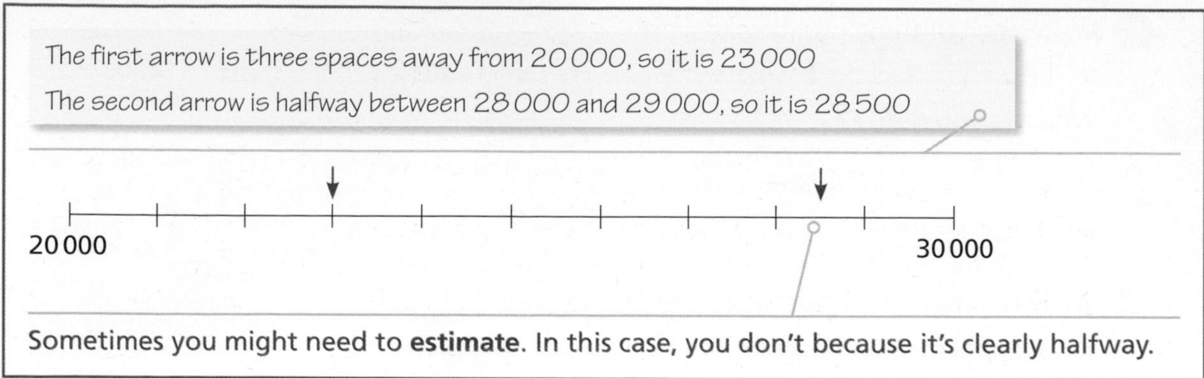

The first arrow is three spaces away from 20 000, so it is 23 000
The second arrow is halfway between 28 000 and 29 000, so it is 28 500
20 000
30 000

Sometimes you might need to **estimate**. In this case, you don't because it's clearly halfway.

Practice 4.1C

1. Write down the number that each arrow is pointing to.

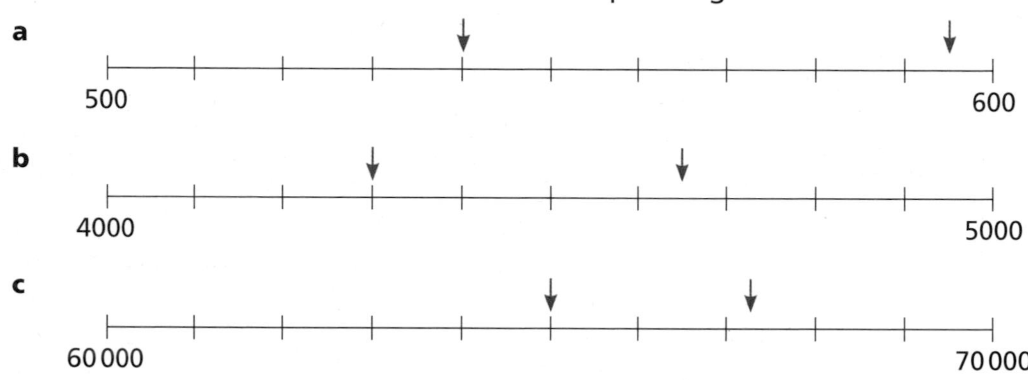

2. Write down the intervals that each of these number lines go up in (for example, 10s, or 20s).

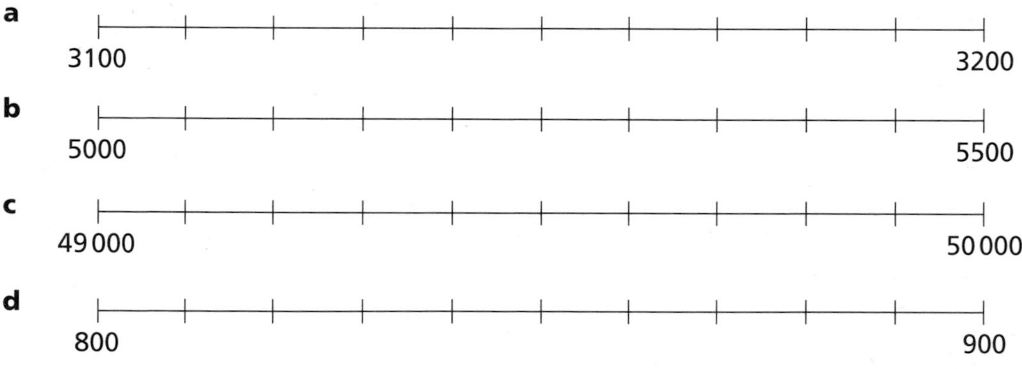

3. **a** Write down the number that each arrow is pointing to.

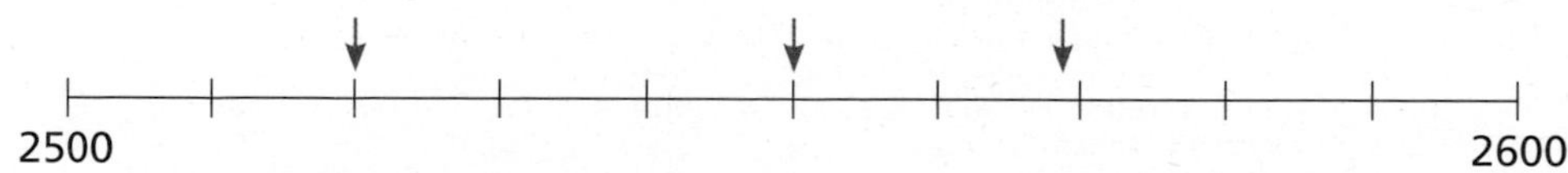

b One of your answers is an estimate. Which answer, and why?

4. Mark each of these numbers on a copy of the number line.

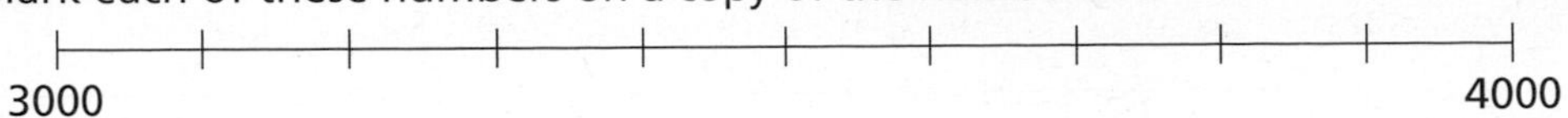

a 3500 | b 3100 | c 3750 | d 3998

5. Mark each of these numbers on a copy of the number line.

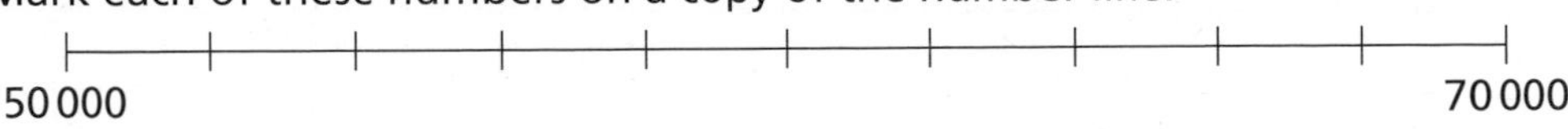

a 60 000 | b 51 000 | c 66 000 | d 68 500

6. Estimate the number that each arrow is pointing to.

a

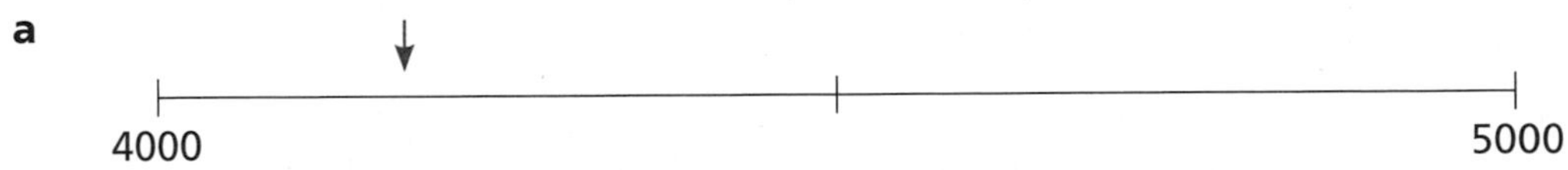

b

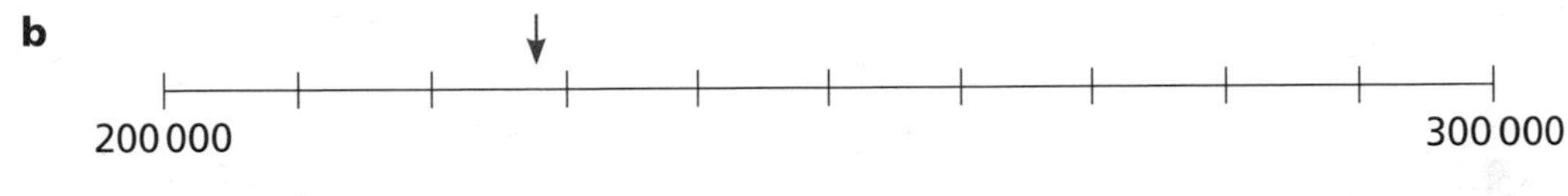

c

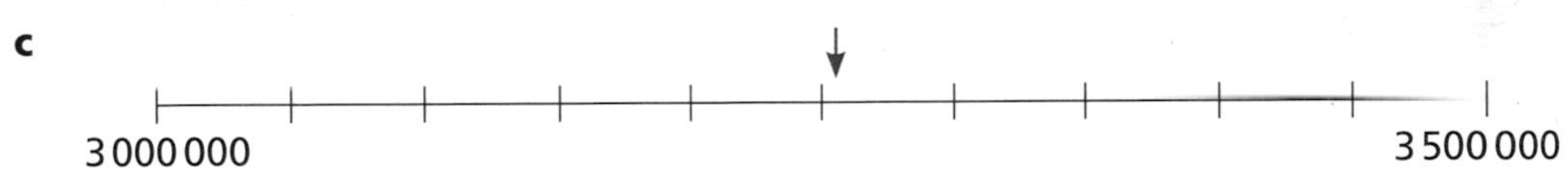

What do you think?

1. Abdullah is asked to mark 3350 on the number line.

a How do you know Abdullah is wrong?

b Between which two letters should the arrow be drawn?

c Describe a method for Abdullah so that he could mark the point as accurately as possible.

Consolidate – do you need more?

1. What is the value of the 5 in each of these numbers?

a 75 | b 350 | c 523 | d 4205 | e 15 900

2. Write these numbers in words.

a 32 | b 635 | c 2400 | d 10 740 | e 12 503

3 What is the arrow pointing to on each number line?

a

50 60

b

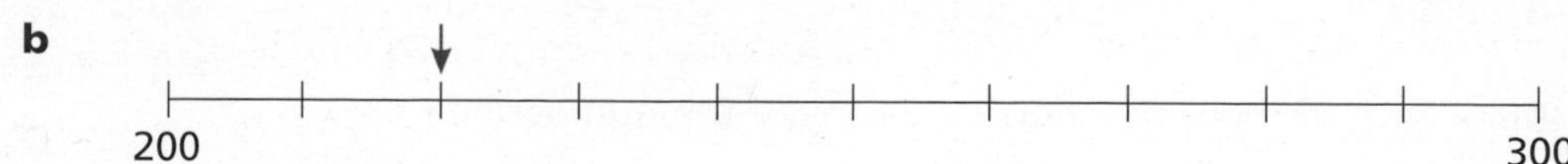

c

320 340

Stretch – can you deepen your learning?

1 Rhys makes a number on a place value grid.

He uses exactly 12 counters.

The number lies between 5000 and 6000.

The number has no hundreds.

Write down all the numbers he could make.

Explain your strategy.

2 Mr Patel writes a number on the board.

15 732 918

Harry starts reading it out loud and says

"a hundred and fifty-seven million…"

Mr Patel stops him and says he is reading it wrong.

Help Harry say the number.

3 Label 1300 on a copy of each of these number lines.

a

b

0 10 000

c

0 5000

d

500 1500

Reflect

1 How can you use a place value grid to help you write and read a number?

2 How do you work out the intervals that a number line goes up in?

4.2 Comparing and ordering integers

Small steps

- Compare two numbers using =, ≠, <, >
- Order a list of integers

Key words

Ascending – increasing in size

Descending – decreasing in size

Inequality symbol – a symbol comparing values showing which is greater and which is smaller

Are you ready?

1 Which is the greater number in each pair?

a 13 or 30 **b** 220 or 190 **c** 103 or 57 **d** 4000 or 999

2 Use the words **equal to, greater than**, or **less than** to complete the statements.

a 42 is ____________ 24

b 39 is ____________ 309

c One thousand is ____________ nine hundred.

d 5 tens is ____________ 50.

3 Put the numbers in order, starting with the smallest.

a 17, 7, 72 **b** 105, 95, 100

c 91, 19, 109 **d** 1500, 1005, 150

e Are your numbers in ascending or descending order?

4 Put the numbers in order, starting with the greatest.

a 79, 97, 96 **b** 30, 200, 1000

c 117, 109, 190 **d** 1111, 11, 111

e Are your numbers in ascending or descending order?

Models and representations

You can use cubes to visualise which way around to use the equality and **inequality symbols**.

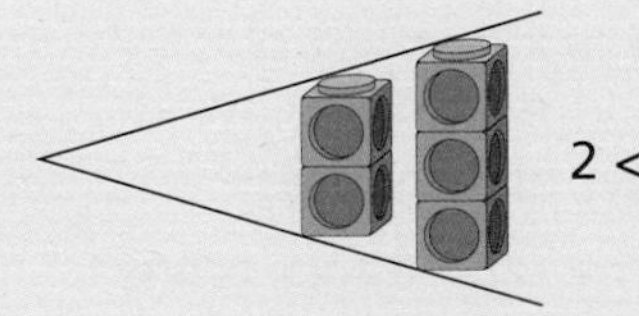

2 is less than 3

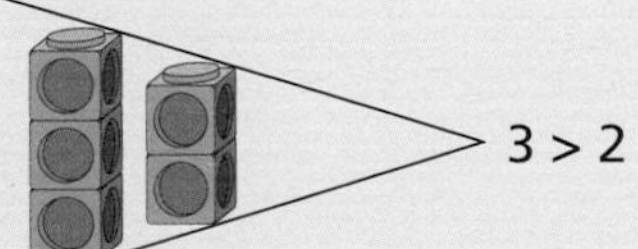

3 is greater than 2

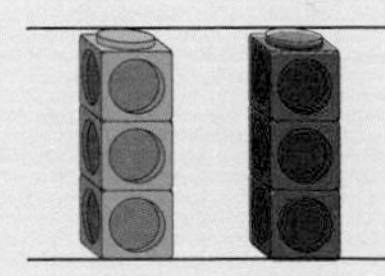

3 is equal to 3

4.2 Comparing and ordering integers

A place value grid is useful to compare numbers.

Thousands	Hundreds	Tens	Ones
3	2	0	5
3	3	0	5
3	2	5	0
	3	5	5

The following symbols will be used throughout

= equal to **≠ not equal to** **> greater than** **< less than**

Example 1

Use >, < or = to compare 205 and 199

Write both numbers in a place value grid.

Highest place value

Hundreds	Tens	Ones
2	0	5
1	9	9

Hundreds are worth more than tens and ones so look at that column first.

205 has more hundreds than 199 so 205 is greater than 199

The greater than symbol is >

205 > 199

Example 2

Use >, < or = to compare 3162 and 3181

Highest place value

Thousands	Hundreds	Tens	Ones
3	1	6	2
3	1	8	1

Write both numbers in a place value grid.

Thousands is the highest place value. The numbers have the same number of thousands.

Look at the next highest place value, which is hundreds. They have the same number of hundreds.

So look at the tens. **6** tens is less than **8** tens, so 31**6**2 is less than 31**8**1

The less than symbol is <

3162 < 3181

Practice 4.2A

1 Write each statement in words.

a 56 > 46 **b** 90 = 9 tens **c** 104 < 105 **d** 90 ≠ 9 ones

Could you write each statement in a different way?

2 Use < or > to complete the statements.

a 562 ◯ 381

Hundreds	Tens	Ones
5	6	2
3	8	1

b 1098 ◯ 2000

Thousands	Hundreds	Tens	Ones
1	0	9	8
2	0	0	0

c 7350 ◯ 7305

Thousands	Hundreds	Tens	Ones
7	3	5	0
7	3	0	5

d 10 799 ◯ 11 003

Ten thousands	Thousands	Hundreds	Tens	Ones
1	0	7	9	9
1	1	0	0	3

3 Use < or > to copy and complete the statements.

You might find it useful to draw a place value grid.

a 650 ◯ 560 **b** 4062 ◯ 4026 **c** 5327 ◯ 6327

d 5103 ◯ 5099 **e** 780 ◯ 7800 **f** 8512 ◯ 10 374

4 Use <, > or = to copy and complete the statements.

a 3 ones ◯ 3 tens **b** 7 tens ◯ 69

c fifty thousand ◯ 10 000 **d** 1 million ◯ 1 000 000

e 4.5 million ◯ 4 000 500 **f** 650 ◯ 65 hundreds

5 Use = or ≠ to copy and complete the statements.

a half a million ◯ 500 000 **b** 57 tens ◯ 57

c sixty-two thousand ◯ 6200 **d** 7 900 000 ◯ 7.9 million

e 9 thousand ◯ 90 hundreds **f** 36 000 250 ◯ $36\frac{1}{4}$ million

6 Write down three numbers that are greater than 104 000 but less than 104 107. Compare your answers with a partner.

7 Find the missing digits to make the statements correct.

a 3☐5 < 324 **b** ☐781 > 7916

c 1264☐ > 12 648 **d** 4302 < 43☐5

What do you think?

1

Marta: There are 11 whole numbers that are greater than 1.3 million but less than 1 300 010

Flo: There are 9 whole numbers that are greater than 1.3 million but less than 1 300 010

Who do you agree with? Explain your answer.

2 Find a missing digit to make the statement correct.

14☐10 < 15 629

Ed: The missing digit has to be less than 6

Faith: The missing digit can be anything.

Who do you agree with? Explain your answer.

Example 3

Write the numbers in **ascending** order.

3417, 2521, 3186, 3452

Thousands	Hundreds	Tens	Ones
3	4	1	7
2	5	2	1
3	1	8	6
3	4	5	2

Ascending means increasing in size, so write the numbers from smallest to largest.

You could write the numbers in a place value grid to compare the place value of each digit before ordering.

↓

Thousands	Hundreds	Tens	Ones
3	4	1	7
2	5	2	1
3	1	8	6
3	4	5	2

First check the thousands column as this has the highest place value.

They all have the same number of thousands apart from 2521 which makes this the smallest number.

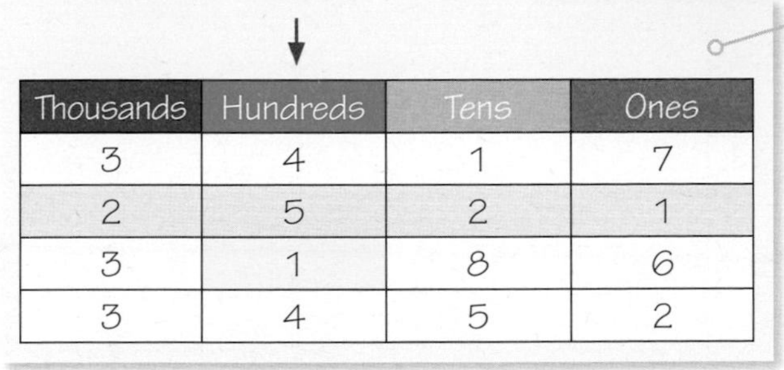

Thousands	Hundreds	Tens	Ones
3	4	1	7
2	5	2	1
3	1	8	6
3	4	5	2

For the other three numbers move to the hundreds column.

The number with the fewest hundreds is 3186 so that is the next smallest number.

The other two numbers have the same amount of hundreds so move onto to comparing the tens.

Thousands	Hundreds	Tens	Ones
3	4	1	7
2	5	2	1
3	1	8	6
3	4	5	2

3417 has fewer tens.

So the numbers in ascending order are 2521, 3186, 3417, 3452

Practice 4.2B

1 Here are some numbers in a place value grid.

TTh	Th	H	T	O
7	4	3	0	2
7	4	2	2	1
6	4	3	0	2
7	3	4	5	2

a Which is the smallest number? How do you know?

b Write the numbers in

i ascending order **ii** descending order.

2 Write each set of numbers in ascending order.

a 3282, 3611, 3265, 3350

b 1062, 2655, 1521, 999

c 5700, 5699, 5701, 5007, 5770

d 12 350, 9000, 10 420, 10 402, 9118

3 Write each set of numbers in descending order.

a 7200, 6980, 7105, 7220

b 5487, 5467, 5497, 5427, 5477

c 33 810, 3381, 33 801, 338 100

d 45 030, 8950, 12 832, 979, 29 760

4 Here are the sales of a computer game each month.

Month	Number of games sold
January	1200
February	1098
March	1850
April	1600
May	2320
June	1755

a Put the **months** in order, starting with the month that sold the fewest games.

b In December there were three times as many games sold as there were in January.

Why do you think this happened?

How many games were sold in December?

5 The table shows the heights, in feet, of the world's six tallest mountains.

Mountain	Height (feet)
Kangchenjunga	28 169
Makalu	27 838
Cho Oyu	26 906
K2	28 251
Lhotse	27 940
Mount Everest	29 029

Write the mountains in height order, starting with the tallest.

6 Here are the prices of four footballers' houses.

A

£1.75 million

B

£1 500 000

C

£1.3 million

D

£1$\frac{1}{2}$million

a Which is the most expensive house?

b Which is the least expensive house?

c Why would it be difficult to write the house prices in order?

What do you think?

Put one digit in each box so that the numbers are in ascending order from the top of the table to the bottom of the table.

Is there more than one solution? Compare with a partner.

	M	HTh	TTh	Th	H	T	O
a		7	5	2	4	6	0
b	2	3	5	0	0		2
c	2		4	6	2	0	5
d	2	4		8	9	1	1
e	2	4	7		5	7	0

Consolidate – do you need more?

1 Copy and complete the statements with "greater than", "less than" or "equal to".

a 82 is ______________ 85

b 341 is ______________ 314

c 6 ones is ______________ 6 tens

d 7590 is ______________ 7 thousand

e fifty thousand is ______________ 50 000

f 800 500 is ______________ 850 000

2 Copy and complete the statements using <, > or =

a 82 ◯ 85

b 341 ◯ 314

c 6 ones ◯ 6 tens

d 7590 ◯ 7 thousand

e fifty thousand ◯ 50 000

f 800 500 ◯ 850 000

3 Write each set of numbers in order, starting with the smallest.

a 835, 7299, 661, 974, 4365

b 8500, 9500, 8200, 8050, 9010

c 73, 301, 67, 550, 703, 113

d 12 320, 12 355, 12 341, 12 095, 12 300

4 Write each set of numbers in order, starting with the greatest.

a 407, 411, 470, 401, 417

b 512, 5012, 5112, 502, 5222

c 57 888, 56 999, 57 000, 57 399, 57 900

d 100 400, 110 400, 101 400, 100 040, 140 000

Stretch – can you deepen your learning?

1 Here are the top five countries when ranked by population size.

Country	Population in 2020	Expected population in 2050
United States	331 002 651	398 328 349
Pakistan	220 892 340	290 847 790
China	1 439 323 776	1 301 627 048
India	1 380 004 385	1 656 553 632
Indonesia	273 523 615	300 183 166

a Write the countries in order, starting with the country with the highest population in 2020

b Write the countries in order, starting with the country with the highest expected population in 2050

c Put the countries in descending order based on their expected increase in population from 2020 to 2050

2 a, b and c are positive integers less than 10

$a > b > c$

Put the numbers represented by the expressions in ascending order.

$100b + 10a + c$, $100c + 10a + b$, $100a + 10b + c$, $100c + 10b + a$

3 $x > 2$

Put the numbers represented by the expressions in descending order.

x, $100x$, $10x$, $x + 17$

Reflect

Always, sometimes or never true?

When comparing and ordering numbers, you start by looking at the digit in the ones position.

Use some examples to support your reasoning.

4.3 Using ordered lists: the range and the median

Small steps

- Find the range of a set of numbers
- Find the median of a set of numbers

Key words

Range – the difference between the greatest value and the smallest value in a set of data

Median – the middle number in an ordered list

Measure of spread – shows how similar or different a set of values are

Are you ready?

1 Write down the greatest number in each list.

a 34, 72, 38 **b** 51, 9, 83, 112

c 1050, 150, 500, 950

2 Write down the smallest number in each of the lists in question 1.

3 Work out

a 95 – 76 **b** 140 – 83 **c** 295 – 273 **d** 1000 – 940

4 Put each set of numbers in order from smallest to largest.

a 345, 712, 341, 502 **b** 423, 71, 2010, 5

5 Here are some number lines.

Work out the number that is halfway between the two numbers marked.

a

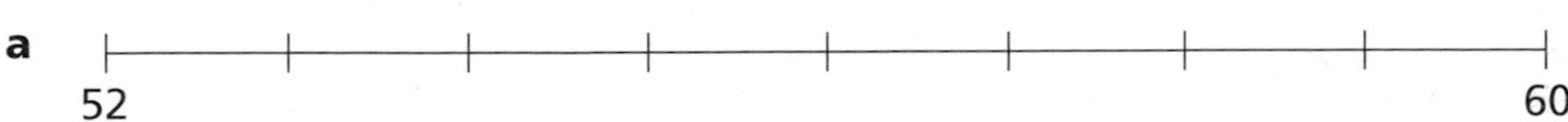

b

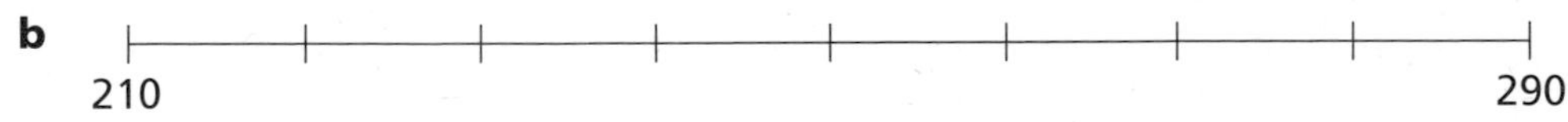

c

d

Describe your method.

4.3 Using ordered lists: the range and the median

Models and representations

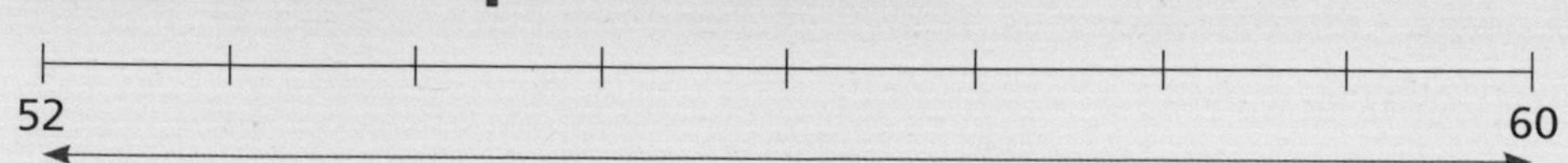

Number lines are useful for ordering data and working out the difference between a pair of numbers.

The **range** is a **measure of spread**. To find the range of a set of data, subtract the smallest value from the largest value.

The **median** is a type of average. To find the median of a set of data

- arrange the numbers in order (usually from smallest to largest),
- then find the middle of the numbers.

Example 1

Here are the shoe sizes of six people.

6 10 10 5 11 7

What is the range of the shoe sizes?

First identify the smallest and largest values. You can do this by putting the data in order.

To work out the range, subtract 5 from 11.

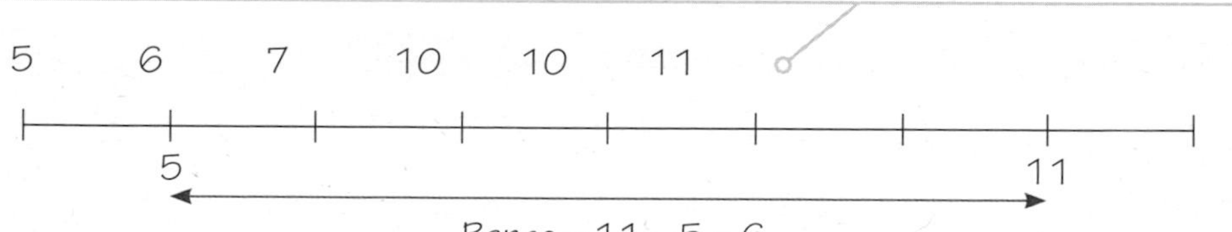

The range of the shoe sizes is 6

Example 2

Rob goes shopping to buy some clothes. The range of the cost of his items is £13.
The cheapest item he bought was £21. How much did the most expensive item cost?

Least expensive

£21

Most expensive

£21 + £13 = £34

Range = £13

The most expensive item cost £34

Sketch a number line

Practice 4.3A

1. Work out the range of the masses of each set of parcels.

   a 148 g, 220 g, 385 g, 431 g, 500 g

   b 220 g, 385 g, 148 g, 431 g, 500 g

   What do you notice?

2. Here are the number of miles Beth drives each day in a week.

   31, 84, 120, 43, 17, 35, 102

   a Beth says "The range of the miles I drive is 17 to 120"

   Explain the mistake Beth has made.

   b Find the range of the number of miles she drives in a week.

3. The table shows the number of copies of a magazine sold per month.

Month	Number of copies sold
January	1200
February	1098
March	1850
April	1600
May	2320
June	1755

   Work out the range of the number of copies sold.

4. Here is the temperature in six cities.

   23°C, 31°C, 8°C, 18°C, 18°C, 28°C

   Work out the range of the temperatures.

5. What mistake has been made each time?

   a

64 cm	111 cm	93 cm	48 cm	180 cm

   The range = 180 – 64 = 116 cm

   b

2 kg	180 g	3 kg	520 g	80 g

   The range = 520 – 2 = 518 g

6. The range of a set of numbers is 12.

   The largest number is 63.

   What is the smallest number?

7 The range of the times taken to complete a puzzle is 26 seconds.

The fastest time was 1 minute 40 seconds.

What was the slowest time?

8 The table shows the number of loaves of bread sold each day in a bakery.

Day	Mon	Tues	Wed	Thurs	Fri
Number of loaves sold	28	39	45	48	37

a What is the range of number of loaves of bread sold?

b The bakery makes 50 loaves of bread each day. Find the range of the number of loaves not sold.

What do you think?

1 Below are the heights of eight children in a line.

143 cm	138 cm	151 cm	139 cm
145 cm	155 cm	152 cm	140 cm

a Another child joins the line.
The range of the heights stays the same.
What can you say about the height of the extra child?

b Another child joins the line.
The range increases by 2 cm.
What could the height of the child be?

Compare your answer with a friend.

Example 3

Here are the number of cups of coffee sold in a coffee shop for 5 days.

120, 143, 210, 192, 245

Find the median number of coffees sold.

120 143 192 210 245

~~120~~ ~~143~~ (192) ~~210~~ ~~245~~

The median number of coffees sold is 192

First put the numbers in order, starting with the smallest.

Then identify the middle number. You can do this by crossing one off from each end of the list until you have one number left.

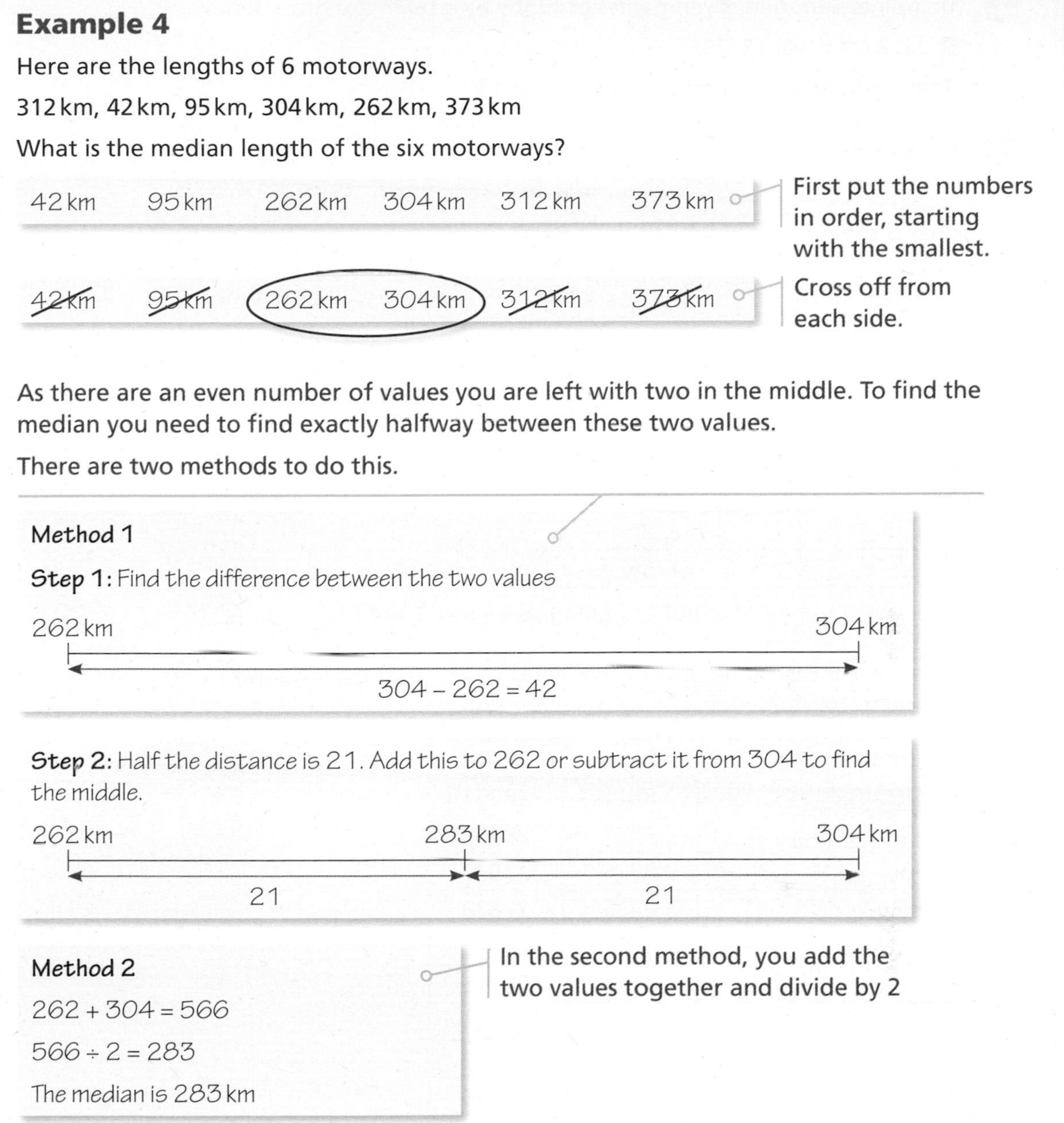

Example 4

Here are the lengths of 6 motorways.

312 km, 42 km, 95 km, 304 km, 262 km, 373 km

What is the median length of the six motorways?

42 km 95 km 262 km 304 km 312 km 373 km

First put the numbers in order, starting with the smallest.

~~42 km~~ ~~95 km~~ 262 km 304 km ~~312 km~~ ~~373 km~~

Cross off from each side.

As there are an even number of values you are left with two in the middle. To find the median you need to find exactly halfway between these two values.

There are two methods to do this.

Method 1

Step 1: Find the difference between the two values

262 km 304 km

304 − 262 = 42

Step 2: Half the distance is 21. Add this to 262 or subtract it from 304 to find the middle.

262 km 283 km 304 km

21 21

Method 2

In the second method, you add the two values together and divide by 2

262 + 304 = 566

566 ÷ 2 = 283

The median is 283 km

Practice 4.3B

1 Work out the median of each set of data.

a £21, £35, £17, £32, £25

b 104 mm, 151 mm, 96 mm, 110 mm, 67 mm, 125 mm, 12 mm

c

d 80 g, 120 g, 50 g, 100 g, 60 g, 90 g

2. The points scored in seven games of rugby by a team are given below.

   12, 24, 32, 17, 10, 15, 26

   What is the median number of points scored?

3. Here are the heights of 10 plants.

   48 cm, 92 cm, 38 cm, 55 cm, 84 cm, 106 cm, 73 cm, 98 cm, 45 cm, 63 cm

   What is the median height of the plants?

4. The table shows the number of copies of a magazine sold per month.

Month	Number of copies sold
January	1200
February	1098
March	1850
April	1600
May	2320
June	1756

   Work out the median number of magazines sold.

5. The table shows the heights of the world's 10 highest mountains.

Mountain	Metres	Feet
Mount Everest	8848	29 029
K2	8611	28 251
Kangchenjunga	8586	28 169
Lhotse	8516	27 940
Makalu	8485	27 838
Cho Oyu	8201	26 906
Dhaulagiri	8167	26 795
Manaslu	8163	26 781
Nanga Parbat	8126	26 660
Annapurna	8091	26 545

   a Find the median height, giving your answer in metres.

   b Find the median height, giving your answer in feet.

6. The masses of five parcels are shown.

   What is the median mass?

7 **a** These five cards have a median of 12

7	20	30	12	

What could the value of the missing card be?

b These six cards have a median of 12

7	20	30	14	9	

What is the value of the missing card?

What do you think?

1 Here are the heights of five basketball players.

200 cm 195 cm 201 cm 197 cm 194 cm

Here are the heights of six gymnasts.

154 cm 152 cm 160 cm 158 cm 146 cm 158 cm

Discuss what you notice about the median height of the basketball players compared to the median height of the gymnasts.

Explain why this might be the case.

2 **a** Write down five numbers that have a median of 7 and a range of 6

b Write down four numbers that have a median of 7 and a range of 6

Compare your answers with a friend.

Consolidate – do you need more?

1 Work out the range of each set of data.

a 12, 15, 14, 17

b 50, 80, 20, 35, 70, 55, 90

c £200, £250, £190, £210, £192, £290

d 93 g, 57 g, 72 g, 18 g, 33 g

2 The range of some numbers is 10. The smallest number is 4. What is the largest number?

3 The range of some numbers is 25. The largest number is 100. What is the smallest number?

4 Work out the median of each set of data.

a 10, 7, 9, 4, 8

b 29, 46, 12, 18, 32, 16, 59, 41, 27

c 15 kg, 20 kg, 18 kg, 21 kg

d 160 m, 190 m, 250 m, 200 m

Stretch – can you deepen your learning?

1 Here are five cards.

26	28	30	36	40

a Another two cards are added.

The median is now 33 and the range is now 20

What are the values of the two cards?

b Another card is added.

The range and median both increase.

What is the new median?

Reflect

What's the same and what's different about finding the median and finding the range?

4.4 Rounding: powers of 10

Small steps

- Round integers to the nearest power of ten

Key words

Round – to give an approximate value of a number that is easier to use

Powers of 10 – the result of multiplying 10 by itself a number of times to give a value such as 10, 100, 1000, 10000 and so on

Are you ready?

1 Between which two multiples of ten does each number lie?

The first one has been done for you.

a 46 lies between 40 and 50

b 73 lies between ______ and ______

c 95 lies between ______ and ______

d 276 lies between ______ and ______

2 Between which two multiples of one hundred does each of these numbers lie?

The first one has been done for you.

a 355 lies between 300 and 400

b 784 lies between ______ and ______

c 1740 lies between ______ and ______

d 3918 lies between ______ and ______

3 Use the number lines to help you.

a Which multiple of 10 is 148 closer to?

b Which multiple of 100 is 148 closer to?

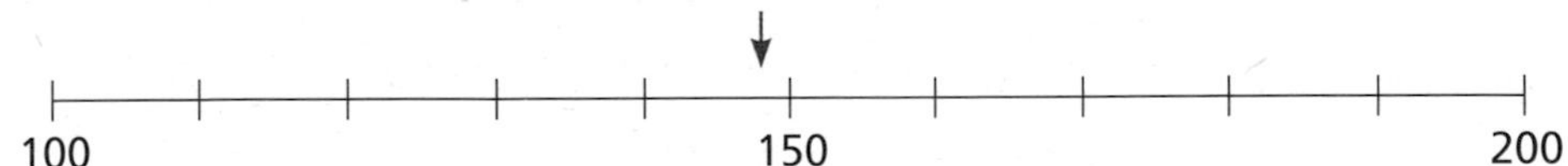

Models and representations

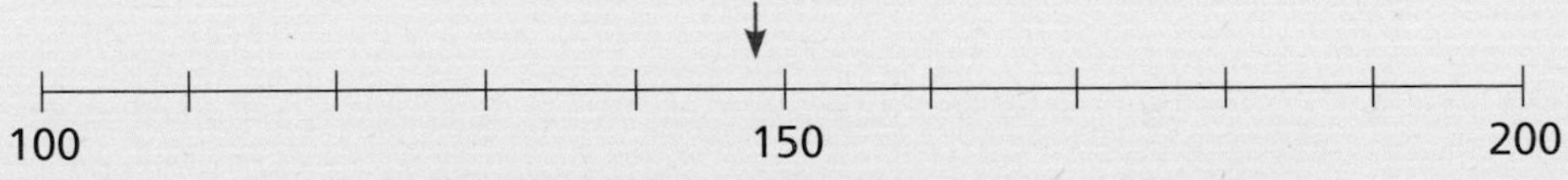

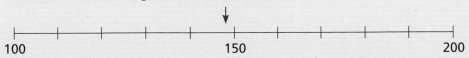

Number lines can help you with **rounding** to any **power of 10**.

Imagine a concert review that reads: "There were 40000 people at the concert."

It is unlikely there were exactly 40000 people at the concert. The exact figure could have been 36841 or 43294 or many other values. The number of people has been rounded to the nearest ten thousand.

4.4 Rounding: powers of 10

Example 1

Ben spends £816 on a holiday. How much money did he spend, to the nearest hundred pounds?

800
816
900
816 is closer to 800 than 900
So £816 rounded to the nearest hundred is £800

Draw a number line.

You want to round to the nearest hundred so think about the two multiples of one hundred that 816 lies between.

Example 2

Round 3500 kg to the nearest thousand.

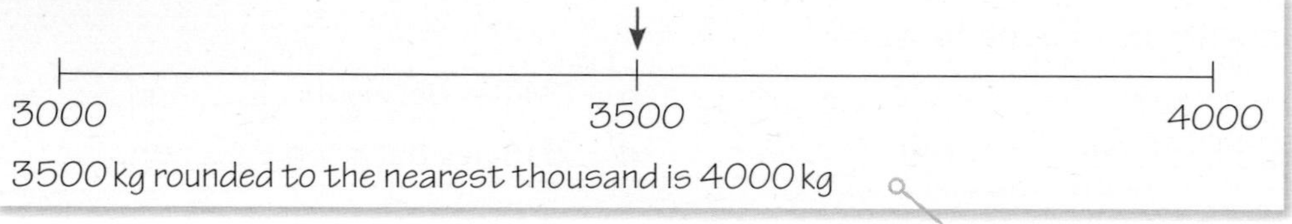

When a number is exactly halfway between two other numbers you always round to the greater number.

3500 kg is exactly halfway between 3000 kg and 4000 kg

Example 3

There are 47 563 people at a football game. Round this number to the nearest

a 10 **b** 100 **c** 1000 **d** 10 000

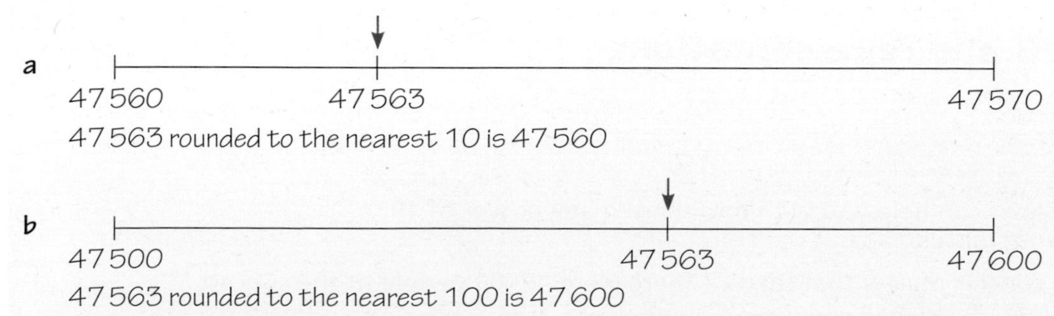

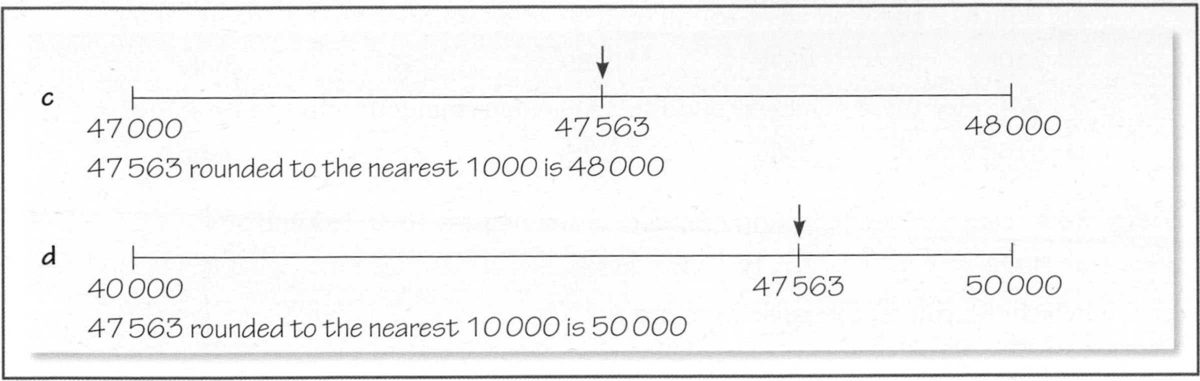

Practice 4.4A

1. Use the number lines to help you round.

   a Round 382 to the nearest 10.

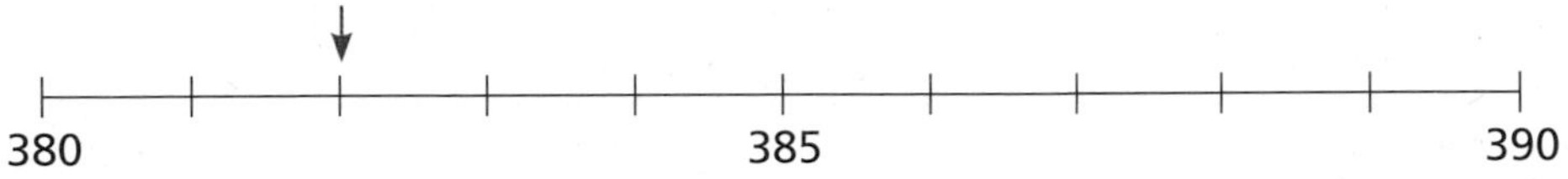

   b Round 382 to the nearest 100.

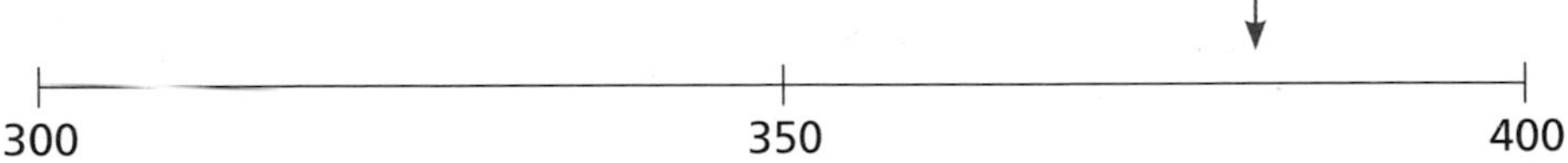

2. Use the number lines to help you round.

   a Round 5783 to the nearest 1000.

   b Round 5783 to the nearest 100.

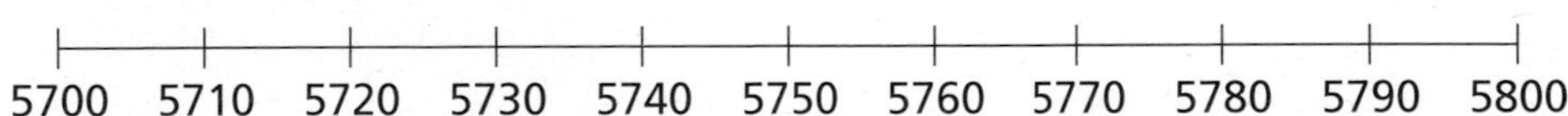

   c Round 5783 to the nearest 10.

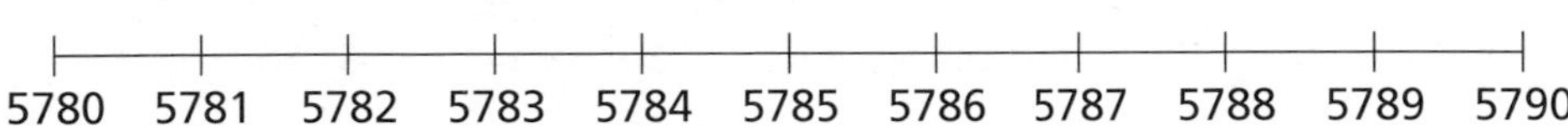

3. Ed says "When you round a number to the nearest 100, you look at the number of 100s there are. If it is equal to or greater than 5, you round up to the next 100.

   For example 37 748, there are 7 hundreds, so you round up to 37 800"

   What mistake has Ed made?

4. a Which of these numbers round to 200 when rounded to the nearest 100?

   176 139 218 150 250

   b Which of these numbers round to 720 when rounded to the nearest 10?

   722 703 709 715 734

c Which of these numbers round to 3000 when rounded to the nearest 1000?

3106 3095 2785 2961 3500

d Which of these numbers round to 3000 when rounded to the nearest 100?

3106 3095 2785 2961 3500

5 Round each of the following numbers to the nearest 1000, 100 and 10

a 1728 **b** 26 380 **c** 172 539

Which did you find it easier to round to?

6 Copy and complete the table

Number	Rounded to nearest 1000	Rounded to nearest 10 000	Rounded to nearest 100 000
265 380			
412 119			
76 289			

7 What could the missing digits be?

a 173☐ to the nearest 10 is 1730 **b** 173☐ to the nearest 100 is 1700

c 26☐09 to the nearest 100 is 26 500 **d** 26☐09 to the nearest 1000 is 27 000

8 Beth has made a six-digit number, but she does not know the last two digits.

Beth says: "I can still round my number to the nearest 1000"

Is Beth correct? Explain your answer with an example.

9 Here are some digit cards.

2 3 5 6 9

Use the cards to

a Make a 3-digit number that rounds to 600 to the nearest 100

b Make a 3-digit number that rounds to 600 to the nearest 10

c Make a 4-digit number that rounds to 3000 to the nearest 1000

d Make a 4-digit number that rounds to 3000 to the nearest 100

e Make a 5-digit number that rounds to 60 000 to the nearest 10 000

How many different answers can you find?

10 Copy and complete the table

Number	Rounded to nearest 100	Rounded to nearest 1000	Rounded to nearest 10 000
	26 700		
	27 900	28 000	
	46 500	46 000	50 000

What do you think?

1 Here is a newspaper article.

> News
>
> **60 000 people attended United versus Rovers football match yesterday**

a Do you think exactly 60 000 people attended the football match?

b What is the greatest number of people that could have attended the football match?

c What is the least number of people that could have attended the football match?

2 Here are some digit cards.

2	8	1	0	3	7

a Use all of the cards to find five different numbers that round to 700 000 when rounded to the nearest 100 000

b Find the range of your answers to part **a**.

Consolidate – do you need more?

1 Round each number to the nearest ten.

a 68 **b** 75 **c** 87 **d** 95

e 381 **f** 212 **g** 759 **h** 899

i 7228 **j** 997 **k** 35 246 **l** 254 884

2 Round each number to the nearest hundred.

a 830 **b** 750 **c** 68 **d** 997

e 3412 **f** 8248 **g** 10 336 **h** 735 670

3 Round each number to the nearest thousand.

a 3911 **b** 6505 **c** 783 **d** 9999

e 34 620 **f** 71 299 **g** 44 498 **h** 230 865

4 Round each number to the nearest ten thousand.

a 20 090 **b** 73 882 **c** 18 567 **d** 9500

e 534 622 **f** 96 510 **g** 1 385 200 **h** 999 999

Stretch – you can deepen your learning?

1 The masses of these boxes have all been rounded to the nearest 100 g.

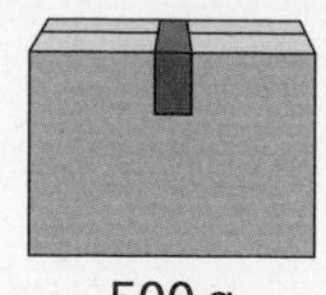
500 g

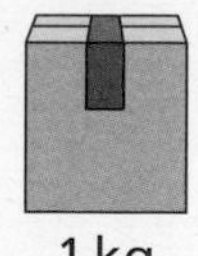
1 kg

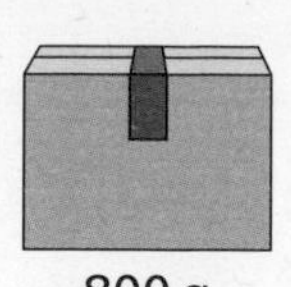
800 g

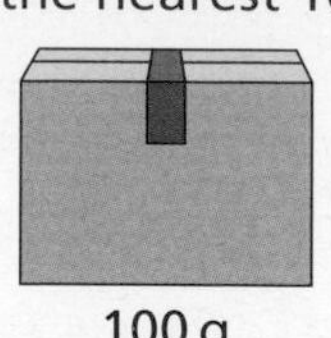
100 g

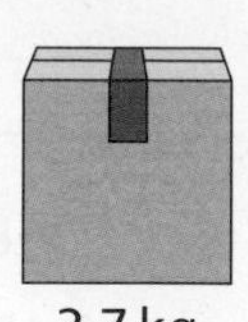
2.7 kg

Find the minimum possible combined mass of all five boxes.

2 Here is a number where two of the digits have been replaced by letters:

67 A4B

For what values of A and B will this number round to 67 000 to the nearest thousand?

Does the value of B matter?

3 The number ABCDEF rounds to 310 000 to the nearest thousand.

What values can each letter take?

Reflect

When rounding 125 736 to the nearest thousand, which digits are the important digits to consider?

Does the value of the other digits matter?

What if you were rounding the number to the nearest ten, hundred or ten thousand?

4.5 Understanding decimals

Small steps

- Understand place value for decimals
- Position decimals on a number line

Key words

Tenth – one of the parts when a whole is split into 10 equal parts

Hundredth – one of the parts when a whole is split into 100 equal parts

Thousandth – one of the parts when a whole is split into 1000 equal parts

Are you ready?

1 What is the value of the 7 in each of these numbers?

a 17 **b** 127 **c** 706 **d** 93.7

2 Copy and complete the following

a 36.2 = 30 + 6 + □ **b** 195.3 = 100 + □ + □ + □

3 **a** Write down 3 tenths as a decimal.

b Write down 3 hundredths as a decimal.

4 How many tenths are there in 1 whole?

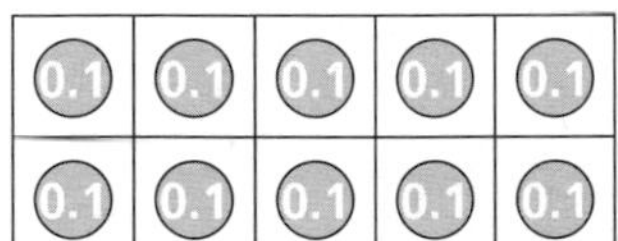

5 Faith is counting in tenths

5.7, 5.8, 5.9, 5.10, 5.11

Explain the mistake that Faith has made.

Models and representations

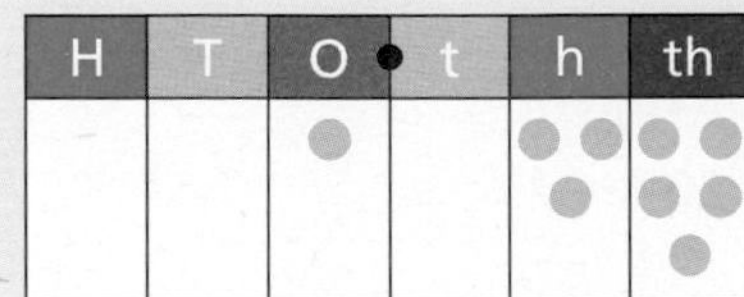

Place value grids and counters are useful for showing decimals.

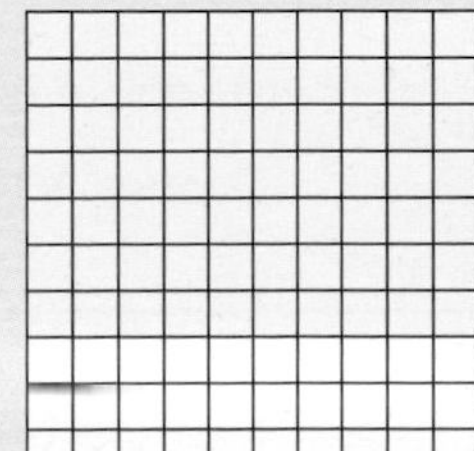

The hundred square shows 0.7

4.5 Understanding decimals

Think about where you might see numbers with a decimal point in real life.

You have already used number lines to position whole numbers.

Number lines can be seen in many places, but you may not have realised they are number lines.

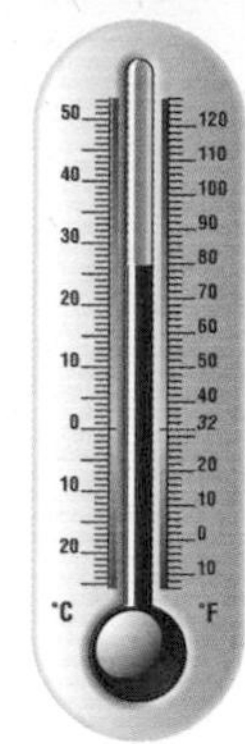

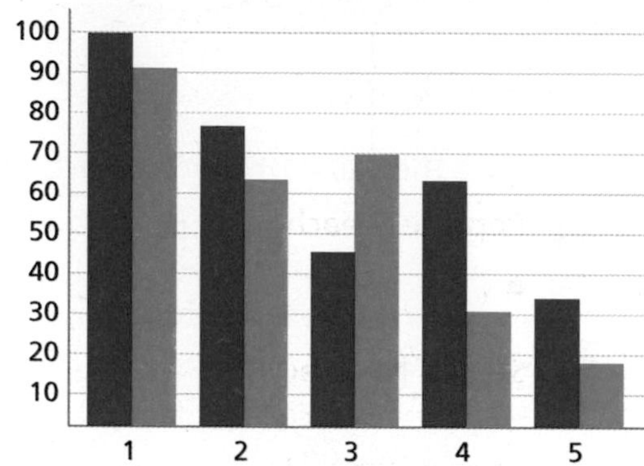

Can you see where the number lines are in each picture?

Example 1

What is the value of the 6 in the number 45.169?

Tens	Ones	tenths	hundredths	thousandths
4	5	1	6	9

The 6 has a value of 6 **hundredths**.

To help you might want to write the number in a place value table.

The 6 is the second digit after the decimal point.

This means it is in the hundredths column.

Example 2

How many **tenths** are there in the number 2.7?

2.7 is 2 ones (or wholes) + 7 tenths

2.7 has 27 tenths.

1 whole is made up of 10 tenths

2 ones is the same as 20 tenths

Practice 4.5A

1 What numbers are represented in each place value grid?

a

H	T	O	t	h
	●	●	●●●●●●●	●

b

H	T	O	t	h
		●		●●●●●

c

H	T	O	t	h	th
	●●	●●●	●	●●●●●	●●●●●●

d

H	T	O	t	h
●●		●●●	●●●●●●	

2 Represent each of these numbers using place value counters.

a 1.5 **b** 1.52 **c** 26.73 **d** 13.035

3 Samira has made this number on a place value grid.

H	T	O	t	h
		●●	●●●	

a What number has Samira made?

b Samira adds 7 hundredths to her table. What number has Samira made now?

c Samira adds 5 tenths to her table. What number has Samira made now?

4 a Which of these numbers have 7 in the tenths position?

13.7 15.72 12.07 m 72.35

b Write down two more numbers that have 7 in the tenths position.

5 a Which of these numbers have 3 hundredths?

300 1.35 kg 61.13 0.137

b Write down two more numbers that have 3 in the hundredths position.

6 The number 213.65 has 5 tenths as this is the second digit after the decimal point.

Is Flo correct?

7 Write down the value of the underlined digit in each of these numbers.

a 15.<u>8</u> **b** 9.<u>6</u>45 **c** 503.<u>6</u>1

d <u>6</u>34 **e** <u>6</u>3.4 **f** <u>6</u>.34

g 0.<u>6</u>34 **h** 0.6<u>3</u>4 **i** 0.63<u>4</u>

8 Copy and complete the following

a 200 + 60 + 9 + 0.3 = ☐

b 7000 + 200 + 5 + 0.8 = ☐

c 900 + 20 + 0.1 + 0.04 = ☐

d 1000 + 720 + 0.9 = ☐

e 60 + 0.06 = ☐

f 205.62 = 200 + 5 + ☐ + ☐

g 19.2 = 10 + 9 + ☐

h 19.2 = 10 + 7.2 + ☐

9 Huda has made this number.

0.1 0.1 0.1 0.1 0.1 0.1 0.1 0.1 0.1 0.1 0.1 0.1 0.1

Rob has made this number.

Huda says she has made the same number as Rob. Explain why Huda is correct.

10 **a** How many tenths are there in each of these numbers?

i 5 **ii** 7 **iii** 7.2 **iv** 19.5

b How many hundredths are there in each of these numbers?

i 3 **ii** 8 **iii** 8.1 **iv** 8.15

c How many hundredths have the same value as 7 tenths?

What do you think?

1 Abdullah is investigating numbers that have more digits after the decimal point.

He types a number into his calculator.

0.5081923

a Tell a partner the value of each of the digits after the decimal point.

For example, the 5 is 5 tenths.

How far can you get along the number together?

b Discuss with your partner what you noticed.

2 Benji has £25.68 in his money box.

Describe the value of the 6 in this amount.

Explain how it relates to tenths.

Example 3

What numbers are the arrows pointing to?

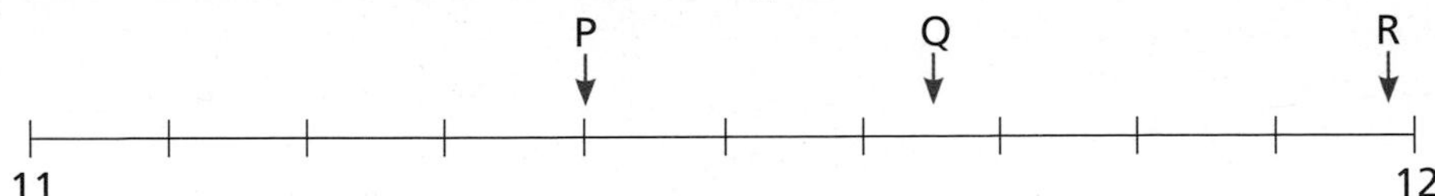

The number line goes from 11 to 12

It has 10 equal intervals.

One whole divided into ten means that this is going up in 0.1s

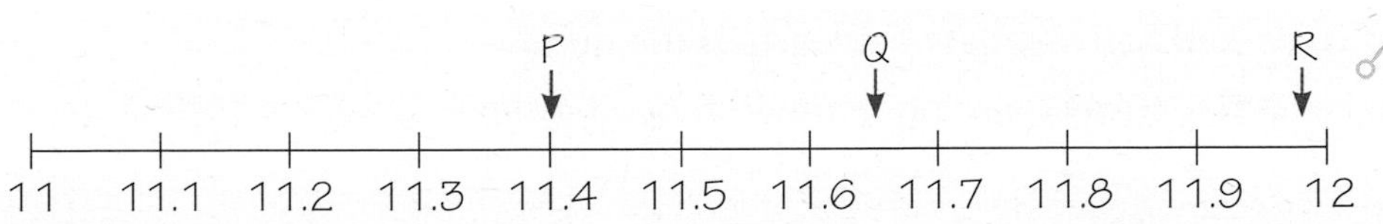

P is equal to 11.4

Q lies halfway between 11.6 and 11.7 so is 11.65

R lies more than halfway between 11.9 and 12. It is much closer to 12 than 11.9, so it is about 11.98

This is an **estimate**.

Example 4

What number is the arrow pointing to?

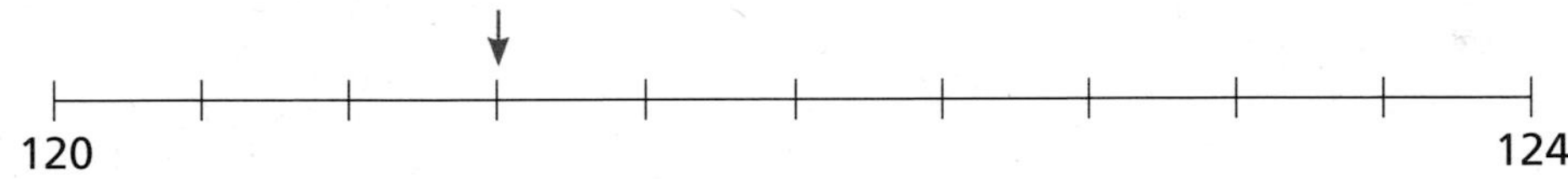

There are 10 equal intervals.

The number line goes from 120 to 124, a distance of 4.

$\frac{4}{10} = 0.4$

120 120.4 120.8 121.2 124

The arrow is pointing to 121.2

Write down the number of equal intervals.

Work out the distance of the number line.

Divide 4 by 10 to work out how much it increases by each time.

$\frac{4}{10}$ is equal to 4 tenths which is 0.4

4.5 Understanding decimals

Practice 4.5B

1 Find the missing values on each of the number lines.

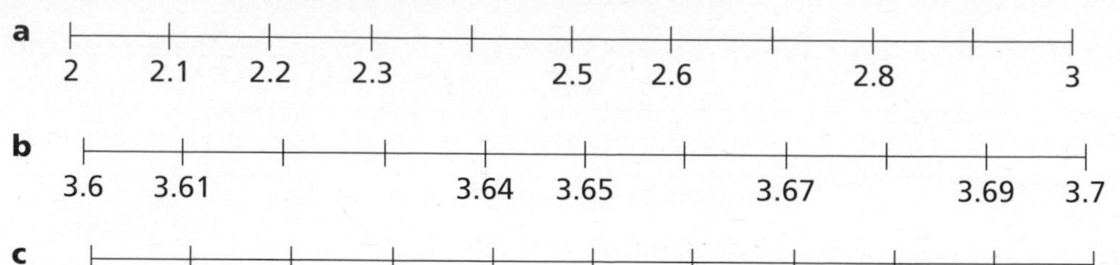

2 The arrow is marked halfway between each of the numbers.

What number is the arrow pointing to on each number line?

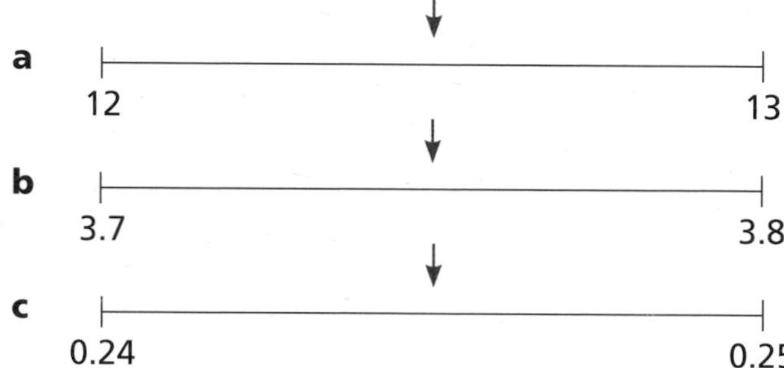

3 What numbers are the arrows pointing to?

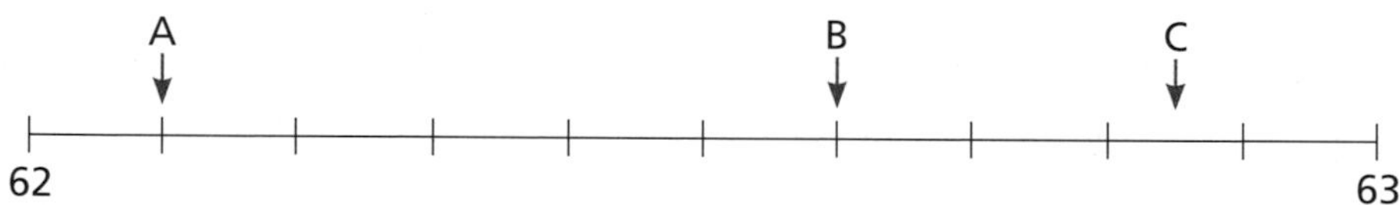

4 What numbers are the arrows pointing to?

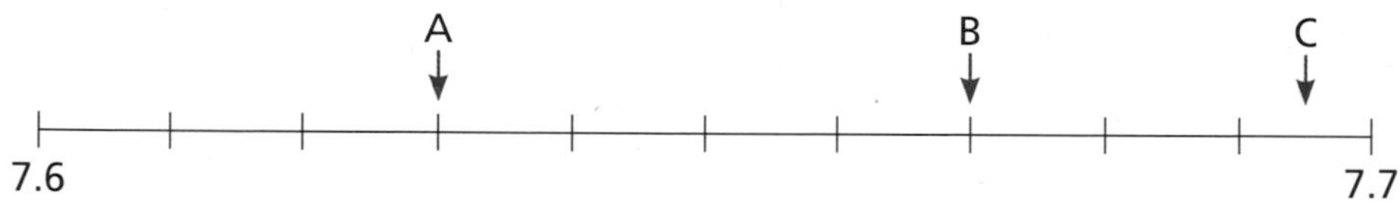

5 Emily says the arrow is pointing to about 128.27

Is Emily correct? Explain your answer.

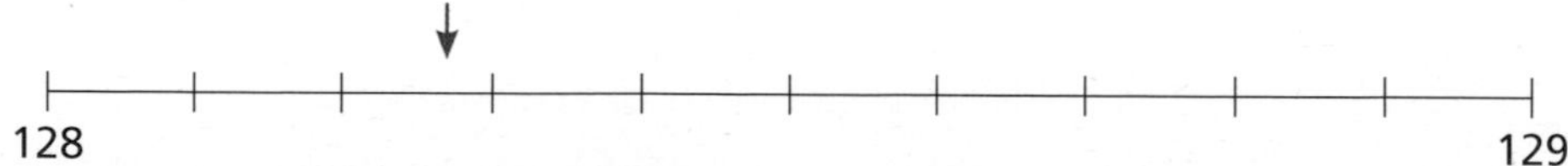

6 Copy the number line into your book.

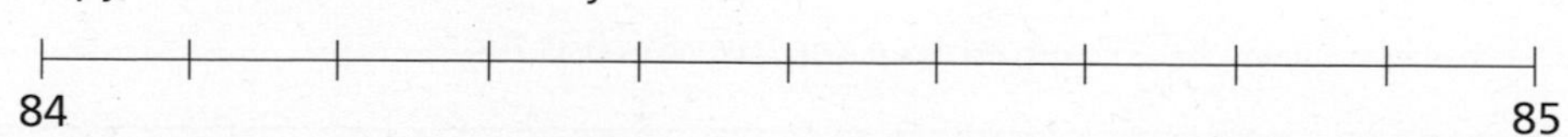

Label each of these numbers on the number line.

A 84.8 B 84.25 C 84.48

7

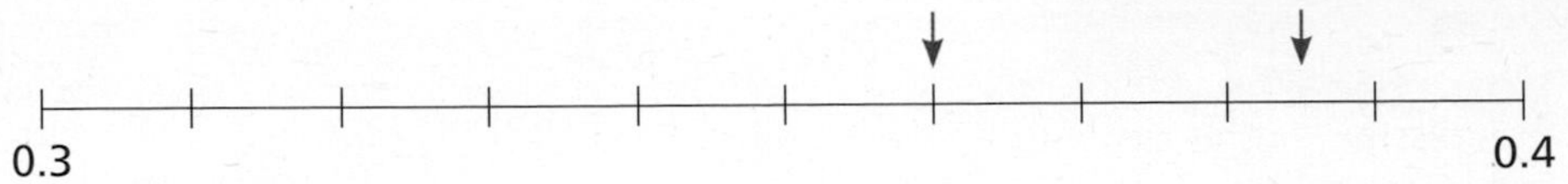

a Write down three numbers that lie between the two arrows marked on the number line.

b Can you write down a number with five digits after the decimal point that lies between the two arrows?

8 What is the mass (in kg) of each box?

a

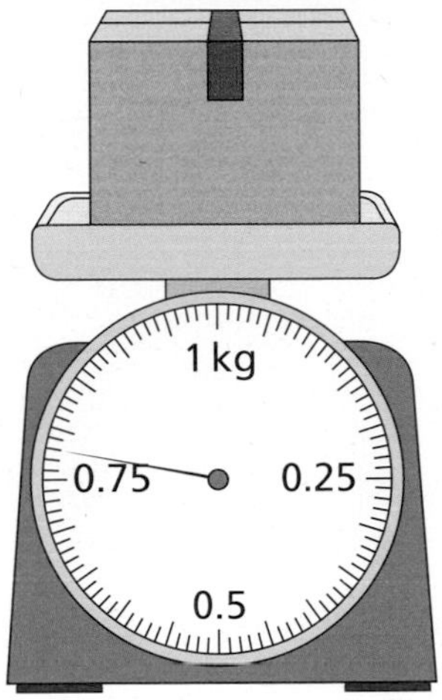

b

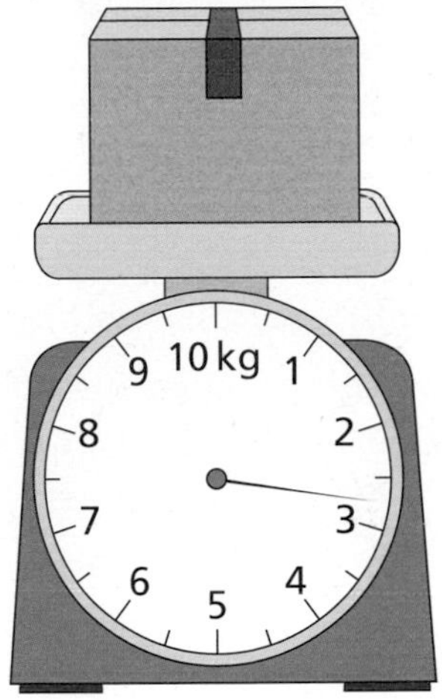

9 Here is part of a thermometer.

a What temperature is shown?

b The temperature decreases by 0.5 degrees.

What is the new temperature?

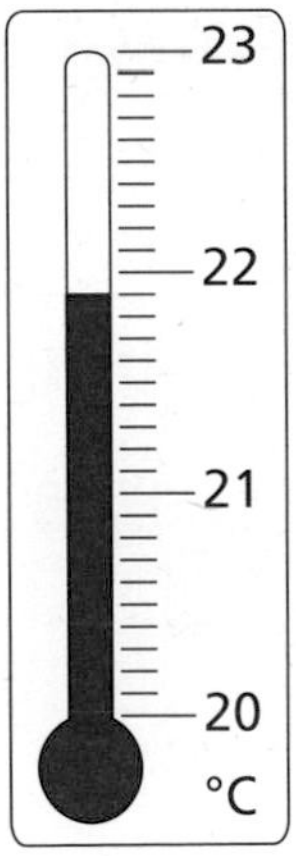

What do you think?

1 Here is a number line.

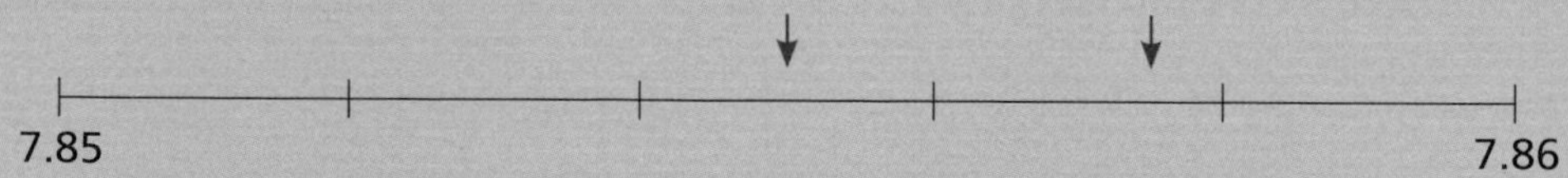

a What two numbers are the arrows pointing to?

b Amina says 7.856 lies between these two numbers.

Show that Amina is correct.

c Amina adds three digits to the end of her number.

Does Amina's number still lie between the two arrows?

Is it possible to tell?

2 Two ferries are sailing from Dover to Calais.

The diagram shows the positions of the two ships.

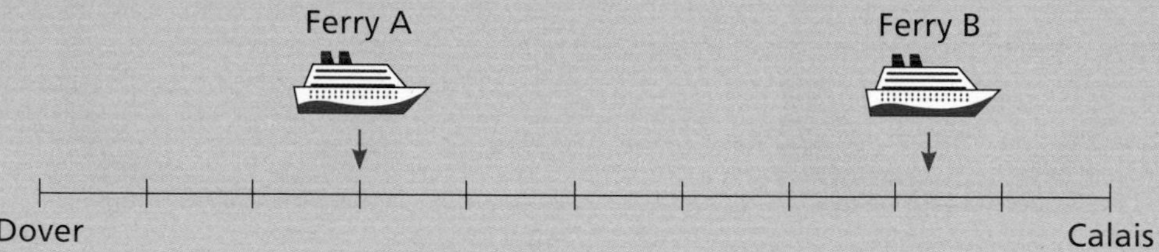

a What fraction of the distance has Ferry A travelled?

b Roughly what fraction of the distance has Ferry B travelled?

Consolidate – do you need more?

1 **a** What number is represented in the place value grid?

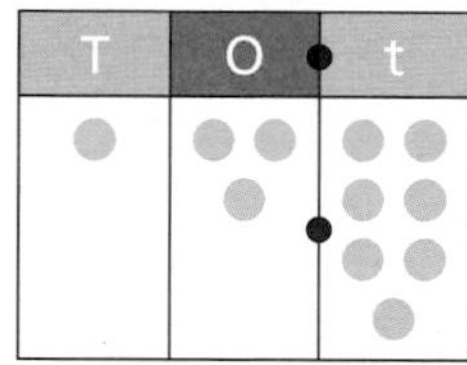

b Two more counters are added to one of the columns in the table.

What numbers could now be made?

2 What is the value of each digit in these numbers?

a 3.7 **b** 4.581 **c** 287.35

3 What is the value of the underlined digit in each number?

a $8.4\underline{3}$ **b** $1\underline{9}.72$ **c** $3\underline{4}07.6$

d $0.\underline{9}05$ **e** $2\underline{7}\,350.18$ **f** $0.00\underline{9}\,13$

4 Explain why 8 tenths is equal to 80 hundredths.

5 What is each arrow pointing to?

a

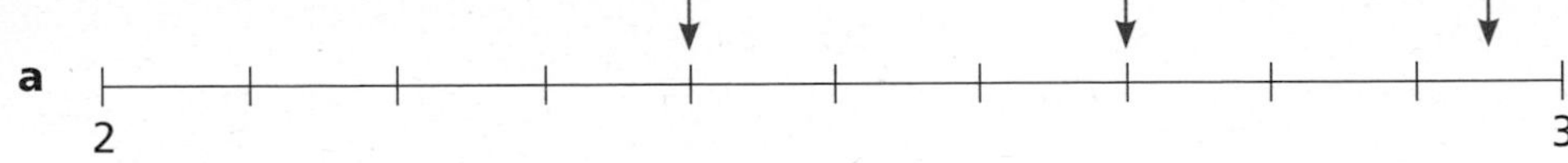

b

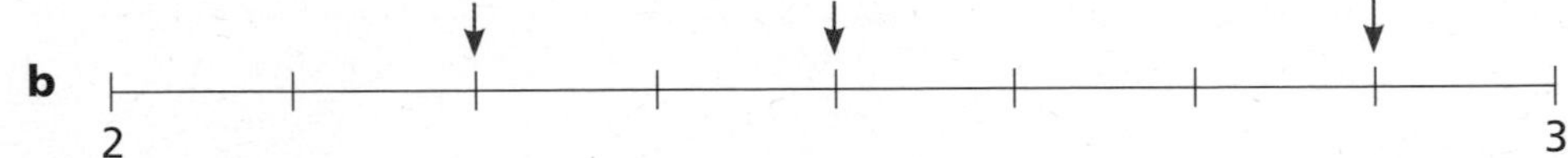

Stretch – can you deepen your learning?

1 A number is made up of four different digits.

There are two digits before the decimal point and two after the decimal point.

The sum of digits is 10

How many numbers between 10.5 and 12.9 can you make?

2 Show where 37.5 lies on each of these number lines.

a

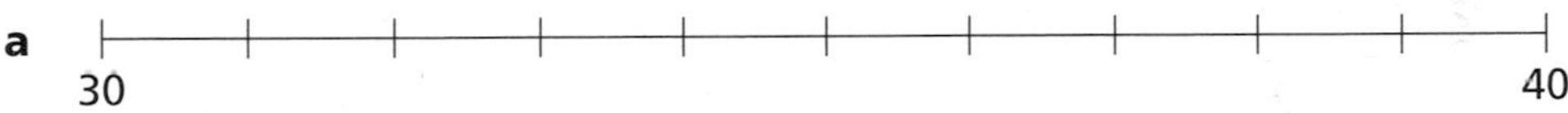

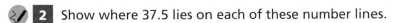

b

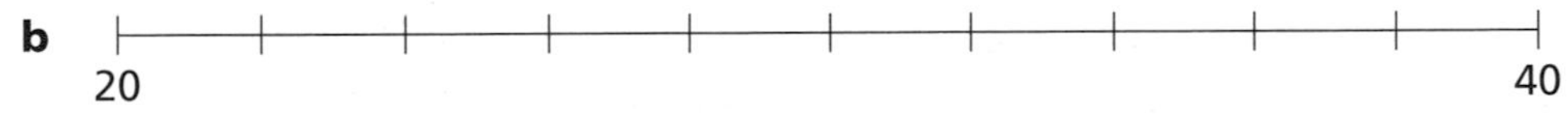

c

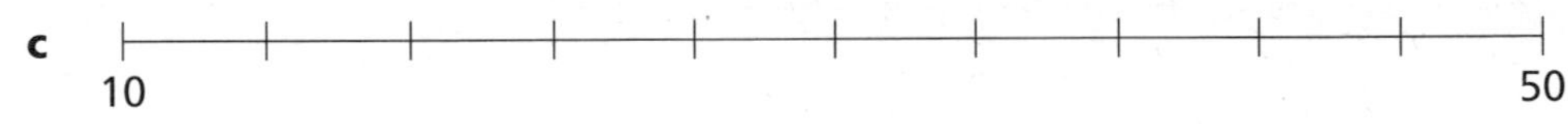

d

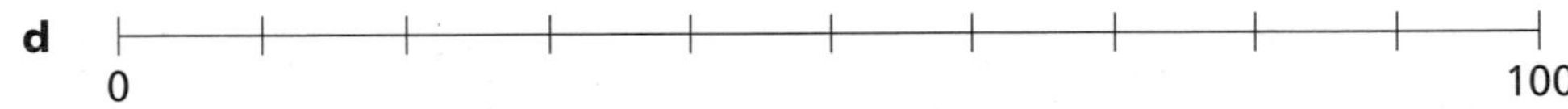

Reflect

1 True or false? A number with 7 hundredths is greater than a number with 4 hundredths. Write down some examples to support your reasoning.

2 How can you compare the size of decimal numbers? How is this the same and how is it different from comparing integers?

4.6 Rounding to one significant figure

Small steps

- Round a number to 1 significant figure

Key words

Significant figures – the most important digits in a number that give you an idea of its size

Order of magnitude – size of a number in powers of 10

Are you ready?

1 Round each of these numbers to the nearest ten.

a 738 **b** 9115 **c** 135.8 **d** 995

2 Round each of these numbers to the nearest hundred.

a 723 **b** 48 799 **c** 68.39 **d** 1999

3 72☐65 rounds to 72 000 to the nearest thousand.

What could the missing digit be?

Is there more than one answer?

4 Ali is working out a calculation. Here is his calculator answer.

Write the number that is displayed

a to the nearest whole number

b to the nearest ten

c to 2 decimal places

d to 5 decimal places.

Models and representations

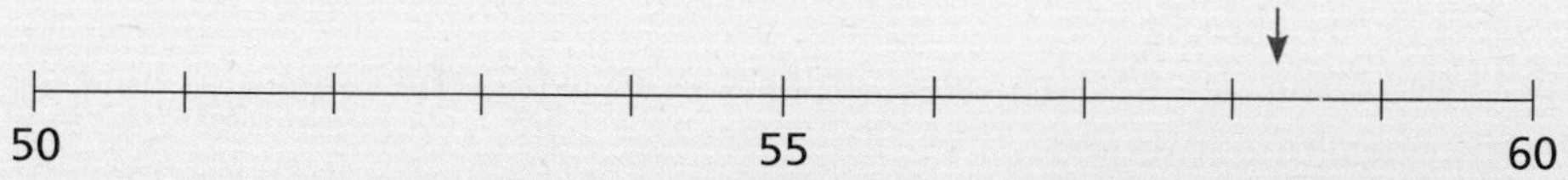

Can you see how the number line helps you to understand rounding?

Which 10 is 58.3 closer to? Is it 50 or 60?

Some numbers need rounding to make them easier to understand.

For example, the number of views of an online video might be 187 319 726

187 000 000 views

This number has been **approximated**.

Example 1

Write down the first **significant figure** in each of these numbers

a 907 312 **b** 0.007 583 6

a 9 is the first significant figure

1st non zero digit from left

$$907\,312$$

2nd significant figure

b 7 is the first significant figure

1st non zero digit from left.
So 7 is the 1st significant figure

$$0.0075836$$

2nd significant figure 3rd significant figure

Example 2

Round the number 56 718 to 1 significant figure

60 000

First identify the 1st significant figure

⑤6 718

The 1st significant figure is 5

This is in the 10 000s column, so rounding to 1 significant figure will be the same as rounding to the nearest 10 000

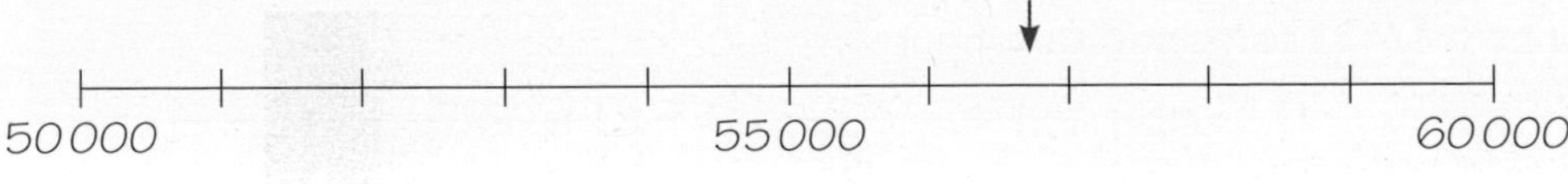

56 718 is closer to 60 000

Notice the answer must be of the same **order of magnitude** as the number itself – when we are approximating the answer we give must represent the original number.

Another way to represent this is by drawing a line on the number

5¦6718

The first significant figure is 5 and the second is a 6. We know that the number is more than halfway on the number line so the rounded number will be greater than the original. So the first significant figure of the answer will round to 6

The answer must be of the same order of magnitude as the number itself, so it is 60 000

Example 3

Round 0.006 956 3 to 1 significant figure

0.006¦9563

To 1 significant figure 0.006 956 3 is 0.007

The first significant figure is 6. We need to look at the number after the 6

This is a 9 so we need to round the 6 to a 7

So to 1 significant figure 0.006 956 3 is 0.007

We don't need to include any zeros after the 7 as they will not change the place value of the number.

Practice 4.6A

1. In each number, which digit is the first significant figure?

   a 923 **b** 1.803 **c** 0.002 58

   What is the place value of each of those digits?

2. In each number, which digit is the second significant figure?

   a 923 **b** 1.803 **c** 0.002 58

   What is the place value of each of those digits?

3. Use the number lines to help you round each number.

   a Round 37.483 to 1 significant figure

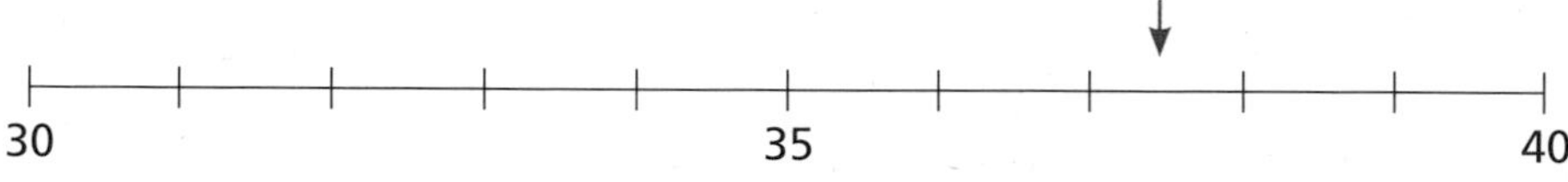

   b Round 37 483 to 1 significant figure

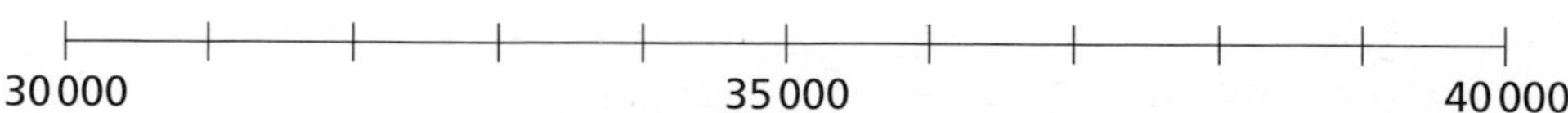

   c Round 0.374 83 to 1 significant figure

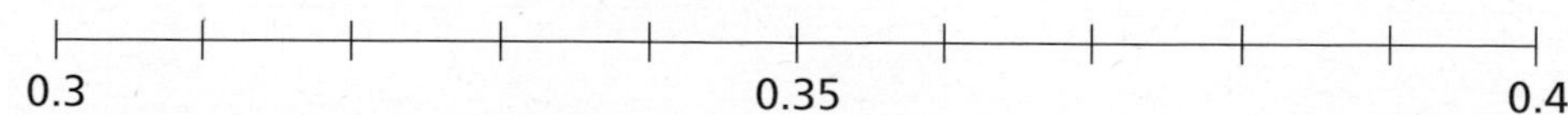

4 Emily is rounding 263.8 to 1 significant figure.

The digit 6 is the 2nd significant figure.
As 6 is greater than 5 it must round up.
The first significant figure is 2, so this will round up to 3. The answer is 3

What mistake has Emily made? Explain what she should have done.

5 Round each number to 1 significant figure.

a 36.192 **b** 1.853

c 2760 **d** 312 840

e 0.381 593 178 9 **f** 206.511 156 77

g 1.9599

Discuss your answer to **g** with your partner.

6 Seb rounds 7382 to 1 significant figure.

The answer is 7000. It was the same as rounding to the nearest 1000

Explain why rounding to 1 significant figure is not always the same as rounding to the nearest 1000

7 Round each of these numbers to 1 significant figure.

a 47 381 **b** 4738.1 **c** 473.81

d 47.381 **e** 4.7381 **f** 0.473 81

g What patterns did you notice?

8 Work out each of these calculations on your calculator.

Round each answer to 1 significant figure.

a 1.89×319.175 **b** $\sqrt{3} \times \sqrt{2}$

9 By rounding each number to 1 significant figure, estimate the answer to each of these calculations.

a 195×4.3 **b** $\frac{2355 \times 978}{1.95}$

Check your estimates using a calculator.

10 The number 5☐3 is rounded to 1 significant figure.

The answer is 600.

What could the original number have been?

4.6 Rounding to one significant figure

What do you think?

1 The Earth is 149 597 870 km from the Sun.

A cell is 0.00876 mm long.

Why might you round a numbers like these to one significant figure rather than the nearest 10, 100, 1000 and so on?

2 Jakub rounds the number of people at a football match to 1 significant figure.

I get 30 000 when I round.

What are the smallest and largest possible numbers of people at the match?

3 The length and width of this rectangle have been rounded to 1 significant figure.

8 cm

2 cm

What is the smallest possible area of the rectangle?

Consolidate – do you need more?

1 Round each of these numbers to 1 significant figure.

a 482 **b** 3057 **c** 5.961 361 **d** 0.000 106 5

2 Work out each calculation and round the answers to 1 significant figure.

a 36.71 × 283.7 **b** 5.7^3 **c** 250 ÷ 0.0175

3 Work out the area of the square.

Give your answer to 1 significant figure.

9.56 m

Stretch – can you deepen your learning?

1 **a** Investigate rounding numbers to 2 and 3 significant figures. What's the same and what's different?

b Round 19 998 to 1 significant figure and 2 significant figures. What do you notice?

c Round 0.199 98 to 1 significant figure and 2 significant figures. What do you notice?

2 A number rounded to 1 significant figure is 100 000.

The same number rounded to 2 significant figures is 100 000.

The same number rounded to 3 and to 4 significant figures is also 100 000.

What could the number be?

Reflect

Think of examples in real life where you might see rounded numbers.

What have they been rounded to?

Why do you think we often round to a given number of significant figures, rather than decimal places and nearest 10, 100, 1000 and so on?

(H) 4.7 Standard form: large numbers

Small steps

- Write 10, 100, 1000 and so on, as powers of 10 (H)
- Write positive integers in the form $A \times 10^n$ (H)

Key words

Power – This is written as a small number to the right and above the base number, indicating how many times to use the number in a multiplication. For example, the 5 in 2^5

Standard form – a number written in the form $A \times 10n$ where A is at least 1 and less than 10, and n is an integer

Are you ready?

1 Work out the following.

a 10^2 **b** 7^2 **c** 2^5 **d** 5^3

2 Write each of these numbers in numerals.

a One million **b** Two hundred thousand

c Ninety thousand **d** Thirty million

3 Write either the symbol < or the symbol > to compare each pair of numbers.

a 70 000 ◯ 7000 **b** 9000 ◯ 6000

c 100 000 ◯ 900 000 **d** 6 million ◯ 600 000 000

4 Work out the following without using a calculator.

a 300 × 300 **b** 20 × 4000 **c** 9000 × 100

Discuss your method with your partner.

Models and representations

To compare numbers you could write them in a place value grid.

Ten Thousands	Thousands	Hundreds	Tens	Ones
	9	0	0	0
5	0	0	0	0

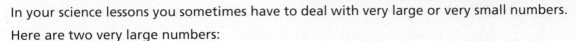

In your science lessons you sometimes have to deal with very large or very small numbers.

Here are two very large numbers:

700 000 000 000 000 60 000 000 000 000

It is quite easy to make a mistake when reading or comparing these numbers.

There is a way to write these numbers that avoids these issues. This is normally known as **standard form** or standard index form.

A number that is a multiple of a **power** of 10 written in standard form is $A \times 10^n$, where A is a number that is at least 1 and less than 10 and n is an integer.

Example 1

Write 100 000 000 as a power of 10

$10 \times 10 \times 10 \times 10 \times 10 \times 10 \times 10 \times 10$

$= 100\,000\,000$

Also $10 \times 10 \times 10 \times 10 \times 10 \times 10 \times 10 \times 10 = 10^8$

So 100 000 000 as a power of 10 is 10^8

You know that $10 \times 10 = 100$

and $10 \times 10 \times 10 = 1000$

Just continue this pattern.

Example 2

Write 800 000 in the form $A \times 10^n$

800 000 is equal to $8 \times 100\,000$

Also 100 000 is equal to $10 \times 10 \times 10 \times 10 \times 10 = 10^5$

So $800\,000 = 8 \times 10^5$

Rewrite it as a number between 1 and 10 multiplied by a power of 10

In this example the number 8×10^5 is written in standard form.

Example 3

Which of these numbers is the greatest?

3×10^5 7×10^4

$3 \times 10^5 = 3 \times 10 \times 10 \times 10 \times 10 \times 10 = 300\,000$

$7 \times 10^4 = 7 \times 10 \times 10 \times 10 \times 10 = 70\,000$

300 000 is greater than 70 000

So 3×10^5 is the greatest number.

As each number is in standard form, 3×10^5 is the greatest number as it has the greater power.

One method for answering this is to write each number in ordinary form.

Another method is to compare the powers.

You know that 10^5 is greater than 10^4

The numbers must be in standard form to make this comparison.

Practice 4.7A

1 Write each of the following numbers as a power of 10.

a 100 **b** 1000 **c** 10 000 **d** 100 000

e What patterns do you notice?

2 Say whether you agree with this statement:

> The digit 1 followed by 8 zeros is equal to 10^8

Give a reason for your response.

3

10^6 is equal to 1 million, that means 10^7 is the same as one billion.

Is Benji correct?

4 Write these as ordinary numbers.

a 10^8 **b** 10^3 **c** 10^{11}

5 **a** Write 100 million as a power of 10 **b** Write 10 billion as a power of 10

6

> **Googol –** the digit 1 followed by one hundred zeros

Write a googol as a power of 10

7 Write each of the following numbers in the form $A \times 10^n$

a 100 000 **b** 200 000 **c** 300 000 **d** 700 000

What do you notice?

8 Write each of these numbers in standard form.

a 2000 **b** 300

c 90 000 **d** 5 000 000 000

e 3 million **f** 80

g Seven hundred million **h** 60 000 000 000 000 000 000 000

9 Write each of these as ordinary numbers.

a 3×10^4 **b** 3×10^5 **c** 3×10^6 **d** 3×10^7

e What is the same about each answer? What is different?

10 Write each of these as ordinary numbers.

a 1×10^4 **b** 6×10^3 **c** 9×10^7 **d** 8×10^4 **e** 9×10^{11}

8 × 10^4 is greater than 3 × 10^5 because 8 is greater than 3

Explain the mistake that Faith has made.

12 Write <, > or = to compare each pair of numbers.

a 3×10^5 ◯ 2×10^5

b 3×10^5 ◯ 7×10^8

c 3×10^5 ◯ 340 000

d 3×10^5 ◯ 4×10^5

e 3×10^5 ◯ 6×10^2

What strategy did you use? Discuss your method with a friend.

13 Which of the following numbers is the greatest? Show your working out.

4 million	4×10^7	400 × 1000

What do you think?

1 The number 10 100 can be written as $10^4 + 10^2$.

Write these numbers as powers of 10 added together.

a 101 000

b 1 000 100

c 1110

2 One definition of a trillion is a million million. Write a trillion as a power of 10

3 10^4 is the same as 1×10^4. Do you agree? Explain your answer.

4 **a** 5×10^7 is greater than 5×10^x. What can you say about the value of x?

b 5×10^7 is greater than $A \times 10^7$. What can you say about the value of A?

c 5×10^7 is greater than $B \times 10^4$. What can you say about the value of B?

5 Write the answers to these in standard form.

a 200 × 300

b 7000×10^3

c 40 × 2 million

How did you approach each question?

Consolidate – do you need more?

1 Write each of the following numbers in the form $A \times 10^n$

a 30 000 **b** 10 000
c 9000 **d** 7 000 000
e 5 million **f** 6 000 000 000 000
g 70 000 000 000 000 000 **h** 900 000 000

2 Write each of these numbers in ordinary form.

a 7×10^4 **b** 8×10^5 **c** 3×10^6
d 5×10 **e** 7×10^{10} **f** 9×10^{12}

Stretch – can you deepen your learning?

1 A number starts with the digit 1 and is followed by 19 zeros.

What is this number as a power of 10?

Were you able to write down the answer straight away? How did you go about it?

2 $7 \times 10^4 > 7 \times 10^x$

Write down the set of values that x could be.

3 Round each of these numbers to 1 significant figure and then write your answer in standard form.

a The mass of the Earth is 5 972 000 000 000 000 000 000 000 kg.

b On average the total hours watched on YouTube each month is 3.25 billion hours.

4 Work out the missing numbers.

a $200 \times \square = 8 \times 10^7$

b $3000 \times \square = 9 \times 10^{15}$

c $200 \times 500 \times \square =$ 3 billion

Reflect

Why do people write numbers in standard form?

Ⓗ 4.8 Standard form: small numbers

Small steps

- Investigate negative powers of 10 Ⓗ
- Write decimals in the form $A \times 10^n$ Ⓗ

Key words

Power – This is written as a small number to the right and above the base number, indicating how many times to use the number in a multiplication. For example, the 5 in 2^5

Standard form – a number written in the form $A \times 10n$ where A is at least 1 and less than 10, and n is an integer

Are you ready?

1 Work out the following powers of 10

a 10^2 **b** 10^3 **c** 10^5

2 Write each of these numbers as a power of 10

a 10 000 **b** 1 000 000 **c** 100 000 000

3 Write each of these numbers as decimals.

a one tenth **b** one hundredth **c** one thousandth

4 Write the following numbers in descending order.

10^4 10 10^3 10^6

5 **a** Explain why 3×10^4 is equal to 30 000

b Work out 4×10^5

c $5\,000\,000 = 5 \times 10^n$. What is the value of n?

Models and representations

In order to compare numbers you might want to consider writing them in a place value grid.

For negative powers the numbers are going to be less than 1

Ones	tenths	hundredths	thousandths	ten thousandths
0	0	0	1	

Using a calculator you can investigate powers of 10

What do you notice as the powers decrease?

10^3	1000
10^2	100
10^1	10
10^0	1
10^{-1}	0.1
10^{-2}	0.01
10^{-3}	0.001

4.8 Standard form: small numbers (H)

What do you notice about the pattern of zeros?

A positive power of 10 means you multiply that number of tens together.

$10^3 = 10 \times 10 \times 10 = 1000$

For negative powers of 10 you divide 1 by that number of tens.

10^{-3} is equal to $\frac{1}{10 \times 10 \times 10}$ or $\frac{1}{1000}$ which is one thousandth or 0.001

Example 1

Explain why 10^{-2} is equal to 0.01

10^{-2} is equal to $\frac{1}{10^2}$ which is equal to $\frac{1}{100}$

$\frac{1}{100}$ is one hundredth or 0.01

So $10^{-2} = 0.01$

Your explanation should be clear and concise

Example 2

Write 0.0001 as a power of 10

$0.0001 = \frac{1}{10000}$

$= \frac{1}{10^4}$

$= 10^{-4}$

Think about the place value of the number.

Ones	•	tenths	hundredths	thousandths	ten thousandths
0	•	0	0	0	1

Example 3

Write 5×10^{-3} as a decimal.

$10^{-3} = 0.001$

So $5 \times 10^{-3} = 5 \times 0.001 = 0.005$

Remember $10^{-3} = \frac{1}{1000} = 0.001$

Practice 4.8A

1 Match the powers of 10 with the number.

Use your calculator if you need to.

10^1

10^0

0.1

10^{-1}

10

10^{-2}

1

2 Copy and complete the table of values.

	10^{-1}	10^{-2}	10^{-3}	10^{-4}	10^{-5}
Fraction	$\frac{1}{10}$				
Decimal					

What patterns do you notice?

Discuss your answer with your friend.

3 Jackson is working out powers of 10

$10^{-3} = -1000$

What mistake has Jackson made? What should the answer be?

4 Write each of the following as decimals.

a 10^{-3} **b** 10^{-4} **c** 10^{-5} **d** 10^{-6}

e 10^{-7} **f** What is the same about each number? What is different? How does the number of zeros relate to the power?

5 Write each of these numbers as a power of 10

a 10 000 **b** 100 000 **c** 100 000 000

6 Write each of these numbers as a power of 10

a $\frac{1}{1000}$ **b** $\frac{1}{10\,000}$ **c** $\frac{1}{100\,000}$ **d** $\frac{1}{1\,000\,000}$

7 Write each of these numbers as a power of 10

a 0.001 **b** 0.000 01

c 0.000 000 01 **d** One millionth

8 Here are some numbers

One hundredth 10^3 10^{-5}

10^1 A thousandth 10^{-1}

Copy the table and write each of the numbers in the correct place.

Numbers greater than 1	Numbers between 0 and 1	Numbers less than 0

9 Put the following numbers in order, starting with the smallest.

10^2 10^{-2} one tenth 10^5

10 Write each of these numbers as a decimal.

a 2×10^{-3} **b** 3×10^{-3} **c** 4×10^{-3} **d** 5×10^{-3}

Explain to your friend how you answered this question.

11 Write each of these numbers as a decimal.

a 7×10^{-2} **b** 7×10^{-3} **c** 7×10^{-4} **d** 7×10^{-5}

e 7×10^{-8}

12 Put the following numbers in descending order.

3×10^2 5×10^{-4} 6×10^{-2} 3×10^{-4}

What do you think?

1 This table shows powers of 10.

10^5	100 000
10^4	10 000
10^3	1000
10^2	100
10^1	10
10^0	1

10^{-5}	0.000 01
10^{-4}	0.000 1
10^{-3}	0.001
10^{-2}	0.01
10^{-1}	0.1

What do you notice as the powers decrease?

What do you notice about the pattern of zeros?

2 Emily and Abdullah are writing 10^{-5} as a decimal.

Both have got the answer incorrect.

10^{-5} is equal to 0.00005

I think the answer is 0.000005

Why do you think they made their mistakes?

Did you get the same reasons as your partner?

3 **a** Write one thousand as a power of 10.

b Write one thousandth as a power of 10.

c What do you notice about your answers to parts **a** and **b**?

Consolidate – do you need more?

1 Write each of the following numbers as powers of 10.

a 0.001 **b** 0.000 001 **c** 0.000 000 1

2 Write each of the following numbers as fractions.

a 10^{-2} **b** 10^{-4} **c** 10^{-5} **d** 10^{-7} **e** 10^{-10}

Write each of your numbers as decimals.

3 Explain why one ten thousandth is equal to 10^{-4}

Stretch – can you deepen your learning?

1 Write each of these numbers in the form $A \times 10^n$

a 0.008 **b** 0.000 04 **c** 0.7

d 0.0009 **e** 0.000 000 007 **f** 0.000 08

2 **a** Write one million as a power of 10

b Write one millionth as a power of 10

c What is the same about your answers? What is different?

3 Write 7 billionths in the form $A \times 10^n$

4 Why is 10 to the power 0 not equal to 0?

Reflect

What is the same about positive and negative powers of 10?

What is different about positive and negative powers of 10?

4 Place value and ordering
Chapters 4.1–4.8

I have become fluent in...

- Recognising the place value of any number
- Comparing and ordering numbers
- Finding the range and median of a set of numbers
- Rounding numbers
- Converting to and from standard form (H)

I have developed my reasoning skills by...

- Working out intervals on a number line
- Exploring what happens to the range and median as numbers change
- Considering how different digits affect a number

I have been problem-solving through...

- Finding more than one answer in missing digit questions
- Finding missing numbers given the median or range
- Considering the bounds of a number when rounding

Check my understanding

1 Write down the value of the 7 in each of these numbers. Give your answers in words.

a 7 286 901 **b** 4789 **c** 627 301 **d** 67 105 293 **e** 765 841 299

2 Use <, > or = to make the statements correct.

a 599 ◯ 1021 **b** 3 million ◯ 3 000 000

c 70 000 ◯ 9999 **d** 4 ones ◯ 4 tens

3 Calculate the range of each set of numbers.

a 1, 5, 11, 15 **b** 47, 38, 29, 26, 20, 19 **c** 30, 47, 19, 28, 100, 51, 14

4 Calculate the median of each set of numbers.

a 9, 17, 3, 25, 21 **b** 45, 30, 25, 28 **c** 507, 500, 395, 627, 510, 599

5 Write down the value of the 3 in each of these numbers.

a 3.71 **b** 19.3 **c** 28.043 **d** 306.547 **e** 27.238

6 Round each number to the nearest 10, 100 and 1000

a 8732 **b** 96 407 **c** 123 685 **d** 9999 **e** 76 595

7 Round each number to 1 significant figure.

a 8732 **b** 96 407 **c** 123 685 **d** 9999 **e** 76 595

8 Write each number in standard form. (H)

a 500 000 **b** 30 000 000 **c** 0.005 **d** 0.000 03

5 Fractions, decimals and percentages

In this block, I will learn...

about tenths and hundredths

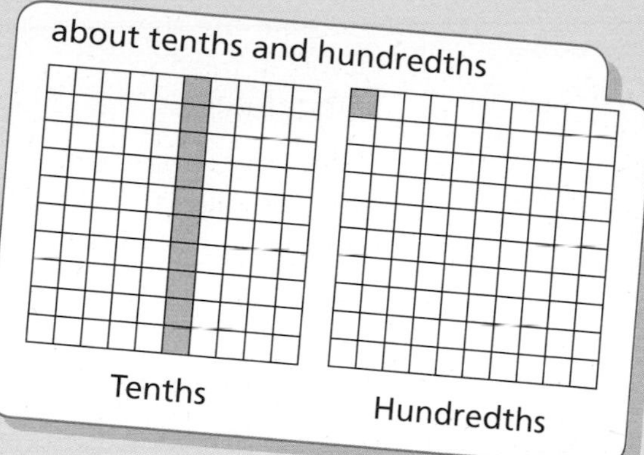

how to use and interpret basic pie charts

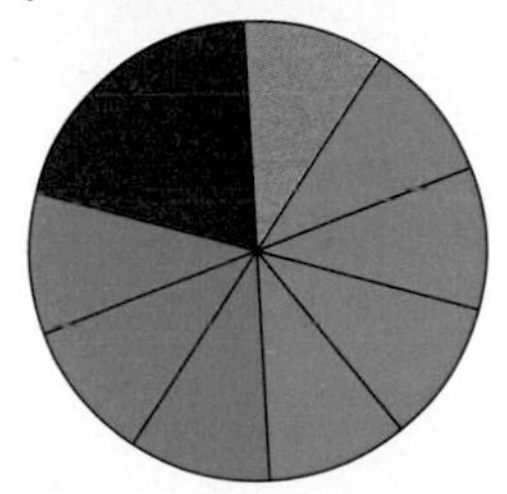

10% is shaded orange

$\frac{2}{5}$ is shaded blue

$\frac{3}{10}$ is shaded green

0.2 is shaded purple

how to convert between simple fractions, decimals and percentages

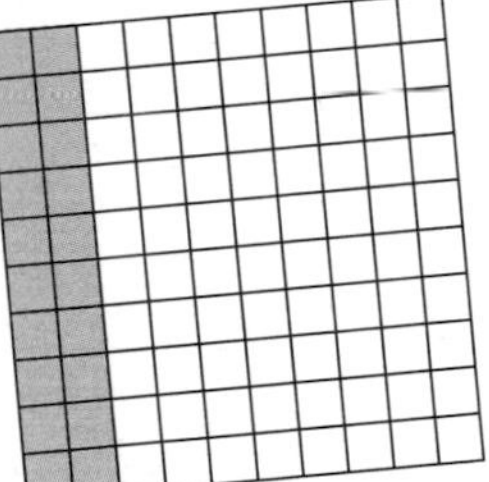

$\frac{1}{5}$ of the squares are shaded

20 out of 100 squares are shaded

20% of the squares are shaded

$\frac{2}{10}$ of the squares are shaded

0.2 of the squares are shaded

how to link fractions with division

$\frac{4}{5} = 4 \div 5$

means divide

about fractions greater than 1 Ⓗ

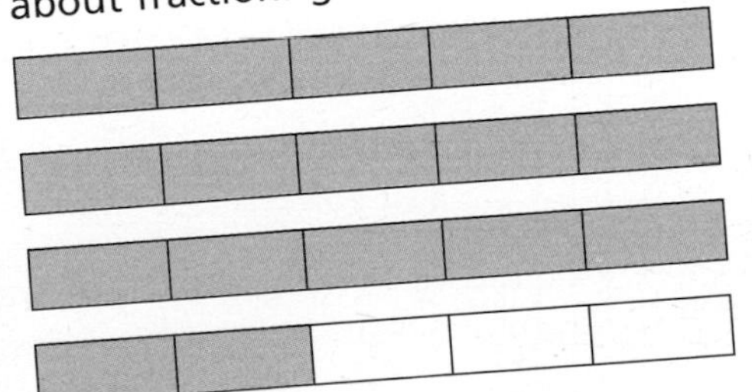

3 wholes and 2 fifths = 17 fifths

3 wholes and 2 fifths = 340%

3 wholes and 2 fifths = 3.4

5.1 Tenths and hundredths

Small steps

- Represent tenths and hundredths as diagrams
- Represent tenths and hundredths on number lines
- Interchange between fractional and decimal number lines
- Convert between fractions and decimals – tenths and hundredths

Key words

Tenths – when one whole is split into 10 equal parts, each part is one tenth

Hundredths – when one whole is split into 100 equal parts, each part is one hundredth

Are you ready?

1 Write these as decimals.

a one tenth **b** one hundredth **c** eight tenths **d** three hundredths

2 What fraction of each diagram is shaded?

a 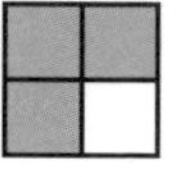 **b**  **c** 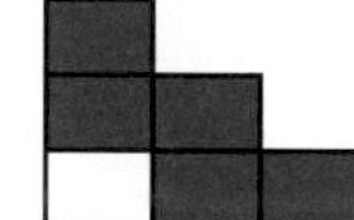

3 Write down the value of the 6 in each number.

a 362 **b** 6031 **c** 0.6 **d** 27.6

4 Write down a number that has seven tenths.

5 Write down a number that has one tenth and three hundredths.

Models and representations

Hundred square

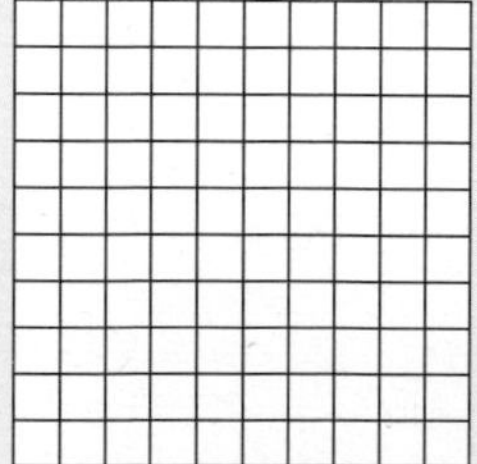

Number line

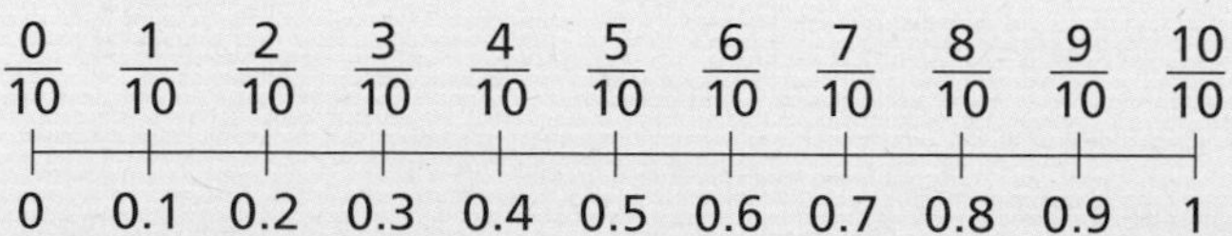

Place value chart

Tens	Ones	tenths	hundredths

Bar model

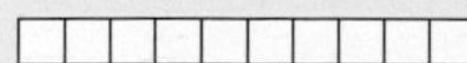

Example 1

On a hundred square, shade

a $\frac{1}{100}$ **b** $\frac{1}{10}$

a

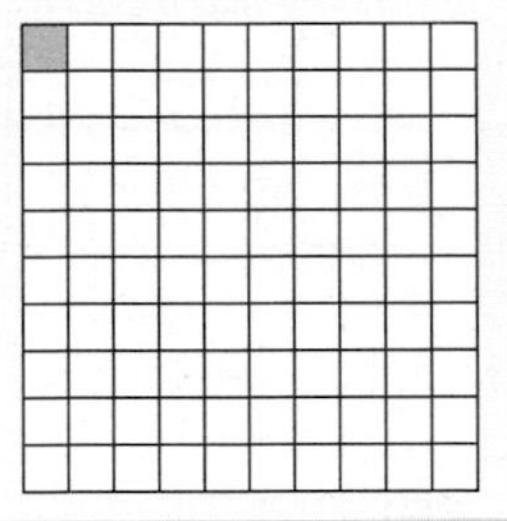

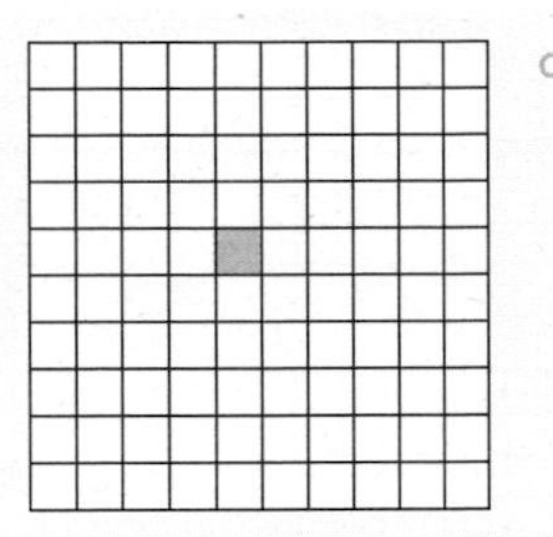

$\frac{1}{100}$ means one **hundredth**.

Each small square represents one hundredth. The whole has been split into 100 equal parts and one of those parts is shaded. This is the same as the decimal 0.01. It doesn't matter which square is shaded.

b

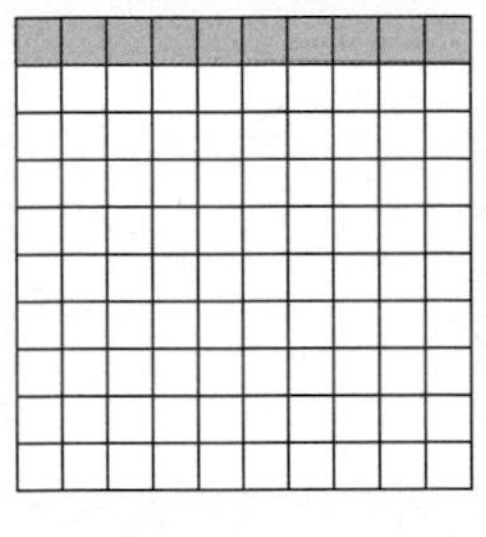

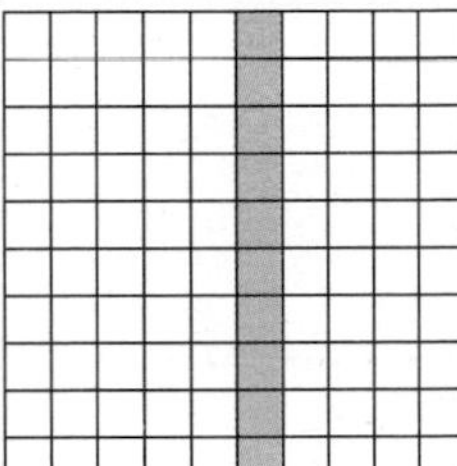

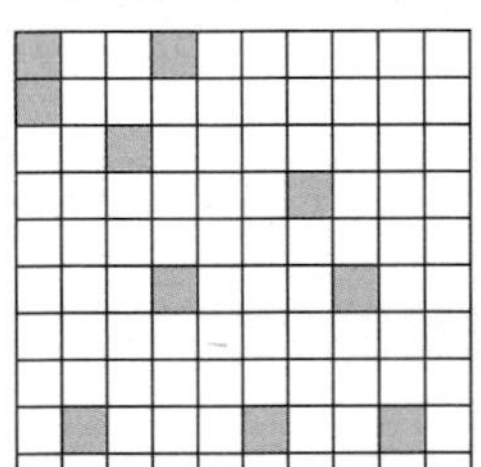

$\frac{1}{10}$ means one **tenth**.

When one row or column of a hundred square is shaded it represents one tenth. The whole has been split into ten equal parts with one part shaded.

You can also think of one tenth as ten hundredths. 10 squares make up one tenth so one tenth is the same as 10 hundredths. Any ten squares shaded would represent one tenth.

$\frac{1}{10} = \frac{10}{100}$ This is the same as 0.1

Practice 5.1A

1 How many tenths are shaded in each hundred square? Write your answers in words and as a fraction.

a

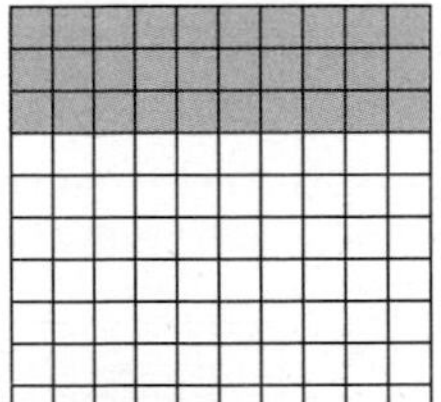

b

c

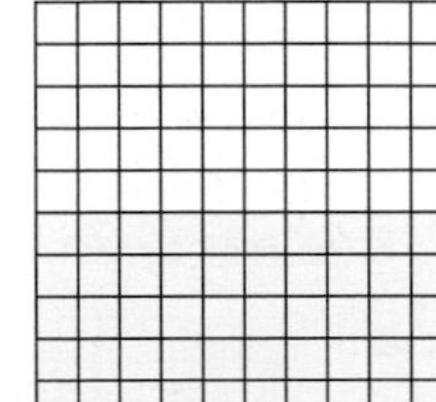

d

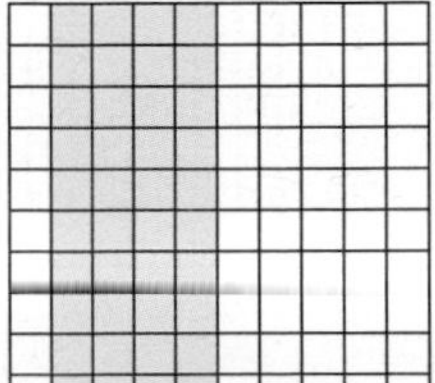

e

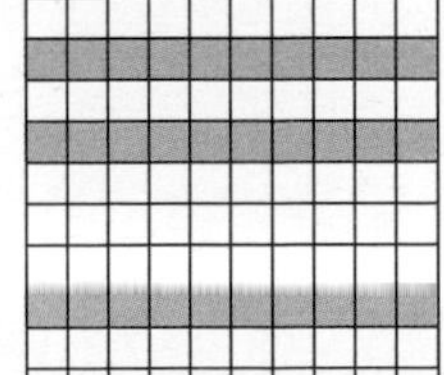

f

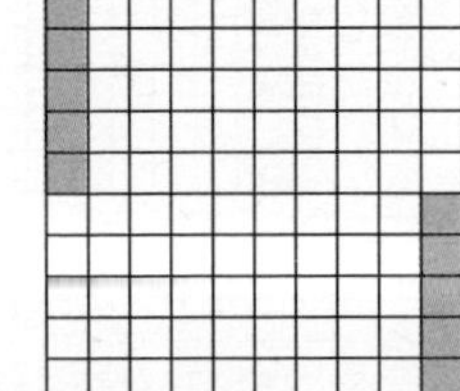

2 How many hundredths are shaded in each hundred square? Write your answers in words and as a fraction.

a 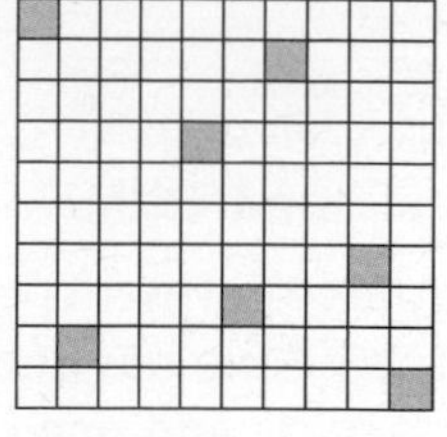

b 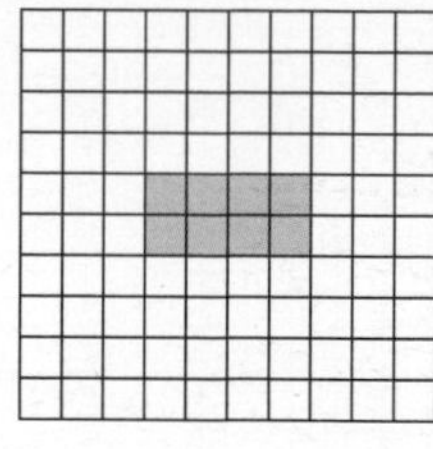

c 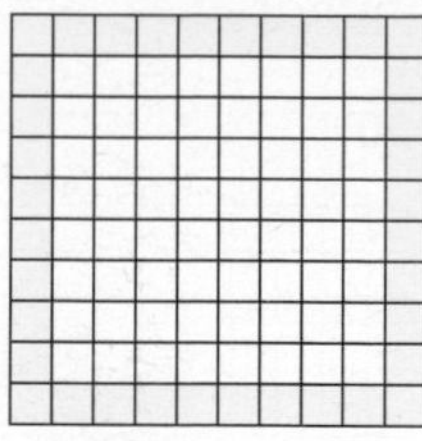

d 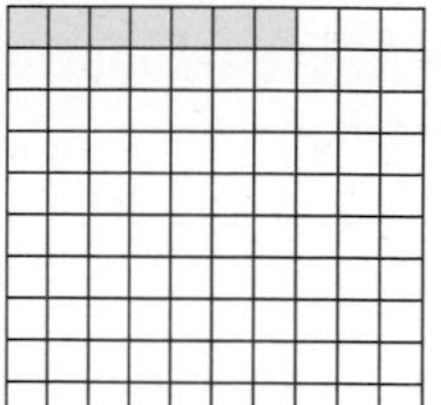

e 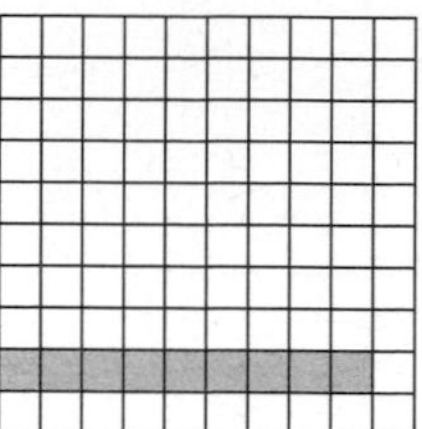

f 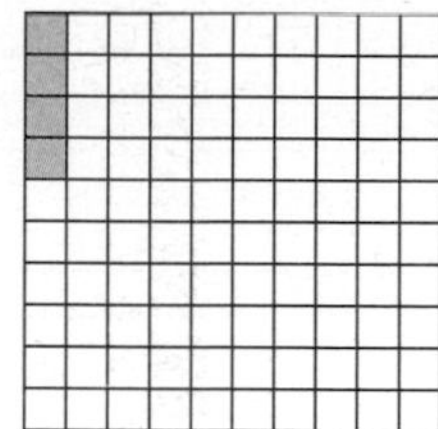

3 Here are two hundred squares.

A 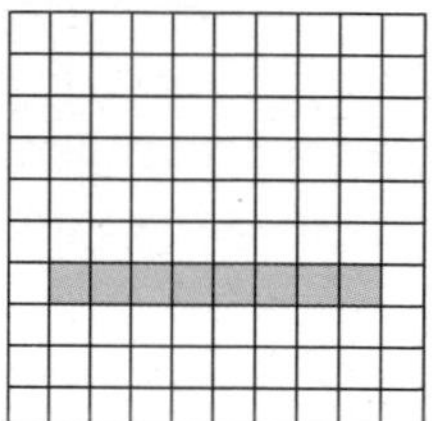

B 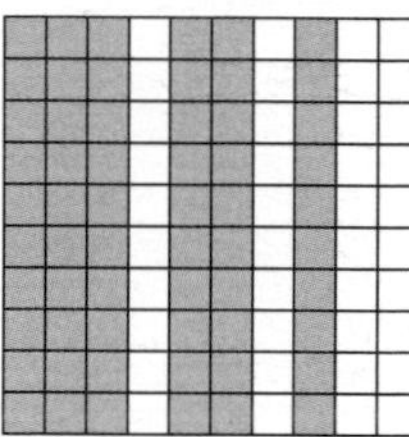

a What fraction of each hundred square is shaded?

b What fraction of each hundred square is not shaded?

c What do you notice about your answers to parts **a** and **b** for each hundred square?

4 Represent each of these fractions on a hundred square.

a 2 tenths **b** 2 hundredths **c** $\frac{7}{100}$

d $\frac{4}{10}$ **e** 10 hundredths **f** 100 hundredths

Compare your diagrams with a partner's. Did you shade the same squares?

5 Write the fraction shaded in each of these hundred squares in two different ways. One has been done for you.

a 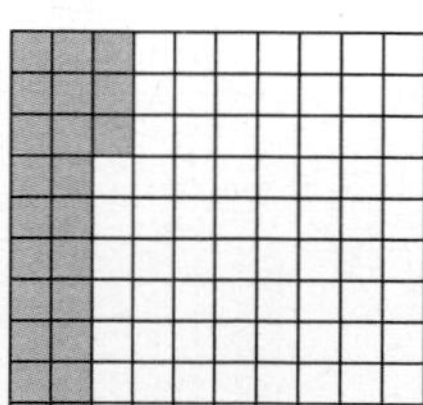

2 tenths and 3 hundredths

23 hundredths

b 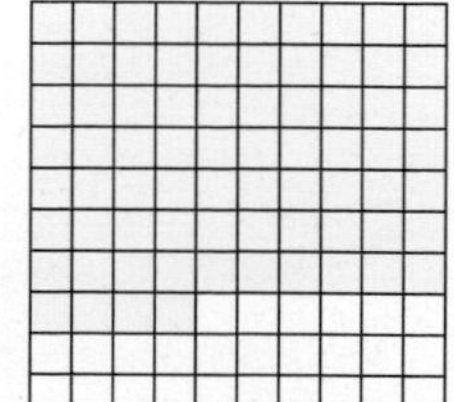

c 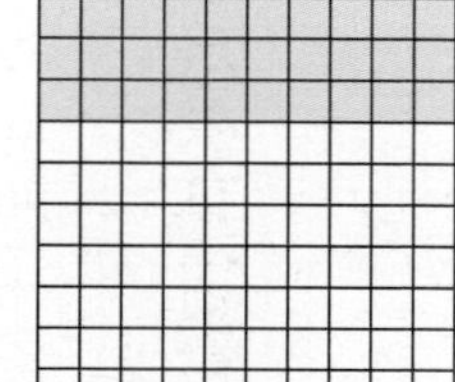

 **6** Represent each of these on a hundred square.

a

Ones	tenths	hundredths
	0.1 0.1 0.1 0.1	

b

Ones	tenths	hundredths
		0.01 0.01 0.01 0.01 0.01 0.01

c

Ones	tenths	hundredths
	0.1 0.1 0.1	0.01 0.01 0.01 0.01 0.01 0.01 0.01 0.01 0.01

What do you think?

1 Sven and Rhys have been asked to represent 99 hundredths on a hundred square.

Here are their answers.

Sven

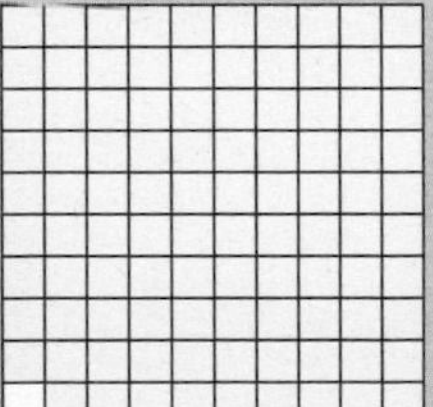

Rhys

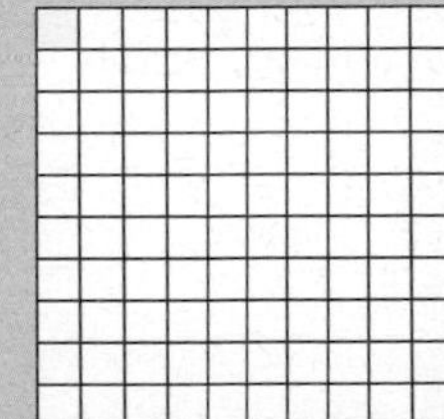

Who is correct? Explain your answer.

2 Kate says, "One row is worth one tenth and one column is worth one tenth so the diagram shows 2 tenths."

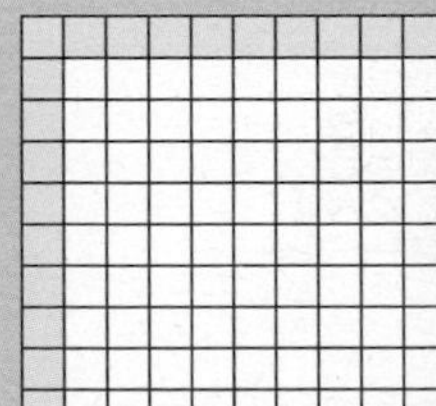

Do you agree with Kate? Explain your reasoning.

3 Samira has represented the calculation $\frac{7}{10} + \frac{9}{100}$ on this hundred square.

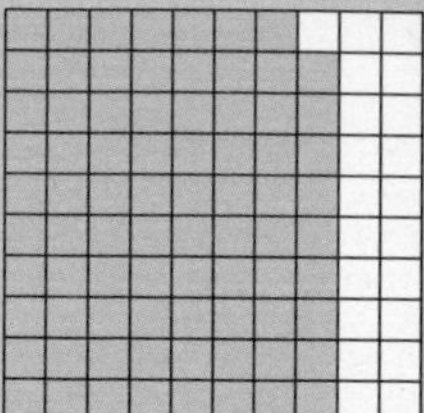

a Explain why Samira is correct.

b Represent the calculation $\frac{3}{10} + \frac{11}{100}$ on a hundred square.

Example 2

What numbers are the arrows pointing to? Write each answer as a fraction and a decimal.

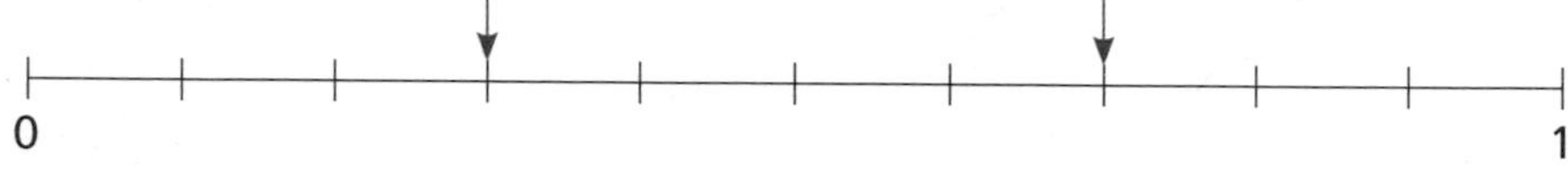

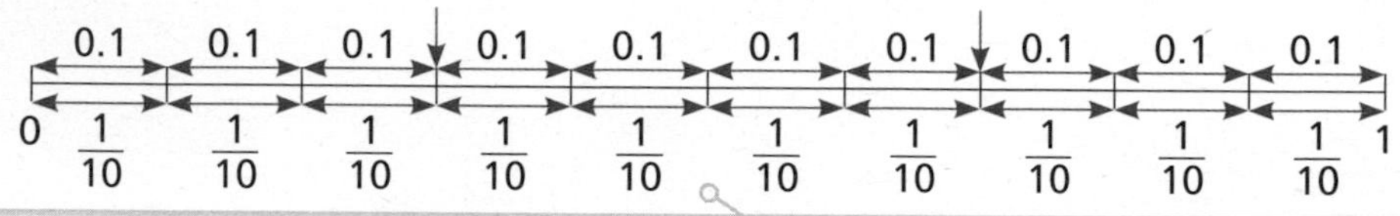

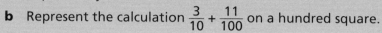

The number line is split into ten equal parts going from 0 to 1, so each part must be worth 0.1 or $\frac{1}{10}$

You can work this out by doing 1 ÷ 10

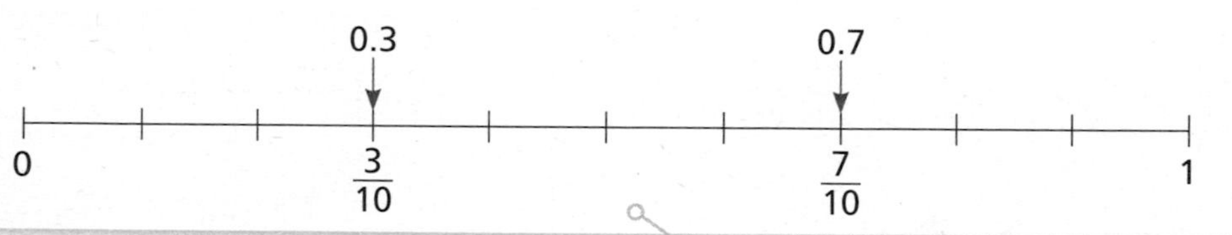

The arrows are pointing to 0.3 or $\frac{3}{10}$ and 0.7 or $\frac{7}{10}$

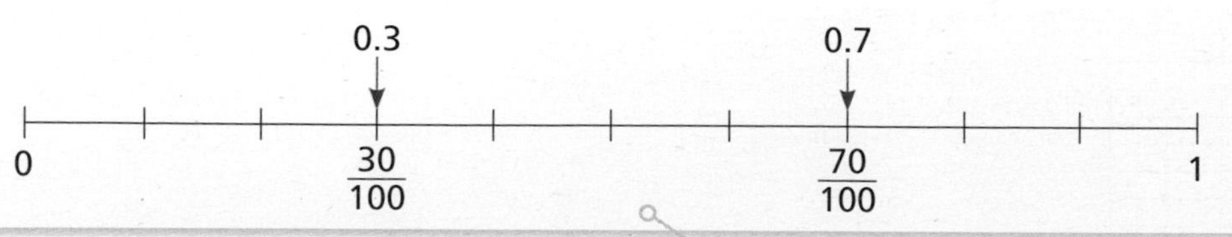

You know that 1 tenth is equal to 10 hundredths so you could complete the number line like this instead.

If you change the end point of the number line, the arrows would be pointing to different numbers.

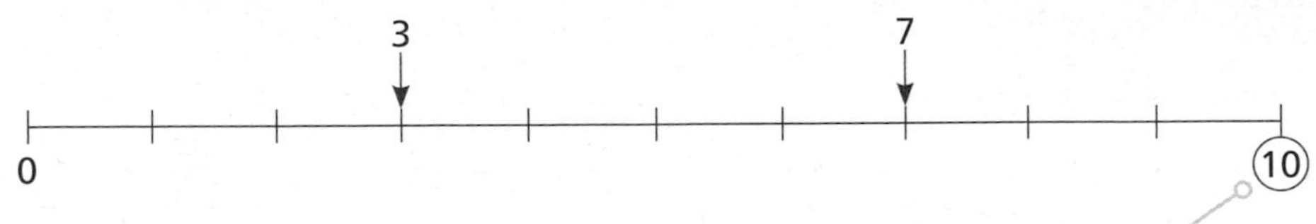

The end point is now 10 so each part on the number line must now be 1

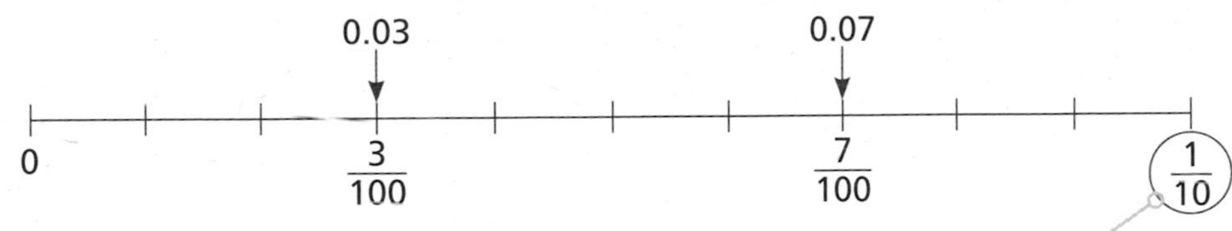

The end point is now $\frac{1}{10}$ so each part on the number line must now be $\frac{1}{100}$

Practice 5.1B

1 Copy and complete each of these number lines.

a

b

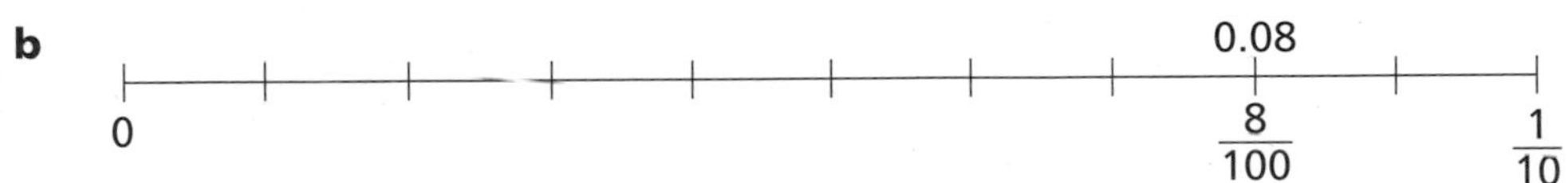

2 What numbers are the arrows pointing to?

a

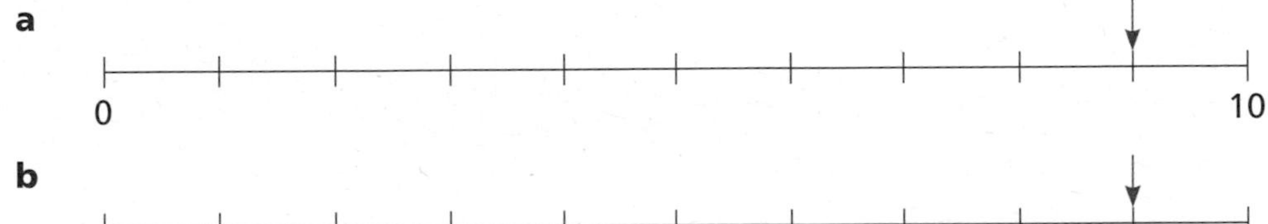

b

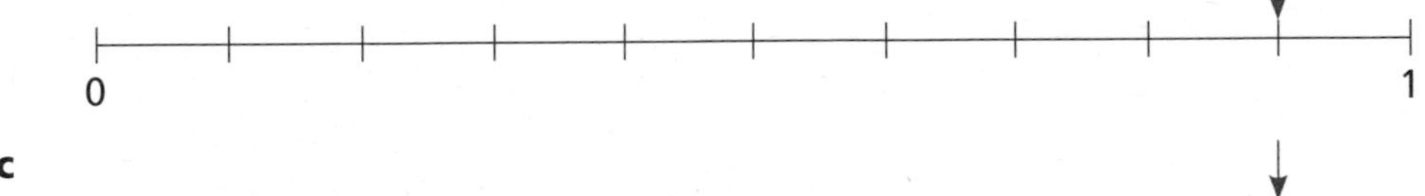

c

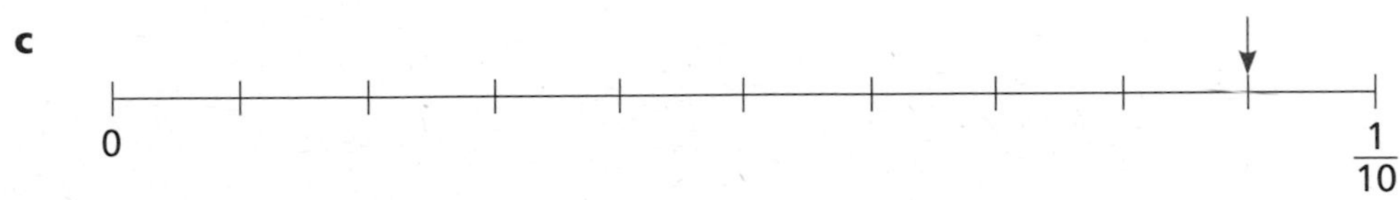

3 Copy and complete the number line.

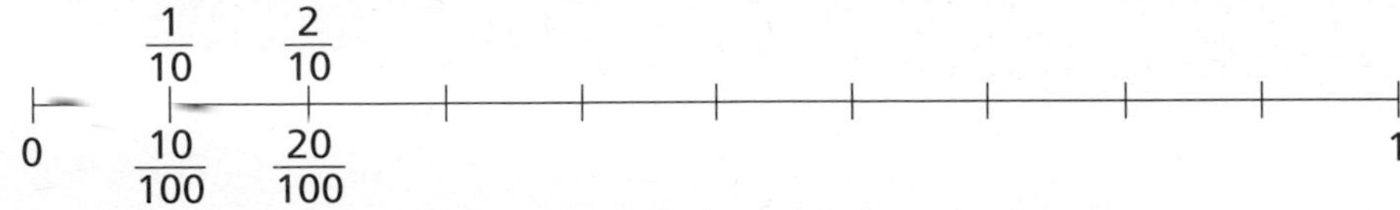

4 What numbers are the arrows pointing to? Write each answer as a fraction and a decimal.

a

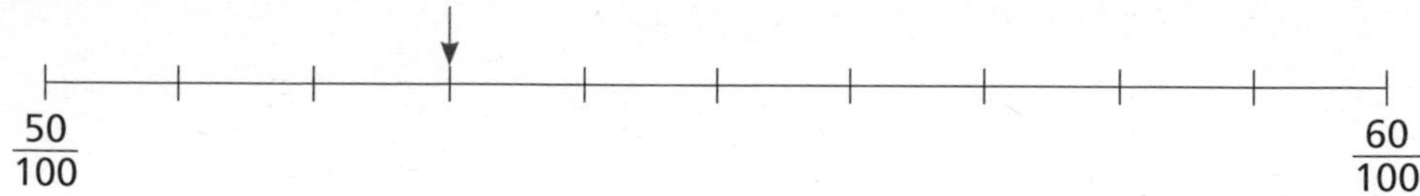

b

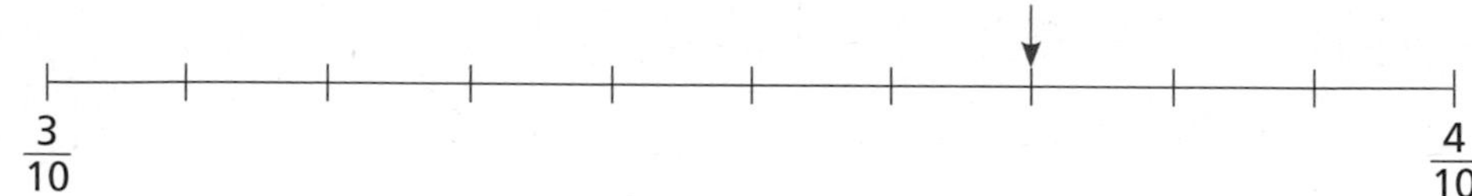

c

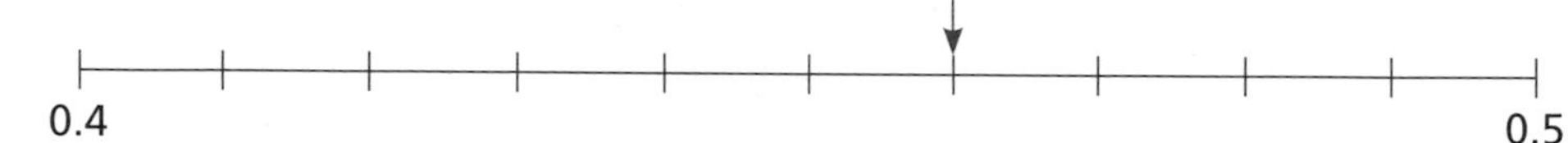

d

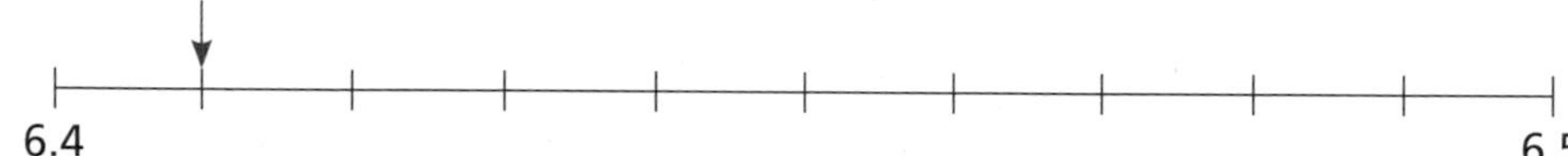

5 What are the missing numbers?

a $\frac{3}{10} = \frac{\square}{100}$

b $0.3 = \frac{\square}{10}$

c $0.3 = \frac{30}{\square}$

6 Copy and complete these statements.

a $0.7 = \frac{\square}{10}$

b $0.07 = \frac{\square}{100}$

c $0.\square = \frac{9}{10}$

d $0.17 = \frac{17}{\square}$

e $\frac{\square}{\square} = 0.8$

f $0.04 = \frac{\square}{\square}$

What do you think?

1 Marta says, "The arrow is pointing to $\frac{14}{10}$"

Jakub says, "The arrow is pointing to 1.4"

Abdullah says, "The arrow is pointing to $\frac{140}{100}$"

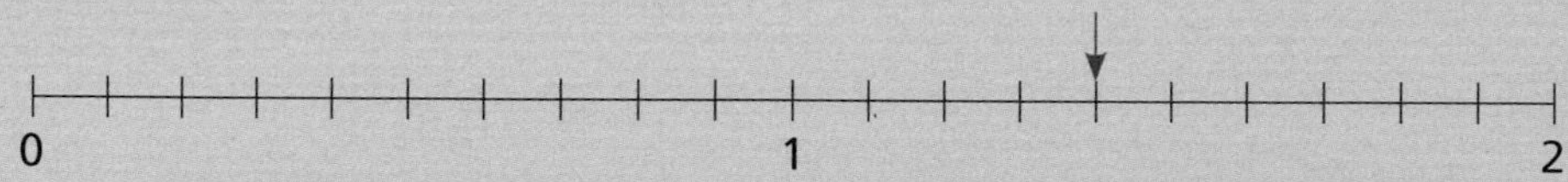

Who is correct?

Use hundred squares to support your reasoning.

2 Huda says, "The arrow is pointing to 3.1"

What mistake has Huda made?

Consolidate – do you need more?

1 Write each of these fractions as a decimal.

a $\frac{4}{10}$ **b** $\frac{9}{10}$ **c** $\frac{5}{10}$

2 Write each of these decimals as a fraction.

a 0.8 **b** 0.1 **c** 0.2

3 Write each of these fractions as a decimal.

a $\frac{3}{100}$ **b** $\frac{6}{100}$ **c** $\frac{7}{100}$

4 Write each of these decimals as a fraction.

a 0.01 **b** 0.09 **c** 0.06

5 Write each of these tenths as hundredths. The first one is done for you.

a $\frac{2}{10} = \frac{20}{100}$ **b** $\frac{6}{10}$ **c** $\frac{3}{10}$ **d** 0.7

6 **a** Partition each of these fractions into tenths and hundredths. The first one has been done for you.

i $\frac{23}{100} = \frac{2}{10} + \frac{3}{100}$ **ii** $\frac{47}{100}$ **iii** $\frac{51}{100}$ **iv** $\frac{99}{100}$ **v** $\frac{28}{100}$ **vi** $\frac{74}{100}$

b Write all of the answers to part **a** as decimals.

Stretch – can you deepen your learning?

1 Write each fraction as a decimal.

a $\frac{129}{100}$ **b** $\frac{147}{100}$ **c** $\frac{28}{10}$ **d** $\frac{250}{100}$

2 Write down the next three terms in each of these linear sequences. Write each term as a fraction and a decimal.

a $\frac{21}{100}, \frac{25}{100} \ldots$ **b** $\frac{99}{100}, \frac{91}{100} \ldots$ **c** $\frac{24}{100}, \frac{27}{100} \ldots$ **d** $\frac{44}{100}, \frac{33}{100} \ldots$

3 Solve these equations. Give each of your answers as a fraction and a decimal.

a $x + \frac{3}{10} = \frac{4}{10}$ **b** $x + \frac{3}{100} = \frac{34}{100}$ **c** $x + \frac{3}{10} = \frac{34}{100}$ **d** $x - \frac{12}{100} = \frac{5}{10}$

4 Work out the value of x and y.

$x + y = \frac{43}{100}$ $x - y = \frac{3}{100}$

Reflect

Explain whether you can

a always write tenths as hundredths

b always write hundredths as tenths

5.2 Converting fractions and decimals

Small steps

- Convert between fractions and decimals – fifths and quarters
- Convert between fractions and decimals – eighths and thousandths (H)

Key words

Convert – to change from one form to another

Proportion – part of something when compared to the whole

Are you ready?

1 Explain why this shape shows one quarter.

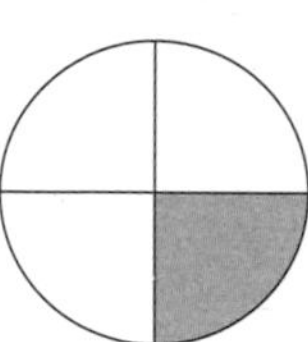

2 What fraction of each shape is shaded?

a 

b

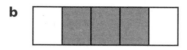

c 

d

e 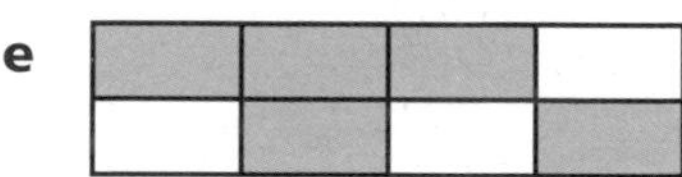

f 

3 Write each fraction as a decimal.

a $\frac{3}{10}$ b $\frac{3}{100}$ c $\frac{7}{10}$ d $\frac{43}{100}$ e $\frac{80}{100}$

4 Write each decimal as a fraction.

a 0.29 b 0.9 c 0.07 d 0.99 e 0.1

Models and representations

Hundred square

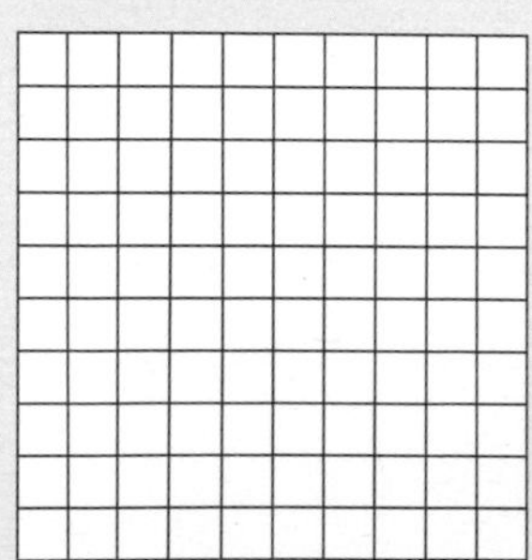

A hundred square is useful when working with fifths and quarters.

Thousand square

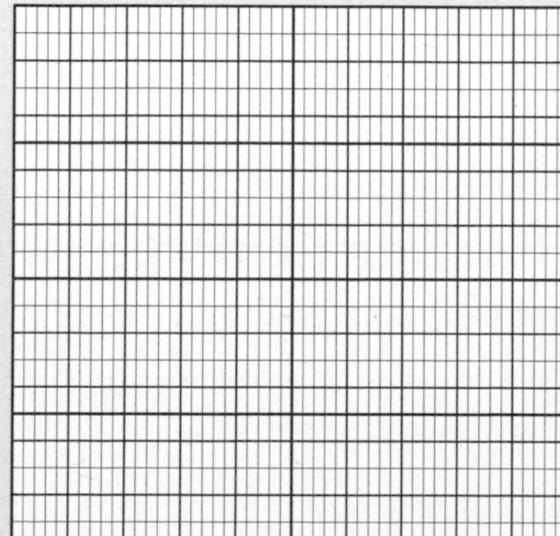

A thousand square is useful when working with eighths and thousandths.

Example 1

On a hundred square, shade

a **i** $\frac{1}{5}$ **ii** $\frac{1}{4}$

b Write the **proportion** of each hundred square shaded in part **a** as a decimal.

a i

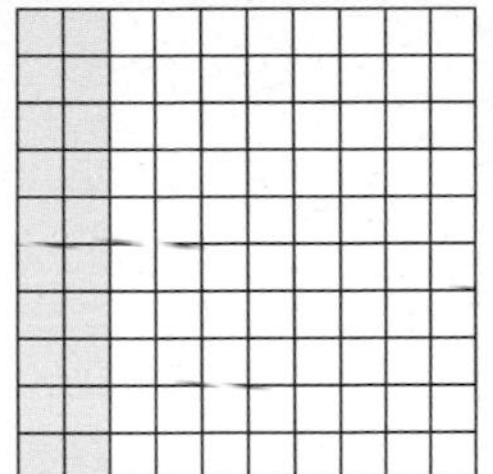

Split the hundred square into five equal parts. One fifth of the hundred square is shaded.

You can divide the hundred square into strips vertically or horizontally, or you could shade any 20 small squares.

a ii

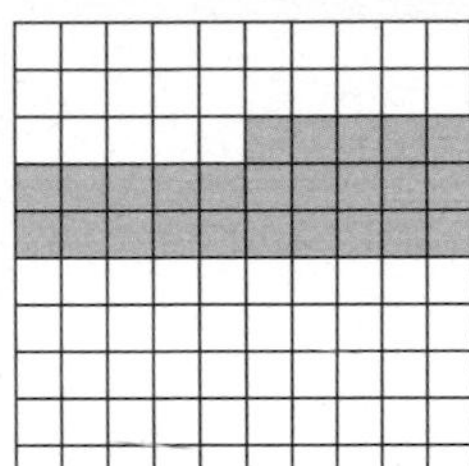

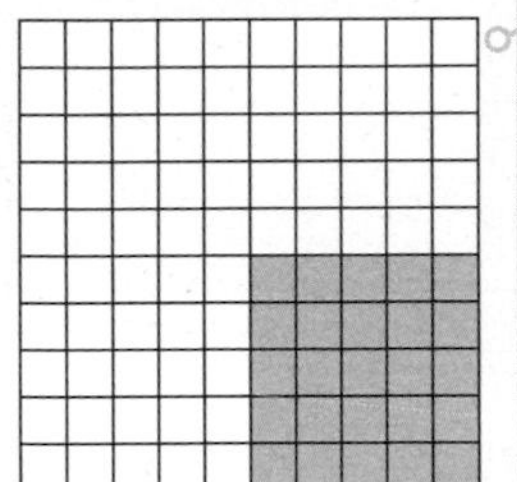

Split the hundred square into four equal parts. One quarter of the hundred square is shaded.

You can divide the hundred in several different ways as long as you shade 25 small squares.

b i $\frac{1}{5} = \frac{20}{100} = 0.2$

The diagram shows that one fifth is equivalent to two tenths or twenty hundredths, which can be written as 0.2

b ii $\frac{1}{4} = \frac{25}{100} = 0.25$

The diagram shows that one quarter is equivalent to 25 hundredths, which can be written as 0.25

Practice 5.2A

1 Each hundred square has been split into fifths.

A

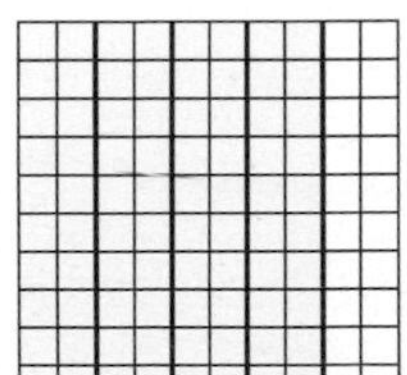

B

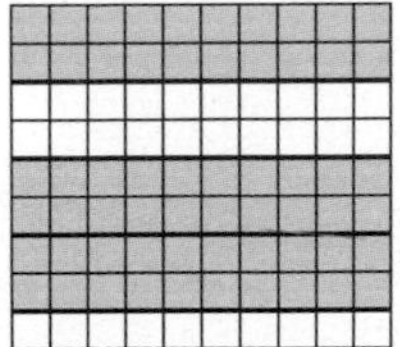

C

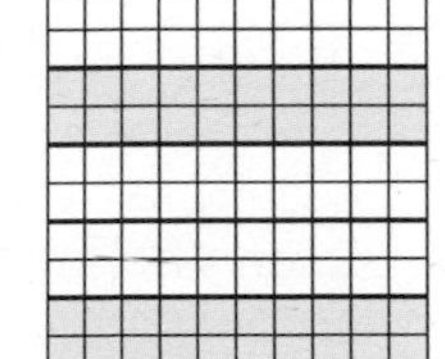

D

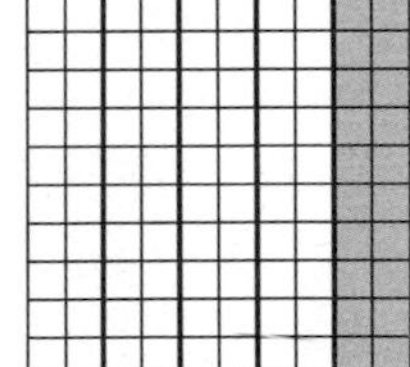

a How many fifths of each hundred square have been shaded?

b How many tenths of each hundred square have been shaded?

c How many hundredths of each hundred square have been shaded?

d Write your answers to parts **a**, **b** and **c** as decimals.

5.2 Converting fractions and decimals

2 Each hundred square has been split into quarters.

A 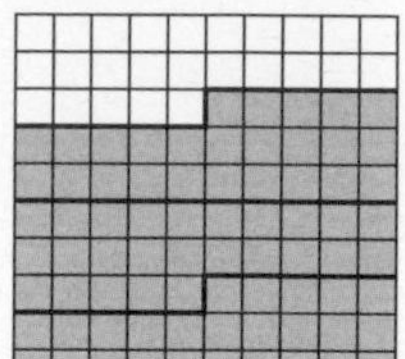 **B** 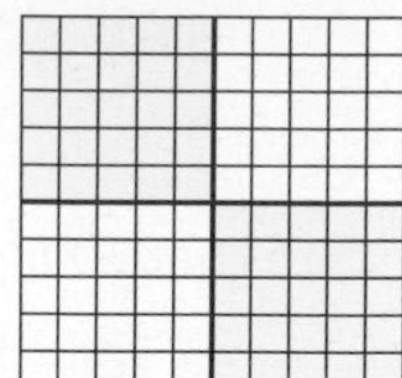 **C** 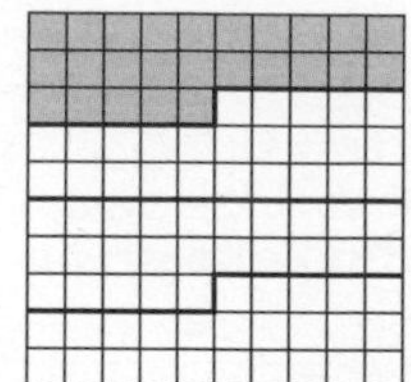 **D** 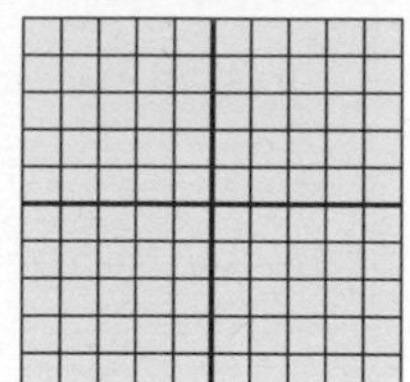

a How many quarters of each hundred square have been shaded?

b How many hundredths of each hundred square have been shaded?

c Write your answers to parts **a** and **b** as decimals.

3 Here are two hundred squares.

A 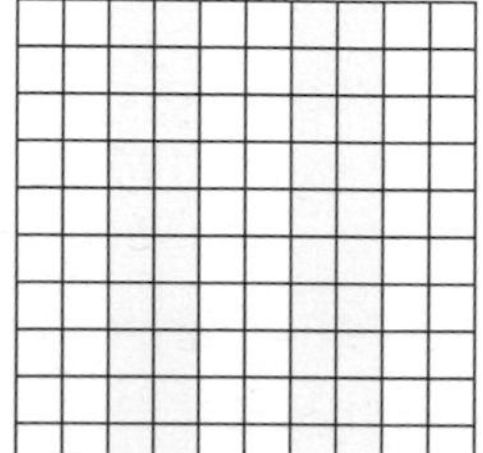 **B** 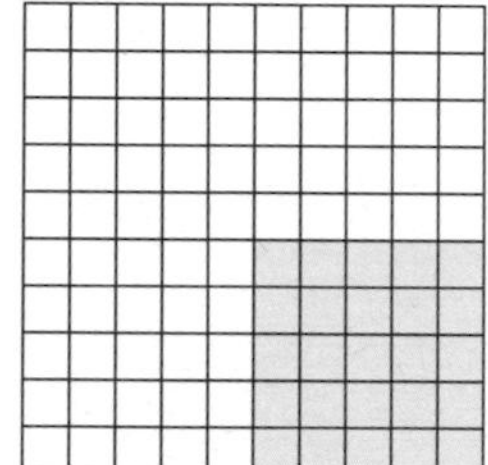

a What proportion of each hundred square is shaded?

b What proportion of each hundred square is not shaded?

c What do you notice about your answers to parts **a** and **b** for each diagram?

4 Write each number as either fifths or quarters. Then write it as a decimal. You can use a hundred square to help you.

a 2 tenths **b** 25 hundredths **c** $\frac{50}{100}$

d $\frac{4}{10}$ **e** 75 hundredths **f** 100 hundredths

Is there more than one answer to part **f**? Discuss this with a partner.

5 The bar model shows that $\frac{4}{4} = \frac{5}{5} = \frac{10}{10}$

$\frac{1}{10}$	$\frac{1}{10}$	$\frac{1}{10}$	$\frac{1}{10}$	$\frac{1}{10}$	$\frac{1}{10}$	$\frac{1}{10}$	$\frac{1}{10}$	$\frac{1}{10}$	$\frac{1}{10}$
$\frac{1}{5}$		$\frac{1}{5}$		$\frac{1}{5}$		$\frac{1}{5}$		$\frac{1}{5}$	
$\frac{1}{4}$			$\frac{1}{4}$		$\frac{1}{4}$			$\frac{1}{4}$	

Copy and complete these statements. Use the bar model to help you.

a $\frac{1}{5} = \frac{?}{10} = 0.\square$ **b** $\frac{2}{4} = \frac{?}{10} = 0.\square$ **c** $\frac{8}{10} = \frac{?}{5} = 0.\square$

6 Use a blank hundred square.

a Shade one fifth of the hundred square. How many squares did you shade?

b Shade one quarter of the hundred square. How many squares did you shade?

c Which is greater, one fifth or one quarter? Explain your answer.

7 The bar model shows that $\frac{4}{4} = \frac{5}{5} = \frac{10}{10}$

$\frac{1}{10}$	$\frac{1}{10}$	$\frac{1}{10}$	$\frac{1}{10}$	$\frac{1}{10}$	$\frac{1}{10}$	$\frac{1}{10}$	$\frac{1}{10}$	$\frac{1}{10}$	$\frac{1}{10}$
$\frac{1}{5}$		$\frac{1}{5}$		$\frac{1}{5}$		$\frac{1}{5}$		$\frac{1}{5}$	
$\frac{1}{4}$			$\frac{1}{4}$			$\frac{1}{4}$		$\frac{1}{4}$	

Use <, > or = to complete each of these statements. You can use the bar model to help you.

a $\frac{1}{5} \bigcirc \frac{3}{4}$ **b** $\frac{1}{4} \bigcirc \frac{1}{5}$ **c** $\frac{1}{4} \bigcirc \frac{2}{5}$

d $\frac{3}{5} \bigcirc \frac{6}{10}$ **e** $\frac{7}{10} \bigcirc \frac{4}{5}$ **f** $\frac{2}{4} \bigcirc \frac{5}{10}$

What do you think?

1 Filipo and Lydia have been asked to represent one fifth on a hundred square.

Here are their answers.

Filipo

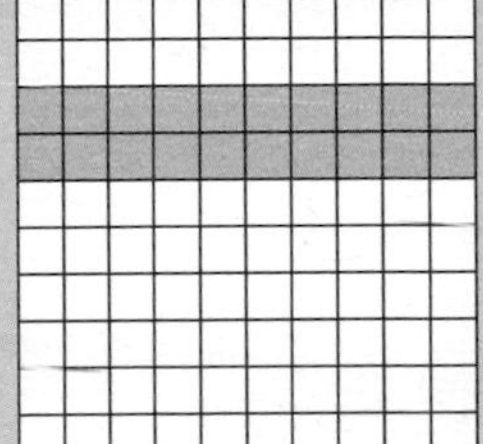

Lydia

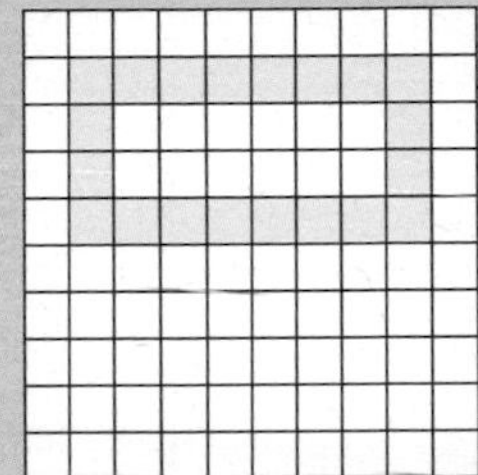

Who is correct? Explain your answer.

2 Mario and Amina are comparing $175\frac{3}{4}$ and $175\frac{2}{5}$.

Mario says, "I can't do it as I haven't looked at fifths and quarters in big numbers yet."

Amina says, "I don't think that matters."

Who do you agree with?

Discuss it with a partner.

Example 2 Ⓗ

a On a thousand square, shade

i $\frac{1}{1000}$ **ii** $\frac{1}{100}$ **iii** $\frac{1}{10}$ **iv** $\frac{1}{8}$

b Write each fraction from part **a** as a decimal.

a i

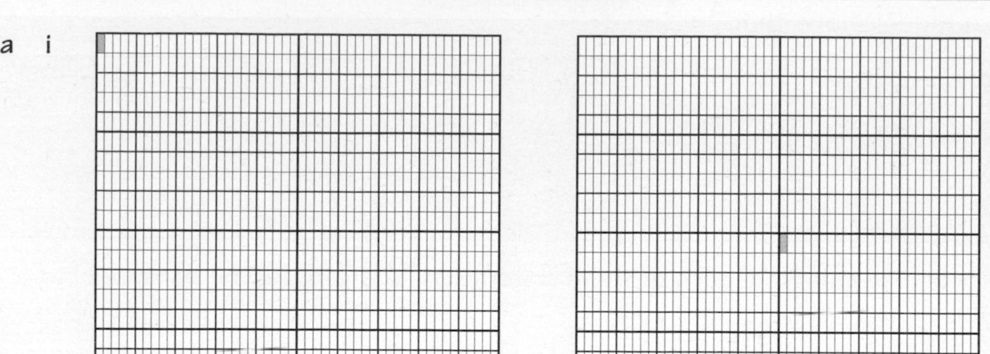

A thousand square is split into one thousand equal parts.
You can shade any one part.

a ii

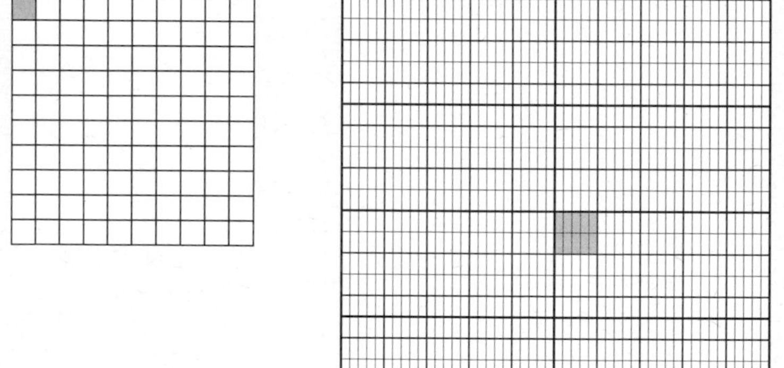

One of the bigger squares in a thousand square is the same as one hundredth. This is the same as 10 small rectangles.
You can shade any of the bigger squares or any 10 small rectangles.

a iii

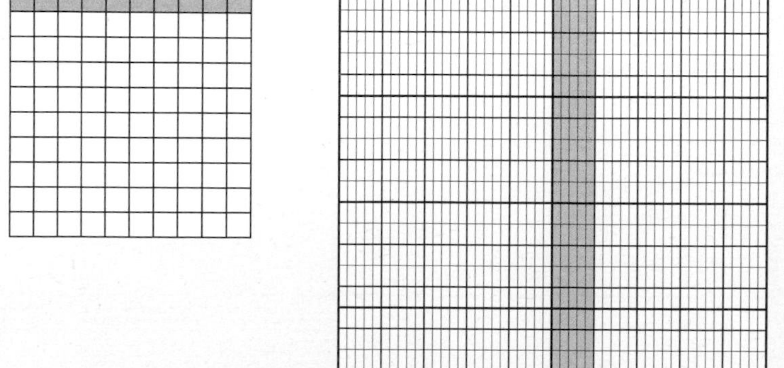

You can shade one full row or column of the thousand square. This is the same as 100 small squares.
You can shade any full row or column or any 100 small rectangles.

a iv

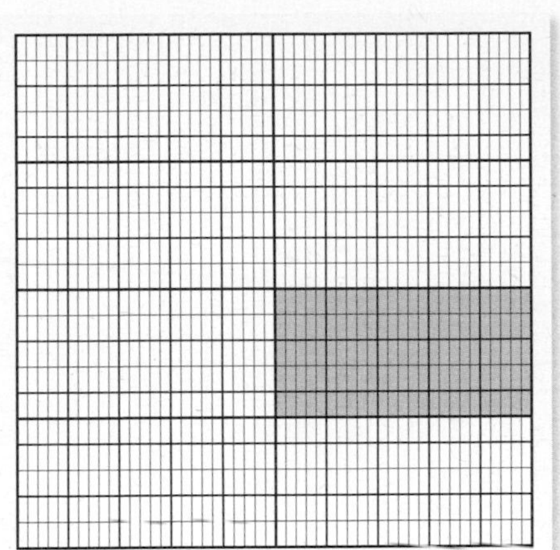

Divide the thousand square into eight equal parts. Shade one of these eight parts.

You can shade any one of the 8 parts or any 125 small rectangles.

b i *0.001* — $\frac{1}{1000}$ is equivalent to 0.001.
You can see this from a place value chart.

b ii *0.01* — The diagram shows that ten thousandths is equivalent to one hundredth.
$\frac{10}{1000} = \frac{1}{100}$ or 0.01

b iii *0.1* — The diagram shows that one hundred thousandths is equivalent to one tenth.
$\frac{100}{1000} = \frac{1}{10}$ or 0.1

b iv *0.125* — The diagram shows that one eighth is equivalent to one hundred and twenty-five thousandths so
$\frac{1}{8} = \frac{125}{1000}$ or 0.125

Practice 5.2B Ⓗ

1 What proportion of each thousand square is shaded? Give each answer as a fraction and a decimal.

a

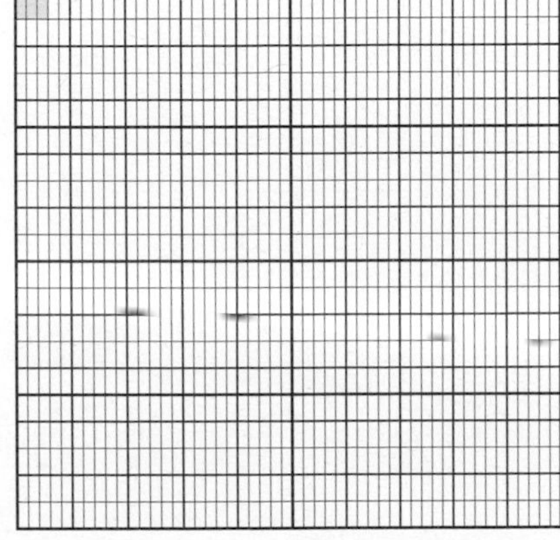

b

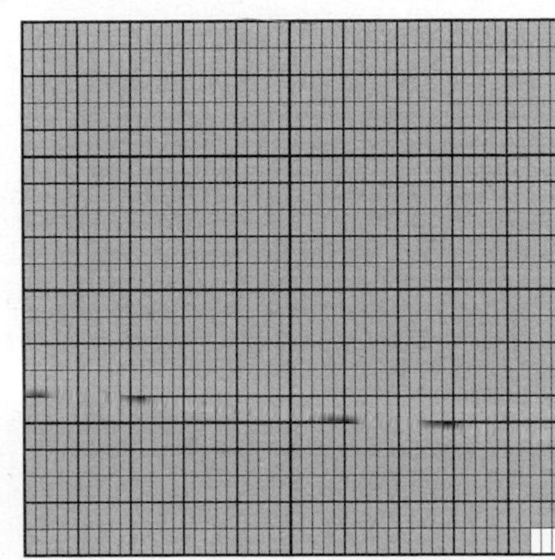

c

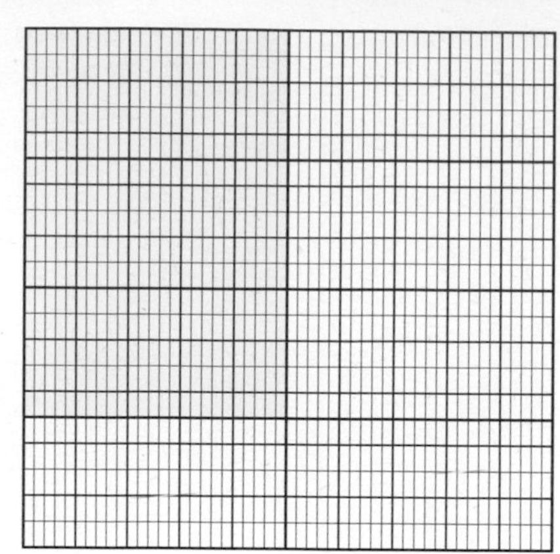

d

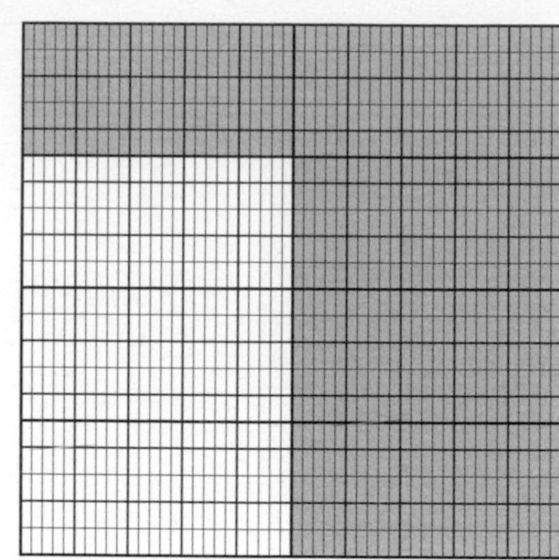

2 Use a thousand square to help you.

a How many thousandths are equivalent to 7 hundredths?

b How many thousandths are equivalent to 7 tenths?

3 Convert each of these fractions to a decimal.

a $\frac{389}{1000}$ b $\frac{273}{1000}$ c $\frac{961}{1000}$ d $\frac{511}{1000}$

e $\frac{800}{1000}$ f $\frac{80}{1000}$ g $\frac{8}{1000}$ h $\frac{19}{1000}$

4 Convert each of these decimals to a fraction.

a 0.176 b 0.999 c 0.004 d 0.076

e 0.345 f 0.763 g 0.800 h 0.750

5 Work out the missing numbers.

a $\frac{1}{4} = \frac{?}{8} = 0.\square$ b $\frac{3}{4} = \frac{?}{8} = 0.\square$ c $\frac{4}{8} = \frac{?}{2} = 0.\square$ d $\frac{1}{8} = 0.\square$

e $\frac{3}{8} = 0.\square$ f $\frac{5}{8} = 0.\square$ g $\frac{7}{8} = 0.\square$ h $\frac{8}{8} = \square$

What do you think?

1 Abdullah has been asked to write 72 thousandths in numerals. Here is his answer.

72 000

Do you agree with Abdullah? Explain your answer.

2 Write down the numbers that are missing from each of these number lines. Write the decimal and fraction for each.

a

b 0 0.001

$\frac{1}{1000}$ $\frac{6}{1000}$

3 Here are 7 fractions.

A $\frac{1}{10}$ B $\frac{1}{100}$ C $\frac{1}{1000}$ D $\frac{1}{4}$ E $\frac{1}{5}$ F $\frac{1}{8}$ G $\frac{1}{2}$

a Write each of these fractions as a decimal.

b Hence, write the fractions in ascending order.

c What do you notice?

Consolidate – do you need more?

1 Write each of these fractions as a decimal.

a $\frac{1}{5}$ **b** $\frac{2}{5}$ **c** $\frac{3}{5}$ **d** $\frac{4}{5}$

2 Write each of these fractions as a decimal.

a $\frac{1}{4}$ **b** $\frac{2}{4}$ **c** $\frac{3}{4}$

3 **a** Explain how the bar model shows that $\frac{2}{8}$ is equivalent to $\frac{1}{4}$.

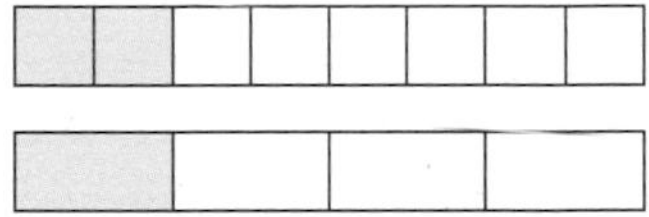

b Write $\frac{2}{8}$ as a decimal.

4 Write each of these fractions as a decimal.

a $\frac{1}{8}$ **b** $\frac{3}{8}$ **c** $\frac{6}{8}$ **d** $\frac{7}{8}$

5 Write each of these decimals as a fraction.

a 0.145 **b** 0.807 **c** 0.035 **d** 0.001

Stretch – can you deepen your learning?

1 For each number line, work out the value at A, the value at B, and the difference between A and B.

Give all answers as fractions and decimals.

a

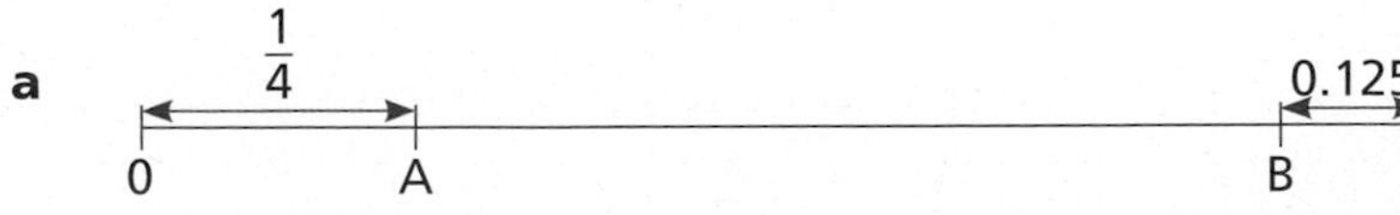

b

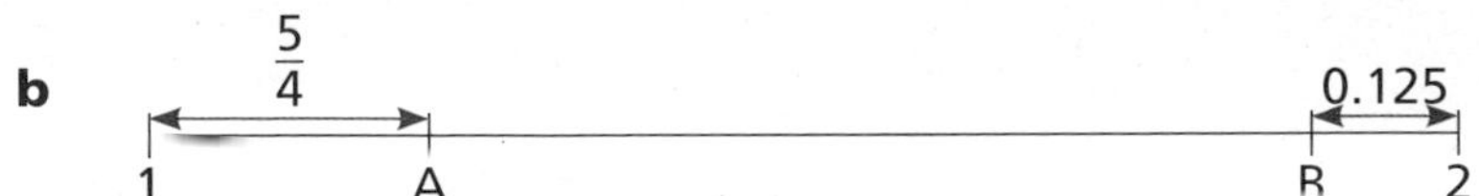

c

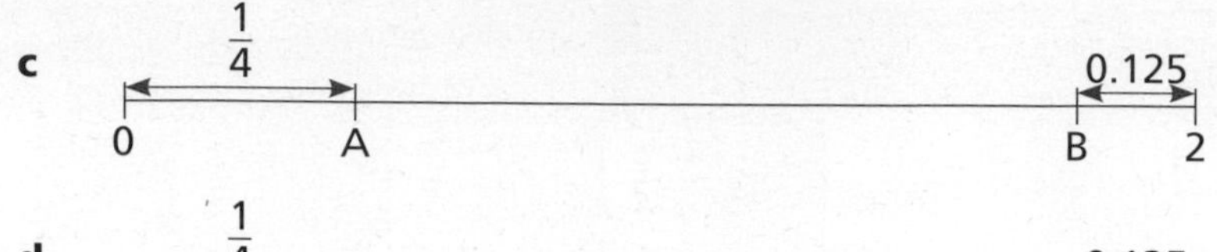

d

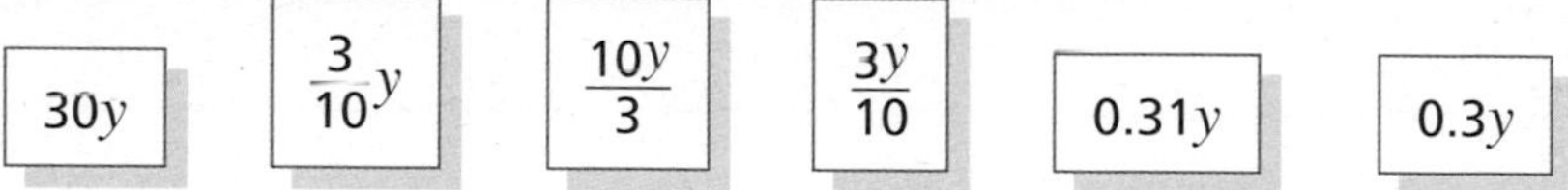

2 Which expressions are equivalent to three tenths of y?

$30y$	$\frac{3}{10}y$	$\frac{10y}{3}$	$\frac{3y}{10}$	$0.31y$	$0.3y$

3 Write down four expressions that are equivalent to two fifths of z.

4 Write each of these decimals as a fraction.

a 3.7 **b** 12.09 **c** 1.871 **d** 5.375

e 6.25 **f** 19.009 **g** 21.08 **h** 184.397

Reflect

$\frac{1}{100}$ is equal to 0.01. Explain how you can use this fact to work out the decimal equivalent of each of these fractions.

a $\frac{1}{10}$ **b** $\frac{1}{5}$ **c** $\frac{1}{4}$ **d** $\frac{3}{4}$

5.3 Using percentages

Small steps

- Understand the meaning of percentage using a hundred square
- Convert fluently between simple fractions, decimals and percentages

Key words

Per cent – parts per hundred

Convert – to change from one form to another

Are you ready?

1 How many squares out of 100 are shaded in each hundred square?

a 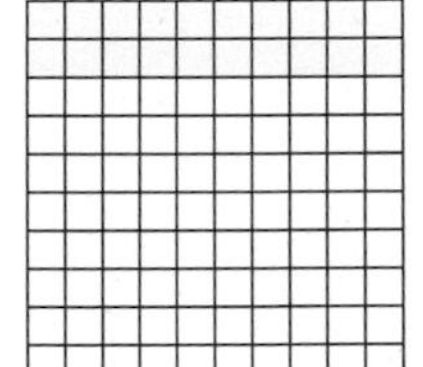 **b** 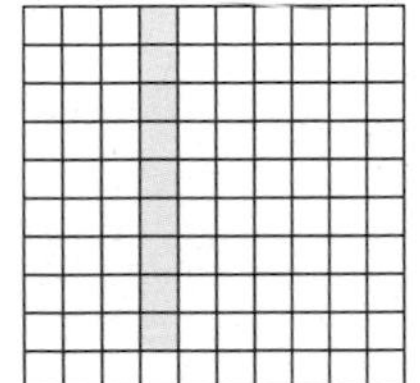 **c** 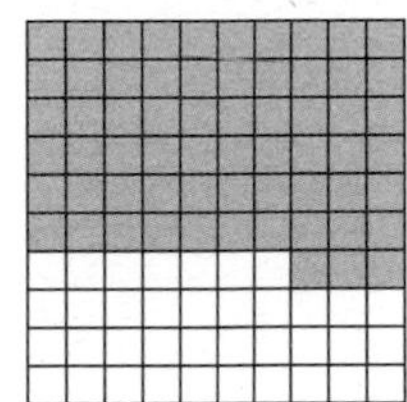 **d** 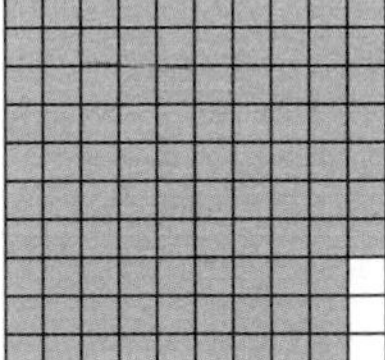

How did you count the squares?

2 Write each decimal as a fraction.

a 0.2 **b** 0.7 **c** 0.9 **d** 0.1

3 Write each decimal as a fraction.

a 0.31 **b** 0.73 **c** 0.97 **d** 0.49

4 Write each fraction as a decimal.

a $\frac{3}{10}$ **b** $\frac{19}{100}$ **c** $\frac{99}{100}$ **d** $\frac{5}{10}$

5 Write each fraction as a decimal.

a $\frac{1}{4}$ **b** $\frac{1}{5}$ **c** $\frac{3}{4}$ **d** $\frac{3}{5}$

6 Write each decimal as a fraction.

a 0.8 **b** 0.4 **c** 0.25 **d** 0.75

Models and representations

Hundred square

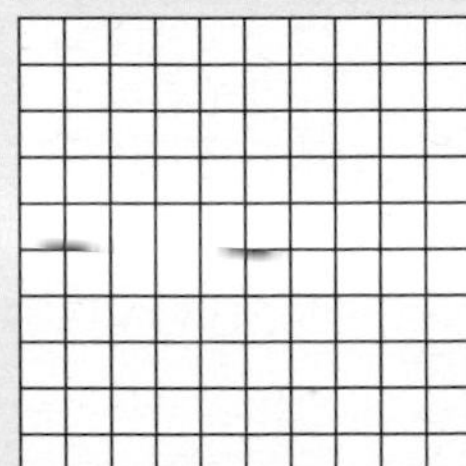

Bar model

Number line

0 10% 20% 30% 40% 50% 60% 70% 80% 90% 100%

0 $\frac{1}{10}$ $\frac{2}{10}$ $\frac{3}{10}$ $\frac{4}{10}$ $\frac{5}{10}$ $\frac{6}{10}$ $\frac{7}{10}$ $\frac{8}{10}$ $\frac{9}{10}$ $\frac{10}{10}$

In this chapter you will use a hundred square to represent 100%. You will learn how to **convert** between fractions, decimals and percentages.

Example 1

Shade hundred squares to show each of these percentages.

Reflect on what you have already learned about fractions and decimals and how you used a hundred square to convert between them.

a 1% **b** 10% **c** 100%

a

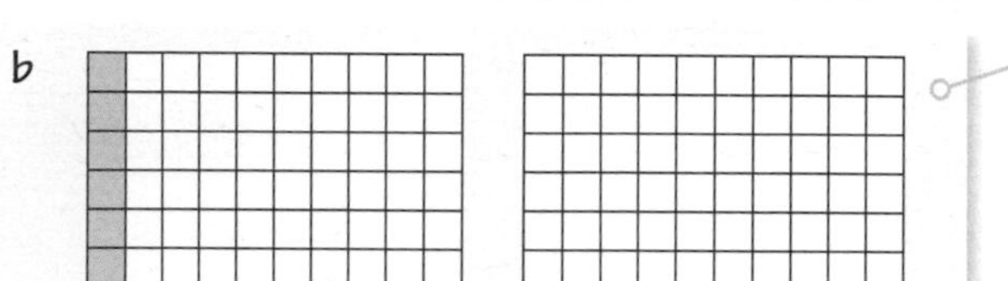

'Per cent' means parts per hundred, so you need to shade 1 square.

1 out of 100, or $\frac{1}{100}$ is equivalent to 1%.

Each individual square on the hundred square, regardless of its position, is worth 1%.

b

10 parts out of 100 is the same as 10%, so you need to shade one full row or column.

c

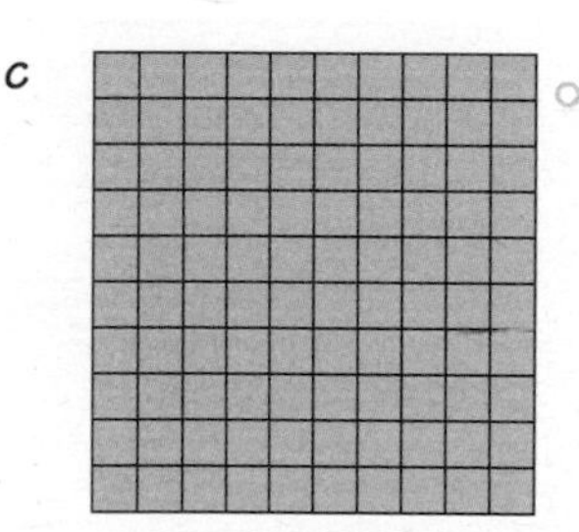

One whole is represented by 100%.

100 out of the 100 equal parts are shaded.

Practice 5.3A

1 Here are four hundred squares.

A

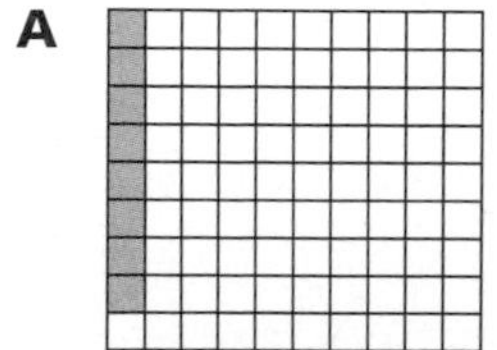

B

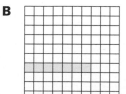

C

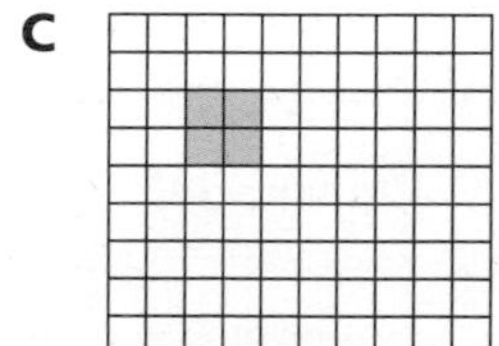

D

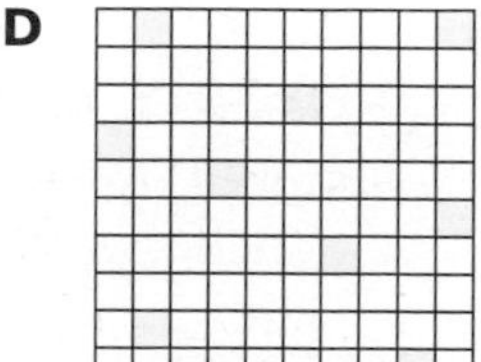

a How many squares are shaded in each hundred square?

b What fraction of each hundred square is shaded? Give your answer as a fraction out of 100.

c What percentage of each hundred square is shaded?

2 Here are four hundred squares.

A 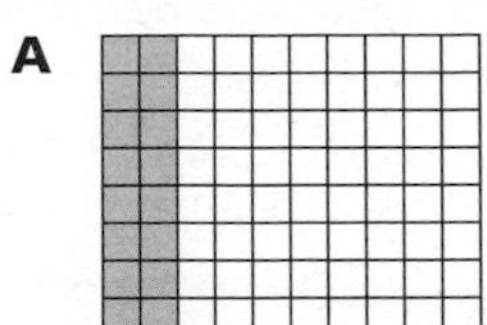 B C D

a How many squares are shaded in each hundred square?

b What fraction of each hundred square is shaded? Give your answer as a fraction out of 100.

c What percentage of each hundred square is shaded?

3 Here are four hundred squares.

A 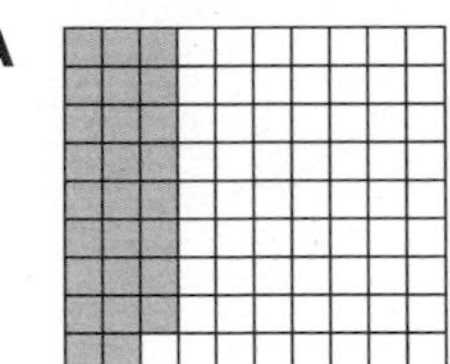 B C 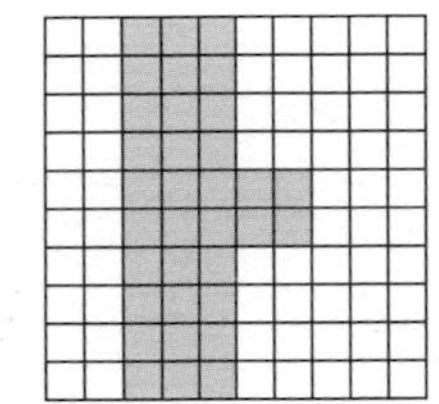 D 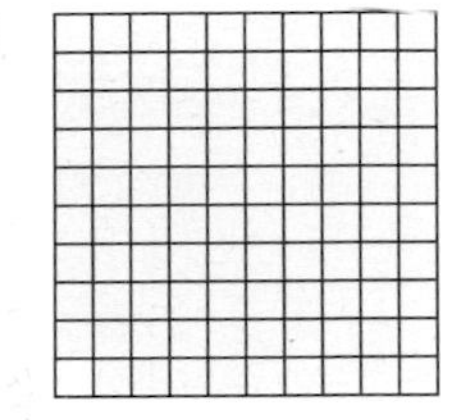

a How many squares are shaded in each hundred square?

b What fraction of each hundred square is shaded? Give your answer as a fraction out of 100.

c What percentage of each hundred square is shaded?

4 Shade hundred squares to show each of these percentages.

a 35% b 71% c 8% d 19% e 25%

5 Chloe shades in 49 squares on a hundred square.

a What percentage of the hundred square has she shaded?

b What percentage of the hundred square has she not shaded?

c Add together your answers from parts **a** and **b**.

d What do you notice? Why does this happen?

6 For each question part, you are given the percentage of a hundred square that is shaded. Write down the percentage of each hundred square that is not shaded.

a 71% b 11% c 53% d 1% e 100%

What do you think?

1 Here is a hundred square.

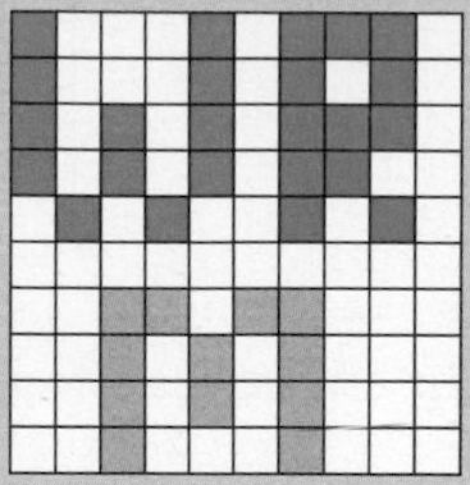

a What percentage of the hundred square is shaded blue (the 'W')?

b What percentage of the hundred square is shaded green (the 'R')?

c What percentage of the hundred square is shaded yellow (the 'M')?

d What percentage of the hundred square is not shaded?

e What could you do to check that you are right?

2 Rob says that he has shaded 10% of the hundred square.

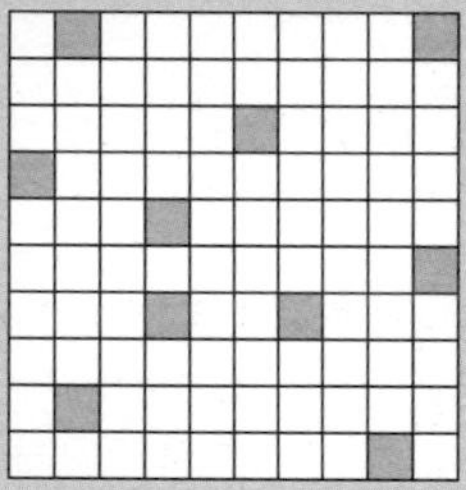

Lydia says that can't be true because he hasn't shaded a full row or column.

Who do you agree with?

Explain your answer.

3 Each bar model is split into 10 equal parts. What percentage of each bar model is shaded?

a

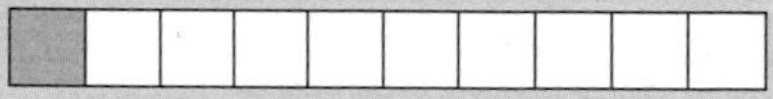

b

c

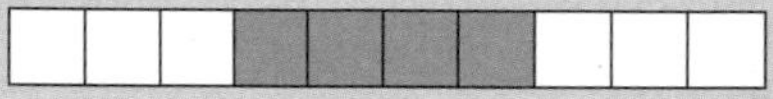

d

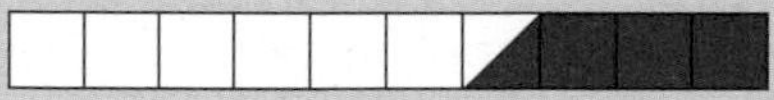

Example 2

Shade hundred squares to show each of these percentages.

a 25% **b** 20%

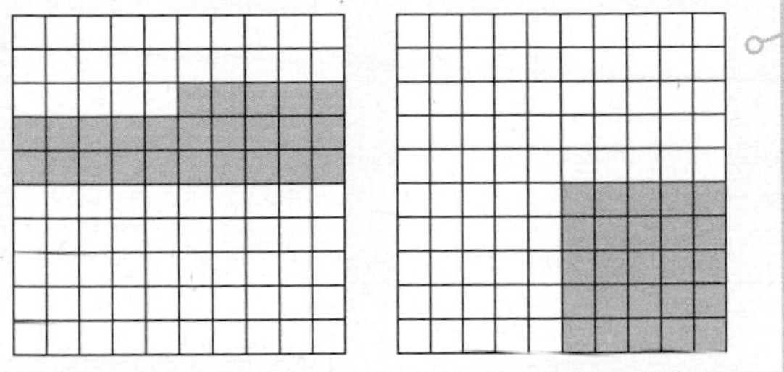

25% is the same as 25 out of 100 so you need to shade 25 squares.

This is the same as $\frac{1}{4}$ of the hundred square.

You can divide the hundred square into 4 equal parts and shade 1 part or you can shade any 25 squares.

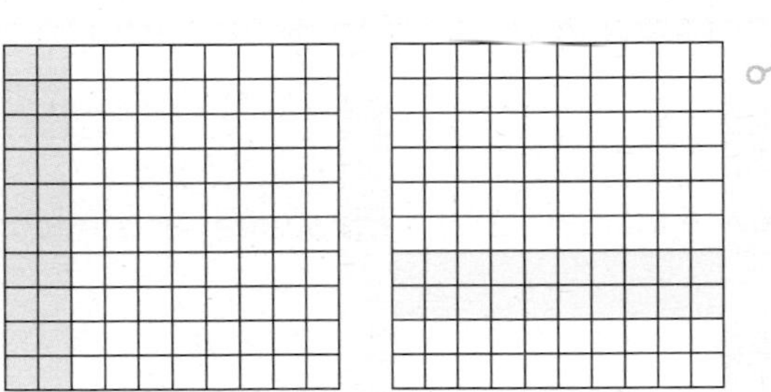

20% is the same as 20 out of 100, so you need to shade 20 squares.

This is the same as $\frac{1}{5}$ of the hundred square.

You can divide the hundred square into 5 equal parts and shade 1 part or you can shade any 20 squares.

Practice 5.3B

1 What proportion of each hundred square is shaded? Give each answer as a fraction, a decimal and a percentage.

a 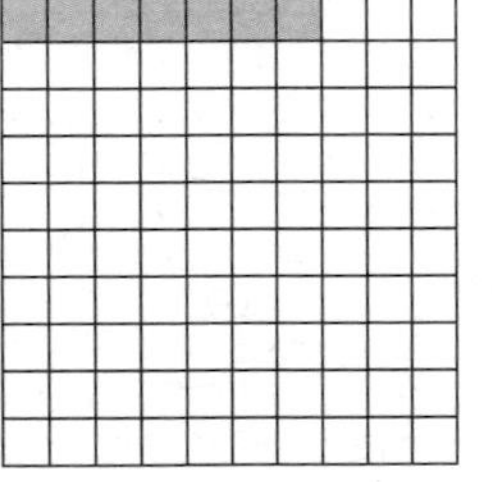 **b** 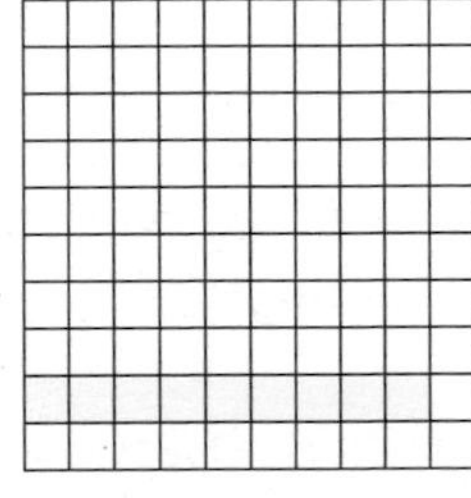 **c** 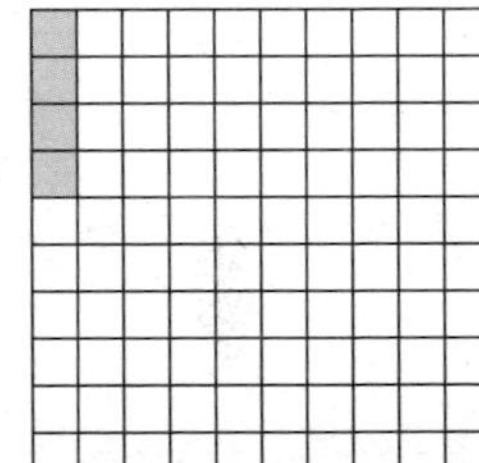

d 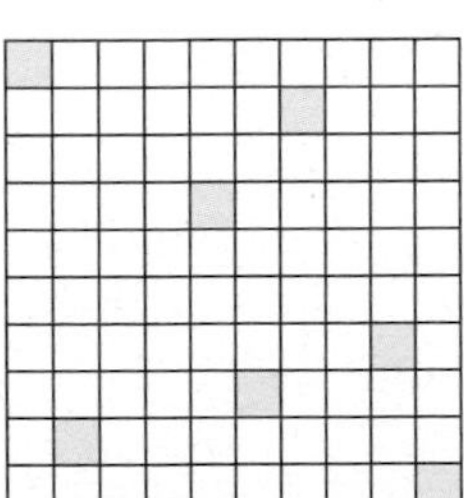 **e** 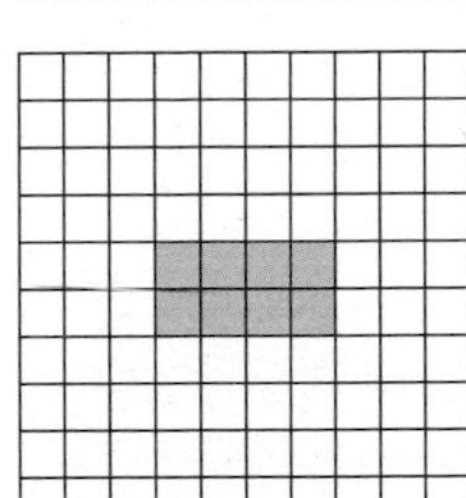 **f** 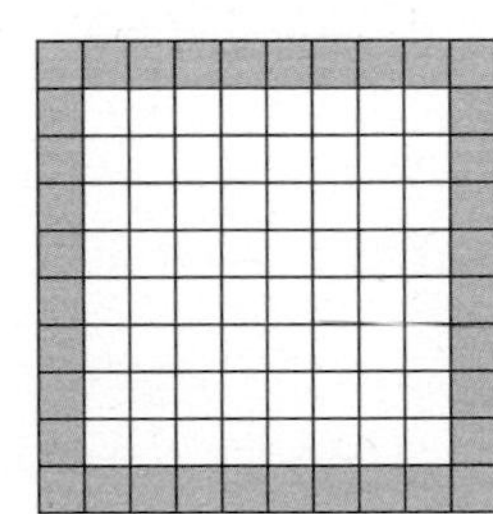

2 What proportion of each hundred square is shaded? Give each answer as a fraction, a decimal and a percentage.

a 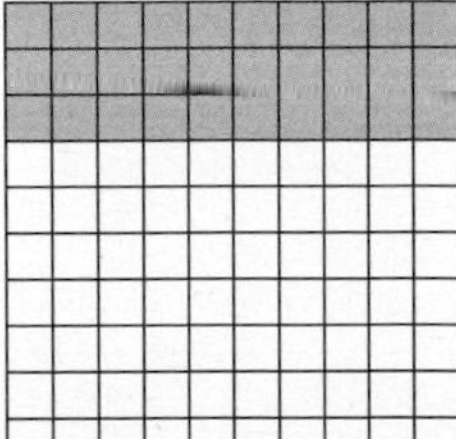 **b** 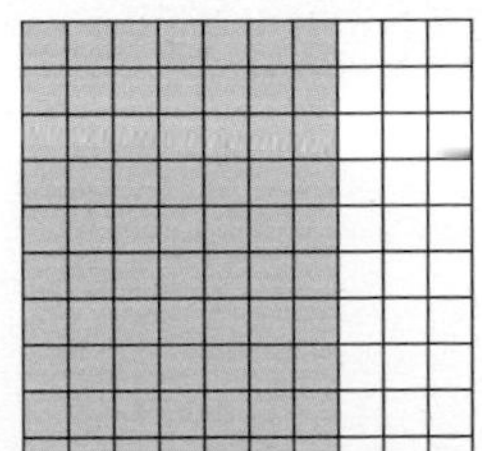 **c** 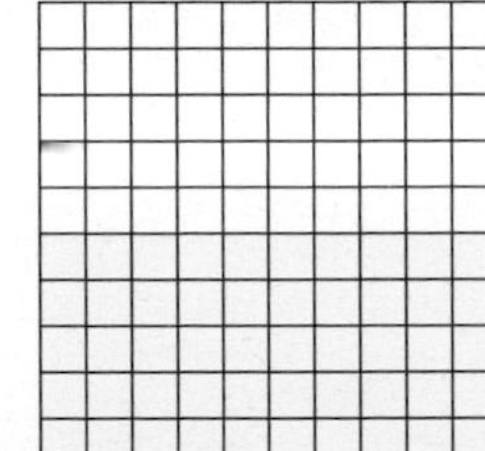

d 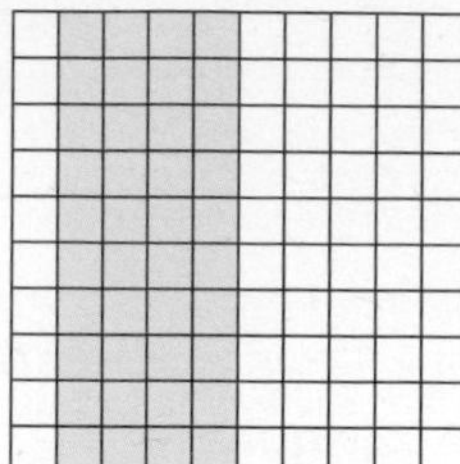
e 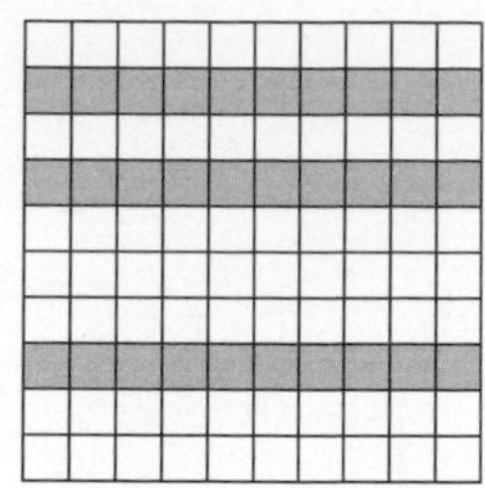
f 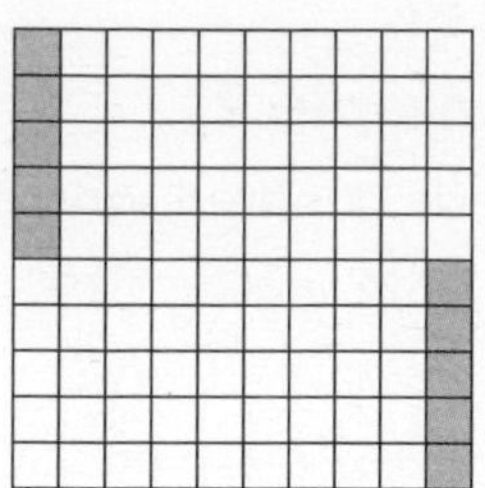

3. What proportion of each hundred square is shaded? Give each answer as a fraction, a decimal and a percentage.

a 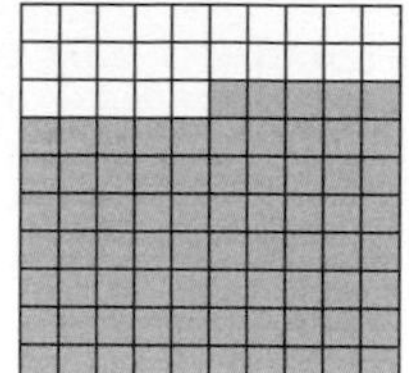
b 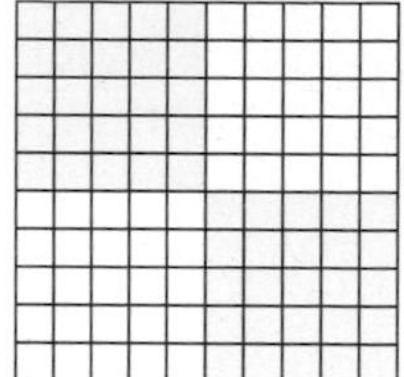
c 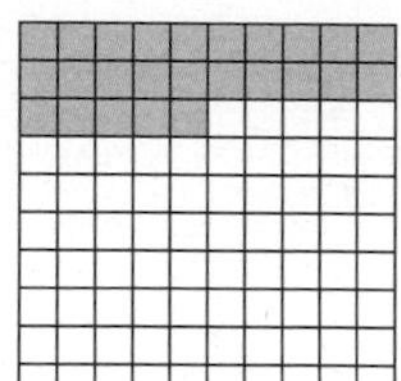
d 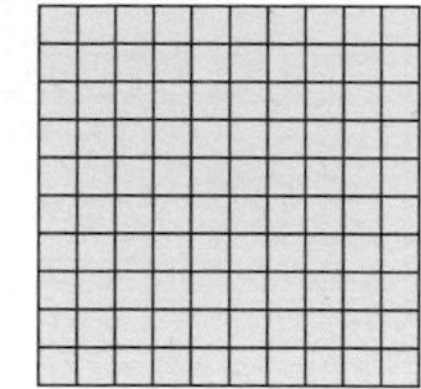

4. What proportion of each hundred square is shaded? Give each answer as a fraction, a decimal and a percentage.

a 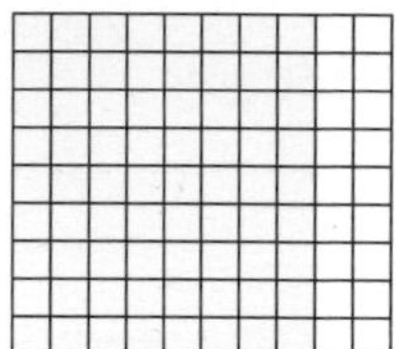
b 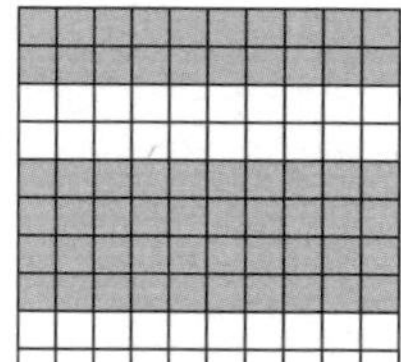
c 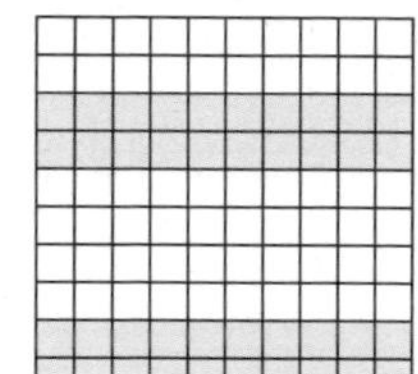
d 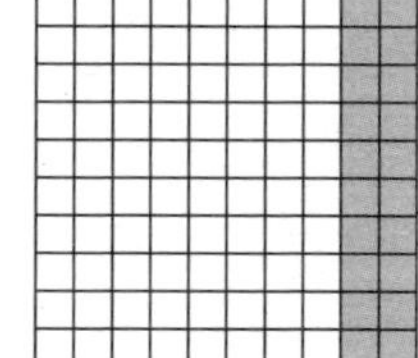

5. Write each of these fractions as a decimal and as a percentage.

a $\frac{23}{100}$ **b** $\frac{7}{10}$ **c** $\frac{4}{5}$ **d** $\frac{3}{4}$ **e** $\frac{1}{2}$

6. Write each of these decimals as a fraction and as a percentage.

a 0.71 **b** 0.03 **c** 0.25 **d** 0.6 **e** 0.9

7. Write each of these percentages as a fraction and as a decimal.

a 99% **b** 12% **c** 7% **d** 20% **e** 75%

What do you think?

1. Decide whether each statement is true or false. Explain your answers.

a 0.5 is equivalent to $\frac{1}{5}$

b $\frac{3}{4}$ is equivalent to 34%

c 100% is equivalent to $\frac{5}{5}$

For any statements that are false, rewrite them so that they are true.

2 Jackson has a pizza.

He eats $\frac{1}{5}$ of the pizza.

He gives 24% of the pizza to his sister.

He gives 0.16 of the pizza to his brother.

How much of the pizza is left?

Write your answer as a fraction, a decimal and a percentage.

Consolidate – do you need more?

1 Write each of these percentages as a fraction and as a decimal.

a 31% **b** 72% **c** 80% **d** 90% **e** 1%

2 Write each of these fractions as a percentage and as a decimal.

a $\frac{11}{100}$ **b** $\frac{27}{100}$ **c** $\frac{93}{100}$ **d** $\frac{3}{10}$ **e** $\frac{1}{4}$

3 Write each of these decimals as a fraction and as a percentage.

a 0.57 **b** 0.61 **c** 0.09 **d** 0.1 **e** 0.8

Stretch – can you deepen your learning?

1 The grid is split into 50 equal pieces.

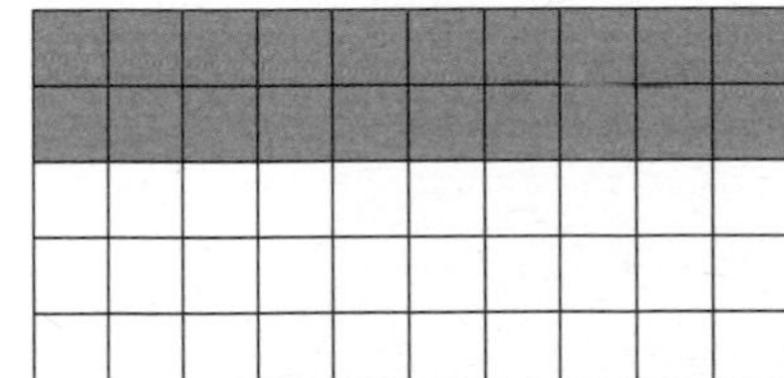

a What proportion of the grid is shaded? Give your answer as a fraction, a decimal and a percentage.

b Ed shades 18% of the grid in another colour. How many squares does he shade?

c What proportion of the grid is now not shaded? Give your answer as a fraction, a decimal and a percentage.

2 Decide whether or not each of these sequence is linear.

For the sequences that are linear, write down the next three terms as fractions, decimals and percentages.

a $\frac{1}{10}, \frac{15}{100}, \frac{1}{5}\ldots$ **b** $\frac{1}{4}, \frac{2}{4}, 1\ldots$ **c** $\frac{99}{100}, \frac{9}{10}, \frac{81}{100}\ldots$ **d** $1, \frac{4}{5}, \frac{3}{5}\ldots$

3 Write each of these decimals as a percentage.

a 1.35 **b** 2.72 **c** 6.09 **d** 8.4

When are percentages greater than 100% used in real life?

4 Here is a bar model.

94.5%	
p	0.71

a Write down the fact family for the bar model.

b Use your fact family to work out the value of p. Give your answer as a fraction, a decimal and a percentage.

5 $\frac{4}{5} < x < 82\%$

Darius says that there is only one possible value of x.

Beth says that there are an infinite number of possible values of x.

Who do you agree with? Explain your answer.

Reflect

Decide whether each statement is true or false. Explain your answers.

a To convert a decimal to a percentage you multiply by 100.

b Only fractions with a denominator of 100 can be written as percentages.

c $\frac{a}{100} = a\%$

5.4 Pie charts

Small steps

- Use and interpret pie charts

Key words

Pie chart – a graph in which a circle is divided into sectors that each represent a proportion of the whole

Proportion – a part, share or number considered in relation to a whole

Sector – a part of a circle formed by two radii and a fraction of the circumference

Are you ready?

1 What fraction of each diagram is shaded?

a

b

c

d 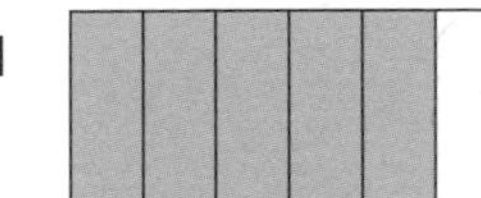

e 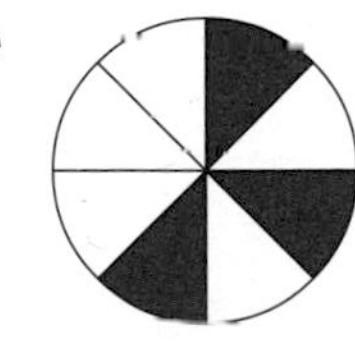

f 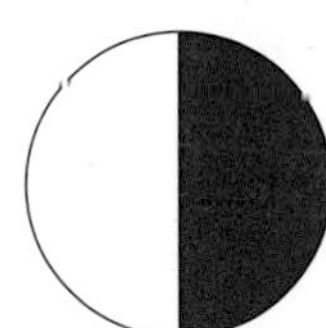

2 What percentage of each diagram is shaded?

a 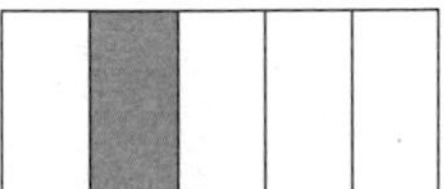

b 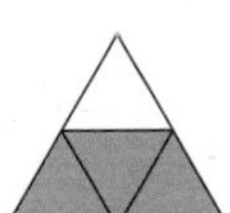

c 

d 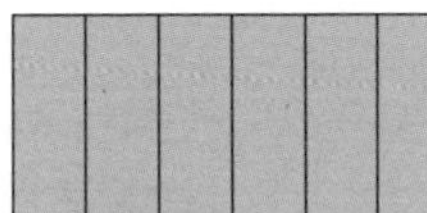

e 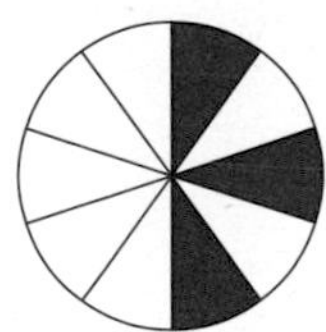

f 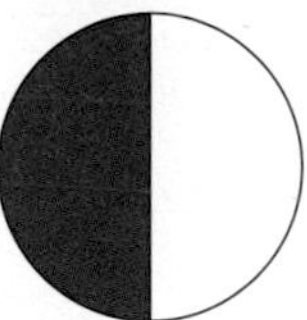

3 Write each of your answers to question 2 as a decimal.

Models and representations

Pie charts

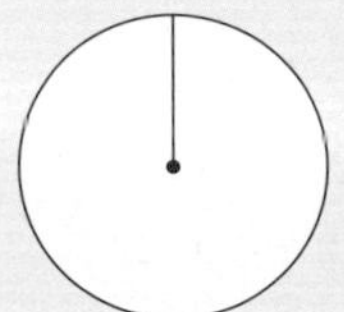 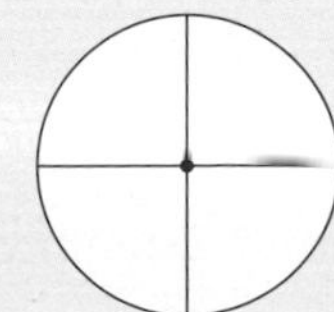 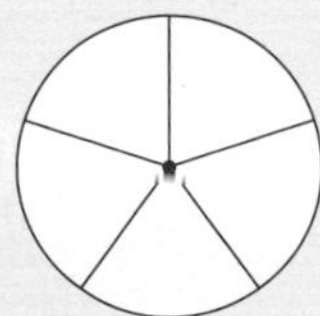 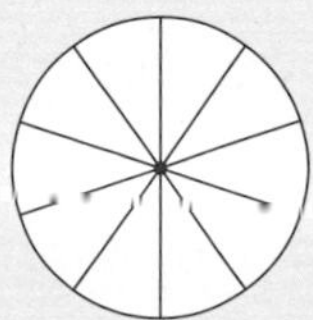

The circles are divided into equal sectors

5.4 Pie charts

In this chapter you will use a pie chart (or circle) to represent one whole, rather than a hundred square or bar model. You will look at how to construct pie charts in chapter 11.5.

A pie chart represents **proportions** of the whole, which can be expressed as fractions, decimals or percentages.

Example 1

A group of people were asked to choose their favourite colour. Their results are shown in the pie chart.

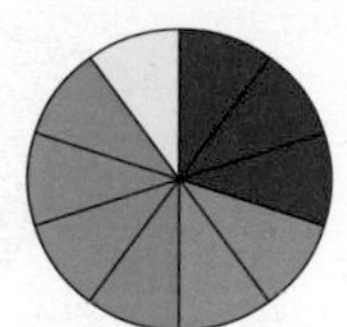

a What proportion of the people chose red?

b What proportion of the people chose blue?

Give each answer as fraction, a decimal and a percentage.

a $\frac{3}{10} = 0.3 = 30\%$

The pie chart is split into 10 equal parts, and 3 of them are red. This means that $\frac{3}{10}$ of the group of people chose red as their favourite colour.

$\frac{3}{10}$ is equivalent to 0.3 and 30%

b $\frac{4}{10} = \frac{2}{5} = 0.4 = 40\%$

4 of the 10 **sectors** are blue. So $\frac{4}{10}$ of the group of people chose blue as their favourite colour.

This can also be written as $\frac{2}{5}$, 0.4 or 40%

Practice 5.4A

1 A group of people were asked how many pets they have. The results are shown in the pie chart.

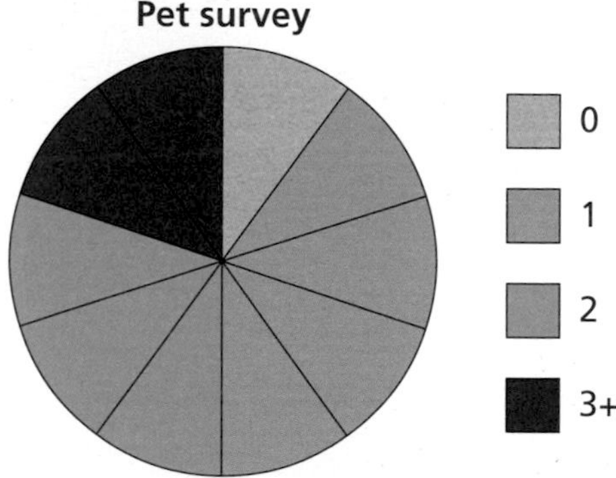

Write your answers to parts **a–d** as a fraction, a decimal and a percentage.

a What proportion of the people had no pets?

b What proportion of the people had 1 pet?

c What proportion of the people had 2 pets?

d What proportion of the people had 3 or more pets?

e Is it possible to tell exactly how many people have 3 pets? Explain your answer.

2 Some students were asked how they travel to school.

The results are shown in the pie chart. The pie chart has markings on the circumference at intervals of 10%.

How students travel to school

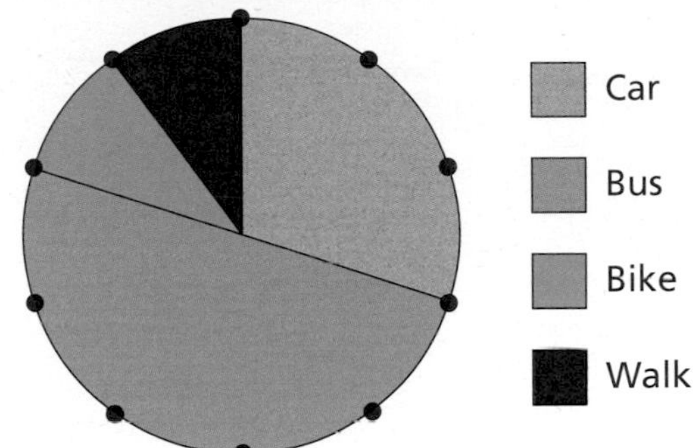

a What is the most common way to travel to school? What proportion of the students use this method? Give your answer as a fraction, a decimal and a percentage.

b Benji says that three students travel to school by car. Do you agree with Benji? Explain your answer.

c What does the pie chart tell you about the number of students who travel by bike and the number of students who walk to school?

3 A group of people were asked their favourite type of food. Their results are shown in the pie chart. The markings around the circumference of the pie chart are equally spaced.

Favourite food

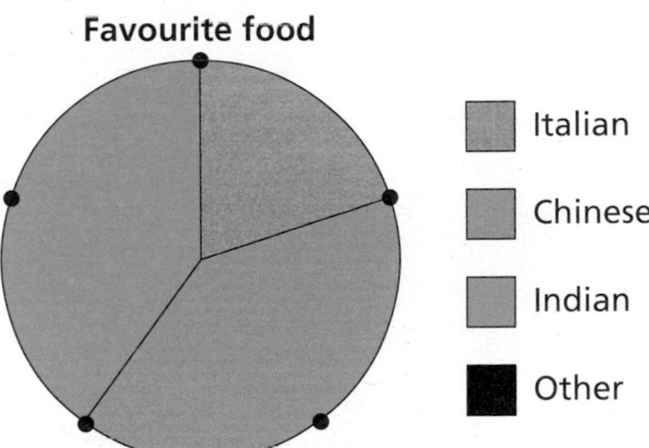

a What proportion of people chose Italian? Give your answer as a fraction.

b What proportion of people chose Chinese? Give your answer as a decimal.

c What proportion of people chose Indian? Give your answer as a percentage.

d What proportion of people chose other?

4 The staff from an office were asked whether or not they have a driving licence. The results are shown in the pie chart.

The markings on the circumference are equally spaced.

Staff who have a driving licence

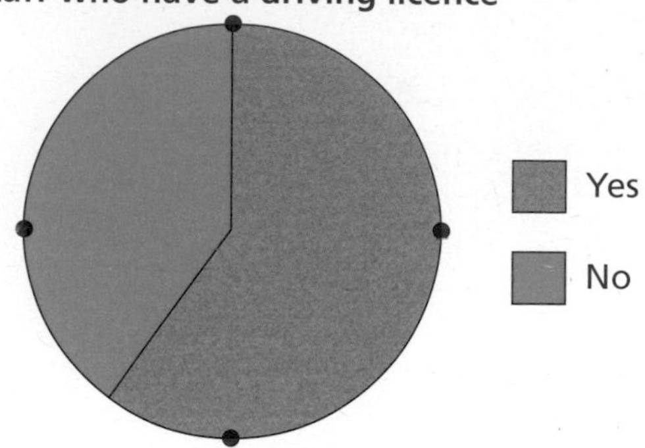

a Estimate the percentage of the staff who have a driving licence.

b Estimate the percentage of the staff who do not have a driving licence.

c Explain why your answers are only estimates.

d Rachel estimates that 43% of the staff do have a driving licence and 47% do not. Give three criticisms of Rachel's estimates.

What do you think?

1 The pie charts show the proportion of males and females in two offices.

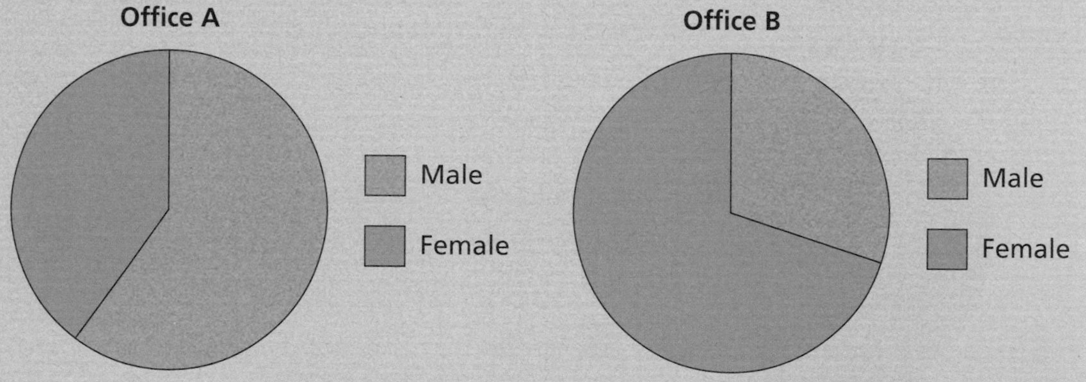

a What proportion of the people in each office are male? Give each answer as a fraction, a decimal and a percentage.

b What proportion of the people in each office are female? Give each answer as a fraction, a decimal and a percentage.

c Decide whether each statement is true, false or you cannot tell. Explain your answer.

i There are more males than females in office A.

ii There are more males in office A than in office B.

iii There is a greater proportion of females in office B than in office A.

iv There are twice as many males as females in office A.

2 Students in Year 7 and Year 8 were asked how many after-school clubs they attend each week.

The results are shown in these pie charts.

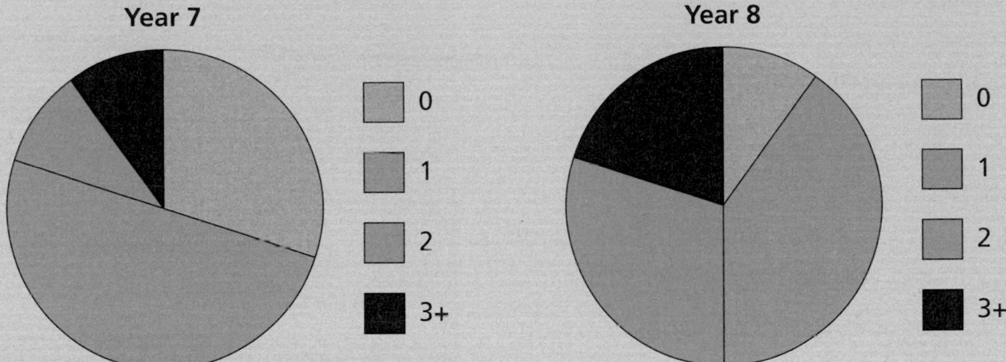

a Copy and complete the table to show the percentage of people in Year 7 and Year 8 who attend each number of after-school clubs.

Number of clubs	Year 7	Year 8
0		
1		
2		
3+		

b Copy and complete the table again, this time giving your answers as fractions.

c Look carefully at the pie charts. Use them to make two comparisons between attendance of after-school clubs in Year 7 and Year 8.

d Beca says that the number of students in Year 7 who attend two after-school clubs is the same as the number of students in Year 8 who attend none.

Do you agree with Beca? Explain your answer.

Consolidate – do you need more?

1 The pie charts show the number of wins, losses and draws for 4 football teams.

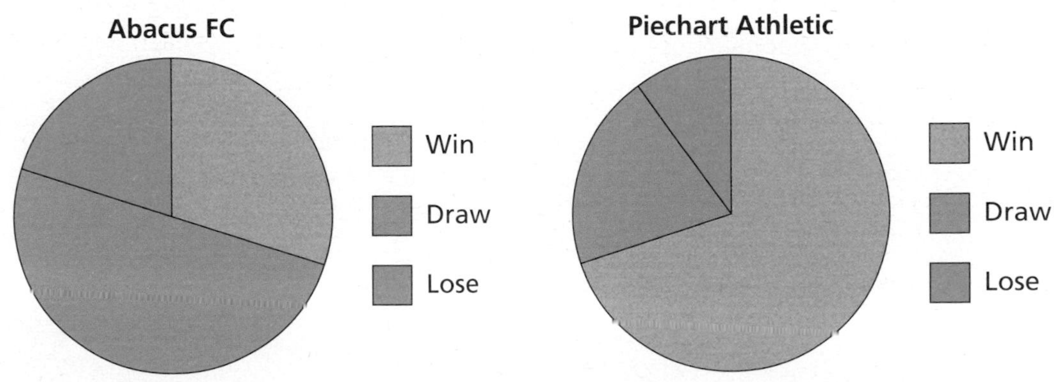

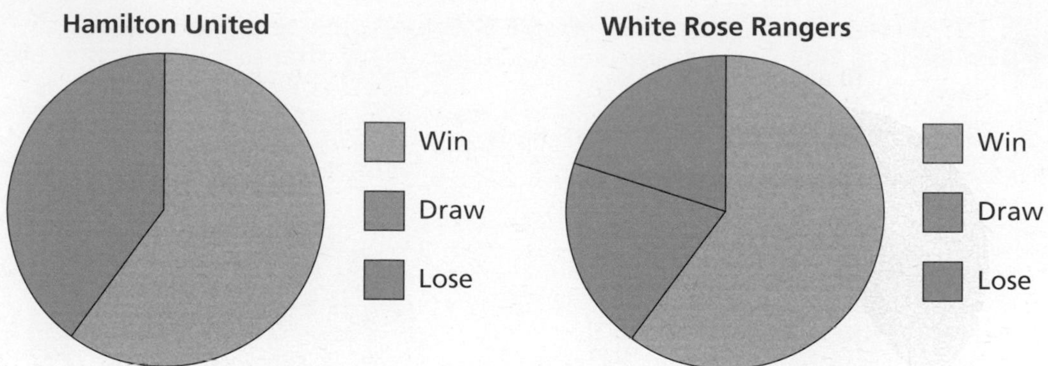

a Which team won the greatest proportion of their games? Explain how you know.

b Which team did not lose any games? Explain how you know.

c What proportion of their games did each team win? Give each answer as a fraction, a decimal and a percentage.

d What proportion of their games did each team draw? Give each answer as a fraction, a decimal and a percentage.

e What proportion of their games did each team lose? Give each answer as a fraction, a decimal and a percentage.

Stretch – can you deepen your learning?

1 The pie chart shows the favourite sports of a group of people. The key is missing.

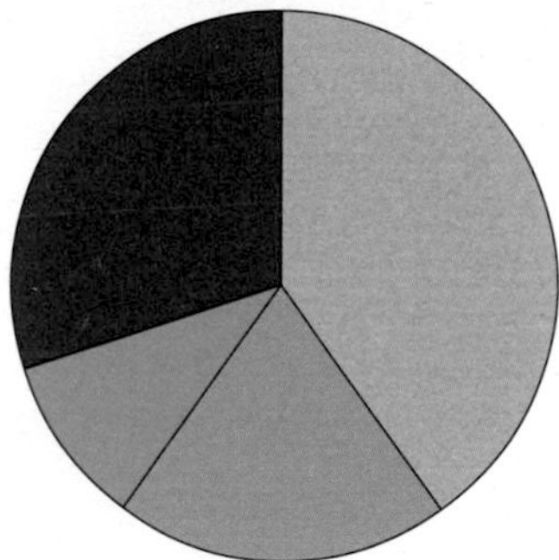

Twice as many people chose football as chose netball.

Twice as many people chose netball as chose rugby.

30% of the people surveyed chose other.

What proportion of the people surveyed is represented by each sector on the pie chart? Give each answer as a fraction, a decimal and a percentage.

2 The pie charts show the ages of people in a gym at two times in the same day.

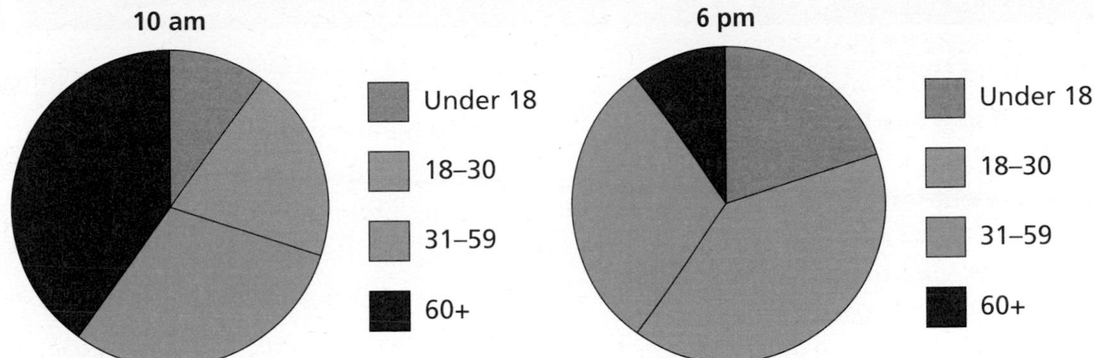

a Estimate the proportion of people at the gym who are in each age group at each of the times. Compare your estimates with a partner.

b What do you notice about the ages of the people in the gym at 10am compared with 6pm? Make two comparisons. Why do you think this is the case?

Reflect

1 Explain why you cannot compare exact values from a pie chart if you are not given the total.

2 Explain what you can interpret from a pie chart.

5.5 Understanding fractions below 1

Small steps

- Represent any fraction as a diagram
- Represent fractions on number lines
- Identify and use simple equivalent fractions

Key words

Fraction – a number that compares equal parts of a whole

Diagram – a simplified drawing showing the appearance, structure, or workings of something

Number line – a line on which numbers are marked at intervals

Equivalent – equal in value

Are you ready?

1 Decide whether or not each diagram represents one half. Explain each answer.

a 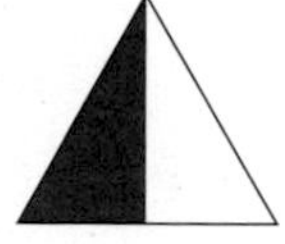

b 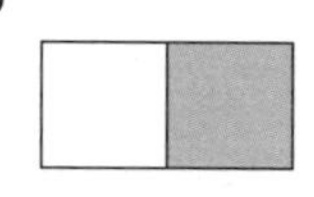

c 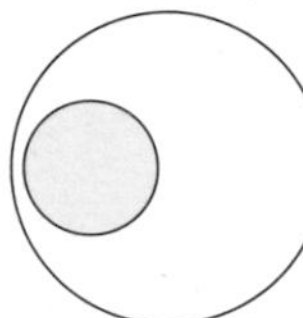

d 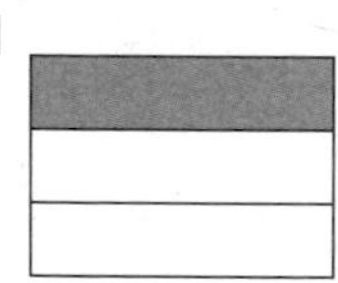

e 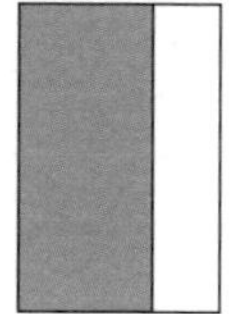

f 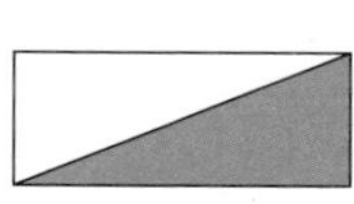

2 Write down the size of the intervals on each number line.

a 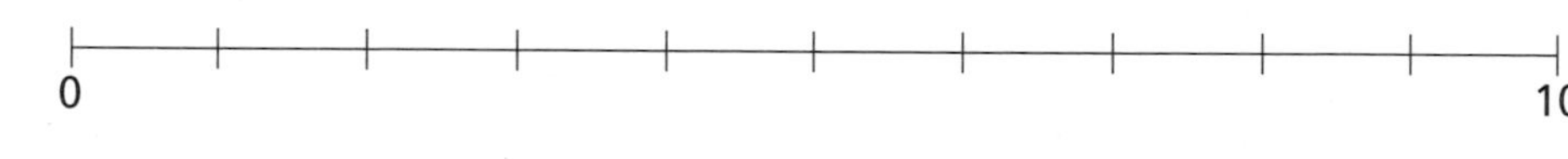

b 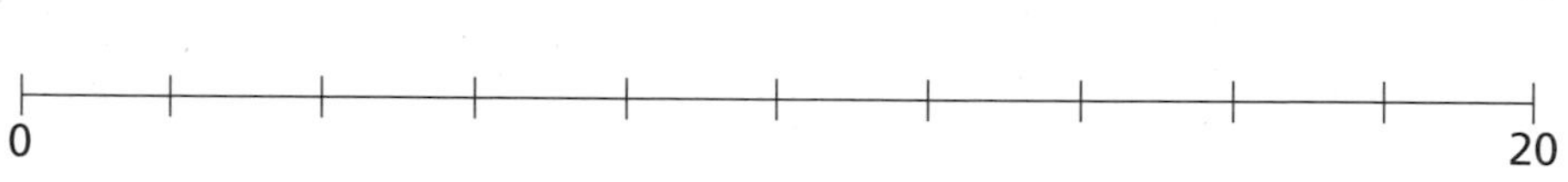

c 

d 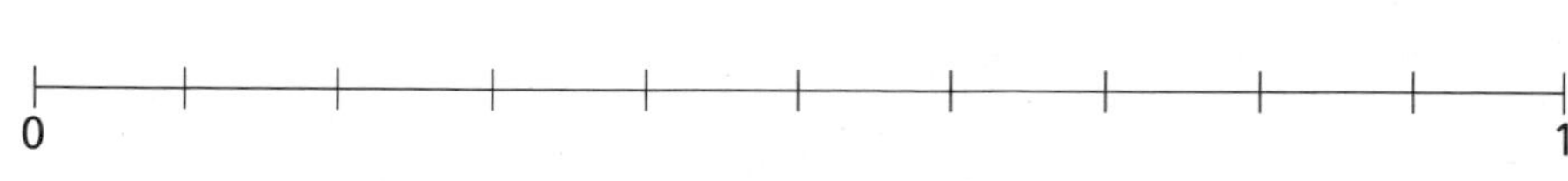

e 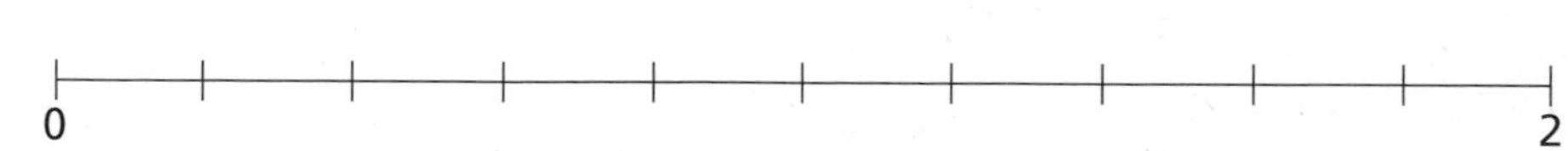

3 Work out these multiplications.

a 3×5 **b** 7×5 **c** 18×7 **d** 21×7 **e** 15×4 **f** 28×4

Models and representations

Bar model

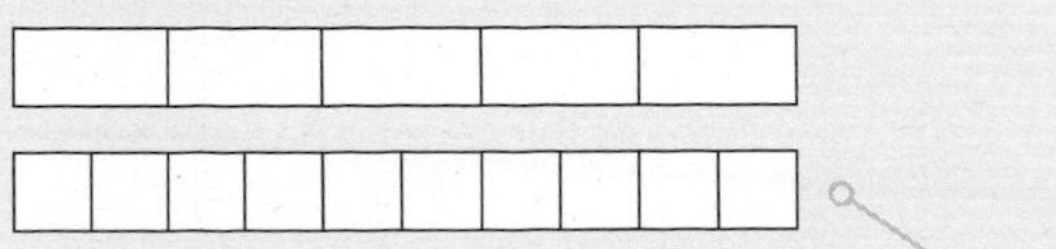

These bar models show the equivalence of fifths and tenths.

When working with equivalent **fractions**, the two bars must be equal in length to show that the whole is the same.

Number line

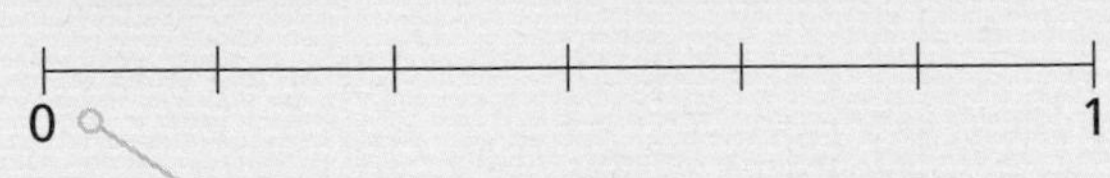

This number line is going up in sixths, and shows that six sixths is **equivalent** to one whole.

When representing a fraction using a **diagram**, you must make sure that the diagram is split into parts of equal size.

Example 1

What fraction of each diagram is shaded?

a 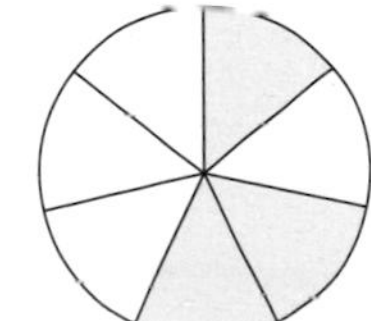

b 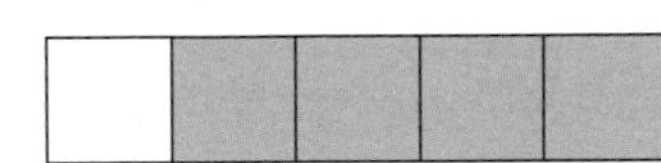

c 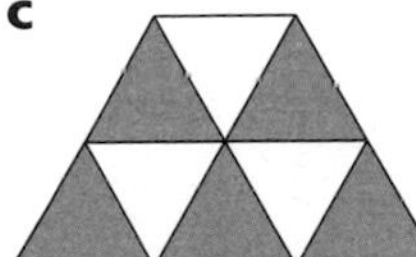

a $\frac{3}{7}$

The whole circle has been split into 7 equal parts so each part represents one seventh.

Three out of the seven equal parts are shaded.

This represents $\frac{3}{7}$.

b $\frac{4}{5}$

The whole bar has been split into 5 equal parts so each part represents one fifth.

Four out of the five equal parts are shaded.

This represents $\frac{4}{5}$.

c $\frac{5}{8}$

The whole has been split into 8 equal parts so each part represents one eighth.

Five out of the eight equal parts are shaded.

This represents $\frac{5}{8}$.

5.5 Understanding fractions below 1

Practice 5.5A

1 Decide whether each diagram shows **equal** or **unequal** parts.

a  **b** 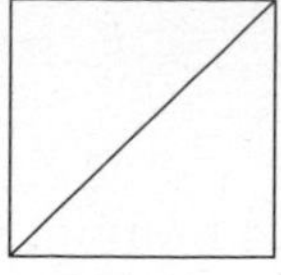 **c** 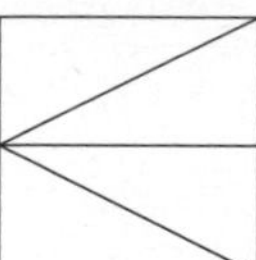

d 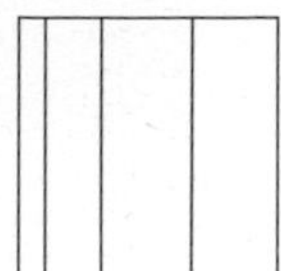 **e** 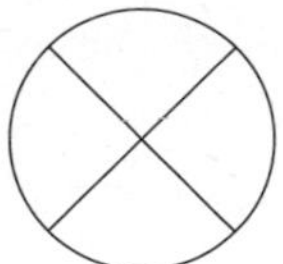 **f** 

2 What fraction of each diagram is shaded?

a 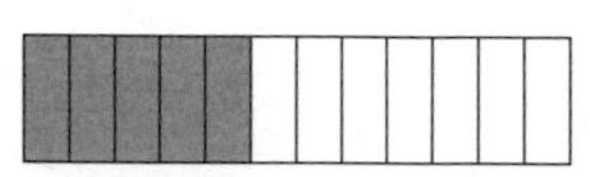 **b** 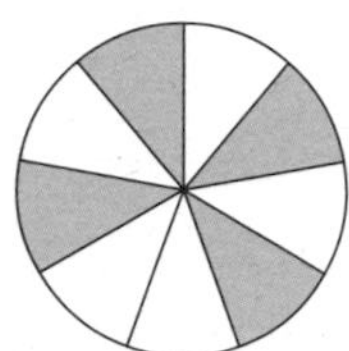 **c** 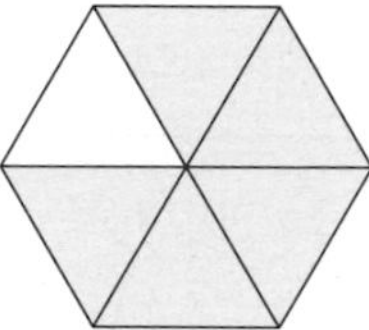

d 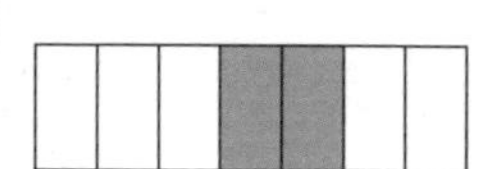 **e** 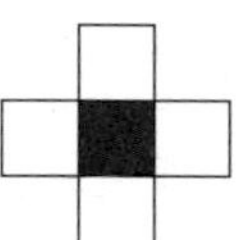 **f** 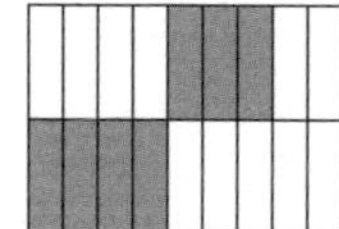

3 What fraction of each diagram is shaded?

a 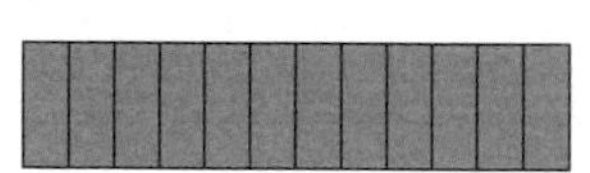 **b** 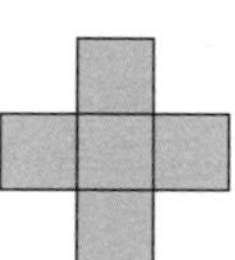

c 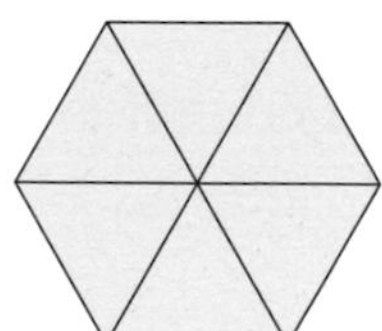 **d** 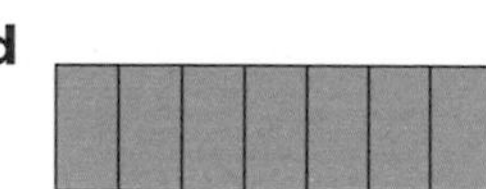

What do you notice?

4 What fraction of each set of counters is red?

a ●●●●●○○○○○○

b ●●●●●●●○

c ●●●●●●●●●

d ●○○○○○○○○○○○○○

5 Draw a diagram to represent each of these fractions.

a $\frac{1}{5}$ **b** $\frac{3}{11}$ **c** $\frac{9}{13}$ **d** $\frac{7}{7}$ **e** $\frac{17}{20}$

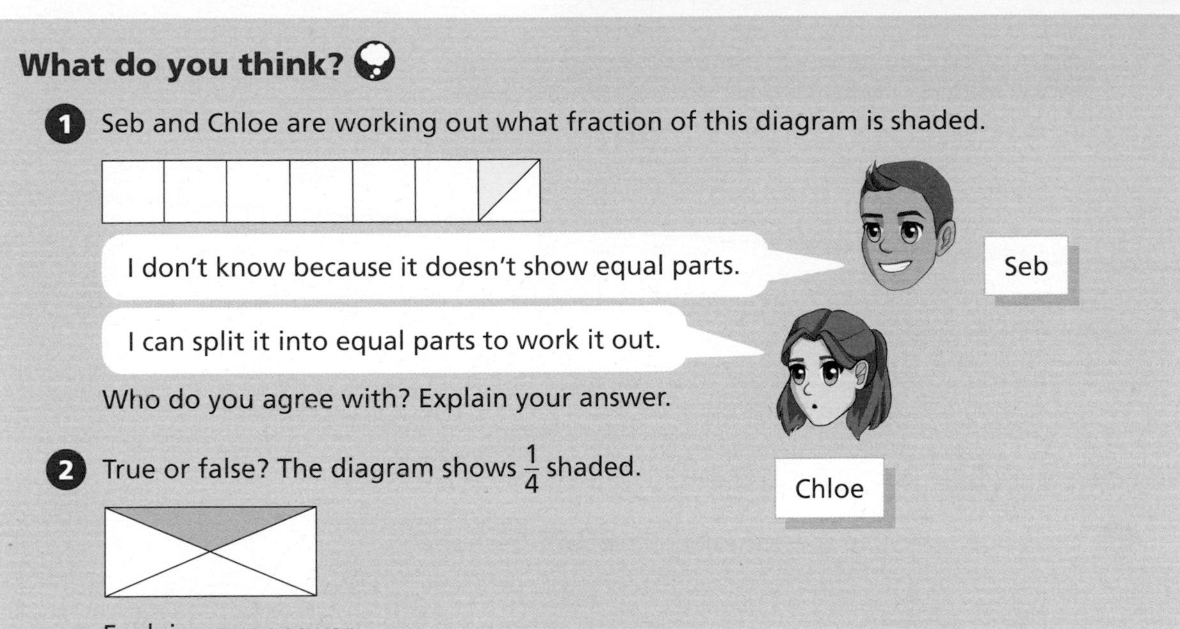

I can split it into equal parts to work it out.

When you are working with fractions, it is important that number lines show equal parts. The intervals should be equally spaced.

Example 2

Here is a number line from 0 to 1

0 1

a What fraction is the arrow pointing to?

b Mark the fraction $\frac{8}{11}$ on the number line.

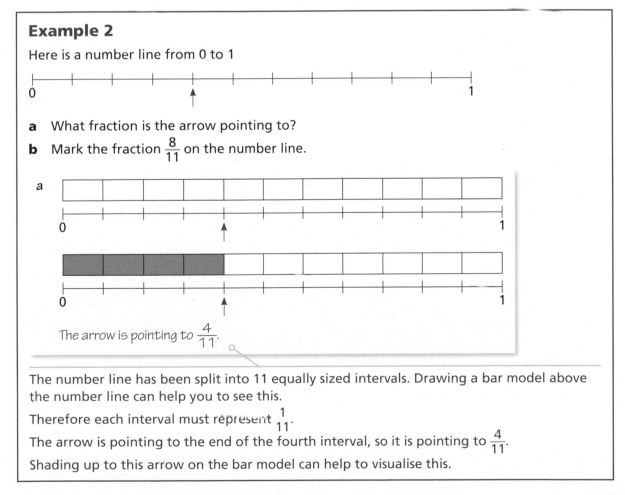

a

0 1

0 1

The arrow is pointing to $\frac{4}{11}$.

The number line has been split into 11 equally sized intervals. Drawing a bar model above the number line can help you to see this.

Therefore each interval must represent $\frac{1}{11}$.

The arrow is pointing to the end of the fourth interval, so it is pointing to $\frac{4}{11}$.

Shading up to this arrow on the bar model can help to visualise this.

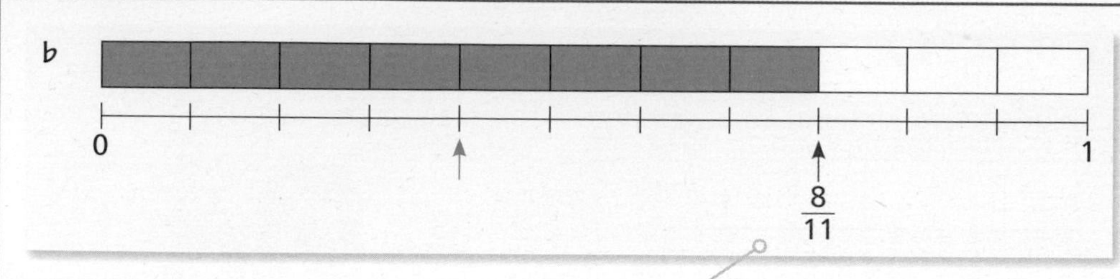

The number line already shows elevenths, so to represent eight elevenths you need to go to the end of the eighth interval.

You can shade $\frac{8}{11}$ on the bar model to show this.

Practice 5.5B

1 Copy and complete these number lines with fractions.

a

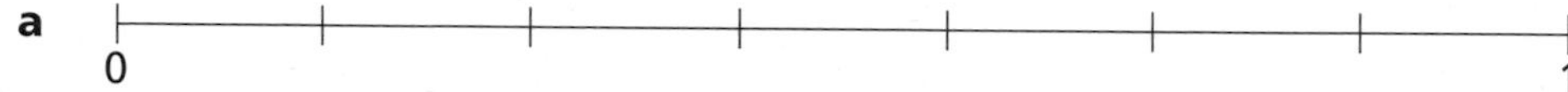

b

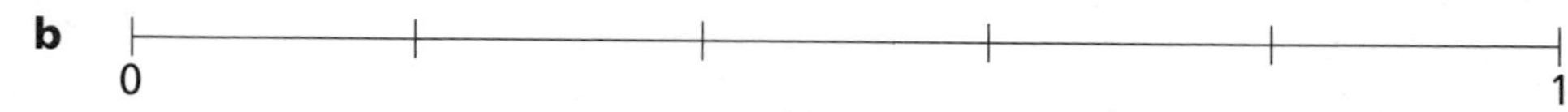

c

0 1

d

0 1

e

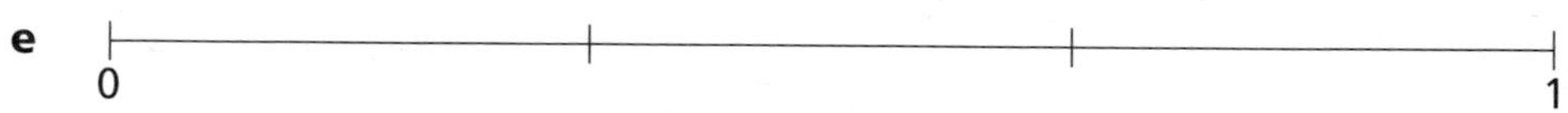

f

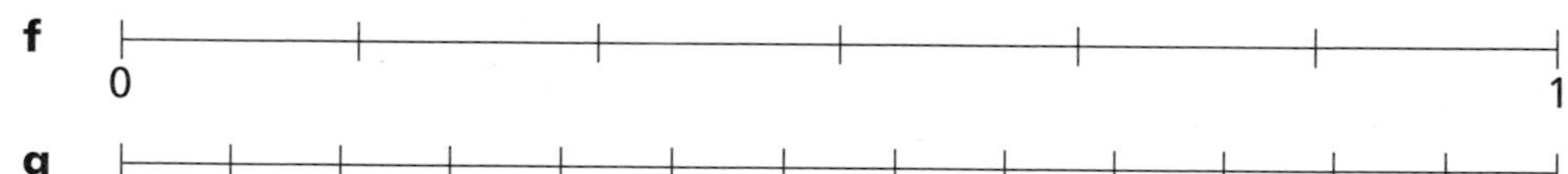

g

0 1

2 What fraction is each arrow pointing to?

a

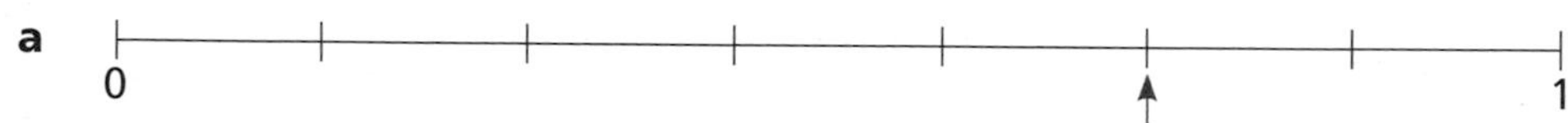

b

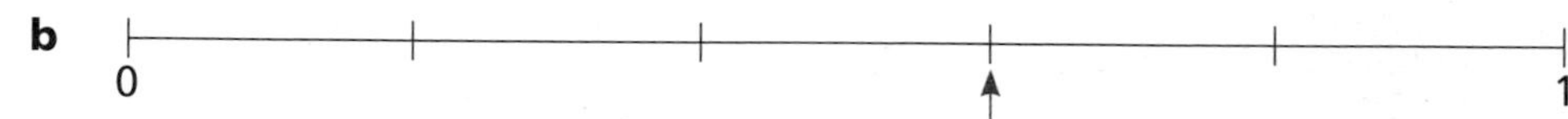

c

d

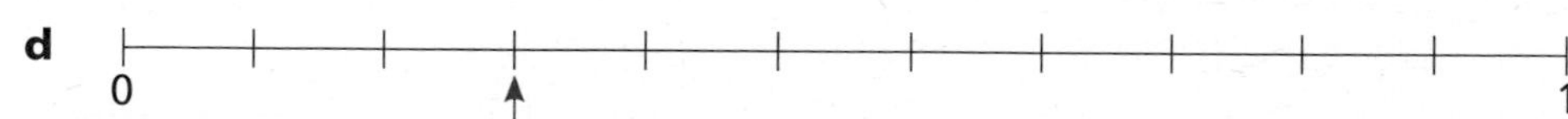

e

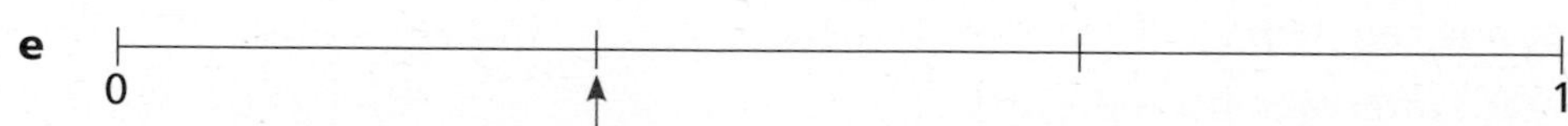

f

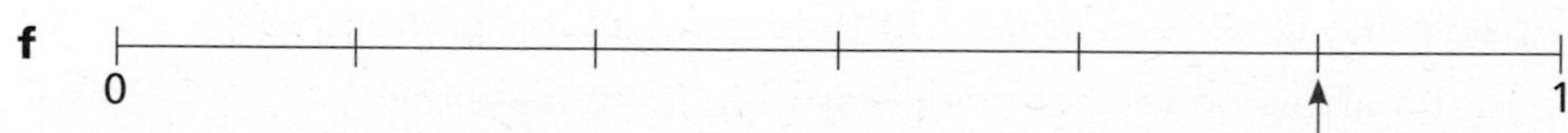

3 **a** Draw a number line from 0 to 1 going up in twelfths.

b Draw an arrow pointing to $\frac{5}{12}$ on your number line.

4 Mario walks 1 km from his home to the park.

a Mario stops at a shop, which is $\frac{1}{4}$ of the way to the park. Copy and complete the number line to show the approximate position of the shop.

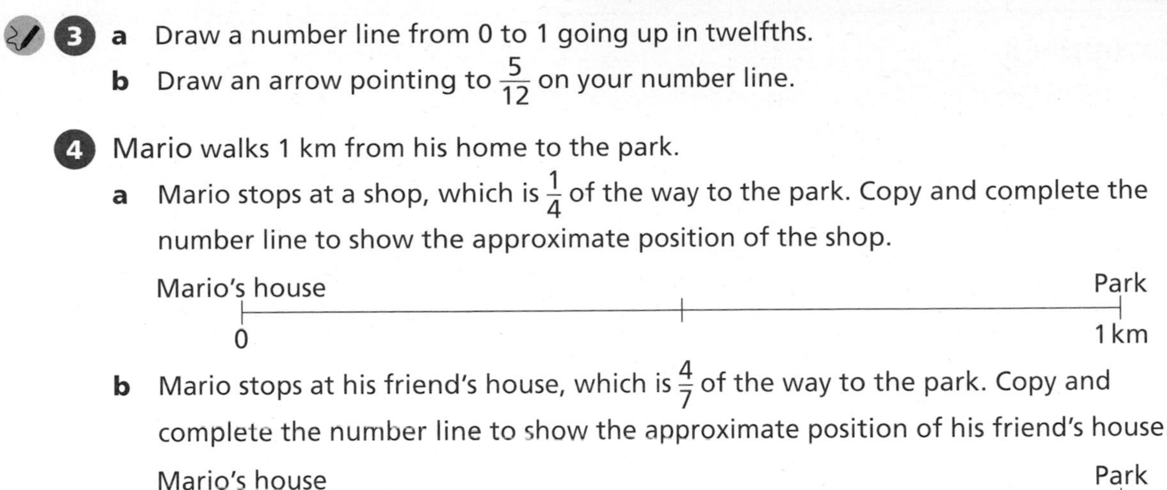

b Mario stops at his friend's house, which is $\frac{4}{7}$ of the way to the park. Copy and complete the number line to show the approximate position of his friend's house.

Mario's house   Park

0   1 km

Which fraction did you find easier to mark? Why?

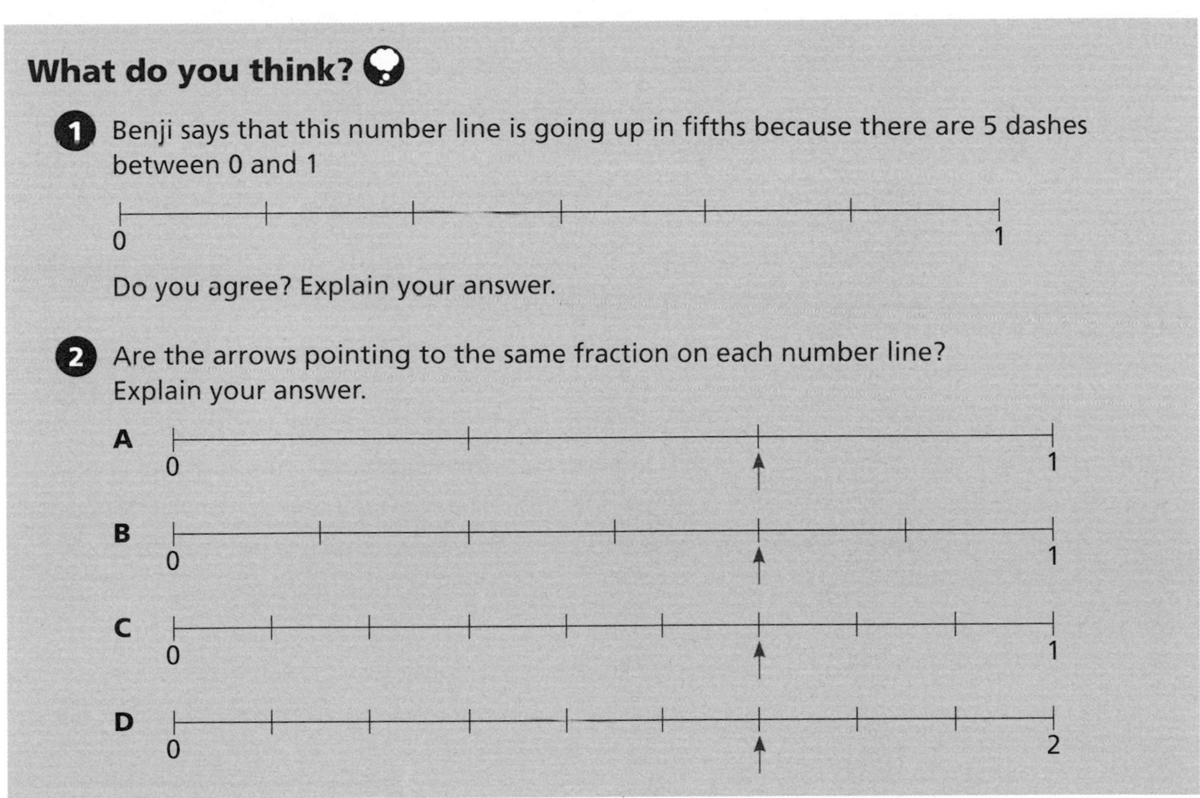

What do you think?

1 Benji says that this number line is going up in fifths because there are 5 dashes between 0 and 1

0   1

Do you agree? Explain your answer.

2 Are the arrows pointing to the same fraction on each number line? Explain your answer.

A 0   1

B 0   1

C 0   1

D 0   2

Representing fractions as diagrams and on number lines will help you to see the link between equivalent fractions.

Example 3

Find the missing numbers in these fractions.

a $\frac{1}{2} = \frac{?}{8}$ **b** $\frac{3}{5} = \frac{9}{?}$ **c** $\frac{2}{8} = \frac{3}{?}$

Method A

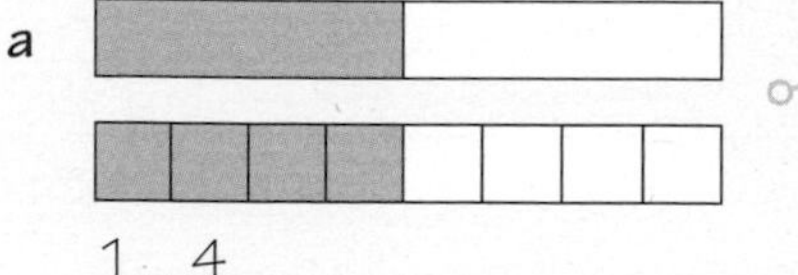

Draw two equally sized bars. Split the first bar into halves and the second bar into eighths.

Shade one half of the first bar, and shade in the same amount of the second bar.

You can see that one half is equivalent to four eighths.

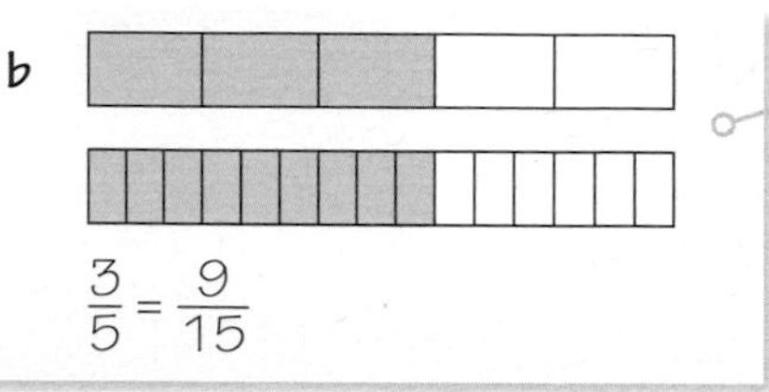

Draw two equally sized bars. Shade three fifths on the first bar. Shade the same amount of the second bar.

In the second bar, the shaded section needs to be split into 9 equal parts.

Each fifth from the first bar is equal in size to three parts in the second bar.

You need to split the second bar into 15 equal parts altogether.

Comparing the bars shows that $\frac{3}{5}$ is equivalent to $\frac{9}{15}$.

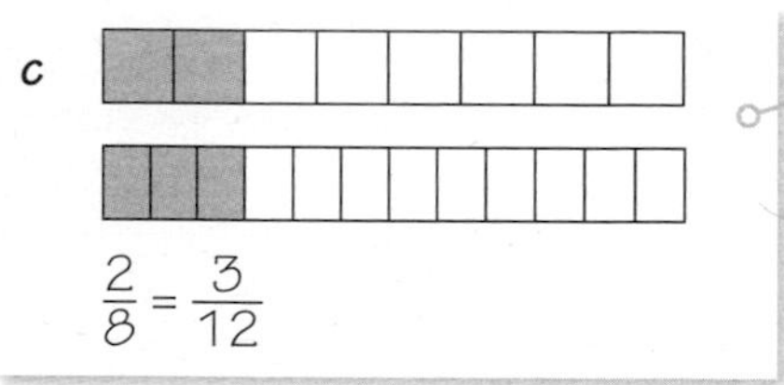

Draw two equally sized bars. Split the first bar into eight equal parts and shade in two parts.

Shade in the same amount of the second bar. This shaded section needs to be split into three equal parts. Split the rest of this bar into equal parts.

You need to split the second bar into 12 equal parts altogether.

Comparing the bars shows that $\frac{2}{8}$ is equivalent to $\frac{3}{12}$.

Method B

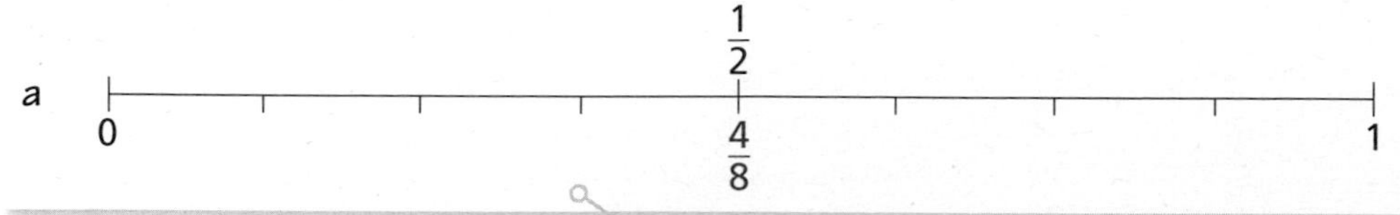

Draw a number line and split it into 2 equal parts on top and 8 equal parts on the bottom.

Mark $\frac{1}{2}$ on the number line.

Look at the corresponding fraction on the bottom.

You can see that $\frac{1}{2}$ is equivalent to $\frac{4}{8}$.

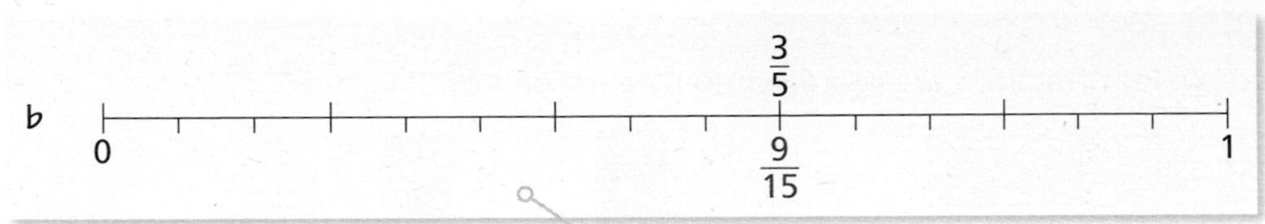

Draw a number line and split it into 5 equal parts on top. Mark $\frac{3}{5}$ on the number line.

On the bottom of the line, draw a marker to show where $\frac{9}{?}$ will go. You know this needs to be at the end of the ninth interval.

Draw in the equally spaced intervals and you will see that $\frac{3}{5}$ is equivalent to $\frac{9}{15}$.

Draw a number line and split it into eight equal parts on top and mark $\frac{2}{8}$ on the number line.

On the bottom, draw a marker that aligns with $\frac{2}{8}$; you know this needs to be at the end of the third interval.

Draw in the remaining intervals on the bottom and you will see that $\frac{2}{8}$ is equivalent to $\frac{3}{12}$.

Method C

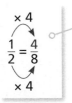

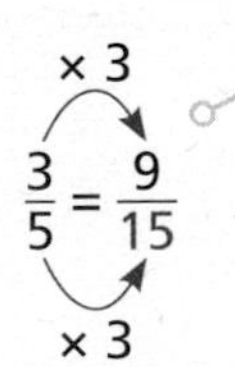

The denominator has been multiplied by 4 so you also need to multiply the numerator by 4.

$\frac{1}{2}$ is equivalent to $\frac{4}{8}$.

$$\times 3: \quad \frac{3}{5} = \frac{9}{15} \quad \times 3$$

The numerator has been multiplied by 3 so the denominator also needs to be multiplied by 3.

$\frac{3}{5}$ is equivalent to $\frac{9}{15}$.

$$\times 4 \quad \frac{2}{8} = \frac{3}{12} \quad \times 4$$

2 isn't a factor of 3, so instead you can consider the relationship between the numerator and the denominator. The denominator of the first fraction is 4 times the numerator, so the same must be true in the second fraction. $\frac{2}{8}$ is equivalent to $\frac{3}{12}$.

Practice 5.5C

1. What equivalent fractions are represented by each pair of diagrams?

   a

   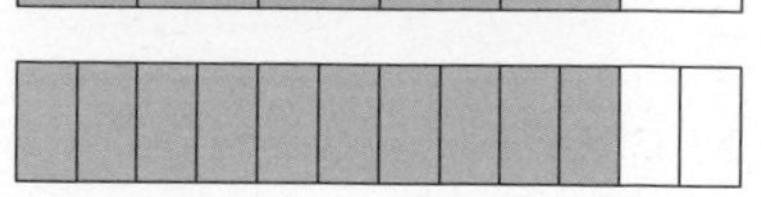

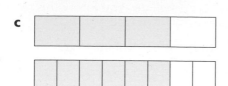

   b

   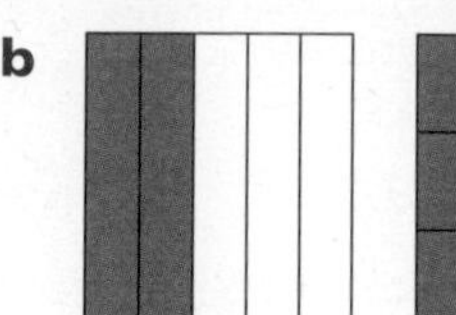

   c

   d

   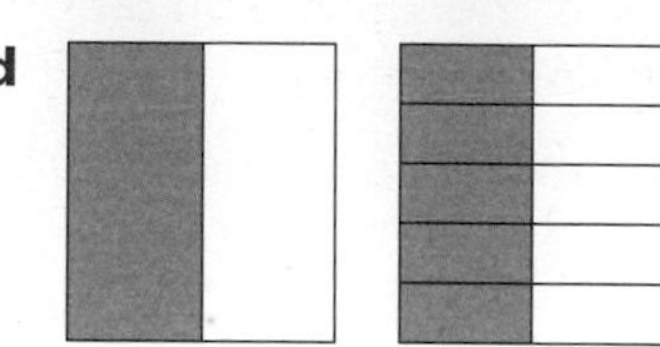

2. What equivalent fractions can be seen from the number line?

   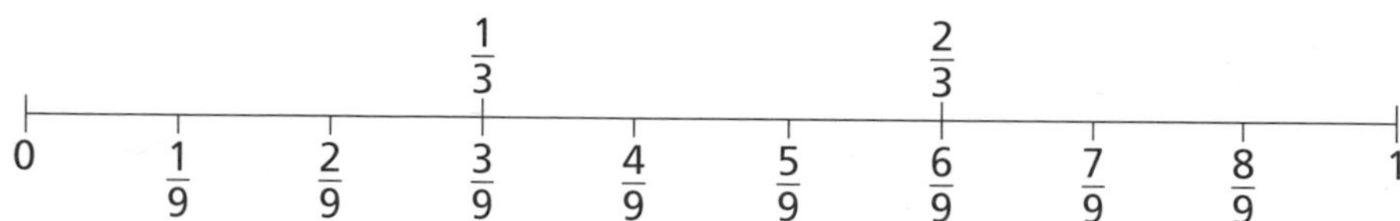

3. Draw a diagram to show the equivalence of each pair of fractions.

   a $\frac{4}{5} = \frac{8}{10}$ **b** $\frac{1}{3} = \frac{5}{15}$ **c** $\frac{2}{7} = \frac{6}{21}$ **d** $\frac{5}{6} = \frac{10}{12}$ **e** $\frac{5}{9} = \frac{20}{36}$

4. Copy and complete these equivalent fractions.

   a $\frac{1}{4} = \frac{?}{36}$ **b** $\frac{7}{18} = \frac{?}{36}$ **c** $\frac{3}{9} = \frac{?}{3}$ **d** $\frac{5}{7} = \frac{?}{35}$ **e** $\frac{6}{11} = \frac{?}{110}$

   f $\frac{3}{7} = \frac{12}{?}$ **g** $\frac{12}{24} = \frac{15}{?}$ **h** $\frac{6}{19} = \frac{12}{?}$ **i** $\frac{17}{21} = \frac{170}{?}$ **j** $\frac{5}{45} = \frac{11}{?}$

5. Write down three fractions that are equivalent to each of these.

   a $\frac{1}{2}$ **b** $\frac{1}{4}$ **c** $\frac{5}{7}$ **d** $\frac{5}{8}$ **e** $\frac{11}{13}$ **f** $\frac{19}{20}$

   Compare your answers with a partner. Did you get the same?

What do you think?

1. A whole cake is shared equally between 8 people. What fraction of the cake do they each get? Give your answer in three different ways.

2. Samira has drawn two bar models.

   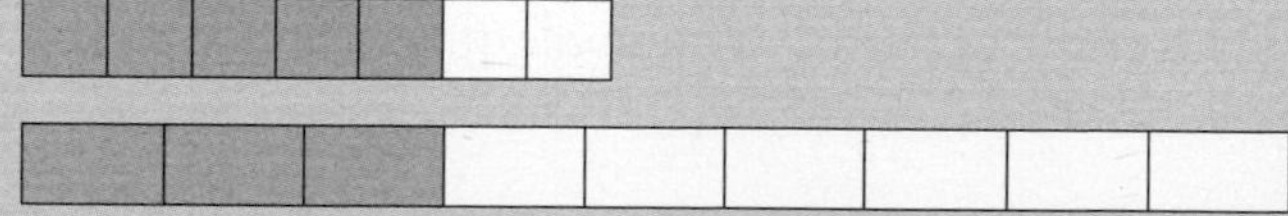

   She says, "My bar models show that $\frac{5}{7}$ is equivalent to $\frac{3}{9}$"

   Do you agree with Samira? Explain your answer.

3. Seb says, "$\frac{11}{13}$ is equivalent to $\frac{12}{14}$ because I've added 1 to both the numerator and the denominator." Show that Seb is wrong.

Consolidate – do you need more?

1 Draw a diagram to represent each fraction.

a $\frac{1}{5}$ **b** $\frac{2}{3}$ **c** $\frac{5}{7}$ **d** $\frac{9}{10}$ **e** $\frac{5}{6}$

2 Represent each fraction on a number line.

a $\frac{1}{5}$ **b** $\frac{2}{3}$ **c** $\frac{5}{7}$ **d** $\frac{9}{10}$ **e** $\frac{5}{6}$

3 Write down three fractions that are equivalent to each of these.

a $\frac{1}{5}$ **b** $\frac{2}{3}$ **c** $\frac{5}{7}$ **d** $\frac{9}{10}$ **e** $\frac{5}{6}$

Stretch – can you deepen your learning?

1 Draw an arrow to estimate the position of $\frac{4}{9}$ on this number line.

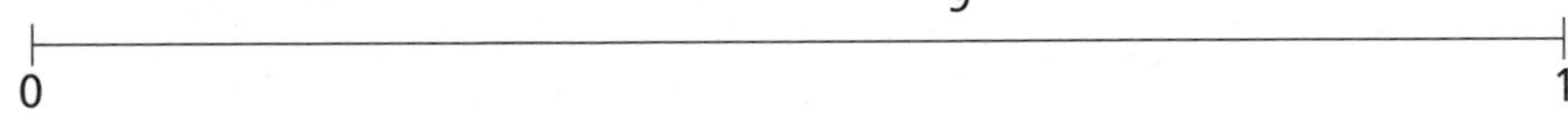

Why is your answer an estimate?

2 Show that $\frac{1}{2}$ of this diagram is shaded.

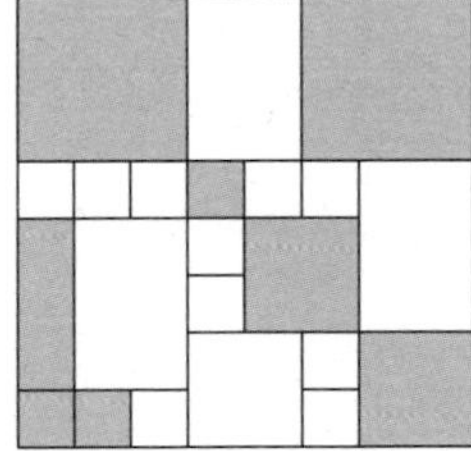

3 True or false? All fractions equivalent to $\frac{1}{2}$ have an even denominator. Explain your answer.

4 Flo got $\frac{3}{4}$ of the questions on an online quiz correct. She then got $\frac{21}{28}$ of the questions on a written test correct. Flo thinks that she did better on the written test than on the online test because 21 is greater than 3. Do you agree? Explain your answer.

Reflect

1 Describe three ways you can show two fractions are equivalent. Use examples to illustrate your answers.

2 In your own words, explain what it means for two fractions to be equivalent.

3 If you do the same to the numerator and the denominator, you'll find an equivalent fraction. Is this always true, sometimes true or never true?

5.6 Linking fractions and division

Small steps

- Understand fractions as division
- Convert fluently between fractions, decimals and percentages

Key words

Fraction – a number that represents part of a whole; used to represent numbers that are not integers

Division – the process of splitting a number into equal parts

Are you ready?

1 Use a calculator to work out these divisions. Give each answer as a decimal. Write down all the figures on your calculator display.

a $1 \div 5$ **b** $1 \div 10$ **c** $1 \div 100$ **d** $1 \div 1000$ **e** $7 \div 100$ **f** $1 \div 4$

2 Write down the decimal equivalent of each fraction.

a $\frac{1}{5}$ **b** $\frac{1}{10}$ **c** $\frac{1}{100}$ **d** $\frac{1}{1000}$ **e** $\frac{7}{100}$ **f** $\frac{1}{4}$

Compare your answers to question 1 and 2. What do you notice?

3 Write each of these fractions as a decimal and a percentage.

a $\frac{27}{100}$ **b** $\frac{49}{100}$ **c** $\frac{3}{10}$ **d** $\frac{3}{4}$ **e** $\frac{4}{5}$

4 Write each of these decimals as a fraction and a percentage.

a 0.79 **b** 0.03 **c** 0.6 **d** 0.9 **e** 0.25

5 Write each of these percentages as a fraction and a decimal.

a 81% **b** 7% **c** 70% **d** 99% **e** 37.1% **f** 80%

The line separating the numerator from the denominator in a **fraction** represents a **division**. For example, two-fifths ($\frac{2}{5}$) represents two out of five equal parts, but it also represents the calculation 2 divided by 5

Example 1

a Shade a bar model to show that 1 shared equally between 5 is the same as $\frac{1}{5}$.

b Work out $1 \div 5$ using a calculator.

c How do parts **a** and **b** show that $\frac{1}{5} = 1 \div 5$?

a [bar model of five equal parts]

The whole has been split into five equal parts.

Each part represents $\frac{1}{5}$.

So 1 shared between 5 is the same as $\frac{1}{5}$.

b $1 \div 5 = 0.2$ — Type [1] [÷] [5] [=] into a calculator

c From part **a**, 1 shared between $5 = \frac{1}{5}$, which is equivalent to 0.2
From part **b**, $1 \div 5 = 0.2$
Therefore $\frac{1}{5} = 1 \div 5$

Remember that $\frac{1}{5}$ is equivalent to the decimal 0.2

Sharing is a type of division.

Practice 5.6A

1 Work out each division on a calculator. Write each answer as a decimal.

a $3 \div 100$ **b** $3 \div 10$ **c** $17 \div 100$ **d** $17 \div 10$ **e** $17 \div 1000$

2 Convert each fraction to a decimal.

a $\frac{3}{100}$ **b** $\frac{3}{10}$ **c** $\frac{17}{100}$ **d** $\frac{17}{10}$ **e** $\frac{17}{1000}$

Compare your answers to questions **1** and **2**. What do you notice?

3 Match each of these divisions to the equivalent fraction.

$9 \div 5$ $7 \div 4$ $17 \div 20$ $3 \div 8$ $19 \div 25$

$\frac{3}{8}$ $\frac{17}{20}$ $\frac{19}{25}$ $\frac{7}{4}$ $\frac{9}{5}$

4 Write each division as a fraction.

a $5 \div 9$ **b** $7 \div 11$ **c** $6 \div 7$ **d** $19 \div 20$ **e** $20 \div 19$

5 Write each fraction as a division.

a $\frac{5}{13}$ **b** $\frac{1}{12}$ **c** $\frac{27}{5}$ **d** $\frac{19}{4}$ **e** $\frac{4}{19}$

6 Write each fraction as a division. Then work out the decimal equivalent using a calculator. Write down all the figures on your calculator display.

a $\frac{9}{13}$ **b** $\frac{7}{11}$ **c** $\frac{15}{27}$ **d** $\frac{134}{25}$ **e** $\frac{25}{134}$

f $\frac{8}{9}$ **g** $\frac{11}{7}$ **h** $\frac{1}{3}$ **i** $\frac{19}{20}$ **j** $\frac{39}{50}$

k Which of your answers had a recurring pattern in the digits?

7 Use a calculator to convert each fraction to a decimal. Then decide which fraction of each pair is greater.

a $\frac{5}{8}$ $\frac{11}{20}$ **b** $\frac{17}{25}$ $\frac{4}{5}$ **c** $\frac{49}{50}$ $\frac{24}{25}$

8 Use a calculator to convert each fraction to a decimal. Then write each set of fractions in ascending order.

a $\frac{17}{20}$ $\frac{11}{10}$ $\frac{4}{5}$ $\frac{1}{20}$ $\frac{14}{28}$ **b** $\frac{23}{25}$ $\frac{13}{20}$ $\frac{7}{8}$ $\frac{11}{8}$ $\frac{1}{6}$

Was it possible to find the smallest and greatest fraction in each set without working out the division?

9 5 pizzas are shared equally between 7 people.

What fraction of a pizza does each person receive?

What do you think?

1 Emily has converted $\frac{4}{25}$ into a decimal using division.

Her answer is 6.25.

a How do you know without doing a calculation that Emily's answer is wrong?

b What mistake has Emily made?

c Convert $\frac{4}{25}$ into a decimal using division.

2 Which of these fractions are equivalent to $\frac{1}{3}$?

$\frac{17}{54}$ $\frac{19}{57}$ $\frac{126}{378}$ $\frac{1249}{3741}$ $\frac{99}{297}$

Compare your method with a partner's.

3 Marta says that $\frac{15}{105}$ and $\frac{19}{133}$ are not equivalent because 15 isn't a factor of 19.
Zach says that they are equivalent because both divisions give the same answer.
Who do you agree with? Explain your answer.

You can use your understanding of fractions as division to convert between any fraction, decimal and percentage.

Example 2

Zach is doing a 10 km walk.

He is currently 7 km into his walk.

a What fraction of his walk has Zach completed?

b Write your answer to part **a** as a decimal.

c What percentage of his walk has Zach completed?

a $\frac{7}{10}$ — The whole walk is 10 km. He has walked 7 km so he is seven tenths of the way through his walk.

b $7 \div 10 = 0.7$ — Seven tenths is the same as seven divided by ten.

c $\frac{7}{10} = \frac{70}{100} = 70\%$	Per cent means per 100. 7 out of 10 is equivalent to 70 out of 100 which is 70%.
or $0.7 \times 100 = 70$ so $\frac{7}{10} = 70\%$	Alternatively, you could use your answer to part **b**. To convert from a decimal to a percentage you multiply by 100

Practice 5.6B

1 Write each of these fractions as a decimal.

a $\frac{11}{20}$ **b** $\frac{21}{25}$ **c** $\frac{39}{50}$ **d** $\frac{7}{8}$ **e** $\frac{3}{8}$

2 Write each of these decimals as a percentage.

a 0.55 **b** 0.84 **c** 0.78 **d** 0.875 **e** 0.375

3 Write each of these fractions as a percentage.

a $\frac{11}{20}$ **b** $\frac{21}{25}$ **c** $\frac{39}{50}$ **d** $\frac{7}{8}$ **e** $\frac{3}{8}$

4 Use the fact that $\frac{1}{20}$ is 5% to write each of these fractions as a percentage.

a $\frac{3}{20}$ **b** $\frac{11}{20}$ **c** $\frac{14}{20}$ **d** $\frac{19}{20}$ **e** $\frac{21}{20}$

5 Write each of these fractions as a decimal and a percentage.

a $\frac{9}{20}$ **b** $\frac{7}{25}$ **c** $\frac{7}{10}$ **d** $\frac{63}{100}$ **e** $\frac{817}{1000}$

f $\frac{3}{4}$ **g** $\frac{4}{5}$ **h** $\frac{1}{8}$ **i** $\frac{13}{25}$ **j** $\frac{79}{1000}$

6 Write each of these decimals as a fraction and a percentage.

a 0.1 **b** 0.6 **c** 0.25 **d** 0.71 **e** 0.22

f 0.379 **g** 0.99 **h** 0.85 **i** 0.625 **j** 0.007

7 Write each of these percentages as a fraction and a decimal.

a 37% **b** 41% **c** 83% **d** 95% **e** 15%

f 13% **g** 20% **h** 75% **i** 37.5% **j** 93.1%

8 Charlie is shading in a grid. $\frac{3}{5}$ of the grid is shaded red. 12% of the grid is shaded blue. The rest of the grid is shaded green. What proportion of the grid is shaded green? Give your answer as a fraction, a decimal and a percentage.

What do you think?

1 Convert each of these fractions to a decimal.

a $\frac{1}{7}$ $\frac{2}{7}$ $\frac{3}{7}$ $\frac{4}{7}$ $\frac{5}{7}$ $\frac{6}{7}$

b $\frac{1}{3}$ $\frac{2}{3}$

What do you notice?

2 Decide whether each statement is true or false. Explain your answers.

a $\frac{3}{5} = 60\%$ b $\frac{4}{7} = 47\%$ c $\frac{7}{8} = 88\%$ d $38\% = \frac{19}{50}$

Compare your answers with a partner's.

3 Flo and Faith are putting these values into descending order.

$\frac{1}{5}$ 72% 0.5

I'm going to use a calculator to write all the values as decimals.
Flo

Faith
I don't think you need to do that.

a Who do you agree with? Discuss it with a partner.

b Write the values in descending order.

Consolidate – do you need more?

1 Work out each division using a calculator. Give each answer as a decimal.

a 7 ÷ 20 b 19 ÷ 25 c 77 ÷ 100 d 9 ÷ 10 e 2 ÷ 5 f 1 ÷ 4

2 Write each of these fractions as a decimal.

a $\frac{7}{20}$ b $\frac{19}{25}$ c $\frac{77}{100}$ d $\frac{9}{10}$ e $\frac{2}{5}$ f $\frac{1}{4}$

3 Work out the missing numbers to complete each of these equivalent fractions.

a $\frac{3}{4} = \frac{?}{100}$ b $\frac{3}{10} = \frac{?}{100}$ c $\frac{9}{20} = \frac{?}{100}$

d $\frac{17}{25} = \frac{?}{100}$ e $\frac{4}{5} = \frac{?}{100}$ f $\frac{37}{50} = \frac{?}{100}$

4 Write each of these fractions as a percentage.

a $\frac{3}{4}$ b $\frac{3}{10}$ c $\frac{9}{20}$ d $\frac{17}{25}$ e $\frac{4}{5}$ f $\frac{37}{50}$

5 Write each of these fractions as a decimal and a percentage.

a $\frac{1}{4}$ b $\frac{17}{20}$ c $\frac{21}{25}$ d $\frac{31}{50}$ e $\frac{5}{8}$ f $\frac{13}{10}$

Stretch – can you deepen your learning?

1 Work out the missing number in each part-whole model. Give each answer as a fraction, a decimal and a percentage.

a

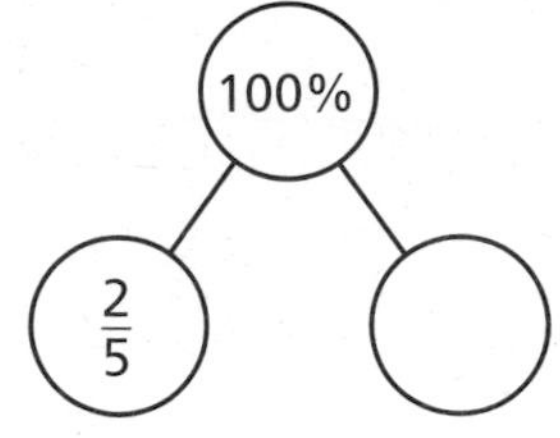

b

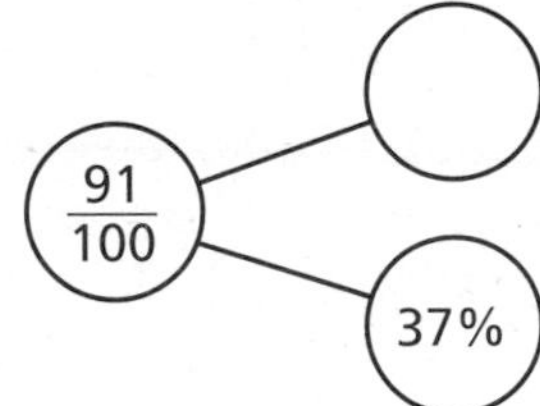

c

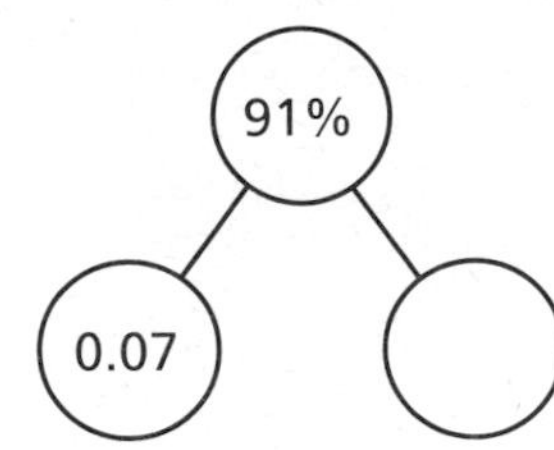

d

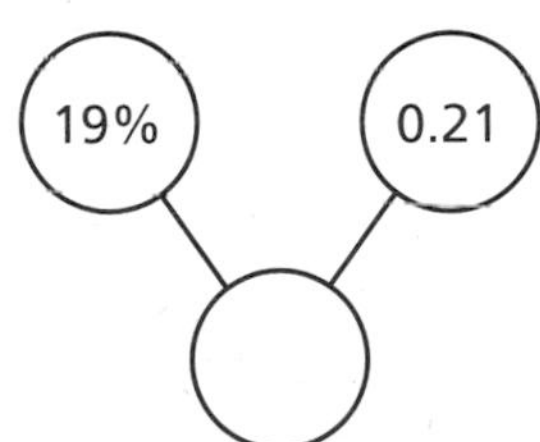

e

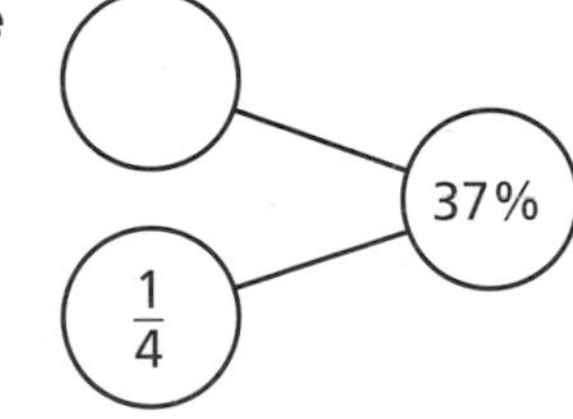

f

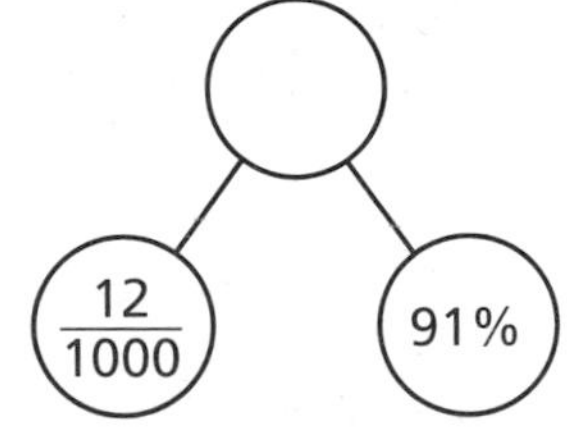

2 What is the term-to-term rule for each linear sequence? Give each answer as a fraction, a decimal and a percentage.

a $\frac{1}{10}$ 20% 0.3 **b** 65% $\frac{3}{5}$ 0.55

c $\frac{1}{20}$ 0.35 65% **d** 100% $\frac{7}{8}$ 75%

3 Write down the next three terms in each linear sequence. Give each term as a fraction, a decimal and a percentage.

a $\frac{1}{10}$ 20% 0.3 **b** 65% $\frac{3}{5}$ 0.55

c $\frac{1}{20}$ 0.35 65% **d** 100% $\frac{7}{8}$ 75%

4 In a Fibonacci sequence, each term is the sum of the previous two terms. Decide whether or not each of these sequences could be Fibonacci. Explain your answers.

a 0 0.1 10% $\frac{1}{5}$ **b** 2% 0.02 $\frac{1}{25}$

c 5% 5% $\frac{1}{5}$ 0.5

For any sequence that is Fibonacci in style, work out the next three terms. Give the terms as fractions, decimals and percentages.

5 Given that $a = 23$, $b = 20$, $c = 25$ and $d = 50$, write the value of each fraction as a decimal and a percentage.

a $\frac{a}{b}$ **b** $\frac{a}{c}$ **c** $\frac{a}{d}$ **d** $\frac{b}{a}$ **e** $\frac{b}{c}$ **f** $\frac{b}{d}$

g $\frac{c}{a}$ **h** $\frac{c}{b}$ **i** $\frac{c}{d}$ **j** $\frac{d}{a}$ **k** $\frac{d}{b}$ **l** $\frac{d}{c}$

Reflect

Decide whether each statement is true or false. Explain your answers.

a Any fraction can be written as a decimal and a percentage.

b Percentages always have to be integers.

c When a fraction has a denominator of 100, the equivalent percentage is given by the numerator.

H 5.7 Exploring fractions greater than 1

Small steps

- Explore fractions above 1, decimals and percentages (H)

Key words

Improper fraction – a fraction in which the numerator is greater than the denominator

Mixed number – a number presented as an integer and a proper fraction

Are you ready?

1 Which of these fractions are less than one whole? How do you know?

$\frac{3}{5}$ $\frac{7}{9}$ $\frac{15}{17}$ $\frac{21}{20}$ $\frac{15}{15}$ $\frac{4}{3}$ $\frac{7}{11}$

$\frac{11}{7}$ $\frac{3}{3}$ $\frac{15}{14}$ $\frac{37}{37}$ $\frac{1}{8}$ $\frac{9}{7}$ $\frac{4}{4}$

2 Which of these fractions are equal to one whole? How do you know?

$\frac{3}{5}$ $\frac{7}{9}$ $\frac{15}{17}$ $\frac{21}{20}$ $\frac{15}{15}$ $\frac{4}{3}$ $\frac{7}{11}$

$\frac{11}{7}$ $\frac{3}{3}$ $\frac{15}{14}$ $\frac{37}{37}$ $\frac{1}{8}$ $\frac{9}{7}$ $\frac{4}{4}$

3 Which of these fractions are greater than one whole? How do you know?

$\frac{3}{5}$ $\frac{7}{9}$ $\frac{15}{17}$ $\frac{21}{20}$ $\frac{15}{15}$ $\frac{4}{3}$ $\frac{7}{11}$

$\frac{11}{7}$ $\frac{3}{3}$ $\frac{15}{14}$ $\frac{37}{37}$ $\frac{1}{8}$ $\frac{9}{7}$ $\frac{4}{4}$

4 Which of these fractions, decimals and percentages are greater than one whole? How do you know?

99% $\frac{7}{6}$ 0.107 2.56 123% $\frac{4}{19}$ 101%

1.001 $\frac{15}{8}$ $\frac{12}{99}$ 17.001 3076% 62.5% 0.879

Models and representations

Bar model

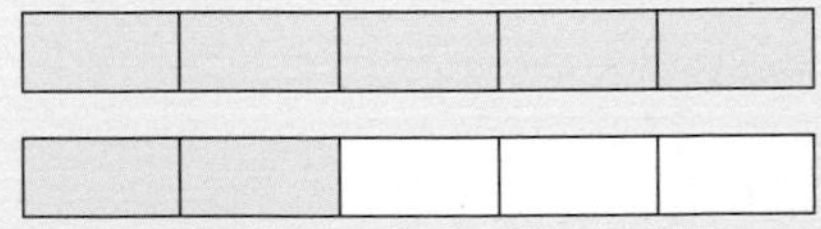

This shows that $\frac{7}{5}$ is equivalent to $1\frac{2}{5}$. Each whole is split into five equal parts representing fifths, and there are seven parts shaded. One whole bar and $\frac{2}{5}$ of the second bar are shaded.

Number line

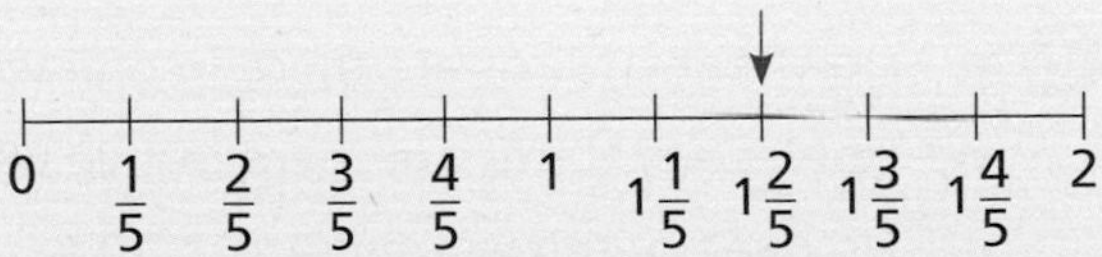

You can also use a number line to help to convert between **improper fractions** and **mixed numbers**.

5.7 Exploring fractions greater than 1

In this lesson you will look at fractions greater than 1

You will use prior learning about fraction, decimal and percentage equivalents.

- If a fraction is greater than one whole, the numerator is greater than the denominator.
- If a decimal is greater than one whole, there will be an integer greater than 0 to the left of the decimal point.
- If a percentage is greater than one whole, the percentage will be greater than 100%.

Example 1

a Convert $\frac{17}{5}$ to a mixed number.

b Write $\frac{17}{5}$ as a decimal.

c Write $\frac{17}{5}$ as a percentage.

a

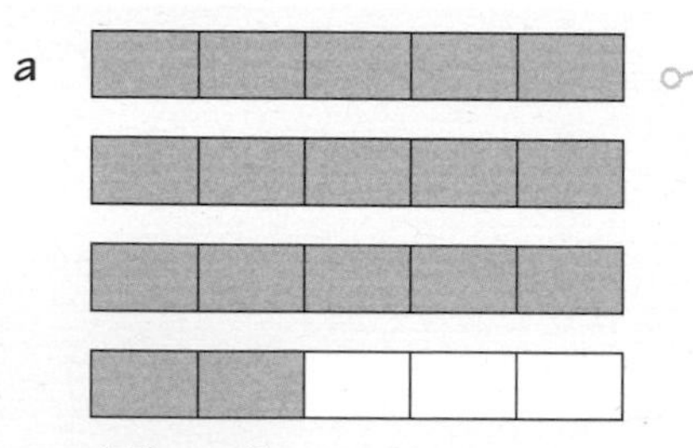

$\frac{17}{5} = 3\frac{2}{5}$

The bar model shows that $\frac{17}{5}$ is equal to three wholes and two fifths.

b $\frac{17}{5} = 3\frac{2}{5}$

$\frac{2}{5} = \frac{4}{10} = 0.4$

$\frac{17}{5} = 3.4$

When converting to a decimal, you can ignore the wholes and focus on the fraction.

You know that $\frac{2}{5} = \frac{4}{10}$ so $\frac{2}{5} = 0.4$

You can then put the wholes and the decimal together to give 3.4

c $\frac{17}{5} = 3\frac{2}{5}$

3 wholes = 300%

$\frac{2}{5} = \frac{4}{10} = 40\%$

$\frac{17}{5} = 340\%$

One whole is equal to 100%, so three wholes must be equal to 300%.

$\frac{2}{5}$ is equivalent to $\frac{4}{10}$ or 40%.

Therefore $\frac{17}{5}$ is equal to 340%.

Practice 5.7A

1 **a** How many thirds are there in one whole?

b How many thirds are there in five wholes?

c How many thirds are there in 12 wholes?

d How many thirds are there in 20 wholes?

e How did you work out your answers?

2 Convert these improper fractions to mixed numbers. Use the bar models to help you.

a $\frac{12}{5}$

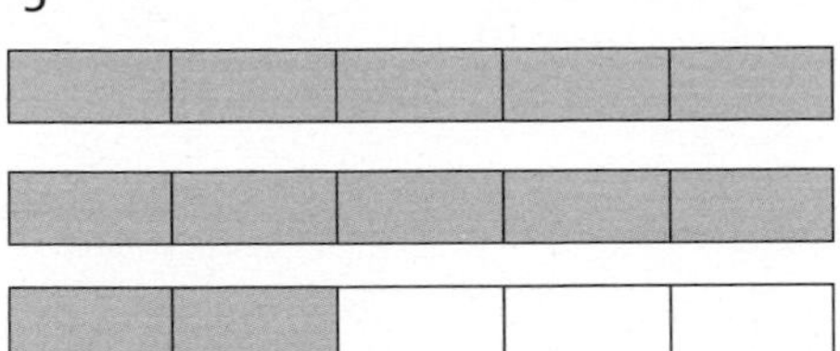

b $\frac{12}{7}$

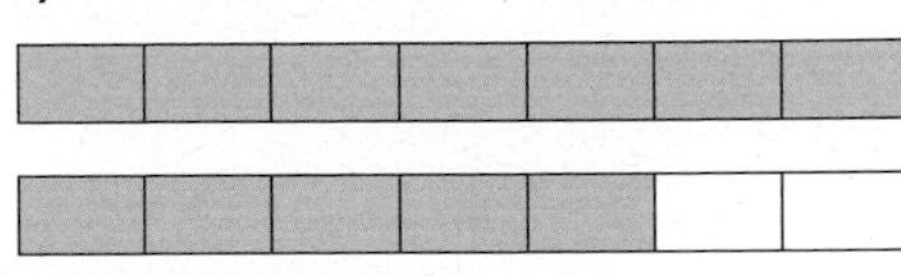

c $\frac{14}{3}$

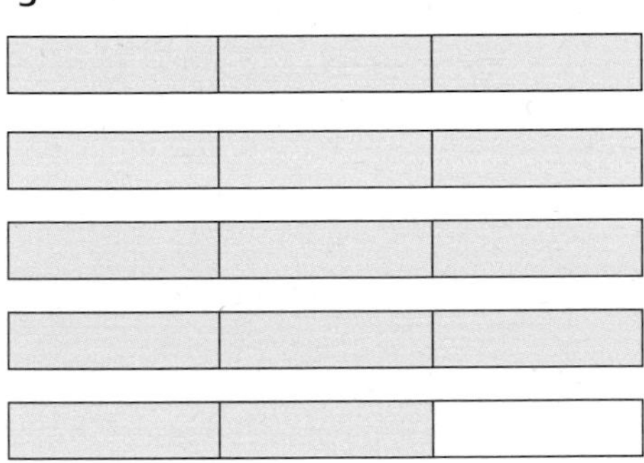

d $\frac{23}{6}$

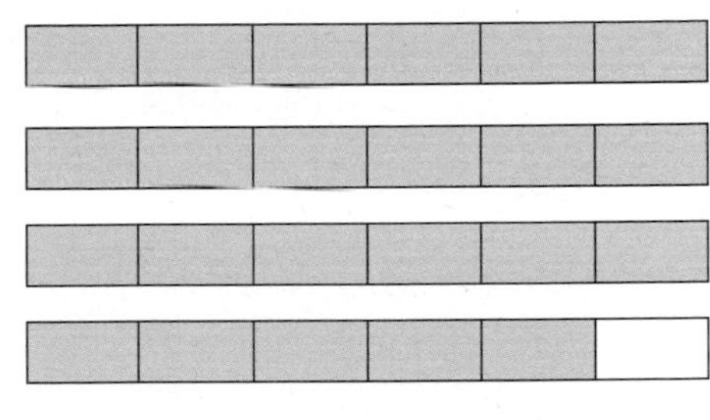

3 Convert these improper fractions to mixed numbers.

a $\frac{24}{5}$ **b** $\frac{17}{4}$ **c** $\frac{14}{3}$ **d** $\frac{19}{9}$ **e** $\frac{31}{10}$ **f** $\frac{77}{8}$

Compare your method with a partner.

4 Convert these mixed numbers to improper fractions.

a $5\frac{1}{3}$ **b** $2\frac{6}{7}$ **c** $9\frac{3}{4}$ **d** $4\frac{4}{9}$ **e** $10\frac{12}{15}$ **f** $15\frac{2}{3}$

Compare your method with a partner.

5 Copy and complete these number lines.

a

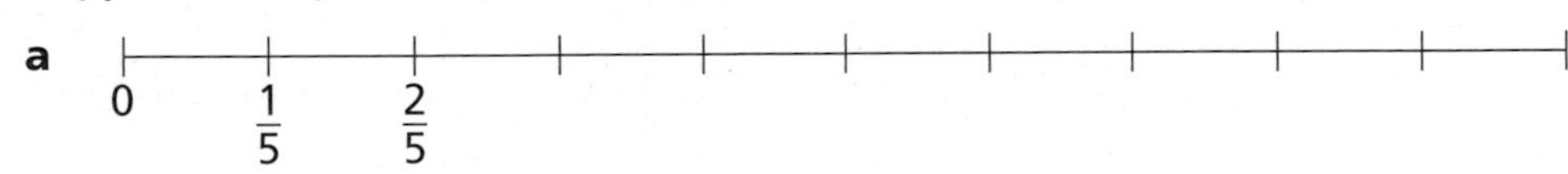

b

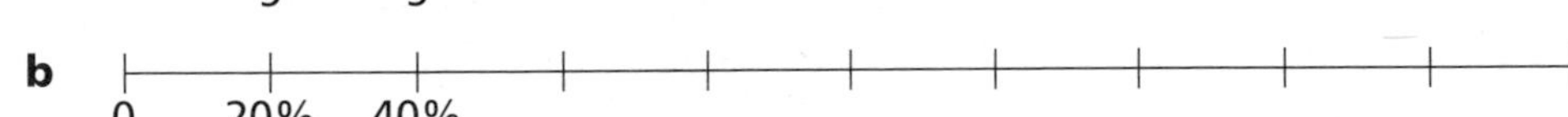

c

0 0.2 0.4

What do you notice about the number lines?

6 Write these fractions as percentages.

a $\frac{1}{5}$ $\frac{5}{5}$ $\frac{6}{5}$

b $\frac{3}{4}$ $\frac{8}{4}$ $\frac{11}{4}$

c $\frac{7}{10}$ $\frac{80}{10}$ $\frac{87}{10}$

d $\frac{71}{100}$ $\frac{300}{100}$ $\frac{371}{100}$

What do you notice?

7 Write these fractions as decimals.

a $\frac{1}{5}$ $\frac{5}{5}$ $\frac{6}{5}$

b $\frac{3}{4}$ $\frac{8}{4}$ $\frac{11}{4}$

c $\frac{7}{10}$ $\frac{80}{10}$ $\frac{87}{10}$

d $\frac{71}{100}$ $\frac{300}{100}$ $\frac{371}{100}$

What do you notice?

8 Copy and complete these number lines.

a 0, $\frac{3}{10}$, $\frac{6}{10}$

b 0, 25%, 50%

c 0, $\frac{2}{7}$, $\frac{4}{7}$

d 0, 0.75, 1.5

e 0, $\frac{5}{9}$, $1\frac{1}{9}$

What do you think?

1 Convert these improper fractions to mixed numbers.

a $\frac{17}{4}$ $\frac{18}{4}$ $\frac{19}{4}$ $\frac{20}{4}$ $\frac{21}{4}$

b $\frac{17}{5}$ $\frac{18}{5}$ $\frac{19}{5}$ $\frac{20}{5}$ $\frac{21}{5}$

c $\frac{17}{3}$ $\frac{18}{3}$ $\frac{19}{3}$ $\frac{20}{3}$ $\frac{21}{3}$

What do you notice?

2 Flo wants to convert $21\frac{7}{10}$ into an improper fraction, a decimal and a percentage.

a Flo says, "I'm going to draw bar models to convert it into an improper fraction, and that will help me to convert it into a decimal and a percentage." Explain why this is not the most efficient way.

b Write $21\frac{7}{10}$ as an improper fraction, a decimal and a percentage.

Consolidate – do you need more?

1 What improper fraction is represented by each diagram?

a

b

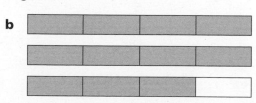

c

d

2 What mixed number is represented by each diagram?

a b

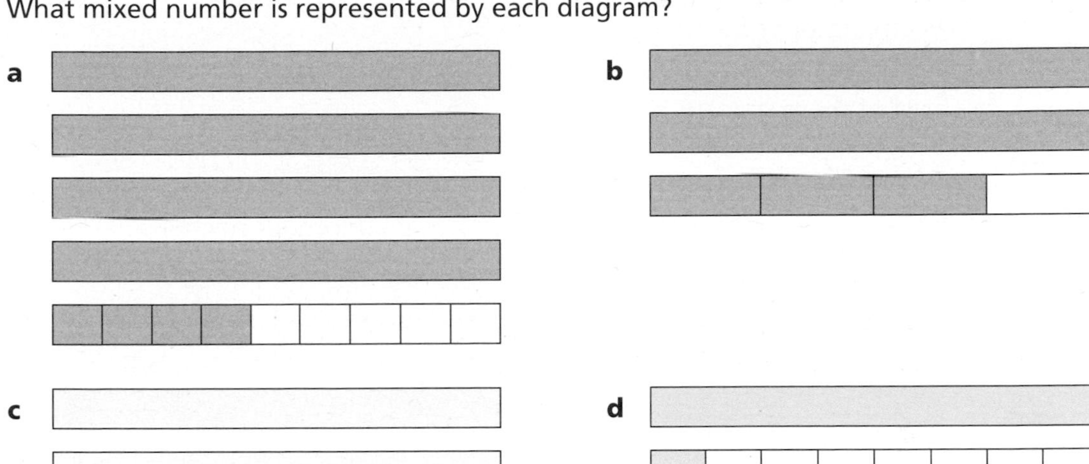

c d

3 Convert these improper fractions to mixed numbers.

a $\frac{12}{5}$ **b** $\frac{13}{4}$ **c** $\frac{19}{6}$ **d** $\frac{21}{8}$ **e** $\frac{25}{7}$ **f** $\frac{25}{8}$

4 Convert these mixed numbers to improper fractions.

a $2\frac{1}{7}$ **b** $1\frac{8}{9}$ **c** $3\frac{4}{5}$ **d** $2\frac{5}{6}$ **e** $5\frac{2}{7}$ **f** $4\frac{9}{10}$

5 Which of these values are the same as $1\frac{2}{5}$?

12.5% $\frac{12}{5}$ $\frac{7}{5}$ 1.25 1.4 140%

How do you know?

Stretch – can you deepen your learning?

1 Work out the range of these fractions.

Give your answer as an improper fraction and as a mixed number.

$\frac{17}{4}$ $2\frac{1}{4}$ $5\frac{3}{4}$ $7\frac{1}{4}$ $\frac{31}{4}$

2 Work out the median of these fractions. Give your answer as an improper fraction and as a mixed number.

$\frac{17}{4}$ $2\frac{1}{4}$ $5\frac{3}{4}$ $7\frac{1}{4}$ $\frac{31}{4}$

3 Which of these improper fractions are equivalent to integers?

$\frac{17}{7}$ $\frac{51}{5}$ $\frac{70}{8}$ $\frac{64}{8}$ $\frac{100}{25}$ $\frac{49}{7}$ $\frac{63}{9}$ $\frac{42}{4}$

How do you know?

4 A regular hexagon is made up of six equilateral triangles.

The area of one equilateral triangle is $\frac{2}{7}$ m^2.

a What is the area of the regular hexagon? Give your answer as a mixed number and as an improper fraction.

b What other shapes can be made from equilateral triangles? What would be the area of these shapes?

5 $a = 17$, $b = 23$, $c = 5$, $d = 11$ and $e = 2$

a Convert each of these mixed numbers into an improper fraction.

i $a\frac{c}{b}$ **ii** $b\frac{c}{d}$ **iii** $c\frac{e}{d}$ **iv** $e\frac{c}{a}$ **v** $d\frac{a}{b}$ **vi** $b\frac{a}{e}$

b Convert each of these improper fractions into a mixed number.

i $\frac{a}{c}$ **ii** $\frac{b}{c}$ **iii** $\frac{d}{c}$ **iv** $\frac{c}{e}$ **v** $\frac{b}{e}$ **vi** $\frac{a}{e}$

6 Convert each of these mixed numbers into an improper fraction.

Write down each step of your working.

a $14\frac{3}{7}$ **b** $14\frac{x}{7}$ **c** $14\frac{3}{y}$ **d** $14\frac{x}{y}$ **e** $z\frac{x}{y}$

Reflect

1 How do you know when a fraction is improper?

2 In your own words, explain how to convert **a** an improper fraction into a mixed number and **b** a mixed number into an improper fraction.

3 True or false?

A positive improper fraction represented as a decimal will always be greater than 1.
Explain your answer.

4 True or false?

A percentage greater than 100% can be represented as an improper fraction and a mixed number.
Explain your answer.

5 Fractions, decimals and percentages

Chapters 5.1–5.7

I have become fluent in…

- Converting simple fractions, decimals and percentages
- Understanding percentages
- Representing fractions as diagrams
- Identifying and using equivalent fractions

I have developed my reasoning skills by…

- Making connections between different representations
- Linking fractions and division
- Interpreting pie charts
- Exploring common misconceptions

I have been problem-solving through…

- Representing fractions, decimals and percentages in different forms
- Using fractions, decimals and percentages in different contexts
- Spotting and describing patterns

Check my understanding

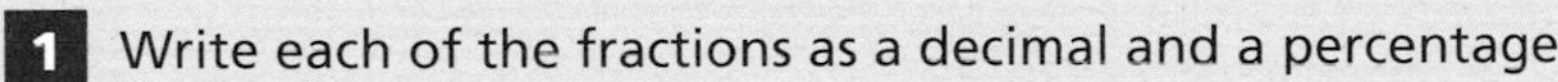

1 Write each of the fractions as a decimal and a percentage.

a $\frac{1}{10}$ **b** $\frac{71}{100}$ **c** $\frac{81}{100}$ **d** $\frac{3}{4}$ **e** $\frac{2}{5}$

2 Write each of the decimals as a fraction and a percentage.

a 0.3 **b** 0.25 **c** 0.79 **d** 0.6 **e** 0.5

3 Write each of the percentages as a fraction and a decimal.

a 27% **b** 40% **c** 90% **d** 99% **e** 75%

4 The pie chart shows the favourite colours of a group of people.

Red
Blue
Green
Other

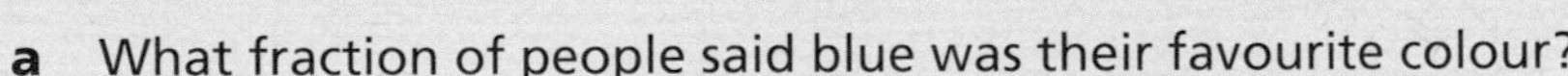

a What fraction of people said blue was their favourite colour?

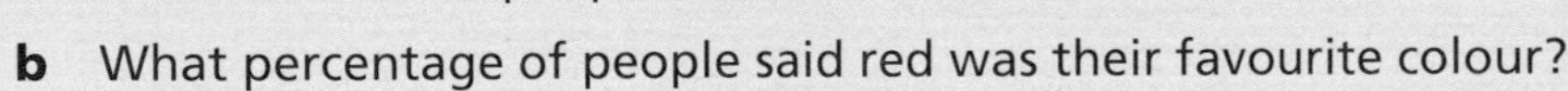

b What percentage of people said red was their favourite colour?

c Write the proportion of people who chose 'other' as a decimal.

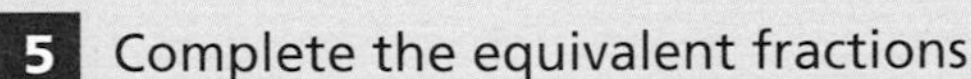

5 Complete the equivalent fractions.

a $\frac{2}{7} = \frac{?}{21}$ **b** $\frac{5}{9} = \frac{25}{?}$ **c** $\frac{?}{3} = \frac{20}{60}$ **d** $\frac{4}{?} = \frac{24}{66}$

6 Write the fractions as decimals. (H)

a $\frac{1}{8}$ **b** $\frac{3}{1000}$ **c** $\frac{5}{8}$ **d** $\frac{731}{1000}$ **e** $\frac{7}{8}$

7 Write each improper fraction as a mixed number and then as a decimal and a percentage. (H)

a $\frac{191}{100}$ **b** $\frac{33}{10}$ **c** $\frac{19}{4}$ **d** $\frac{11}{5}$ **e** $\frac{17}{8}$

6 Addition and subtraction

In this block, I will learn...

about the relationship between addition and subtraction

173		
87	49	37

how to use the column methods for addition and subtraction

	H	T	O
	4	6	8
+	3	8	7
	8	5	5
		1	1

when to use mental methods for addition and subtraction

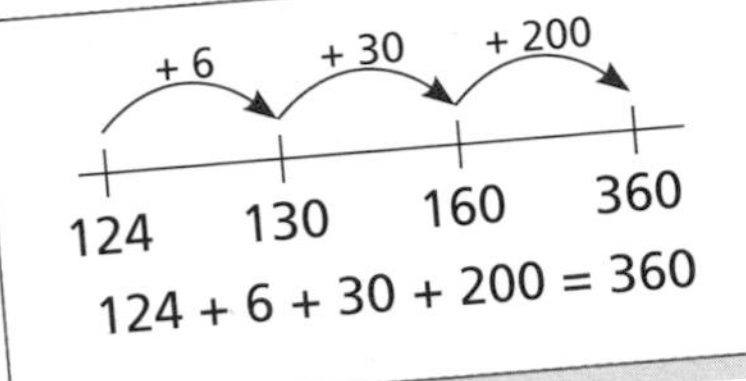

how to solve problems involving perimeter, charts and graphs

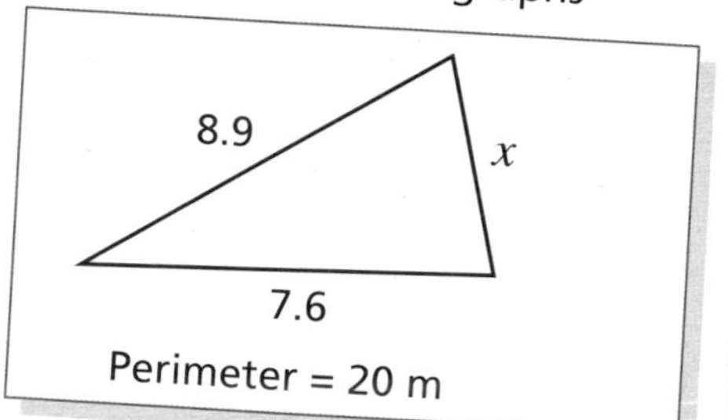

how to solve problems in financial mathematics

Date	Description	Credit (£)	Debit (£)	Balance (£)
May 1	Opening balance			546.23
May 2	Wages	600.00		A
May 6	Phone bill		18.65	B
May 7	Shopping		C	1038.86

how to solve problems with tables, frequency trees and timetables

Marketville	08:05	10:23	11:06	14:30	17:45
Old Town	08:26	10:42	11:23	14:47	18:06
Naze	08:44	–	–	–	18:22
New Town	08:56	11:03	11:40	15:08	18:42
Upper Brough	09:22	11:20	11:58	15:25	18:56
Lower Brough	09:30	11:26	12:05	15:32	19:02

how to add and subtract numbers in standard form (H)

$8 \times 10^9 + 2 \times 10^9$

$8 \times 10^9 - 2 \times 10^9$

6.1 Linking addition and subtraction

Small steps

- Properties of addition and subtraction
- Mental strategies for addition and subtraction

Key words

Total – the result of adding two or more numbers

Difference – the result of subtracting a smaller number from a larger number

Partition – to break up a number into smaller parts

Commutative – relating to an operation on two quantities, when the order of the quantities does not affect the result

Associative law for addition – when you add numbers it does not matter how they are grouped

Are you ready?

1 Complete the addition square.

+	5	7	8
4	9		
6			14
7		14	

2 Work out the missing numbers.

+		4	
	9	11	
		12	15
			16

3 Find the difference between the numbers to complete the table.

–	8	3	6
9	1		
7			
4			2

4 Use the fact that 4 + 7 = 11 to work out the following.

a 40 + 70 **b** 400 + 700 **c** 11 000 – 7000

d 0.4 + 0.7 **e** 0.04 + 0.07 **f** 1.1 – 0.4

Models and representations

Part-whole model

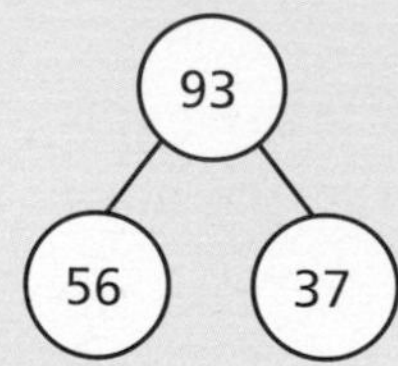

Bar model

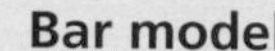

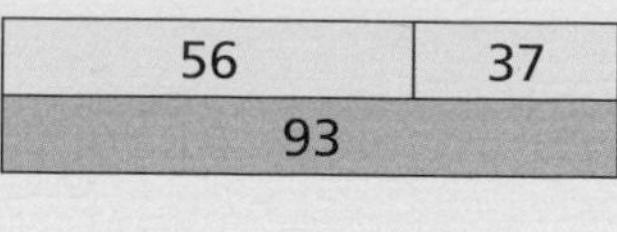

Number line

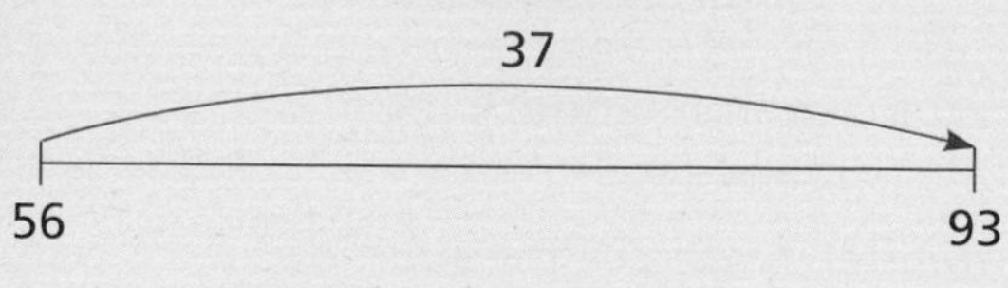

Base 10 blocks

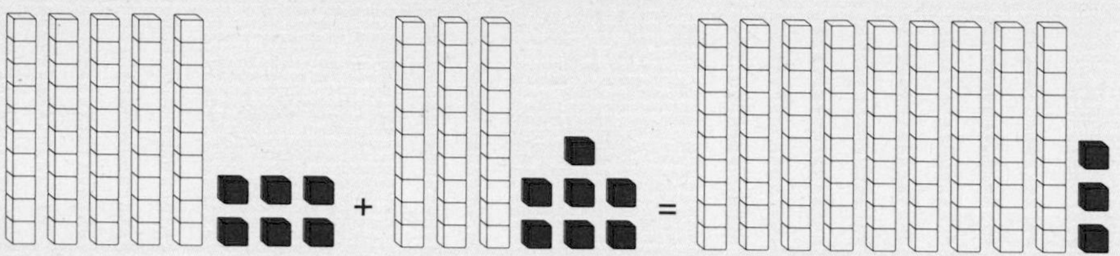

Place value counters

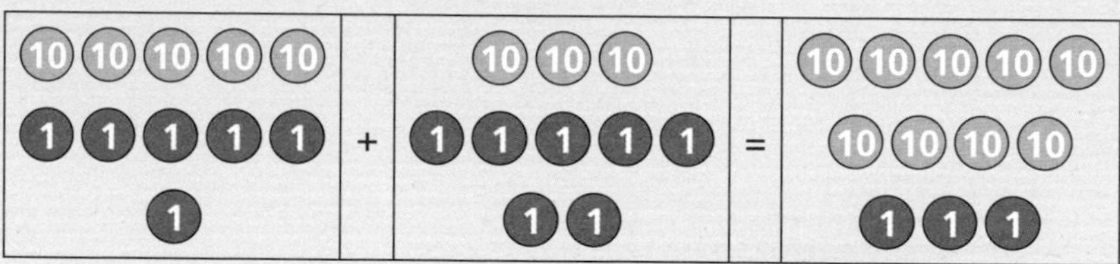

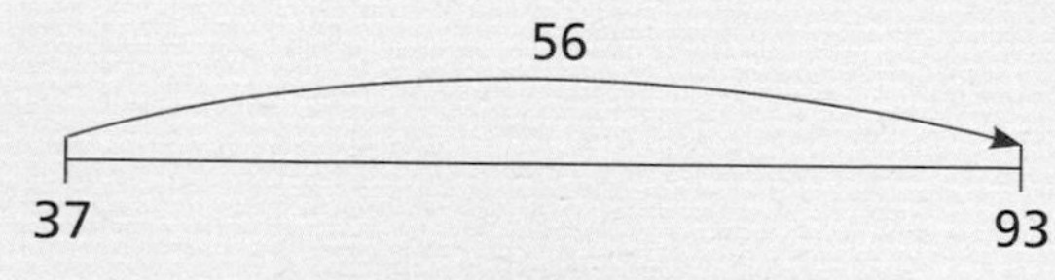

This number line shows 37 + 56 = 93

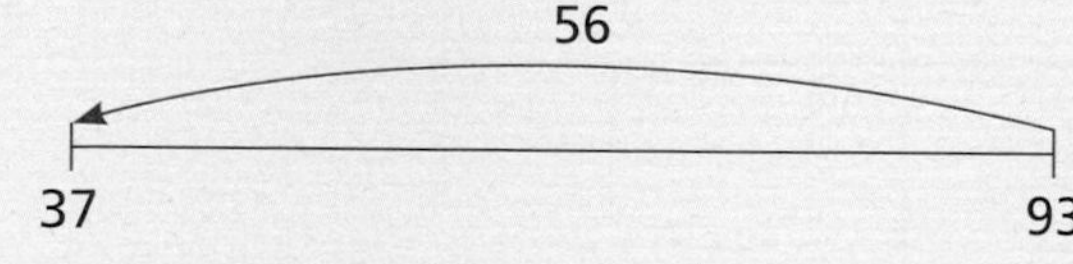

while this shows 93 – 56 = 37

Example 1

a What addition does this number line show?

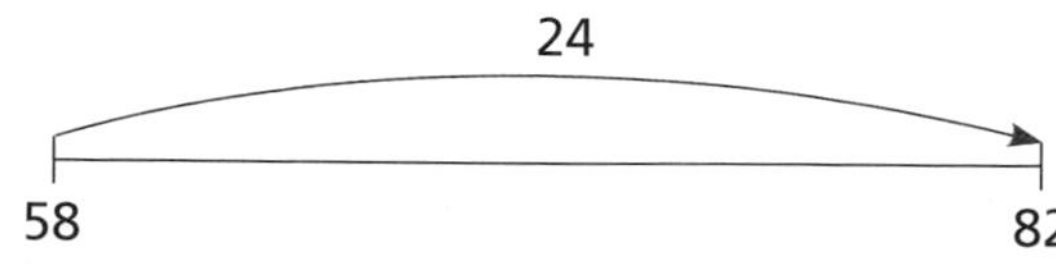

b Show this addition as

i a bar model

ii a part-whole model.

c What other addition and subtraction facts can you write using these numbers?

a 58 + 24 = 82 — Read the number line from left to right.

b i

58	24
82	

82	
58	24

58	24	}82

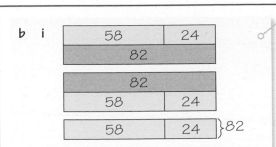

You can show this as a bar model in several ways.

ii

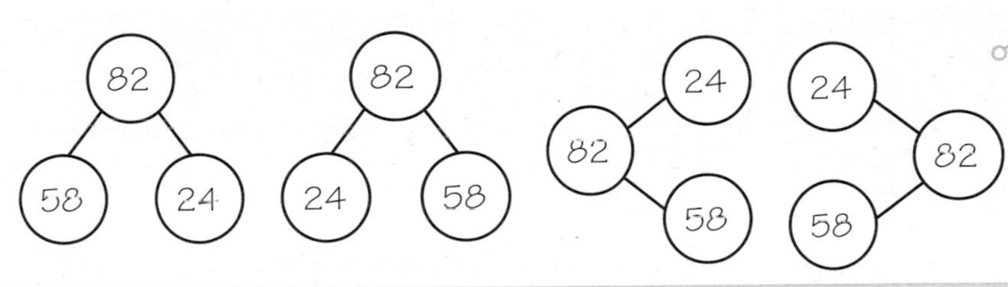

The part-whole model could look like these.

c $24 + 58 = 82$ — Addition is **commutative**.

$82 - 24 = 58$ and $82 - 58 = 24$ — You know these from the fact family.

Example 2

What relationships can you deduce from this bar model?

173		
87	49	37

$173 = 87 + 49 + 37$ — From the bar model.

$173 = 87 + 37 + 49$ — As addition is commutative, there are several ways of ordering the three smaller numbers.

In how many ways can you order three different numbers?

$173 - 37 = 87 + 49$ — If you subtract 37 from 173 you are left with a bar of length 87 + 49

173 − 37		~~37~~
87	49	~~37~~

$173 - 87 = 49 + 37$ — If you subtract 49 from 173 you are left with a bar of length 87 + 37

$173 - 49 = 87 + 37$ — If you subtract 87 from 173 you are left with a bar of length 49 + 37

$173 - 37 - 49 = 87$ — If you subtract both 37 and 49 from 173 you are left with a bar of length 87

173 − 37 − 49	~~49~~	~~37~~
87	~~49~~	~~37~~

173 – (37 + 49) = 87

You can add 37 and 49 first, and them subtract the **total** from 173.

173 – (37 + 49)	~~49 + 37~~
87	~~49 + 37~~

173 – 87 – 49 = 37 or
173 – (87 + 49) = 37

173 – 37 – 87 = 49 or
173 – (37 + 87) = 49

You could draw similar bar models to show these subtractions.

Practice 6.1A

1 **a** Which of these facts are shown by the part-whole model?

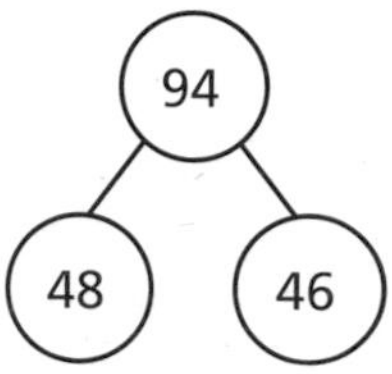

48 + 94 = 46	48 + 46 = 94	48 – 94 = 46	46 + 94 = 48
94 – 46 = 48	94 = 48 – 46	94 – 48 = 46	46 – 94 = 48

b Redraw the part-whole model as

i a bar model

ii a number line.

2

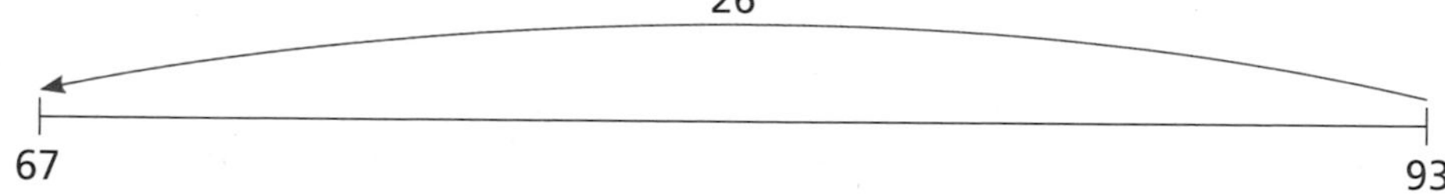

The number line shows 93 – 26 = 67

a What addition does the number line also represent?

b Draw a bar model to show this information.

c What other subtraction fact can you write using these numbers?

3

30	40	50
120		

Which of these facts are shown by the bar model?

30 + 40 + 50 = 120	120 – 50 = 30 + 40	120 – 30 = 40 + 50
120 – 30 – 40 = 50	120 + 50 = 40 + 30	120 – 40 = 30 + 50

4 **a** Draw a number line to show 160 + 80 = 240

b Write down another addition and two subtractions that can be deduced from your number line.

5 **a** Draw a bar model to show that $80 + x = 40 + y$

b Copy and complete the diagram to show $300 = 140 + x + y$

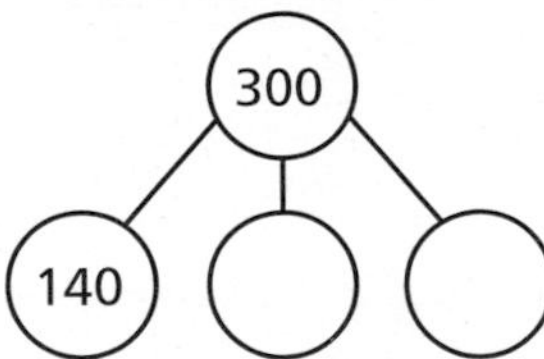

c Use your completed diagram from **b**. Write down whether each of these statements is true or false.

i $x + y = 300 - 140$ **ii** $300 - x = 140 + y$

iii $300 - y = 140 - x$ **iv** $300 - x - y = 140$

v $x = 300 - 140 + y$ **vi** $x = 300 - (140 + y)$

d Use your completed diagram from **b**. Copy and complete this statement:

$y =$ ________

6

20		
13		7
7	13	

a Amina says, "The bar models show that addition is commutative". Explain why Amina is right.

b Is subtraction commutative? Explain your answer.

7

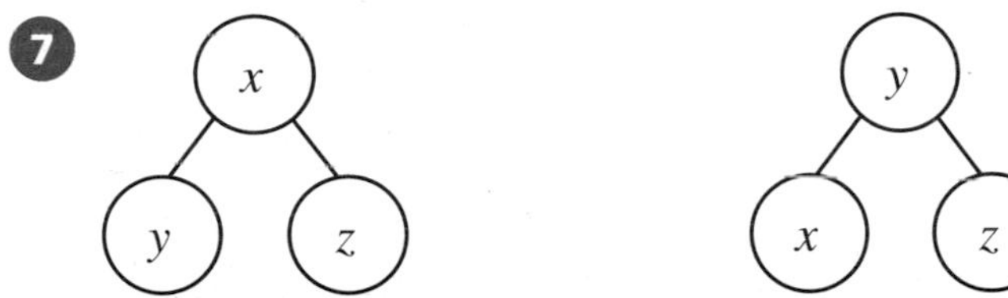

Emily says, "These part-whole models show the same information". Explain why Emily is wrong.

What do you think?

1

a	b	c
x		

a Which of these statements are true?

$a + b + c = x$ $x - a = b + c$ $x - (a + b) = c$

$x - a + b = c$ $x - a - b = c$ $a + b - c = x$

b Copy and complete the statements.

i $x - b =$ ________ **ii** $a + b =$ ________ **iii** $x - (a + b + c) =$ ________

2

a	b
c	d

a Which of these statements are true?

$b + a = d + c$ $a + b - c = d$ $a + b - c - d = 0$

b Copy and complete the statements.

i $c + d - b =$ ________ **ii** $c + d - a =$ ________ **iii** $a - d + b =$ ________

3 **a** Which of these calculations will not have the same answer as the others? Why not?

49 + 63 – 29 49 – 29 + 63 63 + 49 – 29

63 – 49 + 29 63 – 29 + 49

b Which calculation will be the easiest to work out? Why?

Example 3

There are many different ways of working out calculations mentally.

It is important to know number bonds, especially those to 10

The **associative law for addition** means that you can regroup calculations to make them easier,

For example,

86 + 47 + 53 — Look for numbers where the ones digits add to 10

= 86 + (47 + 53) — Notice that 47 + 53 = 100

= 86 + 100

= 186

This was easier to work out than (86 + 47) + 53 = 133 + 53 = 186

Example 4

"Reordering" is another useful strategy. For example:

6 + 7 + 9 + 3 + 4

= 6 + 4 + 7 + 3 + 9 — Addition is commutative, so you can change the order.

= 10 + 10 + 9 — Reorder to use the number bonds to 10

= 29

Here are some ways of working out 39 + 39

Doubling

Double 30 is 60

Double 9 is 18

So 39 + 39 = 60 + 18

= 78

Compensation

39 + 39 = 39 + 40 – 1

= 79 – 1

= 78

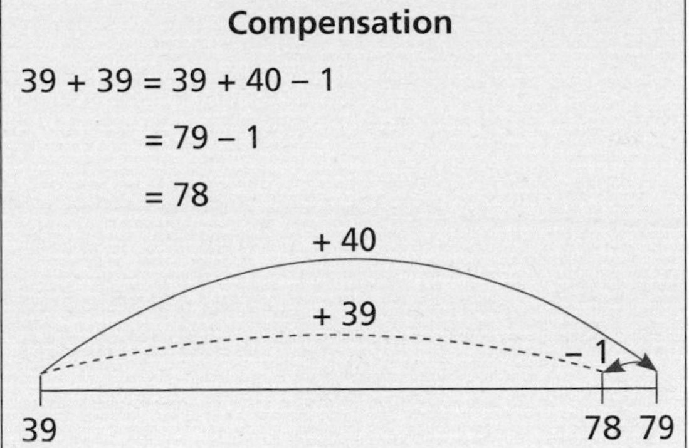

Partitioning

39 + 39 = 39 + 30 + 9

= 69 + 9

= 78

Adjusting

39 + 39

+ 1   – 1

= 40 + 38

= 78

Example 5

Work these out using a mental method.

a 146 + 37   **b** 360 – 124   **c** 513 + 299   **d** 856 – 147 – 256

Remember: there are lots of different methods. You should choose the one that you find easiest and most efficient.

a 146 + 37 = 146 + 30 + 7 — You can **partition** the second number.

= 176 + 7 — Work out 146 + 30 first as you only need to add the tens.

= 183 — Now use number bonds to work out the final answer.

You could partition the 7 this way

176 + 7 = 176 + 4 + 3 = 180 + 3 = 183,

using the fact that 6 + 4 = 10

b 360 – 124 — You can partition the 124 and subtract each part separately.

360 – 124 = 360 – 100 – 20 – 4 — 360 – 100 = 260

= 260 – 20 – 4 — 260 – 20 = 40

= 240 – 4

= 236

Or

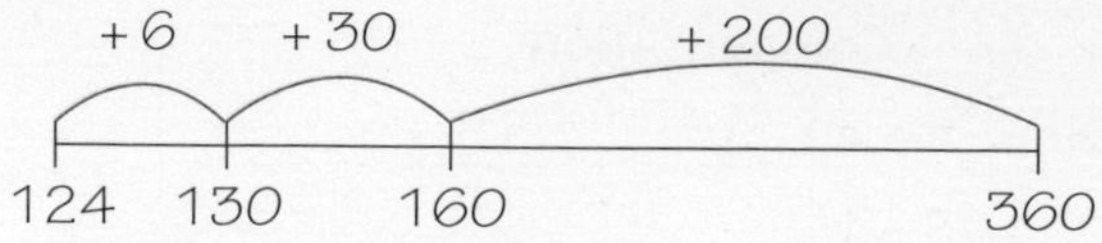

124 + 6 + 30 + 200 = 236

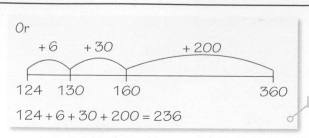

You could "count on" from 124 to 360

c 513 + 299

513 + 299

−1 +1

= 512 + 300

= 812

You could adjust by adding 300 and subtracting 1 from 513

Or

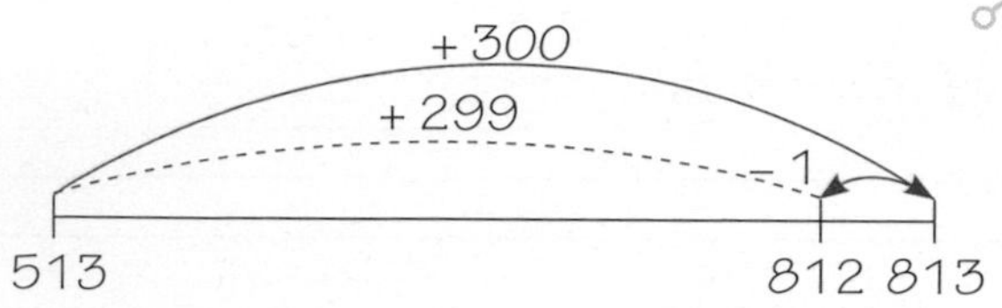

Alternatively, you could compensate.

Or

513 + 299 = 513 + 200 + 90 + 9

= 713 + 90 + 9

= 803 + 9

= 812

You could partition, but this might take longer and still have more difficult calculations.

d 856 − 147 − 256

= 856 − 256 − 147

= 600 − 147

= 600 − 100 − 40 − 7

= 500 − 40 − 7

= 460 − 7

= 453

The first and last numbers have the same numbers of tens.

Reorder.

You can work this out by partitioning or counting on.

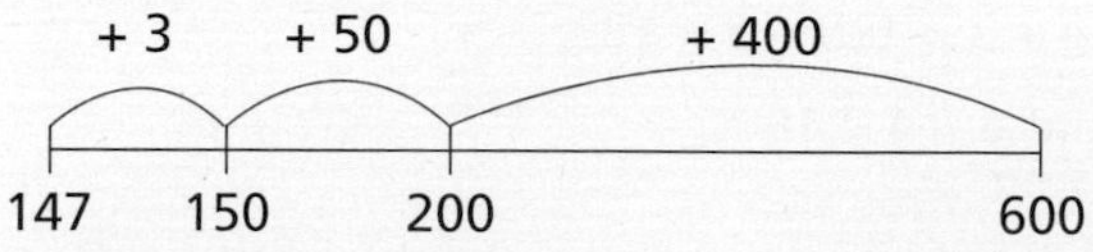

3 + 50 + 400 = 453

Practice 6.1B

1 **a** Which calculations will have the same answer as 63 + 49?

64 + 50   62 + 50   63 + 50 + 1   63 + 50 − 1

b Which calculations will have the same answer as 93 − 29?

94 − 30   94 − 28   93 − 20 − 9   93 − 30 − 1   93 − 30 + 1

2 Check that all of these methods give the same answer to the calculation 83 + 97

83 + 97 = 80 + 3 + 90 + 7

83 + 97 = 83 + 7 + 90

83 + 97 = 83 – 3 + 97 + 3 = 80 + 100

83 + 97 = 83 + 100 – 3

83 + 97 = 97 + 83 = 97 + 3 + 80 = 100 + 80

Which method do you prefer?

3 Find the answers to these calculations. Explain the method you use.

a 268 + 99 **b** 316 – 99 **c** 232 + 125

d 600 – 100 **e** 600 – 240 **f** 600 – 428

g 34 + 77 + 6 **h** 500 – 27 + 37 **i** 76 + 76

j 99 + 98 **k** 800 – 297

4 Rob finds 48 + 99 by working out 48 + 100 and then subtracting 1

a How could Rob adapt his method to work out 4.8 + 9.9?

b Work these out. Explain the method you use for each calculation.

i 26.8 + 0.9 **ii** 26.8 + 9.9 **iii** 26.8 + 19.9

iv 26.8 – 0.9 **v** 26.8 – 9.9 **vi** 26.8 – 19.9

5 Find the missing numbers using a mental method.

a

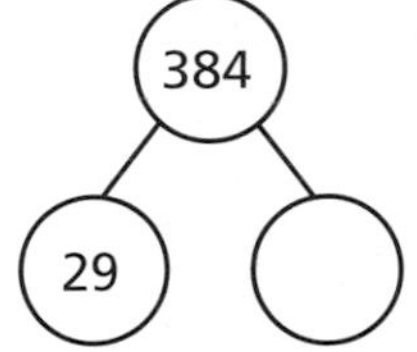

b

648	
127	

c

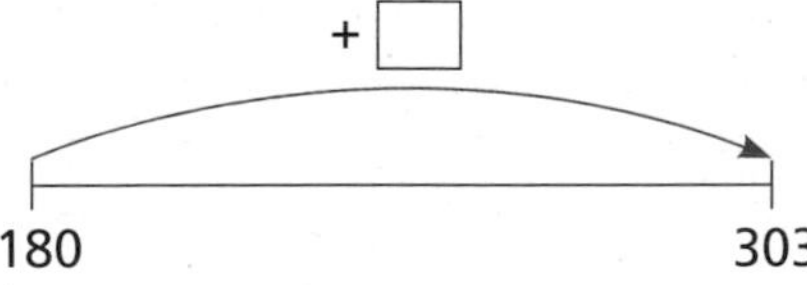

d

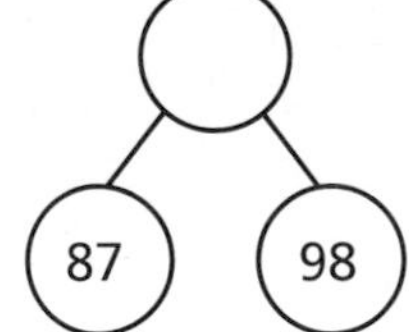

e

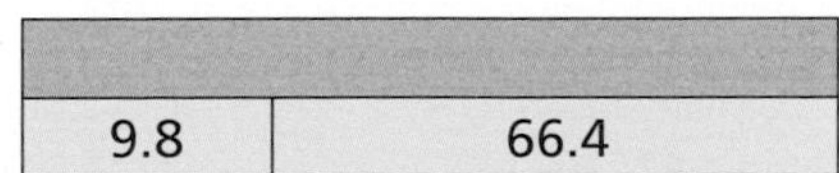

f

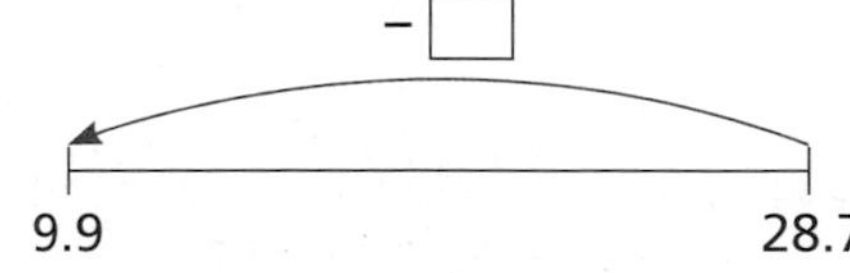

What do you think?

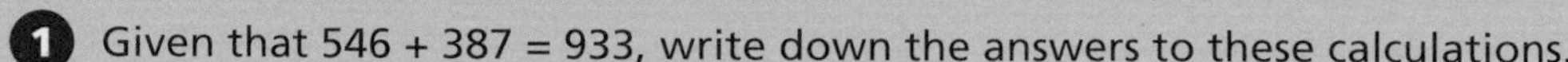

1 Given that 546 + 387 = 933, write down the answers to these calculations.

a 546 + 388 **b** 548 + 387 **c** 547 + 388 **d** 546 + 390

e 933 – 545 **f** 933 – 388 **g** 933 – 390 **h** 9.33 – 5.47

2 Solve each of these equations using a mental method.

a $a + 99 = 842$ **b** $b - 99 = 842$ **c** $c - 10.1 = 36.8$ **d** $10.1 + d = 36.8$

e $e - 19.9 = 47.3$ **f** $19.9 + f = 27.6$ **g** $384 = g + 99$ **h** $384 = h + 29$

i $384 = i + 2.9$ **j** $384 = j - 99$ **k** $384 = k - 29$ **l** $384 = l - 2.9$

Consolidate – do you need more?

1

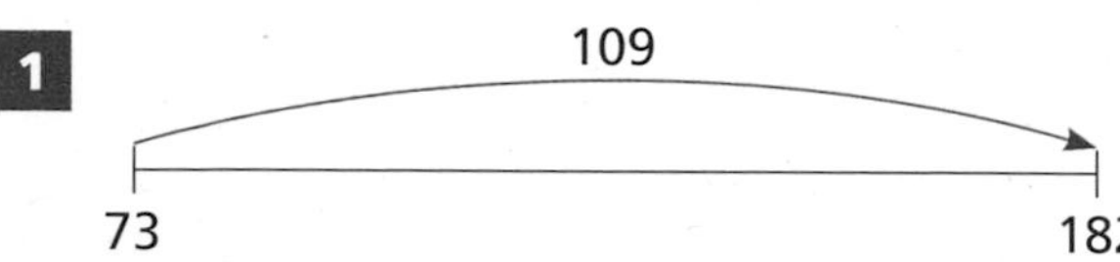

The number line shows 73 + 109 = 182

a Use the number line to copy and complete 182 – ☐ = ☐.

b Draw a bar model to show this information.

c What other subtraction fact can you write using these numbers?

2

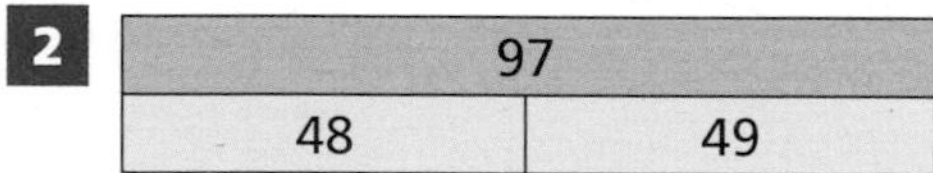

a Which of these facts does the bar model show?

48 + 49 = 97 48 – 97 = 49 97 – 48 = 49 97 = 49 + 48

b Write down the answers to these calculations.

i 97 – 47 **ii** 49 + 49 **iii** 48 + 48

3 **a** Which of these facts can you deduce from the part-whole model?

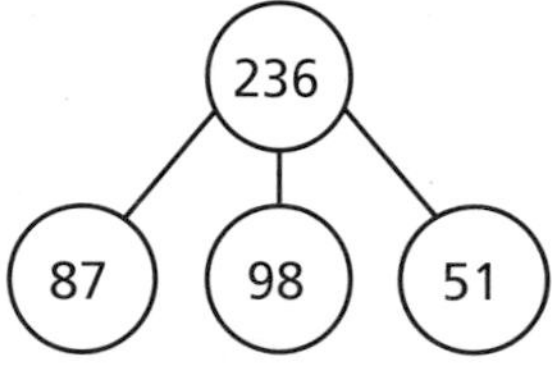

236 – 51 = 87 + 98 236 – 98 + 51 = 87 236 – 98 – 51 = 87

236 – (98 + 51) = 87 236 – 98 = 87 + 51 98 – 51 = 236 – 87

b Redraw the part-whole model as

i a bar model

ii a number line.

4 Find the answers to these calculations. Explain the method you have chosen.

a 399 + 476 **b** 476 – 399 **c** 1000 – 297

d 4.6 + 9.8 **e** 9.9 – 3.1 **f** 10 – 3.1

g 4.7 + 4.8 + 0.3 **h** 564 + 472 – 264 **i** 8.4 + 8.4

j 200 – 9.9 **k** 200 – 10.1

Stretch – you can deepen your learning?

1 Is subtraction associative? Use 6 – 3 – 1 to justify your answer.

2 Ed says that the final digit of the missing number in the calculation 578 + ☐ = 934 must be 6

Is Ed right? How do you know?

3 Find the missing numbers in these calculations. Explain how you work them out.

a 627 – 7 – ☐ = 600 **b** 432 + 28 – ☐ = 442

c 608 – ☐ + 47 = 618

Reflect

1 What do you look for when deciding how to solve an addition or subtraction mentally?

2 Explain why addition is commutative but subtraction is not.

6.2 The column method of addition

Small steps

- Use formal methods for addition of integers
- Use formal methods for addition of decimals
- Solve addition problems in the context of perimeter
- Solve financial maths problems
- Choose the most appropriate method: mental strategies, formal written or calculator

Key words

Exchange – when we change numbers for others with equal value, for example replacing 1 ten with 10 ones or replacing 10 tens with 1 hundred

Perimeter – the total distance around a two-dimensional shape

Credit – an amount of money paid into an account

Debit – an amount of money taken out of an account

Balance – an amount of money in an account

Profit – if you buy something and then sell it for a higher amount,
profit = amount received – amount paid

Loss – if you buy something and then sell it for a smaller amount,
loss = amount paid – amount received

Are you ready?

1 Use the fact that 4 + 9 = 13 to work out

a 400 + 900 **b** 0.4 + 0.9 **c** $\frac{9}{100} + 0.04$

d $\frac{9}{10} + 0.4$ **e** 40 + 90

2 Write these numbers in figures.

a Four hundred and four. **b** Four hundred and forty.

c Four thousand and forty. **d** Four thousand and four.

e Four and four hundredths.

3 Solve these equations.

a $a - 7 = 12$ **b** $b - 8 = 24$ **c** $c - 12 = 356$

4 Match the equivalent decimals and fractions.

$\frac{3}{4}$	$\frac{34}{100}$	$\frac{1}{2}$	$\frac{7}{100}$	$\frac{70}{100}$
0.7	0.75	0.34	0.5	0.07

Models and representations

Base 10 blocks

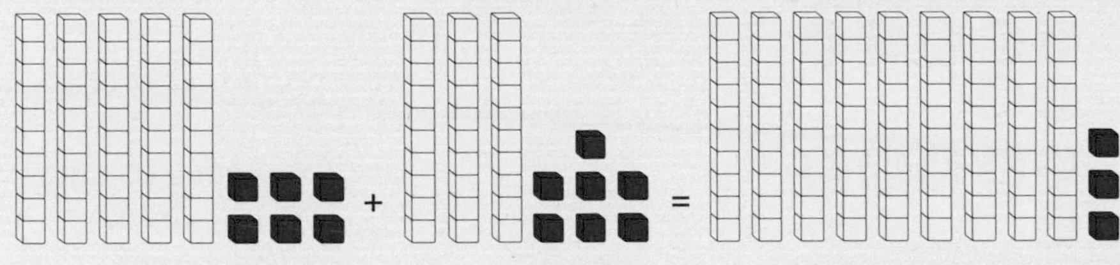

Column method

	T	O
	5	4
+	3	5
	8	9

Place value counters

1 1 1   0.1 0.1 0.1 0.1 0.1 0.1   0.01 0.01 = 3.62

Place value charts

	Tens	Ones
	10 10 10 10 10	1 1 1 1
+	10 10 10	1 1 1 1 1
	10 10 10 10 10 10 10 10	1 1 1 1 1 1 1 1 1

Example 1

Compare these calculations.

a 47 + 32   **b** 47 + 36   **c** 4.7 + 3.2   **d** 0.47 + 0.36

a 47 + 32

	Tens	Ones
	10 10 10 10	1 1 1 1 1 1 1
+	10 10 10	1 1
	10 10 10 10 10 10 10	1 1 1 1 1 1 1 1 1

	T	O
	4	7
+	3	2
	7	9

This calculation does not involve exchanging or regrouping.

Both the place value representation and the column method show that the total in any column does not go above 9

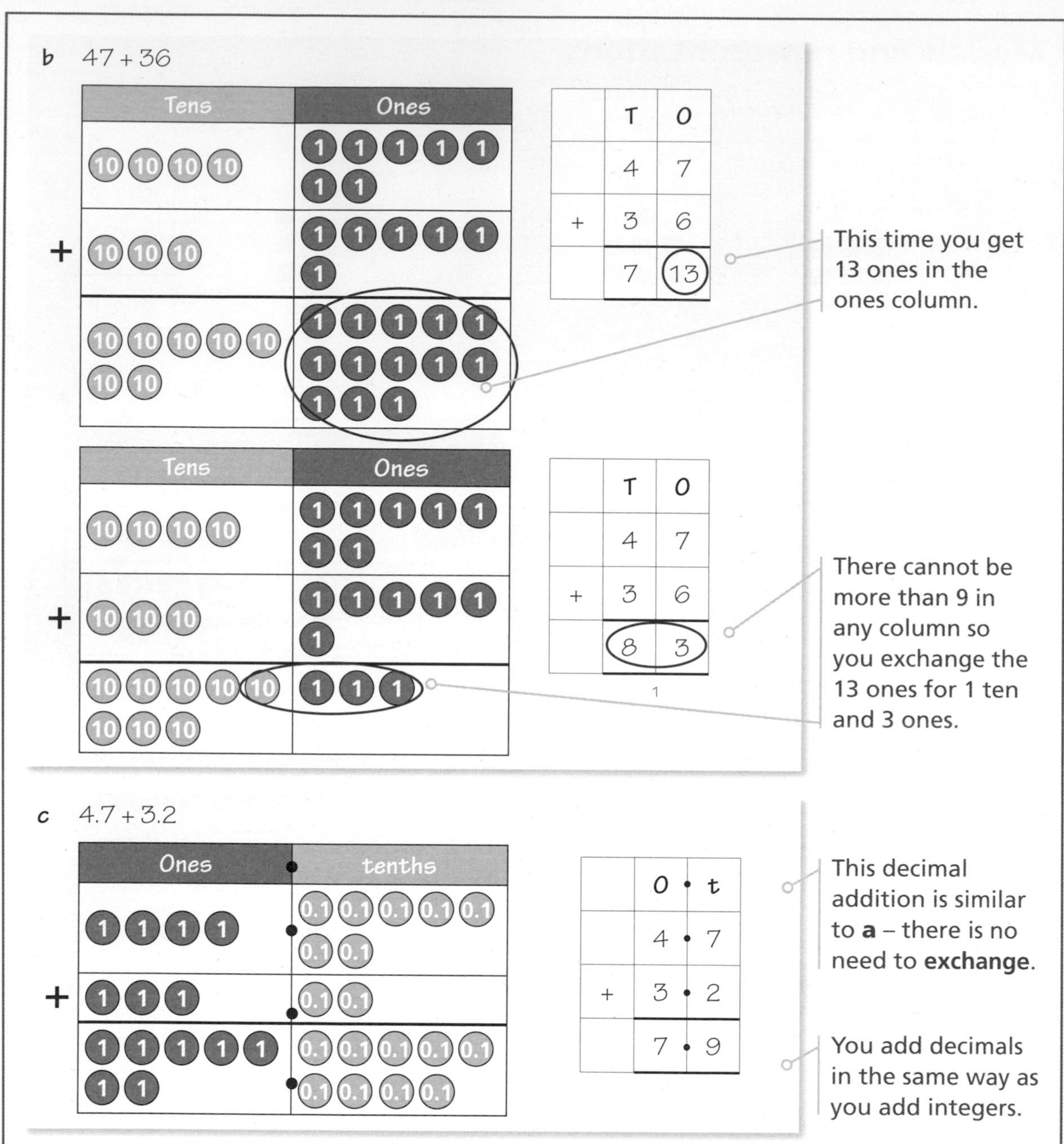
b 47 + 36
Tens
Ones
T O
4 7
+ 3 6
7 13
This time you get 13 ones in the ones column.
T O
4 7
+ 3 6
8 3
1
There cannot be more than 9 in any column so you exchange the 13 ones for 1 ten and 3 ones.
c 4.7 + 3.2
Ones
tenths
O • t
4 • 7
+ 3 • 2
7 • 9
This decimal addition is similar to a – there is no need to exchange.
You add decimals in the same way as you add integers.

d $0.47 + 0.36$

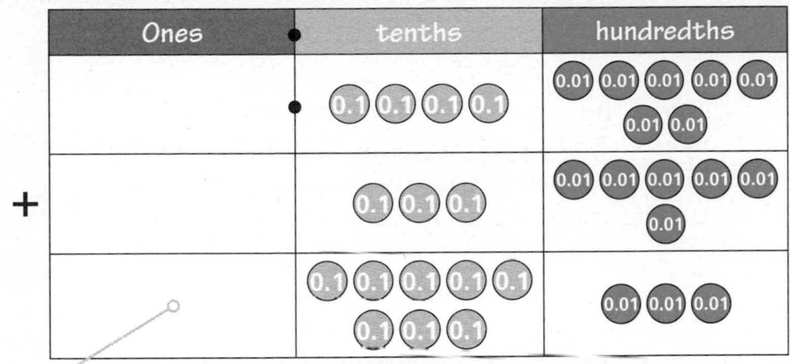

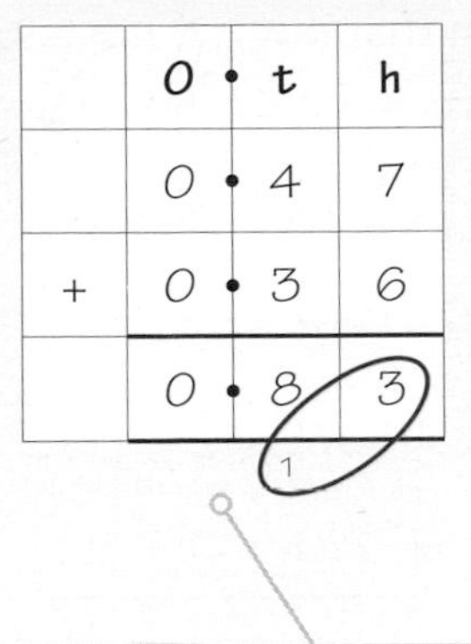

	O	•	t	h
	0	•	4	7
+	0	•	3	6
	0	•	8	3
			1	

The relationship between adjacent columns is the same as for integers, so you can use the same process.

This decimal addition is similar to **b** – you need to make an exchange but this time it is between the hundredths and tenths columns.

Example 2

Find the sum of 843 and 757.

Some people call all calculations in maths "sums" but this is not correct. "Sums" are only additions.

Remember that "find the sum of" means add.

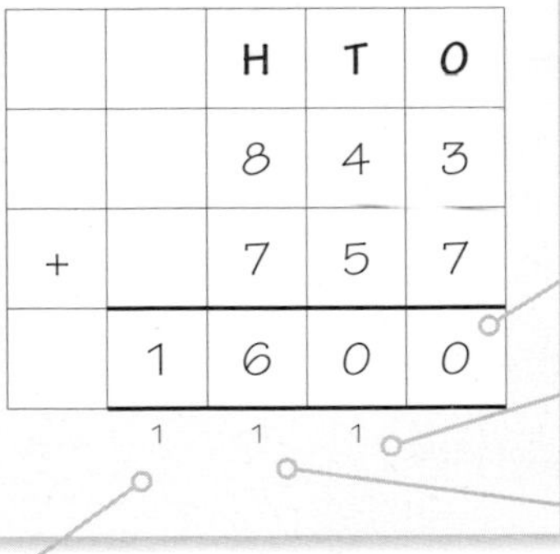

		H	T	O
		8	4	3
+		7	5	7
	1	6	0	0
	1	1	1	

You need to use zeros as place holders.

Add this to the other numbers in the tens column: $1 + 4 + 5 = 10$

Add this to the other numbers in the hundreds column: $1 + 8 + 7 = 16$

You need an extra column for thousands that was not needed for the original numbers.

Example 3

Given that $a = 7.2$, $b = 0.93$, $c = 18$ and $d = 0.68$, show how to work out

a $a + b$ **b** $a + c$ **c** $a + d$ **d** $b + d$

a $a + b = 7.2 + 0.93$

	O	•	t	h
	7	•	2	
+	0	•	9	3
	8	•	1	3
	1			

The column method is efficient here, but you need to be careful to line up the digits with the same place value. You could add a zero to 7.2 to help you as 7.2 and 7.20 are equal in value.

b $a + c = 7.2 + 18$

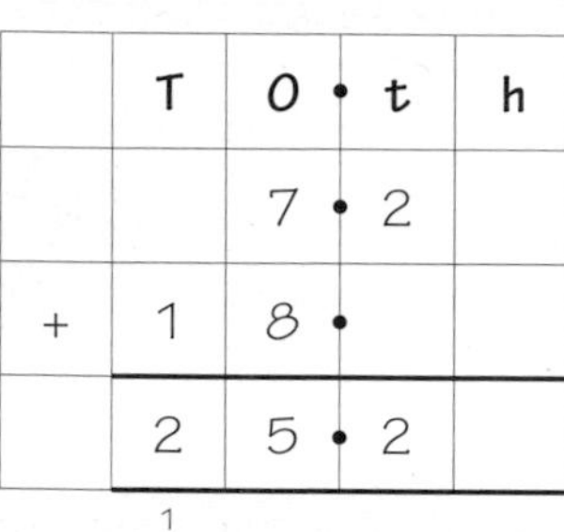

	T	O	•	t	h
		7	•	2	
+	1	8	•		
	2	5	•	2	
	1				

Again the column method works well, but you could work this out mentally if you know that 7 + 18 = 25

c $a + d = 7.2 + 0.68$

	O	•	t	h
	7	•	2	
+	0	•	6	8
	7	•	8	8

The column method is suitable here, but you could choose to do this mentally.

d $b + d = 0.93 + 0.68$

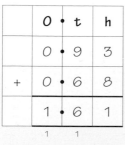

	O	•	t	h
	0	•	9	3
+	0	•	6	8
	1	•	6	1
	1		1	

It is sensible to use the column method here.

Practice 6.2A

1. **a** Use the column method to find the total of each pair of numbers.

   i 123 and 456   **ii** 123 and 486

   iii 124 and 456   **iv** 124 and 476

   What's the same and what's different?

   b Work out

   i 87 + 124   **ii** 193 + 46

   iii 207 + 68   **iv** 70 + 365

2. Work out

   a 5.7 + 3.4   **b** 19.6 + 8.5

   c 3.67 + 4.38   **d** 0.167 + 0.853

3. Find the missing numbers in these diagrams.

   a

   b

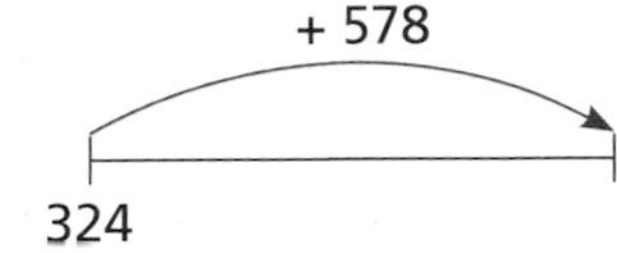

   c

0.84	0.76

4. Work out

   a $6.72 + \frac{8}{100}$ **b** $6.72 + \frac{8}{10}$ **c** $6.72 + 8$ **d** $6.72 + \frac{3}{4}$

5. Given that $w = 2.8$, $x = 0.28$, $y = 2.08$ and $z = 8.2$ find the value of

   a $z + w$ **b** $w + x$ **c** $w + y$

   d $x + y$ **e** $x + z$ **f** $y + z$

6. **a** Add one hundred and thirty-seven thousand to five hundred and twenty-three thousand.

   b What number is eight thousand and eighty more than 8888?

   c What number is sixteen thousand and fifty-six greater than 8794?

What do you think?

1. Solve these equations.

   a $a - 307 = 432$ **b** $b - 614 = 887$ **c** $c - 4.8 = 14.08$ **d** $d - 0.707 = 0.707$

2. Describe how you would use the column method to add three numbers, for example 567 + 345 + 242

3 Work out

a $7.36 + \frac{17}{10}$ **b** $7.36 + \frac{17}{100}$ **c** $7.36 + \frac{17}{1000}$ **d** $7.36 + \frac{17}{10000}$

4 A linear sequence starts 8.2, 9.5 …

Which term in the sequence is the first integer?

You often need to use addition to solve mathematical and real-life problems.

Example 4

Find the **perimeter** of this triangle.

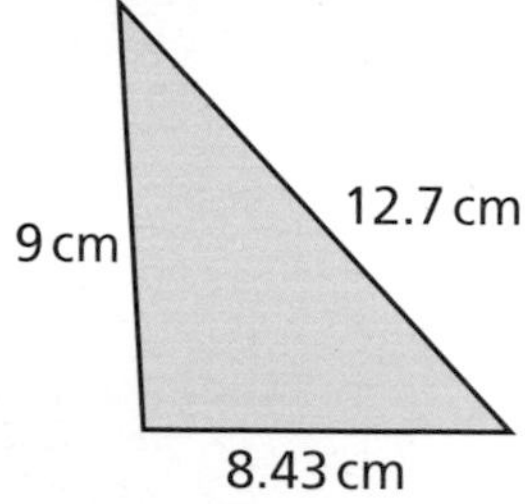

	T	O •	t	h
	1	2 •	7	0
		9 •	0	0
+		8 •	4	3
	3	0 •	1	3
	2	1		

It can make it easier to work with decimals if you add zeros to make the number of decimal places equal.

So the perimeter is 30.13 cm.

State the answer with the unit.

Example 5

Here is part of a bank statement. Copy and complete the statement by working out the values of A and B.

Date	Description	Credit (£)	Debit (£)	Balance (£)
Sep 1	Opening balance			283.15
Sep 2	Electricity bill		67.86	215.29
Sep 6	Wages	443.88		A
Sep 7	Refund	29.99		B

	H	T	O •	t	h
	2	1	5 •	2	9
+	4	4	3 •	8	8
	6	5	9 •	1	7
			1	1	

You can use the column method to find **balance** A after the wages are added.

£659.17 + £29.99
= £659.17 + £30 − £0.01
= £689.16

To find balance B, you need to add £29.99 to £659.17

You can work this out mentally using compensation as you are adding a number of pounds that is close to a multiple of £10

You could also use the column method.

Practice 6.2B

1 Find the perimeter of each of these triangles.

a 57 mm, 78 mm, 68 mm

b 8.6 cm, 6.9 cm, 7.4 cm

c

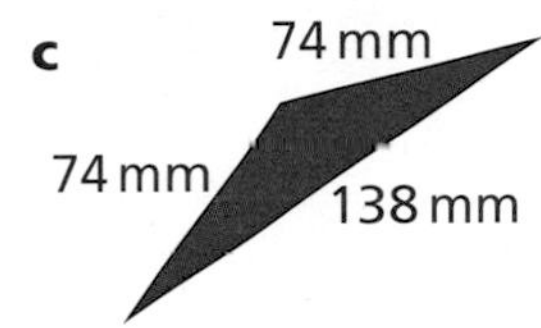

2 Find the perimeter of each of these rectangles.

a 12.7 cm, 8.6 cm, 8.6 cm, 12.7 cm

b 87 mm, 92 mm

3 Find the missing balances in this bank statement.

Date	Description	Credit (£)	Debit (£)	Balance (£)
Nov 1	Opening balance			65.28
Nov 2	Wages	740.00		A
Nov 3	BACS payment	81.73		B
Nov 5	Deposit	150.00		C
Nov 6	Expenses	73.99		D
Nov 7	Bank transfer	99.00		E

Are there any values you can work out mentally?

4 **a** Filipo buys some shares for £386. When he sells the shares, he makes a profit of £276. How much did he sell the shares for?

b Faith sells her old bike for £85. She makes a loss of £147. How much did she pay for the bike?

c Three trains arrive at a station. The first train has 168 passengers, the second has 356 passengers and the third has 275 passengers. How many passengers in total arrive at the station?

d Lydia is 146.3 cm tall. Her mother is 18.5 cm taller. How tall is Lydia's mother?

5 Decide whether it is easier to use a mental or a written method to find the answers to these calculations. Work them out, using your chosen methods.

a 5 billion + 2.6 billion

b £137.56 + £19.99

c £45.23 + £74.68

d 3.61 + 2.05

e 13 568 + 20 000

f 657 + 2400

What do you think?

1 What mistakes have been made in these calculations?

a
$$\begin{array}{r} 5\ 6\ 2 \\ +\ \ 4\ 7\ 5 \\ \hline 9\ 1\ 3\ 7 \\ \hline \end{array}$$

b
$$\begin{array}{r} 8\ 7\ 2 \\ +\ \ 6\ 2\ 8 \\ \hline 1\ 4\ 0\ 0 \\ \hline {}_{1\ \ 1}\ \ \ \end{array}$$

c
$$\begin{array}{r} 1\ 8.7 \\ +\ \ \ 1\ 6 \\ \hline 2\ 0.3 \\ \hline {}_{1\ \ 1}\ \ \end{array}$$

d
$$\begin{array}{r} 4.7\ 2 \\ +\ 1\ 5.8 \\ \hline 6.3.0 \\ \hline {}_{1\ \ 1}\ \ \end{array}$$

e
$$\begin{array}{r} 3\ 1\ 6 \\ +\ \ 6\ 9\ 2 \\ \hline 9\ 1\ 0\ 8 \\ \hline \end{array}$$

f
$$\begin{array}{r} 3\ 2.6 \\ +\ 4\ 8\ \ \ \ \\ \hline 7\ 0.6 \\ \hline {}_{1}\ \ \ \ \ \ \end{array}$$

2 Find the missing numbers in these additions. There may be more than one possible answer.

a
$$\begin{array}{r} 2\ 5\ 3 \\ +\ \square\square\ 1 \\ \hline 6\ 0\ 4 \\ \hline \end{array}$$

b
$$\begin{array}{r} 5\ \square\ 7 \\ \square\ 4\ 8 \\ +\ \ 4\ 9\ \square \\ \hline 2\ \square\ 1\ 1 \\ \hline \end{array}$$

c
$$\begin{array}{r} 4\ \square\ 6 \\ \square\ 1\ 6 \\ +\ \ 9\ 4\ \square \\ \hline 2\ \square\ 7\ 1 \\ \hline \end{array}$$

3 Use the digits 1 to 9 once only to fill the spaces.

$\square\square\square + \square\square\square + \square\square\square$

How close can you get to a total of 1000?

Consolidate – do you need more?

1 **a** Find the sum of each pair of numbers.

- **i** 306 and 423
- **ii** 356 and 433
- **iii** 356 and 72
- **iv** 94 and 306
- **v** 6.7 and 4.5
- **vi** 16.7 and 34.5
- **vii** 16 and 34.5
- **viii** 1.64 and 0.75

b Work out

- **i** 73 + 48
- **ii** 73 + 108
- **iii** 657 + 326
- **iv** 572 + 609
- **v** 6.2 + 2.6
- **vi** 6.2 + 5.6
- **vii** 16.2 + 5.6
- **viii** 20.7 + 12.6

2 Find the missing numbers in these diagrams.

a

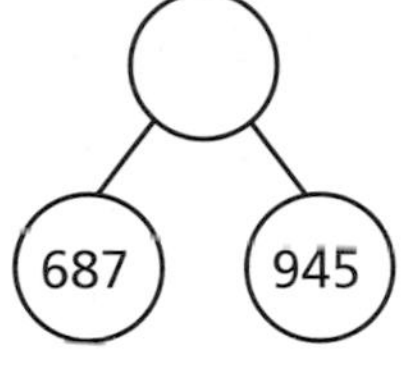

b

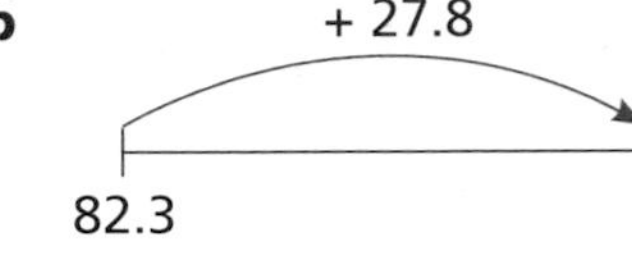

c

12.3	7.86	9.4

3 Find the sum of each pair of numbers.

- **a** 486 and 404
- **b** 486 and 40.4
- **c** 486 and 4.04
- **d** 48.6 and 40.4
- **e** 48.6 and 4.04
- **f** 4.86 and 0.404
- **g** 0.0486 and 0.404

4 **a** Add thirty-five to 6887

b Add three hundred and five to 6887

c Add three thousand and five to 6887

d Add thirty-five thousand to 6887

5 **a** Find the perimeter of an equilateral triangle with sides of 43.8 cm.

b The width of a rectangle is 6.7 cm. Its length is 4.6 cm longer than its width. Find the perimeter of the rectangle.

6 Find the missing balances in this bank statement.

Date	Description	Credit (£)	Debit (£)	Balance (£)
May 23	Opening balance			567.32
May 25	BACS payment	89.99		A
May 26	Wages	1500.00		B
May 28	Bank transfer	137.89		C
May 31	Refund	281.44		D

Are there any values you can work out mentally?

Stretch – can you deepen your learning?

1 Here are the first two terms of some linear sequences. How often will these sequences produce terms that are integers?

a 7.4, 8.1, … **b** 6.3, 6.6, … **c** 5.75, 5.9, …

Investigate for some sequences with other differences. Can you find a way of producing sequences with integers every other term, every third term, every fourth term and so on?

2 Using the digits 1 to 9 exactly once, how many three-digit sums can you produce?

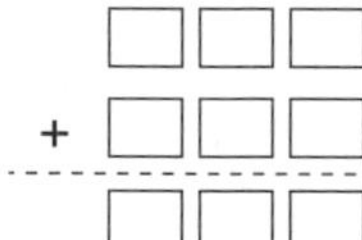

Reflect

1 How do you decide whether to use a mental method, a written method or a calculator to work out an addition?

2 What's the same and what's different about adding integers and decimals?

6.3 The column method of subtraction

Small steps

- Use formal methods for subtraction of integers
- Use formal methods for subtraction of decimals
- Solve problems in the context of perimeter
- Solve financial maths problems
- Choose the most appropriate method: mental strategies, formal written or calculator

Key words

Difference – the result of subtracting a smaller number from a larger number

Debit – an amount of money taken out of an account

Range – the difference between the greatest value and the smallest value in a set of data

Are you ready?

1 Use the fact that 8 – 5 = 3 to work out

a 80 – 30 **b** 0.8 – 0.3

c $0.08 - \frac{3}{100}$ **d** thirty thousand less than 80 000

2 Work out these subtractions. Compare the answers.

a 7 – 1 **b** 7 – 0.1 **c** 7 – 0.01 **d** 7 – 0.001 **e** 7 – 0.0001

3 Solve these equations.

a $a + 8 = 20$ **b** $0.6 + b = 0.9$ **c** $600 = c + 200$

4 Use a mental method or jottings to work out

a 600 – 100 **b** 600 – 10 **c** 600 – 1 **d** 600 – 0.1

e 600 – 200 **f** 600 – 250 **g** 600 – 251 **h** 600 – 251.1

Models and representations

Base 10 blocks

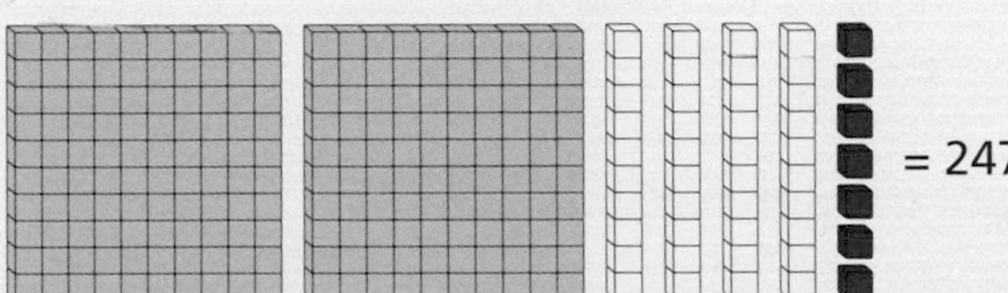

Place value counters

Place value charts

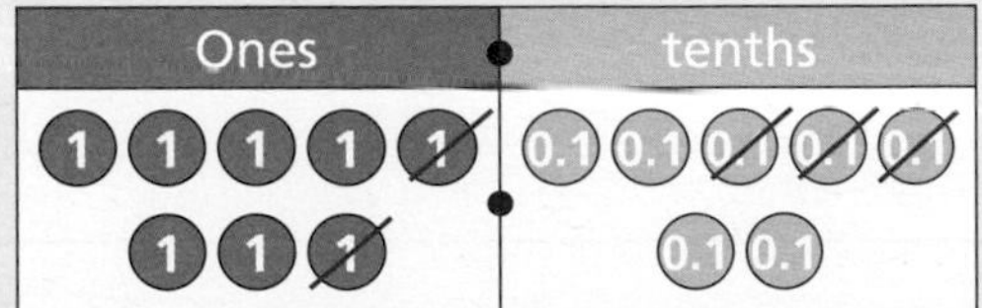

Column method

	O • t
	8 • 7
–	2 • 3
	6 • 4

6.3 The column method of subtraction

Example 1

Work out these subtractions. Compare the calculations.

a 67 – 32 **b** 67 – 38 **c** 6.7 – 3.2 **d** 0.67 – 0.38

a 67 – 32

	T	O
	6	7
–	3	2
	3	5

So 67 – 32 = 35

This calculation does not involve exchanging or regrouping.

There are enough tens and enough ones to subtract the numbers from each column

b 67 – 38

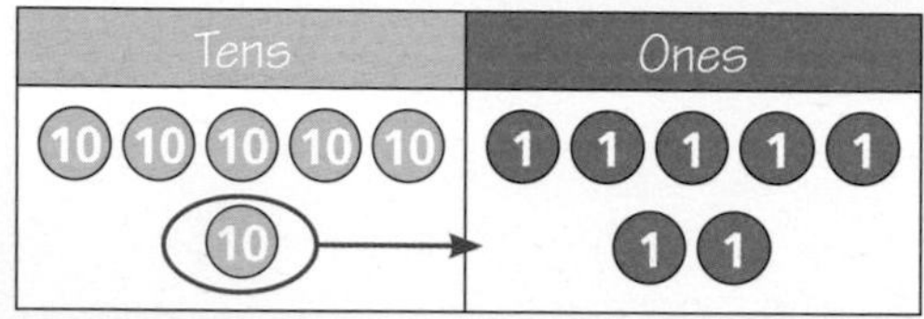

This time you want to subtract 8 ones but there are only 7 ones in the ones column.

You need to exchange one of the tens for 10 ones.

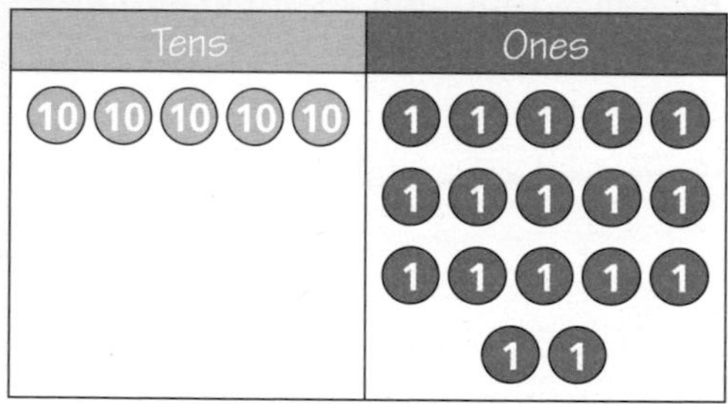

This uses the fact that you can partition 67 as 50 + 17 instead of 60 + 7

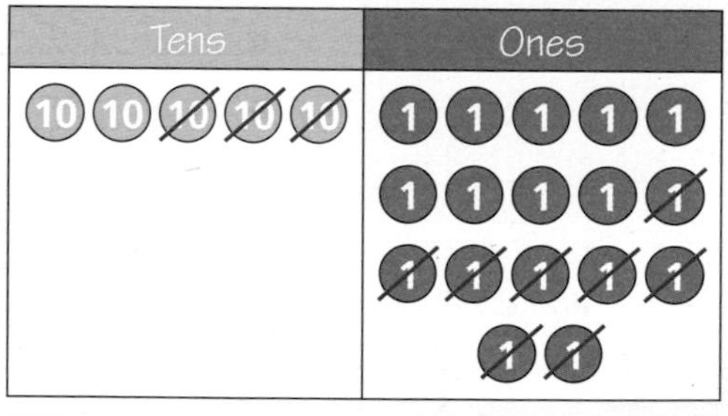

You can now subtract 38 by removing 3 tens and 8 ones.

So 67 – 38 = 29

You can also work this out using the column method.

	T	O
	~~6~~ 5	17
–	3	8
	2	9

67 is partitioned into 50 and 17

c $6.7 - 3.2$

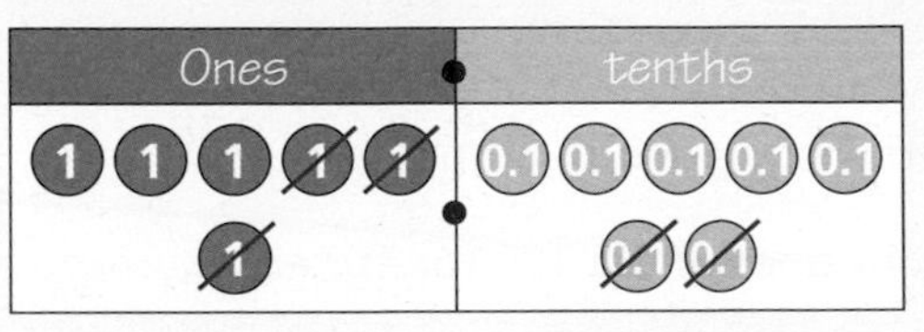

This decimal subtraction is similar to **a** – there is no need to exchange.

You subtract decimals in the same way as you subtract integers.

So $6.7 - 3.2 = 3.5$

You can also work this out using the column method.

	O • t
	6 • 7
–	3 • 2
	3 • 5

d $0.67 - 0.38$

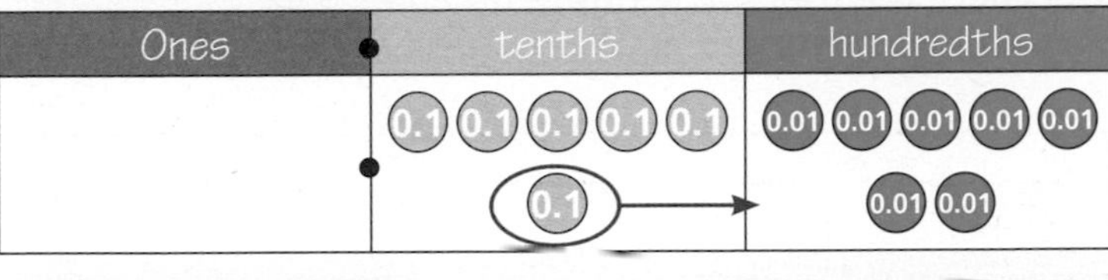

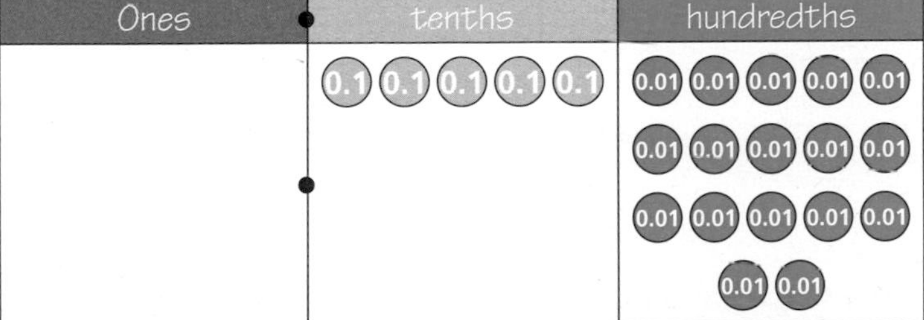

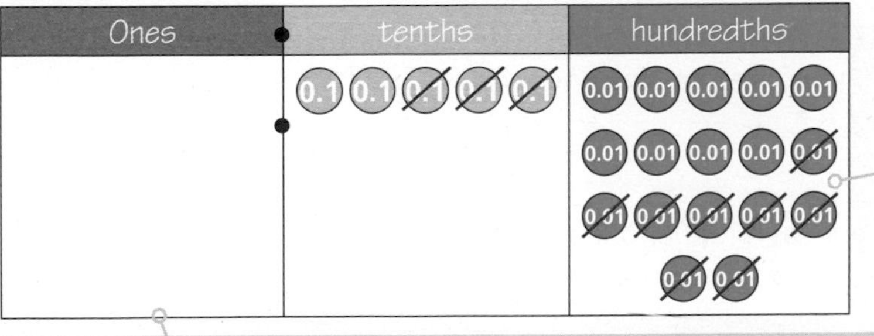

The relationship between adjacent columns is the same as for integers, so you can use the same the process.

This decimal subtraction is similar to **b**. You need to make an exchange between the tenths and the hundredths columns.

So $0.67 - 0.38 = 0.29$

You can also work this out using the column method.

	O • t	h
	0 • ${}^{5}\not{6}$	${}^{1}7$
–	0 • 3	8
	0 • 2	9

6 tenths and 7 hundredths is partitioned into 5 tenths and 17 hundredths.

Example 2

Find the **difference** between 812 and 357.

Remember that "difference" means the result of subtracting the two numbers.

	H	T	O
	7 8̸	10 1̸	12
–	3	5	7
	4	5	5

You need to exchange a ten for 10 ones and a 100 for 10 tens.

	H	T	O
	4	5	5
+	3	5	7
	8	1	2
	1	1	

You can check the answer by adding 455 to 357

As 455 + 357 = 812, it means that 812 – 357 = 455, so the answer is correct.

You can always check addition by subtraction and subtraction by addition as these are inverse operations.

Example 3

Work out

a 47 – 23 **b** 47 – 2.3 **c** 47 – 0.23

Use the column method for each calculation.

a

$$\begin{array}{r} 47 \\ -\ 23 \\ \hline 24 \end{array}$$

You do not need to exchange as 7 > 3

b

$$\begin{array}{r} 4\,{}^{6}\not{7}.{}^{1}0 \\ -\ \ \ 2.3 \\ \hline 44.7 \end{array}$$

Write 47 as 47.0 and line up the decimal points.
As 0 < 3, you need to exchange 1 one for 10 tenths.

c

$$\begin{array}{r} 4\,{}^{6}\not{7}.{}^{9}\not{0}\,{}^{1}0 \\ -\ \ \ 0.23 \\ \hline 46.77 \end{array}$$

Write 47 as 47.00 and line up the decimal points.
Then make two exchanges.

Practice 6.3A

1 Use the column method to find the difference between each pair of numbers.

a 63 and 21 **b** 63 and 27

c 603 and 201 **d** 603 and 210

e 603 and 207 **f** 630 and 7

g 630 and 27 **h** 630 and 57

i 630 and 217 **j** 630 and 347

Which questions were most difficult? Why? Would it have been easier to use a mental method for any of the questions?

2 Work out

a 8.7 – 3.4 **b** 8.7 – 3.8 **c** 8.17 – 3.4 **d** 8.07 – 3.4 **e** 8.7 – 3.04

3 Find the missing numbers in these diagrams.

a

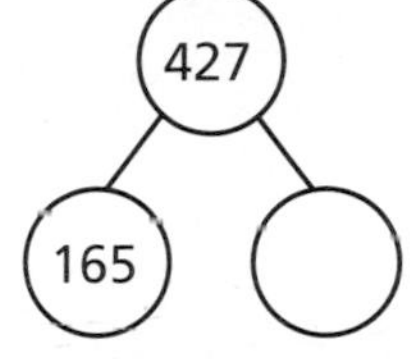

b

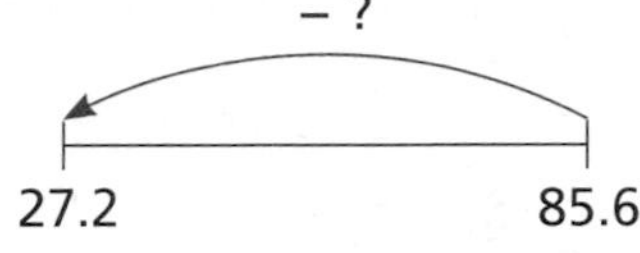

c

76.5	
28.7	

d

472	386
	584

4 Work out

a $6.72 - \frac{8}{10}$ **b** $6.72 - \frac{8}{100}$ **c** $6.72 - \frac{3}{5}$ **d** $6.72 - \frac{3}{4}$

5 Given that $w = 2.8$, $x = 0.28$, $y = 2.08$ and $z = 8.2$, find the value of

a $z - w$ **b** $w - x$ **c** $w - y$

d $y - x$ **e** $z - x$ **f** $z - y$

6 **a** Subtract seven thousand five hundred from 9236.

b Subtract seven thousand and fifty from 9236.

c Subtract seven thousand and five from 9236.

What do you think?

1 

You cannot find the difference between 672 and 921 because 921 is greater than 672

Explain why Jackson is wrong.

2 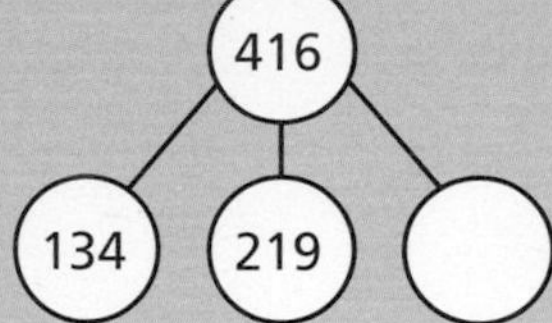

Compare these methods for finding the missing number.

- Subtract 134 from 416 and then subtract 219 from the result.
- Add 134 and 219 and then subtract the result from 416.

Which method do you prefer? Give a reason why.

3 Solve these equations.

a $a + 307 = 410$ **b** $b + 370 = 410$ **c** $c + 37 = 410$ **d** $d + 3.7 = 410$

4 Find the value of each letter.

a $w + 35.2 = 71.7 - 18.3$ **b** $x - 294 = 2000 - 378$

c $582 - y = 217 + 159$ **d** $13.8 - z + 6.7 = 17.4 - 2.53$

Many real-life problems involve subtractions, or additions and subtractions.

Example 4

Complete this bank statement by filling in the balances.

Date	Description	Credit (£)	Debit (£)	Balance (£)
Mar 1	Opening balance			846.72
Mar 2	Rent		520.00	A
Mar 6	Car service		88.53	B
Mar 7	Electricity bill		74.87	C

Remember: **debits** are amounts that are taken out of an account.

	H	T	O	• t	h
	8	4	6	• 7	2
−	5	2	0	• 0	0
	3	2	6	• 7	2

To find the balance A after the rent is taken out of the account, subtract £520.00 from £846.72

You can use the column method.

As no exchange is needed, you could also do this mentally.

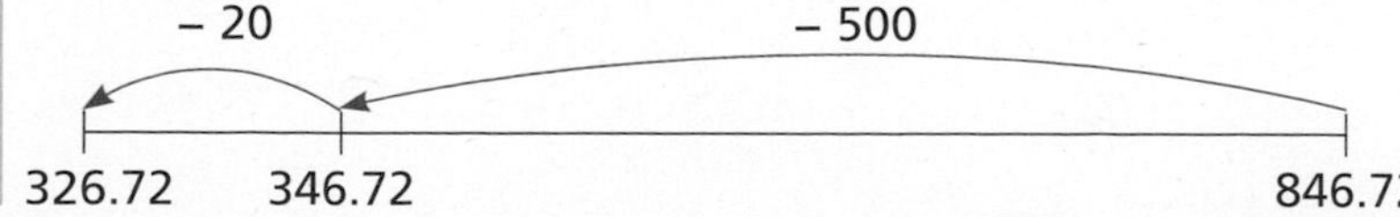

	H	T	O •	t	h
	2 ~~3~~	1 ~~2~~ 1	16 •	6 ~~7~~	12
−		8	8 •	5	3
	2	3	8 •	1	9

To find the balance at B, subtract £88.53 from £326.72

	H	T	O •	t	h
	1 ~~2~~	13	7 ~~8~~ •	11	9
−		7	4 •	8	7
	1	6	3 •	3	2

To find the balance at C, subtract £74.87 from £238.19

Example 5

Find the length of the unknown side of this triangle.

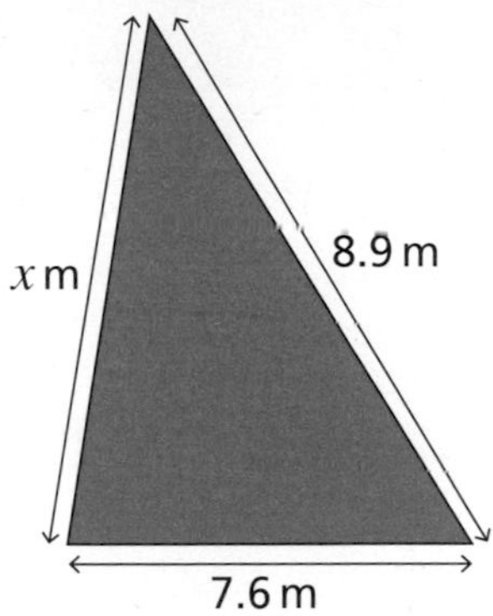

Perimeter = 20 m

$7.6 + 8.9 = 7.6 + 9 - 0.1$
$= 16.6 - 0.1$
$= 16.5$

First, find the total of the two sides you know.

You could use the column method, but because the number you are adding is nearly a whole number, compensation is quicker.

$20 - 16.5 = 3.5$

Then subtract the total of the first two sides from the perimeter.

You could use the column method or counting on.

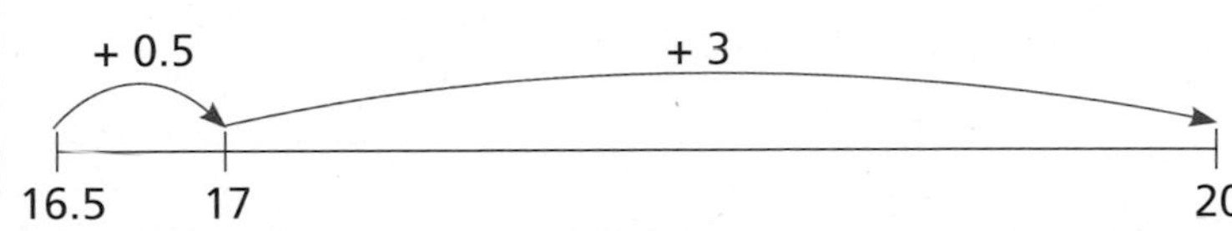

The missing side is 3.5 m

Give the unit with the final answer.

Practice 6.3B

1. Ali buys old furniture and then resells it. The table shows how much he paid for some items and how much he sold them for. Find the profit or loss he makes on each item.

	Item	Bought for	Sold for
a	Table	£35	£80
b	Chest of drawers	£20	£60
c	Sideboard	£110	£47
d	Chair	£12	£40

2. All of these shapes have perimeter 50 cm. Find the lengths of the missing sides. All lengths are given in centimetres.

a

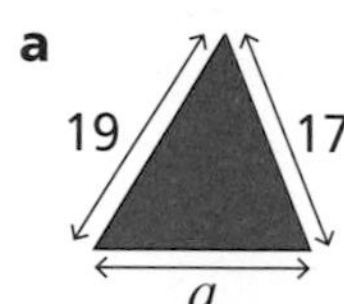

b

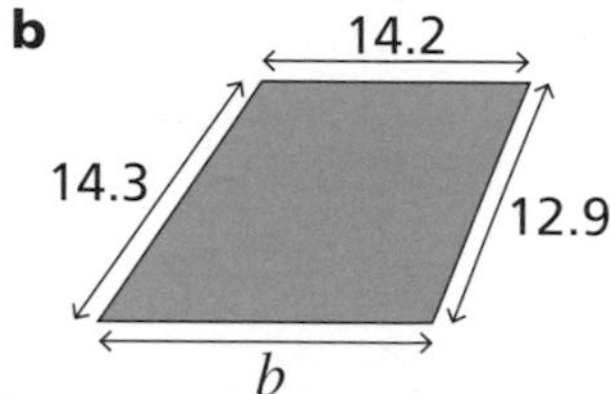

c

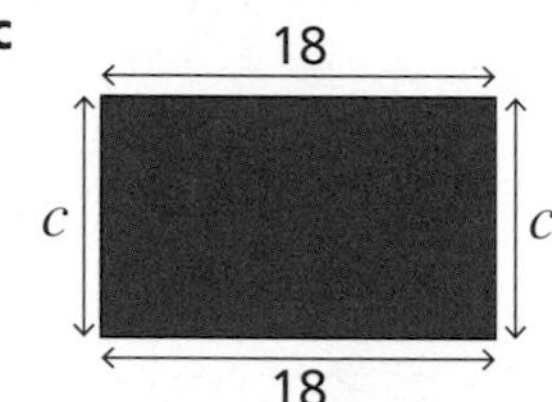

d

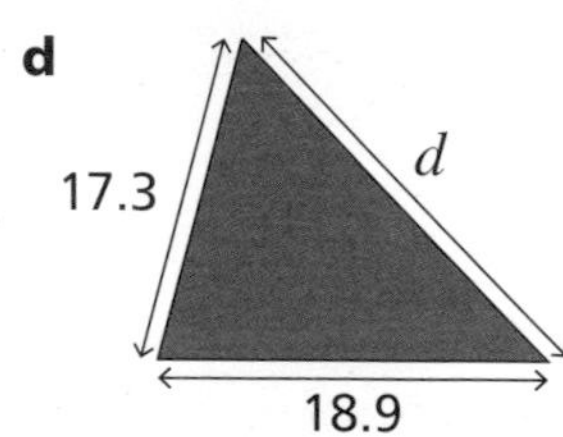

e

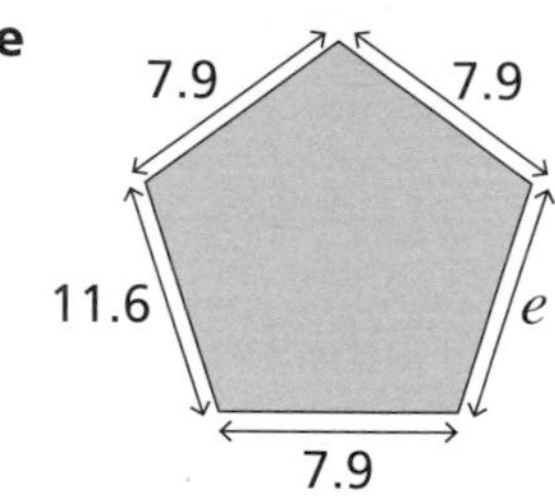

3. Copy and complete this bank statement by filling values A to E.

Date	Description	Credit (£)	Debit (£)	Balance (£)
May 1	Opening balance			546.23
May 2	Wages	600.00		A
May 6	Phone bill		18.65	B
May 7	Shopping		C	1038.86
May 11	TV subscription		39.99	D
May 13	Refund	E		1043.87

Think carefully about whether you need to add or subtract, and whether you can work the amounts out mentally.

4. Find the range of each set of values.

You studied the range in chapter 4.3.

a 12 15 15 18 23 27 31 40

b 127 cm 141 cm 183 cm 152 cm 160 cm

c 8.13 7.2 6.9 9.8 12.4 18.7 21.3

d 3.07 5 2.98 1.64 3 1.92

5 **a** Beca is 146.3 cm tall. Her sister is 12.5 cm shorter. How tall is Beca's sister?

b Samira has £1200 to spend on a holiday. The flights cost £385.28 and accommodation costs £468. How much money does Samira have left?

c A book has 314 pages. Jackson reads 168 pages of the book. How many more pages does he have to read?

d June 20th is the 171st day of 2022. How many days are there after 20th June in 2022?

6 Decide whether to use a mental or written method for these calculations. Then work out each answer using your chosen method.

a 5 billion – 2.6 billion

b £137.56 – £19.99

c £74.68 – £45.23

d 3.61 – 2.05

e 20 000 – 13 568

f 2400 – 657

What do you think?

1 Explain the mistakes that have been made in each of these calculations.

a

$$\begin{array}{r} 4\cdot\overset{8}{\cancel{9}}\,{}^{1}2 \\ -\quad 3\cdot5 \\ \hline 4\cdot5\,7 \end{array}$$

b

$$\begin{array}{r} 9\cdot4 \\ -\;2\cdot6\,3 \\ \hline 7\cdot2\,3 \end{array}$$

c

$$\begin{array}{r} 6 \\ -\;3\cdot0\,7 \\ \hline 3\cdot0\,7 \end{array}$$

2 Find the missing numbers in these subtractions.

a

$$\begin{array}{r} 4\,\square\,5 \\ -\;\square\,2\,\square \\ \hline 2\,8\,6 \end{array}$$

b

$$\begin{array}{r} \square\,3\,\square \\ -\;3\,\square\,6 \\ \hline 2\,3\,5 \end{array}$$

c

$$\begin{array}{r} \square\,\square\,3 \\ -\;8\,2\,\square \\ \hline 8\,7 \end{array}$$

3 $a = 82$, $b = 39$ and $c = 14.5$

Work out the value of

a $a - b - c$ **b** $a - c - b$ **c** $a + b - c$ **d** $a - b + c$

e $a - c + b$ **f** $a + c - b$ **g** $a - (b + c)$

Which calculations give the same answers? Why is this?

Consolidate – do you need more?

1 **a** Find the difference between each pair of numbers.

i

Hundreds	Tens	Ones
100 100 100 100	10 10 10 10 10 10 10 10	1 1 1 1 1 1
100 100 100		1 1 1 1

ii

Hundreds	Tens	Ones
100 100 100 100	10 10 10 10 10 10 10 10	1 1 1 1 1 1
100 100	10	1 1 1 1 1 1 1 1

iii 486 and 391

iv 486 and 52

v 486 and 74

vi 486 and 658

b Work out

i 45 – 23

ii 45 – 28

iii 435 – 218

iv 435 – 83

v 485 – 263

vi 405 – 263

2 Find the missing number in each of these diagrams.

a

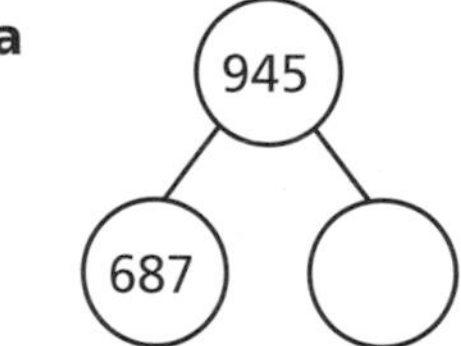

b

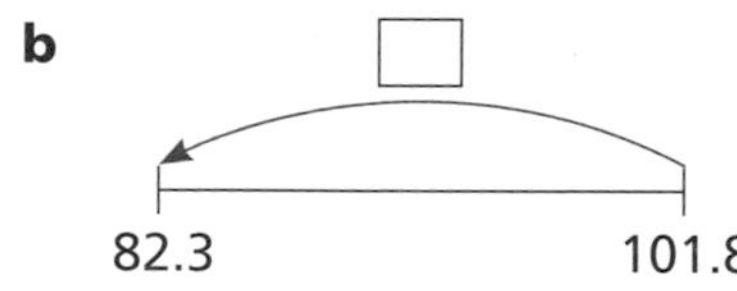

c

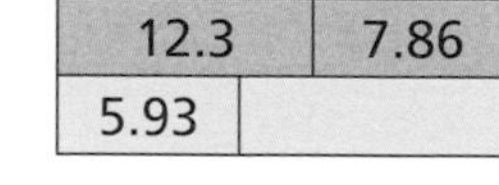

3 **a** Subtract eighty-five from 6837

b Subtract eight hundred and five from 6837

c Subtract five thousand and eight from 6837

d Subtract 6837 from eight thousand and five.

e Subtract 6837 from eight thousand and fifty.

f Subtract 6837 from eight thousand five hundred.

4 **a** An isosceles triangle has two equal sides that are 47 mm long. The perimeter of the triangle is 125 mm. Find the length of the third side.

b The length of a rectangle is 6.7 cm. Its perimeter is 22.6 cm. Find the width of the rectangle.

5 Find the missing values A to F in this bank statement.

Date	Description	Credit (£)	Debit (£)	Balance (£)
Aug 18	Opening balance			136.75
Aug 23	BACS transfer	A		386.75
Aug 25	Bookshop		9.99	B
Aug 27	Withdrawal		75.00	C
Aug 29	Interest	1.47		D
Aug 30	Gas bill		56.88	E
Aug 31	Wages	F		1011.80

Are there any values you can work out mentally?

Stretch – can you deepen your learning?

1 **a** Huda says, "10 – 7.43 is the same as 9.99 – 7.42". Explain why Huda is right.

Use Huda's method to find

b 10 – 7.43

c 20 – 15.72

d the change from a £50 note when you spend £34.83

2 In how many ways can you use the digits 0 to 9, once only, to produce a correct subtraction with this format?

```
   □□□□
 –  □□□
 ───────
    □□□
```

(You cannot use 0 as the first digit of a number.)

Reflect

1 How can you tell whether a problem involves addition or subtraction?

2 How do you decide whether to use a mental method, a written method or a calculator to work out the answer to a subtraction?

3 What's the same and what's different about subtracting integers and decimals?

6.4 Addition and subtraction problems

Small steps

- Solve problems involving tables and timetables
- Solve problems with frequency trees
- Solve problems with bar charts and line charts
- Choose the most appropriate method: mental strategy, column method or calculator

Key words

Frequency – the number of times something happens

Frequency tree – a diagram showing a number of people/objects grouped into categories

Two-way table – this displays two sets of data in rows and columns

Timetable – a table showing times

Bar chart – this uses horizontal or vertical rectangles to show frequencies

Line graph – this has connected points and shows how a value changes over time

Are you ready?

1 Find these totals.

a 47 + 78 **b** 3.7 + 4.4 **c** 164 + 359 **d** 35.4 + 28.7

2 Work out

a 91 – 37 **b** 7.6 – 4.8 **c** 412 – 157 **d** 23.6 – 5.8

3 Use a mental method to work out

a 183 + 99 **b** 183 – 99 **c** £45 + £9.99 **d** £45 – £9.99

4

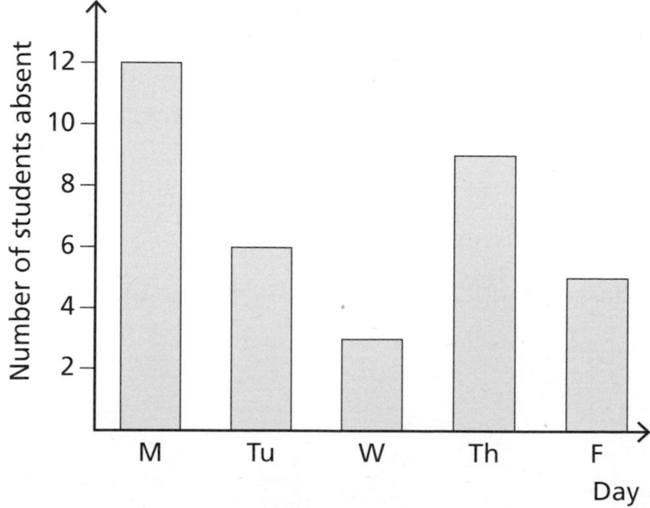

How many students were absent on

a Monday? **b** Wednesday?

Models and representations

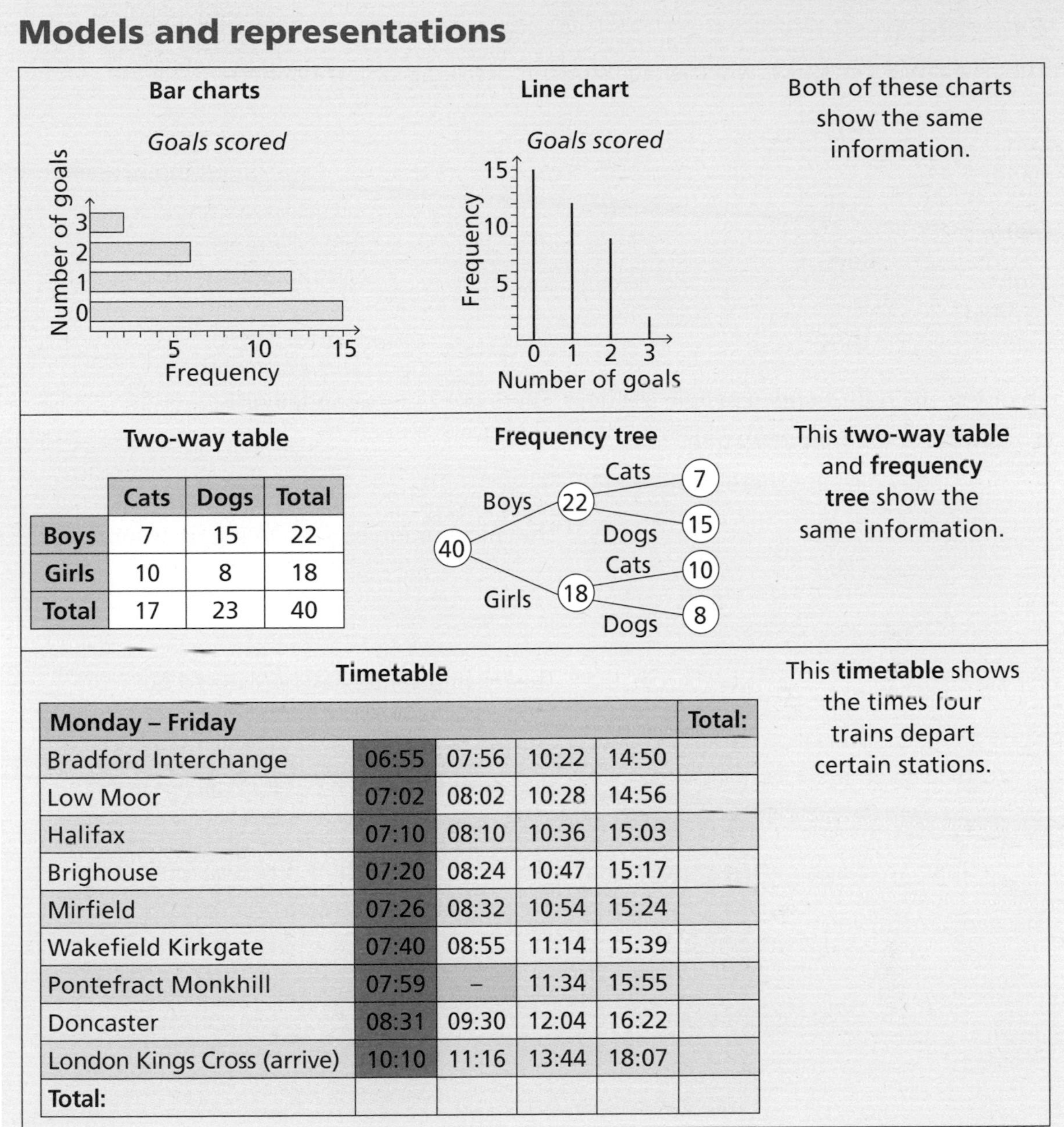

Two-way table

	Cats	Dogs	Total
Boys	7	15	22
Girls	10	8	18
Total	17	23	40

Timetable

Monday – Friday					**Total:**
Bradford Interchange	06:55	07:56	10:22	14:50	
Low Moor	07:02	08:02	10:28	14:56	
Halifax	07:10	08:10	10:36	15:03	
Brighouse	07:20	08:24	10:47	15:17	
Mirfield	07:26	08:32	10:54	15:24	
Wakefield Kirkgate	07:40	08:55	11:14	15:39	
Pontefract Monkhill	07:59	–	11:34	15:55	
Doncaster	08:31	09:30	12:04	16:22	
London Kings Cross (arrive)	10:10	11:16	13:44	18:07	
Total:					

In this section, you will read information from tables, charts and graphs. You will use the information to solve problems.

6.4 Addition and subtraction problems

Example 1

The frequency tree shows whether 200 students are right- or left-handed.

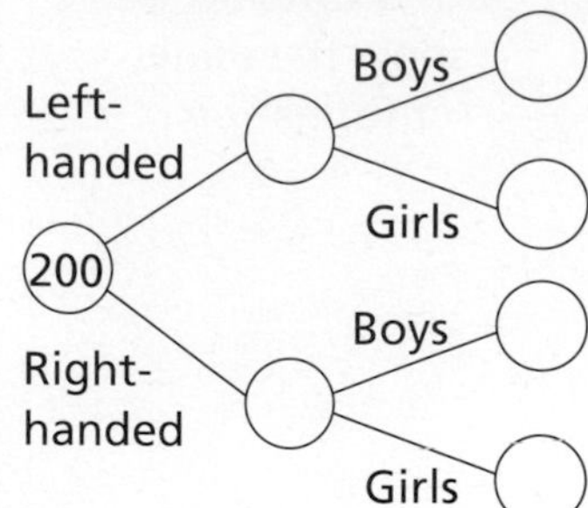

78 of the 164 right-handed people are boys. There are 11 left-handed girls.

a Complete the frequency tree.

b How many girls are there altogether?

a

164 — Boys — 78

This information is given in the question.

Number left-handed = 200 − 164

= 36

Left-handed boys = 36 − 11 = 25

Right-handed girls = 164 − 78 = 86

The missing values can be worked out from the facts you already know.

200 — Left-handed — 36 — Boys — 25; Girls — 11

200 — Right-handed — 164 — Boys — 78; Girls — 86

You can now complete the frequency tree.

b 86 + 11 = 97

The total number of girls is found by the adding the number of left-handed girls to the number of right-handed girls.

Example 2

A group of people were asked to name their favourite colour. The **bar chart** shows the results.

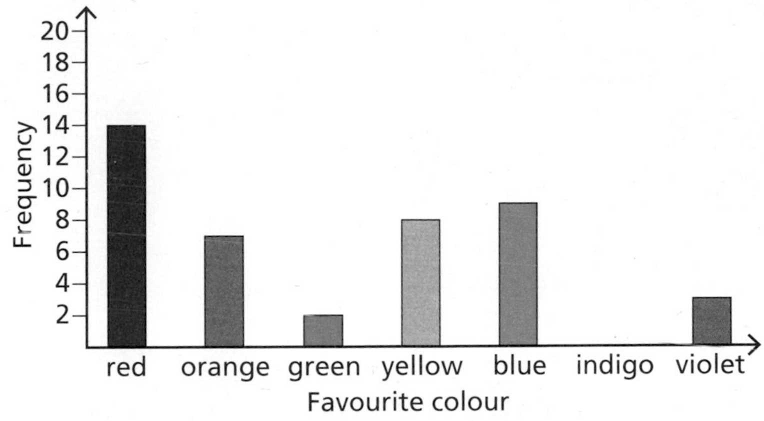

a How many more people chose red than yellow?

b How many people were asked altogether?

a $14 - 8 = 6$ — The bar chart shows that 14 people chose red and 8 chose yellow. To find how many more chose red than yellow, you subtract to find the difference.

b $14 + 7 + 2 + 8 + 9 + 0 + 3 = 43$ — Read off the heights of the bars to see how many people chose each colour. Then add to find the total.

Practice 6.4A

1 The line chart shows the colours of the cars in the school car park.

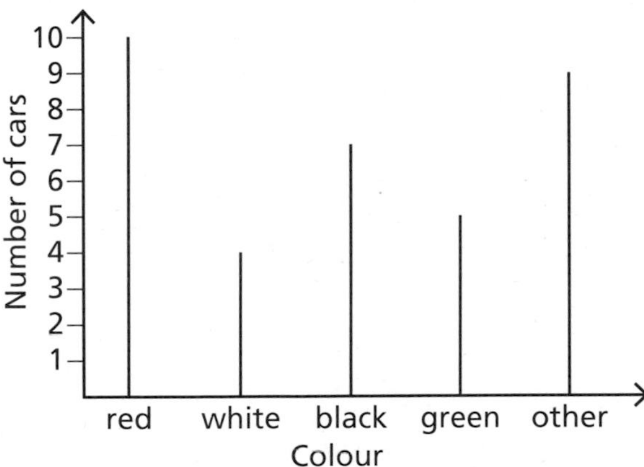

a How many cars are black?

b How many cars are there altogether in the car park?

c How many more red cars are there than white cars?

2 The bar chart shows Year 7 choices of sport for the Summer term.

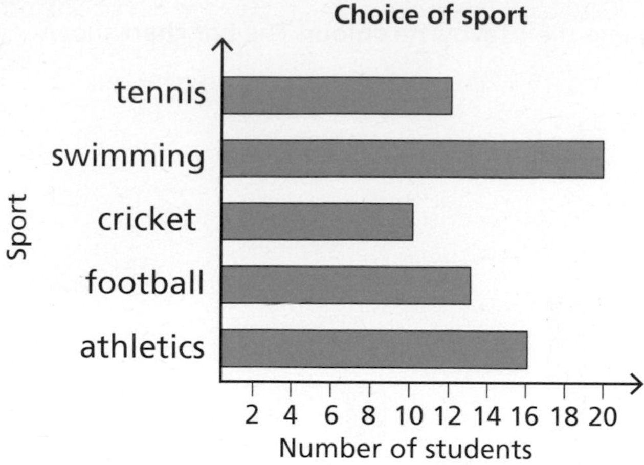

- **a** How many Year 7 students chose athletics?
- **b** How many more Year 7 students chose swimming than chose football?
- **c** How many students are there in Year 7?

3 The line graph shows average temperatures in Aberdeen over a year.

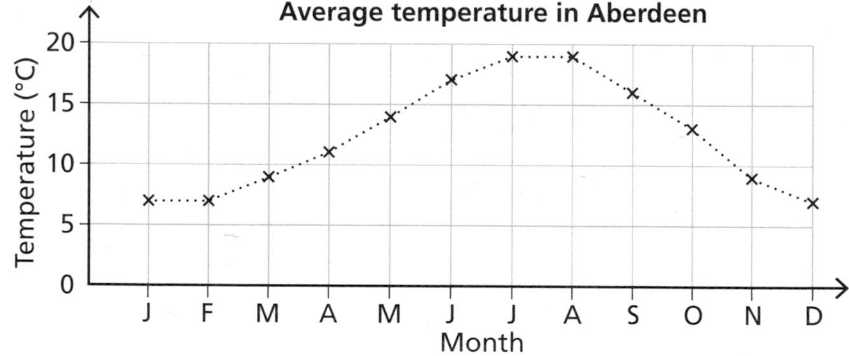

- **a** In how many months was the temperature lower than 15°C?
- **b** Find the difference between the highest and lowest temperatures.

4 In a class of 30 pupils there are 16 boys. Copy and complete the frequency tree.

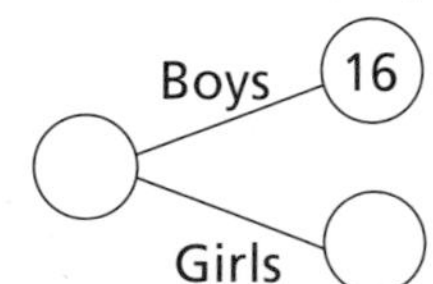

5 150 students go on a trip. 47 travel in coach A, 49 travel in coach B and the rest travel in coach C. Draw a frequency tree to show this information.

6 The frequency tree shows how many students in Year 7 and Year 8 study French and Spanish. Every student in Year 7 and Year 8 studies either French or Spanish (but not both).

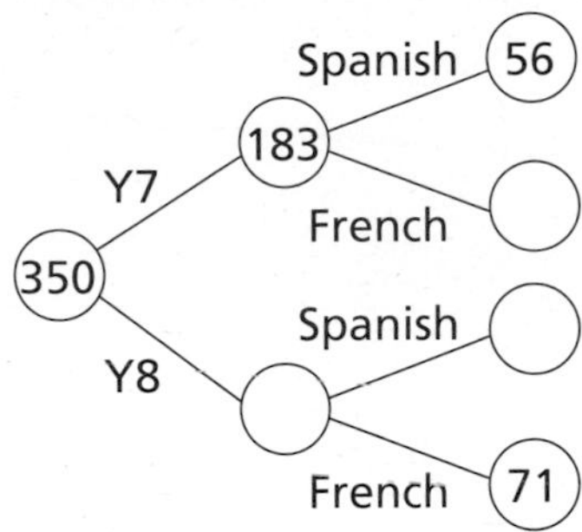

a Copy and complete the frequency tree.

b Find the difference between the number of students in Year 7 and the number of students in Year 8.

c Find the total number of students in Year 7 and Year 8 who study Spanish.

7 The two-way table shows information about some shapes in a bag.

	Green	**Yellow**	**Total**
Triangles	46		
Circles		53	110
Total			200

Draw a completed frequency tree to show this information.

8 A plane holds 200 passengers.

35 travel first class and the rest travel standard class.

68 of the standard-class passengers and 12 of the first-class passengers have hand luggage only.

The rest of the passengers have luggage in the hold.

Show this information on a frequency tree.

What do you think?

1 50 people took a maths test or a spelling test. Some of the results are shown on the frequency tree.

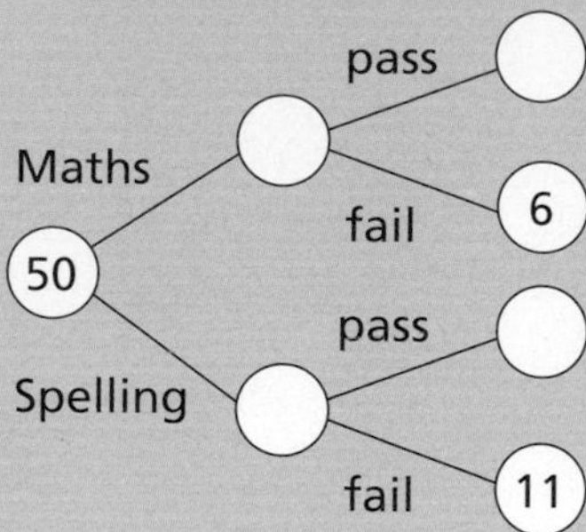

Half of the people who took the spelling test passed. How many people passed the maths test?

2 80 adults took part in a survey on reading.

23 men said they liked reading.

8 out of the 43 women who took part said they did not like reading.

Complete a frequency tree to show this information.

3 The line graph shows the temperature, taken at hourly intervals, in a school one morning.

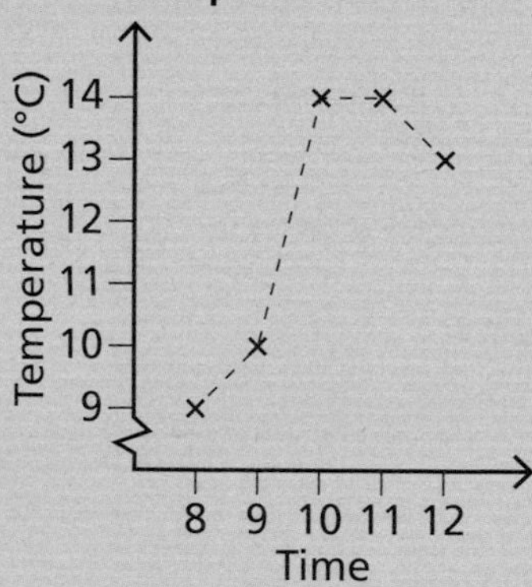

Beth says that the temperature at 9:30 a.m. was 12°C

a Explain why Beth may be wrong.

b Suggest a reason why the temperatures are joined with dotted lines and not solid lines.

Any two-way table can be redrawn as a frequency tree

Do you agree with Seb? Explain your answer.

Example 3

This is part of a bus timetable.

Marketville	08:05	10:23	11:06	14:30	17:45
Old Town	08:26	10:42	11:23	14:47	18:06
Naze	08:44	–	–	–	18:22
New Town	08:56	11:03	11:40	15:08	18:34
Upper Brough	09:22	11:20	11:58	15:25	18:56
Lower Brough	09:30	11:26	12:05	15:32	19:02

Remember: there are 60 minutes in an hour so you cannot use a calculator in decimal mode to work out additions and subtractions involving time.

a How many buses do not stop at Naze?

b How long does the 08:05 take to get from Marketville to Lower Brough?

c How long does the 14:47 bus from Old Town take to get to New Town?

a 3 — Places where a bus or train do not stop are often labelled with a dash on timetables.

b 1 hour and 25 minutes — The bus leaves Marketville at 08:05 and arrives at Brough at 09:30.

You need to find the difference between these times.

You can work this out by counting on, either in your head or using a number line.

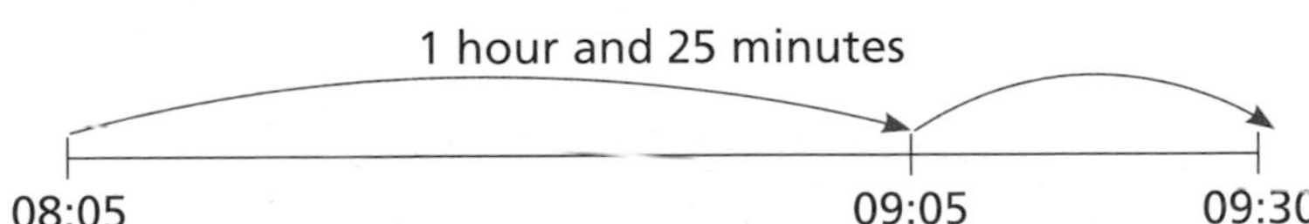

c 21 minutes — The bus leaves Old Town at 14:47 and arrives at New Town at 15:08.

You need to find the difference between these times.

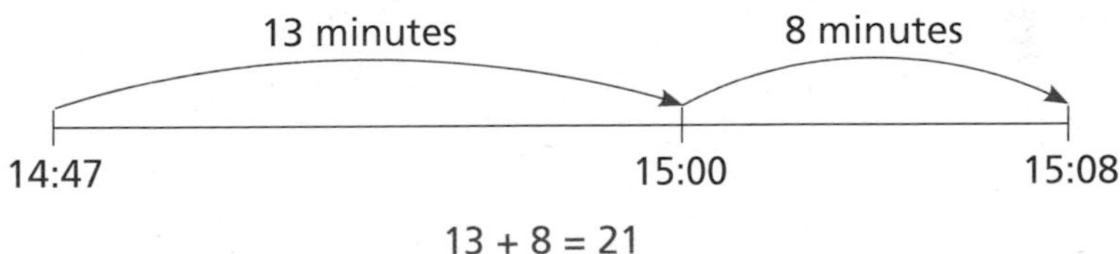

13 + 8 = 21

Example 4

This table shows the distances, in miles, between some UK towns and cities.

Aberdeen					
430	Birmingham				
468	101	Cambridge			
149	372	349	Glasgow		
354	88	153	214	Manchester	
690	278	368	563	358	Penzance

a How far is it from Birmingham to Manchester?

b How much further is it from Aberdeen to Penzance via Birmingham than the direct route?

a *88 miles*

Read off the number where the column for Birmingham meets the row for Manchester.

Aberdeen					
430	Birmingham				
468	101	Cambridge			
149	372	349	Glasgow		
354	88	153	214	Manchester	
690	278	368	563	358	Penzance

b Total distance = 430 + 278
= 708 miles

The direct route is 690 miles

Difference = 708 − 690
= 18 miles

Aberdeen to Birmingham is 430 miles

Aberdeen					
430	Birmingham				
468	101	Cambridge			
149	372	349	Glasgow		
354	88	153	214	Manchester	
690	278	368	563	358	Penzance

and Birmingham to Penzance is 278

Aberdeen					
430	Birmingham				
468	101	Cambridge			
149	372	349	Glasgow		
354	88	153	214	Manchester	
690	278	368	563	358	Penzance

The direct distance from Aberdeen to Penzance is 690 miles.

Aberdeen					
430	Birmingham				
468	101	Cambridge			
149	372	349	Glasgow		
354	88	153	214	Manchester	
690	278	368	563	358	Penzance

Practice 6.4B

1 This table shows the distances, in miles, between some UK towns and cities.

London					
119	Bristol				
401	373	Edinburgh			
260	120	398	Fishguard		
56	74	362	214	Oxford	
414	386	133	411	371	Stranraer

a Which two places are closest to each other?

b How far is Edinburgh from Oxford?

c How much further away from London is Stranraer than Fishguard?

d Rob drives from London to Fishguard and then from Fishguard to Stranraer. How far does he drive altogether?

2 Find the length of time between

a 9 a.m. and 11 a.m.

b 9 a.m. and 11:05 a.m.

c 8:55 a.m. and 11:05 a.m.

d 8:35 a.m. and 11:25 a.m.

e 8:50 a.m. and 12:10 p.m.

f 8:55 a.m. and 2:05 p.m.

g 8:25 a.m. and 2:17 p.m.

3 Find the time that is:

a 20 minutes after 13:15

b 40 minutes after 13:15

c 50 minutes after 13:15

d 25 minutes after 4:40 p.m.

e 35 minutes after 4:40 p.m.

f 37 minutes after 17:50.

4 A train arrives in Leeds at 12:05. The train departed from Glasgow 4 hours and 19 minutes earlier. At what time did the train leave Glasgow?

5 Here is part of a bus timetable.

Bishop's Stortford Interchange	06:00	06:39	08:15	10:30	12:45	15:00	17:15	19:30
Hockerill	06:10	06:53	08:29	10:40	12:55	15:14	17:29	19:40
Birchanger	06:16	06:59	08:35	10:46	13:01	15:20	17:35	19:46
Stansted Mountfitchet	06:28	07:11	08:47	10:58	13:13	15:32	17:47	19:58
Elsenham Station Road	06:33	07:16	08:52	11:03	13:18	15:37	17:52	20:03
Henham	06:41	07:24	09:00	11:11	13:26	15:45	18:00	20:11
Molehill Green Road	06:52	07:35	09:11	11:22	13:37	15:56	18:11	20:22
Stansted Airport	06:57	07:40	09:16	11:27	13:42	16:01	18:16	20:27

a At what time does the 08:15 from Bishop's Stortford arrive at Henham?

b Sven is at the bus stop in Birchanger at 12:35. How long does he have to wait before the next bus is due?

c How long does the 15:14 from Hockerhill take to get to Molehill Green Road?

d How long does the bus take to get from Stansted Mountfitchet to Stansted Airport?

e Which buses take the shortest time to travel from Bishop's Stortford to Stansted Airport? In each case, write down the time that it leaves Bishop's Stortford Interchange.

6 Here is part of a train timetable for some trains from Manchester to Leeds.

Train	A	B	C	D	E	F
Manchester	06:54	07:36	09:37	10:21	11:34	12:21
Halifax	07:38	–	–	11:04	–	13:04
Leeds	08:14	09:04	11:03	11:37	13:03	13:47

a How many of these trains stop at Halifax?

b How long does train D take to travel from Manchester to Leeds? Give your answer in hours and minutes.

c How many minutes does it take train A to travel from Halifax to Leeds?

d How much longer does train A take to travel from Halifax to Leeds than train D?

e Which train takes the shortest amount of time to travel from Manchester to Leeds?

7 The table shows the distances, in miles, between some towns and cities in Jordan.

Amman							
88	Irbid						
50	38	Jerash					
69	50	42	Mafraq				
22	88	45	45	Zarqa			
32	115	77	101	54	Madaba		
118	202	164	188	141	86	Kerak	
183	267	229	252	205	151	63	Tafila

a Which two places are furthest apart?

b How far is Madaba from Ibrid?

c What is the total distance from Amman to Kerak to Zarqa and then back to Amman again?

What do you think?

1 The diagram shows how long a train takes to travel between five towns.

A	17 minutes	B	12 minutes	C	28 minutes	D	14 minutes	E

a Complete this table to show the time the train takes to travel between each pair of towns.

A				
17	B			
		C		
		28	D	
	54	42		E

b Copy and complete the timetable to show the times when the three trains will be at each of the five towns

A	08:40		
B			
C		10:15	
D			
E			12:02

2 The diagram and table show the distances between some towns on the Coast to Coast walking path in England. Copy and complete both diagrams.

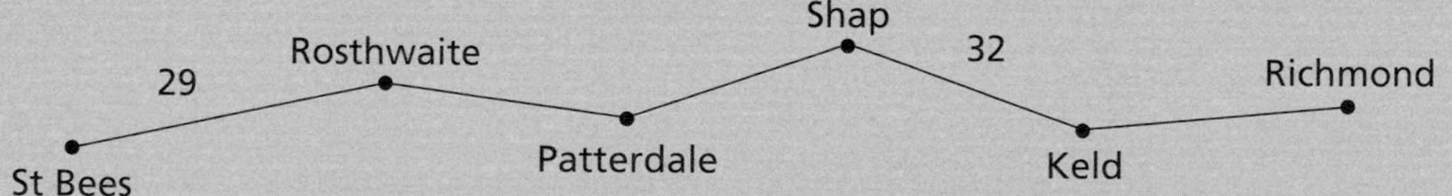

					St Bees
				Rosthwaite	
			Patterdale		
		Shap	17	34	
	Keld				
Richmond		54			

Consolidate – do you need more?

1 Complete these part-whole models.

a 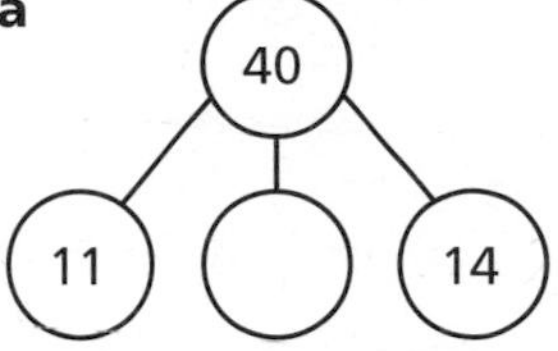

b 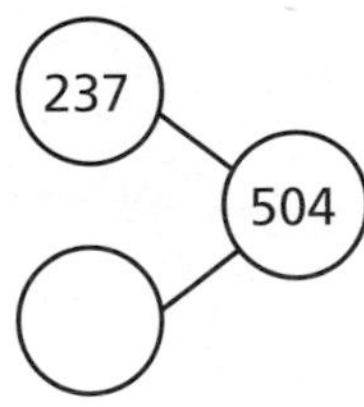

c 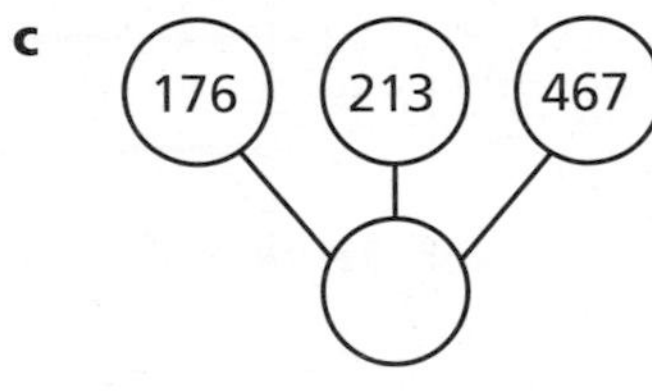

2 There are 180 students in Year 7 at a school.

92 of the students are boys and the rest are girls.

12 of the 28 students who wear glasses are girls.

Show this information in

a a two-way table

b a frequency tree.

3 The table shows the number of adults and children sitting in different parts of a theatre.

	Stalls	Circle	Balcony	Total
Adults	57	46		140
Children		19	42	
Total	88			

a Copy and complete the two-way table.

b There are 410 seats altogether in the theatre. How many of the seats are empty?

4 The bar chart shows the number of customers in a café at lunchtime on Monday to Friday one week.

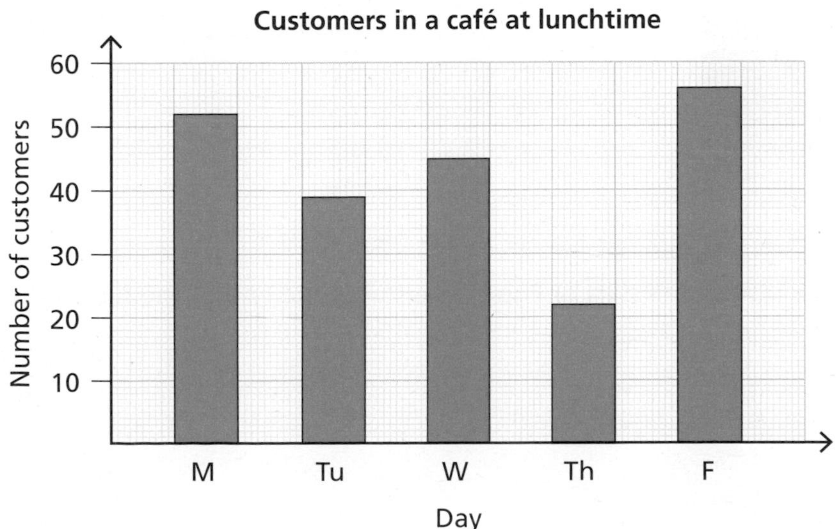

a Find the difference between the highest and lowest number of customers.

b Find the total number of lunchtime customers during this week, from Monday to Friday.

5 The line chart shows how many text messages Beca sent one week.

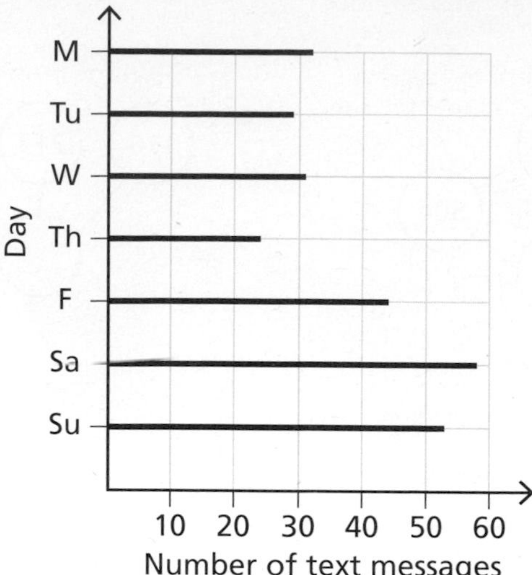

How many more text messages did Beca send from Monday to Friday compared with at the weekend, on Saturday and Sunday?

6 Find the time that is

a half an hour before 17:17

b three quarters of an hour after 11:23

c two and a quarter hours after 15:56

d 3 hours 50 minutes before 17:02

7 Here is an extract from a TV guide.

6:10	Daily Quiz
6:45	Pop Masters
7:35	Science Alive
8:15	Film
10:40	Late News
11:05	Comedy Time
12:15	Close

a How long does the film last?

b How long is it from the start of Pop Masters until the end of the film?

Stretch – can you deepen your learning?

1 **a** What is the minimum amount of information you need to complete all nine cells in a two-way table, as shown below? Does it matter which cells you are given information for?

	Set A	Set B	Total
Category 1	?	?	?
Category 2	?	?	?
Total	?	?	?

b Investigate for other sizes of tables.

2 What would be the most efficient way to work these out?

a The time 55 minutes before a given time.

b The time 119 minutes after a given time.

3 Find a local bus timetable for your area. Investigate whether all the buses take the same length of time between the same two stops or whether the length of time taken varies. Why might this be?

4 In how many ways can you write expressions to fill in the blank cells of this frequency tree?

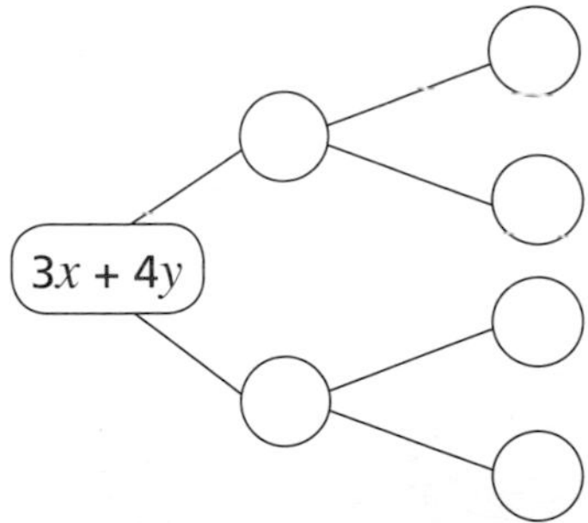

Reflect

1 Describe how you would find missing values in a frequency tree.

2 How can you calculate the difference between two times?

3 What is the difference between a line graph and a bar chart?

6.5 Adding and subtracting in standard form

Small steps

- Write positive integers in the form $A \times 10^n$ (H)
- Write decimals in the form $A \times 10^n$ (H)
- Add and subtract numbers given in standard form (H)

Key words

Standard form – a number written in the form $A \times 10^n$ where A is at least 1 and less than 10, and n is an integer

Are you ready?

Remember you looked at **standard form** in chapters 4.7 and 4.8.

1 Write these numbers in standard form.

- **a** 10000
- **b** 20000
- **c** 200000
- **d** seventy million

2 Write these as ordinary numbers.

- **a** 10^3
- **b** 10^8
- **c** 5×10^4
- **d** 9×10^7

3 Write these as ordinary numbers.

- **a** 10^{-1}
- **b** 5×10^{-1}
- **c** 5×10^{-5}
- **d** 3×10^{-4}

4 Write these numbers in standard form.

- **a** 0.001
- **b** 0.00006
- **c** six thousandths
- **d** seven millionths

Models and representations

Place value grids

Ten thousands	Thousands	Hundreds	Tens	Ones	
	9	0	0	0	= 9000
5	0	0	0	0	= 50000

10^4	10^3	10^2	10^1	10^0	
	9	0	0	0	$= 9 \times 10^3$
5	0	0	0	0	$= 5 \times 10^4$

Ones	tenths	hundredths	thousandths	ten thousandths	
0	0	0	1		= 0.001
0	0	7			= 0.07

10^0	10^{-1}	10^{-2}	10^{-3}	10^{-4}	
0	0	0	1		$= 1 \times 10^{-3}$
0	0	7			$= 7 \times 10^{-2}$

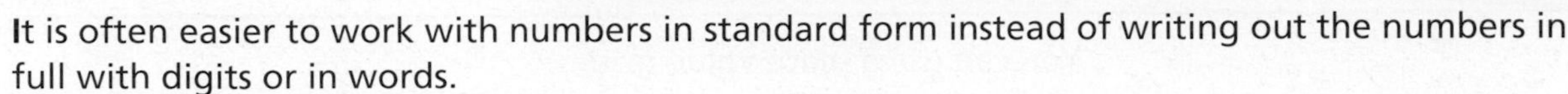

6.5 Adding and subtracting in standard form (H)

It is often easier to work with numbers in standard form instead of writing out the numbers in full with digits or in words.

These represent the same addition:

300 billion + 200 billion = 500 billion

300 000 000 000 + 200 000 000 000 = 500 000 000 000

$3 \times 10^{11} + 2 \times 10^{11} = 5 \times 10^{11}$ The last calculation is easier to write and to read.

Example 1

Are these statements true or false? Explain your answers.

a $3 \times 10^4 + 4 \times 10^4 = 7 \times 10^8$ **b** $3 \times 10^4 + 7 \times 10^4 = 10^5$ **c** $6 \times 10^4 + 6 \times 10^{-4} = 0$

a False

$3 \times 10^4 + 4 \times 10^4 = 7 \times 10^4$

You can check this by converting to ordinary numbers
$3 \times 10^4 + 4 \times 10^4 = 30\,000 + 40\,000$
$= 70\,000 = 7 \times 10^4$

b True

$3 \times 10^4 + 7 \times 10^4$

$= 30\,000 + 70\,000$

$= 100\,000$ and $100\,000 = 10^5$

You could write 100 000 as 1×10^5 but just 10^5 is the same, so we do not usually bother.

This is similar to writing x in algebraic expressions instead of $1x$.

c False

$60\,000 + 0.0006 = 60\,000.0006$

This is not equal to 0.

Remember: negative powers of ten do not represent negative numbers. They represent decimals.

Example 2

Find the answers to these calculations. Give your answers in standard form when appropriate.

a $2 \times 10^6 + 5 \times 10^6$ **b** $3 \times 10^{-3} + 7 \times 10^{-3}$ **c** $5 \times 10^7 + 3 \times 10^1$ **d** $5 \times 10^3 - 2 \times 10^{-3}$

a $2 \times 10^6 + 5 \times 10^6 = 7 \times 10^6$

The power of 10 for both numbers is the same, so you can add the numbers and keep the same power of 10 – the numbers would be in the same column in a place value grid.

	10^6	10^5	10^4	10^3	10^2	10^1	10^0
	2	0	0	0	0	0	0
+	5	0	0	0	0	0	0
	7	0	0	0	0	0	0

b $3 \times 10^{-3} + 7 \times 10^{-3}$
$= 10 \times 10^{-3}$
$= 10^{-2}$

10×10^{-3} is not in standard form.

You can use a place value grid.

	10^0	•	10^{-1}	10^{-2}	10^{-3}
	0	•	0	0	3
+	0	•	0	0	7
	0	•	0	1	0
				1	

You cannot have more than 9 in any column.

So the answer is 1×10^{-2} which you can write as 10^{-2}

c $5 \times 10^7 + 3 \times 10^1$
$= 50\,000\,000 + 30$
$= 50\,000\,030$

The numbers do not have the same power of 10, so 5 and 3 do not have the same place value.

When the numbers in standard form do not have the same power of ten, it is easier to convert them into ordinary numbers before you add or subtract them.

The answer could be written in standard form as $5.000\,003 \times 10^7$, but this is not very neat.

So far you have only written standard form numbers with integer multiples of powers of ten. Numbers like 5.3×10^7 and 5.03×10^7 will be studied in Book 2.

d $5 \times 10^3 - 2 \times 10^{-3}$
$= 5000 - 0.002$
$= 4999.998$

Again it is easier to convert to ordinary numbers.

You can work this out mentally.

As for part **c**, you could write this in standard form, but it is neater not to.

Practice 6.5A

1 Work these out, giving your answers in standard form.

a $5 \times 10^4 + 3 \times 10^4$
b $5 \times 10^{16} + 3 \times 10^{16}$
c $5 \times 10^{-4} + 3 \times 10^{-4}$
d $5 \times 10^{-14} + 3 \times 10^{-14}$
e $5 \times 10^4 - 3 \times 10^4$
f $5 \times 10^{16} - 3 \times 10^{16}$
g $5 \times 10^{-4} - 3 \times 10^{-4}$
h $5 \times 10^{-14} - 3 \times 10^{-14}$

2 Choose the correct answer to each calculation.

a $6 \times 10^4 + 3 \times 10^5$

9×10^9 | 9×10^5 | 63 000 | 360 000

b $6 \times 10^5 - 3 \times 10^4$

3×10^1 | 3×10^5 | 630 000 | 570 000

c $6 \times 10^{-2} - 3 \times 10^{-4}$

3×10^{-6}	3×10^{2}	0.0597	0.005 97

d $6 \times 10^{3} + 3 \times 10^{-3}$

9×10^{0}	6000.003	6000.0003	6.3×10^{3}

e $6 \times 10^{3} - 1 \times 10^{-3}$

5×10^{-3}	6000.001	5999.999	5×10^{0}

f $6 \times 10^{-3} + 3 \times 10^{-3}$

9×10^{-3}	9×10^{-6}	0.006 003	0.0009

3 Work these out. Give your answers as ordinary numbers.

a $7 \times 10^{6} + 8 \times 10^{4}$ **b** $7 \times 10^{6} - 8 \times 10^{4}$ **c** $7 \times 10^{3} + 8 \times 10^{-1}$

d $7 \times 10^{3} - 8 \times 10^{-1}$ **e** $7 \times 10^{-1} + 8 \times 10^{-3}$ **f** $7 \times 10^{-1} - 8 \times 10^{-3}$

4 Work these out. Give your answers in standard form.

a 6 million + 4 million

b Seventy thousand + thirty thousand

c 20 billion added to 80 billion

d One hundredth add nine hundredths

e Five millionths more than five millionths

f $4 \times 10^{7} + 6 \times 10^{7}$

g $3 \times 10^{-4} + 7 \times 10^{-4}$

5 Work out the value of $6 \times 10^{5} - 10^{n}$ when

a $n = 5$ **b** $n = 4$ **c** $n = 3$ **d** $n = 2$

e $n = 1$ **f** $n = 0$ **g** $n = -1$ **h** $n = -2$

i $n = -3$ **j** $n = -4$

6 Work out the value of $6 \times 10^{5} + 10^{n}$ when

a $n = 5$ **b** $n = 4$ **c** $n = 3$ **d** $n = 2$

e $n = 1$ **f** $n = 0$ **g** $n = -1$ **h** $n = -2$

i $n = -3$ **j** $n = -4$

7 Add these pairs of numbers, giving your answers in standard form.

a Two googols and three googols. Remember a googol is 1 followed by 100 zeros.

b Seven googols and two googols.

c Seven googols and three googols.

d Seventeen googols and three googols.

e Seventeen googols and twenty-three googols.

f Seventeen googols and eighty-three googols.

What do you think?

1

$10^3 + 10^3 = 10^6$

Explain why Benji is wrong.

2 $a \times 10^5 + b \times 10^5 = 10^6$

where a and b are both integers.

a List some possible values of a and b

b What can you deduce about $a + b$?

c What is the greatest possible difference between a and b?

3 $6 \times 10^8 + 4 \times 10^8 = 10^9$. Deduce the value of

a $6 \times 10^9 + 4 \times 10^9$

b $6 \times 10^{10} + 4 \times 10^{10}$

c $6 \times 10^{18} + 4 \times 10^{18}$

d $6 \times 10^{-8} + 4 \times 10^{-8}$

e $6 \times 10^{-9} + 4 \times 10^{-9}$

f $10^{12} - 6 \times 10^{11}$

g $10^{20} - 4 \times 10^{19}$

h $10^{-3} - 4 \times 10^{-4}$

4 $10^7, 10^8, 10^9, 10^{10}$ $3 \times 10^{-3}, 4 \times 10^{-3}, 5 \times 10^{-3} \ldots$

Both of these sequences are linear.
Emily

Only one of them is linear.
Jackson

a Who do you agree with? Justify your answer.

b Will the sequences ever have any terms in common? Why or why not?

Consolidate – do you need more?

1 Work these out. Give your answers in standard form.

a $7 \times 10^5 + 2 \times 10^5$
b $7 \times 10^6 - 2 \times 10^6$
c $7 \times 10^{-2} - 2 \times 10^{-2}$
d $7 \times 10^{-10} + 2 \times 10^{-10}$
e $7 \times 10^8 - 2 \times 10^8$
f $7 \times 10^{47} + 2 \times 10^{47}$
g $2 \times 10^{-9} + 7 \times 10^{-9}$
h $7 \times 10^{-9} - 2 \times 10^{-9}$

2 Convert these to ordinary numbers, and then find the total of each set.

a 5×10^4, 6×10^3 and 7×10^2
b 5×10^{-4}, 6×10^{-3} and 7×10^{-2}
c 5×10^4, 6×10^3 and 7×10^{-2}
d 5×10^{-4}, 6×10^3 and 7×10^{-2}
e 5×10^{-4}, 6×10^{-3} and 7×10^2

3 Add each pair of numbers. Give your answers in standard form.

a 3×10^5 and 4×10^5
b 8×10^7 and 10^7
c 6×10^{-2} and 3×10^{-2}
d 5×10^{15} and 4×10^{15}
e 5×10^{15} and 5×10^{15}
f 4×10^{-2} and 6×10^{-2}

4 Find the difference between each pair of numbers. Give your answers in standard form.

a 3×10^5 and 5×10^5
b 3×10^5 and 4×10^5
c 7×10^{-1} and 4×10^{-1}
d 7×10^{-6} and 6×10^{-6}
e 10^5 and 4×10^4
f 10^{-5} and 3×10^{-6}

5 $a = 6 \times 10^5$, $b = 8 \times 10^{-2}$, $c = 2 \times 10^5$ and $d = 2 \times 10^{-2}$

Work out the values of these expressions. Give your answers in standard form.

a $a + c$
b $10a$
c $b - d$
d $a - c$
e $2c$
f $a - 2c$
g $a + 2c$
h $b - 4d$
i $b + d$

Stretch – can you deepen your learning?

1 One billion is one thousand million. Copy and complete these calculations.

a One billion + nine billion = $10^{\square}$
b One billion + nineteen billion = $\square \times 10^{\square}$
c One billion + thirty-nine billion = $\square \times 10^{\square}$
d One million + $\square$ million = 10^7
e One million + $\square$ million = 10^8
f One million + $\square$ million = 10^9

2 **a** One trillion is one thousand billion. Write one trillion in standard form.

b x billion + y billion = one trillion. Find the value of $x + y$ giving your answer in standard form.

3 One micrometre (1 μm) is one millionth of a metre.

Copy and complete

a $1\,\mu\text{m} = 10^{\square}\,\text{m}$

b $6\,\mu\text{m} + 3\,\mu\text{m} = \square \times 10^{\square}\,\text{m}$

c $6\,\mu\text{m} + \square\,\mu\text{m} = 10^{-5}\,\text{m}$

d $6\,\mu\text{m} + \square\,\mu\text{m} = 10^{-4}\,\text{m}$

4 One nanometre (1 nm) is 10^{-9} m.

a How many nanometres are there in one metre?

b p nm + q nm = 1 m. What can you say about $p + q$?

c f nm + g nm = 0.001 m. What can you say about $f + g$?

d How many nanometres are there in a kilometre? Give your answer in standard form and in words.

Reflect

Explain why you do not add or subtract the powers when adding and subtracting numbers written in standard form.

6 Addition and subtraction
Chapters 6.1–6.5

I have become fluent in…

- Using the column method for addition
- Using the column method for subtraction
- Reading and interpreting tables, timetables, charts and graphs
- Calculating with time
- Adding and subtracting with standard form (H)

I have developed my reasoning skills by…

- Extracting relevant information from tables and charts
- Choosing whether to do a calculation using mental or written methods
- Deciding whether a problem needs to be solved using addition, subtraction or both

I have been problem-solving through…

- Designing tables and frequency trees to find required information
- Working out profit, loss, credit and debit
- Working out perimeters and missing lengths in shapes
- Representing equations in a variety of forms

Check my understanding

1 There are 16842 adults and 7548 children at a football match.

a How many people are there altogether at the football match?

b How many more adults are there than children?

2 Find the third term of the linear sequence that starts 36.7, 52.3 …

3 A merchant has £600. She sells her stock, making a profit of £784. She then buys more stock at a cost of £927. How much money does the merchant have now?

4 The perimeter of the triangle is equal to the perimeter of the parallelogram.

Work out the value of a.

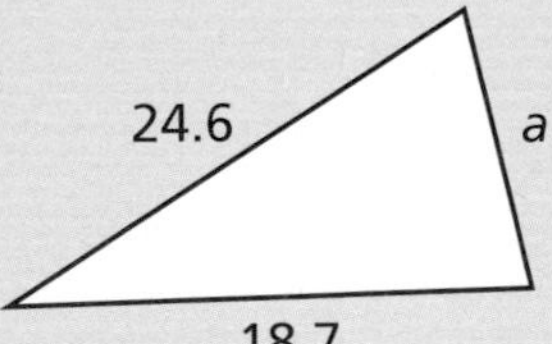

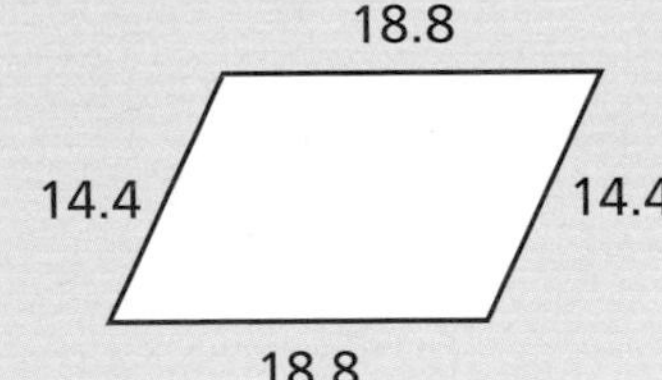

5 Will any of these three calculations be equal in value? Explain your reasoning.

1000 – 567 999 – 566 999 – 568

6 Beca thinks 8 – 5.23 = 3.23. Show that Beca is wrong.

7 Represent the information in this two-way table as a completed frequency tree.

	Prefer tea	Prefer coffee	No preference	Total
Over 50s		27		
Under 50s	46		17	120
Total			40	300

7 Multiplication and division

In this block, I will learn...

about multiples and factors

Factors of 12
1, 2, 3, 4, 6, 12

Multiples of 12
12, 24, 36, 48 ...

how to multiply and divide by powers of 10 and convert units

Ones	•	tenths	hundredths	thousandths	ten thousandths
3	•	7	6		
0	•	0	3	7	6

÷ 100 ÷ 100 ÷ 100

6.4 km = 6.4 × 1000 m = 6400 m

how to use the formal methods of multiplication and division, and deduce new facts from known facts

$$\begin{array}{r} 4\ 7 \\ \times\quad 2\ 6 \\ \hline 2\ 8\ 2 \\ 9\ {}^{1}4\ 0 \\ \hline 1\ {}^{1}2\ 2\ 2 \\ \hline \end{array}$$

47 × 26 = 1222
So
4.7 × 26 = 122.2
1222 ÷ 4.7 = 260

21 ÷ 8

$$\begin{array}{r} 2 \bullet 6\ \ 2\ \ 5 \\ 8\,\overline{)\,2\ \ 1 \bullet {}^{5}0\ {}^{2}0\ {}^{4}0} \end{array}$$

how to work out the areas of shapes

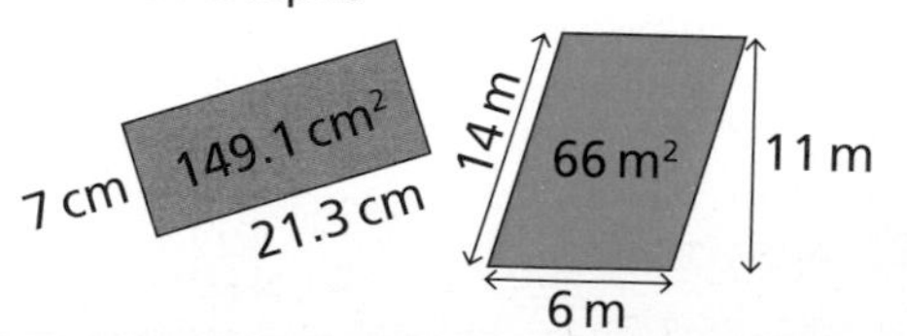

how to find the mean of a set of data

6, 7, 10, 19, 22
Total = 6 + 7 + 10 + 19 + 22 = 64
Mean = 64 ÷ 5 = 12.8

how to use the order of operations

$3 + 4 \times 5^2$
$= 3 + 4 \times 25$
$= 3 + 100$
$= 103$

how to use multiplication and division in complex situations (H)

$5x$ cm
$3x$ cm
$18x^2$ cm²
$7x$ cm

7.1 Linking multiplication and division

Small steps

- Properties of multiplication and division
- Understand and use factors
- Understand and use multiples

Key words

Factor – a positive integer that divides exactly into another positive integer

Factor pair – a pair of numbers that multiply together to give a number

Multiple – the result of multiplying a number by a positive integer

Highest common factor (HCF) – the greatest number that is a factor of every one of a set of numbers

Lowest common multiple (LCM) – the smallest number that is a multiple of every one of a set of numbers

Array – an arrangement of numbers in rows and columns

Are you ready?

1 Copy and complete the table of multiplication facts.

×	5	7	8
2			
4			
6			

2 Copy and complete the table of multiplication facts.

×	4	6	8
3			
7			
9			

3 How can you tell if a number is divisible by

a 10 **b** 5 **c** 2?

4 Here are some whole numbers.

15 18 20 30 24 40 225

Which of the numbers are divisible by

a 10 **b** 5 **c** 2?

Models and representations

Array

These arrays show $10 = 2 \times 5$ and $10 = 5 \times 2$

Bar models can also show relationships between numbers.

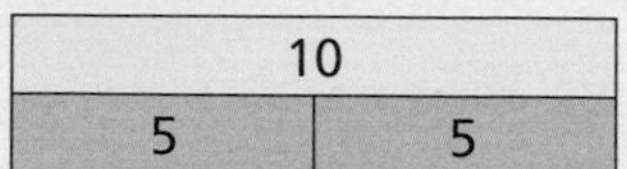

10				
2	2	2	2	2

You could also use a Venn diagram to see relationships between numbers.

Numbers up to 10

6, 8 and 10 are both greater than 5 and even

2 and 4 are even but not greater than 5

1, 5 and 3 are neither greater than 5 nor even

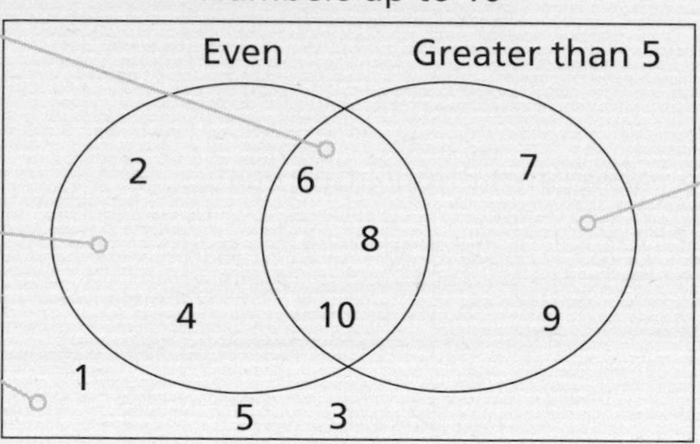

7 and 9 are greater than 5 but not even

You already know that multiplication is commutative, so

$3 \times 4 = 4 \times 3 = 12$

Multiplication is also *associative*, so the order in which numbers are grouped doesn't matter either

$3 \times 8 \times 5 = (3 \times 8) \times 5 = 24 \times 5 = 120$ or $3 \times 8 \times 5 = 3 \times (8 \times 5) = 3 \times 40 = 120$

In this chapter you will look at **factors** and **multiples**.

Factors divide exactly into a number.

The factors of 10 are 1, 2, 5 and 10

1 and 10 is a **factor pair** of 10 as $1 \times 10 = 10$

2 and 5 is also a factor pair of 10 as $2 \times 5 = 10$

You can find multiples by multiplying a number.

The multiples of 6 are 6, 12, 18, 24...

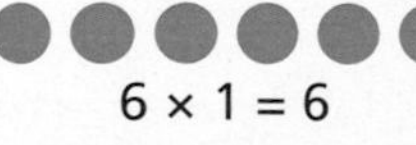

$6 \times 1 = 6$

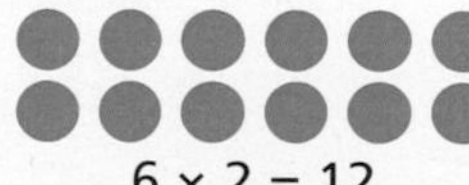

$6 \times 2 = 12$

$6 \times 3 = 18$

$6 \times 4 = 24$

Example 1

a What multiplication and division facts does the array show?

b What factors of 20 does the array show?

c What are the other factors of 20? Draw arrays to illustrate these, and state the fact families for each array.

a $4 \times 5 = 20$ — The array shows that 5 times 4 is 20 and the other members of this fact family

$5 \times 4 = 20$

$20 \div 4 = 5$

$20 \div 5 = 4$

b 5 and 4 — Because the array shows exactly 4 rows and exactly 5 columns, 4 and 5 divide into 20 and so are factors of 20

c

$2 \times 10 = 20$

$10 \times 2 = 20$

$20 \div 2 = 10$

$20 \div 10 = 2$

$1 \times 20 = 20$

$20 \times 1 = 20$

$20 \div 1 = 20$

$20 \div 20 = 1$

So the other factors of 20 are 1, 2, 10 and 20

There are no other ways to arrange the 20 counters into equal rows and columns, so all the factors have been found.

When listing factors, it is useful to write them in order.

The factors of 20 are 1, 2, 4, 5, 10, 20

$1 \times 20 = 20$

$2 \times 10 = 20$

$4 \times 5 = 20$

The factor pairs of 20 are 1 and 20, 2 and 10, 4 and 5

Example 2

What is the highest common factor of 12 and 18?

The factors of 12 are 1, 2, 3, 4, 6 and 12 — You can use arrays or your knowledge of times tables facts to get these lists.

The factors of 18 are 1, 2, 3, 6, 9 and 18

The common factors of 12 and 18 and 1, 2, 3 and 6 — "Common" means that both numbers have this factor.

The highest common factor of 12 and 18 is 6 — The highest common factor is sometimes written as the HCF for short.

Practice 7.1A

1. List the factors of

a 15	**b** 8	**c** 24	**d** 7
e 30	**f** 25	**g** 13	**h** 40

2. In each pair, decide if the first number is a factor of the second number.

a 5, 40	**b** 9, 45	**c** 20, 30
d 6, 56	**e** 75, 75	**f** 1, 38
g 15, 100	**h** 15, 200	**i** 15, 300

3. Find all the factors of 60 that are less than 30

4. Find all the factors of 400 that are greater than 40

5. **a** Copy and complete the Venn diagram.

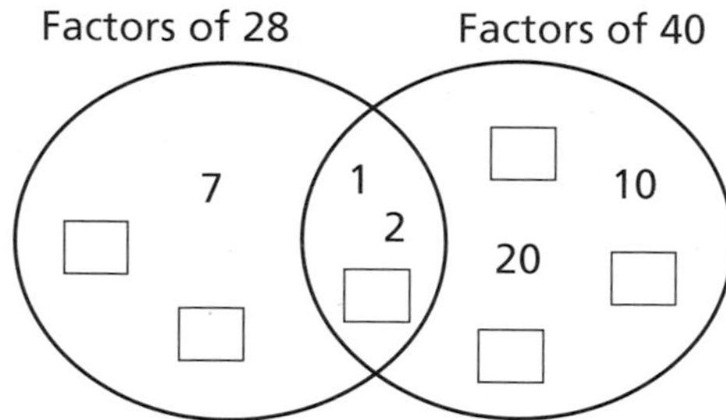

   b What is the highest common factor of 28 and 40?

6. **a** Draw a Venn diagram showing the factors of 20 and 35

   b What is the highest common factor of 20 and 35?

7. **a** List the factors of 15

   b List the factors of 25

   c What are the common factors of 15 and 25?

   d What is the highest common factor of 15 and 25?

8. Find the highest common factor of each set of numbers.

   a 10 and 15 **b** 15 and 30

   c 10, 15 and 30 **d** 17 and 30

   e 30 and 60 **f** 17, 30 and 60

   g 15, 30 and 60 **h** 15, 17, 30 and 60

What do you think?

1. All numbers have at least one factor

   Do you agree with Emily?

2. If 3 is a factor of a number, then 6 is also a factor of the same number

   Give an example to show that Faith is wrong.

3. Investigate each of these statements and decide if they are true or false.

   a If a number ends in a zero, then 10 is a factor of the number.

   b If a number ends in a 5, then 5 is a factor of the number.

   c Only numbers that end in an even number have 2 as a factor.

   d Only numbers that end in 5 have 5 as a factor.

4. **a** The highest common factor of a pair of numbers is 10. What might the numbers be? Is there only one answer?

   b The highest common factor of a set of three numbers is 10. What might the numbers be? Is there only one answer?

5. **a** Which number under 50 has the greatest number of factors?

   b Which number under 100 has the greatest number of factors?

6. A car park space is nine feet wide. Which of these can be marked into an exact number of car park spaces?

   A 36 feet **B** 50 feet **C** 90 feet **D** 120 feet **E** 150 feet

Example 3

a List the first five multiples of 6

b List the first five multiples of 4

c What is the lowest common multiple of 6 and 4?

d List four more common multiples of 6 and 4

a 6, 12, 18, 24, 30...

The first multiple is 6 × 1 = 6
The second multiple is 6 × 2 = 12
The third multiple is 6 × 3 = 18 and so on.

b 4, 8, 12, 16, 20...

The first multiple is 4 × 1 = 4
The second multiple is 4 × 2 = 8
The third multiple is 4 × 3 = 12 and so on.

c 12

12 is the first number that appears in the lists of multiples of both 4 and 6, so it is their lowest common multiple. Lowest common multiple is sometimes written as LCM.

d 24, 36, 48, 60

Any other common multiples must be a multiple of 12 itself.

Practice 7.1B

1 List the first five multiples of

a 7 **b** 8 **c** 10
d 9 **e** 20

2 In each pair, decide if the second number is a multiple of the first number.

a 5, 40 **b** 9, 45 **c** 20, 30
d 6, 56 **e** 75, 75 **f** 1, 38
g 15, 100 **h** 15, 200 **i** 15, 300

3 Find all the multiples of 8 that are less than 50

4 Find all the multiples of 7 that are between 50 and 90

5 Flo is finding the lowest common multiple of 7 and 8. She writes:

Multiples of 7: 7, 14, 21, 28, 35 ...
Multiples of 8: 8, 16, 24, 32, 40 ...

There is no number that is common to both lists, so 7 and 8 don't have a lowest common multiple.

Explain why Flo is wrong.

6 a List the first five multiples of 10
b List the first five multiples of 15
c What is the lowest common multiple of 10 and 15?

7 Find the lowest common multiple of each set of numbers.
a 6 and 7 b 6 and 8 c 6 and 9 d 6 and 12
What's the same and what's different?

8 Find the lowest common multiple of each set of numbers.
a 8 and 10 b 8 and 12 c 10 and 12
d 8, 10 and 12 e 9 and 15 f 9 and 10
g 9, 10 and 15

What do you think?

1 Are the statements always, sometimes, or never true?
a "The LCM of two numbers is greater than both numbers."
b "The LCM of two numbers is found by multiplying the two numbers."

2 a List the first 8 multiples of 5. What pattern do you see in the last digits?
b List the first 8 multiples of 6. What pattern do you see in the last digits?
c Is there a pattern in the final digits of the 8 times table?

3 Can a number be both a factor and a multiple?

4 Here is a set of numbers.
30 31 32 33 34 35 36 37 38 39 40
Which of the numbers are
a multiples of 5 b multiples of 3 c multiples of 10
d factors of 30 e factors of 120?

Consolidate – do you need more?

1 Here is a set of numbers.
12 15 18 20 24 25 30 35 36 40
Write down any of the numbers that
a are factors of 40 b are multiples of 10
c have 4 as a factor d are multiples of both 3 and 5
e are multiples of 9 f are both multiples of 3 and factors of 24

2 List the factors of each number.

a 25 **b** 36 **c** 42

d 28 **e** 50

3 Find the highest common factor of each pair of numbers.

a 10 and 20 **b** 8 and 16 **c** 12 and 24 **d** 20 and 40

What do you notice?

4 Find the lowest common multiple of each pair of numbers.

a 10 and 20 **b** 8 and 16 **c** 12 and 24 **d** 20 and 40

What do you notice?

5 Find the highest common factor of each pair of numbers.

a 15 and 36 **b** 18 and 24 **c** 30 and 45 **d** 60 and 90

6 Find the lowest common multiple of each pair of numbers.

a 6 and 9 **b** 16 and 24 **c** 10 and 15 **d** 25 and 30

7 Find the HCF and LCM of

a 3, 6 and 10 **b** 4, 10 and 12 **c** 10, 12 and 20

Stretch – can you deepen your learning?

1 The digit sum of a number is found by adding all its digits

The digit sum of 145 = 1 + 4 + 5 = 10

Are the statements always, sometimes or never true? If it is sometimes true, give an example of when it is true and when it is false.

a If the digit sum of a number is divisible by 3, then 3 is a factor of the number.

b If 3 is a factor of a number, then the digit sum of the number is divisible by 3.

c If the digit sum of a number is divisible by 9, then 9 is a factor of the number.

d If 9 is a factor of a number, then the digit sum of the number is divisible by 9.

e If the digit sum of a number is divisible by 4, then 4 is a factor of the number.

f If 4 is a factor of a number, then the digit sum of the number is divisible by 4.

2

If 6 is a factor of a number, then 2 and 3 are also factors of the number.

a Investigate Marta's claim.

b If 12 is a factor of a number, what other numbers must be a factor of the number?

c If x is a factor of a number, what other numbers must be a factor of the number?

3 If $a = 5b$ is a definitely in the 5 times table?

4 How many numbers can you find with

a only 1 factor

b exactly 2 factors

c exactly 3 factors

d exactly 4 factors?

5 x and y are two integers. Investigate the claim
"The HCF of a and b multipled by the LCM of a and b is equal to ab."

6 Investigate the claims.

a If x is a factor of y then x is a factor of ay.

b If x is a multiple of y then ax is a multiple of y.

Reflect

1 What is the connection between multiplication and division?

2 What is the difference between a multiple and a factor?

7.2 Multiplying with powers of 10

Small steps

- Multiply and divide integers and decimals by powers of 10
- Convert metric units
- Multiply by 0.1 and 0.01 (H)
- Write positive integers in the form $A \times 10^n$ (H)
- Write decimals in the form $A \times 10^n$ (H)

Key words

Centi – one hundredth

Milli – one thousandth

Kilo – one thousand

Are you ready?

1 What is the value of the figure 4 in each of the numbers?

a 6043 **b** 417 **c** 11.4

d 294.3 **e** 246 017

2 Write the fractions as decimals.

a $\frac{7}{100}$ **b** $\frac{7}{10}$ **c** $\frac{17}{10}$ **d** $\frac{17}{100}$

3 Work out

a 4×10 **b** 7×10 **c** 10×8 **d** 10×7

4 **a** How many centimetres are there in a metre?

b How many metres are there in a kilometre?

Models and representations

Place value charts

Hundreds	Tens	Ones	tenths	hundredths
	1	6	2	4
1	6	2	4	

In a place value chart, the value of each column is 10 times greater than the column to its right.

Base 10 blocks are also very useful.

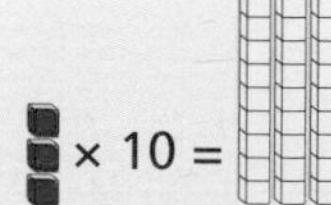

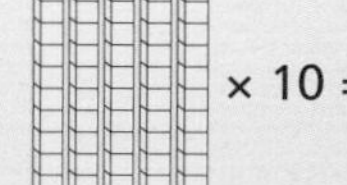

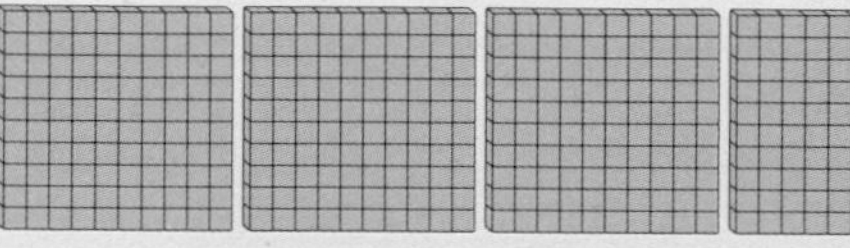

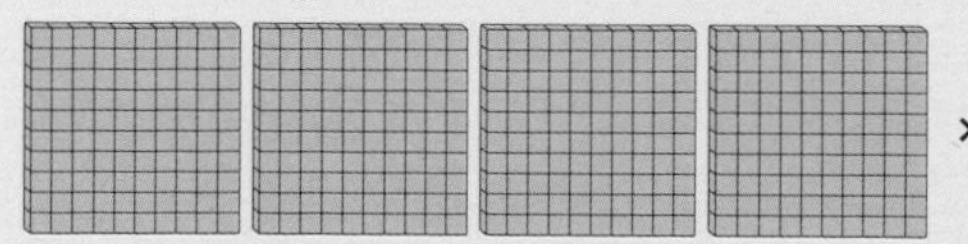

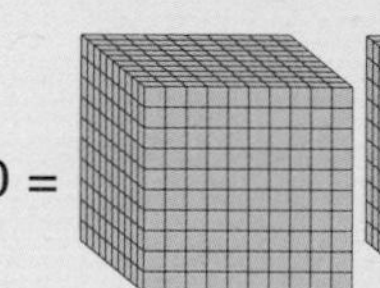
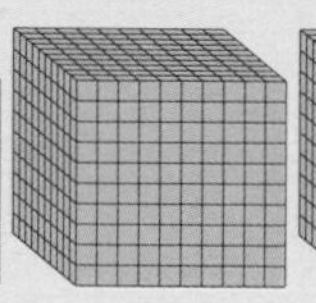
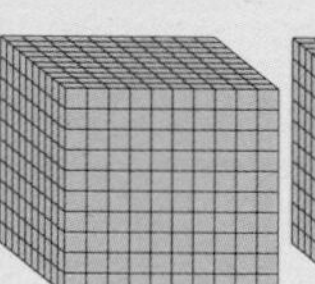
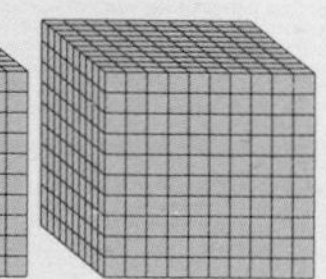

You could also use place value counters to represent multiplying by 10, 100 and so on.

(1) (1) (1) × 10 = (10) (10) (10)   (10) (10) (10) (10) (10) × 10 = (100) (100) (100) (100) (100)

(0.01) (0.01) (0.01) (0.01) × 10 = (0.1) (0.1) (0.1) (0.1)

Example 1

To multiply a number by 10, you just add a zero.

Do you agree with Benji?

No. It is true for integers, for example 23 × 10 = 230, but not for decimals, for example 6.7 × 10 ≠ 6.70

When you multiply a number by 10, each digit in the number is worth 10 times more than before.

For example (10) (10) (1) (1) (1)   23

× 10

(100) (100) (10) (10) (10)   230

Hundreds	Tens	Ones
	2	3
2	3	0

You need to add a zero as without this the number would still look like 23

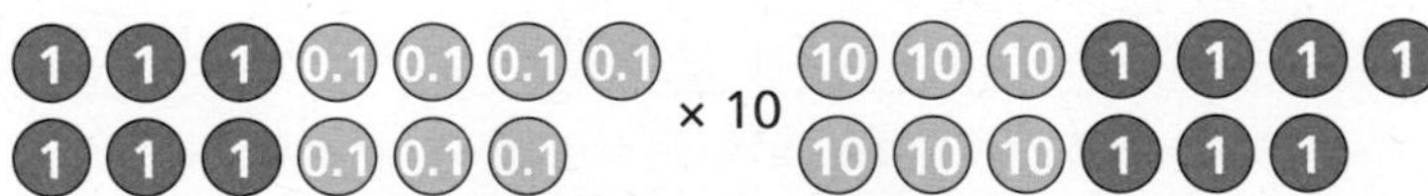

(1) (1) (1) (0.1) (0.1) (0.1) (0.1)
(1) (1) (1) (0.1) (0.1) (0.1)   × 10   (10) (10) (10) (1) (1) (1) (1)
(10) (10) (10) (1) (1) (1)

Tens	Ones	tenths
	6	7
6	7	

You can see that 6.7 × 10 = 67, not 6.70

Example 2

Work out

a 4.63 × 10 **b** 4.63 × 100 **c** 46.3 × 1000

a 4.63 × 10 = 46.3

You can represent this in a place value chart, knowing each digit moves one place to the left when we multiply by 10

Tens	Ones	tenths	hundredths
	4	6	3
× 10	× 10	× 10	
4	6	3	

This can also be done mentally by picturing the numbers moving one place to the left.

b 4.63 × 100 = 463

Multiplying by 100 is the same as multiplying by 10 and multiplying by 10 again.

Hundreds	Tens	Ones	tenths	hundredths
		× 100 4	× 100 6	× 100 3
4	6	3		

Picture the numbers moving two places to the left.

c 46.3 × 1000 = 46 300

In the same way, multiplying by 1000 is the same as multiplying by 10 three times in succession, or moving the digits three places to the left

Ten Thousands	Thousands	Hundreds	Tens	Ones	tenths
			× 1000 4	× 1000 6	× 1000 3
4	6	3	0	0	

This time you need to add in the extra zeros so you do not confuse the answer with 463

Example 3

Work out

a 4.63 ÷ 10

b 46.3 ÷ 1000

a 4.63 ÷ 10 = 0.463

You can represent this in a place value chart, knowing each digit moves one place to the right when we divide by 10

Ones	tenths	hundredths	thousandths
4	6	3	
	÷ 10	÷ 10	÷ 10
0	4	6	3

Picture the numbers moving one place to the right. Remember to include a zero at the start of the number, so you write 0.463 rather than just ".463" b $46.3 \div 1000 = 0.0463$

Dividing by 1000 is the same as dividing by 10 three times in succession.

Tens	Ones	tenths	hundredths	thousandths	ten thousandths
4 ÷ 1000	6 ÷ 1000	3 ÷ 1000			
	0	0	4	6	3

Picture the numbers moving three places to the right. Be careful to include the 0 that represents the tenths.

Practice 7.2A

1 Multiply each number by 10. You may use a place value grid to help you.

a 37 **b** 0.37 **c** 307 **d** 30.7
e 3.07 **f** 0.0307 **g** 370

2 What is the overall effect if you

a multiply a number by 10 and then multiply the result by 10
b multiply a number by 10 and then divide the result by 10
c multiply a number by 10, multiply the result by 10, and then multiply that result by 10?

3 Work out

a 46×100 **b** 4.06×100 **c** 406×10
d 460×100 **e** 4.6×1000 **f** 4.06×1000
g 10×4.006 **h** 100×46 **i** 1000×0.46

4 Divide each number by 10. You may use a place value grid to help you.

a 37 **b** 0.37 **c** 307
d 30.7 **e** 3.07 **f** 0.0307

5 What is the overall effect if you

a divide a number by 10 and then divide the result by 10
b divide a number by 10 three times in succession
c divide a number by 10 three times in succession and then multiply the result by 10?

6 Work out

a $46 \div 100$ **b** $4.06 \div 100$ **c** $406 \div 10$
d $460 \div 100$ **e** $4.6 \div 1000$ **f** $4.06 \div 1000$
g $4.006 \div 10$ **h** $46 \div 1000$ **i** $460 \div 1000$

7 Copy and complete

a $5.8 \times \square = 580$ b $5.8 \times \square = 5800$ c $\square \times 100 = 58$

d $0.58 \times \square = 5.8$ e $\square \times 100 = 508$ f $\square \times 1000 = 50.8$

8 Copy and complete

a $92 \div \square = 0.92$ b $9.2 \div \square = 0.092$ c $\square \div 100 = 9.2$

d $92 \div \square = 0.092$ e $\square \div 100 = 92$ f $\square \div 1000 = 9.02$

9 Find the missing operations and numbers.

a $67 \bigcirc \square = 670$ b $67 \bigcirc \square = 0.67$ c $6.7 \bigcirc \square = 0.067$

d $670 \bigcirc \square = 6.7$ e $670 \bigcirc \square = 6700$ f $0.0067 \bigcirc \square = 6.7$

What do you think?

1 Solve the equations.

a $\frac{a}{10} = 7.8$ b $10b = 7.8$ c $7.8 = \frac{c}{1000}$

d $7.8 = 100d$ e $7.8e = 780$ f $780 = 0.78f$

2 Find the missing numbers.

a $17 \times 100 = 1.7 \times \square$ b $2.6 \times \square = 0.26 \times 100$ c $4 \times \square = 400 \div 10$

d $700 \div \square = 0.7 \div 10$ e $700 \div \square = 0.7 \times 10$ f $53 \times 100 = \square \div 100$

3 Find some possible values for the missing numbers.

$48 \div \square = 480 \div \square$

4

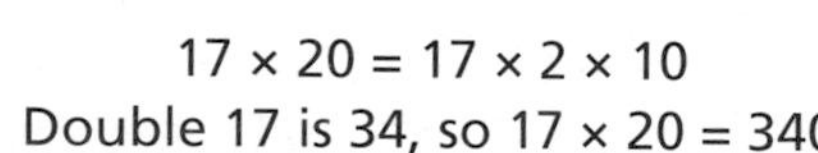

17 × 20 = 17 × 2 × 10
Double 17 is 34, so 17 × 20 = 340

Use Abdullah's strategy to work out:

a 36×20 b 14×30 c 59×200 d 31×40

5

64 ÷ 20 is 64 ÷ 2 ÷ 10
Half of 64 is 32, so 64 ÷ 20 = 3.2

Use Beca's strategy to work out

a $36 \div 20$ b $45 \div 30$ c $28 \div 400$ d $8 \div 4000$

6 A B C

A is 10 times bigger than B and 100 times smaller than C.

What is the relationship between B and C?

You can use multiplying and dividing by powers of 10 to convert between metric units.

"**Centi**" means "One hundredth" so 1 cm = $\frac{1}{100}$ m or 1 m = 100 cm

"**Milli**" means "One thousandth" so 1 mm = $\frac{1}{1000}$ m or 1 m = 1000 mm

$\frac{1}{100}$ is 10 times bigger than $\frac{1}{1000}$, so 1 cm = 10 mm, or 1 mm = $\frac{1}{10}$ cm

"**Kilo**" means "One thousand" so 1 km = 1000 m, or 1 m = $\frac{1}{1000}$ km

These prefixes apply to all metric units. So 1 cg = 10 mg and 1 cl = 10 ml and so on.

Example 4

Convert 36 cm to

a mm **b** m **c** km

a 36 cm = 36 × 10 mm = 360 mm — Use the fact that 1 cm = 10 mm and your knowledge of how to multiply numbers by 10

b 36 cm = 36 ÷ 100 m = 0.36 m — Use the fact that 1 m = 100 cm and your knowledge of how to divide numbers by 100

c 36 cm = 0.36 m = 0.36 ÷ 1000 km = 0.000 36 km — Use the answer to part **b** before using the fact that 1 km = 1000 m and your knowledge of how to divide numbers by 1000

Example 5

Convert 460 g to

a kg **b** cg **c** mg

a 460 g = 460 ÷ 1000 kg = 0.46 kg — Use the fact that 1 kg = 1000 g

b 460 g = 460 × 100 cg = 46 000 cg — Use the fact that 1 g = 100 cg

c 460 g = 460 × 1000 mg = 460 000 mg

Use the fact that 1 g = 1000 mg

You could have used your answer to **b** and the fact that 1 cg = 10 mg. 46 000 cg = 46 000 × 10 mg = 460 000 mg

Practice 7.2B

1 Kate is converting 36 cm to m.

a Will the answer be greater or less than 36?

b Copy and complete the following. 36 cm = 36 ◯ 100 m = ☐ m

2 Convert these lengths to m.

a 700 cm **b** 700 km **c** 700 mm
d 7000 cm **e** 7000 km **f** 7000 mm
g 70 cm **h** 70 km **i** 7 cm

3 Convert these masses to g.

a 8 kg **b** 80 kg **c** 0.08 kg
d 800 cg **e** 80 000 cg **f** 80 cg
g 8000 mg **h** 80 mg **i** 8 mg

4 Convert these capacities to cl.

a 2 litres **b** 20 litres **c** 200 litres
d 200 ml **e** 20 ml **f** 2 ml

5 **a** By converting both lengths to cm, find which is greater, 0.7 m or 70 mm.

b Find the greater length in each pair, justifying your answers.

i 3000 cm, 3 m **ii** 0.08 m, 80 cm **iii** 2 km, 20 000 mm

6 Which metric unit would you use to measure

a the capacity of a small glass
b the length of a classroom
c the thickness of a magazine
d the mass of an orange
e the mass of an adult human
f the capacity of a water tank
g the length of an exercise book
h the mass of a coin
i the mass of an ant
j the distance between two towns?

7 The length and width of a rectangle are 56 mm and 7 cm.

Find the perimeter of the rectangle.

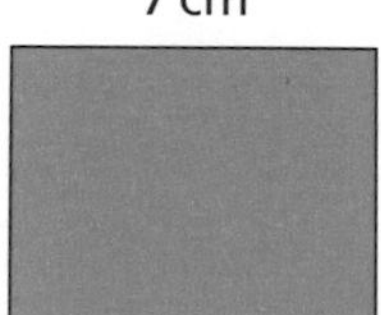

8 Find the total mass of the shopping in the basket.

9 A glass will hold 250 ml of liquid and a cup will hold 150 ml. Find the total capacity of 10 glasses and 100 cups, giving your answer in litres.

What do you think?

1. 1 litre of water has mass 1 kg.

   Find the mass, in grams, of

   a 250 ml of water **b** 35 cl of water **c** 6.5 litres of water

2. **a** How many millimetres are there in a kilometre?

   b How many centigrams are there in a kilogram?

3. 1 decimetre (dm) is one tenth of a metre.

   a How many centimetres are there in a decimetre?

   b How many decimetres are there in a kilometre?

   c Work out how many decigrams (dg) are equivalent to

   i 1 g **ii** 1 kg **iii** 45 g **iv** 60 cg **v** 60 mg

   d Work out how many decilitres (dl) are equivalent to

   i 1 litre **ii** 100 ml **iii** 84 cl **iv** 1 kl (1 kilolitre)

4. 1 tonne = 1000 kg

   How many milligrams are there in a tonne?

(H) Remember $0.1 = \frac{1}{10}$

26×0.1 is $26 \times \frac{1}{10}$

which is the same as $\frac{1}{10} \times 26$ or $\frac{1}{10}$ of 26

So $26 \times 0.1 = \frac{1}{10}$ of $26 = 26 \div 10 = 2.6$

In the same way $26 \times 0.01 = 26 \div 100 = 0.26$

Example 6

Work out

a 3.6×0.1 **b** 0.46×0.01

a $3.6 \times 0.1 = 3.6 \div 10 = 0.36$ — Use the fact that $\times\ 0.1$ is the same as $\div\ 10$

b $0.46 \times 0.01 = 0.0046$ — Use the fact that $\times\ 0.01$ is the same as $\div\ 100$

Practice 7.2C (H)

1. Match the cards that show equivalent calculations.

÷ 1000	× 0.1	× 0.01	÷ 10	× 0.001	÷ 100

2. Multiply each number by 0.1

   a 2000 **b** 204 **c** 24

   d 2.04 **e** 0.24 **f** 0.0204

3 Multiply each number by 0.01

a 70 b 7000 c 78 d 708

e 7.8 f 0.708 g 0.78

4 Match the cards that will have the same answer.

1.8 × 0.01	18 × 0.1	18000 × 0.01	18 ÷ 1000	1800 ÷ 10	180 ÷ 100

5 Find the missing numbers.

a 48 ÷ 100 = 4.8 × ☐ b 2.6 × 0.1 = ☐ ÷ 1000

c 4 × ☐ = 400 ÷ 1000 d 80 ÷ ☐ = 0.8 × 0.01

6 Put in order of size, starting with the smallest.

650 ÷ 10000	6.05 × 0.01	65 ÷ 100	650 × 0.01	65 × 0.1	65 ÷ 10

What do you think?

1 

24 × 0.2 = 24 × 2 × 0.1 = 48 × 0.1 = 4.8

Use Jackson's strategy to work out

a 430 × 0.2 b 12 × 0.3 c 31 × 0.03 d 22 × 0.4

e 13 × 0.04 f 0.2 × 24 g 0.03 × 21 h 0.004 × 12

2 Multiplying by 0.1 is the same as dividing by 10

What multiplication would be equivalent to dividing by 0.1?

Write down the answers to

a 6 ÷ 0.1 b 0.2 ÷ 0.1 c 6.2 ÷ 0.1 d 0.07 ÷ 0.1 e 0.083 ÷ 0.01

3 Work out the calculations, giving your answers in standard form.

a 0.08 × 0.1 b 0.008 × 0.1 c 0.008 × 0.01 d 0.008 × 0.001

e 10^6 × 0.1 f 10^{12} × 0.01 g $(4 \times 10^7) \times 0.1$ h $(4 \times 10^{-5}) \times 0.01$

Consolidate – do you need more?

1 Which is greater, 12.6 × 100 or 1260 ÷ 10? How do you know?

2 In each set, which calculation has a different answer?

a 10 × 37, 100 × 3.07, 3.70 × 100 b 46 × 10, 4.6 × 100, 460 ÷ 10

c 0.04 × 1000, 40 × 100, 4 × 10

3 Chloe writes $1.7 \times 100 = 1.700$

Explain what Chloe has done wrong.

4 Copy and complete the following.

a $38 \times \square = 380$ **b** $38 \div \square = 0.38$ **c** $38 \div \square = 0.038$

d $3.8 \times \square = 380$ **e** $0.38 \times \square = 380$ **f** $3.8 \div \square = 0.0038$

5 Find the difference between

a 2 km and 860 m **b** 0.6 litres and 75 cl

c 30 cg and 20 mg **d** 500 cm and 4000 mm

e 150 g and 200 cg **f** 2 m and 1000 mm

g 20 cm and 500 mm **h** 420 ml and 20 cl

6 Solve the equations.

a $\frac{a}{10} = 65$ **b** $10b = 65$ **c** $6.5 = \frac{c}{100}$

d $6.5 = 100d$ **e** $6.05 = \frac{e}{100}$ **f** $100f = 605$

g $60.5 = 10g$ **h** $\frac{h}{1000} = 6.05$

Stretch – can you deepen your learning?

1 In chemistry, one mole of a substance contains roughly 6×10^{23} atoms.

Work out, giving your answers in standard form, how many atoms are contained in

a 10 moles **b** 100 moles **c** 0.1 moles **d** 0.001 moles

2 A decimetre (dm) is one tenth of a metre. Also, a decametre is 10 m.

a How many decimetres are there in a decametre?

b What fraction of one decametre is one decimetre?

A micrometre (μm) is one millionth of a metre.

c How many μm are the same length as

i a decimetre **ii** a decametre?

3 Research the meaning of other prefixes such as nano and mega. Find the connections between the different measures.

Reflect

1 What's the same and what's different about multiplying and dividing by 10, 100 and 1000?

2 Explain the prefixes used in the metric system. How do they connect to each other?

7.3 Multiplication: short and long

Small steps

- Properties of multiplication and division
- Use formal methods for multiplication of integers
- Use formal methods for multiplication of decimals
- Solve problems using the area of rectangles and parallelograms
- Choose the most appropriate method: mental strategies, formal written or calculator

Key words

Product – the result of a multiplication

Parallelogram – a quadrilateral with two pairs of parallel sides

Base – a side of a shape that is used as the foundation

Perpendicular – at right angles to

Area – the space inside a two-dimensional shape

Are you ready?

1 Write the number that is ten times greater than

a 17 **b** 1.7 **c** 1.07 **d** 10.07

2 Write the number that is ten times smaller than

a 17 **b** 1.7 **c** 1.07 **d** 10.07

3 Copy and complete the table of multiplication facts.

×	6	8	9
3			
4			
5			

4 Write the numbers correct to one significant figure.

a 68 **b** 60.8 **c** 0.68 **d** 608

Models and representations

Place value chart

This place value chart shows that 34 × 5 equals 34 + 34 + 34 + 34 + 34. You can link this to the column method of multiplication.

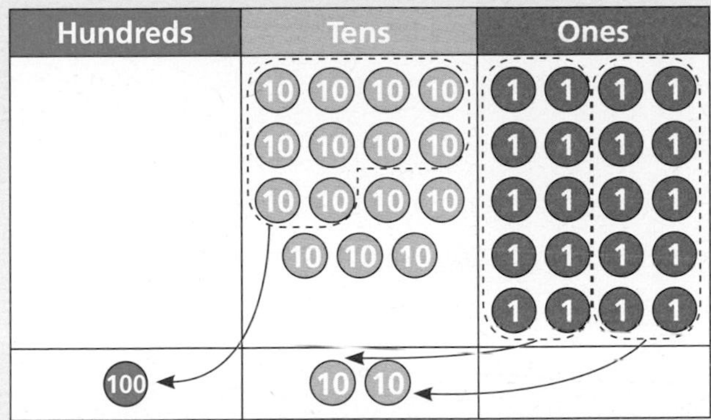

	H	T	O
		3	4
×			5
	1	7	0
		2	

You can also use base 10 blocks to show multiplications. This links to the grid model and the column method.

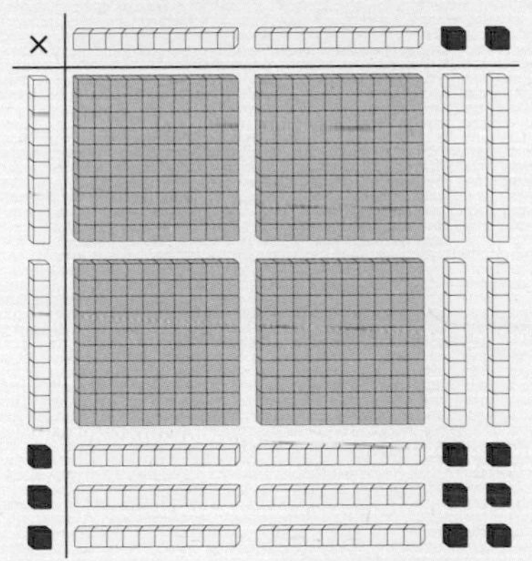

×	20	2
20	400	40
3	60	6

400 + 40 = 440
60 + 6 = 66
506

	H	T	O
		2	2
×		2	3
		6	6
	4	4	0
	5	[1]0	6

You can use your knowledge of multiplying and dividing by powers of 10 to find answers to related calculations.

Example 1

Work out

a 60 × 3 **b** 600 × 300 **c** 0.6 × 0.3

a 60 × 3 = 180

You know 6 × 3 = 18

You also know that 60 × 3 is ten times greater than 6 × 3

60 × 3 = 6 × 10 × 3 = 6 × 3 × 10 = 18 × 10 = 180

b 600 × 300 = 180 000

Again 6 × 3 = 18

You have multiplied both numbers by 100, so

18 × 100 × 100 or 18 × 10 000 = 180 000

c 0.6 × 0.3 = 0.18

Again 6 × 3 = 18

You have divided both numbers by 10 so

18 ÷ 10 ÷ 10 or 18 ÷ 100 = 0.18

Practice 7.3A

1 Use the fact that 7 × 8 = 56 to write down the answers to

a 70 × 8 **b** 7 × 80 **c** 70 × 80 **d** 7 × 800

e 70 × 800 **f** 80 × 7 **g** 800 × 70 **h** 0.7 × 8

i 0.8 × 0.7 **j** 0.08 × 0.07

2 Abdullah says

4 × 5 = 20 so 40 × 50 = 200

Explain why Abdullah is wrong.

3 Work out

a 30 × 400 **b** 60 × 30 **c** 0.7 × 4 **d** 8 × 0.3

e 9 × 0.7 **f** 60 × 7 **g** 0.8 × 9 **h** 0.3 × 0.7

i 20 × 700 **j** 0.07 × 0.6

4 Copy and complete the steps to work out 900 × 0.7

9 × 7 = 63

× 10 ↓ 90 × 7 = ☐ ↓ × 10

× 10 ↓ 900 × 7 = ☐ ↓ × 10

÷ 10 ↓ 900 × 0.7 = ☐ ↓ ÷ 10

5 Use the same method as in question 4 to work out the calculations. Write out the steps if you need to.

a 30 × 0.4 **b** 0.6 × 40 **c** 0.7 × 400

d 800 × 0.04 **e** 9000 × 0.7 **f** 400 × 0.07

g 0.08 × 70 **h** 6000 × 0.3 **i** 0.04 × 6000

j 0.005 × 6000

6 Write down three questions that will have the same answer as 700 × 0.5

7 Given that 67 × 28 = 1876, write down the value of

a 67 × 2.8 **b** 0.67 × 28 **c** 6.7 × 0.28

d 28 × 0.067 **e** 2.8 × 0.067

What do you think?

1 ab = 1200. Suggest some values for a and b if both a and b are integers.

2 Jackson says

800 × 0.04 will have the same answer as 8 × 4

Explain whether or not Jackson is correct.

3 ab = 1200

Only one of a and b is an integer.

Suggest some values for a and b.

4 Given that 38 × 46 = 1748, write down four calculations with the answer 17.48

7.3 Multiplication: short and long

You can think of multiplication in several ways.

Place value counters linked to the column method

Hundreds	Tens	Ones
	10 10 10 10 10 10 10 10 10 10 10 10 10 10 10 10 10 10	1 1 1 1 1 1 1 1 1 1 1 1
100	10 10 10 10 10 10 10 10 10	1 1

	H	T	O
		3	2
×			6
	1	9	2
		1	

Grid method

	40	6
70	2800	420
3	120	18

3	2	2	0
	1	3	8
3	3	5	8

Example 2

Estimate the answers to these calculations and then work out the exact answers.

a 28 × 3 **b** 2.8 × 3 **c** 36 × 24 **d** 0.36 × 2.4

a 28×3

$\approx 30 \times 3 = 90$

	T	O
	2	8
×		3
	8	4
	2	

You estimate by rounding the numbers to one significant figure.

28 ≈ 30

3 is already to one significant figure.

You know 3 × 3 = 9 so 30 × 3 will be 10 times greater.

To work out the exact answer, use the column method as that is the most efficient.

b 2.8×3

$\approx 3 \times 3 = 9$

$2.8 \times 3 = 8.4$

2.8 ≈ 3

Now use the fact that 28 × 3 = 84

2.8 × 3 will be 10 times smaller which is 8.4

This is close to the estimate of 9

You could have used the column method with decimals.

	O •	t
	2 •	8
×		3
	8 •	4
	2	

c 36×24

$\approx 40 \times 20 = 800$

	H	T	O
		4	0
×		2	0
	8	0	0

To work out the exact answer, use the column method as that is the most efficient.

Estimate: $36 \approx 40 \quad 24 \approx 20$

		3	6
×		2	4
	1	$_2$4	4
	$_1$7	2	0
	8	6	4

d 0.36×2.4

$\approx 0.4 \times 2 = 0.8$

Estimate: $0.36 \approx 0.4 \quad 2.4 \approx 2$

0.36×2.4

$= 36 \times 24 \div (100 \times 10)$

$= 864 \div 1000$

$= 0.864$

Now use the answer to 36 × 24 to work out the exact answer

÷ (100 × 10) is the same as ÷ 1000

0.864 is very close to the estimate of 0.8. That suggests the answer is probably correct.

Practice 7.3B

1 Estimate and then find the exact answer to

a 26 × 3 **b** 28 × 4 **c** 62 × 5 **d** 4 × 73
e 7 × 28 **f** 39 × 6 **g** 8 × 42 **h** 346 × 3
i 7 × 229 **j** 513 × 8

2 Use your answers to question 1 to write down the answer to

a 26 × 0.3 **b** 2.8 × 4 **c** 6.2 × 0.5 **d** 4 × 7.3
e 0.7 × 2.8 **f** 3.9 × 6 **g** 0.8 × 4.2 **h** 3.46 × 0.3
i 0.7 × 2.29 **j** 51.3 × 0.8

3 Work out

a 5.7 × 4 **b** 0.3 × 68 **c** 165 × 0.2 **d** 72 × 0.8 **e** 0.03 × 51.6

4 Estimate and then find the exact answer to

a 32 × 47 **b** 26 × 63 **c** 57 × 29 **d** 41 × 72
e 70 × 37 **f** 68 × 45 **g** 58 × 27 **h** 27 × 53
i 82^2 **j** 467 × 62

5 **a** A textbook weighs 635 g. Work out the weight of

i six textbooks **ii** 36 textbooks

b Each textbook costs £7.49. Work out the cost of

i six textbooks **ii** 36 textbooks

6 Estimate and then find the exact answer to

a 1.6 × 37 **b** 2.8 × 7.9 **c** 0.29 × 85
d 6.7 × 8.2 **e** 3.9^2

7 Solve the equations.

a $\frac{a}{34} = 17$ b $4.9 = \frac{b}{30}$ c $\frac{c}{22} = 57$ d $\frac{d}{19} = 0.38$

What do you think?

1 Rob multiplies 75 by 62 and gets the answer 580

$$\begin{array}{r} 7\ 5 \\ \times\quad 6\ 2 \\ \hline 1\ {}^{1}5\ 0 \\ 4\ {}^{3}5\ 0 \\ \hline 5\ 8\ 0 \\ \hline \end{array}$$

a Use estimation to show that Rob is wrong. b What mistake has Rob made?

2 a Huda finds the product of two 2-digit numbers. The last digit of the product is 7. What can you deduce about the numbers she started with?

b Huda finds the product of two different 2-digit numbers. The last digit of the product is 0

One of Huda's numbers must have been a multiple of 10

Zach is wrong. How many different possibilities are there for the last digits of Huda's numbers this time?

3 Find the missing numbers in these calculations.

a
$$\begin{array}{r} 2\ \square \\ \times\quad \square\ 6 \\ \hline 1\ 5\ 6 \\ \square\ 8\ 0 \\ \hline 9\ \square\ 6 \\ \hline \end{array}$$

b
$$\begin{array}{r} \square\ 8 \\ \times\quad 3\ \square \\ \hline 2\ 3\ 2 \\ \square\ 7\ 4\ \square \\ \hline \square\ \square\ 7\ 2 \\ \hline \end{array}$$

c
$$\begin{array}{r} 7\ \square \\ \times\quad \square\ \square \\ \hline \square\ 1 \\ 4\ \square\ 6\ \square \\ \hline \square\ 3\ \square\ \square \\ \hline \end{array}$$

4 Seb and Chloe are working out 358 × 9

I used the column method

I did 358 × 10 = 3580 and subtracted 358

Compare Seb's and Chloe's methods. Which one do you prefer?

5 Adapt Chloe's method in question 4 to work out

a 358 × 99 b 276 × 9.9 c 57 × 1.9

The **area** of a rectangle is found by multiplying its length and its width.

A **parallelogram** can be transformed into a rectangle like this:

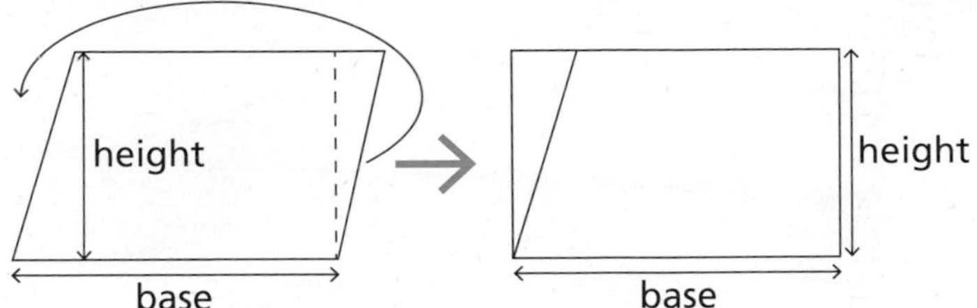

Area of parallelogram = area of rectangle with length equal to the **base** of the parallelogram and width equal to the **perpendicular** height of the parallelogram.

So,

Area of parallelogram = base × perpendicular height.

Example 3

Find the areas of the shapes.

a

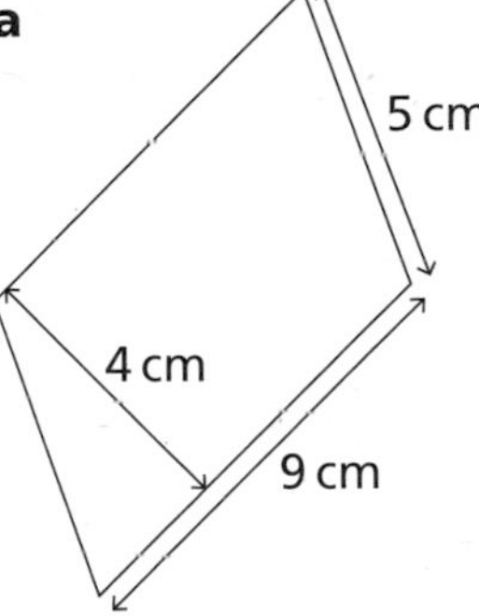

b

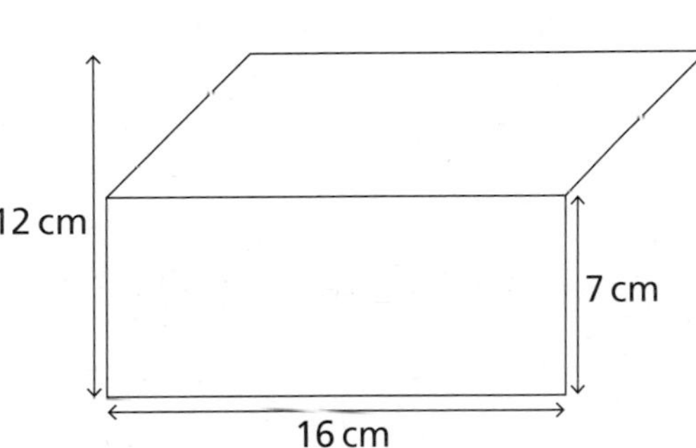

a Area = 9 cm × 4 cm
= 36 cm^2

Remember to multiply the base by the perpendicular height and not the other side of the parallelogram.

Remember to measure area in cm^2, m^2, mm^2 and so on.

b Area of rectangle = 16 cm × 7 cm
= 112 cm^2

You can find this practising the formal method or using a calculator.

Height of parallelogram = 12 cm − 7 cm = 5 cm

Area of parallelogram = 16 cm × 5 cm
= 80 cm^2

Total area = 112 cm^2 + 80 cm^2
= 192 cm^2

The total height of both shapes is 12 cm so you can find the height of the parallelogram by subtraction.

Practice 7.3C

1. Find the areas of the rectangles.

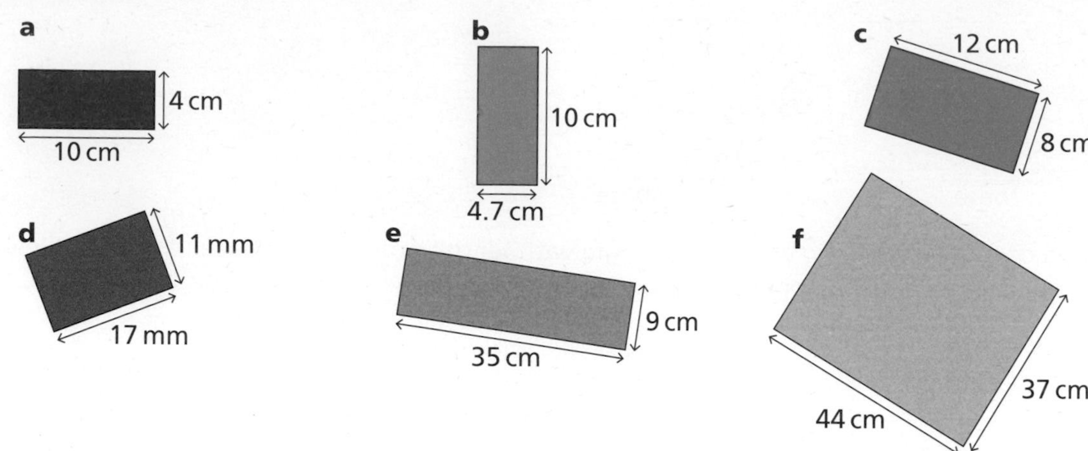

2. Find the areas of the parallelograms.

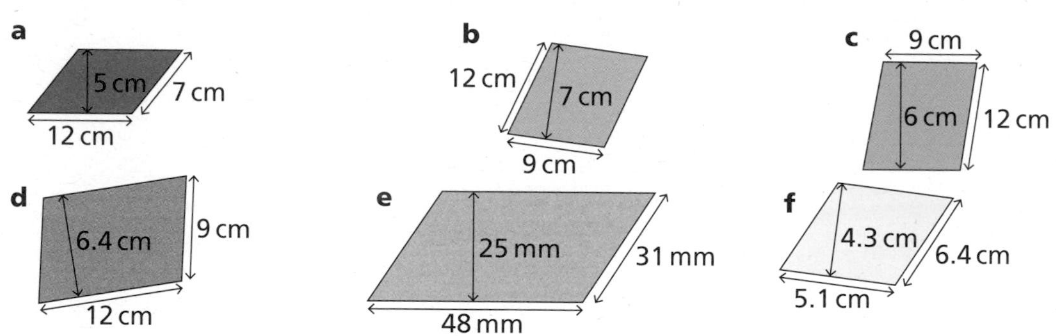

3. Find the areas of the compound shapes.

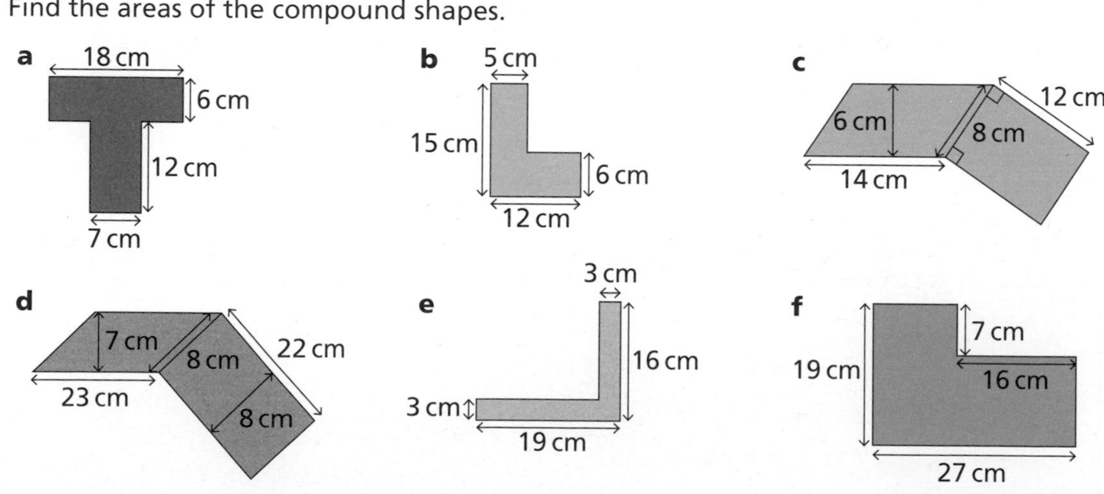

4 Find the shaded areas.

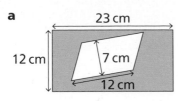

a

23 cm
12 cm
7 cm
12 cm

b

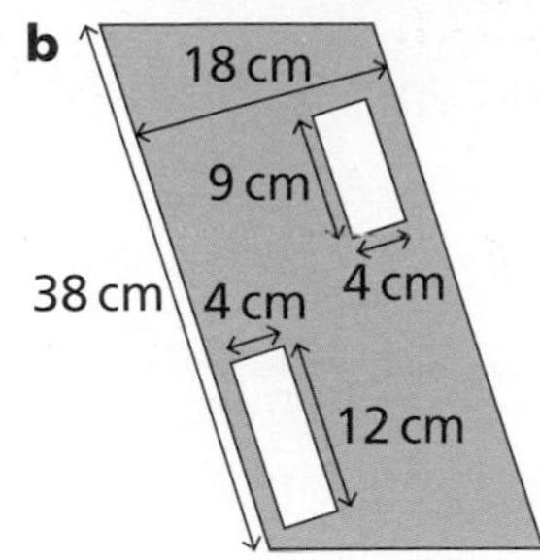

c

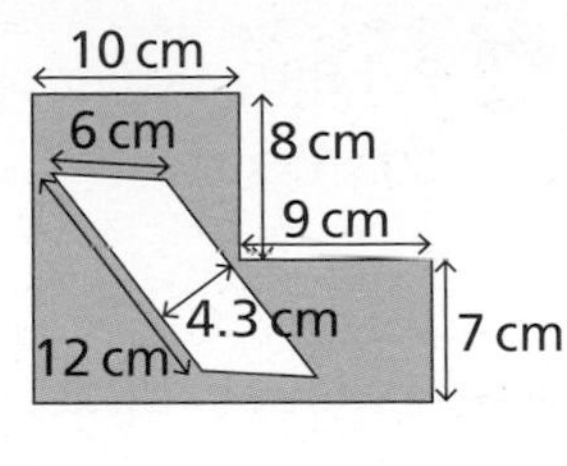

5 **a** A football field is 360 feet long and 160 feet wide. Find the area of the field in square feet.

b A tennis court for singles matches is 78 feet long and 27 feet wide. For doubles matches, the court is the same length but 36 feet wide. Find the difference in area between a court used for a singles match and a court used for a doubles match, giving your answer in square feet.

c A cricket pitch is 66 feet (20.12 m) long and 10 feet (3.05 m) wide. Find the area of a cricket pitch in square feet and in square metres.

What do you think?

1 A rectangle has area 72 cm^2. The lengths of the sides are integers.

a Find the possible lengths of the sides of the rectangle in centimetres.

b What is the connection to the factors of 72?

c Use your answers to parts **a** and **b** to find the possible integer side lengths of a rectangle with area 144 cm^2.

2

The area of the rectangle is 28 × 3 = 84 cm^2

a Explain why Jakub is wrong

b Find the area of the rectangle in cm^2.

c Find the area of the rectangle in m^2.

d Find the area of the rectangle in mm^2.

3

The only rectangle with integer side lengths and area 19 cm^2 measures 1 cm by 19 cm

It's impossible to have a parallelogram with integer side lengths and area 19 cm^2

Do you agree with Marta or Benji? Justify your answers.

7.3 Multiplication: short and long

Consolidate – do you need more?

1 Put the calculations into groups that have the same answers.

80 × 0.3	80 × 3	8 × 3	800 × 0.3	30 × 8
0.8 × 3	80 × 0.03	0.3 × 80	0.8 × 300	3000 × 0.008

2 Find the product of each pair of numbers.

a 48 and 7 **b** 53 and 8 **c** 243 and 5 **d** 316 and 12

e 51 and 17 **f** 38 and 39 **g** 62 and 48 **h** 514 and 46

3 Use your answers to question 2 to find the products of these pairs of numbers.

a 4.8 and 7 **b** 53 and 0.8

c 24.3 and 0.5 **d** 31.6 and 120

e 5.1 and 0.17 **f** 3.8 and 3900

g 0.62 and 4.8 **h** 0.514 and 4600

4 A rule for generating a sequence is

> Multiply the previous term by 8 and then add 14

Find the third term in the sequence if the first term is

a 10 **b** 17 **c** 6.4

5 A rule for generating a sequence is

> Multiply the previous term by 2.7 and then subtract 5.2

Find the third term in the sequence if the first term is

a 10 **b** 17 **c** 6.4

6 Find the areas of the shapes.

a 14 cm, 19 cm

b

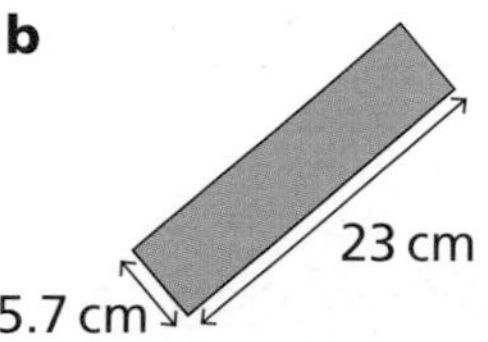

c

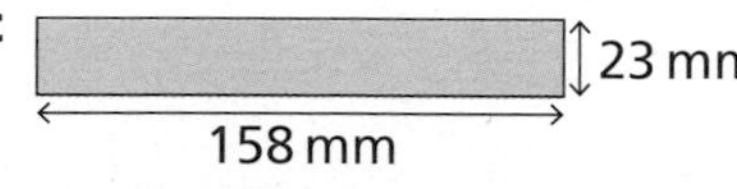

d 9 cm, 11 cm, 16 cm

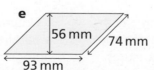

e 56 mm, 74 mm, 93 mm

f 5.2 cm, 6.1 cm, 7.9 cm

7 A rectangular lawn is 8.5 m long and 6.7 m wide.

The lawn contains two square flower beds with sides 1.3 m long.

The rest of the lawn is grass.

Work out the area of grass, in square metres.

Stretch – can you deepen your learning?

1 Given that 67 × 28 = 1876, what divisions can you write the answer to?

2 **a** Show that there is only one way to fill in this multiplication using the digits 1, 2, 3, 4 and 5

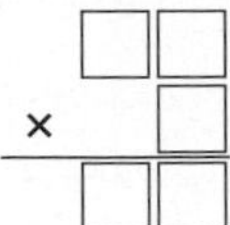

b How many different ways can you fill in the missing numbers for this calculation?

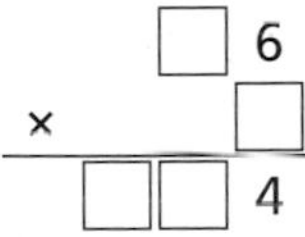

3 Use the fact that multiplication is commutative and associative to find easy ways to work out

a 2 × 37 × 5 **b** 25 × 39 × 4 **c** 8 × 87 × 125

4 Here is a method for working out 75 × 36 using factors and the laws of multiplication

75 × 36

= 25 × 3 × 9 × 4

= 25 × 4 × 3 × 9

= 100 × 27

= 2700

You will explore strategies like this further in Chapter 13.1

Use similar strategies to work out

a 75 × 28 **b** 125 × 16 **c** 32 × 175

5 Explain why the product of two two-digit numbers must be less than 10 000

6 Work out 9 × 9, 99 × 9 and 999 × 9. Describe the pattern and predict the next three terms.

Reflect

Describe how you use the column method to multiply two numbers together. How does it connect to other representations of multiplication?

7.4 Division: short and long

Small steps

- Use formal methods for division of integers
- Use formal methods for division of decimals
- Solve problems using the area of triangles
- Choose the most appropriate method: mental strategies, formal written or calculator

Key words

Dividend – the amount you are dividing

Divisor – the number you are dividing by

Quotient – the result of a division

Remainder – the amount left over when a division is not exact

Are you ready?

1 Calculate

a $24 \div 2$ b $24 \div 3$ c $24 \div 4$

d $24 \div 6$ e $24 \div 8$ f $24 \div 12$

2 Calculate

a $40 \div 4$ b $28 \div 4$ c $16 \div 4$

d $8 \div 4$ e $4 \div 4$ f $2 \div 4$

3 Without finding the answers, decide which will have the greater answer in each pair.

a $800 \div 8$, $800 \div 20$ b $240 \div 8$, $240 \div 10$

4 Write each number to the nearest integer.

a 7.8 b 7.2 c 7.5 d 83.2

e 68.497 f 0.382 g 59.82

Models and representations

Counters and cubes

Counters and cubes can be used to show sharing into equal groups.

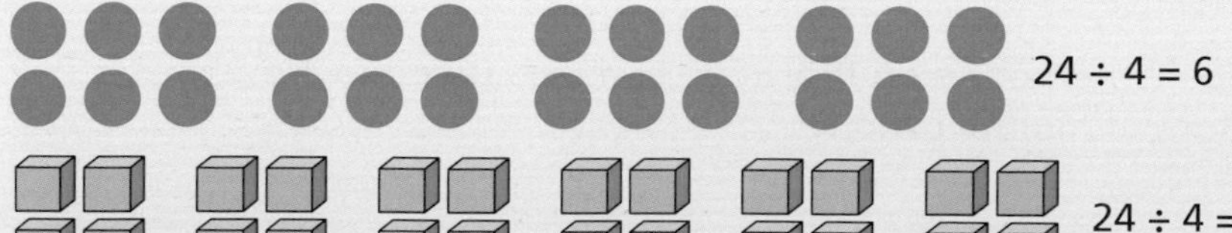

Bead strings

Bead strings can also show sharing into equal groups.

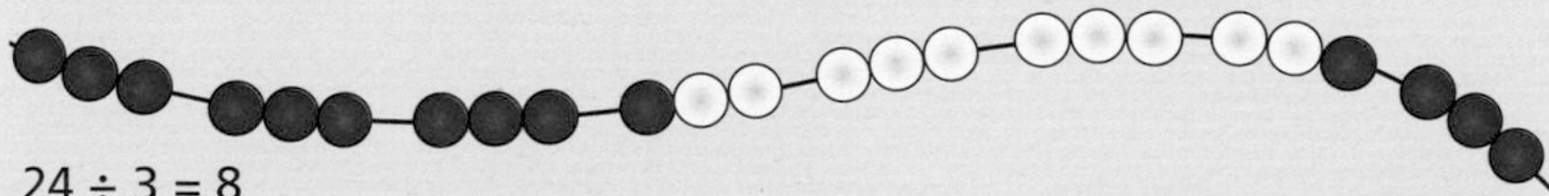

Bar models

60			
15	15	15	15

Bar models can illustrate division facts.

60 ÷ 4 = 15 ⟶ What is 60 shared into 4 equal groups?

60 ÷ 15 = 4 ⟶ How many groups of 15 are there in 60?

You can use place value counters to illustrate the formal method of division you learnt at Key Stage 2.

1

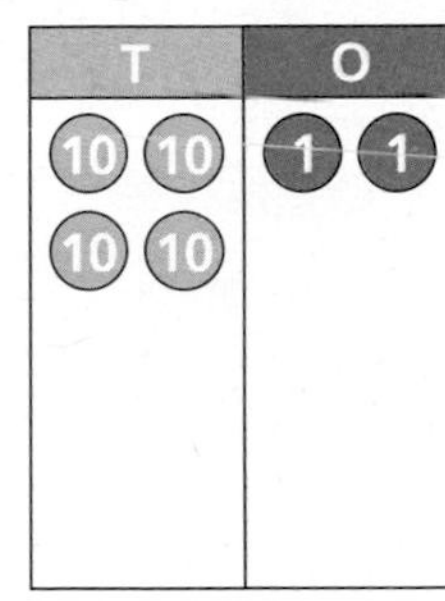

$3\overline{)4\ \ 2}$

To work out 42 ÷ 3, represent 42 with counters and write the calculation as a short division

2

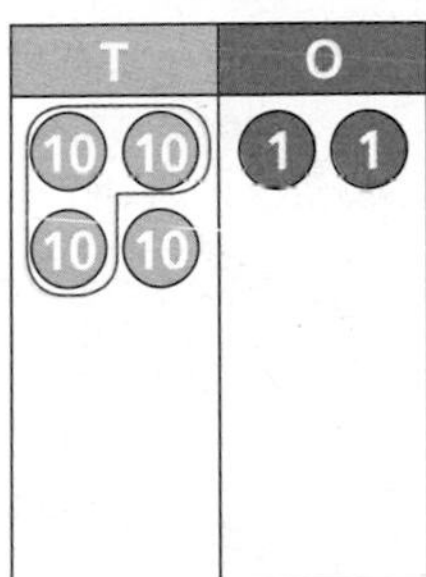

$\begin{array}{r} 1\ \ \ \\ 3\overline{)4\ \ 2} \end{array}$

Firstly, look at how many groups of 3 tens there are in 4 tens. There is one group of 3 tens and one 10 left.

3

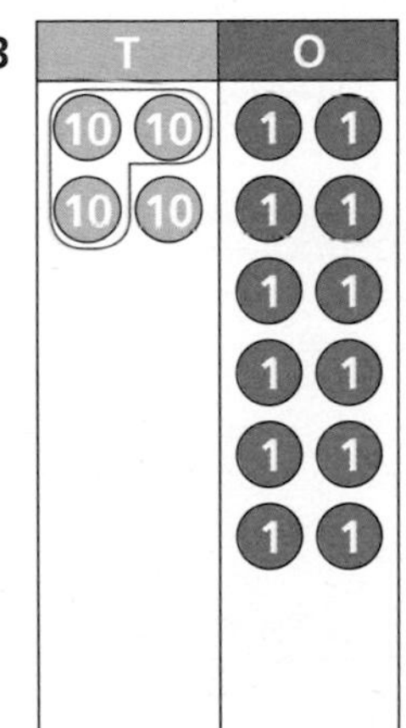

$\begin{array}{r} 1\ \ \ \\ 3\overline{)4\ {}^{1}2} \end{array}$

The one 10 is then exchanged for 10 ones, so there are now 12 ones in the ones column of the place value grid. There is also a 12 in the last column of the short division.

4

$\begin{array}{r} 1\ \ 4 \\ 3\overline{)4\ {}^{1}2} \end{array}$

Now see how many groups of 3 ones there are in 12 ones. The answer is 4 and the division is complete.

So 42 ÷ 3 = 14

The **dividend** is 42, the **divisor** is 3 and the **quotient** is 14

Decimal calculations work in a similar way.

$\begin{array}{r} 1\bullet4 \\ 3\overline{)4\bullet{}^{1}2} \end{array}$

Example 1

Work out 5616 ÷ 24

Method A – long division

```
    234
24 ) 5616
   - 48↓
     81
   - 72↓
      96
    - 96
       0
```

Long division works in the same way as short division but you write the subtractions underneath the calculation to make them easier to see.

You cannot find groups of 24 in 5, so try 56 ÷ 24

24 × 2 = 48 is too small and 24 × 3 = 72 is too large.

So write 2 in the hundreds column of the answer, and subtract 48 from 56 to leave 8

Now look at the 10s column: "bring the 1 down" and there are 81 tens.

24 × 3 = 72 is close, so write 3 in the tens column of the answer, and subtract 72 from 81 to leave 9

Now look at the 1s column: "bring the 6 down" and there are 96 ones.

24 × 4 = 96, so write 4 in the ones column of the answer, subtracting 96 from 96 to show there is no **remainder**.

Method B – using factors

As 24 = 6 × 4, you can divide by 24 by first dividing by 4 and then dividing the result by 6

```
   1 4 0 4
4 )5 ¹6 1 ¹6

   2 3 4
6 )1 4 ²0 ²4
```

Firstly find the quotient of 5616 and 4. As shown, this is 1404

Then find the quotient of 1404 and 6. The answer is 234

You can always check the result of a division by multiplying.

```
     2  3  4
×       2  4
------------
     9  3  6
  4 ¹6 ¹8  0
------------
 ¹5 ¹6  1  6
```

Example 2

Find the quotient of

a 3 and 4 **b** 4 and 3

a

```
   0 • 7  5
4 )3 • ³0 ²0
```

So the quotient of 3 and 4 is 0.75

There are no groups of 4 in 3, so the answer is less than 1. Insert a decimal point and a zero and then find 30 ÷ 4. This is 7 with 2 to carry to another zero.

20 ÷ 4 = 5

b

```
   1 • 3  3 ...
3 )4 • ¹0 ¹0 ¹0
```

So the quotient of 4 and 3 is $1.\dot{3}$

If you work out 4 ÷ 3 in the same way, you repeatedly get the calculation 10 ÷ 3 leading to the answer 1.333333... going on forever. The 3 is "recurring" and the answer is written $1.\dot{3}$.

Notice 3 ÷ 4 Is the same as $\frac{3}{4}$ which you know is 0.75 from Block 5.

Practice 7.4A

1 Use the short division method to work out

a 528 ÷ 3 **b** 52.8 ÷ 3

What's the same and what's different?

2 204 ÷ 6 = 34. Write down the answer to

a 20.4 ÷ 6 **b** 2.04 ÷ 6 **c** 2040 ÷ 6 **d** 408 ÷ 6

3 Work out, showing your method

a 140 ÷ 5 **b** 324 ÷ 6 **c** 324 ÷ 4 **d** 392 ÷ 7

e 255.2 ÷ 4 **f** 164.5 ÷ 7 **g** 390 ÷ 8 **h** 486 ÷ 18

i 754 ÷ 13 **j** 3402 ÷ 27

4 **a** Will the answers to these calculations be greater than or less than 1?

i 5 ÷ 4 **ii** 4 ÷ 5 **iii** 10.8 ÷ 6

iv 3 ÷ 8 **v** 10 ÷ 3

b Work out the answers to the calculations in part **a**.

c Why do you not need to use the short division method to work out 8.4 ÷ 10?

5 Three students are working out 46 ÷ 6

I get 7.6

Ed

Who do you agree with? Why?

6 **a** A school event raises £9360, which is shared equally between 15 charities. How much does each charity receive?

b 12 textbooks are in a pile. The pile is 21.6 cm high. How thick is each textbook?

c 34 rods are laid end to end. The total length is 47.6 m. How long is each rod?

What do you think?

1. In Worked Example 1, Method B, 5616 ÷ 24 is found by dividing 5616 by 4 and then dividing the result by 6
   - **a** Check that you get the same result if you divide 5616 by 6 and then divide the result by 4
   - **b** What other pairs of single digit numbers could you use to work out 5616 ÷ 24? Check you still get the same answer.
   - **c** Use factors to calculate

     i 8766 ÷ 18 **ii** 16 555 ÷ 35 **iii** 2256 ÷ 48
   - **d** What factors could you use to divide a number by

     i 72 **ii** 144 **iii** 84?

     Is there more than one possible answer?

2. What can you say about the quotient of two numbers if the divisor is greater than the dividend?

3. What's the same and what's different about how you round your answers to these problems?
   - **a** 1380 bottles are to be packed into crates containing 18 bottles each. How many full crates will there be?
   - **b** 650 bottles are to be packed into crates containing 18 bottles each. How many crates are needed?

You can use division to solve area problems.

Any triangle is half of a parallelogram.

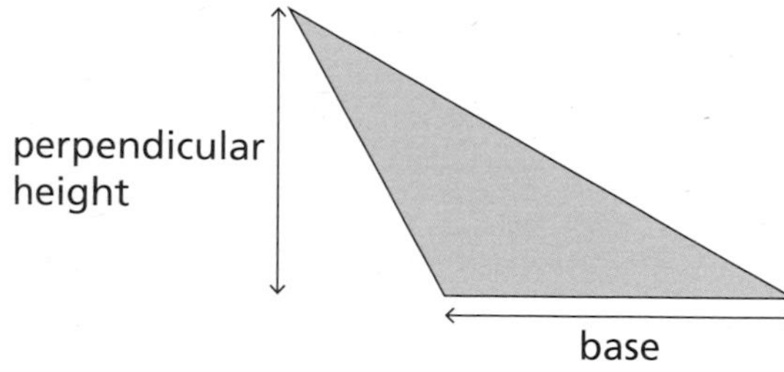

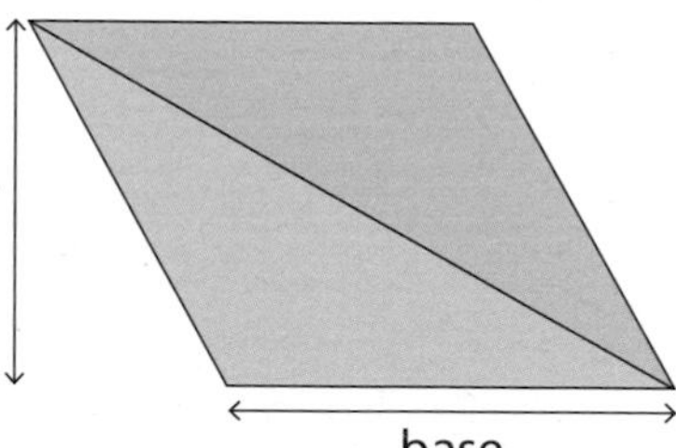

You know the area of a parallelogram = base × perpendicular height

So,

the area of a triangle = $\frac{1}{2}$ × base × perpendicular height, or (base × perpendicular height) ÷ 2

Example 3

The area of the triangle is equal to the area of the parallelogram. Work out the height of the parallelogram.

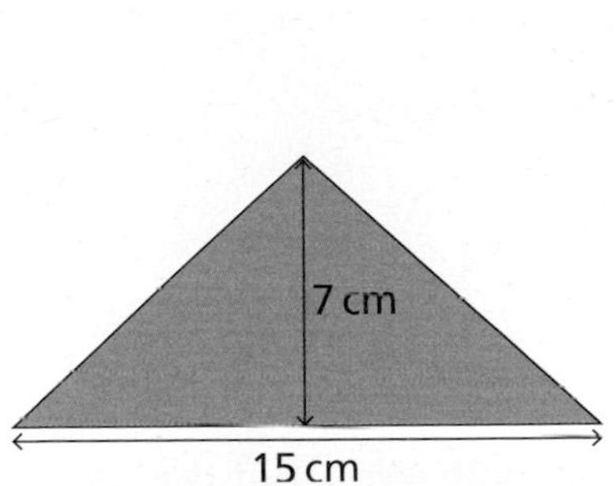

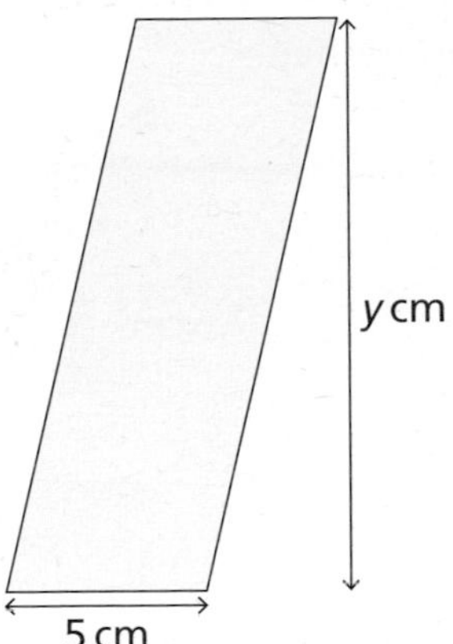

Area of triangle = 15 × 7 ÷ 2
= 105 ÷ 2 = 52.5 cm²

First find the area of the triangle. You might need to write the calculations 25 × 7 and 105 ÷ 2 out in full or you may be able to do them mentally.

For the parallelogram

Base × y = 52.5

5 × y = 52.5

y = 52.5 ÷ 5

Substitute the values into the formula for the area of the parallelogram as it has the same area as the triangle.

Division is the inverse of multiplication.

```
   1 0.5
5 )5 2.²5
```

The height of the parallelogram is 10.5 cm.

Practice 7.4B

1 All these shapes have an area of 108 cm². Work out the missing lengths.

a

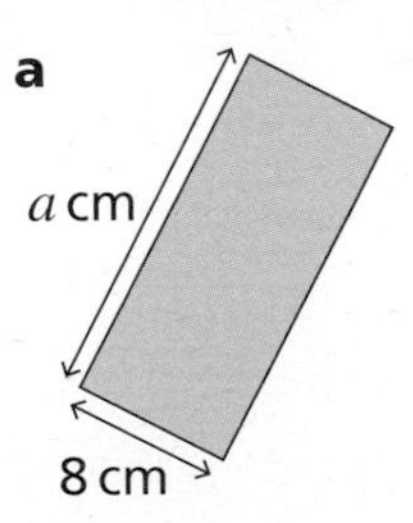

b

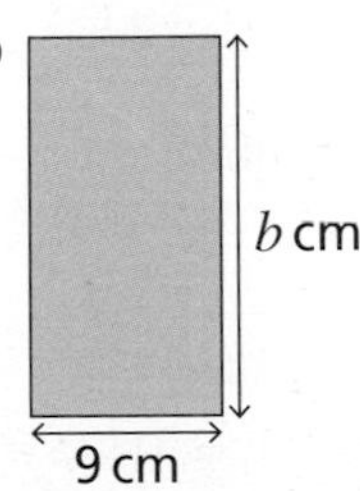

c

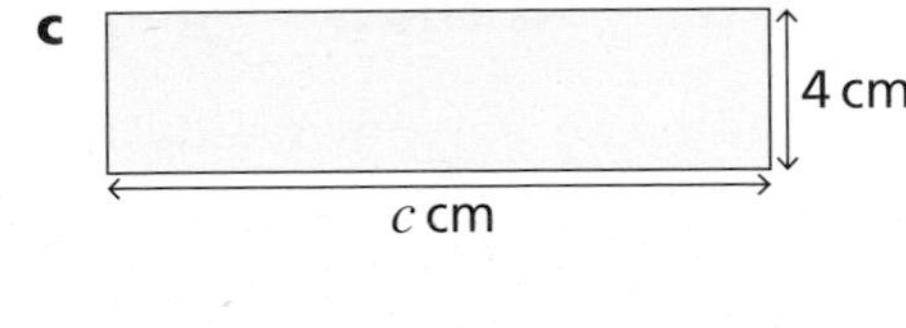

2 Find the areas of the triangles.

a

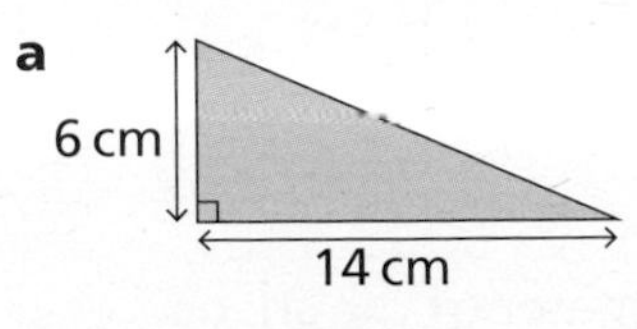

b

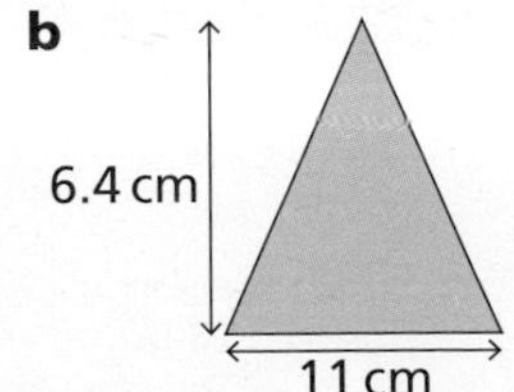

c

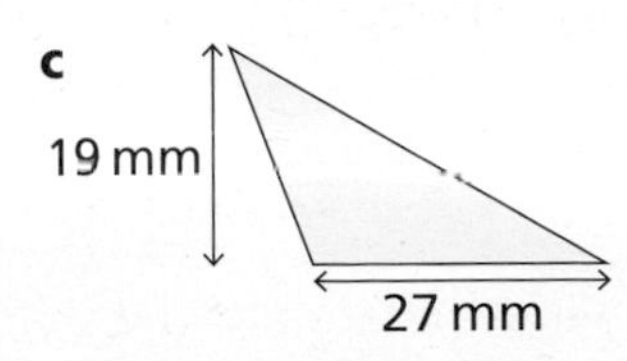

3. The area of the triangle is $112\,mm^2$.

   Find the height of the triangle.

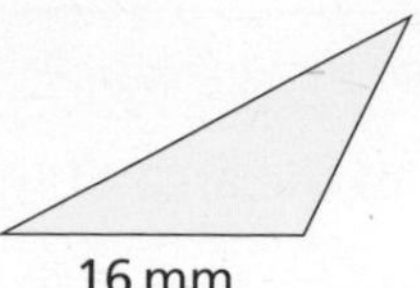

4. Find the area of the compound shapes.

   a

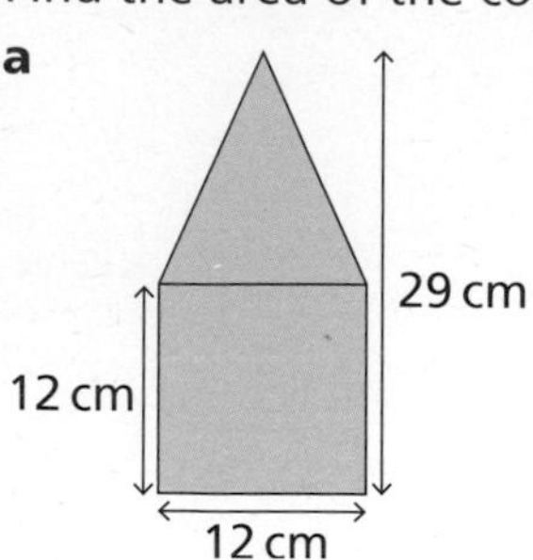

   b

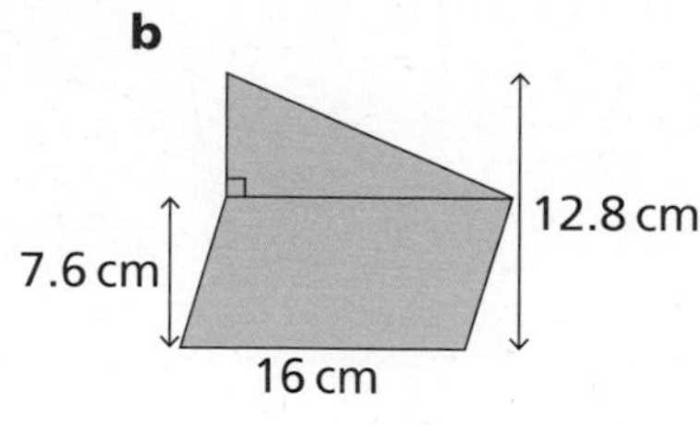

5. The area of the parallelogram is twice the area of the rectangle. Work out the height of the rectangle.

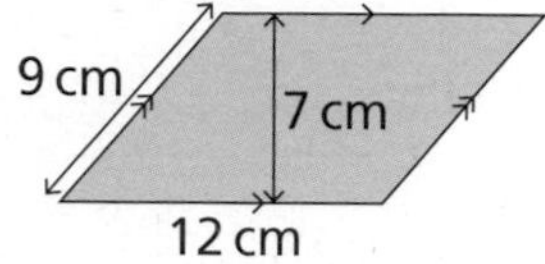

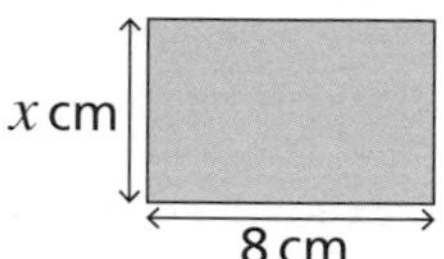

6.

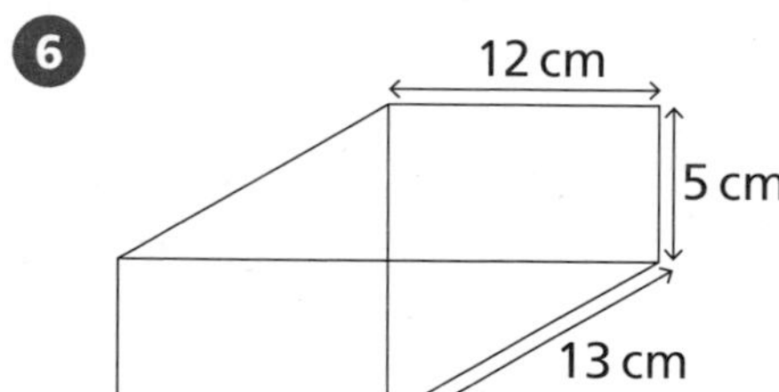

   In this shape, the triangles are identical.

   The rectangles are also identical.

   Identical shapes are also called congruent shapes.

   a Find the area of the shape.

   b Find the perimeter of the shape.

What do you think?

1. The diagram shows a square and two equilateral triangles.

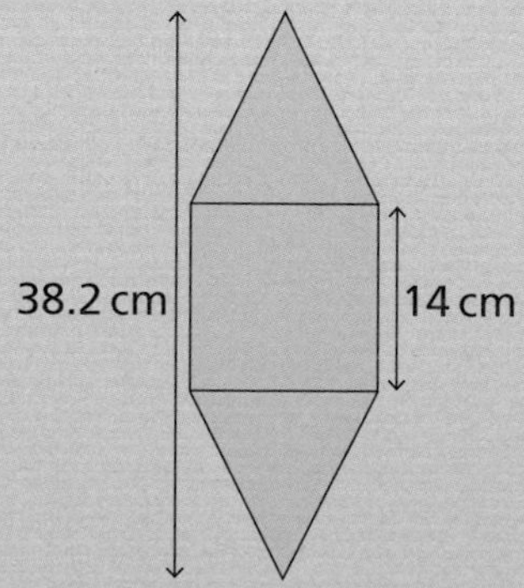

   a Find the area of the shape.

   b Find the perimeter of the shape.

2 Kate estimates the answer to 347 ÷ 9.8 by working out 347 ÷ 10. Is her answer an overestimate or an underestimate?

3 **a** The area of a rectangle is 17 m^2. The length of the rectangle is 25 m.

Chloe, Ed and Faith are trying to work out the width of the rectangle.

Do all of their methods work? Which do you prefer?

I'm going to do 17 ÷ 25 as a long division

Chloe

I'm going to do 68 ÷ 100

Faith

I'm going to do 17 ÷ 5 and then ÷ 5 again

Ed

b The length should have been given as 2.5 m, not 25 m. What is the actual width of the rectangle?

c Find the width of a rectangle with area 12 m^2 and length 2.5 m.

Consolidate – do you need more?

1 Work out, showing your method

a 336 ÷ 4 **b** 3036 ÷ 4 **c** 30 036 ÷ 4 **d** 33.6 ÷ 4

e 33.6 ÷ 8 **f** 8.75 ÷ 7 **g** 8.75 ÷ 8

2 Work out, showing your method

a 912 ÷ 12 **b** 784 ÷ 14 **c** 3248 ÷ 16 **d** 1764 ÷ 36

e 1764 ÷ 49 **f** 552 ÷ 23 **g** 4154 ÷ 31

3 Find the area of a triangle with

a base 12 cm, height 8 cm

b base 12 cm, height 8.4 cm

c base 12.8 cm, height 8.4 cm

4 Work out the height of

a a parallelogram of area 60.3 cm^2 and base 9 cm

b a rectangle of area 79.8 mm^2 and base 7 mm

c a triangle of base 12 cm and area 38.4 cm^2

Stretch – can you deepen your learning?

1 A triangle with integer base and height has area 24 cm^2.

a What might the base and height be?

b How can you be sure you have found them all?

c Investigate for other areas.

2 **a** Use the formal method of division to convert the fractions $\frac{1}{2}, \frac{1}{3}, \frac{1}{4}$ … up to $\frac{1}{11}$ to decimals.

b Which divisions result in terminating decimals and which result in recurring decimals?

Can you give a reason for your answer?

3 **a** Investigate the patterns found when converting $\frac{1}{7}, \frac{2}{7}, \frac{3}{7}$ … up to $\frac{6}{7}$. What do you notice?

b Investigate the patterns found when dividing by 13, 17 and 23

4 **a** Will the answer to $27 \div 2$ be greater than or smaller than the answer to $27 \div 3$?

b Will the answer to $27 \div 0.3$ be greater than or smaller than the answer to $27 \div 3$?

c Compare strategies for working out $27 \div 0.3$

Reflect

Which types of division can you do mentally? Which would you use the written method for, and which would you do on a calculator? Why?

7.5 Using the mean

Small steps

- Solve problems using the mean
- Choose the most appropriate method: mental strategies, formal written or calculator

Key words

Total – the sum of a set of numbers

Mean – the result of sharing the total of a set of data equally between them

1 Find the sums of these sets of numbers.

a 7, 8, 12

b 10, 12, 16, 18, 22, 26

c 100, 94, 103, 97, 110, 108

d 6.3, 6.8, 9.1, 5.7, 6.6, 5.9

2 Work out

a $78 \div 2$ **b** $78 \div 3$ **c** $78 \div 4$ **d** $78 \div 5$ **e** $78 \div 6$

3 Work out

a $107 \div 5$ **b** $107 \div 10$ **c** $107 \div 8$ **d** $114.8 \div 4$ **e** $114.8 \div 7$

Models and representations

Counters

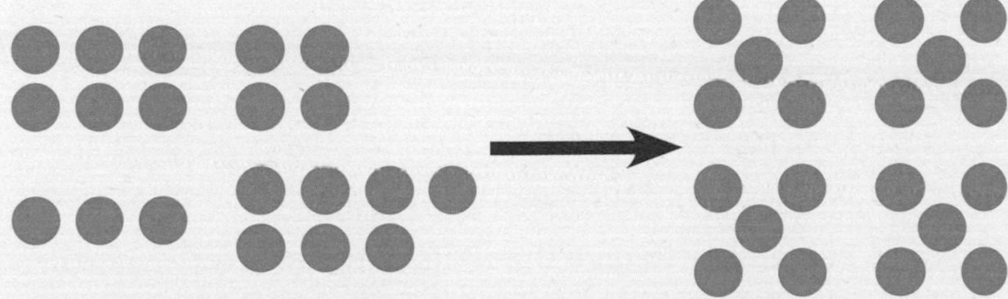

You can use counters to share unequal groups into equal groups.

You can show the same thing with a bar chart.

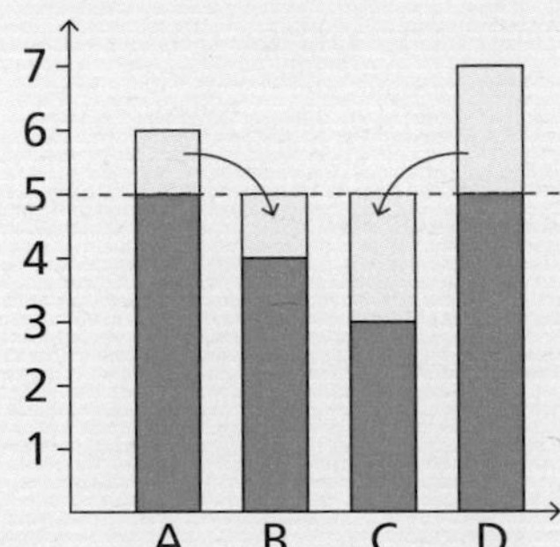

You can also practise using the formal method of division.

$$\begin{array}{r} 6\ 4.8 \\ 5\,\overline{)\,3\ 2\ {}^{2}4.{}^{4}0} \end{array}$$

7.5 Using the mean

The **mean** is an example of an average (remember you used the median, a different average, in Chapter 4.3). The mean of a set of numbers is found by dividing their **total** by the number of items. It's like sharing the total amount equally amongst each member. When people talk about 'the average' they are usually referring to the mean.

Example 1

Here is a set of numbers.

9 7 6 8 6 6

a Find the sum of the set of numbers.

b Hence find the mean of the set of numbers.

a $9 + 7 + 6 + 8 + 6 + 6 = 42$ — Remember "sum" is another word for total.

b mean $= 42 \div 6 = 7$ — Divide the total by the number of items.

Example 2

a Find the mean of these masses.

30 g 52 g 44 g 200 g

b Is the mean representative of the data? Explain your answer.

c The 200 g mass is replaced by a 20 g mass.

How does the mean change?

a $30 + 52 + 44 + 200 = 326g$ — First you find the total.

mean $= 324g \div 4 = 81.5g$ — As there are four weights, you divide the total by 4

Remember to give the units in the answer.

b No, because the mean is not close to any of the values. — To be representative, an average needs to give a sense of what the whole group of data is like. In this case, the mean is quite a lot bigger than three of the masses and quite a lot smaller than the other mass.

c $326 - 200 + 20 = 126g$

new mean $= 126g \div 4 = 31.5g$ — You can adjust the total by subtracting the 200 g and adding on the 20 g. This is quicker than working out 30 + 52 + 44 + 20

This strategy is really useful if you have a large number of items.

Example 3

The mean of five numbers is 40

What might the numbers be? Give two possible answers.

$40 \times 5 = 200$

You know that the mean is 40, which means the total divided by 5 is 40

$\frac{\text{total}}{5} = 40$

Solve the equation by using the inverse operation.

total = 40 × 5 = 200

The numbers could be 40, 40, 40, 40 and 40

This is an obvious set!

The numbers could also be 30, 30, 30, 30 and 80

If you take 10 off each of the first four numbers you would need to add 40 (4 × 10) to the last number. If the total does not change, then the mean will not change either.

There are many more possible answers, for example 39, 41, 40, 40, 40 or 0, 0, 0, 0, 200 and so on.

Practice 7.5A

1 Find the means of these sets of numbers.

a 5, 6 **b** 5, 6, 7 **c** 5, 6, 7, 8 **d** 5, 6, 7, 8, 9 **e** 5, 6, 7, 8, 9, 10

2 **a** Find the mean of the data set 12, 15, 20, 32, 40

b An extra number is added to the data set, and its value is 19. How does the mean change?

c Instead of 19, the value of the extra number is 91. How does the mean change now?

3 Find the means of these sets of numbers.

a 7.1, 9.9, 11.2 **b** 7.4, 3.8, 9.2, 4.7, 6.4

4 Find the means of these lengths.

a 24 cm, 28 cm, 39 cm and 60 cm

b 31 mm, 52 mm, 74 mm, 82 mm and 105 mm

c 8 cm, 12 cm, 9.5 cm, 47 mm and 82 mm

d 95 cm, 85 cm, 1 m 3 cm, 99 cm and 1 m

5 Jackson is finding the mean of this set of numbers.

21 32 34 37 54

a Show that the answer Jackson should get is 35.6

Jackson realises the last number in the list should be 45 not 54

b Is the mean he has found larger or smaller than the mean of the correct set of numbers? Explain your reasoning.

c Find the mean of the correct set of numbers.

6 The mean of a group of numbers is 20. The total of the numbers is 180. How many numbers are in the set?

7 The mean of a set of numbers is 12. There are 15 numbers in the set. What is the total of the set of numbers?

8 I have five number cards. The numbers on all five cards are all different.

?	?	?	?	?

The mean of the numbers on the cards is 8

What might the numbers on the cards be?

What do you think?

1 The mean mass of a team of rugby players is 95 kg.

a Would the mean increase or decrease if a new player of mass 80 kg joined the team?

b Describe how the mean would change if a player of mass 95 kg joins the team.

c Describe how the mean would change if a player of mass 105 kg joins the team.

2 a How would the mean change if the smallest item is removed from a data set?

b How would the mean change if the largest item is removed from a data set?

3 Find the means of these sets of data.

a 1, 2, 3 and 7

b 11, 12, 13 and 17

c 101,102, 103 and 107

d 10, 20, 30 and 70

e 20, 40, 60 and 140

What do you notice about your results?

4 The mean of 1.7, 3.2, 6.8, 6.8 and 11.2 is 5.94

Write down the mean of

a 31.7, 33.2, 36.8, 36.8 and 41.2

b 170, 320, 680, 680 and 1120

c 174, 324, 684, 684 and 1124

d 0.7, 2.2, 5.8, 5.8 and 10.2

In each case, explain how you found your answers.

5 Six friends play a game.

Their mean score in the game is 19

The mean score of four of the friends is 24

Find the mean score of the remaining two friends.

Consolidate – do you need more?

1 What number should you divide the total by to find the mean for each of the sets of data?

a 7, 10, 8, 6

b 7, 10, 8

c 7, 8, 10, 6, 4

d 7, 8, 10, 6, 40

2 Here are Chloe's marks in her end of term tests.

English	73%
Maths	85%
Science	71%

French	88%
History	71%
Geography	56%

a Find Chloe's mean mark.

b Which average of Chloe's marks is greater, the mean or the median?

c Find the range of Chloe's marks.

3 Find the total of the values in a data set if

a there are 10 values and their mean is 6

b there are 6 values and their mean is 10

c there are 15 values and their mean 46

d there are 15 values and their mean is 4.6

e there are 150 values and their mean is 4.6

4 The mean height of the Year 7 students in a school is 147.3 cm. There are 205 Year 7 students in the school. What is their total height?

5 Seb gets seven books for his birthday. The number of pages in each book is

204 187 315 118 46 309 207

Seb says: "My books have an average of nearly 200 pages."

Show that Seb is correct.

6 The school football team plays 10 matches one season. Their score is shown first in each pair. The results were

3 – 0, 3 – 2, 2 – 3, 1 – 0, 0 – 4, 2 – 5, 3 – 1, 4 – 2, 1 – 3, 0 – 3

a Find the mean number of goals scored per game.

b Find the mean number of goals scored by the school's team.

c Find the mean number of goals scored by their opponents.

7 There are 20 students in a class. Their mean mark in a Science test is given as 62%. However, one paper has been marked incorrectly and awarded 58% instead of 85%. What should the correct mean mark in the test be?

Stretch – can you deepen your learning?

1 **a** The mean of the numbers on these cards is 8

5	10	4	?

Find the missing number.

b The mean of another set of four cards is 7. The median of these cards is 10. What could the numbers be if all the cards show integers? What could the numbers be if none of the cards show integers?

2 Beca has four number cards.

The mean of the first two cards is 8

The mean of the first three cards is 9

The mean of all cards is 10

Which cards can you work out the value of?

How much more information do you need to work out the value of all of the cards?

3 How many sets of four integers have a mean of 7, a median of 8 and a range of 10?

4 The mean of a set of numbers is an integer. Their total is 50. How many numbers might there be?

5 Find the mean of each sets of expressions.

a $3a$, $4a$, $5a$, $6a$, $12a$

b $a + 3$, $a + 5$, $a + 13$

6 Chloe says the mean of a set of three consecutive numbers is the middle number. Is Chloe right? Why or why not?

Reflect

1 What's the difference between finding the mean and finding the median of a set of numbers?

2 How can you find missing values in a set of data if you know the mean and some of the values?

7.6 Order of operations

Small steps

- Understand and use order of operations
- Choose the most appropriate method: mental strategies, formal written or calculator

Key words

Operation – a mathematical process such as addition, multiplication, squaring and so on

Priority – a measure of the importance of something

Are you ready?

1 Write down the answers to

a 4×6 **b** 7×3 **c** 9×2

d 6×6 **e** 5×9 **f** 12×3

2 Write down the answers to

a $14 \div 2$ **b** $30 \div 3$ **c** $24 \div 8$

d $40 \div 5$ **e** $28 \div 4$ **f** $30 \div 5$

3 Work out

a $8 + 3 - 4$ **b** $8 - 4 + 3$ **c** $9 + 7 - 2$ **d** $9 - 2 + 7$

What do you notice about your answers to parts **a** and **b** and your answers to parts **c** and **d**?

4 Work out

a $4 \times 5 \div 2$ **b** $4 \div 2 \times 5$ **c** $40 \div 10 \times 3$ **d** $40 \times 3 \div 10$

What do you notice about your answers to parts **a** and **b** and your answers to parts **c** and **d**?

Models and representations

Calculations involving more than one **operation** can be represented using counters.

shows $2 + 5 \times 3$

Altogether there are 17 counters.

shows $(2 + 5) \times 3$

Altogether there are 21 counters this time.

7.6 Order of operations

There are several rules of **priority** in calculations you need to know.

Addition and *subtraction* have equal priority.

If a calculation has only additions and subtractions, you just work from left to right:

$8 + 4 - 7 - 2 + 5$

$= 12 - 7 - 2 + 5$

$= 5 - 2 + 5$

$= 3 + 5$

$= 8$

Multiplication and *division* have equal priority too.

If a calculation has only multiplications and divisions, you just work from left to right:

$8 \times 4 \div 2 \times 5$

$= 32 \div 2 \times 5$

$= 16 \times 5$

$= 80$

Multiplication and division take priority over *addition and subtraction*

In $3 + 4 \times 5$, do the multiplication first

$3 + 4 \times 5$

$= 3 + 20$

$= 23$

Powers and *roots* take priority over *multiplication and division*

$3 \times 5^2 = 3 \times 25 = 75$

If you want the calculations in a different order, put brackets around the part you want to do first.

$(3 \times 5)^2$ means do the multiplication first.

$(3 \times 5)^2 = 15^2 = 225$

This is different from 3×5^2

$(3 + 4) \times 5$ means do the addition first.

$(3 + 4) \times 5 = 7 \times 5 = 35$

This is different from $3 + 4 \times 5$

The order can be summarised in this pyramid.

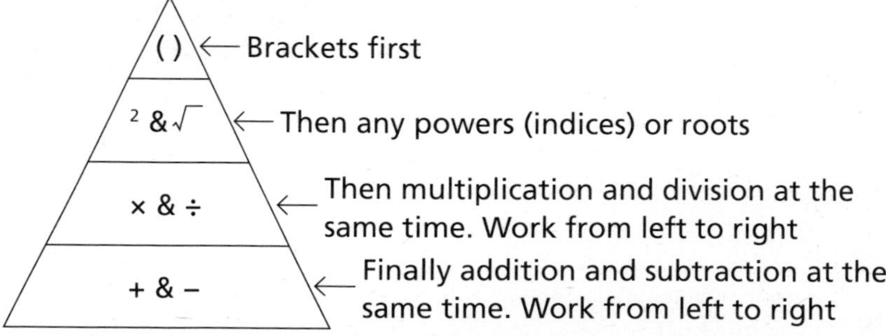
( )
Brackets first
² & √
Then any powers (indices) or roots
× & ÷
Then multiplication and division at the same time. Work from left to right
+ & –
Finally addition and subtraction at the same time. Work from left to right

This is called the "order of operations".

Example 1

Work out

a $6 \div 2 + 4$ **b** $6 \div (2 + 4)$ **c** $6^2 \div 2 + 4$

d $6 \div 2^2 + 4$ **e** $6 \div (2^2 + 4)$ **f** $(6 \div 2)^2 + 4$

It can be helpful to underline the parts you are doing first as you do them.

a $\underline{6 \div 2} + 4$ — Division takes priority over addition.
$= \underline{3 + 4}$
$= 7$

b $6 \div \underline{(2 + 4)}$ — Calculations in brackets take priority, so you do the addition first.
$= \underline{6 \div 6}$
$= 1$

c $\underline{6^2} \div 2 + 4$ — The squaring takes priority.
$= \underline{36 \div 2} + 4$ — The division is next.
$= \underline{18 + 4}$ — Finally the addition.
$= 22$

d $6 \div \underline{2^2} + 4$ — The squaring takes priority.
$= \underline{6 \div 4} + 4$ — The division is next.
$= \underline{1.5 + 4}$ — Finally the addition.
$= 5.5$

e $6 \div (\underline{2^2} + 4)$ — Calculations in brackets take priority, and in the brackets, squaring takes priority.
$= 6 \div \underline{(4 + 4)}$ — Next do the addition in the brackets.
$= \underline{6 \div 8}$ — Finally you do the division.
$= 0.75$

f $\underline{(6 \div 2)}^2 + 4$ — Calculations in brackets take priority, so do the division first.
$= \underline{3^2} + 4$ — Next, squaring takes priority over addition.
$= \underline{9 + 4}$ — Finally the addition.
$= 13$

Practice 7.6A

1 Work out

a $20 + 6 - 8$ **b** $20 - 8 + 6$ **c** $6 + 20 - 8$

d $20 - 8 - 6$ **e** $20 - (8 - 6)$ **f** $20 - (8 + 6)$

2 Work out

a $48 \div 4 \times 3$ **b** $48 \times 3 \div 4$ **c** $3 \times 48 \div 4$

d $4 \times 48 \div 3$ **e** $48 \div 3 \times 4$ **f** $48 \div (3 \times 4)$

3 Work out

a $18 - 4 \times 2$ **b** $(18 - 4) \times 2$ **c** $18 \times 4 - 2$

d $(18 \times 4) - 2$ **e** $18 - 4 \div 2$ **f** $(18 - 4) \div 2$

4 Flo works out $4 \times 5 + 6 \times 10$

She gets the answers 260

a Explain what Flo has done wrong

b Work out the correct answer to $4 \times 5 + 6 \times 10$

5 A taxi charges £5 for a journey plus £2 for every mile travelled. Which of these calculations shows the cost of a 20-mile journey?

$(5 + 2) \times 20$ $5 + 2 \times 20$ $5 \times 20 + 2 \times 20$ $5 \times 20 + 2$

6 Work out

a $5 \times 2 + 4$ **b** $5 \times 2^2 + 4$ **c** $5 \times 2 + 4^2$ **d** $5^2 \times 2 + 4$

e $(5 \times 2)^2 + 4$ **f** $(5 \times 2 + 4)^2$ **g** $5^2 - 2 + 4$ **h** $5^2 - (2 + 4)$

i $5 - 2^2 + 4$ **j** $(5 - 2)^2 + 4$ **k** $5 - 2 + 4^2$ **l** $(5 - 2 + 4)^2$

7 Insert brackets, if needed, to make each calculation correct.

a $7 + 3 \times 4 + 2 = 42$ **b** $7 + 3 \times 4 + 2 = 25$

c $7 + 3 \times 4 + 2 = 21$ **d** $7 + 3 \times 4 + 2 = 60$

What do you think?

1

In the rule BIDMAS A comes before S so the calculation 10 – 4 + 5 = 10 – 9 = 1

a Explain why Ed is wrong.

b Find the correct answer to 10 – 4 + 5

2 Beca says you do not need the brackets in the calculation 3 + (4 × 5). Is she correct? Why or why not?

3 You can make 24 using 5, 3, 2 and 1 in several ways, like this

5 × (3 + 2) – 1   (5 + 3) × (2 + 1)   (5 + 2 + 1) × 3   (5 – 1) × 3 × 2

Write a calculation with the answer 24 using each set of numbers, including brackets where necessary.

a 5, 4, 2 and 1   **b** 10, 4, 2 and 1   **c** 7, 4, 2 and 3   **d** 5, 6, 7 and 8

There are many possible answers for each one.

Compare your answers with a partner's. What strategies did you use?

4 Investigate other sets of numbers to make 24. Can you find sets that do not need brackets?

5 Find the difference between $3a^2$ and $(3a)^2$ when $a = 5$

Consolidate – do you need more?

1 Work out

a 12 + 4 – 2   **b** 12 + 4 × 2   **c** 12 + 4 ÷ 2   **d** (12 + 4) ÷ 2

e 12 + (4 × 2)   **f** 12 + 2 ÷ 4   **g** (12 + 2) ÷ 4   **h** 12 × 2 ÷ 4

i 12 × 4 ÷ 2   **j** 12 × 4 + 2   **k** 12 × (4 + 2)   **l** 12 ÷ (4 + 2)

m 12 ÷ 4 + 2   **n** 12 ÷ 2 + 4   **o** 12 ÷ (2 + 4)   **p** 12 ÷ 2 – 4

q 12 ÷ 4 – 2   **r** 12 ÷ 2 – 4

2 I start with 7, add on 5 and then multiply the answer by 6

Which calculation shows this: 7 + 5 × 6 or (7 + 5) × 6?

3 Work out

a $4^2 + 5$ **b** $4 + 5^2$ **c** $(4 + 5)^2$ **d** $4^2 + 5^2$

4 Beca has fifteen marbles. She gives away seven of them and then loses half of the rest.

Write a calculation that shows this, and find out how many marbles Beca has now.

5 By inserting brackets, how many different answers can you find for each of these calculations?

a $6 \times 4 + 2 \times 5$ **b** $120 \div 40 - 16 \div 2$ **c** $72 \div 12 - 3 \times 2$

Do any of the questions give the same answer with or without brackets?

Stretch – can you deepen your learning?

You can use the numbers 1, 4, 5 and 9 together with +, −, ×, ÷, √ and brackets to make other numbers in many ways, for example

$1 = 1^{459}$ or $1 = (1 \times 4 + 5) \div 9$ or $1 = 1 + 4 + 5 - 9$

$20 = 1^9 \times 4 \times 5$ or $20 = (1^5 + 9) \times \sqrt{4}$

Investigate how many numbers up to 50 you can make. Remember to include brackets where they are needed.

Reflect

1 Why do you think there is an order for operations?

2 Why do you think multiplication and division take priority over addition and subtraction?

H 7.7 More complex multiplication and division

Small steps

- Solve problems using the area of trapezia H
- Explore multiplication and division in algebraic expressions H

Key word

Trapezium – a quadrilateral with one pair of parallel sides

Are you ready?

1 Find the area of each rectangle, giving your answer in appropriate units.

a

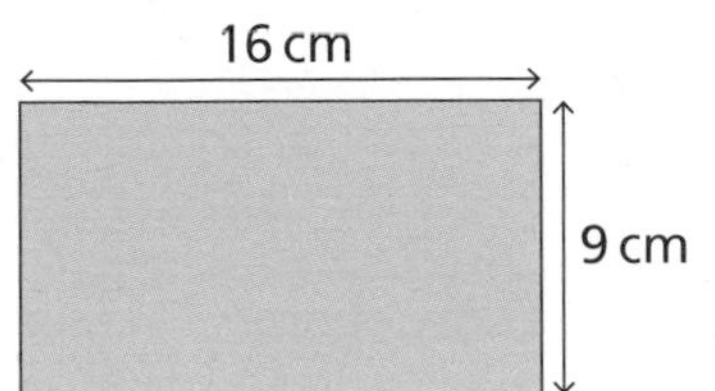

b

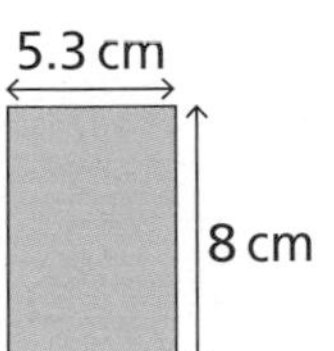

c

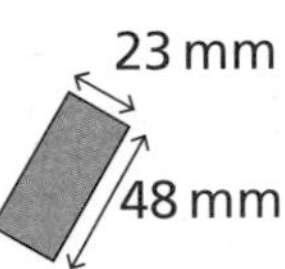

2 Find the areas of the parallelograms, giving your answer in appropriate units.

a 6 cm 4 cm 5 cm

b

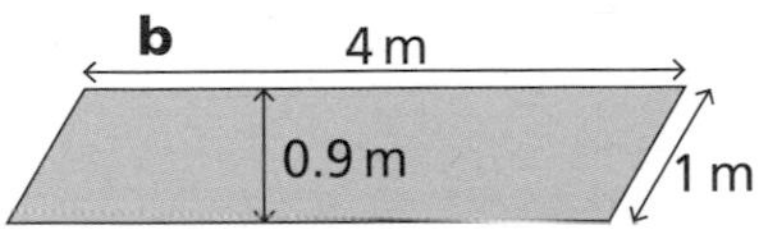

c

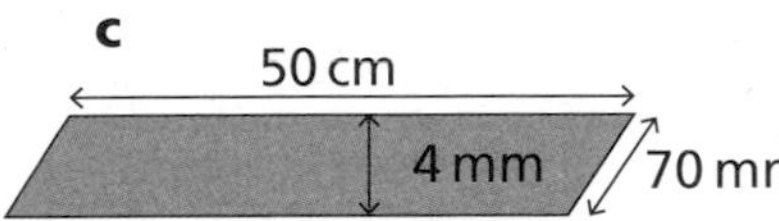

not drawn to scale

3 Find the areas of the triangles, giving your answer in appropriate units.

a

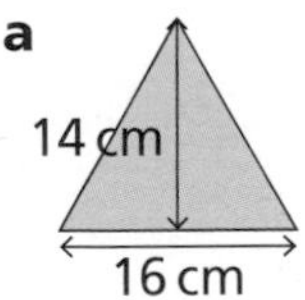

b

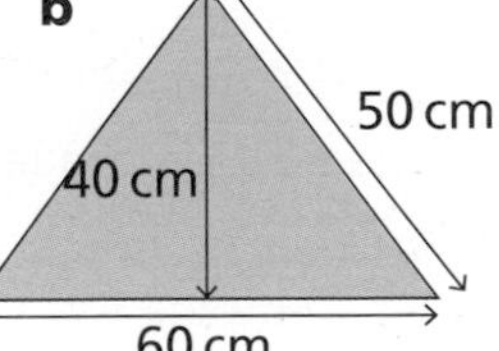

c

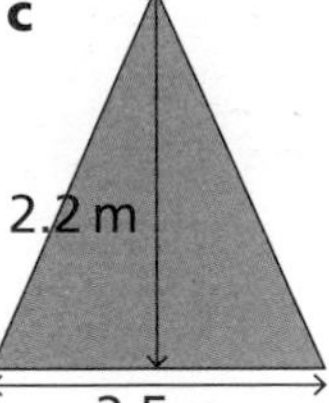

4 Given that $a = 10$, $b = 16$ and $h = 6$ find the values of

a $a + b$ **b** $\frac{1}{2}(a + b)$ **c** $(a + b) \times h$ **d** $\frac{1}{2}(a + b) \times h$

5 Simplify the expressions.

a $2 \times a$ **b** $a \times 2$ **c** $a \times b$ **d** $b \times a$

e $a \times a$ **f** $b \times b \times b$ **g** $a \times 4 \times b$

Models and representations

You will need to recognise trapezia in various orientations – look for a pair of parallel sides, which will usually (but not always) be indicated by arrows.

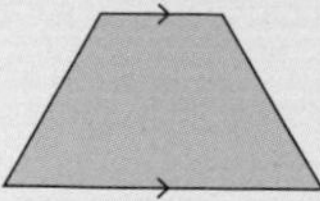
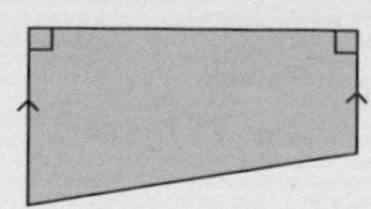
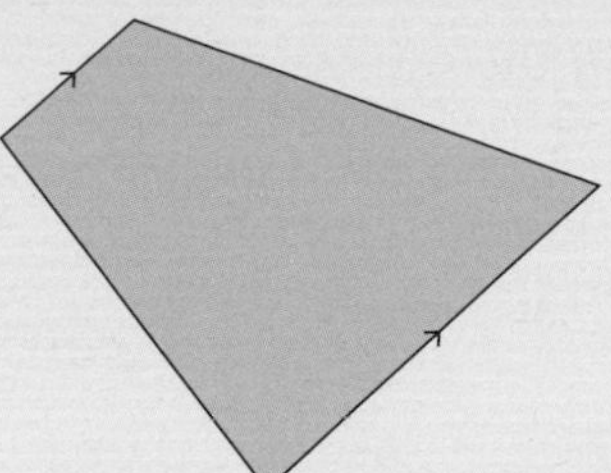
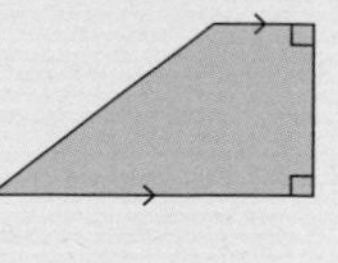

You also need to be able to work fluently with algebraic expressions, including those that involve brackets.

$c(a + b)$

$\frac{x}{y}(p + q^2)$

It is not obvious how to find the area of a **trapezium** just by looking at the shape.

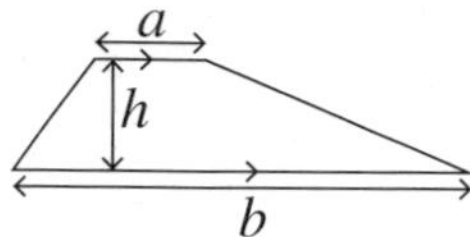

However, if you make two copies of the shape and then put them together…

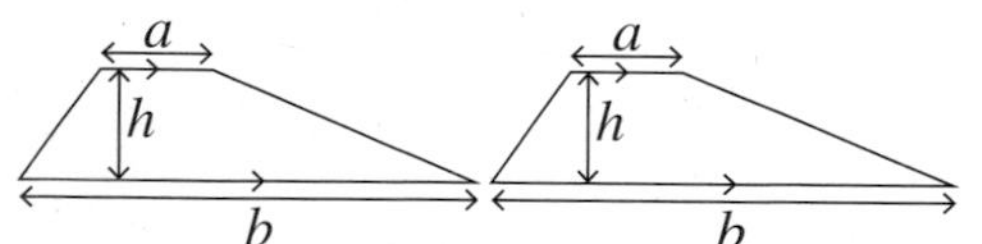

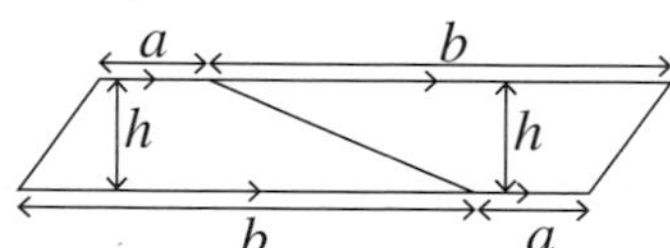

You see that they make a parallelogram.

The area of the parallelogram is base × perpendicular height, or $(a + b) \times h$.

The area of one trapezium is half the area of the parallelogram.

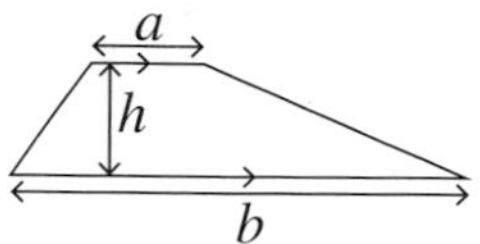

Area of a trapezium $= \frac{1}{2}(a + b) \times h$

$= \frac{1}{2}(a + b)\,h$

You might see this written without the multiplication sign as $\frac{1}{2}(a + b)h$, or you might find it helpful to think of the formula as "half the sum of the parallel sides multiplied by the distance between them".

Example 1

Find the areas of the trapezia, giving your answers in appropriate units.

a

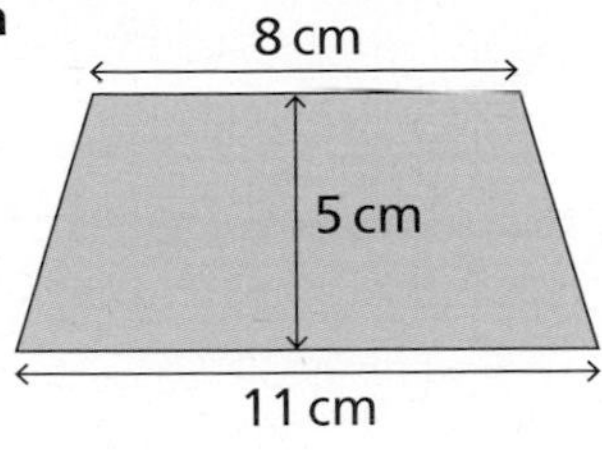

b

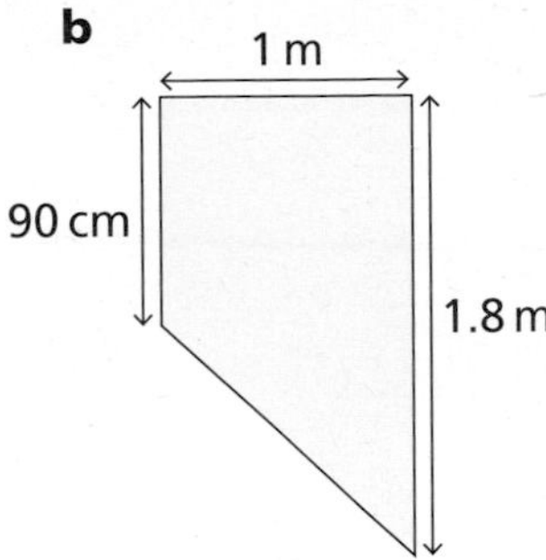

a $\text{Area} = \frac{1}{2}(a+b)h$ — Use the formula for the area of a trapezium.

$= \frac{1}{2}(11+8)5$ — a and b are the parallel sides – it does not matter which is 11 and which is 8. h is the perpendicular height, which is 5

$= \frac{1}{2} \times 19 \times 5$ — Do the calculation in brackets first, and then multiply the numbers together.

$= 47.5\ \text{cm}^2$

b $\text{Area} = \frac{1}{2}(a+b)h$ — Use the formula for the area of a trapezium.

$= \frac{1}{2}(1.8+0.9)1$ — a and b are the parallel sides. You should change 90 cm to 0.9 m so all three lengths are in the same units.

$= \frac{1}{2} \times 2.7 \times 1$ — Do the calculation in the brackets first, and then multiply the numbers together.

$= 1.35\ \text{m}^2$

Practice 7.7A

1 Which of these shapes are trapezia?

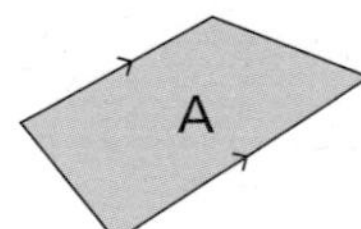

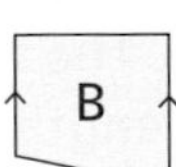

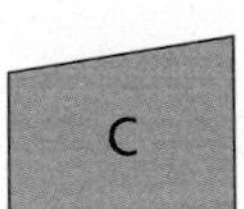

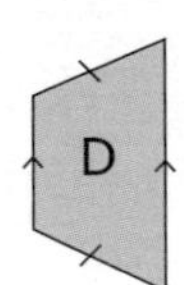

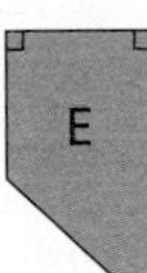

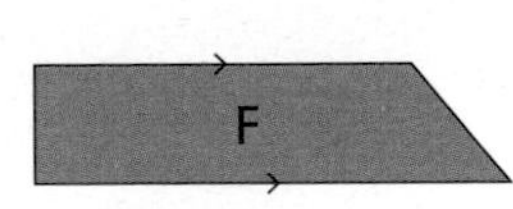

2 Find the area of each trapezium.

a 9 cm, 8 cm, 15 cm

b

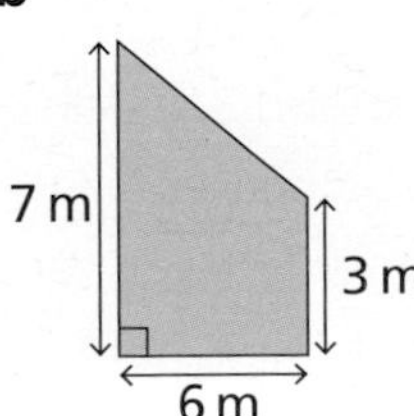

c

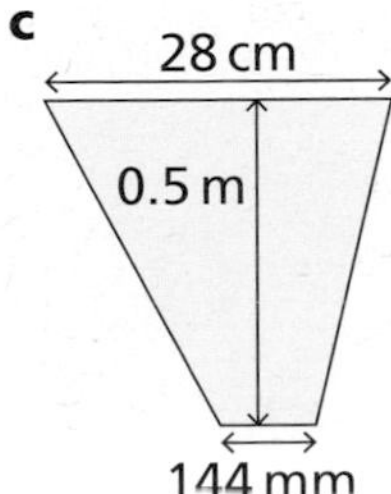

3 The diagram shows the cross-section of a squash court. Find the area of the cross-section.

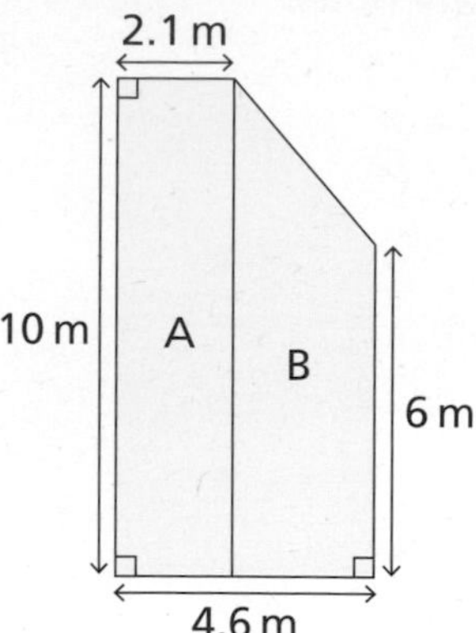

4 Find the area of the compound shapes.

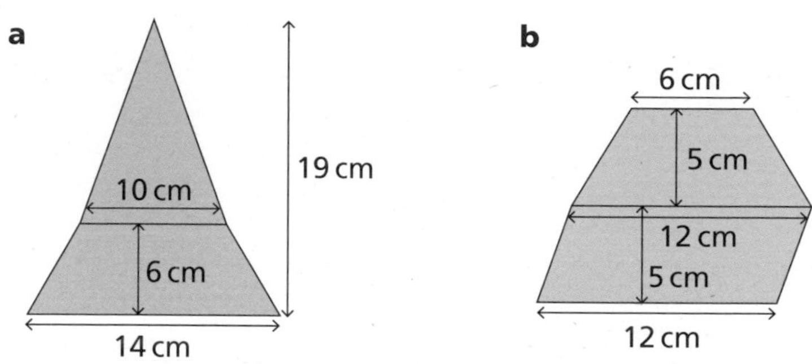

5 A logo consists of a parallelogram on top of a trapezium, as shown.

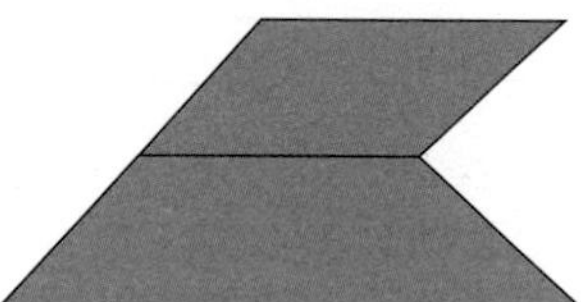

The parallelogram has height 16 mm and base 32 mm. The trapezium is the same height as the parallelogram and its base is twice the length of its top. Find the area of the logo, in square millimetres.

What do you think?

1 A trapezium has area 40 cm^2.

a Its parallel sides are 7 cm and 3 cm long. Find the height of the trapezium.

b Another trapezium with area 40 cm^2 has one parallel side 10 cm long and height 5 cm. Find the length of the second parallel side.

2 How many trapezia can you find with an area of 24 cm^2, given that the lengths of the parallel sides and the height are all integer numbers of centimetres?

You can use the commutative and associative laws of multiplication to simplify algebraic expressions.

$$2x \times 3y = 2 \times x \times 3 \times y$$
$$= 2 \times 3 \times x \times y$$
$$= 6xy$$
$$7ab \times 4a \times 2b = 7 \times 4 \times 2 \times a \times a \times b \times b$$
$$= 56a^2b^2$$

You look at the coefficients and each letter in turn.

You can do the same with division.

$$18a^3b^2 \div 6ab = (18 \div 6) \times (a^3 \div a) \times (b^2 \div b)$$
$$= 3a^2b$$

Example 2

a Find the product of $12ab$ and $7b^2$. **b** Find the quotient of $30pq$ and $5q$.

a $12ab \times 7b^2$
$\equiv 84ab^3$

Find the product means multiply.

$12 \times 7 = 84$

There is only a single a in the original terms, so there will only be a single a in the product.

$b^2 \times b = b^3$

The answer is written with the coefficient first and then the letters in alphabetical order.

b $30pq \div 5q$
$\equiv 6p$

$30 \div 5 = 6$

$q \div q = 1$

So the quotient is $6 \times p \times 1 = 6p$

Practice 7.7B

1 Given that $a = 5$ and $b = 3$, work out the values of

a a^b **b** $a + b^2$ **c** $(a + b)^2$

d $\dfrac{(a + b)^2}{a}$ **e** $\dfrac{(a + b)^3}{a - b}$ **f** $ab(a^2 - 7b)$

2 Simplify the expressions.

a $9m + m$ **b** $9m \times m$ **c** $9m \div m$

d $9m + 3m$ **e** $9m \times 3m$ **f** $9m \div 3m$

3 Find the result of multiplying $5p$ by

a 5 **b** p **c** $5p$ **d** q

e $5q$ **f** $5pq$ **g** $5p^2q$ **h** $5pq^2$

4 Find the result of dividing $48x^2y^3$ by

a 4 b 8 c $4x$ d $8y$

e $12xy$ f $6x^2$ g $6y^2$ h $16x^2y^2$

5 Simplify the expressions.

a $4ab \times 2ab$ b $4ab \div 2ab$ c $4a^2b \times 2ab^2$

d $4a^2b + 2a^2b$ e $24a^3b \div 6ab$ f $24a^3b \times 6ab$

6 a Which of these expressions is equivalent to $(5x)^2$?

$10x^2$ $5x^2$ $25x^2$

b Write each expression as a single term.

i $(4y)^2 - y^2$ ii $(4y)^2 - 2y^2$ iii $(4y)^2 - 4y^2$

iv $(4y)^2 - (2y)^2$ v $(4y)^2 - 4 \times 4y^2$

What do you think?

In terms of p and q means the answer will be an expression involving the letters p and q

1 Find, in terms of p and q, the area of

a a rectangle of length $8pq$ cm and width $4p$ cm

b a parallelogram of base $8pq$ cm and height $4q$ cm

c a triangle base $8pq$ cm and height $4pq$ cm

2 Beca simplifies an expression involving an operation and gets the answer $12ab$. How many possible expressions can you find if the operation was

a a multiplication b a division?

3 Why is it possible to simplify $3m \times 4n$ but not $3m + 4n$?

Consolidate – do you need more?

1 Find the area of the trapezia described.

a parallel sides 9 cm and 11 cm, and height 12 cm

b height 10 cm, and parallel sides 2.4 cm and 5.3 cm

2 Find, in terms of x and y, the area of the trapezia described.

a parallel sides x cm and $3x$ cm, and height y cm

b height $2x$ cm, and parallel sides y cm and $6y$ cm

3 Simplify the expressions.

a $7 \times c$ b $7b \times c$ c $7b \times 7c$

d $7bc \times c$ e $7bc \times 7b$ f $30fg \div 3$

g $30fg \div 3f$ h $30fg \div 3g$ i $30fg \div 3fg$

j $30fg \div (3f \times 2g)$

4 Put the cards in ascending order of value when $a = 4$ and $b = 9$

A	$a + b$	**B**	ab	**C**	$\frac{ab}{2}$	**D**	$\frac{a^2b}{3}$
E	$\frac{1}{2}ab^2$	**F**	$\sqrt{ab}$	**G**	$b^2 - \sqrt{a}$	**H**	$(a - \sqrt{b})^2$

5 Simplify

a $12pq + 6pq$ **b** $12pq \div 6$ **c** $12pq \times 6q$

d $12pq \times 6p$ **e** $12pq \times 6pq$ **f** $12pq \div 6p$

g $12pq \div 6pq$ **h** $12p^2q \times 6pq$ **i** $12pq^2 \div 6pq$

j $12(pq)^2 \div 6q$

Stretch – can you deepen your learning?

1 A set of shapes have lengths and widths represented by multiples of x, y and z. Decide whether each expression represents a perimeter or an area.

a $3x$ **b** $4y + 6z$ **c** $4x^2$

d $x^2 + xy$ **e** $x^2 - y^2$ **f** $8x + 2y + 7z$

g $x(2z - y)$

2 The parallel sides of a trapezium are $(x + y)$ cm and $(2x + 3y)$ cm long. The height of the trapezium is $6z$ cm. Find an expression for the area of the trapezium.

3 A trapezium has area $18x^2$ cm. The parallel sides and the height are all multiples of x. Suggest some possible expressions for the parallel sides and the height of the trapezium.

4

$(4x)^2$ is greater than the value of $4x^2$

Show that Zach's claim is only sometimes true.

Reflect

1 What is the difference between a trapezium and a parallelogram? How would you explain this to someone who had never seen either shape?

2 Explain the steps you take in multiplying and dividing algebraic expressions involving more than one letter.

7 Multiplication and division
Chapters 7.1–7.7

I have become fluent in…

- Identifying factors and multiples of a number
- Using the formal methods for multiplication and division
- Finding the areas of triangles and parallelograms
- Using the order of operations
- Working out the mean of a data set

I have developed my reasoning skills by…

- Identifying the correct side lengths to calculate the areas of shapes
- Using the result of one calculation to find other results
- Choosing whether to do a calculation using mental or written methods
- Deciding whether a problem needs to be solved using addition, subtraction or both

I have been problem-solving through…

- Representing multiplication and division in a variety of forms
- Finding unknown lengths given areas
- Work out missing numbers in data sets
- Converting and working with different metric units

Check my understanding

1 **a** Explain why 6 is both a factor and a multiple of 6.

b Are there any other numbers that are both factors and multiples of 6?

2 Work out the calculations. Explain your choice of a written or mental method.

a 27×10 **b** 27×16 **c** 2.7×1.6 **d** $34.8 \div 100$ **e** $34.8 \div 6$

3 Find the missing numbers in the calculations.

a $\square \div 100 = 7.6$ **b** $\square \div 1000 = 7.06$ **c** $\square \div 12 = 40.8$ **d** $\square \times 12 = 40.8$

4 **a** Find the area of a parallelogram of base 9 cm and perpendicular height 37 mm, giving your answer in cm^2.

b A triangle with base 9 cm has area $33.3\,cm^2$. Find the perpendicular height of the triangle, giving your answer in cm.

5 The mean of four numbers is 7.5. Two of the numbers are 10. What might the other two numbers be?

6 Write down the calculation that will have the greatest answer. Justify your answer.

$19 - 4 \times 3$ $(19 - 4) \times 3$ $19 \times (4 - 3)$ $19 \times 4 - 3$ $19 + 4^3$

7 Find the area of the trapezium, giving your answer in cm^2. (H)

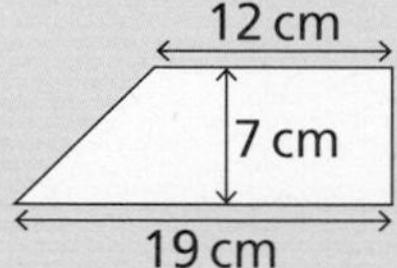

8 Amina says $48cd \div 8d = 4c$.
Explain why Amina is wrong. (H)

$$\frac{4\cancel{8}c\cancel{d}}{\cancel{8}\cancel{d}}$$

8 Fractions and percentages of amounts

In this block, I will learn...

how to work out a fraction of an amount

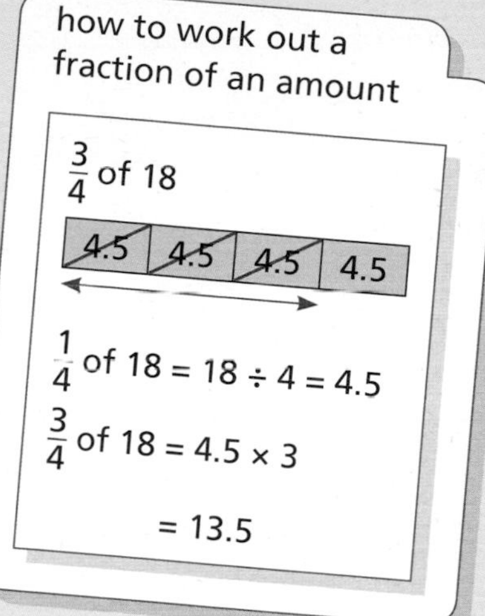

how to work out fractions and percentages using a calculator

$\frac{3}{4} \times 72$ 0.75 × 72

how to solve problems with fractions greater than 1 (H)

$\frac{7}{5}y = 490$

y

70	70	70	70	70	70	70

490

$\frac{1}{5}y = 490 \div 7 = 70$

$y = 70 \times 5 = 350$

how to work out the whole given a fraction of an amount

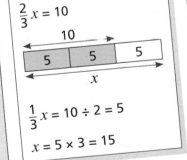

$\frac{2}{3}x = 10$

10

5	5	5

x

$\frac{1}{3}x = 10 \div 2 = 5$

$x = 5 \times 3 = 15$

how to work out a percentage of an amount

45% of 80

50% of 80 = 40

5% of 80 = 4

45% of 80 = 40 − 4 = 36

how to solve problems with percentages greater than 1 (H)

160% of a number is 72

10% of the number is 72 ÷ 16 = 4.5

the number is 4.5 × 10 = 45

8.1 Working with fractions of amounts

Small steps

- Find a fraction of a given amount
- Use a given fraction to find the whole and/or other fractions

Key words

Unit fraction – a fraction in which the numerator is 1

Are you ready?

1 What fraction of each bar is shaded?

a 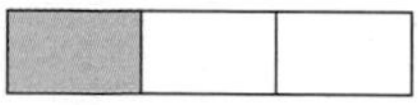 **b**  **c** 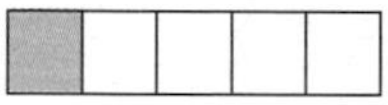 **d** 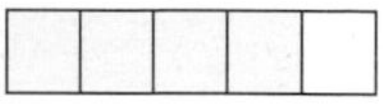

2 Work out

a $30 \div 2$ **b** $30 \div 3$ **c** $30 \div 4$ **d** $30 \div 5$ **e** $30 \div 10$ **f** $30 \div 12$

3 Work out

a 2.5×2 **b** 2.5×3 **c** 2.5×5 **d** 2.5×7 **e** 2.5×11

4 Work out

a $20 \div 4 \times 3$ **b** $30 \div 6 \times 5$ **c** $60 \div 12 \times 7$

d $12 \div 6 \times 5$ **e** $400 \div 10 \times 7$

Models and representations

Arrays

Arrays are useful for showing fractions of amounts.

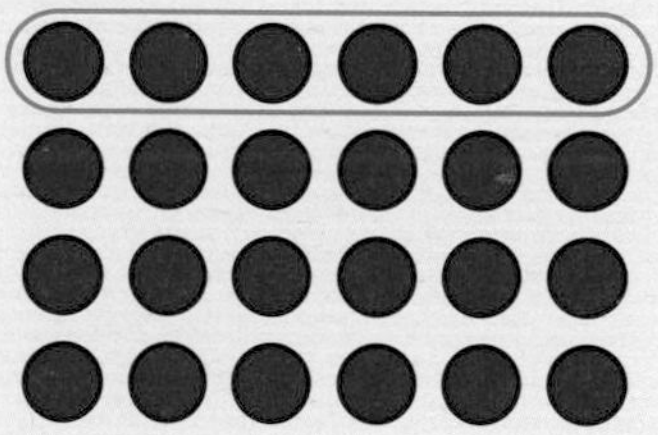

$\frac{1}{4}$ of 24 = 6

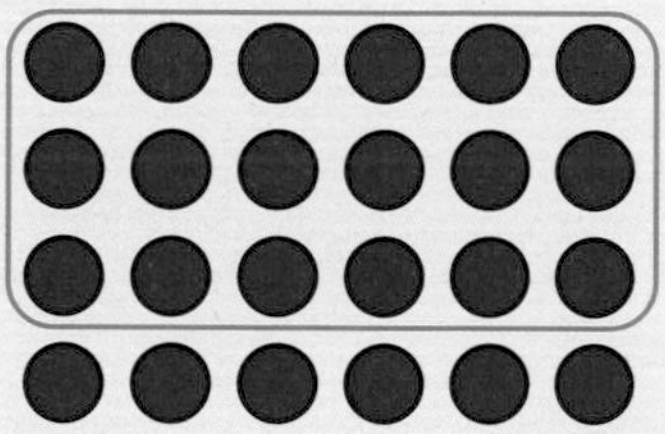

$\frac{3}{4}$ of 24 = 18

$\frac{5}{6}$ of 24 = 20

Bar models

Bar models are also useful for showing fractions of amounts.

one whole				
$\frac{1}{5}$	$\frac{1}{5}$	$\frac{1}{5}$	$\frac{1}{5}$	$\frac{1}{5}$

$\frac{1}{5}$ is an example of a **unit fraction**

60				
12	12	12	12	12

$\frac{3}{5}$ of 60 = 36

Example 1

Work out

a $\frac{1}{5}$ of 85 **b** $\frac{3}{5}$ of 85 **c** $\frac{7}{5}$ of 85

a $\frac{1}{5}$ of $85 = 85 \div 5 = 17$

85

17	17	17	17	17

You find one fifth of a number by dividing the number into 5 equal parts.

b $\frac{3}{5}$ of $85 = 17 \times 3 = 51$

17	17	17	17	17

Three fifths is three of the equal parts, so multiply $\frac{1}{5}$ by 3

c $\frac{7}{5}$ of $85 = 17 \times 7 = 119$

17	17	17	17	17		17	17

Seven fifths is seven of the equal parts, so multiply $\frac{1}{5}$ by 7

The answer is greater than 85, because $\frac{7}{5}$ is greater than one whole.

Practice 8.1A

1 Work out

a $\frac{1}{2}$ of 600 **b** $\frac{1}{3}$ of 600 **c** $\frac{1}{4}$ of 600 **d** $\frac{1}{5}$ of 600

e $\frac{1}{6}$ of 600 **f** $\frac{1}{10}$ of 600 **g** $\frac{1}{12}$ of 600 **h** $\frac{1}{15}$ of 600

i $\frac{1}{20}$ of 600 **j** $\frac{1}{25}$ of 600 **k** $\frac{1}{30}$ of 600 **l** $\frac{1}{40}$ of 600

m $\frac{1}{50}$ of 600 **n** $\frac{1}{60}$ of 600 **o** $\frac{1}{100}$ of 600 **p** $\frac{1}{120}$ of 600

q $\frac{1}{150}$ of 600 **r** $\frac{1}{200}$ of 600 **s** $\frac{1}{300}$ of 600

2 Faith notices her answer to question 1 part **d** is double her answer to part **f**, and 4 times as big as her answer to **i**. What other connections can you find between your answers?

3 **a** Work out $\frac{1}{7}$ of 140

b Use your answer to part **a** to work out:

i $\frac{3}{7}$ of 140 **ii** $\frac{5}{7}$ of 140 **iii** $\frac{6}{7}$ of 140 **iv** $\frac{2}{7}$ of 140

v $\frac{4}{7}$ of 140 **vi** $\frac{9}{7}$ of 140

4 Work out

a $\frac{3}{5}$ of 20 **b** $\frac{5}{9}$ of 27 **c** $\frac{6}{7}$ of 28 **d** $\frac{2}{5}$ of 30

e $\frac{4}{5}$ of 45 **f** $\frac{7}{10}$ of 160

5 You can't find $\frac{1}{5}$ of 28 because 28 isn't in the 5 times table.

Explain why Seb is wrong.

6 Work out

a $\frac{1}{4}$ of 30 b $\frac{3}{4}$ of 50 c $\frac{2}{5}$ of 16 d $\frac{3}{8}$ of 100

e $\frac{5}{6}$ of 15 f $\frac{7}{5}$ of 24

7 Choose one of these three symbols to go between each pair of calculations.

<

a $\frac{3}{4}$ of 60 ◯ $\frac{1}{2}$ of 90 b $\frac{2}{5}$ of 30 ◯ $\frac{3}{10}$ of 50 c $\frac{3}{7}$ of 42 ◯ $\frac{5}{6}$ of 24

d $\frac{5}{3}$ of 21 ◯ $\frac{7}{5}$ of 25 e $\frac{11}{8}$ of 48 ◯ $\frac{5}{11}$ of 99

8 How can you use the fraction keys on your calculator to find a fraction of an amount?

What do you think? 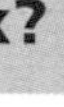

1

Flo: I find $\frac{4}{5}$ of a number by dividing the number by 5 then multiplying my answer by 4

Jakub: I find $\frac{4}{5}$ of a number by dividing the number by 5 then subtracting my answer from the original number.

Compare the two strategies by finding

a $\frac{4}{5}$ of 150 b $\frac{4}{5}$ of 1000 c $\frac{4}{5}$ of 1280 d $\frac{4}{5}$ of 27

Whose strategy do you prefer?

Use your preferred strategy to work out

e $\frac{9}{10}$ of 150 f $\frac{7}{8}$ of 240 g $\frac{29}{30}$ of 600

2

I can find $\frac{1}{20}$ of a number by dividing the number by 10 then dividing my answer by 2

Find similar strategies to work out

a $\frac{1}{15}$ of a number

b $\frac{1}{40}$ of a number

c $\frac{1}{25}$ of a number

d $\frac{1}{500}$ of a number

You can use similar methods to work out other fractions, or even to work out the whole if you are given a fraction.

For example, if half a number is 16,

16	

then the other half must be 16 as well.

16	16

So the whole must be 16 + 16 or 16 × 2 = 32

Example 2

Amina spends three fifths of her money on a magazine costing £6

How much money does Amina have left?

$\frac{3}{5}$ = £6

Amina's money

£6 spent

Three of the five equal parts is £6

$\frac{1}{5}$ = £6 ÷ 3 = £2

So each one of the parts must be £6 ÷ 3

£2	£2	£2	£2	£2

amount left

The bar shows there are two equal parts remaining. Each equal part is £2, so Amina has 2 × £2 left.

Amina has $\frac{2}{5}$ left, which is £2 × 2 = £4

Example 3

$\frac{2}{3}$ of a number is 80. What is $\frac{3}{5}$ of the number?

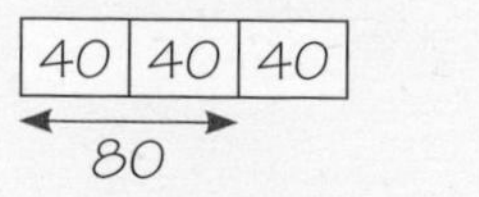

Two of the three equal parts is 80, so each one of the parts must be $80 \div 2 = 40$

The number is $40 \times 3 = 120$

$\frac{1}{5}$ of the number $= 120 \div 5 = 24$

24	24	24	24	24

$\frac{3}{5}$ of the number $= 24 \times 3 = 72$

Now you only need to find $\frac{3}{5}$ of the number by finding $\frac{1}{5}$ and multiplying the result by 3

Practice 8.1B

1 Work out the number if

a $\frac{1}{2}$ of the number is 80 **b** $\frac{1}{3}$ of the number is 80 **c** $\frac{1}{5}$ of the number is 80

d $\frac{1}{7}$ of the number is 80 **e** $\frac{1}{10}$ of the number is 80 **f** $\frac{1}{9}$ of the number is 80

You might find it helpful to draw diagrams like these.

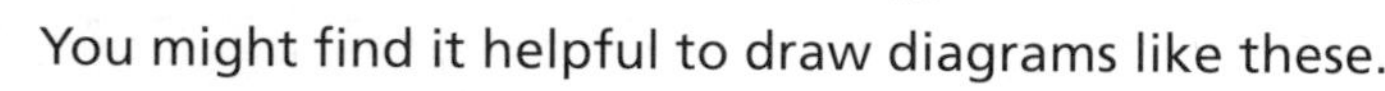

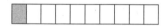

2 $\frac{3}{4}$ of a number is 60

a What is $\frac{1}{4}$ of the number? **b** What is $\frac{5}{8}$ of the number?

c What is the number?

3 Find the number if

a $\frac{3}{7}$ of the number is 9 **b** $\frac{5}{8}$ of the number is 40 **c** $\frac{4}{9}$ of the number is 28

d $\frac{4}{5}$ of the number is 100 **e** $\frac{7}{10}$ of the number is 21 **f** $\frac{5}{11}$ of the number is 35

4 **a** Bobbie spends $\frac{2}{5}$ of his money on a book. He spends £12 on the book. How much money did he start with?

b Lydia spends $\frac{2}{5}$ of her money on a book. She has £12 left. How much money did she start with?

5 20 students in a class have a dog. This is $\frac{5}{8}$ of the class. $\frac{3}{4}$ of the class have a cat. How many students have a cat?

6 Find $\frac{5}{8}$ of the number if

a $\frac{1}{2}$ the number is 40 **b** $\frac{1}{4}$ of the number is 60 **c** $\frac{3}{4}$ of the number is 90

d $\frac{7}{5}$ of the number is 56 **e** $\frac{10}{3}$ of the number is 7.5

What do you think?

1 Solve the equations.

a $\frac{1}{5}a = 22$ **b** $\frac{3}{7}b = 18$ **c** $\frac{4}{5}c = 160$ **d** $\frac{9}{7}d = 81$

2 Find the expression if

a $\frac{1}{2}$ of the expression is $20ab$ **b** $\frac{1}{5}$ of the expression is $20ab$

c $\frac{2}{5}$ of the expression is $20ab$ **d** $\frac{4}{5}$ of the expression is $20ab$

e $\frac{10}{7}$ of the expression is $20ab$

3 What is different about these questions?

What is $\frac{1}{3}$ of 18?

$\frac{1}{3}$ of what number is 18?

Consolidate – do you need more?

1 Work out

a $\frac{1}{2}$ of 90 **b** $\frac{1}{3}$ of 60 **c** $\frac{1}{4}$ of 24 **d** $\frac{1}{5}$ of 100

e $\frac{1}{6}$ of 42 **f** $\frac{1}{10}$ of 900 **g** $\frac{1}{12}$ of 36 **h** $\frac{1}{15}$ of 30

i $\frac{1}{20}$ of 80 **j** $\frac{1}{25}$ of 100 **k** $\frac{1}{30}$ of 60 **l** $\frac{1}{40}$ of 8000

2 Use your answers to question 1 to work out

a $\frac{3}{2}$ of 90 **b** $\frac{2}{3}$ of 60 **c** $\frac{3}{4}$ of 24 **d** $\frac{9}{5}$ of 100

e $\frac{5}{6}$ of 42 **f** $\frac{9}{10}$ of 900 **g** $\frac{7}{12}$ of 36 **h** $\frac{23}{15}$ of 30

i $\frac{11}{20}$ of 80 **j** $\frac{19}{25}$ of 100 **k** $\frac{41}{30}$ of 60 **l** $\frac{17}{40}$ of 8000

3 There are 1200 students at a school in New Town.

a $\frac{11}{20}$ of the students are boys. How many boys are there?

b $\frac{5}{6}$ of the students travel to school by bus. How many students travel to school by bus?

c $\frac{2}{3}$ of the children who live in New Town attend this school. How many children live in New Town?

4 Work out the whole number if

a $\frac{1}{2}$ of the number is 120 **b** $\frac{5}{6}$ of the number is 120 **c** $\frac{3}{4}$ of the number is 120

d $\frac{6}{7}$ of the number is 120 **e** $\frac{10}{11}$ of the number is 120 **f** $\frac{3}{20}$ of the number is 120

5 Find the missing values in these function machines.

a 80 → $\frac{4}{5}$ of the number → ☐ **b** ☐ → $\frac{4}{5}$ of the number → 80

Stretch – can you deepen your learning?

1 **a** Marta thinks $\frac{2}{5}$ of 60 = $\frac{1}{5}$ of 30. Show that Marta is wrong.

b Find the missing numbers.

i $\frac{2}{5}$ of 60 = $\frac{1}{5}$ of ☐ **ii** $\frac{4}{7}$ of 28 = $\frac{2}{7}$ of ☐ **iii** $\frac{4}{5}$ of 90 = $\frac{1}{5}$ of ☐

2 Copy and complete.

a $\frac{2}{5}$ of x = $\frac{1}{5}$ of ☐ **b** $\frac{3}{5}$ of x = $\frac{3}{10}$ of ☐ **c** $\frac{1}{4}$ of x = $\frac{1}{2}$ of ☐

Illustrate your answers using bar models.

3 I know that $\frac{7}{15}$ of £600 is less than $\frac{11}{20}$ of £600 without working out either of the calculations

How does Benji know this?

4 The second term of a linear sequence is $\frac{3}{4}$ of the fourth term. The fourth term of the sequence is $\frac{5}{8}$ of 64. What is the fifth term of the sequence?

Reflect

1 Use clearly labelled diagrams to show your method to work out $\frac{3}{5}$ of an amount.

2 Use clearly labelled diagrams to show your method to work out a number if you know $\frac{3}{5}$ of the number.

3 How are your approaches to questions 1 and 2 the same and how are they different?

8.2 Working with percentages of amounts

Small steps

- Find a percentage of a given amount – mental methods
- Find a percentage of a given amount using a calculator

Key words

Percentage – the number of parts per hundred

Convert – change from one form to another, for example a percentage to a decimal

Are you ready?

1 Write these percentages as fractions.

a 47% **b** 81% **c** 70% **d** 7%

2 Write these fractions as percentages.

a $\frac{1}{2}$ **b** $\frac{1}{4}$ **c** $\frac{3}{4}$

3 Write these fractions as percentages.

a $\frac{1}{10}$ **b** $\frac{3}{10}$ **c** $\frac{7}{10}$ **d** $\frac{1}{5}$

4 Write these percentages as decimals.

a 64% **b** 60% **c** 6%

Models and representations

Bar models can be used to emphasise the links between **percentages** and fractions.

100%	
50%	50%

50% is the same as a half

100%			
25%	25%	25%	25%

$\frac{1}{4}$ = 25%, $\frac{2}{4}$ (or $\frac{1}{2}$) = 50% and $\frac{3}{4}$ = 75%

100%									
10%	10%	10%	10%	10%	10%	10%	10%	10%	10%

$\frac{1}{10}$ = 10%, $\frac{2}{10}$ = 20%, $\frac{3}{10}$ = 30%, and so on up to $\frac{10}{10}$ = 100%

100%				
20%	20%	20%	20%	20%

$\frac{1}{5}$ = 20%, $\frac{2}{5}$ = 40%, and so on up to $\frac{5}{5}$ = 100%

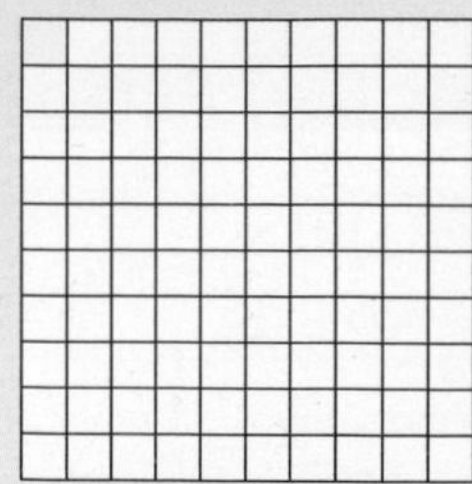

Remember also that $1\% = \frac{1}{100}$

Another useful fact to know is that $\frac{1}{3} = 33\frac{1}{3}\%$.

100%		
$33\frac{1}{3}\%$	$33\frac{1}{3}\%$	$33\frac{1}{3}\%$

$\frac{1}{3} = 33\frac{1}{3}\%$ and $\frac{2}{3} = 66\frac{2}{3}\%$

Notice $33\frac{1}{3}\% + 33\frac{1}{3}\% + 33\frac{1}{3}\% = 99\frac{3}{3}\% = 99\% + 1\% = 100\%$

Example 1

Work out

a 50% of 90 **b** 25% of 90 **c** 75% of 90 **d** $33\frac{1}{3}\%$ of 90

e $66\frac{2}{3}\%$ of 90

a $50\% \text{ of } 90 = 90 \div 2 = 45$

Use the knowledge that 50% is $\frac{1}{2}$, and to find $\frac{1}{2}$ you divide by 2

b $25\% \text{ of } 90 = 90 \div 4 = 22.5$

Use the knowledge that 25% is $\frac{1}{4}$, and to find $\frac{1}{4}$ you divide by 4

You could also find 25% by "halving and halving again"

$90 \div 2 = 45$, $45 \div 2 = 22.5$

c $75\% \text{ of } 90 = 25\% \text{ of } 90 \times 3$
$= 22.5 \times 3 = 67.5$

Use the knowledge that 75% is equal to 3 × 25% or $\frac{3}{4}$, so you can multiply the answer to part **b** by 3

d $33\frac{1}{3}\% \text{ of } 90 = 90 \div 3 = 30$

Use the knowledge that $33\frac{1}{3}\%$ is $\frac{1}{3}$, and to find $\frac{1}{3}$ you divide by 3

e $66\frac{2}{3}\% \text{ of } 90 = 33\frac{1}{3}\% \text{ of } 90 \times 2$
$= 30 \times 2 = 60$

Use the knowledge that $66\frac{2}{3}\%$ is equal to $2 \times 33\frac{1}{3}\%$ or $\frac{2}{3}$, so you can multiply the answer to part **d** by 2

Example 2

Work out

a 10% of 80   **b** 30% of 80   **c** 15% of 44   **d** 95% of 70

e 27% of 3000

a 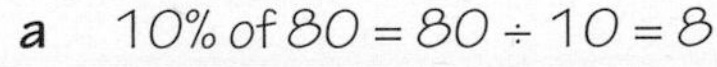

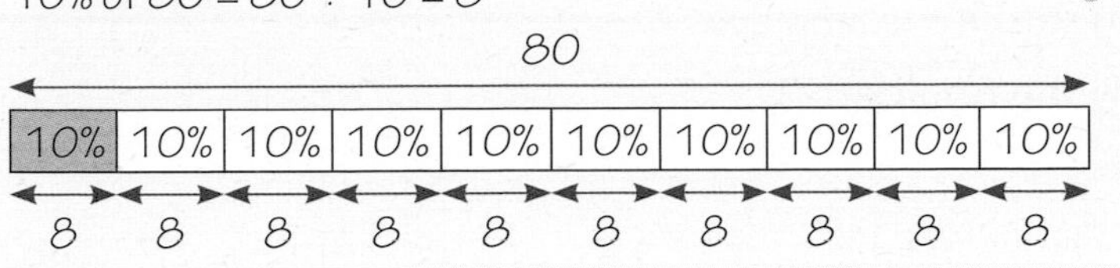

Use the knowledge that 10% is $\frac{1}{10}$, and to find $\frac{1}{10}$ you divide by 10

b 30% of $80 = 3 \times 10\%$ of $80 = 3 \times 8 = 24$

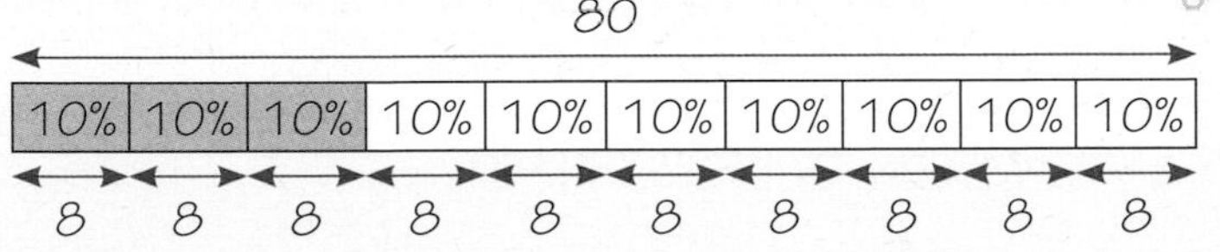

Use the answer to part **a** and the fact that 30% is 3 × 10%.

c $15\% = 10\% + 5\%$ — Partition 15% into 10% and 5%.

10% of $44 = 44 \div 10 = 4.4$ — Find 10% as before.

5% of $44 = \frac{1}{2}$ of 10% of 44 — Find 5% as half of 10%.

$= 4.4 \div 2 = 2.2$

15% of $44 = 10\%$ of $44 + 5\%$ of $44 = 4.4 + 2.2 = 6.6$ — Add 10% and 5% together to give 15%.

d 10% of $70 = 70 \div 10 = 7$ — Find 10% as before.

5% of $70 = 10\%$ of $70 \div 2 = 7 \div 2 = 3.5$ — Find 5% as half of 10%.

95% of $70 = 100\%$ of $70 - 5\%$ of $70 = 70 - 3.5 = 66.5$ — Find 95% as 100% – 5%.

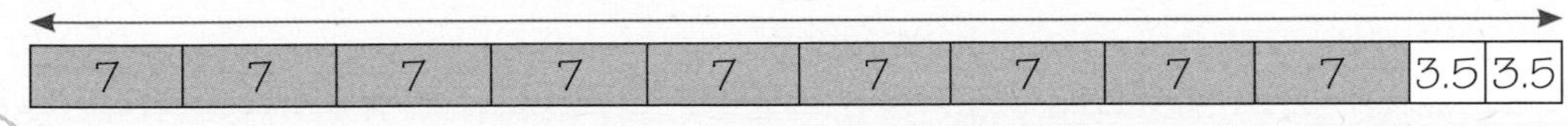

You could find 90% of 70 = 9 × 10% of 70 = 9 × 7 = 63 and then add 90% and 5% together, but it is more efficient just to take 5% away from 100%.

e 1% of 3000 = 3000 ÷ 100 = 30

20% of 3000 = 30 × 20 = 600

7% of 3000 = 30 × 7 = 210

27% of 3000 = 20% of 3000 + 7% of 3000 = 600 + 210 = 180

There are lots of ways of finding this, for example

- Find 1% of 3000 and multiply the result by 27
- Find 25% and add on 2 × 1%.
- Add together 10% + 10% + 5% + 1% + 1%.

You should consider the options and choose the one you find most efficient and reliable.

Practice 8.2A

You should try and do this exercise using mental methods if you can.

1 Work these out. Use bar models to help you if you like.

a 50% of 8 **b** 25% of 12 **c** 75% of 24

d 50% of 30 **e** 25% of 30 **f** 75% of 30

2 Work these out. Use bar models to help you if you like.

a $33\frac{1}{3}$% of 12 **b** $66\frac{2}{3}$% of 12 **c** $33\frac{1}{3}$% of 27

d $66\frac{2}{3}$% of 30 **e** $33\frac{1}{3}$% of 90 **f** $66\frac{2}{3}$% of 180

3 Work out

a 10% of 40 **b** 10% of 42 **c** 30% of 40 **d** 30% of 42

e 10% of 60 **f** 5% of 60 **g** 15% of 60 **h** 15% of 80

i 60% of 80 **j** 20% of 31 **k** 25% of 50 **l** 80% of 300

m 1% of 900 **n** 11% of 900 **o** 23% of 60 **p** 41% of 130

4 **a** Ms Edwards buys a house for £250 000. She sells it five years later making a 20% profit. How much profit does she make?

b Mr Singh earns £45 000 a year. He gets a 2% pay rise. How much is his pay rise?

5 What percentages of 600 can you easily work out mentally?

6 **a** Find the difference between 30% of 600 and $33\frac{1}{3}$% of 600

b Write down $3\frac{1}{3}$% of 600

c Work out $3\frac{1}{3}$% of 90

d Work out $6\frac{2}{3}$% of 450

What do you think?

1. Mario is working out 25% of 60. He writes

   25% of 60 = 60 ÷ 4 = 15%

   What mistake has Mario made?

2. Show that 60% of 40 = $\frac{2}{3}$ of 36

3. Discuss strategies you would use to work out

   a 35% of a number **b** 49% of a number **c** 23% of a number

4. **a** Work out

   i 50% of 30 **ii** 30% of 50

   iii 90% of 40 **iv** 40% of 90

   b What do you notice about your results in part **a**? Find a general rule.

   c Use your rule to work out

   i 68% of 50 **ii** 84% of 25 **iii** 90% of $33\frac{1}{3}$

Although you can work out most percentages mentally by building up from multiples of 10% and 1%, it is more efficient to use a calculator when the numbers become difficult.

Some calculators have a % key you can use, or you can **convert** percentages to decimals, as in the examples below.

Example 3

Work out

a 39% of 6000 **b** 18% of 2.3 kg **c** 17% of 2.8 m

a 39% of 6000
= 0.39 × 6000 = 2340

First convert 39% to a decimal: 39% = $\frac{39}{100}$ = 0.39
Then enter 0.39 × 6000 on your calculator.

b 18% of 2.3 kg
= 0.18 × 2.3 = 0.414 kg

You could give this answer as 414 g

c 17% of 2.8 m
= 0.17 × 2.8 = 0.476 m

You could give this answer here as 47.6 cm

Practice 8.2B

You should use your calculator for this exercise.

1. Write these percentages as decimals.

   a 62% **b** 91% **c** 40% **d** 4% **e** 34% **f** 34.2%

2. Work out

   a 62% of 450 **b** 91% of 7000 **c** 40% of 132

   d 4% of 88 **e** 34% of 6200 **f** 34.2% of 17 000

3. Work out, giving your answers in cm

   a 56% of 90 cm **b** 18% of 4.5 m **c** 43% of 85 mm

4. 70% of the human body is water.

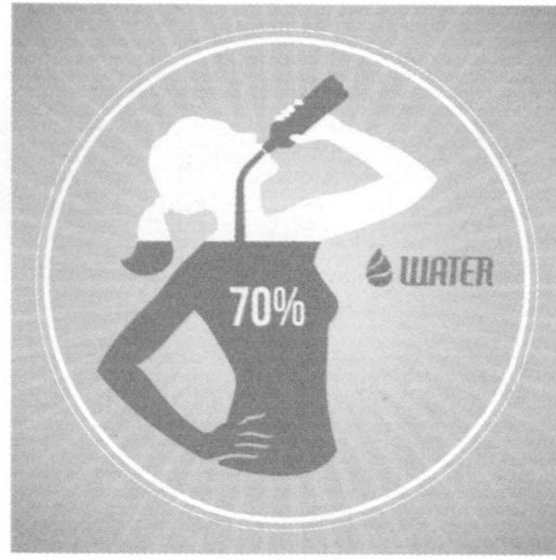

   Flo weighs 43.4 kg. What is the mass of water in Flo's body?

5. Decide whether you would you use a mental method or a calculator to work out each calculation and find the answers.

   a 60% of £400 **b** 60% of £423 **c** 23% of £100 **d** 50% of £23

   e 1% of 900 kg **f** 3.2% of 600 g **g** 7.6% of 840 mg

6. How does the % key work on your calculator? Compare different models of calculators.

What do you think?

1. 23% of a number is 1334

   a What is 1% of the number? **b** What is the number?

   c Explain how you would find

   i 46% of the number **ii** 81% of the number

2. $\frac{4}{7}$ of a number is 60. Explain the steps you would take to work out 37% of the number.

3. Work out

   a 17% of 6000 **b** 1.7% of 6000 **c** 0.17% of 6000

4 In 2005, it was calculated that the average mass of each adult on the planet was 62 kg.

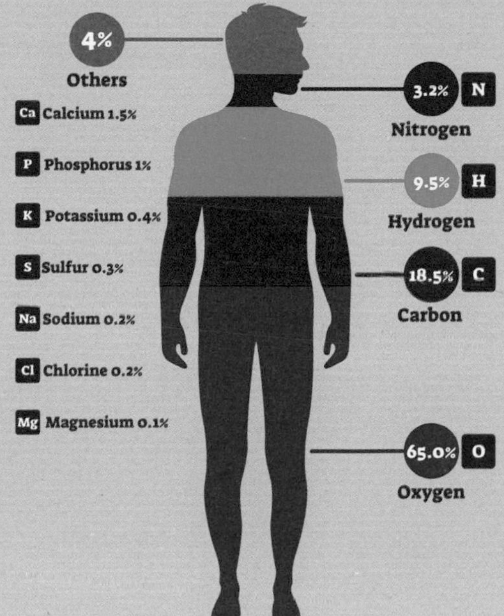

Source: Walpole, S.C., Prieto-Merino, D., Edwards, P. *et al*. The weight of nations: an estimation of adult human biomass. *BMC Public Health* **12**, 439 (2012). https://doi.org/10.1186/1471-2458-12-439

Use the chart to find the mass of these elements in an average adult.

a Carbon **b** Calcium **c** Potassium

Consolidate – do you need more?

Try to do these without a calculator unless the question asks you to use one.

1 Work out

a 25% of 48 **b** 75% of 48 **c** $33\frac{1}{3}$% of 30 **d** $66\frac{2}{3}$% of 15

e 50% of 64 **f** 75% of 200

2 Work out

a 10% of 90 g **b** 50% of 90 g **c** 60% of 90 g **d** 90% of 90 g

e 10% of £240 **f** 5% of £240 **g** 15% of £240 **h** 85% of £240

i 60% of 42 **j** 45% of 80 **k** 35% of 120 **l** 55% of 90

m 1% of 400 **n** 1% of 40 **o** 7% of 40 **p** 21% of 600

Compare your methods with a partner's. Did you take the same approach to all the questions?

3 Use the fact that 17.5% = 10% + 5% + 2.5% to work out 17.5% of

a £800 **b** £300 **c** £220 **d** £64

4 Use a calculator to work out

a 37% of 820 **b** 29% of 800 **c** 4% of 650 **d** 16% of 204

e 73% of 1800 **f** 4.2% of 310

5 **a** Add together 25% of 40 and $33\frac{1}{3}$% of 60

b Find the difference between 50% of 82 and $66\frac{2}{3}$% of 90

Stretch – can you deepen your learning?

1 **a** Write $\frac{1}{8}$ as a percentage.

b Write $\frac{3}{8}$, $\frac{5}{8}$ and $\frac{7}{8}$ as percentages.

c Why might $\frac{2}{8}$, $\frac{4}{8}$ and $\frac{6}{8}$ not be included in a list to convert to percentages?

d Which is greater, 37.5% of 48 or 62.5% of 32? Show calculations to justify your answer.

2 **a** Show that $\frac{1}{6} = 16\frac{2}{3}$% **b** Work out $16\frac{2}{3}$% of 48 **c** Work out $83\frac{1}{3}$% of 48

3 Find the number if

a $33\frac{1}{3}$% of the number is 60 **b** 75% of the number is 60

c 12% of the number is 60 **d** 12.5% of the number is 60

e 80% of the number is 60

4 FInd the missing number: 25% of 40 = $33\frac{1}{3}$% of $\square$

5 **a** Filipo thinks that finding 10% of a number is the same as multiplying the number by 0.1. Is he correct? Why?

b How is multiplying a number by 0.01 connected to working out percentages?

Reflect

1 What's the same and what's different about working out percentages of amounts and working out decimals of amounts?

2 When would you use a mental or written method, and when would you use a calculator, to work out a percentage of an amount?

H 8.3 Over 100%

Small steps

- Solve problems with fractions more than 1 and percentages over 100% (H)
- Find a percentage of a given amount – mental methods
- Find a percentage of a given amount using a calculator

Key word

Original – Referring to a number, the number that you started with

Are you ready?

1 Work out

a $\frac{1}{5}$ of 90 **b** $\frac{3}{5}$ of 90 **c** $\frac{7}{5}$ of 90

2 $\frac{3}{5}$ of a number is 90. What is the number?

3 Without a calculator, work out

a 10% of 50 **b** 30% of 50 **c** 90% of 50 **d** 65% of 50

4 Use a calculator to find

a 37% of £240 **b** 56% of 12 kg

Models and representations

Hundred squares

Hundred squares can be used to represent percentages below and above 100%

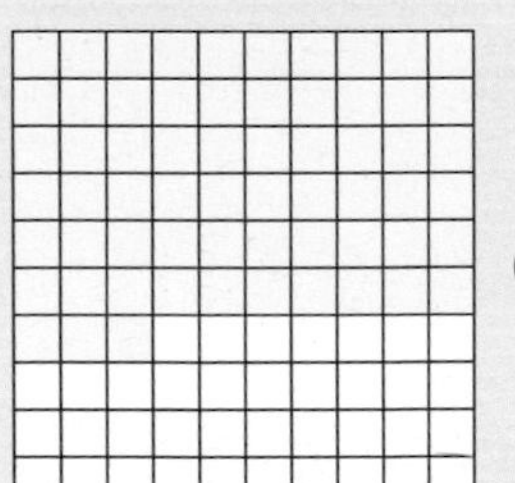

63%

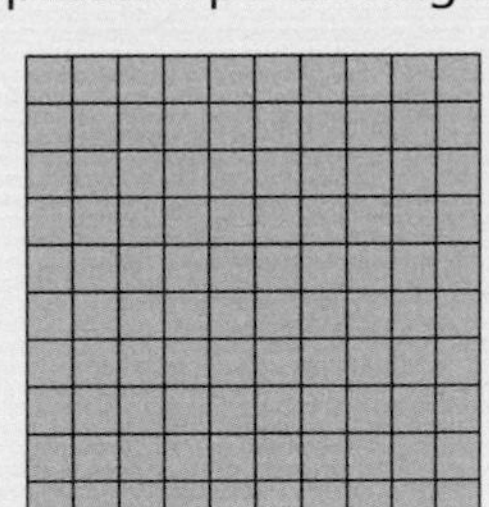

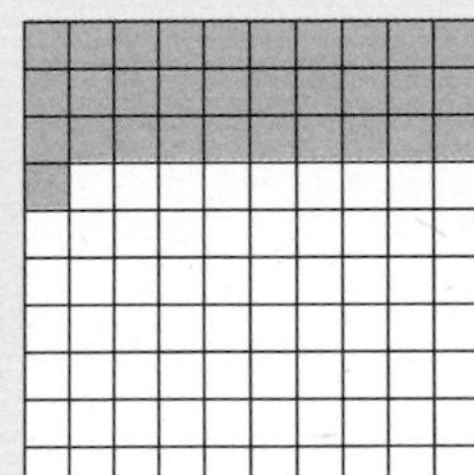

131%

You can also use bar models to show fractions below and over 1

This represents $\frac{4}{5}$

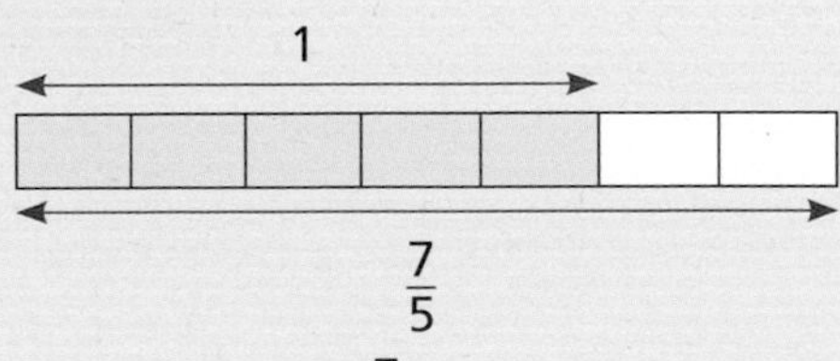

This represents $\frac{7}{5}$

You can convert between fractions and percentages with numbers greater than 1.

Example 1

Write as a percentage

a $\frac{13}{10}$ **b** $\frac{7}{5}$

a $\frac{13}{10} = 130\%$

You know $\frac{1}{10} = 10\%$

So $\frac{13}{10} = 13 \times 10\% = 130\%$

b $\frac{7}{5} = 7 \times 20\% = 140\%$

You know $\frac{1}{5} = 20\%$ and you multiply this by 7 to find $\frac{7}{5}$

Alternatively you could write $\frac{7}{5} = \frac{14}{10}$ and use the fact that $\frac{1}{10} = 10\%$ so $\frac{14}{10} = 14 \times 10\% = 140\%$

Example 2

Write as a fraction

a 180% **b** 102% **c** 250%

a $\frac{180}{100} = \frac{18}{10} = \frac{9}{5}$

Per cent means out of 100, so write 180% as $\frac{180}{100}$ and simplify.

b $\frac{102}{100} = \frac{51}{50}$

c $\frac{250}{100} = \frac{25}{10} = \frac{5}{2}$

You can also see results like this using bar models.

50%	50%

50%	50%

50%	50%

$250\% = 2\frac{1}{2} = \frac{5}{2}$

Example 3

Work out

a $\frac{8}{5}$ of 20 **b** 175% of 60

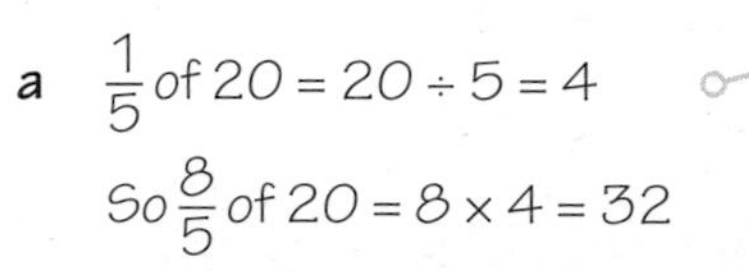

a $\frac{1}{5}$ of $20 = 20 \div 5 = 4$

So $\frac{8}{5}$ of $20 = 8 \times 4 = 32$

You can use a bar model to help see this.

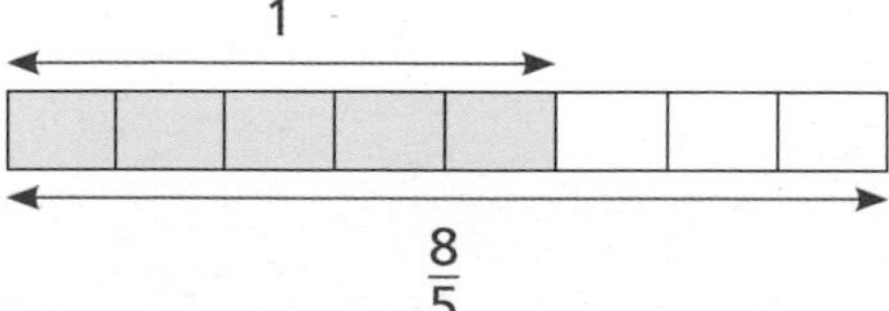

b $175\% = \frac{7}{4}$

$\frac{1}{4}$ of $60 = 60 \div 4 = 15$

So $\frac{7}{4}$ of $60 = 7 \times 15 = 105$

Think of $175\% = 100\% + 75\% = 1 + \frac{3}{4} = \frac{4}{4} + \frac{3}{4} = \frac{7}{4}$

You can then work the answer out in the same way as part **a**.

There are often lots of different ways to solve problems like this. For example, you could also think of 175% as 200% – 25%

$2 \times 60 - \frac{1}{4}$ of $60 = 120 - 15 = 105$

Or you could find 1% of 60 = 60 ÷ 100 = 0.6

So 175% of 60 = 175 × 0.6 = 105

Or you could write 175% as a decimal and use a calculator.

175% = 1.75

1.75 × 60 = 105

Practice 8.3A

1 The bar model shows $\frac{7}{4}$

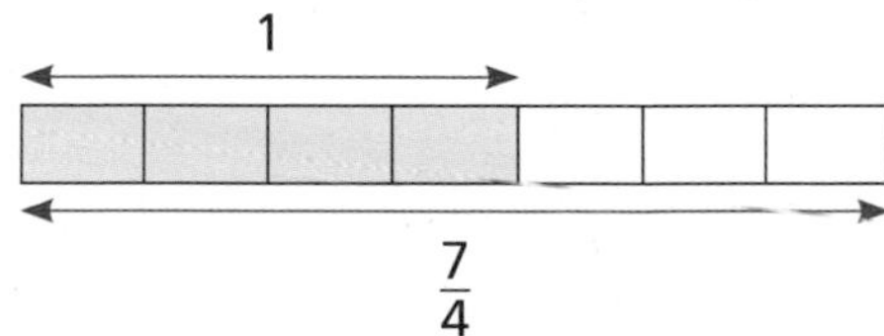

Draw bar models to show

a $\frac{8}{5}$ **b** $\frac{7}{3}$ **c** $\frac{5}{2}$ **d** $1\frac{3}{5}$

2 **a** What percentage does the bar model show?

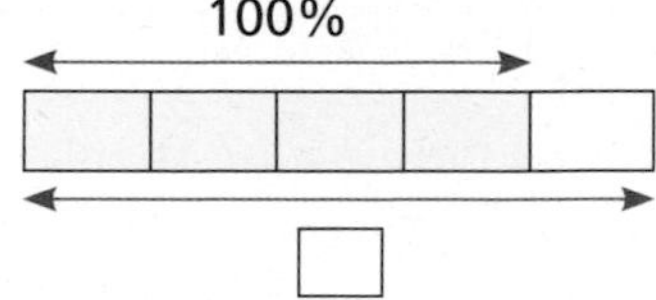

b Draw bar models to show

i 120% **ii** 200% **iii** 175% **iv** 250%

3 Copy and complete the workings to find $\frac{5}{3}$ of 90

$\frac{1}{3}$ of $90 = 90 \div 3$

$= \square$

So $\frac{5}{3}$ of $90 = \square \times 5$

$= \square$

4 Work out these fractions. You may draw bar models if you wish.

a $\frac{5}{4}$ of 60 **b** $\frac{7}{5}$ of 15 **c** $\frac{7}{2}$ of 40 **d** $1\frac{1}{8}$ of 40

5 Work out

a 150% of 50 b 200% of 30 c 120% of 90 d $166\frac{2}{3}$% of 120

e 350% of 12

6 Which of these claims make sense?

A

I scored 150% in a test.

B

My test mark this term is 150% of my test mark last term.

C

In the last year my test mark has improved by 150%

D

I scored 70% in Test 1 and 80% in Test 2. Overall, I scored 150%

7 Sven's height is 120% of Huda's height. Emily's height is 110% of Sven's height. Huda is 115 cm tall. How tall is Emily?

What do you think?

1 a Amina thinks that increasing a number by 100% is the same as multiplying the number by 2. Explain why Amina is right.

b What percentage increase would be the same as

i multiplying by 3 ii multiplying by 5 iii multiplying by 2.5?

2 Marta thinks that because you find 99% of a number by multiplying the number by 0.99, you can find 101% of a number by multiplying the number by 0.101

a Explain why Marta is wrong.

b What would you multiply a number by to find 101% of the number?

You will study this in Book 2

3 If you increase a number by $\frac{1}{5}$ and then decrease your answer by $\frac{1}{5}$ you get back to the original number.

Do you agree with Ed? Give an example to support your view.

4 Work out the missing numbers.

a ☐ % of 80 = 100 b ☐ % of 150 = 250 c 250% of ☐ = 35

Consolidate – do you need more?

1 Match the equivalent cards.

200%	180%	150%	110%	240%

$\frac{12}{5}$	$\frac{12}{6}$	$\frac{3}{2}$	$\frac{9}{5}$	$\frac{11}{10}$

2 Work out these fractions of 150

a $\frac{1}{3}$ **b** $\frac{7}{3}$ **c** $\frac{5}{2}$ **d** $\frac{11}{10}$

e $\frac{13}{10}$ **f** $\frac{9}{5}$ **g** $\frac{71}{50}$

3 Work out these percentages of 180

a 75% **b** 175% **c** 200% **d** 250%

e 220% **f** $133\frac{1}{3}$% **g** $266\frac{2}{3}$%

4 A special offer cereal box is labelled "$\frac{1}{3}$ extra free". What percentage of the weight of the normal box is the special offer box?

5 **a** What percentage is there now if a number has been increased by 80%?

b What percentage is there now if a number has been decreased by 80%?

6 Solve the equations.

a $\frac{7}{5}a = 28$ **b** $\frac{8}{3}b = 16$ **c** $\frac{15}{7}c = 45$

Stretch – can you deepen your learning?

1 Given $\frac{6}{5}$ of a number, what else can you find?

2 Filipo is thinking of two numbers, x and y. He decreases x by a quarter and increases y by 20%. The new numbers are equal.

a Write this information as an equation of the form $\square x = \square y$

b If $y = 60$, find the value of x

3 Mrs A earns £60000 a year after a 20% pay rise. Mrs B earns £60 000 a year after a 20% pay cut. How much did they earn altogether before the changes to their pay?

You will study questions like this in detail in Books 2 and 3

Reflect

1. When can you and when can you not have percentages over 100%?

2. What's the same and what's different about converting a fraction below 1 and a fraction above 1 to a percentage?

3. What's the same and what's different about calculating a fraction below 1 of an amount and calculating a fraction above 1 of an amount?

8 Fractions and percentages of amounts

Chapters 8.1–8.3

I have become fluent in…

- Working out a fraction of an amount
- Working out a percentage of an amount
- Using a calculator to work out fractions and percentages
- Finding the whole given a fraction or a percentage

I have developed my reasoning skills by…

- Deciding when to multiply or divide when finding fractions of amounts
- Comparing different methods to find fractions and percentages
- Linking fractions and percentages
- Deciding when to use mental, written or calculator methods

I have been problem-solving through…

- Representing fractions and percentages in a variety of forms
- Using models to represent situations
- Working out fractions and percentages of amounts in real-life contexts

Check my understanding

1 **a** Write down two-fifths of 40

b Two-fifths of what number is 40?

What's the same and what's different about answering these questions?

2 **a** Discuss the different methods you could use to work out

i 35% of 90 **ii** 37% of 90 **iii** 90% of 35

iv 77% of 35 **v** 0.2% of 90 kg

b Find the answers to the calculations in part **a**.

3 There are 300 students in Year 7 at Moortown School. $\frac{3}{5}$ of the Year 7 students live in Moortown and 27% live in Bridge Vale. How many of the Year 7 students live in other places?

4 15% of a company's workforce work from home. This is 120 people. How many people does the company employ altogether?

5 Which of these claims make sense? (H)

A "The population of my town this year is 120% of the population last year."

B "120% of the people in my town think the roads are too busy."

6 Find the difference between $\frac{7}{3}$ of 90 and 180% of 150 (H)

9 Directed number

In this block, I will learn...

how to represent directed numbers

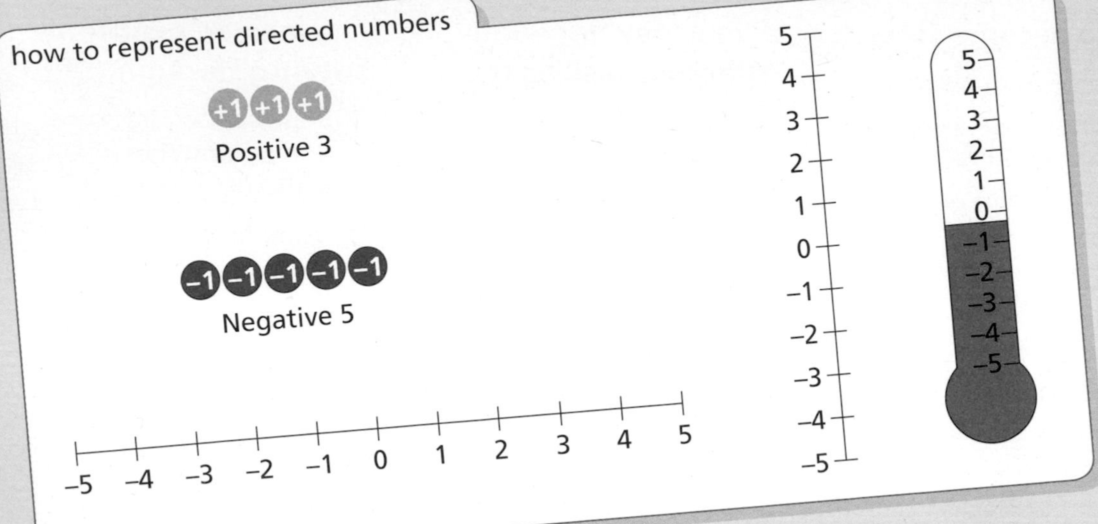

how to add and subtract with directed numbers

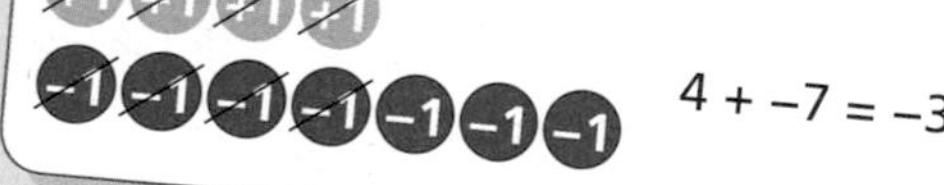

$4 + -7 = -3$

how to multiply and divide with directed numbers

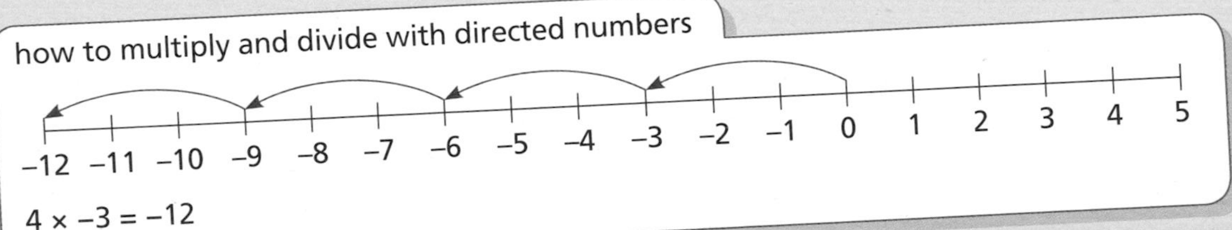

$4 \times -3 = -12$

about substitution and solving equations with directed numbers

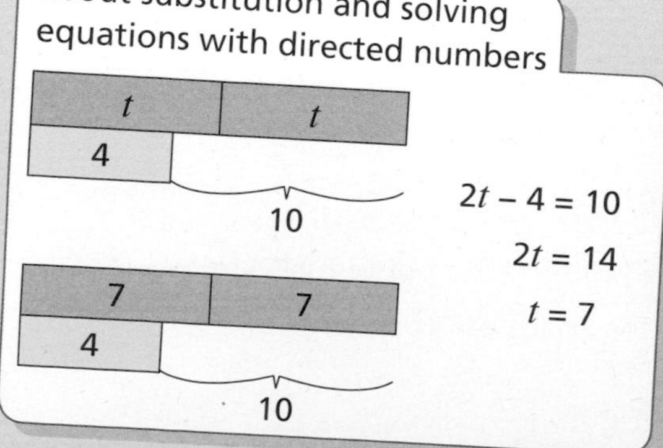

$2t - 4 = 10$

$2t = 14$

$t = 7$

about squaring and cubing negative numbers (H)

$7 \times 7 = 49$ and $-7 \times -7 = 49$

9.1 Understanding directed number

Small steps

- Understand and use representations of directed numbers
- Order directed numbers using lines and appropriate symbols
- Perform calculations that cross zero

Key words

Directed numbers – numbers that can be negative or positive

Negative numbers – numbers less than zero

Are you ready?

1 Work out

a 11 + 2 **b** 11 − 2

2 Which of these calculations will have an answer less than 0?

2 − 8 | 3 − 5 | 1200 − 500 | 140 − 165 | 11 − 63 | 16.67 − 13.45

How do you know? Did you have to work them out?

3 Write down the numbers you say in each case.

a Starting at 10, count down in 1s to −10

b Starting at 10, count down in 2s to −10

c Starting at 10, count down in 5s until you get to −30

4 What numbers are the arrows pointing to?

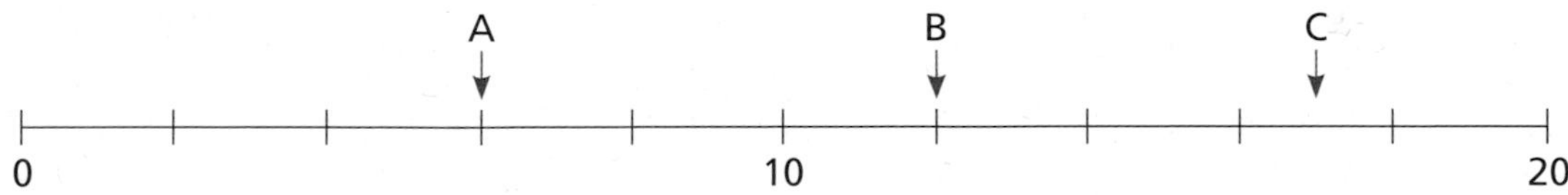

Models and representations

Double sided counters

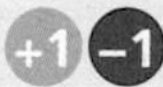

This represents +4

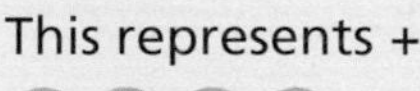

Number lines

−10 −9 −8 −7 −6 −5 −4 −3 −2 −1 0 1 2 3 4 5 6 7 8 9 10

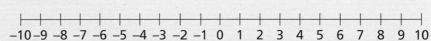

You will use a number line to help work out calculations with **directed numbers** that cross zero.

9.1 Understanding directed number

Numbers can either be positive or negative.

Negative numbers are used in many real-life contexts, for example when giving temperatures.

You will to need to understand negative numbers when plotting coordinates or substituting values into expressions.

Example 1

What numbers are the arrows pointing to?

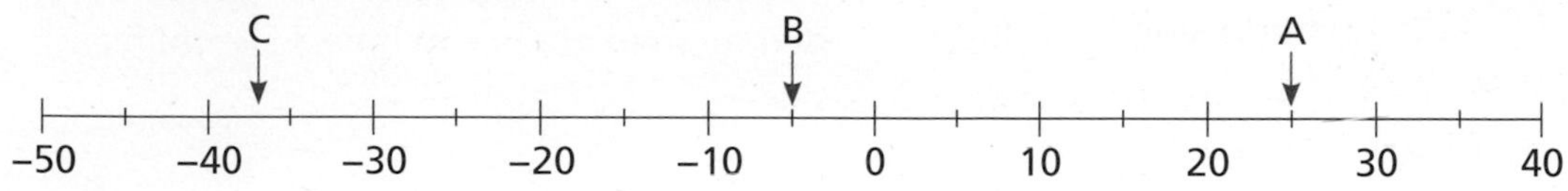

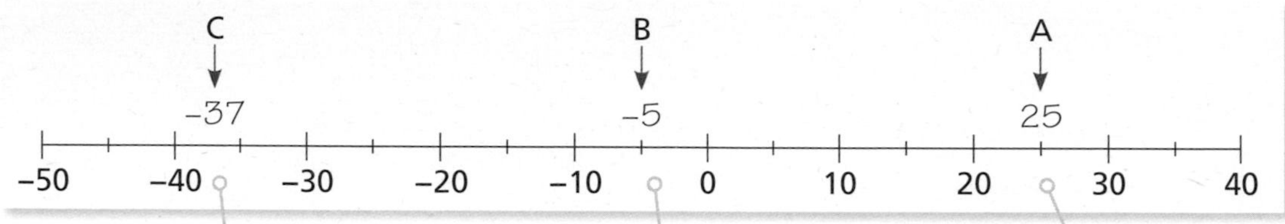

C is closer to –40 than to –30, so C is about –37

B is half way between 0 and –10, so this is –5

A is half way between 20 and 30. So A is pointing to 25

Example 2

Work out 7 – 12

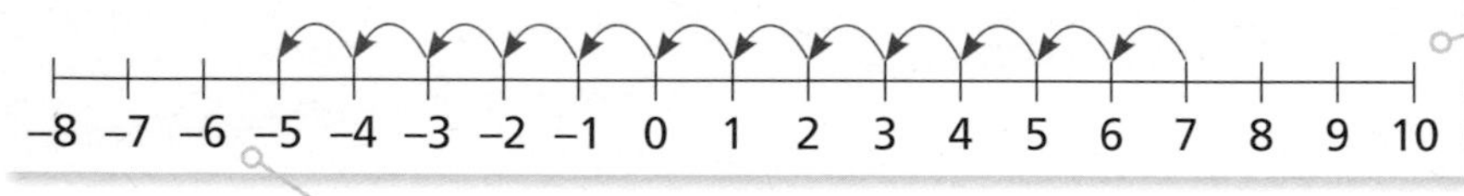

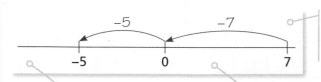

This is not the most efficient way.

You could draw a number line to help you.

You start at 7 and count down 12 through 0.

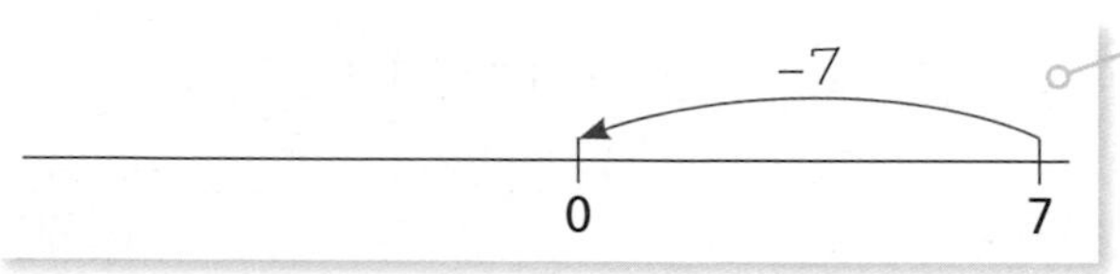

Or you could subtract 7 first to make 0.

–5 –7

–5 0 7

Since 7 + 5 = 12, you still need to subtract 5 more.

So 7 – 12 = –5

You can sketch an open number line like this to help you.

This is a more efficient method. You could imagine the number line in your head so that you do not have to draw it.

Practice 9.1A

1 What numbers are represented here?

a +1 +1 +1 +1 +1 **b** −1 −1 −1 −1 −1 **c** +1 +1 +1

2 Make each of these numbers using +1 and −1 counters.

a −4 **b** 6 **c** −2

3 **a** What numbers are the arrows pointing to?

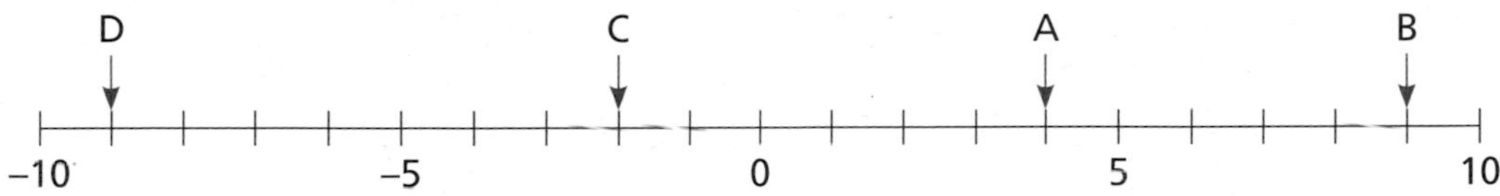

b Put the numbers in order, starting with the lowest.

What did you notice?

4 The thermometers show the temperature in two cities.

a What is the temperature in Leeds?

b What is the temperature in Toronto?

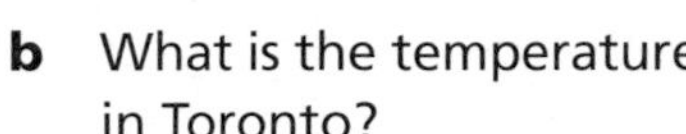

5 What numbers are the arrows pointing to?

a

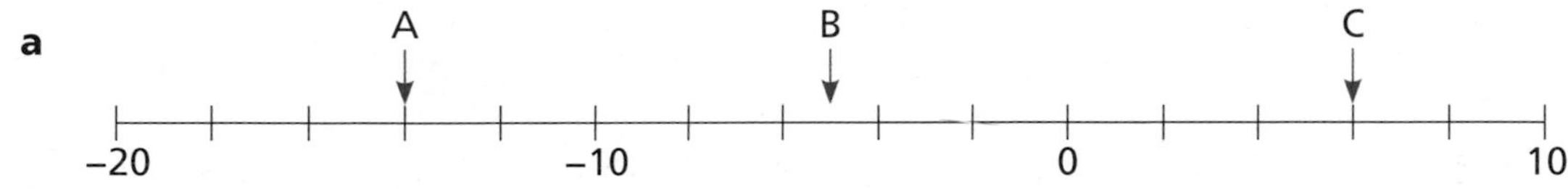

b

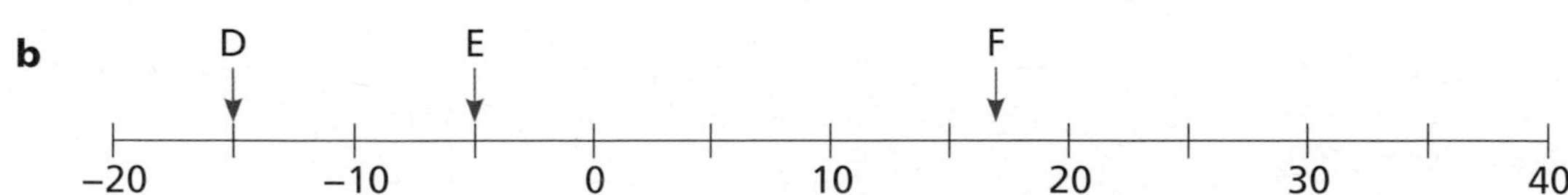

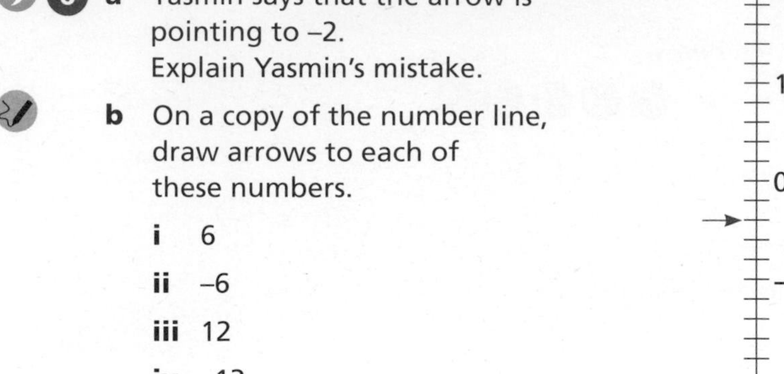

6 **a** Yasmin says that the arrow is pointing to –2.
Explain Yasmin's mistake.

b On a copy of the number line, draw arrows to each of these numbers.

i 6

ii –6

iii 12

iv –13

7 Put these numbers in order, starting with the highest.

a –8, 11, –6, 0

b 197.5, –109.4, –129.6, 14.5, 88.9

8 The table shows the temperature in five cities.

City	Temperature
London	12°C
Moscow	–7°C
Sydney	32°C
Helsinki	–16°C
Toronto	–1°C

a Which city is it the coldest?

b Which city is the warmest?

c The temperature in Warsaw is warmer than Moscow but colder than Toronto.
What could the temperature be in Warsaw?

9 Which symbol, < or >, should go in each circle to make the statements correct?

a –5 ◯ 5

b 3 ◯ –11

c –6 ◯ –10

d –5.6 ◯ –5.9

10 Seb is using a number line to work out 2 – 9

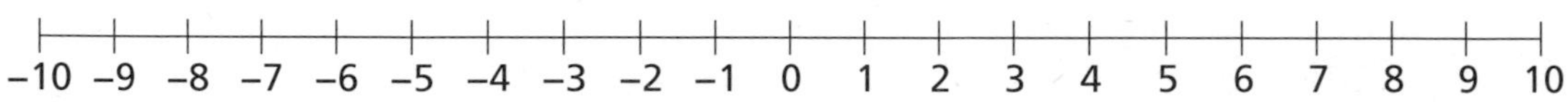

I started at 2 and counted back 9 and got –7

a Sketch a number line to show Seb a more efficient method.

b Compare your method with a partner's.

11 A hotel has several floors.

- Levels 1 to 15 are above ground.
- Ground floor is level 0.
- Levels −1 to −5 are below ground.

a Rob travels from floor 1 to floor −4.
How many floors does he travel through?

b Kate is on the 11th floor.
She travels down 13 floors.
What floor is she on now?

c She then travels up 8 floors.
What floor is she on now?

12 Work these out. You may want to use a number line to help you.

a $4 - 9$ **b** $7 - 2$ **c** $2 - 8$ **d** $-3 + 6$
e $-5 + 18$ **f** $-6 - 3$ **g** $-12 + 5$ **h** $-6 - 6$
i $-8 + 1.5$ **j** $5^2 - 6^2$

13 Find the missing numbers.

a $3 - \square = -6$ **b** $7 - \square = -1$ **c** $-8 + \square = -4$
d $\square - 10 = 3$ **e** $-4 - \square = -15$ **f** $-7 + \square = 7$

14 Zach is counting down in 10s from 42.

What mistake has Zach made?

15 The table shows the temperature in Manchester at different times during one day.

12 midnight	3 a.m.	6 a.m.	9 a.m.	12 noon	3 p.m.	6 p.m.	9 p.m.
−6 °C	−7 °C	−7 °C	−3 °C	2 °C	8 °C	6 °C	3 °C

a By how many degrees does the temperature rise between 12 midnight and 12 noon?

b By how many degrees does the temperature rise between 6 a.m. and 12 noon?

c Explain what happens to the temperature during the day.

d The temperature falls by 6 degrees from 9 p.m. to 12 midnight the next day. What is the temperature at 12 midnight on the next day?

What do you think?

1 The number line has two numbers marked by arrows. Both numbers are the same distance from zero.

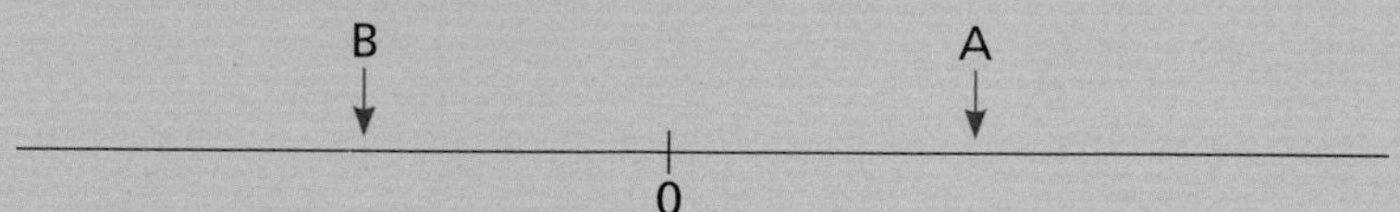

A is the number 17.5

a What number is B pointing to? **b** How do you know?

2 The difference between two numbers is 0

What can you say about the two numbers?

3 **a** Work out 123 – 147 in your head.

Explain your method to your partner.

Does your method always work?

b Use your method to work these out in your head.

i 1052 – 1056 **ii** 113.6 – 114.8

4 **a** Work out

i 3 – 7 = ☐ **ii** $3a - 7a =$ ☐ **iii** $3x - 7x =$ ☐

b What do you notice about your answers?

c Work out

i $5p - 7p$ **ii** $-2m + 5m$ **iii** $-6q - 8q$

iv $3a + 6a - 15a$ **v** $-10b + 2b - 2b$

Consolidate – do you need more?

1 Work out

a 2 – 7 **b** –3 + 8 **c** 19 – 24

d –5 – 7 **e** 3 – 8 **f** 7 – 100

g –5 – 8 **h** 0 – 9 **i** –37 – 50

2 Write the answers to these calculations in descending order.

15 – 18 32 – 38 54 – 52 76 – 77

Explain your method.

3 **a** What numbers are these arrows pointing to?

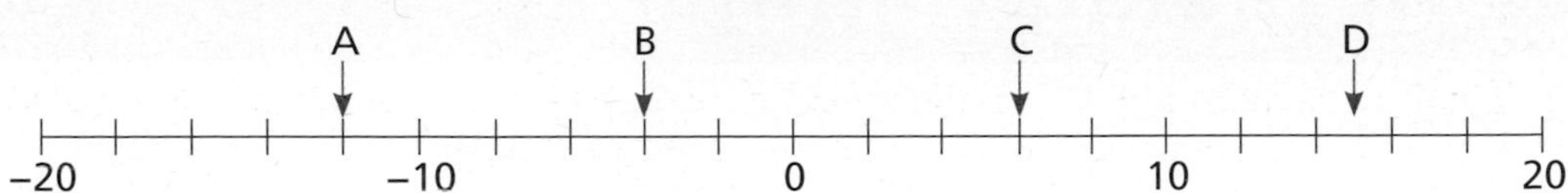

b What do you need to add to A to get to B?

c What do you need to subtract from D to get to B?

Stretch – can you deepen your learning?

1 **a** A number sequence starts at 17

It decreases by 4 each time.

What is the first number less than 0?

b Another number sequence starts at 170

It decreases by 4 each time.

What is the first number less than 0?

2 Here is a number line.

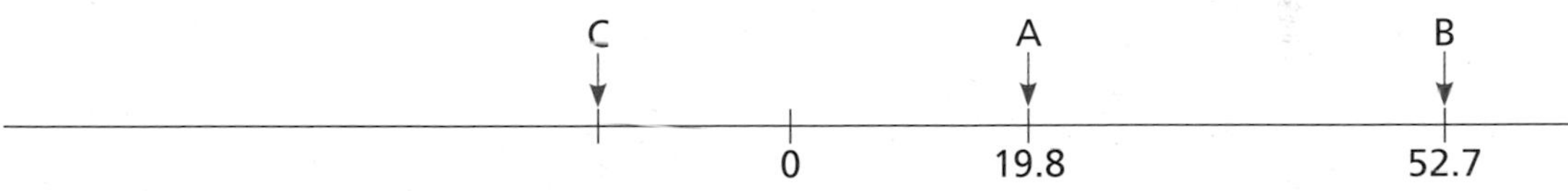

The difference between A and B is the same as the difference between C and A.

Find the value of C.

3 A town is 2.75 metres below sea level.

Another town is 126.8 metres above sea level.

Harry drives from the first town to the second town.

How many metres does the car rise between the two towns?

4 Work these out without using a calculator.

a $-39 - 48 = \square$ **b** $72 - 148 = \square$ **c** $14.5 - 26.4 = \square$

d $-1389 + 765 = \square$ **e** $0.35 - 1.07 = \square$ **f** $-4 + 0.185 = \square$

Explain your strategy to your partner.

5 Work out the missing numbers.

a $7 - 4 - 9 = 7 - \square$ **b** $2 - 8 - 3 = 2 - \square$

c $-5 - 4 - 8 = -5 - \square$ **d** $3 - 5 - 5 - 5 = 3 - \square$

What did you notice about the missing numbers?

Reflect

1 Where might you see negative numbers in real life?

2 Explain why the difference between two numbers is always positive.

9.2 Adding and subtracting directed numbers

Small steps

- Add directed numbers
- Subtract directed numbers
- Using a calculator for directed number calculations

Key words

Zero pairs – for example, +1 and –1 make zero

Demonstrate – to show how to do something

Are you ready?

1 Without working them out, state which calculations will give answers less than 0

10 – 4   4 – 10   2 – 8   3.71 – 5.95   11.5 – 7.7

Explain how you know.

2 Work out

a 7 – 12   **b** –5 – 3   **c** 2 – 15

d –5 + 8   **e** –10 + 18   **f** –18 + 10

3 The thermometer shows the temperature at midnight in New York.

By midday the temperature has risen by 13°C.

What is the temperature at midday?

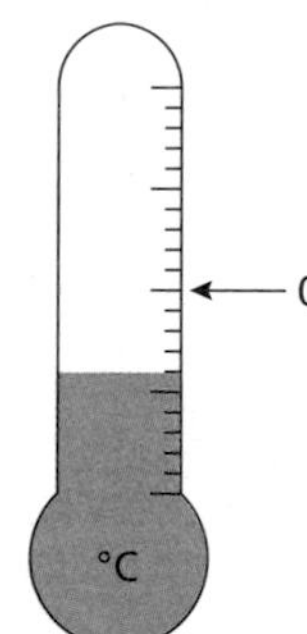

4 Work out the missing numbers.

a ☐ – 12 = 4   **b** ☐ – 12 = –4   **c** 12 – ☐ = 4   **d** 12 – ☐ = –4

What do you notice about the answers?

Models and representations

Double sided counters

Number lines

–10 –9 –8 –7 –6 –5 –4 –3 –2 –1 0 1 2 3 4 5 6 7 8 9 10

This helps you to see the difference between two numbers.

Zero pairs

+1
–1

Why do you think this is called a zero pair?

Example 1

Use counters to work out 4 + –7

Start with four '+1' counters.

You now need to add in seven '–1' counters.

The counters in each zero pair cancel each other out.

So you are left with –3

4 + –7 = –3

Example 2

Explain why 5 – –3 = 8

Start with five '+1' counters.

You need to subtract three '–1' counters. However, you do not have any '–1' counters to subtract.

To subtract three '–1' counters you need to add in three zero pairs. As they are zero pairs they do not change the value of the original number.

You can now subtract three '–1' counters.

You are left with eight '+1' counters.

5 – –3 = 8

Practice 9.2A

1. Beth is working out 5 + –2

   She has these counters.

   She adds the following counters.

   Explain how Beth can use the idea of zero pairs to work out the answer.

2. Demonstrate to your partner how to use counters to work these out.

a 6 + –2	**b** 3 + –7	**c** –3 + 6
d –3 + +6	**e** –4 + –2	**f** –5 + –2

3 Jakub is solving some directed number problems.

He is working out 3 – +6

I need to start with three counters.

I need to subtract six +1 counters. I don't have enough.

Explain what Jakub can do to work out the answer to the problem.

Show your partner.

4 Demonstrate to your partner how to use counters to work these out.

a 5 – +2 **b** 5 – +7 **c** –2 – +3 **d** 7 – –2 **e** –2 – –6

5 Match up the calculations that give the same answer.

You may match more than one calculation on the left to the same calculation on the right.

Left	Right
9 – +3	
9 – –3	3 – 9
9 + –3	9 + 3
–9 + 3	–9 – 3
–9 – +3	9 + –3
–9 – –3	

What was your strategy?

6 Work out

a 7 + –4 **b** 3 – +8 **c** 9 – +2

d 10 – –5 **e** –6 – +3 **f** –12 + –3

g –11 – –7 **h** –7 – –11 **i** 26 + +3

7 Here are some number cards.

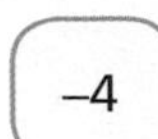

–4   –10   +6   +5   +9

Use the cards to make these statements correct.

a ☐ + ☐ = 14   **b** ☐ – ☐ = 3   **c** ☐ + ☐ = –14

d ☐ – ☐ = –3   **e** ☐ – ☐ = 10   **f** ☐ – ☐ = –1

Do any of the calculations have more than one answer?

8 Copy and complete

a **i** 8 + ☐ = 5   **ii** 8 – ☐ = 5   **iii** 8 + ☐ = 5   **iv** 8 – ☐ = 5

b **i** 20 + ☐ = 12   **ii** 20 – ☐ = 12   **iii** 12 + ☐ = 20   **iv** 12 – ☐ = 20

9 Work out

a 5 + –3 – –7   **b** 6 – –5 – +22   **c** 17.3 – +16.1   **d** $-7\frac{1}{2} + -3\frac{1}{2}$

10 Simplify

a $4x + -3x$ **b** $-7y - +3y$

11 **a** Check your answers to questions 1 to 9 using a calculator. Discuss with a partner how to enter directed number calculations into your calculators.

b Can you use a calculator to help with question 10? Why or why not?

What do you think?

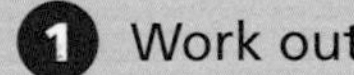

1 Work out

a 5 + –2   **b** 10 + –2   **c** 5 + –3   **d** 3 + –8

What do you notice?

2 Work out

a 8 + –3   **b** 8 – +3   **c** –5 + –2   **d** –5 – +2   **e** 11 + –7   **f** 11 – +7

What do you notice?

3 Jackson is trying to work out the missing number. 7 – ☐ = 12

This is not possible. You cannot subtract a number from 7 so that it increases to 12

Is Jackson correct?

Explain your answer.

Consolidate – do you need more?

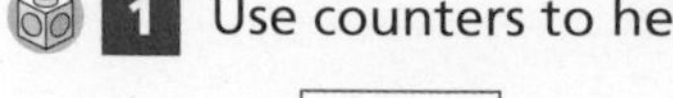

1 Use counters to help you to work these out.

a 3 + 6 **b** 3 + −6 **c** 3 − 6 **d** 3 − +6

2 Work out

a 5 + −8 **b** 6 − +8 **c** 11 − +5

d −15 + −5 **e** −10 + −2 **f** 11 − −4

g −2 − −7 **h** −8 + −3 **i** −9 − +1

j 25 + −100 **k** −73 + −2 **l** 17 − +16 + −1

3 Work out the missing numbers.

a 7 − □ = 1 **b** 7 + □ = 1 **c** □ − 7 = 1 **d** □ + 7 = 1

Stretch – can you deepen your learning?

1 Explain why 3 + −6 is equal to 3 − +6

You can use double sided counters to help you.

2 Here are the temperatures in five cities.

−4.3 °C, −7.5 °C, −12.9 °C, −8.4 °C, −5.9 °C

What is the range of the temperatures?

Write down the calculation you did to work out the answer.

3 Here are some number and symbol cards.

−6 −2 +3 +7 + −

a Use three of the cards to make a calculation with an answer of 4

□ □ □ = 4

Find two different ways.

b Use five of the cards to make a calculation with an answer of −1

□ □ □ □ □ = −1

c What calculation will give the highest answer?

d What calculation will give the lowest answer?

Reflect

Explain how you can use double sided counters to add and subtract directed numbers.

What rules did you notice?

9.3 Multiplying and dividing directed numbers

Small steps

- Multiplication of directed numbers
- Division of directed numbers

Key words

Product – the result of multiplying two or more numbers

Commutative – when an operation can be in any order

Are you ready?

1 Work out

a 7 × 5 **b** 4 × 8 **c** 9 × 19

d 2 × 13 × 5 **e** 36 × 25 **f** 1.7 × 2.1

2 Which parts of question 1 were you able to calculate mentally?

What strategies did you use to answer each question?

3 Work out

a 7 + −3 **b** −7 − +2 **c** 11 − −5

d 3 + −18 **e** −9 − −8 **f** −2 − −12

4 Write the missing number in each calculation.

a 3 + 3 + 3 + 3 + 3 = ☐ × 3

b 12 + 12 + 12 + 12 + 12 + 12 + 12 = 7 × ☐

c 5 × 8 = 8 + 8 + ☐

Models and representations

Double sided counters

This shows four groups of −2

Number lines

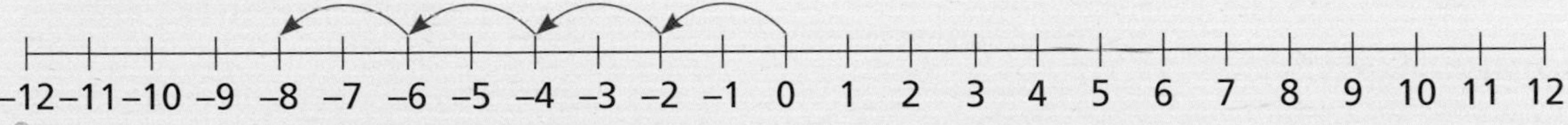

A number line is useful for showing repeated addition or subtraction.

You will use what you already know about multiplication and division, and link this to calculating with directed numbers.

Example 1

Work out

a 4 × 3 **b** 4 × –3 **c** –4 × 3 **d** –4 × –3

Method A

a 4 × 3 = 12

4 × 3 is the same as '4 lots of 3' or 3 + 3 + 3 + 3 = 12

You can show this on a number line.

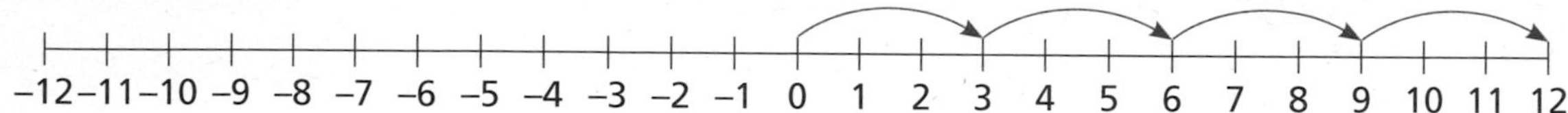

b 4 × –3 = –12

4 × –3 is the same as '4 lots of –3' or (–3) + (–3) + (–3) + (–3) = –12

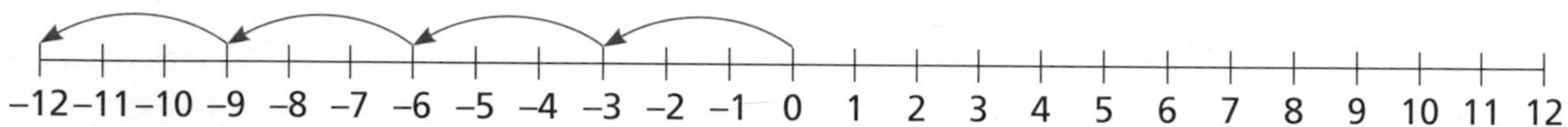

When one of the numbers you are multiplying is negative, the arrows change direction and go to the left on the number line.

c –4 × 3 = –12

What about –4 × 3? Having 'negative 4 groups of 3' does not necessarily make sense.

–4 × 3 is the same as 3 × –4 or '3 groups of –4' which is (–4) + (–4) + (–4) = –12

Multiplication is commutative.

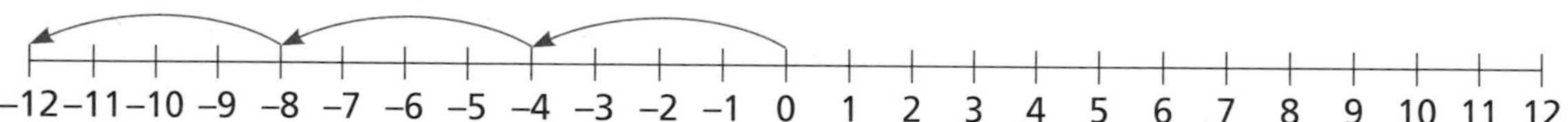

d –4 × –3 = 12

4 × 3

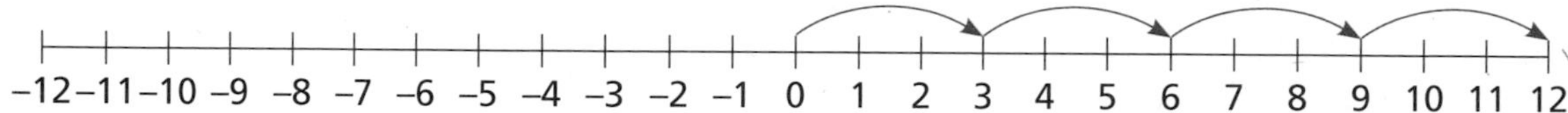

4 × –3

One number being negative means the arrows change direction.

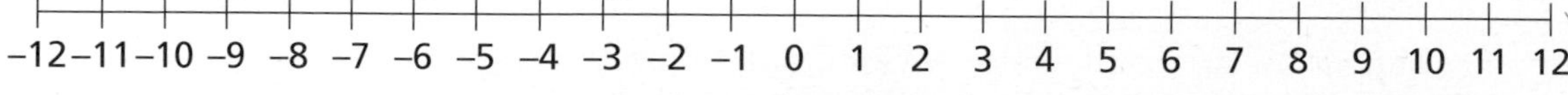

–4 × –3

Both numbers being negative means the arrows need to change direction twice, or not change direction at all.

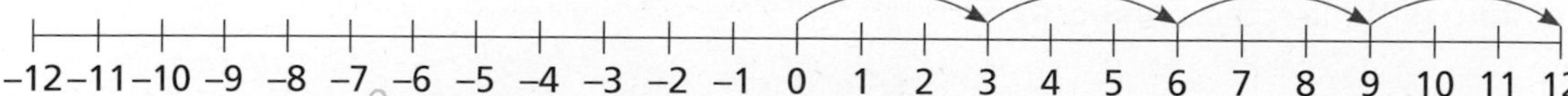

So –4 × –3 must equal positive 12

In general, any negative number multiplied by any other negative number gives a positive answer. It is useful to learn this rule.

Method B

a $4 \times 3 = 12$

4 × 3 is the same as '4 lots of 3' or 3 + 3 + 3 + 3 = 12

You can show this using double sided counters.

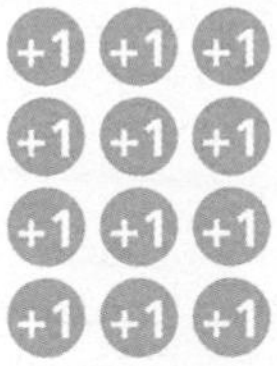

+1 +1 +1
+1 +1 +1
+1 +1 +1
+1 +1 +1

b $4 \times -3 = -12$

4 × –3 is the same as '4 lots of –3' or (–3) + (–3) + (–3) + (–3) = –12

–1 –1 –1
–1 –1 –1
–1 –1 –1
–1 –1 –1

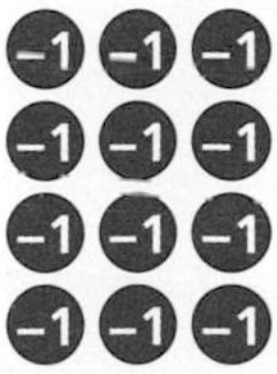

Making one of the numbers negative 'flips' the double sided counters.

c $-4 \times 3 = -12$

–4 × 3 is the same as 3 × –4 or '3 groups of –4' which is (–4) + (–4) + (–4) = –12

–1 –1 –1 –1
–1 –1 –1 –1
–1 –1 –1 –1

Multiplication is commutative.

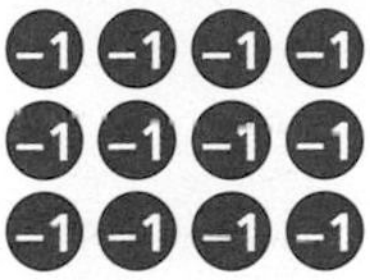

d $-4 \times -3 = 12$

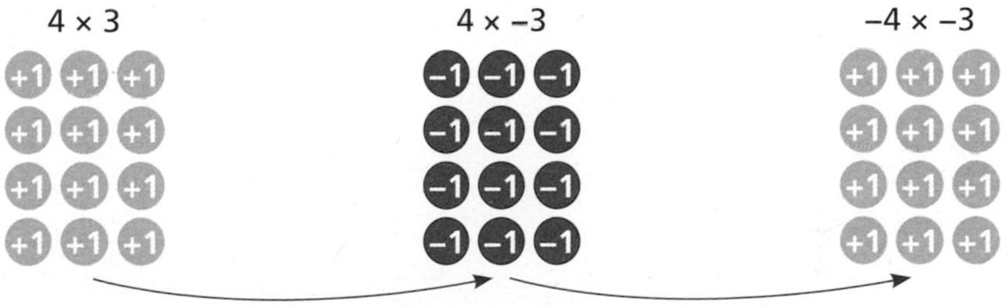

4 × 3	4 × –3	–4 × –3
+1 +1 +1	–1 –1 –1	+1 +1 +1
+1 +1 +1	–1 –1 –1	+1 +1 +1
+1 +1 +1	–1 –1 –1	+1 +1 +1
+1 +1 +1	–1 –1 –1	+1 +1 +1

One number being negative means the counters flipped.

Two numbers being negative means we need to flip them again.

So –4 × –3 must equal positive 12

Practice 9.3A

1. Copy and complete this multiplication grid.

	−3	−2	−1	0	1	2	3
3				0			
2				0			
1				0			
0	0	0	0	0	0	0	0
−1				0			
−2				0			
−3				0			

What patterns do you notice?

2. Here are some multiplication and division calculation cards.

-5×-4	-3×8	10×-5	11×17
$-8 \div 2$	-176×-32	$64 \div -2$	$-132 \div -10$

a Sort the cards into two groups. Do not use a calculator.

Calculations with a positive answer	Calculations with a negative answer

b Discuss your strategy with a partner.

3. Work out

a 9×-2 **b** -12×-3 **c** -5×4

d -5×-3 **e** 7×-5 **f** -11×2

4. Work out

a $10 \div 2$ **b** $15 \div -3$ **c** $-20 \div -10$

d $\frac{18}{-2}$ **e** $\frac{-40}{-5}$ **f** $\frac{-30}{6}$

5. Here are some number cards.

−24	−6	−4	−2	−1	3	4	12

Use two of the number cards to complete these calculations.

a $\square \times \square = 12$ **b** $\square \times \square = -12$

c $\square \div \square = -6$ **d** $\square \div \square = 6$

In how many ways can you complete each calculation?

6 Jakub multiplies two numbers together.

His answer is negative.

This means that at least one of the numbers is negative.

Do you agree with Jakub? Explain your answer.

7 Work out the missing numbers.

a **i** $12 \times \square = -48$ **ii** $-12 \times \square = -48$ **iii** $-12 \times \square = 48$

b **i** $\square \div 7 = -4$ **ii** $\square \div -4 = -7$ **iii** $\square \div -4 = +7$

8 Simplify

a $2a \times -3$ **b** $-2a \times -3$ **c** $3a \times -a$

What do you think?

1 Here are two number cards placed face down.

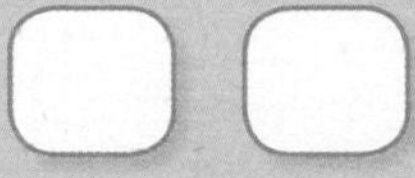

The product of the two cards is positive.

a What do you know must be true about the two cards?

b What do you not know about the cards? Discuss your answers with a partner.

2 Explain how you can work out $-5 \times 3 \times -2$

How many different methods can you find?

3 **a** Samira is working out $-211 \times -76 \times -95 \times 18765$

Her answer is positive.

How do you know that she has made a mistake without doing the calculation?

b Is the answer to this question positive or negative?

$$\frac{12.4 \times 3.5 \times -7.6}{-0.2}$$

How do you know?

Work out the exact answer on your calculator.

Consolidate – do you need more?

1 Work out

a 6×-6 **b** 3×-8 **c** -2×-11

d -5×9 **e** -17×-2 **f** -4×9

2 Complete these calculations.

a **i** $48 \div -6 = \square$ **ii** $48 \div -8 = \square$ **iii** $-48 \div -4 = \square$ **iv** $-48 \div 6 = \square$

b **i** $32 \div -1 = \square$ **ii** $-32 \div 1 = \square$ **iii** $-32 \div -1 = \square$ **iv** $\frac{-32}{-4} = \square$

3 Work out the missing numbers.

a $10 \times \square = -20$ **b** $-6 \times \square = -30$ **c** $\square \times -7 = 21$

d $\square \times -9 = -72$ **e** $12 \div \square = -2$ **f** $-90 \div \square = 10$

4 Here is a multiplication grid.

	−8	−5	−3	2	7
−8					
−5					
−3					
2					
7					

a Which squares will have positive answers?

b Which squares will have negative answers?

c Copy and complete the multiplication grid.

Stretch – can you deepen your learning?

1 The product of two numbers is −30

One of the numbers is −10

What is the sum of the two numbers?

2 **a** The product of two numbers is −20

The sum of the two numbers is 8

What are the two numbers?

b The product of two numbers is −20

The sum of the two numbers is −8

What are the two numbers?

3 Seb and Faith each think of a number.

When we multiply our two numbers, the answer is 36

The sum of our two numbers is −13

Seb thinks that the sum cannot be negative if the product is positive.

Explain why Seb is not correct.

4 $a \times b \times c = -20$

a What can you deduce about the values of a, b and c?

b Give some possible values of a, b and c.

5 Copy this multiplication grid and fill in the missing numbers.

×				
		−77		88
4			−36	32
	39	21		
−2				

Reflect

1 Two numbers multiplied together give a negative answer. What do you know about the two numbers? What do you not know?

2 How do you know that this answer is incorrect, without doing any working out?

$-5.9 \div -3.99 = -1.47\ldots$

9.4 Substitution with directed numbers

Small steps

- Substitute values into two-step expressions
- Evaluating algebraic expressions with directed number
- Use a calculator for directed number calculations

Key words

Evaluate – work out the value of an expression

Substitute – put numbers in place of letters to evaluate an expression

Expression – numbers, symbols and operators (such as ×, +…) grouped together that show the value of something

Are you ready?

1 Given that $x = 3$ and $y = 8$, evaluate

a $2x$ **b** $y - 7$ **c** $x + y$

d $10x + 6$ **e** $\frac{y}{4}$ **f** $x - y$

2 Given that $a = 4$ and $b = 10$, work out the value of

a $2a + b$ **b** a^2 **c** $(a + b)^2$

d $\frac{a + b}{2}$ **e** ab **f** $3ab$

3 Biscuits are sold in small and large packets.

A small packet contains 3 biscuits.

A large packet contains 15 biscuits.

This formula works out the total number of biscuits in s small packets and l large packets.

$T = 3s + 15l$

a Explain why the formula gives the total number of biscuits.

b When $s = 7$ and $l = 4$, what is the value of T?

4 Work out

a $11 - -2$ **b** 8×-4 **c** $3 - 15$

d $18 + -3$ **e** $-7 + -3$ **f** $\frac{16}{-4}$

g -5×-2 **h** $6 \times -3 \times -2$ **i** $-7 - -4$

Models and representations

Bar model

This shows $2p + 5$

p	p	+5

+7	+7	+5
19		

This shows the substitution $p = 7$

If $p = 7$, $2p + 5 = 19$

There are many formulae that help us to solve a wide range of real-life problems. Substitution into expressions and formulae involves replacing a variable with a number.

Example 1

If $p = 3$ and $q = -5$, find the value of

a $p + q$ **b** $p - q$ **c** $2p - 10$ **d** pq **e** $\frac{q}{-10}$

a $p + q = (3) + (-5) = 3 - 5 = -2$ — Adding a negative number is the same as subtracting the number.

b $p - q = (3) - (-5) = 3 + 5 = 8$ — Subtracting a negative number is the same as adding the number.

c $2p - 10 = 2(3) - 10$
$= 2 \times 3 - 10$
$= 6 - 10$
$= -4$

Remember to use the correct priority of operations.

d $pq = (3)(-5) = 3 \times -5 = -15$ — Multiplying a positive number by a negative number gives a negative answer.

e $\frac{q}{-10} = \frac{(-5)}{-10} = \frac{5}{10} = \frac{1}{2}$ — A negative number divided by a negative number gives a positive answer. $\frac{5}{10}$ is the same as $\frac{1}{2}$

Practice 9.4A

1 **a** Flo is working out the value of $c + d$. She is told that $c = 4$ and $d = -9$

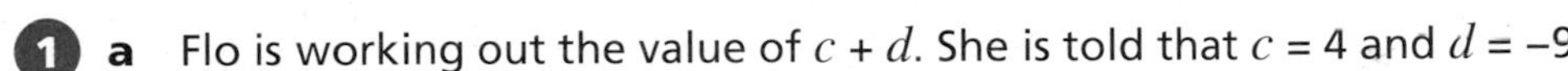

$c + d$ is equal to $4 - 9 = -5$

Is Flo correct?

b Flo is now working out the value of $c - d$ using the same numbers.

$c - d$ is still equal to $4 - 9 = -5$

Is Flo correct?

2 $x = -6$ and $y = 12$

a Show that $x + y$ is equal to $y + x$

b Show that $x - y$ is not equal to $y - x$

3 If $m = -4$ and $n = 20$, find the value of

a $m + 9$ **b** $n - 25$ **c** $7 + m$ **d** $m + n$

e $n - m$ **f** $m - n$ **g** $2m$ **h** $2n$

i $-2m$ **j** mn **k** n^2 **l** m^2

4 $a = 10$, $b = -2$ and $c = -5$. Show that $a + b = 3 - c$

5 $f = -3$, $g = -10$ and $h = 9$. Find the value of

a $2f + h$ **b** $3f + h$ **c** $h - g$ **d** $f - 2g$

e fg **f** gh **g** fgh **h** $fg - 20$

i $2g + 3h$ **j** $g^2 + h^2$

6 The terms of a sequence are given by $7 - 3n$. Find the value of the 20th term.

7 Samira has answered some questions on substitution. She has got the first three correct, but the last one is wrong.

i $x + 2y = 1$ ✓ ii $2z - y = 8$ ✓

iii $xz = 15$ ✓ iv $x + y + z = 0$ ✗

One of the letters that Samira used is equal to –2, one of them is equal to 3 and the other is equal to 5.

a Which letter is represented by which value?

b Explain how you can work out the answer to part **iv**, without knowing which letter is which value.

Sometimes it is necessary to use a calculator to work out the value of an expression.

Most calculators have a button on them that looks like this [(–)]

You can use this button to enter a negative number.

It acts differently to the subtraction button. [–]

Try pressing this button followed by any number to see how it works.

Example 2

Use your calculator to evaluate $\frac{3.9 \times -2.7}{76.5 - -39.8}$

Give your answer to four decimal places.

[fraction] [3] [·] [9] [×] [(–)] [2] [·] [7] [▼] [7] [6] [·] [5] [–] [(–)] [3] [9] [·] [8] [=]

Press the fraction button first.

You can type this into your calculator all in one go using these buttons.

Answer: –0.0905

Practice 9.4B

1 Use your calculator to work out

a -2.3×11.8 **b** $7.8 \div -0.32$

c $\frac{18.4 \times -0.75}{3}$ **d** $11 - (-9.4)^2$

Compare with a partner how you used your calculator.

2 $a = 11.7$, $b = -0.95$ and $c = \frac{3}{8}$

Use your calculator to find the value of

a $3.5a - b$ **b** $5c + 7.2b$ **c** abc

d $0.01(c - 3b)$ **e** $\frac{3b}{c - a}$ **f** $a^2 + b^2 + c^2$

3 This formula converts temperatures in degrees Celsius, C, to degrees Fahrenheit, F

$$F = 32 + \frac{9C}{5}$$

a Convert 15.5 °C into °F **b** Convert −10.6 °C into °F

What do you think?

1 $a = 1$, $b = 4$, $c = -8$ and $d = -2$

Choose letters to make these statements correct.

a □ + □ = 2 **b** □ − □ = 6 **c** □ × □ = 16 **d** □ ÷ □ = −2

Do any of the statements above have more than one answer?

2 $T + S = U - V$

S and T are both negative and U and V are both positive.

S, T, U and V are all whole numbers between −8 and 8

Find possible values for all the letters.

Discuss your strategy with a partner.

Consolidate – do you need more?

1 $a = 5$ and $b = -12$

Find the value of

a $a - 7$ **f** ab **k** a^2

e $2b$ **j** $3a + 2b$ **d** $b - a$

i $100 - 7b$ **c** $a - b$ **h** $\frac{b}{a}$

b $a + b$ **g** $2ab$ **l** b^2

2 $x = 3$, $y = -5$ and $z = -2$

Find the value of

a $x + y + z$ **b** $x - y - z$ **c** $2x + y$

d $xy + z$ **e** xyz

Stretch – can you deepen your learning?

1 $x = \frac{1}{2}$, $y = -\frac{1}{4}$ and $z = -1$

a Show that $x + y = -y$ **b** Show that $x + y + z = 3y$

2 Here are some number cards.

−32.5 −8.7 −5.1 −1.6 7.3

I know that the mean value of these cards is going to be negative without working it out.

a Explain how Ed knows that the mean value will be negative.

b Find the mean value of the cards.

c Find the range of the numbers on the cards.

3 $a = -5.6$, $b = -2.4$ and $a + b = b - c$

Work out the value of c

Show all your reasoning.

4 Explain why, for any two numbers, $ab = -a \times -b$

Reflect

Why do you have to be careful when substituting negative numbers into expressions or formulae?

What do you think are the most common mistakes?

9.5 Two-step equations

Small steps

- Introduction to two-step equations
- Solving two-step equations

Key words

Equation – a mathematical statement showing that two things are equal

Solve – to find the value (or values) of a letter that makes an equation true

Solution – the value (or values) of a letter that makes an equation true

Unknown – a variable (letter), whose value is not yet known

Are you ready?

1 **a** What equation is represented by this bar model?

40	
x	26

b Solve the equation to work out the value of x

2 Solve these equations.

a $3x = 18$ **b** $7m = 28$ **c** $125 = 10g$

d $\frac{y}{5} = 7$ **e** $3 = \frac{k}{7.4}$ **f** $40f = 20$

3 Solve these equations.

a $u + 7 = 11$ **b** $19 + m = 28$ **c** $g - 3 = 11$

d $300 = n - 120$ **e** $10 - y = 4$ **f** $f + 20 = 1$

Models and representations

Bar model

This represents $3p + 5 = 23$

23			
p	p	p	5

A bar model can help you to solve two-step equations.

Cups and counters

Cups and counters can be a fun way of forming equations. Hide an **unknown** under a cup and ask a partner to solve it.

You learnt how to **solve** one-step linear **equations** in Block 3.

Two-step equations require two steps to solve them. This lesson builds on the work you have already done.

Example 1

Marta is solving the equation $3x + 2 = 29$

a Draw a bar model to represent the equation.

b Solve the equation.

a

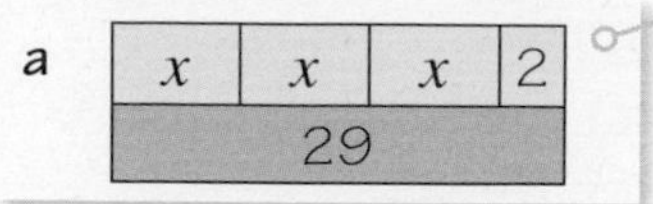

Draw a bar for $3x + 2$

Underneath, draw a bar of the same length showing 29

The bars are the same length because they are equal.

b

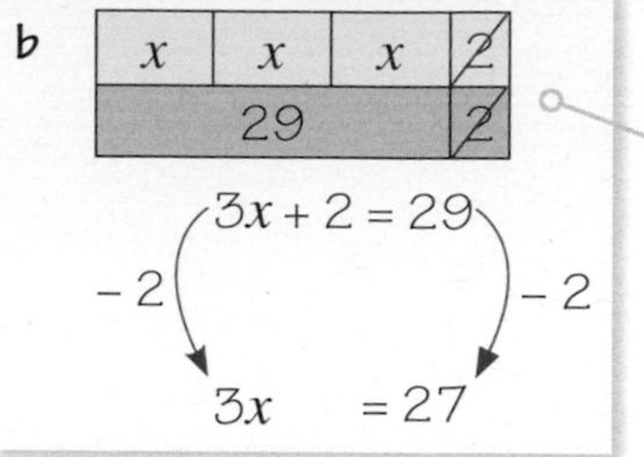

First subtract 2 from both bars.

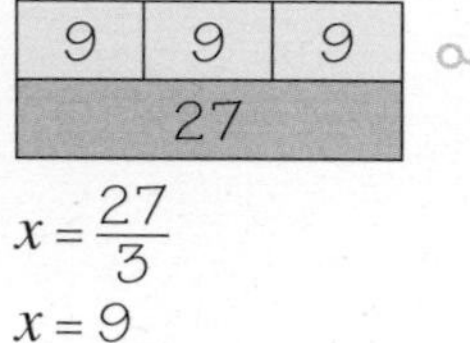

Then you need to divide 27 by 3

Practice 9.5A

1 The bar model shows the equation $2k + 7 = 17$

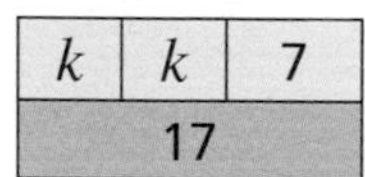

Copy and complete the working to show the steps for solving the equation.

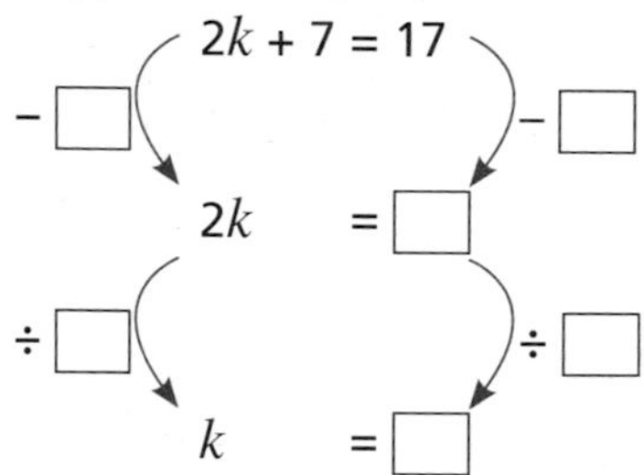

2 **a** What equation does the bar model show?

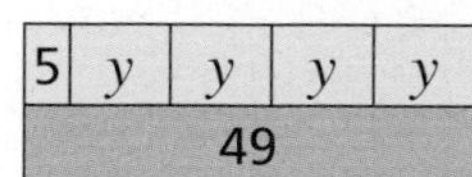

b Solve the equation to find the value of y.

3 Solve these equations.

a $5m + 3 = 33$ **b** $7t + 11 = 39$ **c** $2g + 7 = 12$

d $6 + 4m = 8$ **e** $28 = 3m + 1$ **f** $5k + 20 = 5$

g $3t + 2 = 10$ **h** $2k + 3 = -7$

4 Emily is solving the equation $3x - 4 = 11$

She draws this bar model.

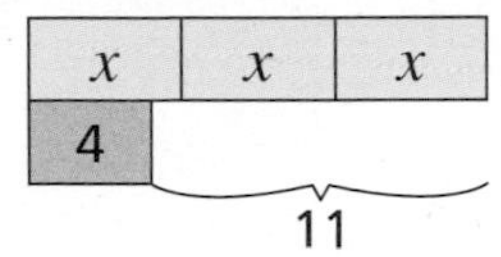

a Explain why Emily has set up her bar model in this way.

b Solve the equation $3x - 4 = 11$

5 Rhys solves the equation $5x - 1 = 11$

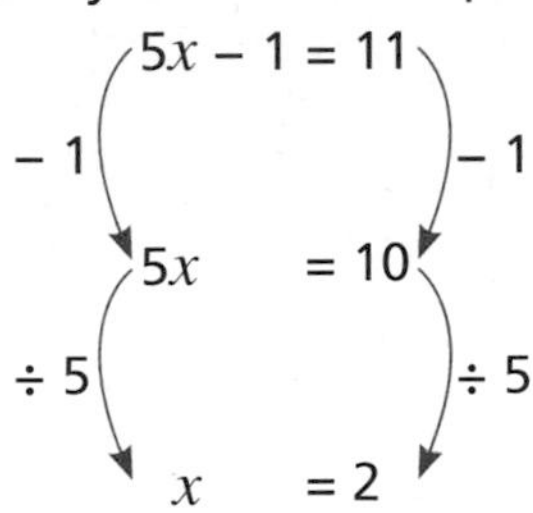

a How can Rhys check if his answer is correct?

b Show that his answer is incorrect.

c Identify the mistake that Rhys made.

6 Solve these equations.

a $6p - 2 = 16$ **b** $4g - 26 = 8$ **c** $11 = 10d - 8$

d $0.5k - 2.3 = 4.4$ **e** $4y - 10 = -30$

7 Discuss with a partner how to solve the equation $10 - 3m = 7$

What mistakes do you think someone might make when solving this equation?

8 Solve these equations.

a $\frac{x}{3} + 1 = 5$ **b** $\frac{x + 1}{3} = 5$

c What's the same and what's different about the way you solve these equations?

9 Solve these equations.

a $\frac{2u}{3} = 5$ **b** $\frac{y}{5} - 7 = 11$ **c** $76 = \frac{n - 2}{2}$ **d** $\frac{10m}{3} = -6$

10 How would you solve this equation?

$a + a + a + 2 + a + a + 3 + 7 = 2$

What do you think?

1. Max is using some cups and counters to help him to solve an equation. There are the same number of counters in each cup.

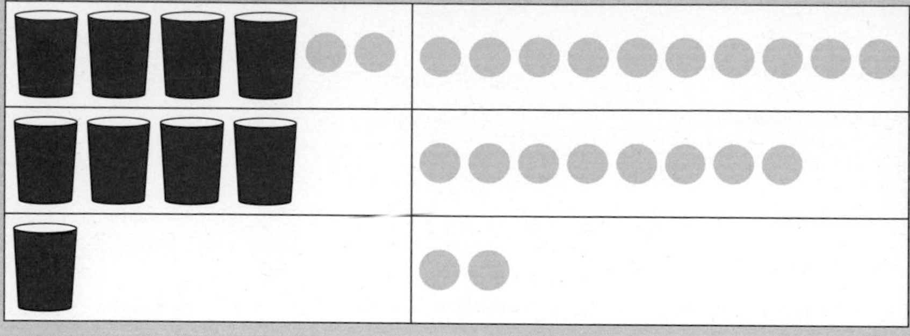

$$4x + 2 = 10$$
$$-2 \quad -2$$
$$4x = 8$$
$$\div 4 \quad \div 4$$
$$x = 2$$

   a How does the diagram match the calculation?

   b How would the diagram change if you represented $4x - 2 = 10$?

2. a Match each function machine to the correct expression.

   A $x \rightarrow$ $\times 2$ $\rightarrow$ $+ 5$ $\rightarrow$

   B $x \rightarrow$ $\times 5$ $\rightarrow$ $+ 2$ $\rightarrow$

   C $x \rightarrow$ $\div 2$ $\rightarrow$ $+ 5$ $\rightarrow$

   D $x \rightarrow$ $+ 2$ $\rightarrow$ $\div 5$ $\rightarrow$

   E $5x + 2$

   F $\frac{x+2}{5}$

   G $2x + 5$

   H $\frac{x}{2} + 5$

   b Explain how you can use one of the function machines to solve $5x + 2 = 22$

3. Solve each pair of equations.

   a i $4x + 3 = 31$ ii $3 + 4x = 31$ b i $5x - 1 = 34$ ii $-1 - 5x = 34$

   c What's the same and what's different about how you solved each pair?

4. Here is a function machine.

   $x \rightarrow$ $+ 2$ $\rightarrow$ $\times 3$ $\rightarrow$?

   Chloe is trying to work out what number she needs to put in to get an output of 72.

   She says:

   I need to solve the equation $3x + 2 = 72$

   Explain why Chloe is wrong.

5. Write down four different two-step equations that have a solution of $w = 5$

   Explain your strategy.

Consolidate – do you need more?

1 Use the bar models to solve these equations.

a $2x + 1 = 9$

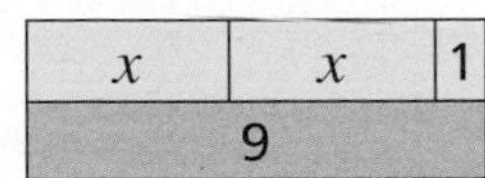

b $2m - 3 = 5$

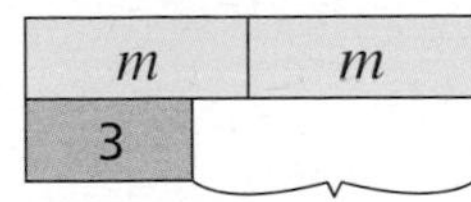

2 Faith solves the equation $2 + 5x = 47$

She gets the answer 9

Show that Faith's answer is correct.

Explain your method to a partner. Did you both check Faith's answer in the same way?

3 Solve these equations.

a $7x - 2 = 12$ **b** $2y + 5 = 16$ **c** $\frac{u}{4} + 1 = 5$

d $4 + 5m = 11$ **e** $\frac{t - 3}{4} = 2$ **f** $7 - 2p = 45$

g $4t + 20 = 12$ **h** $\frac{5y}{2} = 3$ **i** $6h - 2 = -11$

j $-11 = 6h + 4$ **k** $10 - 4x = 15$ **l** $-3x + 2 = 11$

Stretch – can you deepen your learning?

1 Explain why these two equations have the same solution.

$4 - 7x = 11$ $-7x + 4 = 11$

2 Solve these equations.

a $3b + 5 + 2b + 1 = 17$ **b** $f + 2f + 10 - 2 = 2$

c $5g - 2 + 3g - 7 = 24$ **d** $2x + 3 - 5 - 8x = -5$

3 What's the same and what's different about the way you solve these equations?

$\frac{1}{3}x + \frac{1}{2} = 1$ $\frac{x}{3} + 0.5 = 1$

Reflect

1 Explain the difference between a one-step equation and a two-step equation.

2 Explain how you can use a bar model to help solve an equation.

3 How can you check that a solution to an equation is correct?

9.6 Order of operations

Small steps

- Use order of operations with directed numbers
- Use a calculator for directed number calculations
- Choose the most appropriate method: mental strategies, formal written or calculator

Key words

Priority – something which takes priority needs to be done first

Indices – an index number (or **power**) tells you how many times to multiply a number by itself

Operation – a mathematical process such as addition, subtraction, multiplication or division

Are you ready?

1 **a** Explain why $2 \times 3 + 5 = 11$ **b** Explain why $2 + 3 \times 5 = 17$

c Why do you think someone might write 25 as the answer for part **b**?

2 Copy each calculation and underline the part of each calculation that you do first.

a $7 \times 4 + 3$ **b** $6 + 3 \times 6$ **c** $9 - 4 \div 2$

d $2 \times (3 + 5) + 4$ **e** $20 - 3 \times 4 + 6$ **f** 4×5^2

3 Work out the answers to each part of question 2.

4 Zach is working out $10 - 3 + 2$

I have to work out 3 + 2 first as you do additions before subtractions.

Is Zach correct?

Models and representations

Counters or tiles

This shows the calculation $3 + 2 \times 4$

First work out how many counters are in the 2×4 array ($2 \times 4 = 8$) and then add on 3

The answer is 11 not 20

In a calculation, the **operations** need to be carried out in a specific order. The triangle shows the order or **priority** of the operations.

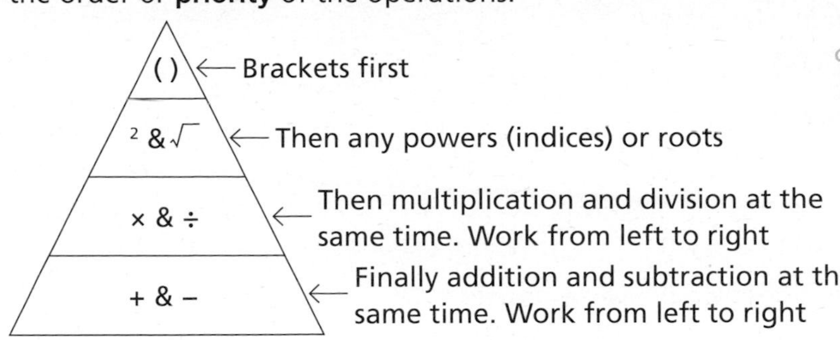
( )
Brackets first
^{2} & $\sqrt{}$
Then any powers (indices) or roots
× & ÷
Then multiplication and division at the same time. Work from left to right
+ & –
Finally addition and subtraction at the same time. Work from left to right

Some people think you do division before multiplication. This is incorrect. If there is both division and multiplication in a question they should be done at the same time – you work from left to right.

Example 1

Work out $4 + 2 \times -3$

$4 + \underline{2 \times -3}$ — From the priority of operations, you first work out 2×-3

$= 4 + -6$ — (–1)(–1) (–1)(–1) (–1)(–1)

You know that $2 \times -3 = -6$

$= 4 - 6$ — Adding a negative number is the same as subtracting.

$= -2$ — Then work out $4 + -6$ (+1)(+1)(+1)(+1) + (–1)(–1) (–1)(–1) (–1)(–1)

$4 - 6 = -2$ — The 4 yellow tiles cancel out 4 red ones (zero pairs). This leaves you with 2 red tiles.

Example 2

Work out $6 \times 3 - 10 \div -2$

$\underline{6 \times 3} - \underline{10 \div -2}$ — There are no brackets or indices so you start with multiplication and division.

$= 18 - -5$ — Do the multiplication and division from left to right.

$6 \times 3 = 18$ $\quad$ $10 \div -2 = -5$

$= 18 + 5$ — Then apply your knowledge of directed numbers.

$= 23$

Subtracting a negative number is the same as adding.

Practice 9.6A

1. Use the counters to explain why $3 + 4 \times -2 = -5$

   +1 +1 +1   −1 −1 −1 −1
−1 −1 −1 −1

2. Copy each calculation and underline the part of it that you need to work out first.

   a $4 + 3 \times -6$ **b** $(4 + 3) \times 6$ **c** $3 \times -6 + 4$ **d** $4 + (3 \times -6)$

   e Which of these calculations give the same answer?
   How can you tell this from what you have underlined?

3. Underline the part of each calculation that you need to work out first. Then work out the answers.

   a **i** $5 + 10 \times -2$ **ii** $5 - 10 \times -2$

   b **i** $-6 - 3 \times 5$ **ii** $6 - 3 \times 5$

   c **i** $7 \times (3 - 5)$ **ii** $-7 \times (3 - 5)$

   d **i** $1 + 8 \times 3 - 4$ **ii** $1 - 8 \times 3 - 4$

   e **i** $11 + 10 \div -2$ **ii** $11 - 10 \div -2$

   f **i** $4 - 10 \div 5$ **ii** $4 - 10 \div -5$

   g **i** $(2 + 5) \times 6 + 2$ **ii** $(-2 + 5) \times -6 + 3$

   h **i** $11 \times (2 - 10)$ **ii** $-11 \times (2 - 10)$

   Check your answers using a calculator.

4. Use the counters to explain why $2 \times 5 + 3 \times -4 = -2$

   +1 +1 +1 +1 +1   −1 −1 −1 −1
+1 +1 +1 +1 +1   −1 −1 −1 −1
−1 −1 −1 −1

5. Faith is working out the value of $7 - 2 + 3 \times -5$

   First I work out 3×-5, which is -15

   a What should Faith do next?

   b What is the answer to the calculation?

6. Copy each calculation and underline all the parts that have equal priority.

   a **i** $4 \times -2 + 3 \times 5$ **ii** $4 \times -2 + 3 \times -5$ **iii** $4 \times -2 - 3 \times 5$

   b **i** $(2 - 10) \div (9 - 7)$ **ii** $(2 - 10) \div (7 - 9)$ **iii** $(2 - 10) - (9 - 7)$

   c **i** $3 + 4 \times -6 \div 2$ **ii** $3 - 4 \times -6 \div 2$ **iii** $3 + 4 \times -6 \div -2$

7 Work out the missing numbers in these calculations.

a **i** $\square \times 5 + 1 = 31$ **ii** $\square \times 5 + 1 = -9$

b **i** $6 + \square \times 3 = 12$ **ii** $6 + \square \times 3 = 0$ **iii** $6 + \square \times 3 = -12$

c **i** $9 - \square + 3 = 5$ **ii** $9 - \square + 3 = 10$ **iii** $9 - \square + 3 = 20$

8 Copy each calculation and insert brackets into the calculations to make the answer correct.

a **i** $4 \times 2 - 9 + 1 = -2$ **ii** $4 \times 2 - 9 + 1 = -27$

b **i** $3 \times 5 - 3 - 10 = -4$ **ii** $3 \times 5 - 3 - 10 = 22$

9 How would you work out each of these calculations?

a $\sqrt{5 \times 4 + 5}$ **b** $2 \times \sqrt{16} - 3$

What is the same? What is different?

10 If $p = 4$, $q = -3$ and $r = 5$, work out the value of

a $pq + qr$ **b** $pq - qr$ **c** $p^2 + q^2$ **d** $(r - q) \div (p - 6)$

What do you think?

1 Flo and Jackson are working out $3 \times 10 \div 5$

I am going to do deal with the 10 and 5 first.

I am going to work from left to right.

a Show that Flo and Jackson get the same answer.

b What if the calculation was $3 \div 10 \times 5$?

2 Why do each pair of calculations give the same answer?

a $6 \times -3 + 2$ $\quad 2 - 6 \times 3$

b $7 \times 2 + 3 \times -4$ $\quad (7 \times 2) - (3 \times 4)$

3 Here are some digit cards.

−2 3 4 8 −9

Here are some operation cards.

+ − × ÷

Choose 3 digits and 2 operations to go into these boxes. You can only use each card once.

a What answer did you get? See if you can work it out mentally. Use your calculator to check.

b In how many different ways can you make the answer 0?

c What is the highest number you can make?

d What is the lowest number you can make?

Consolidate – do you need more?

1 Use counters to explain why $6 + -2 \times 4$ is equal to -2. Now convince your partner.

2 Work out

a $6 + 3 \times -2$ **b** $7 - 4 \times 5$ **c** $3 \times -5 + 2$

d $10 \times 3 + 3 \times -1$ **e** $8 \times 9 - 11 \times 10$ **f** $62 - 3 \times -2$

g $12 - 10 \div 2$ **h** $12 - 10 \div -2$

3 Explain how to work out $\sqrt{3 \times 10 + 5 \times -1}$

4 Insert a pair of brackets into each calculation to make the answers correct.

a $5 + 3 \times -2 = -16$ **b** $5 + 3 \times -2 = -1$ **c** $10 \div -2 + 12 \times 1 = 1$

Stretch – can you deepen your learning?

1 Without working them out, explain why these two calculations will always give the same answer.

A $0.765 \times 3.96 - 5.61 \times 7.134$ **B** $3.96 \times 0.765 + 5.61 \times -7.134$

Use your calculator to check that they do give the same answer.

2 Evaluate the expression on each card.

$(-3)^2$ -3^2

What's the same and what's different about these questions and answers?

3 Work out

a $15 - 50 \div 5 + 3 \times -4$

b $14 \times -2 + \sqrt{25 - 9}$

c $-3 \times 5 + 8 - 5 \times -7$

d $103 - 6 \times 3 + 4$

4 If $p = -10$, $q = -3$ and $r = 5$, work out the value of

a $3 \times p + 2 \times q$

b $\frac{pq}{r}$

c $p - q + r$

d $q + p \div r$

e $r^2 + p \times 2$

f $\sqrt{p^2 - 6r - 17q}$

Reflect

1

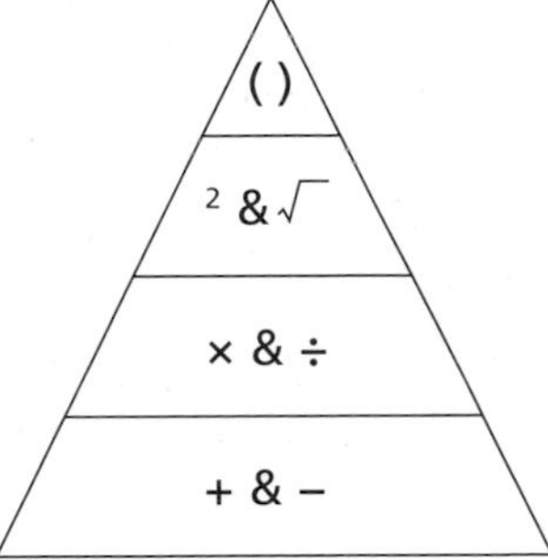

a What operation does the diagram tell you to do first? Which comes next?

b Why are × and ÷ next to each other?

c Why are + and – next to each other?

2 How do you know that these two calculations give the same answer, without working them out?

A $13.6 \times -5.9 + 8.6$

B $8.6 + 13.6 \times -5.9$

Ⓗ 9.7 Powers and roots

Small steps

- Roots of positive numbers Ⓗ
- Explore higher powers and roots Ⓗ

Key words

Power (or exponent) – this tells you how many times to use a number or variable in a multiplication

Root – the nth root of a number x is the number that is equal to x when multiplied by itself n times

Are you ready?

1 Calculate the area of each square.

a

5 cm

b

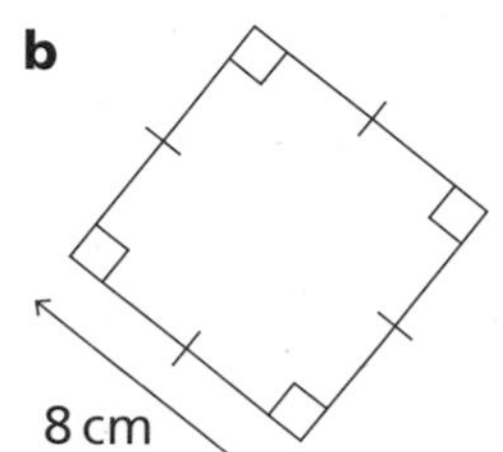

c

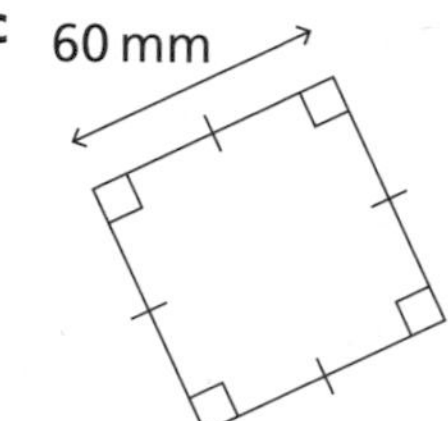

d

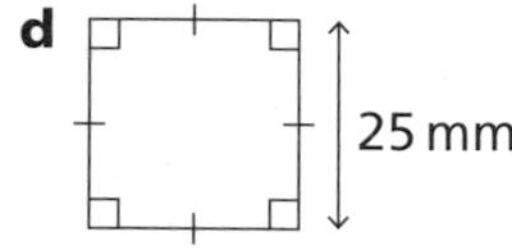

2 How many small cubes are used to make each bigger cube?

a

b

c

3 Complete these calculations.

a 3 × 3 **b** 7 × 7 **c** 9 × 9 **d** 12 × 12

e −3 × −3 **f** −7 × −7 **g** −9 × −9 **h** −12 × −12

Models and representations

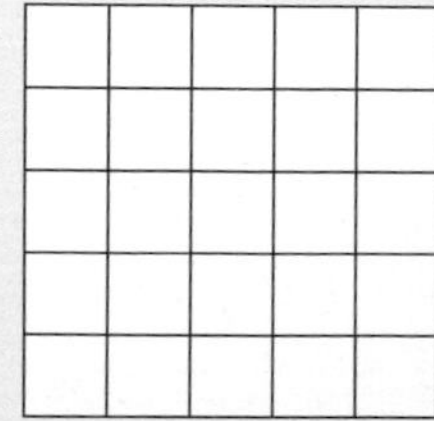

The square of a number is linked to the area of a square. Calculating 5 squared is the same as calculating the area of a square with side length 5

The cube of a number is linked to the volume of a cube. For example, calculating 4 cubed is the same as calculating the volume of a cube with side length 4

In this section, you will look at **roots** of positive numbers.

When you square any number you always get a positive answer. This means that every positive number has two square roots, one positive and one negative. It also means negative numbers don't have any square roots.

The square root symbol ($\sqrt{\ }$) indicates the positive square root of a number, so $\sqrt{25}$ is equal to 5. However, if you were told that $x^2 = 25$, then x could be 5 or –5, which you can write as $x = \pm5$

Example 1

a Work out $\sqrt{81}$

b Solve $x^2 = 81$

a 9 — The square root symbol means that you want the positive square root of the number. Since 9 × 9 = 81, the square root of 81 is 9

b $x = \pm9$ — You know that a number, x, multiplied by itself is equal to 81

From part **a** you know that 9 × 9 = 81

However, –9 × –9 = 81 since the product of two negative numbers is positive. So x could be positive or negative 9

Practice 9.7A

1 Evaluate

a $\sqrt{49}$ **b** $\sqrt{100}$ **c** $\sqrt{144}$ **d** $\sqrt{1}$ **e** $\sqrt{2500}$

2 Solve

a $x^2 = 49$ **b** $y^2 = 100$ **c** $b^2 = 144$ **d** $a^2 = 1$ **e** $p^2 = 2500$

3 **a** Copy and complete the number line.

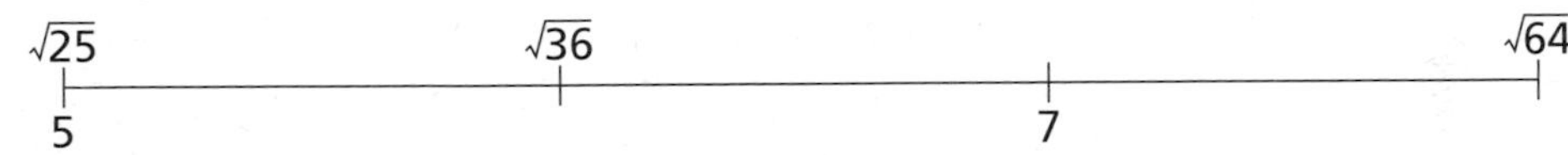

b Between which two integers does each value lie?

i $\sqrt{50}$ **ii** $\sqrt{29}$ **iii** $\sqrt{45}$ **iv** $\sqrt{63}$

4 Estimate

a $\sqrt{51}$ **b** $\sqrt{105}$ **c** $\sqrt{8}$ **d** $\sqrt{17}$ **e** $\sqrt{79}$

Compare your answers with a partner, then check them using a calculator.

9.7 Powers and roots H

What do you think?

1. A square has an area of $225\,cm^2$

   Rob says, "This means that the side length of the square is either 15 cm or –15 cm."

   Do you agree with Rob?

   Explain your answer.

2. Use your calculator to evaluate -4^2 and $(-4)^2$

   What do you notice? Why does this happen?

3. Beth says, "The square root of 25 is 5. This means that the square root of –25 is –5."

   Explain why Beth is incorrect.

In this section, you will look at powers and roots that are greater than 2.

Remember: the **power** of a number tells you how many times to use that number in a multiplication. For example, when calculating 2^5 you need to find the product of five 2s.

Just like the inverse of squaring is finding the square root, the inverse of higher powers also involves finding a root. The inverse of cubing is finding the cube root ($\sqrt[3]{\ }$); the inverse of raising a number to the power of 4 is finding the fourth root ($\sqrt[4]{\ }$), and so on for higher powers.

Example 2

Calculate

a 5^4 **b** $\sqrt[4]{625}$

a 625

To calculate 5 to the power of 4, you need to find the product of four 5s.

$5 \times 5 \times 5 \times 5$

$= 25 \times 5 \times 5$

$= 125 \times 5$

$= 625$ so $5^4 = 625$

b 5

You saw in part **a** that 5 raised to the power 4 is equal to 625. The 4th root is the inverse of raising to the power 4, so the 4th root of 625 is equal to 5

Practice 9.7B

1. Write each calculation out in full and then work out the answer.

   The first one has been done for you.

   a $6^5 = 6 \times 6 \times 6 \times 6 \times 6 = 7776$

   b 10^4

   c 1^8

   d 3^5

2 Work out

a 1^1 1^2 1^3 1^4 1^5

b 2^1 2^2 2^3 2^4 2^5

c 3^1 3^2 3^3 3^4 3^5

d 4^1 4^2 4^3 4^4 4^5

e 5^1 5^2 5^3 5^4 5^5

What do you notice about the numbers in each part?

3 Evaluate

a $\sqrt[3]{27}$ b $\sqrt[4]{16}$ c $\sqrt[5]{100\,000}$ d $\sqrt[3]{125}$ e $\sqrt[3]{343}$

4 a Use the fact that $5^7 = 78\,125$ to work out $\sqrt[7]{78\,125}$

b Use the fact that $\sqrt[4]{1296} = 6$ to evaluate 6^4

c Given that $2^{10} = 1024$, work out $(\sqrt[10]{1024})^2$

5 Work out

a $(-1)^1$ $(-1)^2$ $(-1)^3$ $(-1)^4$ $(-1)^5$

b $(-2)^1$ $(-2)^2$ $(-2)^3$ $(-2)^4$ $(-2)^5$

c $(-3)^1$ $(-3)^2$ $(-3)^3$ $(-3)^4$ $(-3)^5$

d $(-4)^1$ $(-4)^2$ $(-4)^3$ $(-4)^4$ $(-4)^5$

e $(-5)^1$ $(-5)^2$ $(-5)^3$ $(-5)^4$ $(-5)^5$

What do you notice about the numbers in each part?

What do you think?

1 Here are the first 8 powers of −10

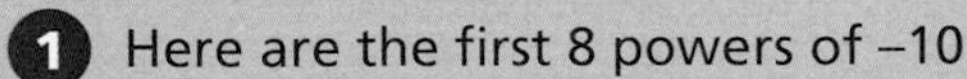

$(-10)^1 = -10$	$(-10)^2 = 100$	$(-10)^3 = -1000$
$(-10)^4 = 10\,000$	$(-10)^5 = -100\,000$	$(-10)^6 = 1\,000\,000$
$(-10)^7 = -10\,000\,000$	$(-10)^8 = 100\,000\,000$	

a What do you notice? Why does this happen?

b In your own words, complete these sentences

i A negative number raised to an odd power …

ii A negative number raised to an even power …

c Decide whether the answer to each calculation will be positive or negative. You do not need to work out the answers.

i $(-9)^{24}$ ii $(5)^{19}$ iii $(-1)^{11}$ iv $(-200)^{298}$

2 Benji and Chloe want to work out the value of $(-27)^4$.

Chloe
The answer will be negative because 27 is odd.
Benji
The answer will be positive because 4 is even.

a Who do you agree with? Explain your answer.

b Work out the value of $(-27)^4$ using a calculator.

c Without further calculation, what is the value of 27^4? How do you know?

3 Given that $12^5 = 248\,832$, what is the value of $(-12)^5$? How do you know?

Consolidate – do you need more?

1 Work out

a 5^2 b 7^2 c 9^2 d 10^2 e 8^2

f $(-5)^2$ g $(-7)^2$ h $(-9)^2$ i $(-10)^2$ j $(-8)^2$

2 Write each calculation in expanded form. Two have been done for you.

a $3^5 = 3 \times 3 \times 3 \times 3 \times 3$ b 7^4 c 2^8

d 10^3 e $(-3)^5 = -3 \times -3 \times -3 \times -3 \times -3$ f $(-7)^4$

g $(-2)^8$ h $(-10)^3$

3 Work out

a 3^5 b 7^4 c 2^8 d 10^3

e $(-3)^5$ f $(-7)^4$ g $(-2)^8$ h $(-10)^3$

4 Work out

a $\sqrt{25}$ b $\sqrt{49}$ c $\sqrt{121}$ d $\sqrt{2500}$

5 Solve each equation.

a $x^2 = 25$ b $q^2 = 49$ c $s^2 = 121$ d $t^2 = 2500$

6 Use a calculator to evaluate

a $\sqrt[3]{64}$ b $\sqrt[5]{-32\,768}$ c $\sqrt[4]{83\,521}$ d $\sqrt[10]{59\,049}$

Stretch – can you deepen your learning?

1 $x^2 = 564$

What is the value of $(-x)^2$? How do you know?

2 The cube root of p is 43

What is the cube root of $-p$? How do you know?

3 If $x^2 = 71$, explain why either $8 < x < 9$ or $-9 < x < -8$

4 Between which pair of integers could the solution to each of these equations lie?

Give your answers as inequalities.

a $b^2 = 103$ **b** $y^2 = 131$ **c** $z^2 = 5$ **d** $x^2 = 387$

5 **a** Explain why $y^2 = 156.25$ has two solutions.

b Use your calculator to work out one of the solutions.

Why does your calculator only work out one solution?

c What is the second solution? How do you know?

6 Given that $a^5 > 0$, what do you know about a?

7 n is a positive integer.

Decide whether each statement is always true, sometimes true or never true.

Explain your answers.

a x^{2n} is positive **b** x^{2n} is negative **c** x^{2n+1} is positive

Reflect

1 Explain why a negative number gives a positive answer when it is squared.

2 Explain why $x^2 = 1$ has two solutions whereas $x^3 = 1$ has only one.

3 Explain why any number raised to an even power will always give a positive answer.

9 Directed number

Chapters 9.1–9.7

I have become fluent in…

- Representing directed numbers
- Ordering directed numbers
- Calculating with directed numbers
- Substituting with directed numbers
- Solving two-step equations

I have developed my reasoning skills by…

- Identifying how to solve equations given in different forms
- Identifying when to use a mental, written or calculator method
- Exploring common misconceptions

I have been problem-solving through…

- Representing directed numbers in different forms
- Using directed numbers in different contexts
- Forming and solving equations from real-life problems

Check my understanding

1 What number is represented in each?

a +1 +1 +1 +1 +1 +1

b −1 −1 −1 −1 −1 −1

c +1 +1 +1
−1 −1 −1

d +1
−1 −1 −1 −1 −1

2 Write each list of numbers in ascending order.

a 17, −24, 15, 4, −1, 0 **b** −265, 127, 10 600, −99, 162, −500

3 Complete the calculations.

a $15 - 29$ **b** $-7 - 12$ **c** $14 - -21$ **d** $122 + -99$
e $-43 + 25$ **f** 17×-4 **g** -6×-8 **h** $125 \div -25$
i $-600 \div -6$ **j** $-280 \div 7$

4 Evaluate each expression when $x = -4$

a $x + 7$ **b** $x - 10$ **c** $3x$ **d** $6x - 5$ **e** $20 - 5x$

5 Solve the equations.

a $2x + 1 = 15$ **b** $3x - 2 = 25$ **c** $2x + 11 = -9$ **d** $5x - 7 = -12$ **e** $\frac{x + 4}{3} = -2$

6 Complete the calculations.

a $3 + 5 \times -2$ **b** $17 \times 2 - 50$ **c** $6 \times (-3 + -7)$ **d** $180 \div -6 + 7$ **e** $-20 - 3 \times -5$

7 Solve the equations. (H)

a $x^2 = 144$ **b** $y^2 = 225$ **c** $z^2 = 36$ **d** $5p^2 = 500$ **e** $\frac{t^2}{2} = 32$

10 Fractional thinking

In this block, I will learn...

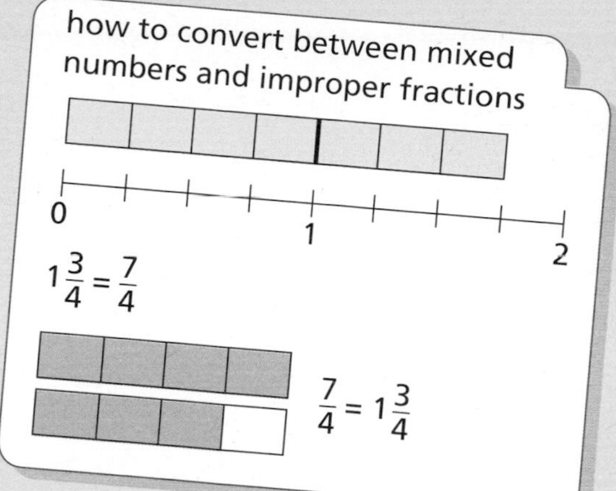

$1\frac{3}{4} = \frac{7}{4}$

$\frac{7}{4} = 1\frac{3}{4}$

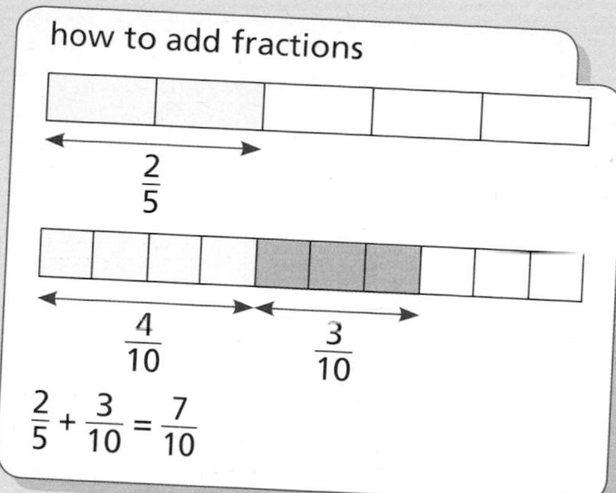

$\frac{2}{5} + \frac{3}{10} = \frac{7}{10}$

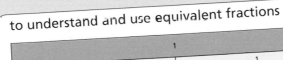

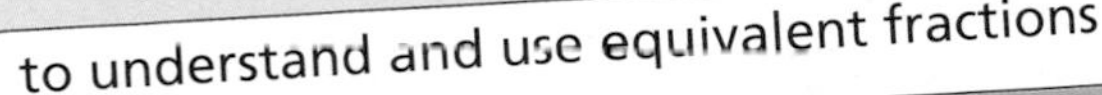

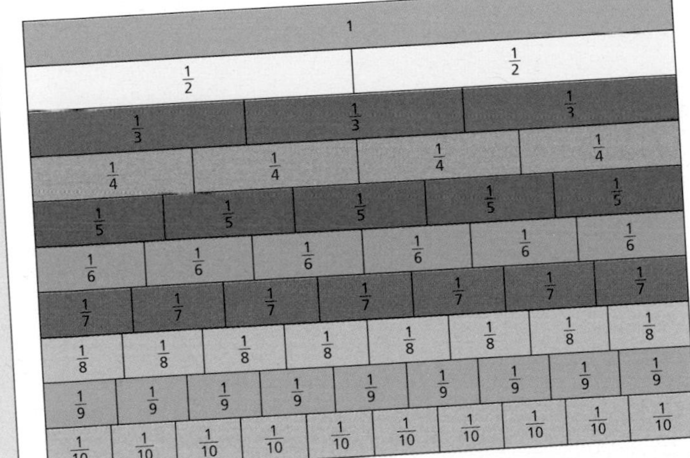

$\frac{1}{2} = \frac{2}{4} = \frac{3}{6} = \frac{4}{8}$

$\frac{1}{3} = \frac{2}{6}$

$\frac{4}{5} = \frac{8}{10}$

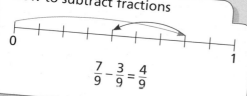

how to subtract fractions

0 1

$\frac{7}{9} - \frac{3}{9} = \frac{4}{9}$

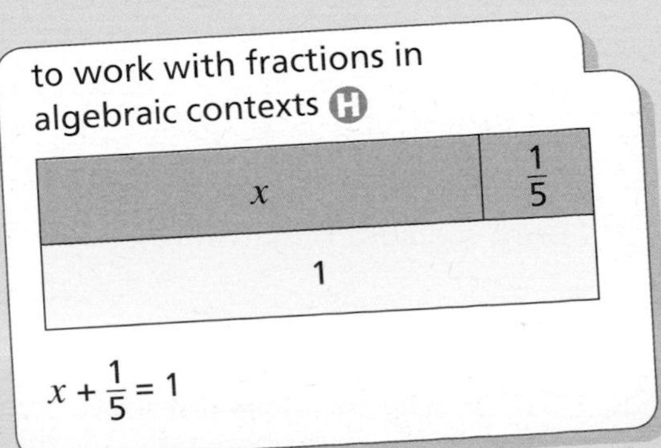

to work with fractions in algebraic contexts H
x
1/5
1

$x + \frac{1}{5} = 1$

10.1 Converting fractions

Small steps

- Understand representations of fractions
- Convert between mixed numbers and improper fractions

Key words

Mixed number – a number made up of a whole number and a fraction

Numerator – the top number in a fraction that shows the number of parts

Denominator – the bottom number in a fraction; it shows how many equal parts one whole has been divided into

Improper fraction – a fraction in which the numerator is greater than the denominator

Convert – Change from one form to another

Are you ready?

1 What fraction is shaded in each diagram?

a

b

c

d

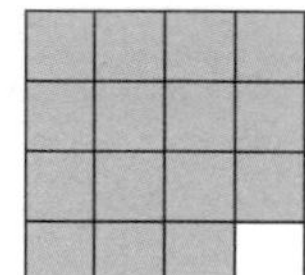

e

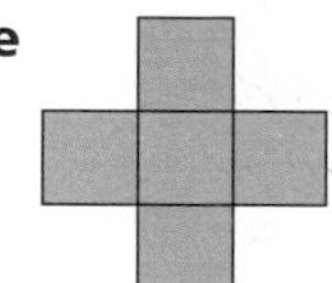

2 What fraction is the arrow pointing to on each number line?

a

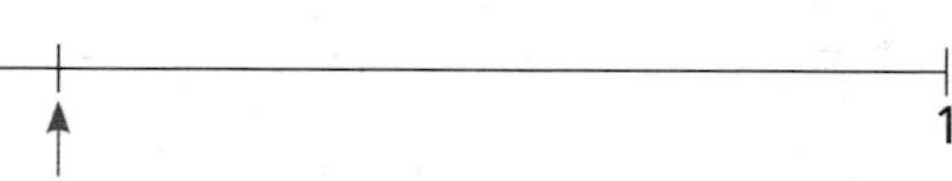

b

c

d

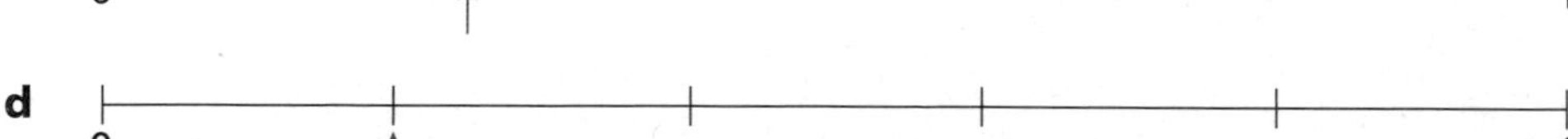

3 Draw a diagram that shows $\frac{1}{3}$

4 Draw a diagram that shows $\frac{3}{4}$

5 Explain why this does not show $\frac{1}{5}$

Models and representations

Both of these diagrams show $1\frac{3}{4}$

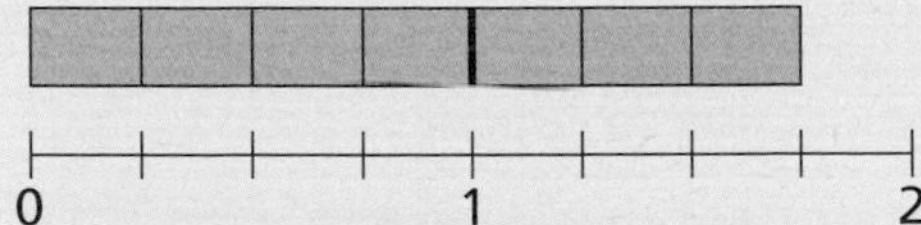

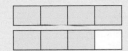

You can use bar models and number lines to help you to **convert** between **mixed numbers** and **improper fractions**.

It is important that you understand that a mixed number and an improper fraction are made up of wholes and fractions. When drawing bar models, you should not draw the wholes touching because this can lead to misconceptions such as $1\frac{2}{5}$ being equal to $\frac{7}{10}$

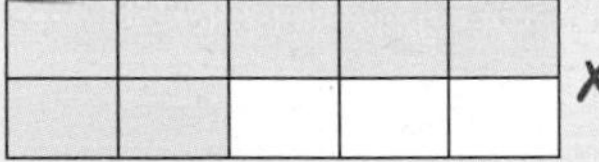 ✗ 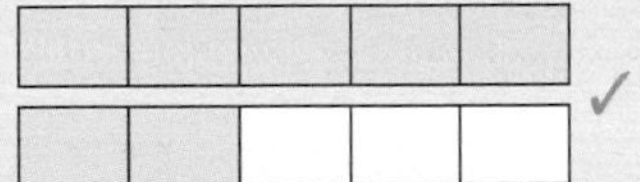 ✓

Example 1

Convert $2\frac{1}{2}$ to an improper fraction.

Remember: in an improper fraction the **numerator** is bigger than the **denominator**.

Method A

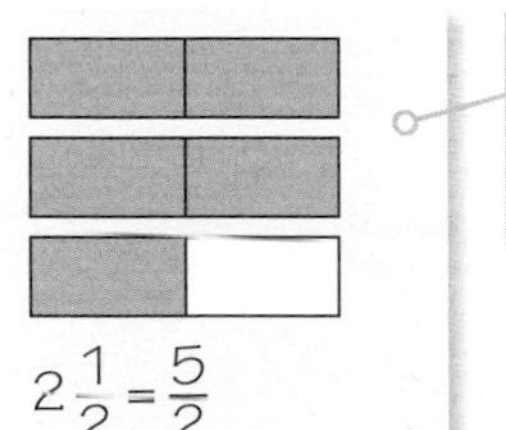

All of the bar models are split into halves.

Two whole bars and one half are shaded to represent $2\frac{1}{2}$

5 of the parts are shaded, so this is equivalent to 5 halves or $\frac{5}{2}$

Method B

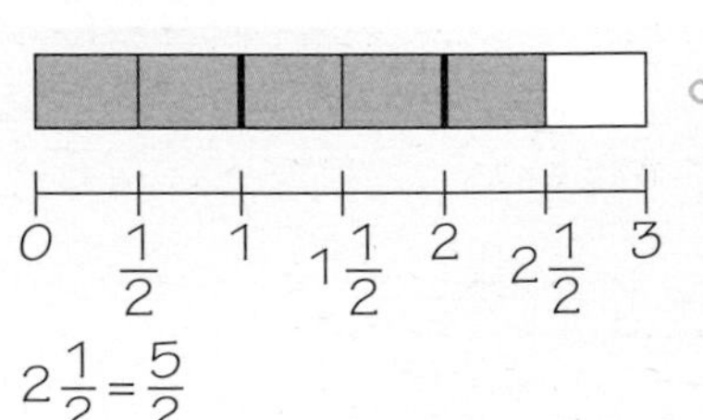

The number line is going up in halves.

The bar model shows that 5 halves are equal to $2\frac{1}{2}$

10.1 Converting fractions

Example 2

Write $\frac{13}{5}$ as a mixed number.

Remember: a mixed number is made up of a whole number and a fraction.

Method A

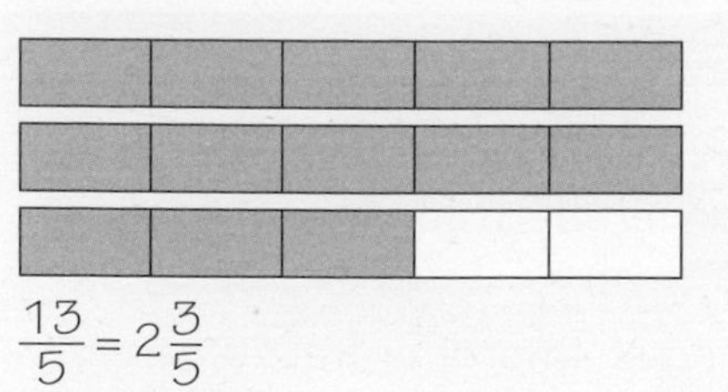

5 fifths make a whole one so $\frac{13}{5}$ makes 2 whole bars and 3 fifths of a bar. You can write this as $2\frac{3}{5}$

Method B

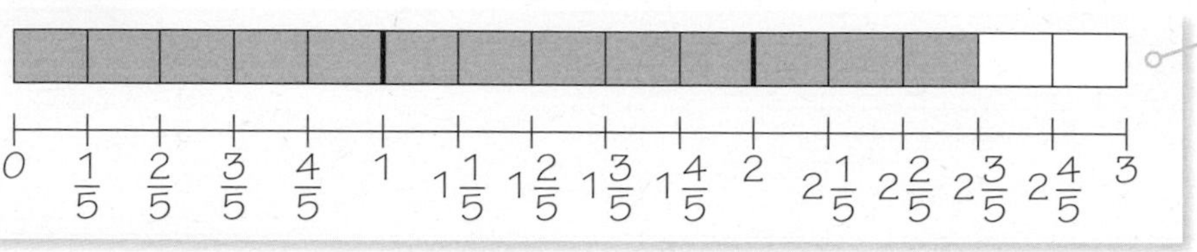

The number line is going up in fifths. The bar model shows that $\frac{13}{5}$ is equivalent to $2\frac{3}{5}$

Practice 10.1A

1. Choose the improper fraction from this list.

   $\frac{5}{6}$ $\quad$ $\frac{90}{91}$ $\quad$ $\frac{7}{5}$ $\quad$ $\frac{1}{12}$

2. Choose the mixed number from this list.

   $1\frac{1}{7}$ $\quad$ 3.5 $\quad$ $\frac{11}{3}$ $\quad$ $\frac{10}{10}$

3. What fraction of each diagram is shaded?

   a

   b

   c

4. How many sixths make

   a one whole $\quad$ **b** two wholes $\quad$ **c** five wholes?

5. This diagram represents $3\frac{1}{4}$

   Write $3\frac{1}{4}$ as an improper fraction.

6. **a** Draw a diagram to represent $2\frac{2}{5}$

   b Write $2\frac{2}{5}$ as an improper fraction.

7 **a** Draw a diagram to represent $\frac{7}{3}$

b Write $\frac{7}{3}$ as a mixed number.

8 Convert these mixed numbers into improper fractions.

a $1\frac{3}{4}$ **b** $3\frac{1}{7}$ **c** $2\frac{5}{6}$ **d** $2\frac{5}{8}$

9 Convert these improper fractions into mixed numbers.

a $\frac{7}{2}$ **b** $\frac{12}{5}$ **c** $\frac{17}{7}$ **d** $\frac{14}{3}$

10 Use <, > or = to compare these pairs of numbers.

a $3\frac{4}{5}$ ◯ $\frac{18}{5}$ **b** $7\frac{2}{3}$ ◯ $\frac{23}{3}$ **c** $\frac{96}{7}$ ◯ 14 **d** $\frac{71}{9}$ ◯ $\frac{35}{3}$

11 Kate says, "To convert $8\frac{4}{9}$ into an improper fraction, just do 8 multiplied by 9, then add 4 to find the numerator. The denominator stays the same. So $8\frac{4}{9} = \frac{76}{9}$"

a Explain why Kate's method works.

b Use Kate's method to convert $12\frac{3}{7}$ into an improper fraction.

What do you think?

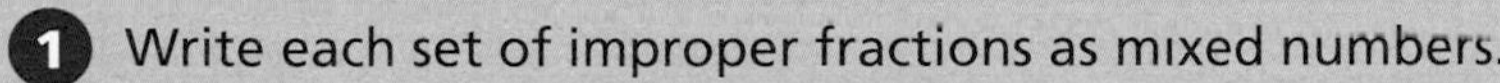

1 Write each set of improper fractions as mixed numbers.

a $\frac{20}{4}$ $\frac{16}{4}$ $\frac{12}{4}$ $\frac{8}{4}$

b $\frac{21}{4}$ $\frac{17}{4}$ $\frac{13}{4}$ $\frac{9}{4}$

c $\frac{21}{8}$ $\frac{22}{8}$ $\frac{23}{8}$ $\frac{24}{8}$

What do you notice about each set?

Why did this happen?

2 Jakub has some number cards.

15 10 9 4 3

He makes a fraction using two of the cards.

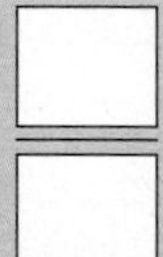

He wants to make a number as close to 4 as he can.

Which two cards should he choose?

Consolidate – do you need more?

1 Use the number lines to convert each mixed number to an improper fraction.

a $1\frac{2}{3}$

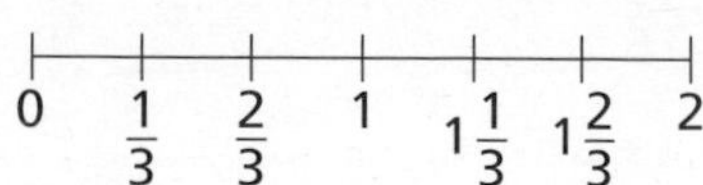

b $2\frac{1}{4}$

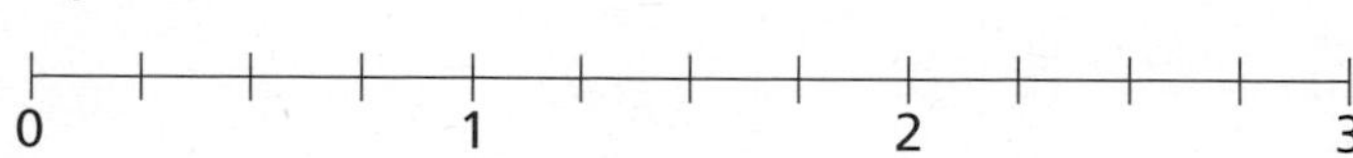

2 Use the number lines to convert these improper fractions to mixed numbers.

a $\frac{9}{5}$

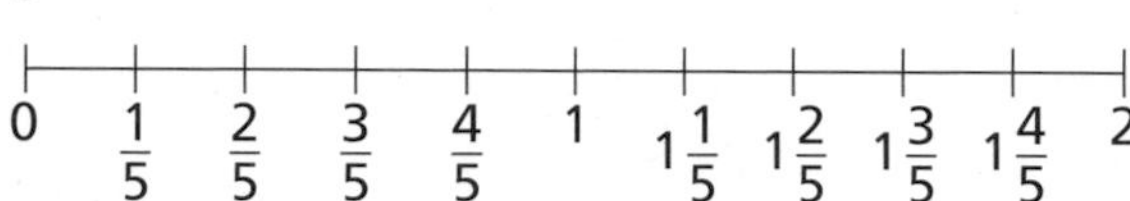

b $\frac{11}{3}$

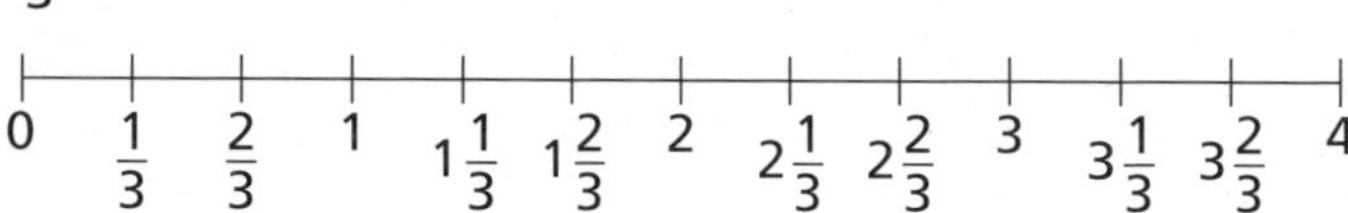

3 Copy and complete.

a $2\frac{\square}{5} = \frac{11}{5}$ **b** $\square\frac{3}{4} = \frac{7}{4}$ **c** $4\frac{1}{\square} = \frac{\square}{2}$ **d** $\frac{24}{\square} = \square\frac{3}{7}$

4 Convert these improper fractions to mixed numbers.

a $\frac{13}{3}$ **b** $\frac{11}{2}$ **c** $\frac{27}{5}$ **d** $\frac{21}{10}$

5 Convert these mixed numbers to improper fractions.

a $1\frac{3}{4}$ **b** $5\frac{1}{3}$ **c** $3\frac{1}{10}$ **d** $2\frac{3}{7}$

Stretch – can you deepen your learning?

1 Write these numbers in descending order.

$\frac{73}{6}$ $12\frac{3}{5}$ $\frac{109}{11}$ $\frac{54}{4}$

2 Copy and complete to show different ways of writing $\frac{39}{7}$

a $\frac{39}{7} = 5\frac{\square}{7}$ **b** $\frac{39}{7} = 4\frac{\square}{7}$ **c** $\frac{39}{7} = 3\frac{\square}{7}$

d $\frac{39}{7} = 2\frac{\square}{7}$ **e** $\frac{39}{7} = 1\frac{\square}{7}$

3 Given that $\frac{c}{a}$ is a proper fraction, find the value of $a + b + c$

$\frac{15}{11} = 1\frac{a}{11}$ $\frac{29}{a} = b\frac{c}{a}$

Reflect

1 Explain how you can use a diagram to change $3\frac{1}{5}$ into an improper fraction.

2 Why would a diagram not be appropriate to show that $15\frac{5}{6} = \frac{95}{6}$?
Explain a method that could be used instead.

10.2 Adding and subtracting fractions (1)

Small steps

- Add and subtract unit fractions with the same denominator
- Add and subtract fractions with the same denominator
- Add and subtract fractions from integers expressing the answer as a single fraction

Key words

Numerator – the top number in a fraction that shows the number of parts

Denominator – the bottom number in a fraction; it shows how many equal parts one whole has been divided into

Unit fraction – a fraction with a numerator of 1

Non-unit fraction – a fraction with a numerator that is not 1

Are you ready?

1 Choose the **unit fraction** from this list.

$\frac{3}{4}$ $\frac{2}{3}$ $\frac{1}{8}$

2 Write down the fraction shaded in each bar model.

a

b

c

d

3 Draw a bar model to show $\frac{3}{5}$

4 From this list, write down the fractions with the same **denominator**.

$\frac{1}{12}$ $\frac{3}{8}$ $\frac{5}{12}$ $\frac{1}{2}$ $\frac{12}{12}$

5 Which of these fractions are equivalent to one whole?

$\frac{1}{9}$ $\frac{3}{3}$ $\frac{7}{2}$ $\frac{4}{4}$ $\frac{11}{11}$ $1\frac{3}{4}$

6 Copy and complete each number line.

a 0, $\frac{1}{4}$, $\frac{2}{4}$, 1

b 0, $\frac{1}{7}$, 1

c 0, 1

Models and representations

Adding fractions

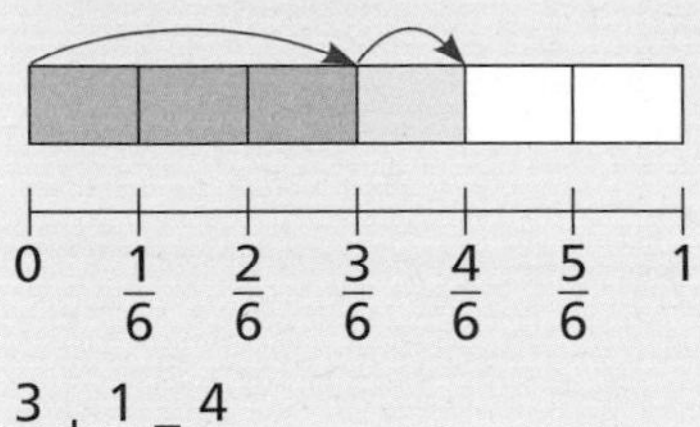

$\frac{3}{6} + \frac{1}{6} = \frac{4}{6}$

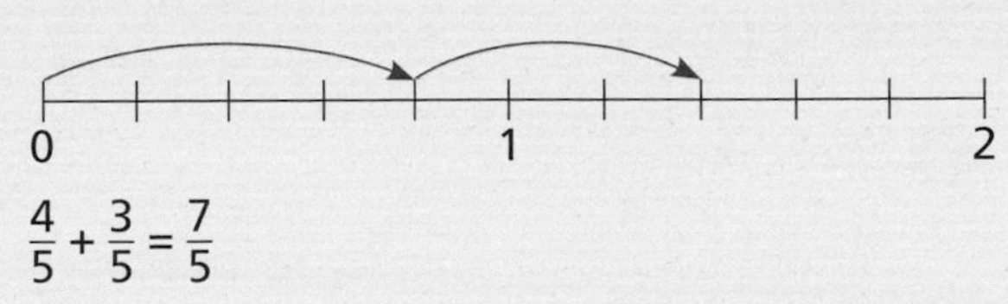

$\frac{4}{5} + \frac{3}{5} = \frac{7}{5}$

Subtracting fractions

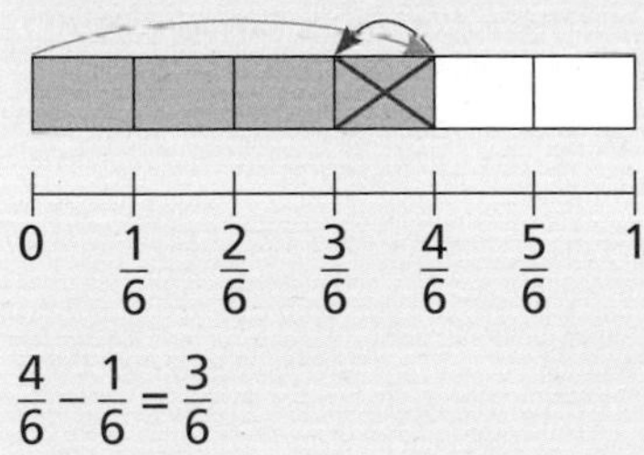

$\frac{4}{6} - \frac{1}{6} = \frac{3}{6}$

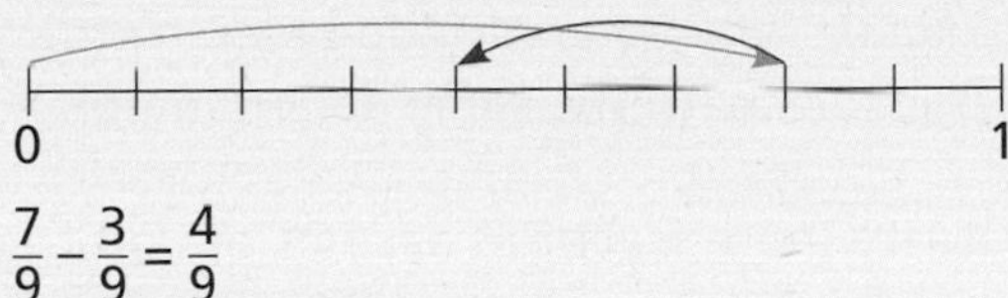

$\frac{7}{9} - \frac{3}{9} = \frac{4}{9}$

Example 1

Samira and Abdullah each eat $\frac{1}{5}$ of a bar of chocolate.

What fraction of the chocolate bar have Samira and Abdullah eaten altogether?

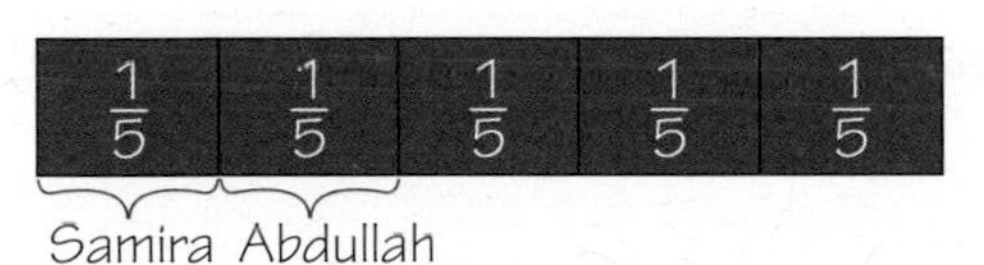

Each part of the chocolate bar is worth $\frac{1}{5}$

Altogether they have eaten $\frac{2}{5}$ of the chocolate bar.

You could show this on a number line.

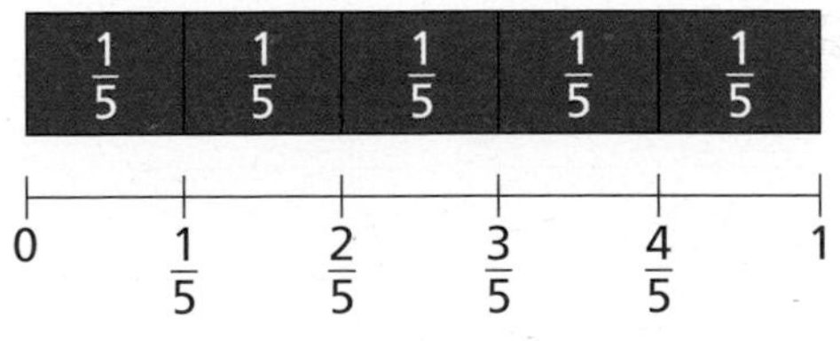

Example 2

Rob has a piece of ribbon that is $\frac{7}{8}$m long. He cuts $\frac{3}{8}$m from the ribbon.

What length of ribbon is left?

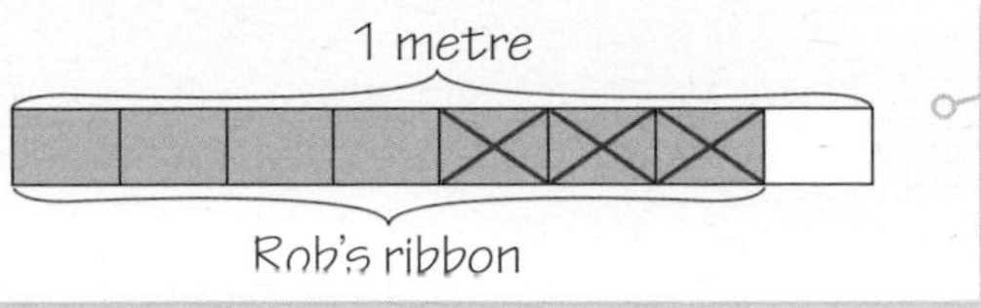

You can show this on a bar model divided into 8 equal pieces.

Start by shading 7 pieces.

Then cross out 3 pieces.

Rob has $\frac{4}{8}$m of ribbon left.

This is equivalent to $\frac{1}{2}$

10.2 Adding and subtracting fractions (1)

Practice 10.2A

1 Draw and shade in bar models to represent each calculation.

a $\frac{1}{3}+\frac{1}{3}=\frac{2}{3}$ **b** $\frac{1}{4}+\frac{1}{4}+\frac{1}{4}=\frac{3}{4}$ **c** $\frac{1}{7}+\frac{1}{7}+\frac{1}{7}+\frac{1}{7}+\frac{1}{7}=\frac{5}{7}$

2 Work out

a $\frac{1}{8}+\frac{1}{8}+\frac{1}{8}$ **b** $\frac{1}{12}+\frac{1}{12}$ **c** $\frac{1}{74}+\frac{1}{74}+\frac{1}{74}$

d $\frac{1}{9}+\frac{1}{9}-\frac{1}{9}$ **e** $\frac{1}{13}+\frac{1}{13}+\frac{1}{13}+\frac{1}{13}-\frac{1}{13}-\frac{1}{13}$

f $\frac{1}{6}+\frac{1}{6}-\frac{1}{6}-\frac{1}{6}$ **g** $\frac{1}{3}+\frac{1}{3}+\frac{1}{3}$

3 Explain why you only add or subtract the **numerators** when adding and subtracting fractions with the same denominator.

4 Write these fractions as sums of unit fractions.

a $\frac{4}{5}$ **b** $\frac{2}{9}$ **c** $\frac{5}{31}$

5 Copy and complete the calculation for each representation.

a $\frac{3}{6}+\frac{2}{6}=\frac{\square}{6}$

b $\frac{4}{11}+\frac{\square}{\square}=\frac{\square}{\square}$

c $\frac{\square}{4}-\frac{\square}{\square}=\frac{\square}{\square}$

d 0 1 $\frac{4}{7}+\frac{\square}{\square}=\frac{\square}{\square}$

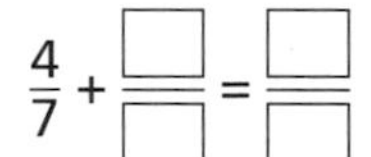

e 0 1 $\frac{\square}{\square}+\frac{\square}{\square}+\frac{\square}{\square}=\frac{\square}{\square}$

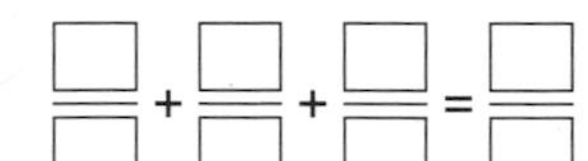

f 0 1 $\frac{\square}{\square}-\frac{3}{9}=\frac{\square}{\square}$

6 Flo says, "$\frac{3}{8}+\frac{1}{8}$ is equal to $\frac{4}{16}$"

Draw a diagram to show that Flo is wrong.

7 **a** Work out

i $\frac{5}{7}+\frac{1}{7}$ **ii** $\frac{2}{15}+\frac{8}{15}+\frac{1}{15}$ **iii** $\frac{5}{9}-\frac{4}{9}$

iv $\frac{5}{6}+\frac{5}{6}$ **v** $\frac{5}{6}-\frac{5}{6}$ **vi** $\frac{4}{11}-\frac{3}{11}+\frac{5}{11}$

b For which question in part **a** was the answer

i a unit fraction **ii** an improper fraction?

8 **a** Work out the missing part in the bar model.

b Write the fact family for the bar model.

$\frac{4}{17}$	
$\frac{15}{17}$	

9 Ed, Marta and Jakub are sharing a large pizza.

Ed eats $\frac{2}{9}$ of the pizza.

Marta eats $\frac{2}{9}$ of the pizza.

Jakub eats $\frac{4}{9}$ of the pizza.

What fraction of the pizza is left?

10 Work out the perimeter of each shape.

a

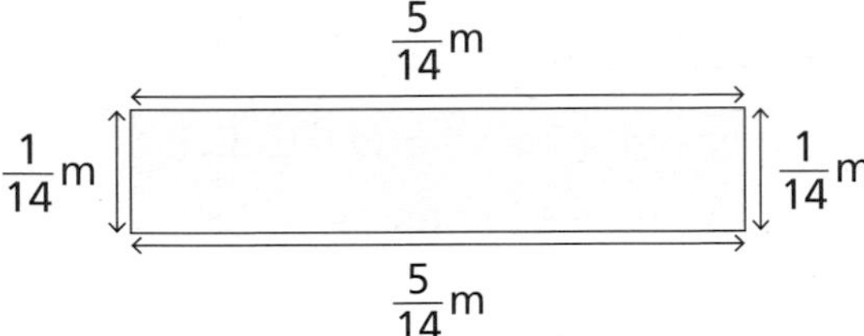

b

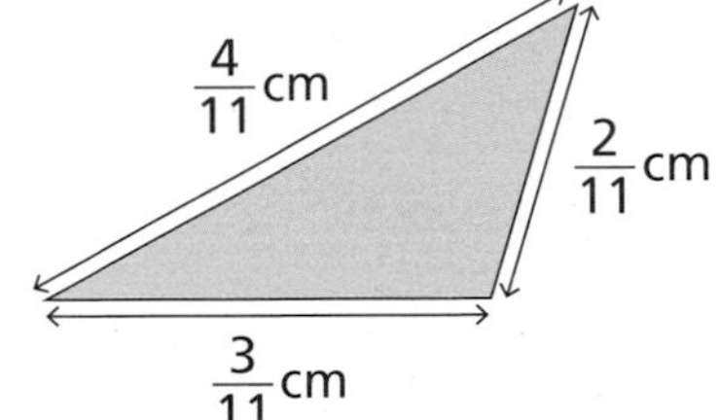

c

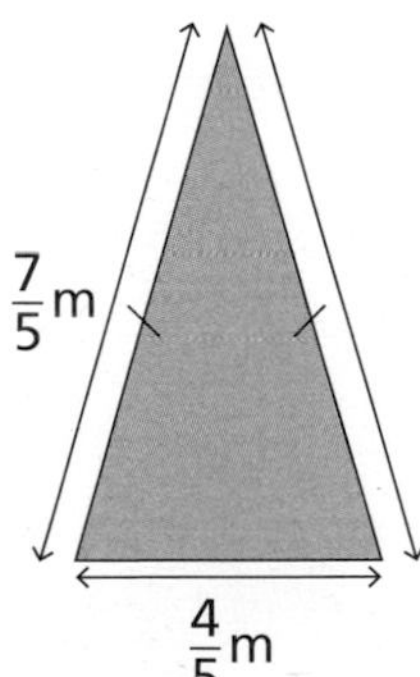

d

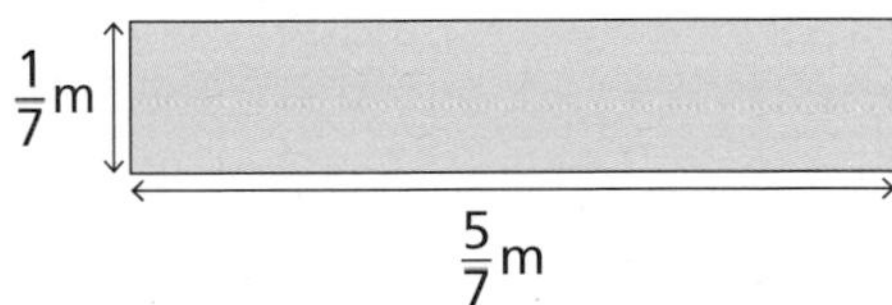

11 Complete these calculations. Write your answers as mixed numbers.

a $\frac{1}{3}+\frac{1}{3}+\frac{1}{3}+\frac{1}{3}$ **b** $\frac{7}{8}+\frac{3}{8}+\frac{1}{8}$ **c** $\frac{33}{35}+\frac{6}{35}$

12 Work out

a $3-5$ **b** £3 – £5 **c** 3 sixths – 5 sixths **d** $\frac{3}{7}-\frac{5}{7}$

What do you think?

1 A linear sequence starts at $\frac{1}{17}$

It decreases by $\frac{3}{17}$ each time.

Find the fifth term of the sequence.

2 $\frac{2}{w} + \frac{a}{w} = \frac{9}{w}$

a Work out the value of a

b Is it possible to work out the value of w? Does it matter?

c w is a positive integer. For what values of w is $\frac{9}{w}$ an improper fraction?

3 Work out

a $\frac{1}{3} - \frac{2}{3}$ **b** $\frac{1}{4} - \frac{2}{4}$ **c** $\frac{1}{5} - \frac{2}{5}$ **d** $\frac{1}{217} - \frac{2}{217}$

What do you notice?

Example 3

Beca and Ed each have a piece of ribbon.

Beca's ribbon is $\frac{2}{7}$ of a metre.

Ed's ribbon is $\frac{5}{7}$ of a metre.

What is the total length of their ribbon?

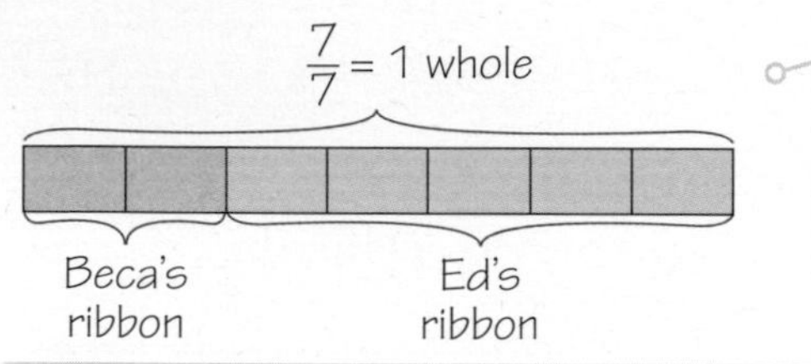

You need to add $\frac{2}{7}$ and $\frac{5}{7}$ to find the total.

You can use a bar model.

$\frac{2}{7} + \frac{5}{7} = \frac{7}{7} = 1$

The total length of their ribbon is 1 metre.

Simplify.

Give the unit with your answer.

Example 4

Ed has another piece of ribbon that is 1 metre long.

He cuts off $\frac{3}{8}$ of a metre and gives it to Beca.

How much ribbon does Ed have left?

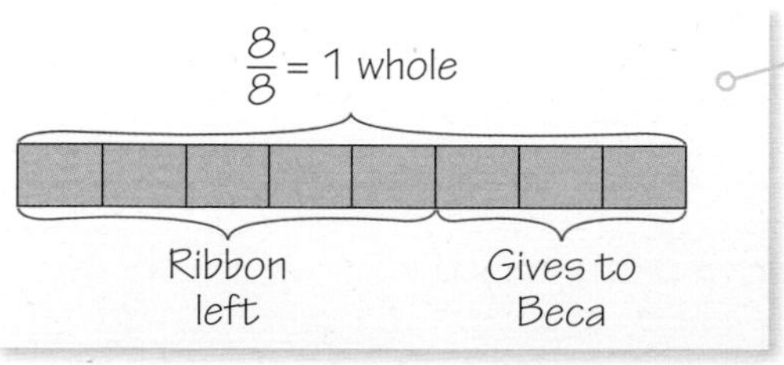

To give Beca $\frac{3}{8}$ of a metre, the whole bar needs to be split into 8 equal parts.

You can rewrite 1 whole as $\frac{8}{8}$

So you can write this as

$1 - \frac{3}{8} = \frac{8}{8} - \frac{3}{8} = \frac{5}{8}$

Ed has $\frac{5}{8}$ of a metre left.

Practice 10.2B

1. Work out

   a $\frac{1}{2}+\frac{1}{2}$ **b** $\frac{1}{6}+\frac{5}{6}$ **c** $\frac{3}{10}+\frac{3}{10}+\frac{4}{10}$ **d** $\frac{98}{91}-\frac{7}{91}$

   What is the same about each answer?

   What is different about each answer?

2. Draw a diagram to show

   a $\frac{1}{5}+\frac{4}{5}=1$ **b** $1-\frac{3}{7}=\frac{4}{7}$

3. Work out

   a $1-\frac{3}{4}$ **b** $1-\frac{3}{10}$ **c** $1-\frac{5}{18}$ **d** $1-\frac{2}{3}-\frac{1}{3}$ **e** $1-\frac{47}{47}$ **f** $1-\frac{63}{100}$

4. How many wholes are represented in each case?

   a $\frac{6}{6}$ **b** $\frac{12}{6}$ **c** $\frac{18}{6}$

5. Complete these calculations.

   a $2-\frac{5}{6}$ **b** $3-\frac{1}{6}$ **c** $2-\frac{2}{3}$ **d** $4-2\frac{5}{6}$

6. Work out the perimeter of each shape.

   a

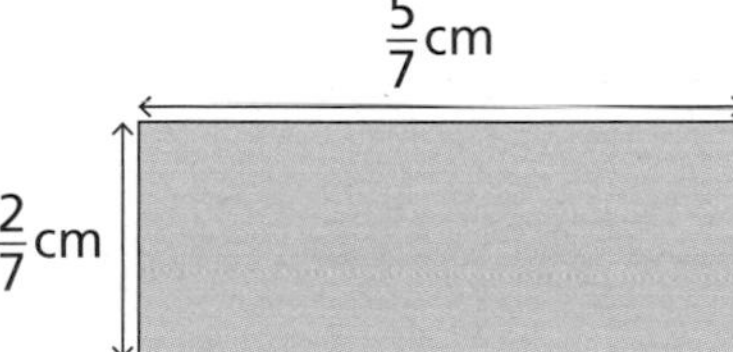

   b

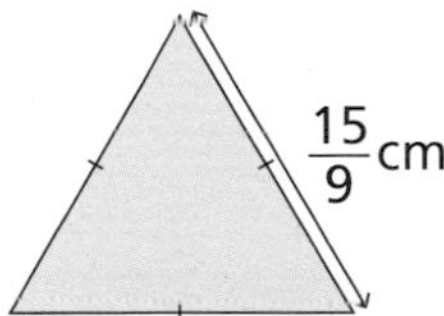

   c

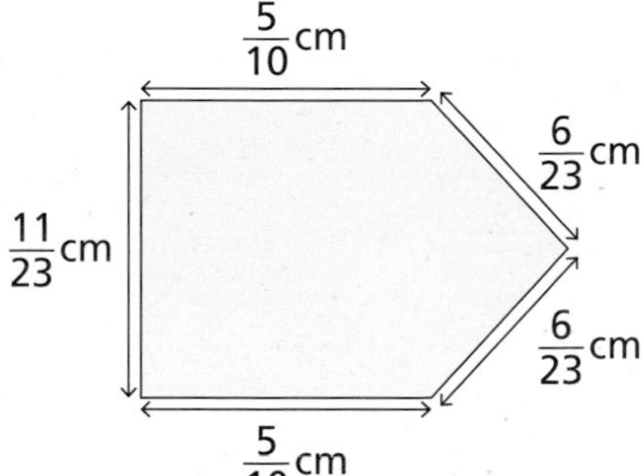

   d

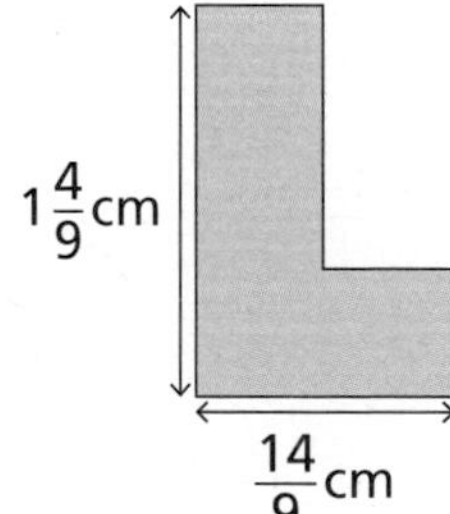

What do you think?

1 Darius says, "I can't work out $\frac{2}{7}+\frac{3}{5}+\frac{5}{7}+\frac{2}{5}$ easily because the fractions have different denominators."

Flo says, "I can see an easy way to work it out."

What do you think Flo has noticed? Explain her strategy.

2 Find the missing integers.

a $\square - \frac{1}{9} = \frac{17}{9}$ b $\square - \frac{5}{12} = \frac{3}{12} + \frac{4}{12}$ c $\frac{2}{3} + \square = 3\frac{2}{3}$

Consolidate – do you need more?

1 Complete these calculations to match each representation.

a $\frac{1}{3}+\frac{1}{3}$

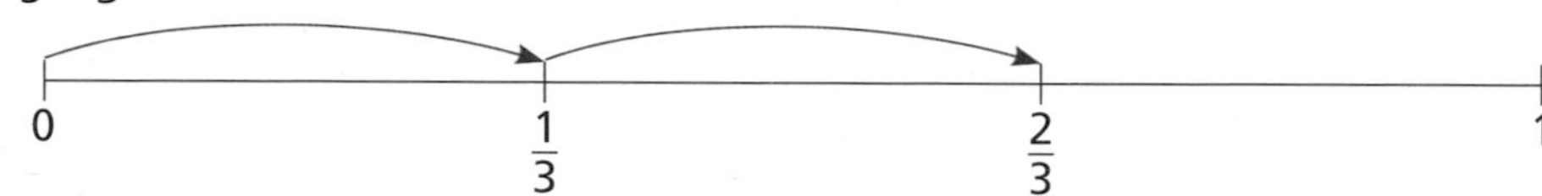

b $\frac{4}{5}+\frac{3}{5}$

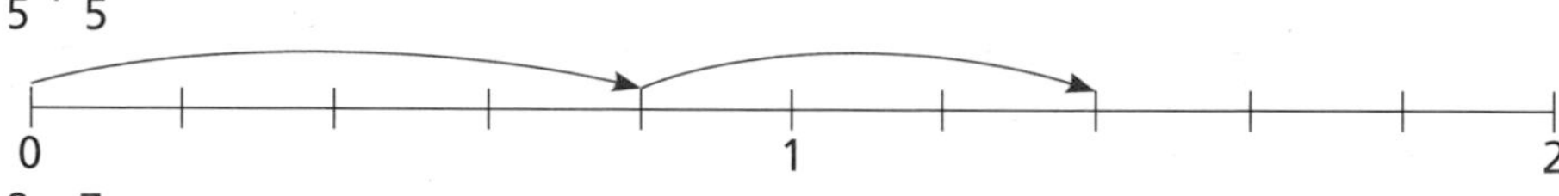

c $\frac{8}{9}-\frac{7}{9}$

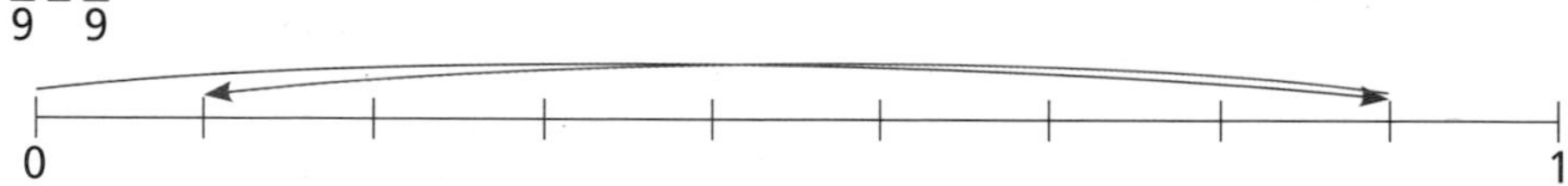

2 Write these fractions as whole numbers.

a $\frac{6}{6}$ b $\frac{3}{3}$ c $\frac{7}{7}$ d $\frac{10}{10}$

e $\frac{68}{68}$ f $\frac{4}{2}$ g $\frac{6}{3}$ h $\frac{9}{3}$

3 Work out

a $\frac{2}{5}+\frac{1}{5}$ b $\frac{2}{7}+\frac{3}{7}$ c $\frac{9}{10}-\frac{6}{10}$ d $\frac{1}{11}+\frac{2}{11}+\frac{3}{11}$

e $\frac{3}{8}+\frac{5}{8}$ f $1-\frac{1}{3}$ g $\frac{6}{7}-\frac{2}{7}-\frac{4}{7}$ h $1-\frac{9}{9}$

i $2-\frac{1}{6}$

4 Emily and Marta each have a piece of ribbon. Emily's piece of ribbon is $\frac{3}{11}$ of a metre long. Marta's piece of ribbon is $\frac{2}{11}$ of a metre long. What fraction of a metre of ribbon do Emily and Marta have in total?

5 Ali is $\frac{4}{27}$ of his way through a journey. What fraction of the journey does he have left?

6 Ed eats $\frac{1}{9}$ of a cake. Flo eats $\frac{4}{9}$ of the same cake. What fraction of the cake is left?

Stretch – can you deepen your learning?

1 The first term of an increasing linear sequence is $\frac{1}{9}$

The constant difference is $\frac{7}{9}$

In what position is the first integer term of the sequence?

2 The first term of a decreasing linear sequence is $\frac{1}{9}$

The constant difference is $\frac{7}{9}$

In what position is the first integer term of the sequence?

3 Work out

a **i** $1 - \frac{11}{12}$ **ii** $\frac{11}{12} - 1$ **iii** $-1 + \frac{11}{12}$ **iv** $\frac{11}{12} - -1$

b **i** $2 - \frac{7}{5}$ **ii** $\frac{7}{5} - 2$ **iii** $-2 - \frac{7}{5}$ **iv** $-2 + \frac{7}{5}$

c **i** $5 - \frac{17}{9}$ **ii** $\frac{17}{9} \quad 5$ **iii** $5 \quad \frac{17}{9}$ **iv** $-\frac{17}{9} - 5$

4 Solve these equations.

a $x + \frac{4}{11} = 2$ **b** $\frac{13}{15} = y - \frac{8}{15}$ **c** $z + \frac{14}{23} = \frac{5}{23}$

d $\frac{27}{39} - a = 1$ **e** $b + \frac{5}{13} = 1 - \frac{15}{13}$ **f** $2c - \frac{3}{51} = -1$

Reflect

In your own words, explain how to add or subtract two fractions with the same denominator.

10.3 Adding and subtracting fractions (2)

Small steps

- Understand and use equivalent fractions
- Add and subtract fractions where denominators share a simple common multiple

Key words

Equivalent – numbers that are written differently but are equal in value

Multiple – a multiple of a number is its product with any integer

Lowest common multiple – the lowest common multiple (LCM) of two or more numbers is the lowest number that is a multiple of them all

Common denominator – two or more fractions have a common denominator when their denominators (bottom numbers) are the same

Are you ready?

1 Copy and complete to show the first 6 multiples of each number.

a 5, 10…

b 3, 6…

c 7, 14…

2 Which of these numbers is a multiple of 8?

1   4   18   40

3 What is the difference between a factor and a multiple?

4 **a** Write the first 5 multiples of each of these numbers.

i 4 and 5   **ii** 4 and 6   **iii** 6 and 12

b Use your answers to part **a** to find the lowest common multiple of each pair.

Models and representations

Fraction wall

1				
$\frac{1}{2}$	$\frac{1}{2}$			
$\frac{1}{3}$	$\frac{1}{3}$	$\frac{1}{3}$		
$\frac{1}{4}$	$\frac{1}{4}$	$\frac{1}{4}$	$\frac{1}{4}$	
$\frac{1}{5}$	$\frac{1}{5}$	$\frac{1}{5}$	$\frac{1}{5}$	$\frac{1}{5}$

You can use a fraction wall to find equivalent fractions. You could make your own and add more rows

Finding equivalent fractions

In this section, you will use diagrams and multiplication and division facts to find **equivalent** fractions.

Example 1

Find two fractions that are equivalent to $\frac{3}{5}$

Method A

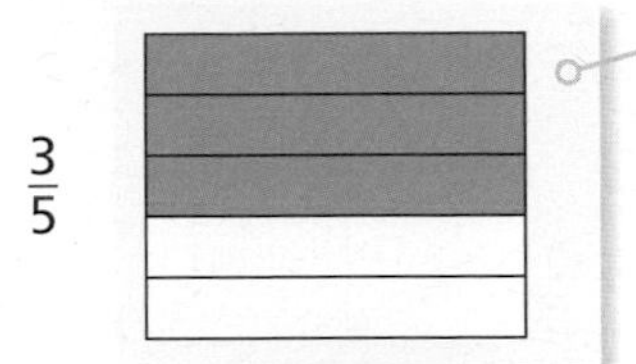

To show $\frac{3}{5}$ you can split a rectangle into 5 equal parts and shade 3 of those parts.

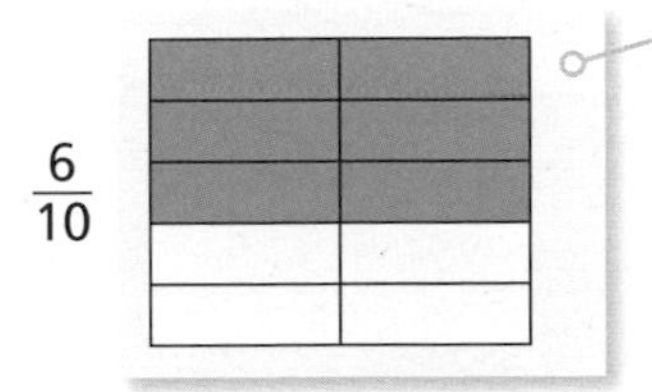

If you split the shape in half vertically, you now have 10 equal parts. 6 of those parts are shaded so $\frac{3}{5}$ must be equivalent to $\frac{6}{10}$

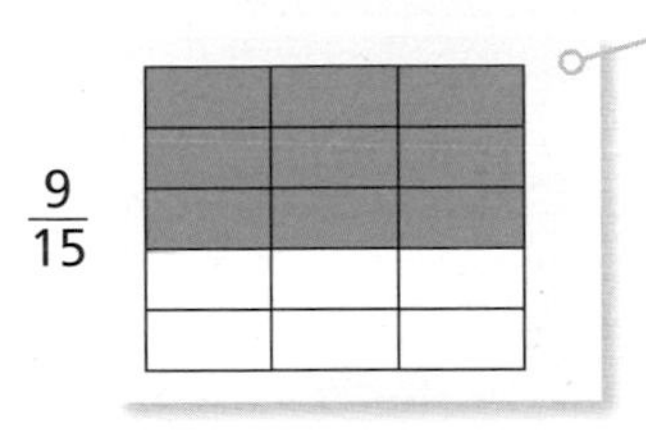

If you split the original diagram into thirds vertically instead of halves, you have 15 equal parts.

9 of these parts are shaded.

So $\frac{3}{5}$ is also equivalent to $\frac{9}{15}$

You could continue this process to find lots of other fractions that are equivalent to $\frac{3}{5}$

Method B

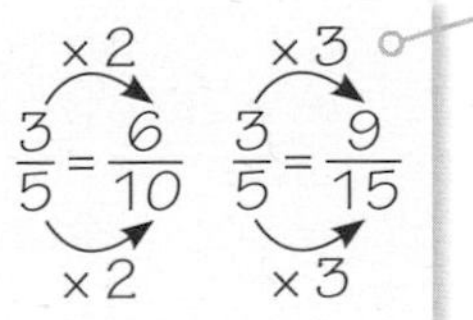

You can multiply the numerator and the denominator by the same number to find equivalent fractions.

There are an infinite number of fractions that are equivalent to $\frac{3}{5}$ because you could multiply the numerator and the denominator by any number.

Practice 10.3A

1 Use the fraction wall to complete the equivalent fractions.

1

$\frac{1}{2}$ $\frac{1}{2}$

$\frac{1}{3}$ $\frac{1}{3}$ $\frac{1}{3}$

$\frac{1}{4}$ $\frac{1}{4}$ $\frac{1}{4}$ $\frac{1}{4}$

$\frac{1}{5}$ $\frac{1}{5}$ $\frac{1}{5}$ $\frac{1}{5}$ $\frac{1}{5}$

$\frac{1}{6}$ $\frac{1}{6}$ $\frac{1}{6}$ $\frac{1}{6}$ $\frac{1}{6}$ $\frac{1}{6}$

$\frac{1}{7}$ $\frac{1}{7}$ $\frac{1}{7}$ $\frac{1}{7}$ $\frac{1}{7}$ $\frac{1}{7}$ $\frac{1}{7}$

$\frac{1}{8}$ $\frac{1}{8}$ $\frac{1}{8}$ $\frac{1}{8}$ $\frac{1}{8}$ $\frac{1}{8}$ $\frac{1}{8}$ $\frac{1}{8}$

$\frac{1}{9}$ $\frac{1}{9}$ $\frac{1}{9}$ $\frac{1}{9}$ $\frac{1}{9}$ $\frac{1}{9}$ $\frac{1}{9}$ $\frac{1}{9}$ $\frac{1}{9}$

$\frac{1}{10}$ $\frac{1}{10}$ $\frac{1}{10}$ $\frac{1}{10}$ $\frac{1}{10}$ $\frac{1}{10}$ $\frac{1}{10}$ $\frac{1}{10}$ $\frac{1}{10}$ $\frac{1}{10}$

a $\frac{1}{2} = \frac{\square}{4}$ **b** $\frac{2}{3} = \frac{\square}{6}$ **c** $\frac{1}{2} = \frac{5}{\square}$

d $\frac{6}{9} = \frac{\square}{3}$ **e** $\frac{6}{\square} = \frac{\square}{4}$ **f** $1 = \frac{\square}{7}$

2 Write down the equivalent fractions shown by each diagram.

a

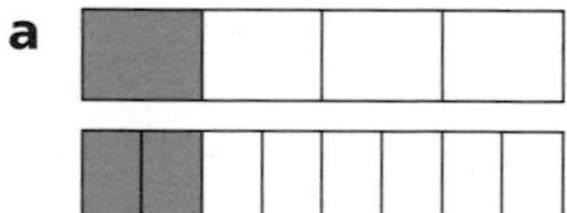

b

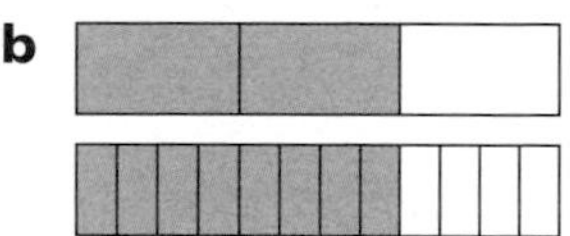

c

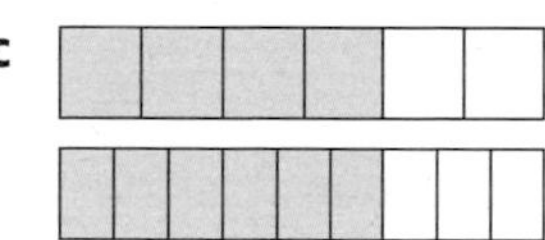

What do you notice about the length of the bars in each question?

3 Draw a bar model to show that $\frac{1}{2} = \frac{3}{6}$

4 Write down five fractions that are equivalent to $\frac{2}{5}$

5 Copy and complete using = or ≠ to say whether or not the fractions in each pair are equivalent.

a $\frac{3}{5} \bigcirc \frac{6}{10}$ **b** $\frac{2}{3} \bigcirc \frac{20}{30}$ **c** $\frac{1}{7} \bigcirc \frac{7}{1}$

d $\frac{8}{8} \bigcirc \frac{2}{2}$ **e** $\frac{3}{4} \bigcirc \frac{3}{5}$ **f** $\frac{1}{3} \bigcirc \frac{11}{31}$

6

$\frac{6}{24}$ and $\frac{7}{28}$ are not equivalent because you cannot multiply 6 by something to get 7 or 24 by something to get 28

$\frac{6}{24}$ and $\frac{7}{28}$ are equivalent because they both simplify to $\frac{1}{4}$

Who do you agree with? Explain your answer.

7 True or false? $\frac{2}{8} = \frac{9}{36}$
Explain your answer.

8 Complete these equivalent fractions.

a $\frac{1}{3} = \frac{5}{\square}$ **b** $\frac{5}{6} = \frac{25}{\square}$ **c** $\frac{3}{4} = \frac{\square}{20}$

d $\frac{1}{\square} = \frac{5}{35}$ **e** $\frac{\square}{20} = \frac{30}{200}$ **f** $\frac{4}{12} = \frac{\square}{15}$

What do you think?

1 These two fractions are equivalent.

$\frac{\square}{20} = \frac{10}{\square}$

Write down five possible pairs of values that could make the statement correct.

Check your answers with a partner. Were their answers the same as yours?

2 Write down an equivalent fraction to $\frac{2}{7}$ where the denominator is a square number.

Is there more than one possible answer?

Adding and subtracting fractions with different denominators

In this section, you will add and subtract fractions where one of the denominators is a **multiple** of the other.

You will use your knowledge of multiples and lowest common multiples.

Example 2

Jakub eats $\frac{2}{5}$ of a bunch of grapes. Faith eats $\frac{3}{10}$ of the same bunch of grapes.
What fraction of the grapes have they eaten altogether?

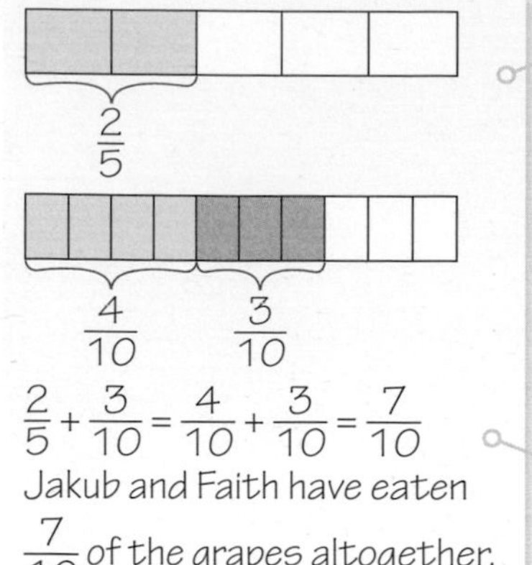

$\frac{2}{5}$
$\frac{4}{10}$
$\frac{3}{10}$
$\frac{2}{5} + \frac{3}{10} = \frac{4}{10} + \frac{3}{10} = \frac{7}{10}$
Jakub and Faith have eaten $\frac{7}{10}$ of the grapes altogether.

To add two fractions, the denominators need to be the same.

Start with a bar model split into 5 equal parts. Shade 2 parts.

Then divide each part in half to get 10 equal parts. Shade 3 more parts.

The bar models show that $\frac{2}{5}$ is equivalent to $\frac{4}{10}$

When the fractions have a **common denominator** you can then add them together.

Remember to add the numerators only – not the denominators as well.

Example 3

Work out $\frac{7}{9} - \frac{1}{3}$

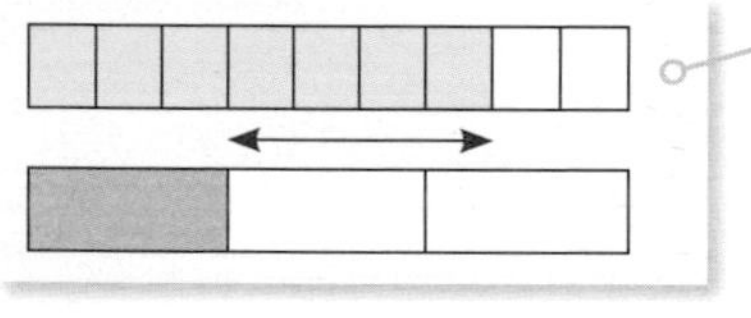

Looking at $\frac{7}{9}$ and $\frac{1}{3}$ on a bar model, you can see you need to make the denominators the same before you can complete the calculation.

The **lowest common multiple** (LCM) of 9 and 3 is 9, so you need each fraction to have 9 as the denominator.

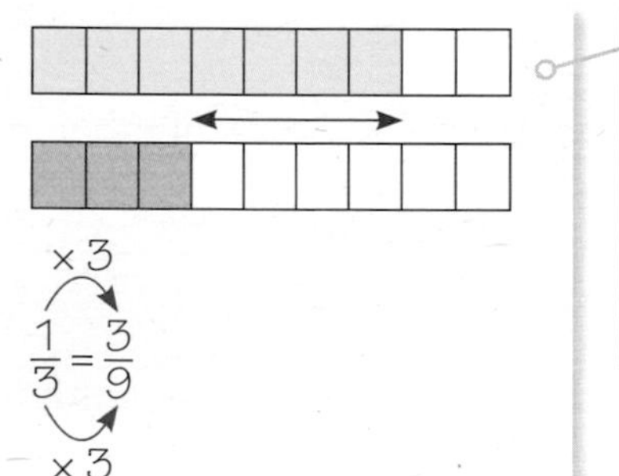

The denominator of $\frac{7}{9}$ is already 9, so this doesn't change.

$\frac{1}{3}$ is equivalent to $\frac{3}{9}$ so you can rewrite the calculation as

$\frac{7}{9} - \frac{3}{9} = \frac{4}{9}$

The answer to the subtraction can be seen by the difference in the bar model.

Practice 10.3B

 **1** Copy the bar models and use them to work out these additions and subtractions.

a $\frac{2}{3} + \frac{1}{6}$

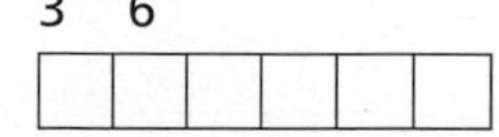

b $\frac{2}{5} + \frac{1}{10}$

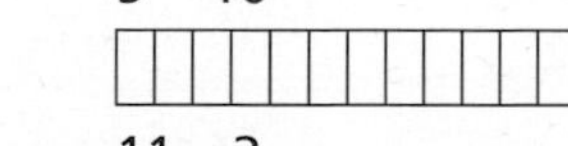

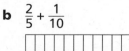

c $\frac{3}{4} - \frac{3}{8}$

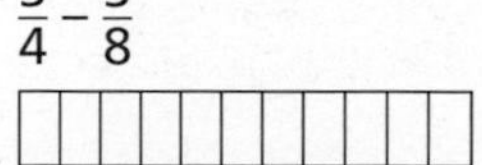

d $\frac{11}{12} - \frac{2}{3}$

2 Find the lowest common multiple of each pair of numbers.

a 2 and 8 **b** 14 and 7 **c** 3 and 9

What do you notice about your answers?

3 Work these out.

a $\frac{1}{2}+\frac{1}{8}$ **b** $\frac{3}{14}+\frac{5}{7}$ **c** $\frac{2}{3}-\frac{4}{9}$

4 Work these out.

Write your answers in their simplest form.

a $\frac{3}{5}+\frac{2}{15}$ **b** $\frac{3}{5}-\frac{2}{15}$ **c** $\frac{3}{5}+\frac{3}{5}+\frac{2}{15}$

d $\frac{7}{20}+\frac{5}{10}$ **e** $\frac{7}{8}-\frac{3}{4}$ **f** $\frac{11}{12}-\frac{1}{3}-\frac{1}{4}$

5 Chloe eats $\frac{1}{10}$ of her birthday cake.

She gives her friends $\frac{3}{5}$ of the cake.

She gives her family $\frac{3}{10}$ of the cake.

Does Chloe have any cake left? Show your working to explain how you know.

6 Beca has some football cards.

She gives $\frac{1}{8}$ of them to her friend Benji and $\frac{1}{2}$ of them to her friend Abdullah.

What fraction of the cards does Beca have left?

What do you think?

1 Find the missing denominators.

a $\frac{1}{5}+\frac{2}{\square}=\frac{1}{3}$ **b** $\frac{3}{\square}+\frac{5}{21}=\frac{2}{3}$ **c** $\frac{5}{12}-\frac{1}{\square}=\frac{1}{6}$

2 Solve these equations.

a $y-\frac{1}{2}=\frac{1}{4}$ **b** $\frac{2}{5}+w=\frac{9}{10}$ **c** $\frac{6}{35}=x-\frac{3}{7}$

Consolidate – do you need more?

1 Ed eats $\frac{1}{4}$ of a cake. Faith eats $\frac{1}{8}$ of the same cake.

What fraction of the cake have they eaten altogether?

Use the diagram to help you.

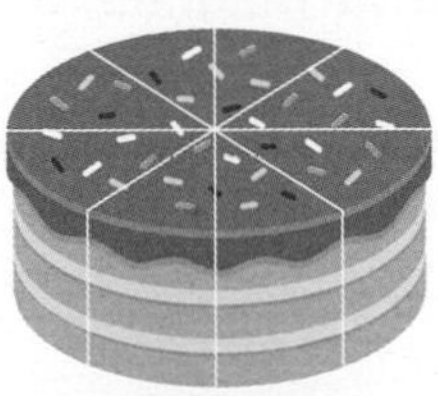

2 Marta has this chocolate bar.

She eats $\frac{2}{3}$ of the chocolate bar and gives her friend $\frac{1}{6}$ of it.

a What fraction of the chocolate bar is eaten?

b What fraction of the chocolate bar is left?

3 Work these out.

a $\frac{1}{4} + \frac{3}{8}$　　b $\frac{2}{5} + \frac{4}{15}$　　c $\frac{7}{10} - \frac{2}{5}$

d $\frac{5}{6} - \frac{1}{3}$　　e $\frac{1}{5} + \frac{3}{10} + \frac{2}{15}$　　f $\frac{1}{3} + \frac{5}{12} - \frac{1}{6}$

Stretch – can you deepen your learning?

1 Work these out. Give your answers as mixed numbers in their simplest form.

a $\frac{2}{3} + \frac{5}{6}$　　b $\frac{5}{14} + \frac{6}{7}$　　c $\frac{11}{16} + \frac{5}{8} + \frac{23}{24}$

2 Jakub cycles $1\frac{3}{4}$ km on Monday.

He cycles $2\frac{1}{8}$ km on Tuesday.

How far does he cycle in total on Monday and Tuesday?

3 A barrel holds $12\frac{1}{4}$ litres of water.

A bucket can hold $3\frac{11}{12}$ litres of water.

Max fills up the bucket with water from the barrel.

How much water is left in the barrel?

4 Three points A, B and C lie on a number line.

A section of the number line is shown.

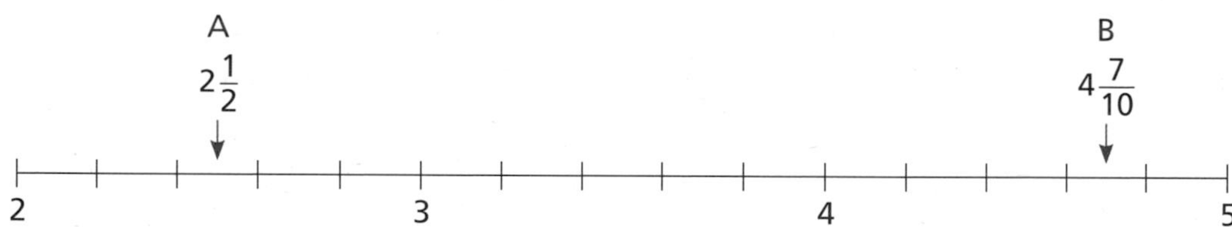

B lies halfway between A and C.

What is the value of C?

Reflect

1 Draw a diagram to show that $\frac{3}{5} + \frac{3}{10} \neq \frac{6}{15}$

2 Draw a diagram to show that $\frac{1}{5} + \frac{3}{15} = \frac{6}{15}$

10.4 Adding and subtracting fractions (3)

Small steps

- Add and subtract fractions with any denominator
- Add and subtract improper fractions and mixed numbers

Key words

Lowest common multiple – the lowest common multiple (LCM) of two or more numbers is the lowest number that is a multiple of them all

Common denominator – two or more fractions have a common denominator when their denominators are the same

Are you ready?

1 Simplify these fractions.

a $\frac{6}{12}$ **b** $\frac{6}{8}$ **c** $\frac{21}{28}$ **d** $\frac{22}{55}$

2 Write these improper fractions as mixed numbers.

a $\frac{11}{7}$ **b** $\frac{4}{3}$ **c** $\frac{15}{2}$ **d** $\frac{27}{5}$

3 Write these mixed numbers as improper fractions.

a $1\frac{2}{3}$ **b** $1\frac{7}{8}$ **c** $2\frac{1}{9}$ **d** $4\frac{3}{5}$

Models and representations

Bar model

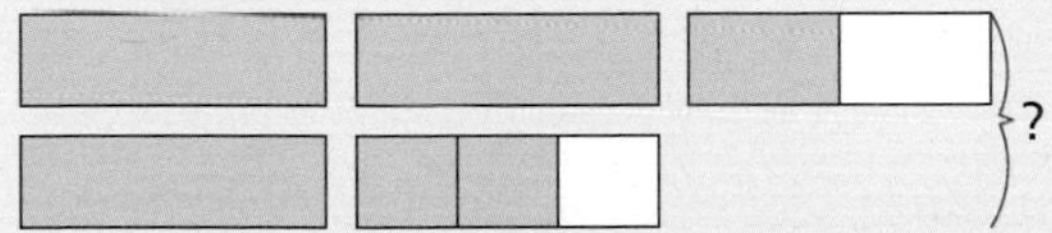

This bar model represents $2\frac{1}{2} + 1\frac{2}{3}$

Part-whole model

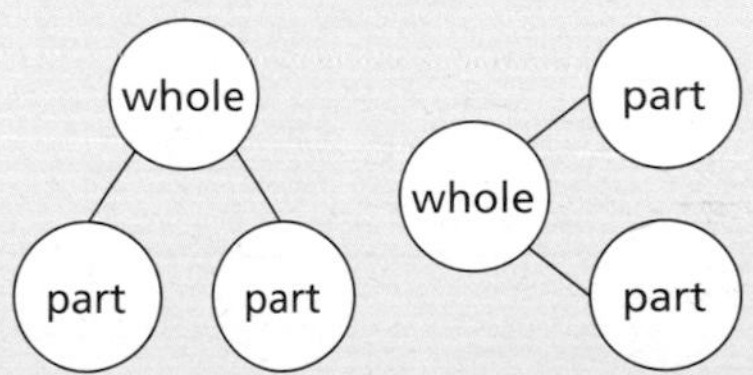

Part-whole models may be drawn in different orientations.

Adding and subtracting fractions with different denominators

In this chapter, you will add and subtract fractions with any denominator.

Always simplify your answers and write any improper fractions as mixed numbers.

Example 1

Work out $\frac{5}{6} + \frac{3}{4}$

Write your answer as a mixed number.

$\frac{5}{6} + \frac{3}{4} = \frac{?}{12} + \frac{?}{12}$

Find the **lowest common multiple** of 6 and 4 to find the lowest **common denominator**.

The fractions must have the same denominator before you can add them.

You could do this by writing out some multiples of 6 and 4

6, (12) 18, 24…

4, 8, (12) 16 …

This mean that your common denominator needs to be 12

Think about what you multiply each denominator by to get 12.

$\overset{\times 2}{\underset{\times 2}{\frac{5}{6}}} + \overset{\times 3}{\underset{\times 3}{\frac{3}{4}}} = \frac{10}{12} + \frac{9}{12}$

Multiply the numerator by the same number as the denominator.

Now change both fractions into equivalent fractions with a denominator of 12

When the fractions have the same denominator you can add them together.

Only add the numerators – the denominator is still 12

$\overset{\times 2}{\underset{\times 2}{\frac{5}{6}}} + \overset{\times 3}{\underset{\times 3}{\frac{3}{4}}} = \frac{10}{12} + \frac{9}{12} = \frac{19}{12}$

$= 1\frac{7}{12}$

$\frac{19}{12}$ is an improper fraction so you need to convert it into a mixed number.

Because $\frac{12}{12} = 1$

$\frac{19}{12} = \frac{12}{12} + \frac{7}{12} = 1\frac{7}{12}$

Practice 10.4A

1 **a** What is the lowest common multiple of 3 and 8?

b Work out $\frac{2}{3} + \frac{1}{8}$

2 Beca is working out $\frac{1}{6} + \frac{1}{8}$

Here is her method.

$\frac{1}{6} + \frac{1}{8} = \frac{8}{48} + \frac{6}{48} = \frac{14}{48}$

a How can Beca's answer be improved?

b Is there a more efficient way of working out $\frac{1}{6} + \frac{1}{8}$?

3 Show how the bar model could be used to work out $\frac{1}{2} + \frac{1}{5}$

4 Work out these additions and subtractions. Write the answers in their simplest form.

a $\frac{1}{4} + \frac{1}{12}$ **b** $\frac{2}{5} + \frac{1}{6}$ **c** $\frac{3}{8} + \frac{1}{6}$

d $\frac{5}{7} - \frac{2}{3}$ **e** $\frac{8}{9} - \frac{5}{6}$ **f** $\frac{1}{6} + \frac{3}{4} - \frac{1}{3}$

5 Work out the missing number in each part-whole model.

a

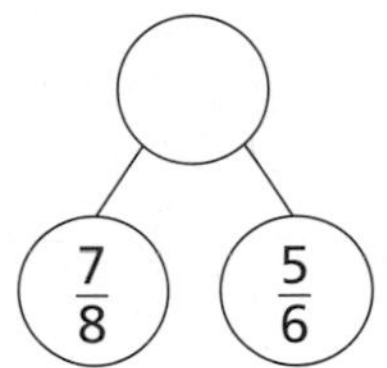

b

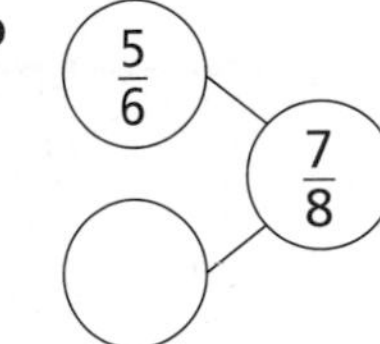

6 Seb reads $\frac{5}{36}$ of his book on Monday.

He reads $\frac{1}{9}$ of his book on Tuesday.

He reads $\frac{1}{6}$ of his book on Wednesday.

He reads the rest over the weekend.

What fraction of the book does he read over the weekend?

7 Work out $\frac{3}{7} + \frac{4}{13} + \frac{5}{8} + \frac{9}{13} + \frac{4}{7} + \frac{3}{8}$

Compare your method with a partner. Did you both use the same method? Whose method is more efficient?

8 Here are the masses of some boxes.

$\frac{4}{9}$kg $\frac{11}{12}$kg $\frac{1}{3}$kg $\frac{3}{4}$kg $\frac{1}{2}$kg

a Work out the range of these masses.

b What strategy did you use to decide which boxes where the heaviest and lightest?

What do you think?

1 The answer to an addition is $1\frac{3}{5}$. What could the question be if

a the fractions in the question have the same denominator

b the fractions in the question have different denominators?

2 The circle has an area of $\frac{17}{25}$m^2

The triangle has an area of $\frac{3}{10}$m^2

Work out the area of the shaded part of the shape.

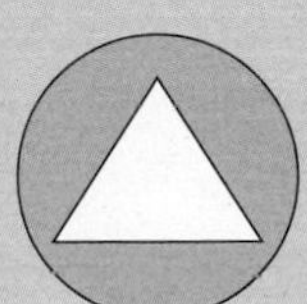

Example 2

Work out $2\frac{4}{5} + 1\frac{7}{10}$

Method A

$2 + 1 = 3$

$\frac{4}{5} + \frac{7}{10} = \frac{8}{10} + \frac{7}{10} = \frac{15}{10}$

$\frac{15}{10} = \frac{10}{10} + \frac{5}{10} = 1\frac{5}{10} = 1\frac{1}{2}$

$3 + 1\frac{1}{2} = 4\frac{1}{2}$

So $2\frac{4}{5} + 1\frac{7}{10} = 4\frac{1}{2}$

Add the wholes first.

Then add the fractions.

Convert your answer to a mixed number and simplify if necessary.

Add the two parts of the answer together to get the final answer.

Method B

$2\frac{4}{5} = \frac{14}{5}$ $\quad 1\frac{7}{10} = \frac{17}{10}$

$\frac{14}{5} + \frac{17}{10} = \frac{28}{10} + \frac{17}{10} = \frac{45}{10}$

$\frac{45}{10} = 4\frac{5}{10} = 4\frac{1}{2}$

Convert the mixed numbers to improper fractions.

Add the fractions by finding a common denominator.

Convert the improper fraction to a mixed number and simplify.

Think about the most appropriate method for each question rather than always using the same method. Choose the method that is more efficient.

Practice 10.4B

1 a Work out $1\frac{1}{6} + 3\frac{3}{4}$ by converting each mixed number into an improper fraction.
Use the diagram to help you.

b Work out $1\frac{1}{6} + 3\frac{3}{4}$ by first adding the wholes and then adding the fractions.
Use the diagram to help you.

c Which method did you prefer? Why?

d Use your preferred method to answer these questions.

i $2\frac{1}{3} + 2\frac{1}{4}$ ii $3\frac{2}{7} + 1\frac{2}{5}$ iii $4\frac{7}{9} + 3\frac{2}{3}$

2 Ed is calculating $3\frac{2}{5} - 1\frac{7}{8}$. He says,

First I will subtract the wholes, then I will subtract the fractions

a Try using Ed's method. What problems did you face?

b Use a different method to work out $3\frac{2}{5} - 1\frac{7}{8}$

3 Work out these subtractions.

a $1\frac{5}{6} - \frac{1}{3}$ **b** $3\frac{7}{12} - 1\frac{3}{8}$ **c** $2\frac{4}{5} - 2\frac{7}{9}$ **d** $10\frac{7}{16} - 7\frac{1}{8}$

Compare your method for each question with a partner. Did you use the same method each time? Whose method was more efficient?

4 Marta runs part of a journey then walks the rest.
She runs $4\frac{6}{7}$ km then walks for $2\frac{3}{4}$ km.

a How far is Marta's journey altogether?

b How much further does she run than walk?

5 Here are the heights, in metres, of some of the tallest buildings in the world.

Building	Height (metres)
Burj Khalifa	$829\frac{4}{5}$
Ping An Finance Centre	$599\frac{1}{10}$
Lotte World Tower	$555\frac{13}{20}$
CN Tower	$553\frac{3}{10}$
One World Trade Center	$546\frac{1}{5}$

a How much taller is the Burj Khalifa than the Ping An Finance Centre?

b What is the total height of all five buildings?

c How much taller is the Ping An Finance Centre than the One World Trade Center?

d What is the range of the heights of the buildings?

What do you think?

1 **a** Show that this hexagon and this triangle have the same perimeter.

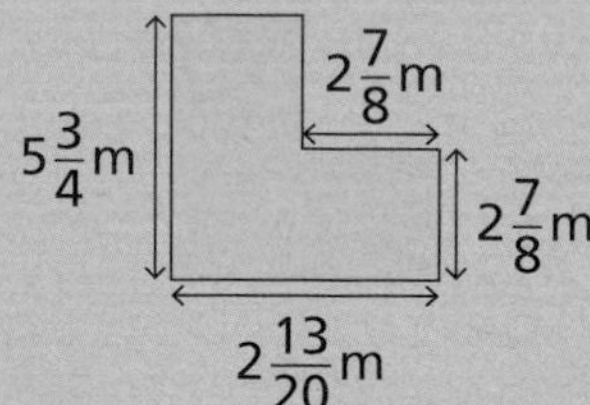

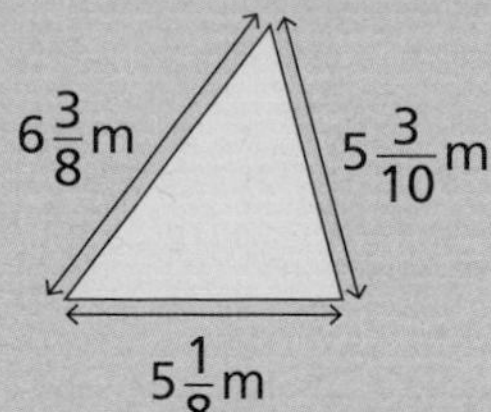

b A square has the same perimeter as the hexagon and the triangle. What is the side length of the square?

2 Is this statement always, sometimes or never true?

Adding two mixed numbers gives a mixed number answer.

Use examples to support your reasoning.

Consolidate – do you need more?

1 Complete these additions.

a $\frac{1}{5}+\frac{3}{10}$ **b** $\frac{1}{5}+\frac{7}{15}$ **c** $\frac{1}{10}+\frac{7}{15}$

d $\frac{3}{10}+\frac{7}{15}$ **e** $\frac{3}{4}+\frac{5}{6}$ **f** $\frac{2}{3}+\frac{3}{5}$

g $\frac{1}{6}+\frac{29}{30}$ **h** $\frac{2}{5}+\frac{3}{4}$ **i** $\frac{4}{7}+\frac{1}{2}$

2 Complete these additions.

a $4\frac{1}{5}+2\frac{3}{10}$ **b** $3\frac{1}{5}+1\frac{7}{15}$ **c** $5\frac{1}{10}+6\frac{7}{15}$

d $12\frac{3}{10}+23\frac{7}{15}$ **e** $99\frac{3}{4}+101\frac{5}{6}$ **f** $4\frac{2}{3}+2\frac{3}{5}$

g $3\frac{1}{6}+1\frac{29}{30}$ **h** $5\frac{2}{5}+6\frac{3}{4}$ **i** $12\frac{4}{7}+23\frac{1}{2}$

3 Complete these subtractions.

a $\frac{3}{10}-\frac{1}{5}$ **b** $\frac{7}{15}-\frac{1}{5}$ **c** $\frac{7}{15}-\frac{1}{10}$

d $\frac{7}{15}-\frac{3}{10}$ **e** $\frac{5}{6}-\frac{3}{4}$ **f** $\frac{2}{3}-\frac{2}{5}$

g $\frac{29}{30}-\frac{1}{6}$ **h** $\frac{3}{4}-\frac{2}{5}$ **i** $\frac{4}{7}-\frac{1}{2}$

4 Complete these subtractions.

a $5\frac{3}{10}-3\frac{1}{5}$ **b** $6\frac{7}{15}-1\frac{1}{5}$ **c** $3\frac{7}{15}-2\frac{1}{10}$

d $12\frac{7}{15}-10\frac{3}{10}$ **e** $19\frac{5}{6}-7\frac{3}{4}$ **f** $25\frac{2}{3}-25\frac{2}{5}$

g $117\frac{29}{30}-99\frac{1}{6}$ **h** $199\frac{3}{4}-199\frac{2}{5}$ **i** $63\frac{4}{7}-27\frac{1}{2}$

5 Seb and Faith are working out $3\frac{1}{6} - 2\frac{3}{4}$

Here are their methods.

Seb's method

$3\frac{1}{6} - 2\frac{3}{4} = \frac{5}{12}$

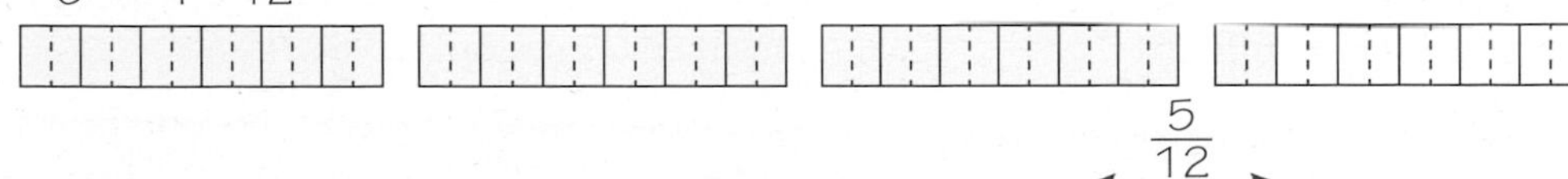

$\frac{5}{12}$

Faith's method

$3\frac{1}{6} - 2\frac{3}{4} = 2\frac{7}{6} - 2\frac{3}{4}$

$= 2\frac{14}{12} - 2\frac{9}{12}$

$= \frac{5}{12}$

Discuss each method with a partner. Which one do you prefer?

6 Use your preferred method to complete these subtractions.

a $2\frac{1}{5} - 1\frac{1}{4}$ **b** $3\frac{1}{2} - 2\frac{7}{10}$ **c** $12\frac{5}{6} - 5\frac{7}{8}$ **d** $15\frac{4}{9} - 11\frac{3}{4}$

Stretch – can you deepen your learning?

1 $a = 3\frac{1}{5}$ $b = 6\frac{1}{3}$ $c = \frac{47}{10}$

Work out the value of

a $a + b$ **b** $b + c$ **c** $a + c$ **d** $a + b + c$

e $b - a$ **f** $b - c$ **g** $a - b$ **h** $a - c$

2 Complete these calculations.

a $3\frac{3}{16} + 4\frac{3}{8}$ **b** $3\frac{x}{16} + 4\frac{x}{8}$ **c** $3\frac{3}{2x} + 4\frac{3}{x}$

3 $x + y = 19$ $5 < x < 6$ $13 < y < 14$

Give five possible pairs of mixed number values of x and y

4 Mario and Emily are working out $7\frac{1}{5} - 3\frac{11}{18}$

They want to subtract the wholes and then the fractions.

Mario says, "I can't do it because $\frac{11}{18}$ is greater than $\frac{1}{5}$"

Emily says, "I can use negative numbers to help me."

Show that Emily is correct.

Reflect

Explain two different methods you could use to find the sum of $2\frac{5}{6}$ and $4\frac{3}{8}$ and find the difference between $2\frac{5}{6}$ and $4\frac{3}{8}$

10.5 Fractions and algebra

Small steps

- Use fractions in algebraic contexts
- Add and subtract simple algebraic fractions (H)

Key words

Expression – a collection of terms involving mathematical operations

Equation – a statement with an equals sign, which states that two expressions are equal in value

Evaluate – to work out the numerical value of an expression

Are you ready?

1 Solve these equations.

a $x + 7.2 = 56.5$ b $x - 0.61 = 3.59$ c $17x = 136$

d $\frac{x}{17} = 136$ e $5.3 + x = 2$ f $17 - x = 12.19$

g $12x = -156$ h $x - 17.64 = -49.12$

2 Solve these equations.

a $2x + 7.2 = 56.5$ b $5x - 0.61 = 3.59$ c $17x + 51 = 136$

d $51 + \frac{x}{17} = 136$ e $5.3 + 3x = 2$ f $17 - 4x = 12.19$

g $12 + 12x = -156$ h $\frac{x}{2} - 17.64 = -49.12$

3 Complete the calculations.

a i $\frac{1}{5} + \frac{3}{5}$ ii $\frac{5}{9} - \frac{2}{9}$ iii $\frac{14}{47} + \frac{29}{47}$ iv $\frac{783}{999} - \frac{317}{999}$

b i $\frac{1}{15} + \frac{3}{5}$ ii $\frac{5}{9} - \frac{1}{18}$ iii $\frac{91}{94} - \frac{29}{47}$ iv $\frac{7}{9} - \frac{317}{999}$

c i $\frac{2}{5} + \frac{3}{4}$ ii $\frac{5}{6} - \frac{3}{8}$ iii $\frac{4}{7} + \frac{1}{9}$ iv $\frac{15}{16} - \frac{7}{12}$

4 Simplify the expressions.

a $x + x$ b $y + y - y$ c $z + z + z + z - z$

d $w + 2w - 4w$ e $2x + 3y + x$ f $4y - z + 3y$

g $7z - 19y + 32z - 18y$ h $5w + 16 - 11w + 24$

Models and representations

Bar models

You can use a bar model to help to solve an equation.

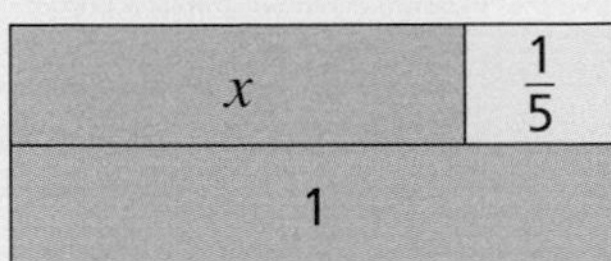

This bar model represents the equation $x + \frac{1}{5} = 1$

When representing fractions, the bar model is split into equal parts.

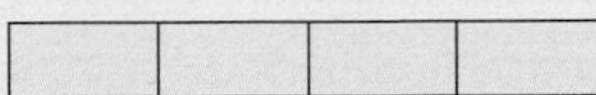

This bar model is split into 4 equal parts and so represents quarters.

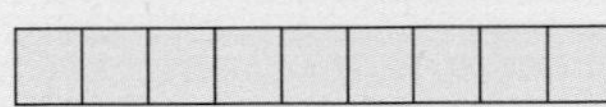

This bar model is split into 9 equal parts and so represents ninths.

If the denominator of a fraction is algebraic, for example x, you do not know how many equal parts there are but you can visualise this using dashed lines.

In this section you will use fractions in algebraic contexts. This will bring together previous learning on algebra including solving equations, substitution and sequences, and further consolidate recent learning on fractions.

Example 1

The nth term of a sequence is $\frac{3n}{2}$

a What is the fifth term of the sequence?

b In which position is the term $13\frac{1}{2}$?

Method A

a $\frac{3n}{2} = 3 \times \frac{n}{2}$

$3 \times \frac{5}{2} = \frac{15}{2} = 7\frac{1}{2}$

The fifth term is $7\frac{1}{2}$

$\frac{3n}{2}$ means 3 multiplied by n divided by 2

Multiply the numerator by 3; do not multiply the denominator as well.

In the expression for the nth term of a sequence, n represents the term number.

For the fifth term, n is equal to 5

Give the answer as a mixed number

$\frac{15}{2} = \frac{14}{2} + \frac{1}{2} = 7\frac{1}{2}$

b $\frac{3n}{2} = 13\frac{1}{2}$

$3n = 27$

$n = 9$

$13\frac{1}{2}$ is the ninth term

Solve this equation to find the value of n.

You know that the nth term is $13\frac{1}{2}$
You can write this as an equation.

Multiply both sides by 2 to get $3n = 27$

Then divide both sides by 3

Method B

a $\frac{3(5)}{2} = \frac{15}{2} = 7\frac{1}{2}$

The fifth term is $7\frac{1}{2}$

Substitute $n = 5$ into the expression for the nth term.

n represents the term number;
for the fifth term, n is equal to 5

Change $\frac{15}{2}$ into a mixed number.

b $\frac{3n}{2} = 13\frac{1}{2}$

$\frac{3n}{2} = \frac{27}{2}$

$3n = 27$

$n = 9$

$13\frac{1}{2}$ is the ninth term

$13\frac{1}{2}$ is the same as $\frac{27}{2}$

You know that the nth term is $13\frac{1}{2}$. You can write this as an equation.

You can convert $13\frac{1}{2}$ into an improper fraction.

Since the denominators are equal, the numerators must also be equal.

Divide both sides of the equation by 3

Practice 10.5A

1 $g = 3$ and $h = 15$

Evaluate each of these expressions.

a $\frac{1}{g} + \frac{1}{g}$ **b** $\frac{14}{h} - \frac{12}{h}$ **c** $\frac{1}{g} + \frac{1}{h}$ **d** $\frac{1}{g} - \frac{1}{h}$

e $\frac{1}{g^2} + \frac{1}{g}$ **f** $\frac{5}{g^2} - \frac{1}{h}$ **g** $h - \frac{3}{g}$ **h** $\frac{g}{h} + \frac{h}{g}$

2 The nth term of a sequence is given by $\frac{n}{5}$

a Work out the first five terms of the sequence.

b What is the term-to-term rule of the sequence?

c Is the sequence linear? Explain your answer.

d What is the 200th term of the sequence?

e In what position is the term with value 25?

3 Work out the missing inputs and outputs of these function machines.

a 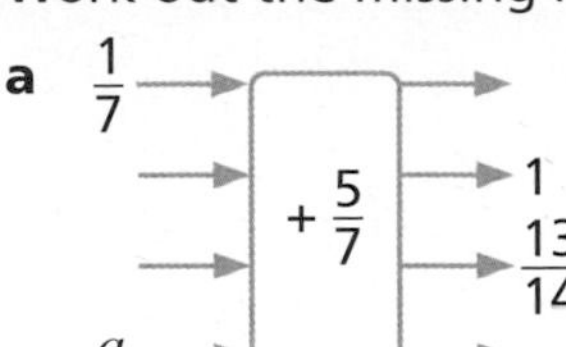

b 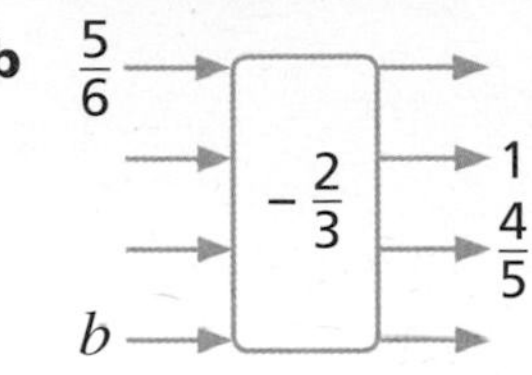

4 Solve these equations.

a $x + \frac{1}{5} = \frac{4}{5}$ **b** $z + \frac{5}{9} = \frac{7}{9}$ **c** $\frac{4}{11} - w = \frac{1}{11}$

d $1 - t = \frac{23}{37}$ **e** $s + \frac{1}{5} - \frac{1}{10} = \frac{3}{5}$ **f** $\frac{3}{7} + q = 1 - \frac{1}{14}$

g $\frac{2}{11} + \frac{1}{2} = y - \frac{1}{4}$ **h** $\frac{21}{40} - p = \frac{1}{5} - \frac{7}{40}$

5 Solve these equations.

a $3\frac{1}{2} + a = 5$ **b** $4 - b = 3\frac{1}{4}$ **c** $7 = c + 5\frac{2}{3}$

d $12\frac{1}{9} - d = 10\frac{5}{9}$ **e** $e + 9\frac{3}{4} = 12\frac{1}{2}$ **f** $f - 11\frac{2}{5} = 23\frac{3}{10}$

g $3\frac{1}{4} = 4\frac{5}{6} - g$ **h** $17\frac{3}{5} + h = 29\frac{3}{4}$

What do you think?

1 The nth term of a sequence is given by $\frac{5n}{7}$

a Work out the first 5 terms of the sequence.

b In what position is the term $46\frac{3}{7}$?

c In what position is the first integer term of the sequence?

d How often does the sequence produce integer terms? Explain your answer.

e How many of the first 100 terms of the sequence will be integers?

2 a and b are positive integers.

a Show that, when $a = 1$ and $b = 2$, $\frac{a}{b} > \frac{a}{b^2}$

b Choose three other pairs of values for a and b, and show that $\frac{a}{b} > \frac{a}{b^2}$ for each.

c Is this statement always true, sometimes true or never true?

$\frac{a}{b} > \frac{a}{b^2}$ for any values of a and b.

Discuss this with a partner.

Adding and subtracting algebraic fractions (H)

You can use the same methods as when adding and subtracting numerical fractions.

Remember:

- when the denominators are the same, you can just add/subtract the numerators
- if the denominators different, you need to convert to equivalent fractions with a common denominator.

Example 2

Work these out.

a $\frac{4}{x} - \frac{1}{x}$ **b** $\frac{3y}{7} + \frac{y}{7}$ **c** $\frac{3}{z} + \frac{1}{2z}$

a

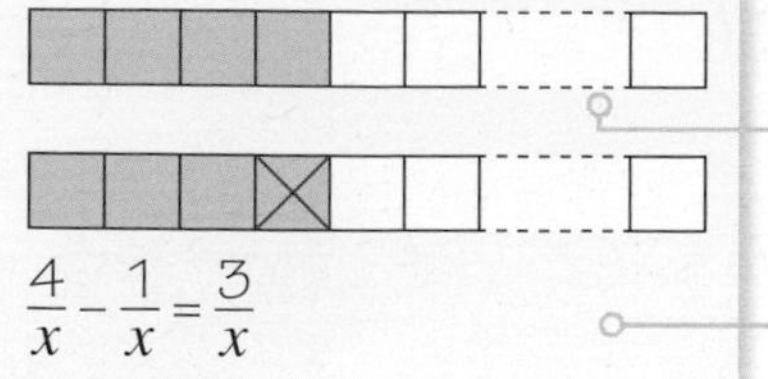

You do not know how many equal parts there are, but you can represent this using a dashed line.

First represent $\frac{4}{x}$ using a bar model.

Four of the parts are shaded.

You are subtracting $\frac{1}{x}$, so cross out one of the parts.

There are now three parts out out of x.

Because the denominators are the same, you can just subtract the numerators.

b As the denominators are the same, you can just add the numerators.

c

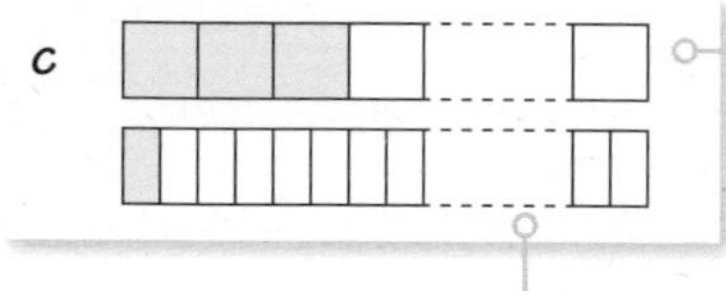

First represent each fraction on a bar model.

The bar model for $\frac{1}{2z}$ must be split into twice as many equal parts as the one for $\frac{3}{z}$ because $2z$ is double z.

You do not know how many parts the bar models are split into but you can use a dashed line.

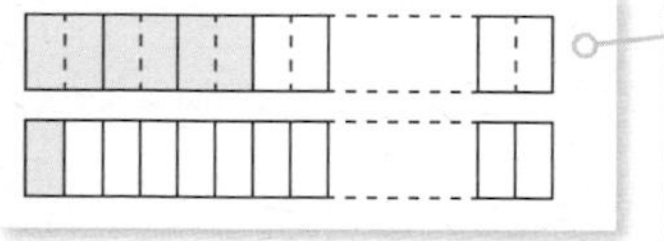

The bar model shows that $\frac{3}{z}$ is equivalent to $\frac{6}{2z}$ so you can rewrite the calculation as $\frac{6}{2z} + \frac{1}{2z}$

The denominators are the same, so you can add the numerators.

Practice 10.5B (H)

1 Complete these calculations.

a **i** $\frac{5}{17} + \frac{9}{17}$ **ii** $\frac{5}{23} + \frac{9}{23}$ **iii** $\frac{5}{x} + \frac{9}{x}$ **iv** $\frac{5}{11z} + \frac{9}{11z}$

b **i** $\frac{9}{17} - \frac{5}{17}$ **ii** $\frac{9}{23} - \frac{5}{23}$ **iii** $\frac{9}{x} - \frac{5}{x}$ **iv** $\frac{9}{11z} - \frac{5}{11z}$

What's the same and what's different about the calculations in each set?

2 Complete these calculations.

a $\frac{13}{x} + \frac{19}{x}$ **b** $\frac{21}{y} - \frac{19}{y}$ **c** $\frac{43}{w} + \frac{17}{w}$

d $\frac{11}{z} - \frac{10}{z}$ **e** $\frac{3}{g} - \frac{3}{g}$ **f** $\frac{24}{b} + \frac{5}{b} - \frac{9}{b}$

g $\frac{3}{c} - \frac{5}{c} + \frac{91}{c}$ **h** $\frac{5}{t} + \frac{2}{s} + \frac{103}{t} - \frac{2}{s}$

3 Complete these calculations.

a **i** $\frac{7}{199}+\frac{3}{199}$ **ii** $\frac{70}{199}+\frac{30}{199}$ **iii** $\frac{7x}{199}+\frac{3x}{199}$ **iv** $\frac{70x}{199}+\frac{30x}{199}$

b **i** $\frac{7}{199}-\frac{3}{199}$ **ii** $\frac{70}{199}-\frac{30}{199}$ **iii** $\frac{7x}{199}-\frac{3x}{199}$ **iv** $\frac{70x}{199}-\frac{30x}{199}$

What's the same and what's different about the calculations in each set?

4 Complete these calculations.

a $\frac{2y}{3}+\frac{5y}{3}$ **b** $\frac{11d}{7}-\frac{5d}{7}$ **c** $\frac{24a}{31}-\frac{13a}{31}$ **d** $\frac{8b}{19}+\frac{5b}{19}$

e $\frac{100k}{21}-\frac{99k}{21}$ **f** $\frac{5g^2}{47}+\frac{12g^2}{47}$ **g** $\frac{a}{2}-\frac{a}{2}$ **h** $\frac{7p}{9}+\frac{p}{9}$

5 Complete these calculations.

a **i** $\frac{3}{5}+\frac{1}{10}$ **ii** $\frac{3}{40}+\frac{1}{20}$ **iii** $\frac{3}{z}+\frac{1}{2z}$ **iv** $\frac{3}{2x}+\frac{1}{x}$

b **i** $\frac{3}{5}-\frac{1}{10}$ **ii** $\frac{3}{40}-\frac{1}{20}$ **iii** $\frac{3}{z}-\frac{1}{2z}$ **iv** $\frac{3}{2x}-\frac{1}{x}$

What's the same and what's different about the calculations in each set?

6 Complete the calculations.

a $\frac{2}{a}+\frac{5}{2a}$ **b** $\frac{7}{b}-\frac{1}{2b}$ **c** $\frac{17}{3c}+\frac{1}{c}$

d $\frac{4}{d}-\frac{1}{3d}$ **e** $\frac{11}{e}+\frac{12}{7e}$ **f** $\frac{17}{100x}-\frac{7}{50x}$

g $\frac{113}{4y}+\frac{1}{2y}-\frac{3}{y}$ **h** $\frac{97}{6k}-\frac{5}{3k}+\frac{1}{k}$

What do you think?

1 Beca and Abdullah are working out $\frac{11a}{4}-\frac{5a}{2}$

The answer is $\frac{a}{4}$

The answer is $\frac{1}{4}a$

Who do you agree with? Explain your answer.

2 Decide whether each statement is always true, sometimes true or never true.

Give reasons for your answers.

a $\frac{b}{x}\equiv\frac{2b}{2x}$ **b** $\frac{m}{2}+\frac{m}{2}\equiv 1$ **c** $\frac{7}{n}>\frac{5}{n}$

3 Find a value of n that would make each statement true.

a $\frac{7}{n}+\frac{3}{n}>1$ **b** $\frac{7}{n}+\frac{3}{n}<1$ **c** $\frac{7}{n}+\frac{3}{n}=1$

Compare your answers with a partner. Is there more than one answer for each question?

Consolidate – do you need more?

1 Complete these calculations.

a $\frac{1}{11}+\frac{3}{11}$ **b** $\frac{1}{17}+\frac{3}{17}$ **c** $\frac{1}{119}+\frac{3}{119}$

d $\frac{1}{1001}+\frac{3}{1001}$ **e** $\frac{1}{y}+\frac{3}{y}$ **f** $\frac{1}{z}+\frac{3}{z}$

g $\frac{1}{2x}+\frac{3}{2x}$ **h** $\frac{1}{k^2}+\frac{3}{k^2}$

What do you notice?

2 Complete these calculations.

a $\frac{3}{11}-\frac{1}{11}$ **b** $\frac{3}{17}-\frac{1}{17}$ **c** $\frac{3}{119}-\frac{1}{119}$

d $\frac{3}{1001}-\frac{1}{1001}$ **e** $\frac{3}{y}-\frac{1}{y}$ **f** $\frac{3}{z}-\frac{1}{z}$

g $\frac{3}{2x}-\frac{1}{2x}$ **h** $\frac{3}{k^2}-\frac{1}{k^2}$

3 Find three equivalent fractions for each of these.

a $\frac{1}{z}$ **b** $\frac{1}{2x}$ **c** $\frac{3}{5a}$ **d** $\frac{7}{4b}$

4 Complete these calculations. Show each step of your working.

a $\frac{1}{2z}+\frac{1}{z}$ **b** $\frac{2}{3z}+\frac{1}{z}$ **c** $\frac{3}{z}+\frac{1}{4z}$ **d** $\frac{4}{5z}+\frac{7}{z}$

e $\frac{4}{3a}+\frac{2}{a}$ **f** $\frac{6}{11c}+\frac{2}{c}$ **g** $\frac{17}{g}+\frac{5}{2g}$ **h** $\frac{1}{x}+\frac{1}{2x}+\frac{1}{4x}$

Stretch – can you deepen your learning?

1 Solve these equations.

a $\frac{x}{5}+\frac{2x}{5}=18$ **b** $\frac{4y}{9}-\frac{y}{9}=21$

c $\frac{8t}{11}+\frac{9t}{11}=17$ **d** $\frac{99x}{100}-\frac{31x}{100}=1.36$

2 Solve these equations.

a $\frac{4}{x}+\frac{5}{x}=1$ **b** $\frac{7}{y}+\frac{11}{y}=2$

c $\frac{21}{z}+\frac{19}{z}=5$ **d** $\frac{21}{2z}+\frac{19}{2z}=5$

3 Solve these equations.

a $\frac{4}{x}+\frac{5}{2x}=1$ **b** $\frac{7}{3y}+\frac{11}{y}=2$

c $2\frac{1}{2z}+\frac{19}{z}=5$ **d** $2\frac{1}{4z}+\frac{19}{2z}=5$

4 The first term of an increasing linear sequence is $\frac{1}{x}$

The constant different is $\frac{7}{2x}$

a Work out the first five terms of the sequence.

b The 6th term of the sequence is 11. Work out the value of x.

5 The first term of a decreasing linear sequence is $\frac{1}{x}$

The constant different is $\frac{7}{2x}$

a Work out the first five terms of the sequence.

b The 6th term of the sequence is 10. Work out the value of x.

6 The nth term of a linear sequence is $\frac{3n}{7}$

The nth term of a different linear sequence is $-\frac{2n}{9}$

The two sequences are added together to form a third linear sequence.

Work out the nth term of the third sequence.

Reflect

1 Decide whether each statement is true or false. Give reasons to support your answers.

a $\frac{3}{x} + \frac{5}{x} = \frac{8}{2x}$

b $\frac{16}{3y} - \frac{4}{3y} = \frac{4}{y}$

c $\frac{19}{abcde} + \frac{43}{abcde} + \frac{81}{abcde} = \frac{143}{abcde}$

d $\frac{5}{a} - \frac{10}{2a} = 0$

2 In your own words, explain how to add or subtract algebraic fractions.

3 Find two different pairs of fractions that have a total of

a $\frac{17}{19x}$

b $\frac{19x}{17}$

c $\frac{17}{19x}$

d $\frac{19x}{17}$

10 Fractional thinking
Chapters 10.1–10.5

I have become fluent in…

- Converting between mixed numbers and improper fractions
- Using equivalent fractions
- Adding and subtracting proper fractions
- Adding and subtracting mixed numbers
- Adding and subtracting fractions from integers

I have developed my reasoning skills by…

- Using equivalence to calculate with decimals and fractions
- Making connections between different representations
- Making and testing patterns spotted when converting improper fractions and mixed numbers

I have been problem-solving through…

- Representing fractions in different forms
- Calculating with fractions in real-life contexts
- Working with fractions in algebraic contexts
- Representing questions in different ways

Check my understanding

1 Convert each improper fraction to a mixed number.

a $\frac{17}{3}$ b $\frac{21}{5}$ c $\frac{19}{17}$ d $\frac{48}{7}$ e $\frac{117}{2}$

2 Convert each mixed number to an improper fraction.

a $5\frac{1}{2}$ b $2\frac{3}{4}$ c $9\frac{1}{6}$ d $3\frac{7}{8}$ e $1\frac{13}{19}$

3 Complete the calculations. Give your answers as a mixed number and an improper fraction.

a $5 - \frac{1}{3}$ b $6 - \frac{2}{5}$ c $8 - \frac{5}{9}$ d $10 - \frac{3}{4}$ e $12 - \frac{9}{10}$

4 Complete the calculations. Give your answers in their simplest form.

a $\frac{1}{5} + \frac{3}{10}$ b $\frac{7}{9} - \frac{2}{3}$ c $\frac{5}{8} + \frac{1}{4}$ d $\frac{7}{10} - \frac{3}{4}$ e $\frac{2}{3} + \frac{1}{7}$

5 Complete the calculations. Give your answers as mixed numbers in their simplest form.

a $2\frac{1}{5} + 5\frac{1}{2}$ b $1\frac{1}{4} - \frac{3}{5}$ c $8\frac{9}{10} - 2\frac{1}{3}$ d $4\frac{2}{9} + 10\frac{5}{6}$

6 Complete the calculations. Give your answers as fractions.

a $0.5 - \frac{1}{10}$ b $0.25 + \frac{3}{4} - \frac{5}{6}$ c $0.6 - \frac{2}{5} + \frac{3}{10}$ d $0.9 - \frac{1}{2} + 0.2$

7 Complete the calculations. H

a $\frac{x}{2} + \frac{x}{4}$ b $\frac{y}{5} - \frac{y}{10}$ c $\frac{5}{x} - \frac{7}{2x}$ d $\frac{3b}{7} + \frac{2b}{5}$ e $\frac{12}{11a} + \frac{10}{a}$

11 Construction and measuring

In this block, I will learn...

how to use letters to name angles and sides

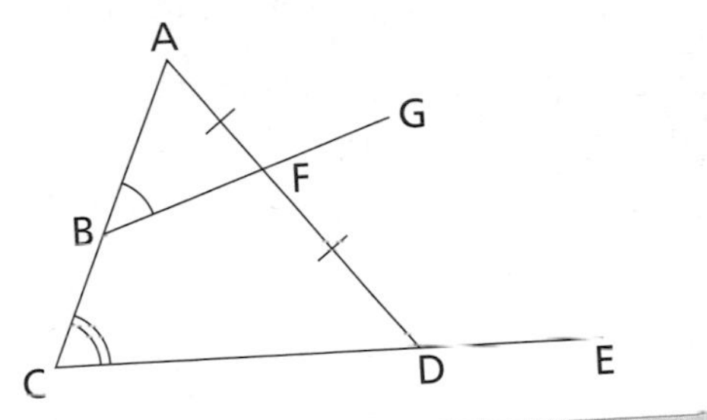

how to interpret geometric diagrams

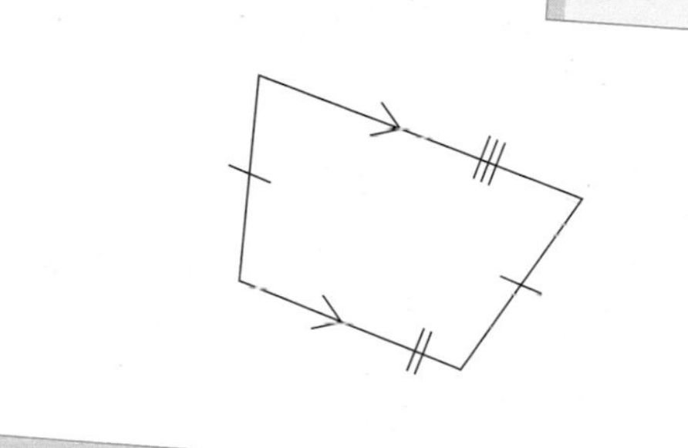

how to measure and draw angles

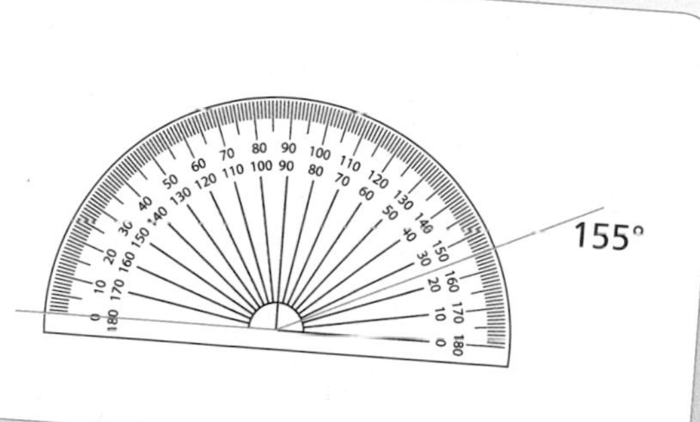

how to identify triangles, quadrilaterals and other shapes

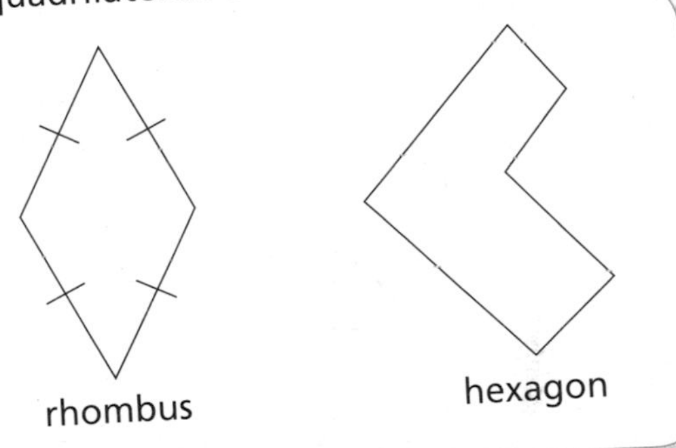

how to construct triangles and other shapes from given information

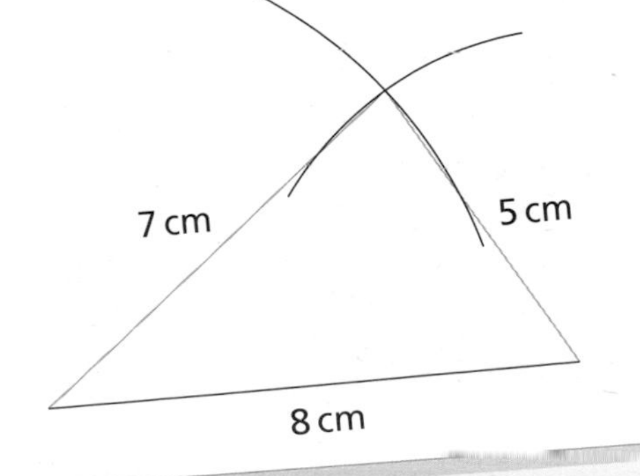

how to construct and interpret pie charts

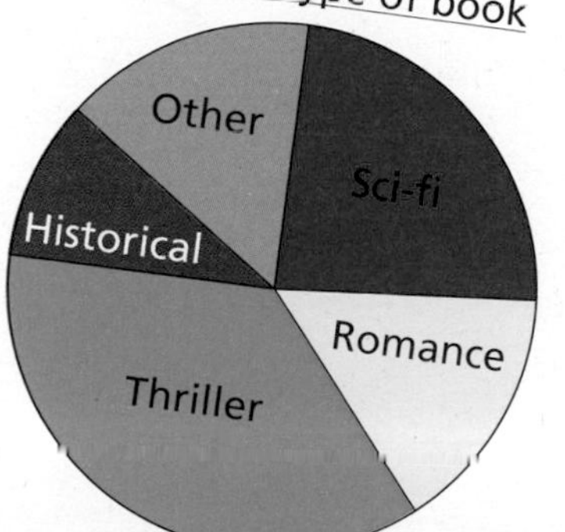

11.1 Geometric notation

Small steps

- Understand and use letter and labelling conventions including those for geometric figures
- Draw and measure line segments including geometric figures
- Understand angles as a measure of turn

Key words

Line segment – a part of a line that connects two points

Angle – a measure of turn between two lines around their common point

Clockwise – in the same direction as the hands of an analogue clock move

Anticlockwise – in the opposite direction to the way the hands of an analogue clock move

Are you ready?

1 How many degrees are there in

a a right angle **b** two right angles?

2 Measure these line segments.

a

b

c

3 Draw a line segment of length 7.4 cm.

4 Which of these is the correct symbol for the unit used to measure angles?

°F ° °C %

Models and representations

Compass

Clock face

Diagrams are labelled with capital letters to identify specific parts of a shape.

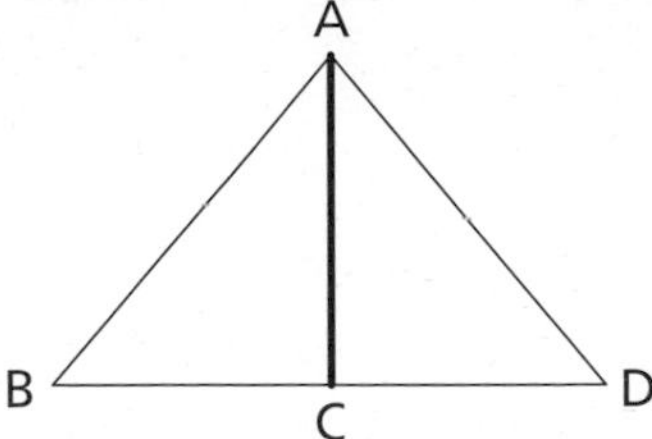

The **line segment** highlighted is called AC.

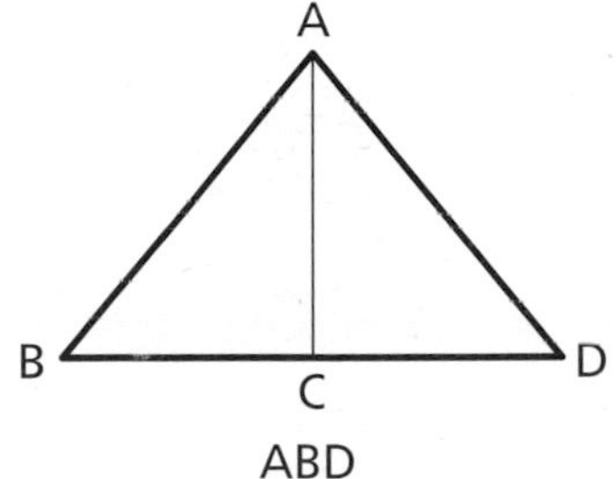

ABD

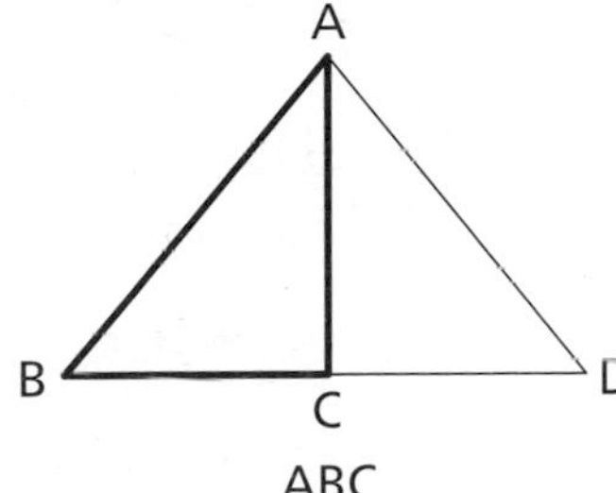

ABC

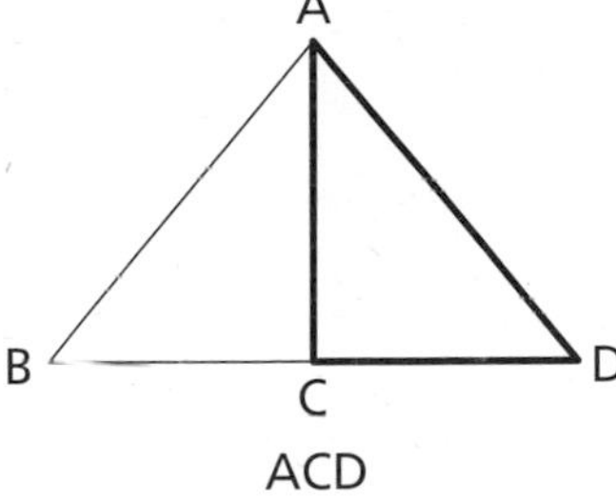

ACD

The larger triangle is called ABD and is made up of two triangles ABC and ACD.

Example 1

Draw and measure the line segments AH and GI.

A B C D

E F G H

I J K L

Make sure you start measuring from 0 and not the end of the ruler.

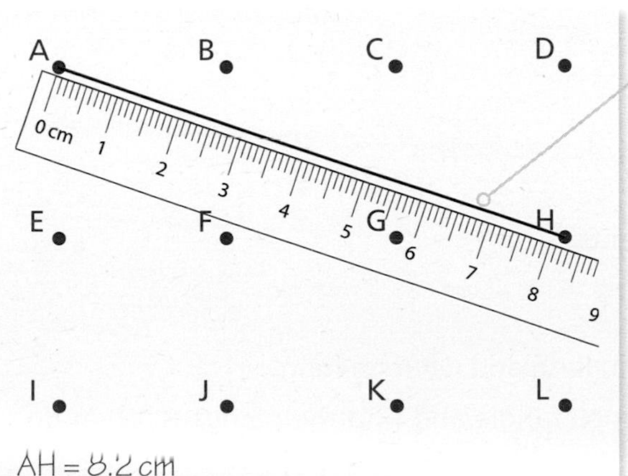

Line segment AH joins the points A and H.

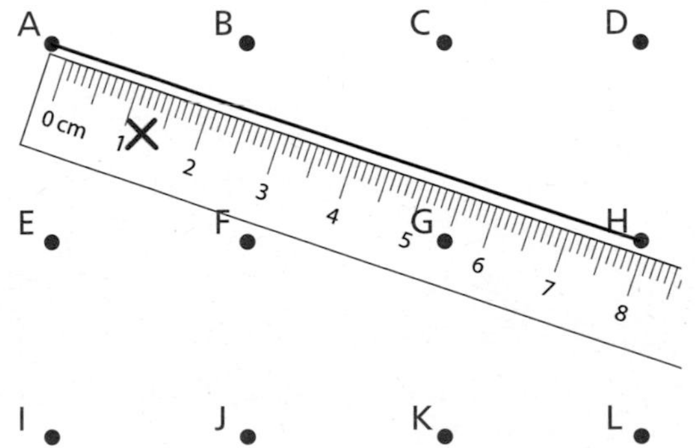

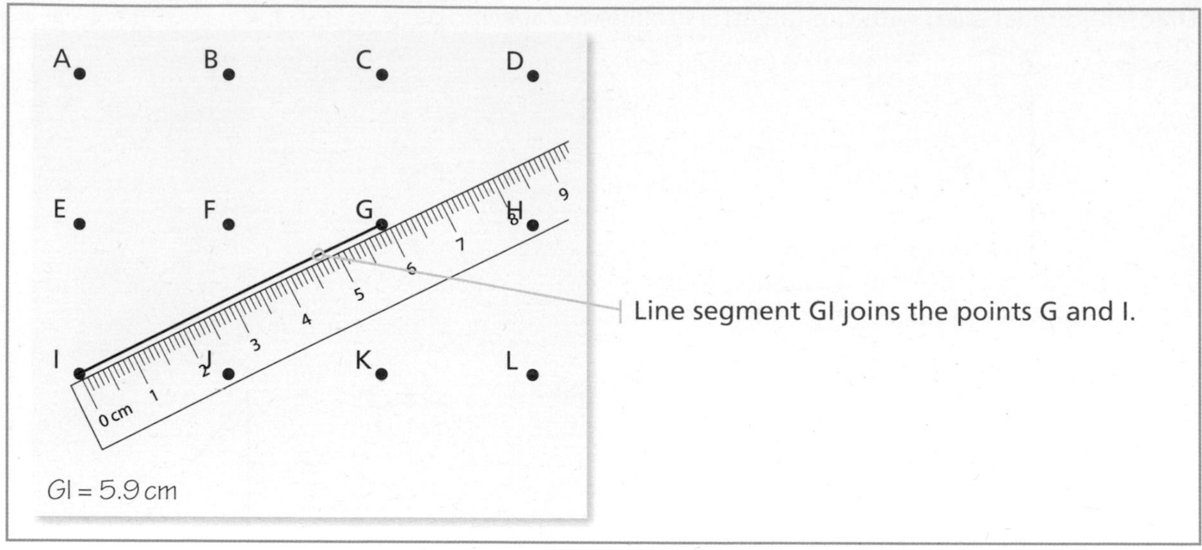

Practice 11.1A

1 Measure these line segments. Give your answers correct to the nearest millimetre (mm).

a

b

c

2

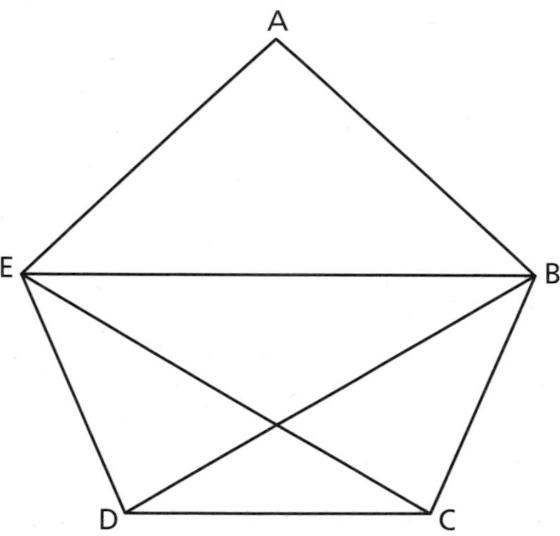

Measure the lengths of these line segments.

a AE **b** BE **c** BD

3 Draw a rectangle PQRS with sides PQ 6 cm long and QR 4 cm long.

Add the line segments PR and QS to your rectangle and find their lengths. What do you notice?

Will your result be true for any rectangle?

4

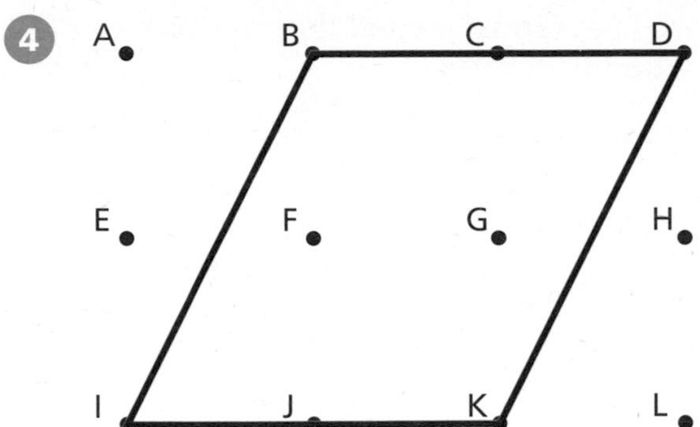

Show that BD ≠ BI

5 In the diagram, triangle DXY is shaded. Name six more triangles that can be found in the diagram.

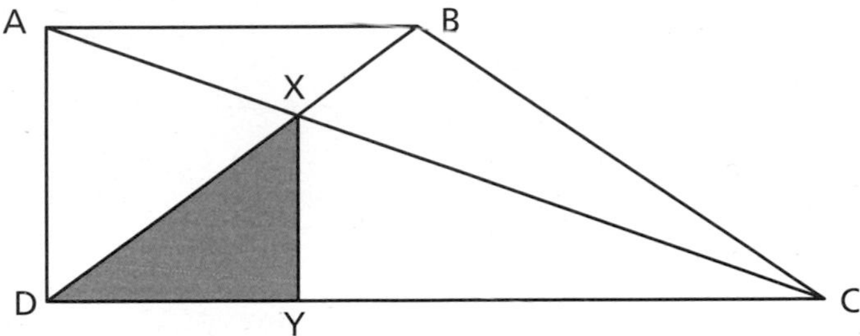

What do you think? 

1

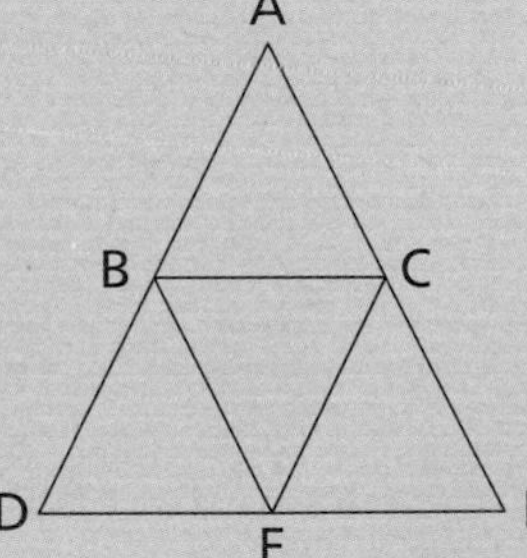

Beca thinks that the diagram shows four triangles, but Ed thinks he can see five.

Who do you agree with?

Name all the triangles in the diagram.

2 The diagram shows rectangles ABFE and CDGH joined by four line segments.

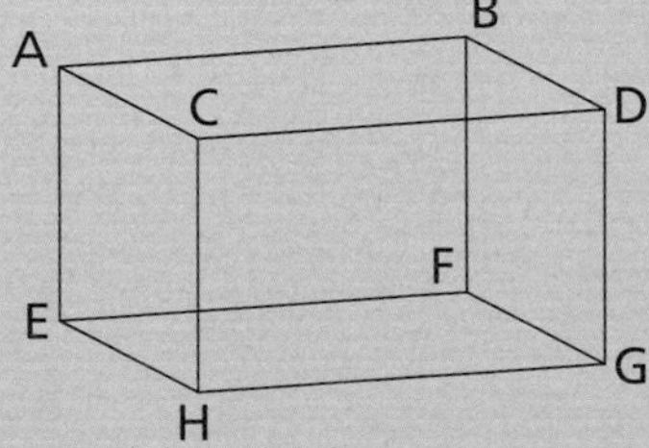

a Name the line segments that join the two rectangles.

b Name all the quadrilaterals, pentagons and hexagons that you can see on the diagram. Discuss your answers with your class.

Angles measure the amount of turn between two lines (arms) around their vertex.

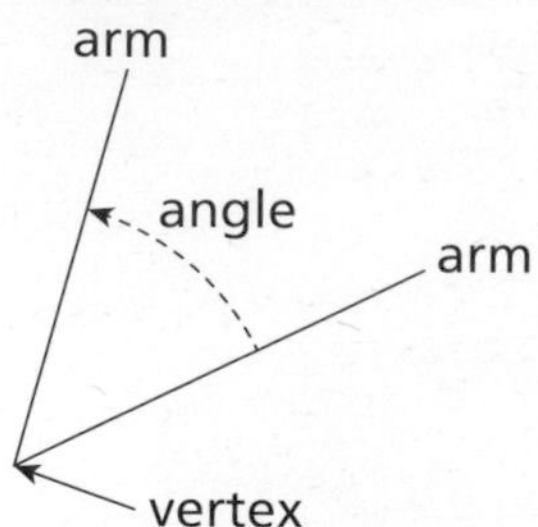

A full turn is split into 360 degrees or 360°, where the symbol ° stands for degrees.

You can name an angle using three letters. The vertex of the angle is the middle letter.

Example 2

Name the angle shaded in each diagram.

a 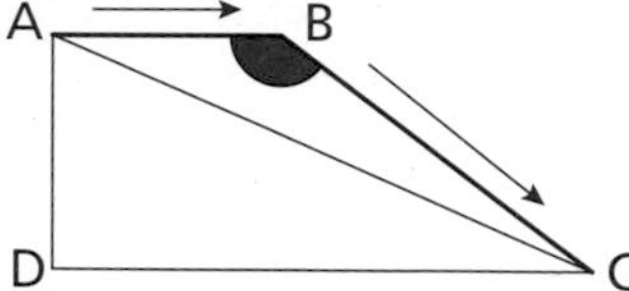

b 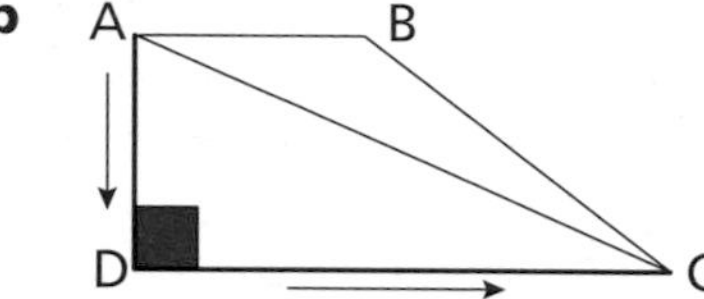

a Angle ABC — The angle is formed by line segments AB and BC. The vertex is B.

"Angle ABC" can be written as ∠ABC or $\widehat{ABC}$ for short.

You could also call this angle CBA but you usually begin with the letter that appears first in the alphabet.

b Angle ADC — The angle is formed by line segments AD and DC. The vertex is D.

There are three angles at vertex C. How could you label them?

Example 3

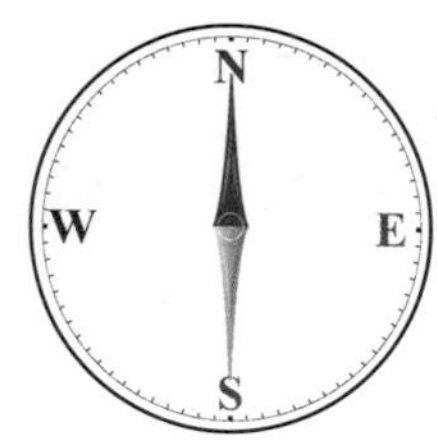

a In what direction will I face if I start facing east and turn $\frac{1}{2}$ turn **clockwise**?

b In what direction will I face if I start facing east and turn $\frac{3}{4}$ turn **anticlockwise**?

a West — The compass points are $\frac{1}{4}$ of a turn from each other. $\frac{1}{2}$ turn clockwise is the same as two quarters: from E to S and then S to W.

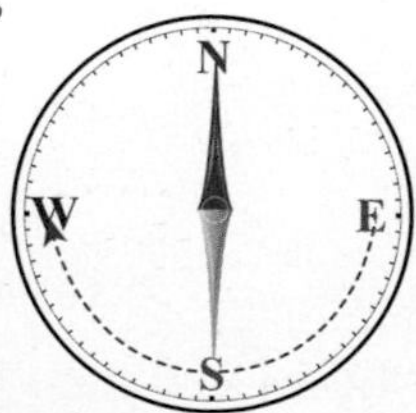

b South

$\frac{3}{4}$ turn anticlockwise goes from E to N, N to W and then W to S.

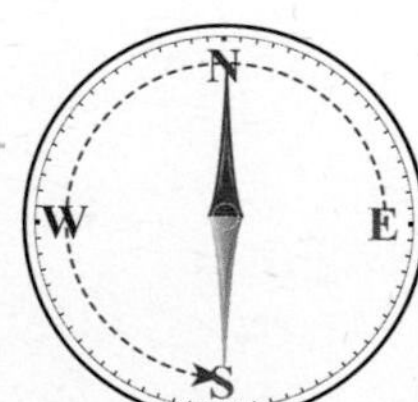

Practice 11.1B

1. Copy the diagram and colour in each of these angles. Label them a, b, c, d.

   a GFI **b** ABF **c** BEF **d** FEH

2. In the diagram for question 1, Jackson says angles BEF and BEG are the same angle. Do you agree? Why or why not?

3. Use three letters to describe each of the angles labelled a, b, c, d and e. There is more than one possibility for each angle.

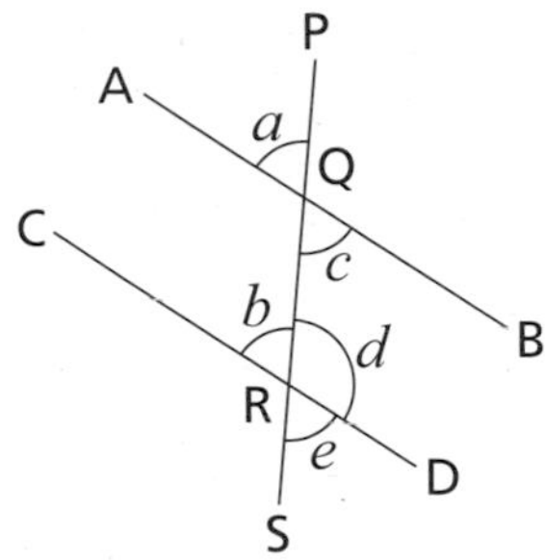

4. **a** Copy the diagram for question 3. Colour each of the angles AQR, CRS and SQB.

   b Beca thinks that the angle AQB does not exist. What do you think?

5. Find the angle, turning clockwise, between each of these pairs of compass points.

   a N and W **b** NE and E **c** SW and N **d** NW and W

6. In the diagram, find the sizes of

   a ∠AOB **b** ∠BOC **c** ∠AOC

   d ∠AOD **e** ∠BOD **f** ∠DOC

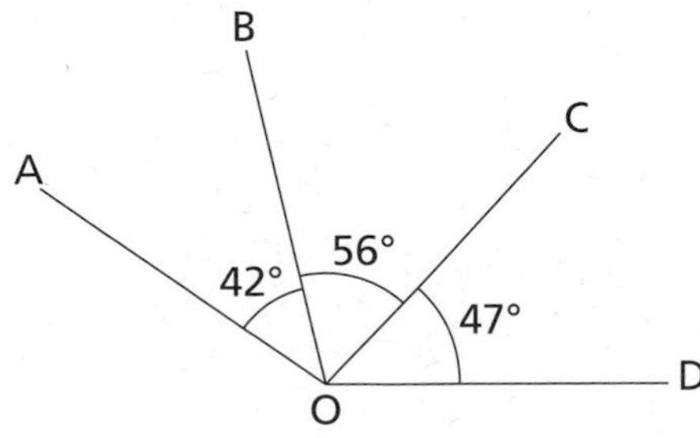

   Note – the diagram is not drawn accurately, so do not measure the angles.

7. **a** Draw a sketch that shows ∠ABC = 104° and ∠CBD = 55°

   b Write down the size of ∠ABD.

What do you think?

1 Is the angle XYZ the same as or different from the angle ZYX? Why or why not?

2 **a** Chloe says that if you turn $\frac{1}{2}$ turn, it does not matter if you go clockwise or anticlockwise.
Do you agree? Explain your answer.

b Which of these turns are equivalent?

i A $\frac{1}{4}$ turn clockwise and a $\frac{3}{4}$ turn anticlockwise.

ii A $\frac{3}{4}$ turn clockwise and a $1\frac{1}{4}$ turn anticlockwise.

3 **a** What fraction of a full turn is 60 degrees?

b Write a turn that would be equivalent to

i 60° clockwise

ii 240° anticlockwise

iii $\frac{1}{5}$ turn clockwise

iv 300° clockwise

v 47° anticlockwise

4 **a** Zach is facing north. He is told to turn 75° anticlockwise but he turns 75° clockwise instead. How many degrees does Zach need to turn to get to where he is meant to be if he turns

i anticlockwise

ii clockwise?

b Repeat part **a** with an angle of 120°.

c Repeat part **a** with an angle of x°.

5 Abdullah says that ∠PQR= 75° but Flo says that ∠PQR= 285°. Who is correct?

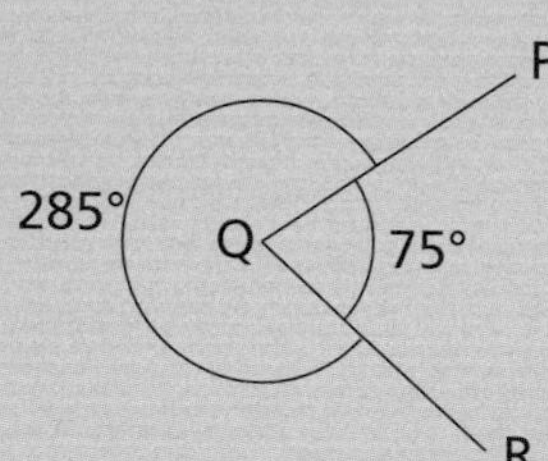

Consolidate – do you need more?

1 Draw a parallelogram. Label the vertices ABCD, as shown.

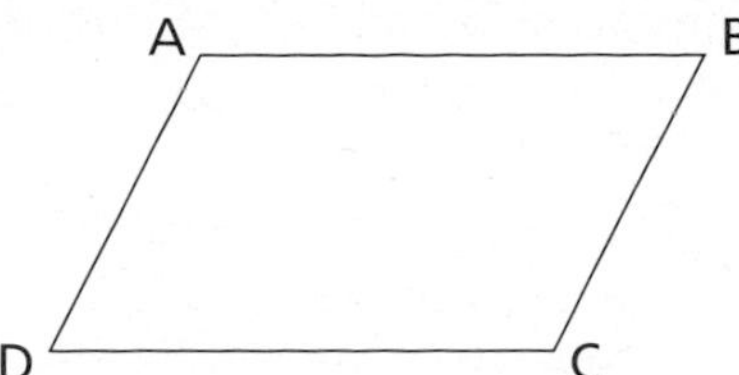

Add the line segment AC to your diagram.

Colour the angles ABC, CAB and ACD on your diagram.

Measure the length of AC. Swap your diagram with a partner. Check each other's measurement and angles.

2 Draw two line segments AB and XY that cross at a point O.

Colour the angles YOA and XOA.

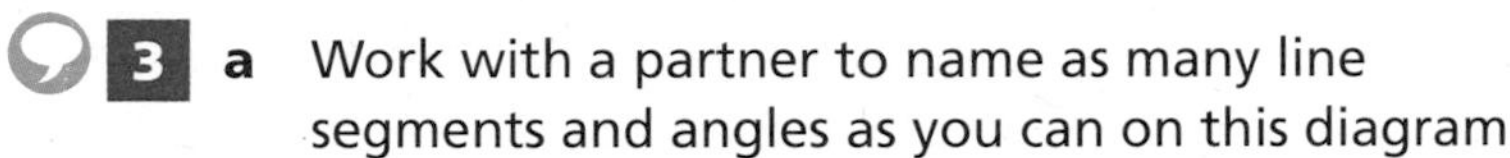

3 **a** Work with a partner to name as many line segments and angles as you can on this diagram.

b Why can you not label angle ZOL or angle XOW?

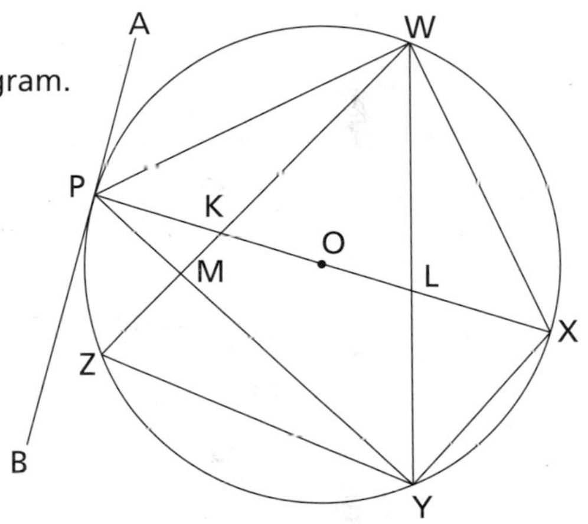

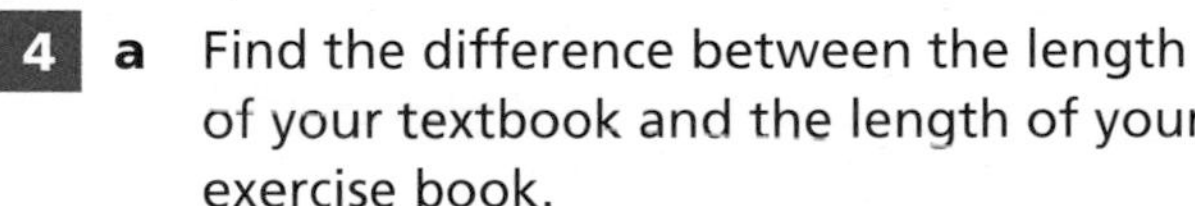

4 **a** Find the difference between the length of your textbook and the length of your exercise book.

b Find the difference between the width of your textbook and the width of your exercise book.

c Find the difference between the perimeter of your textbook and the perimeter of your exercise book.

Stretch – can you deepen your learning?

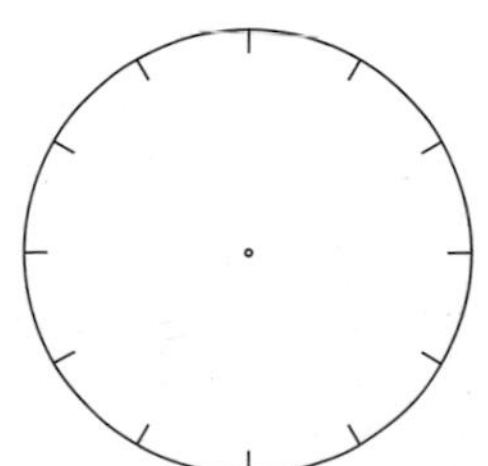

The diagram shows a clock face without numbers or hands. You can use this to help you to answer the questions.

1 Through what angle does the minute hand of a clock turn in

a 1 hour **b** $\frac{1}{2}$ hour **c** $\frac{2}{3}$ hour **d** 45 minutes **e** $\frac{5}{6}$ of an hour?

2 Through what angle do you need to turn to get from one hour on the clock to the next?

3 Through what angle does the hour hand of a clock turn in

a 1 hour **b** $\frac{1}{2}$ hour **c** $\frac{2}{3}$ hour **d** 45 minutes **e** $\frac{5}{6}$ of an hour?

4 The minute hand of a clock turns through 180° from pointing at 12 to pointing at 6. Find five other pairs of numbers that also involve a 180° turn for the minute hand. What do you notice about your answers?

5 **a** Benji thinks that the angle between the hands of a clock at 6:30 pm is 0°. Explain why Benji is wrong and find the correct angle.

b Find a time when the angle between the hands of the clock is

i 0°

ii 180°

iii 90°

c Investigate other angles and times.

Reflect

1 What is an angle?

2 Describe how to use three letter notation to name an angle.

3 What is the difference between a clockwise and an anticlockwise turn?

11.2 Working with angles

Small steps

- Classify angles
- Measure angles up to 180°
- Draw angles up to 180°
- Draw and measure angles between 180° and 360°

Key words

Acute angle – an angle less than 90°

Right angle – an angle of exactly 90°

Obtuse angle – an angle more than 90° but less than 180°

Reflex angle – an angle more than 180° but less than 360°

Are you ready?

1 How many degrees are there in a full turn?

2 Emily says that an angle is the distance between two lines. Explain why Emily is wrong.

3 Give the three letter names of the angles labelled a, b and c.

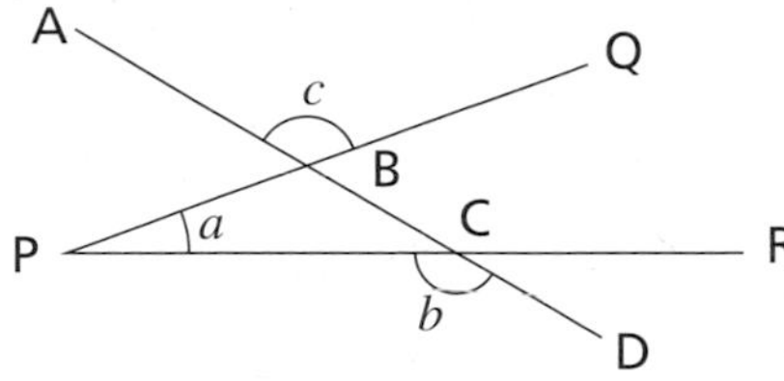

4 Find the angle, turning clockwise, between each of these pairs of compass points.

a E and S **b** E and W **c** W and S **d** N and SE

Models and representations

Geo strips

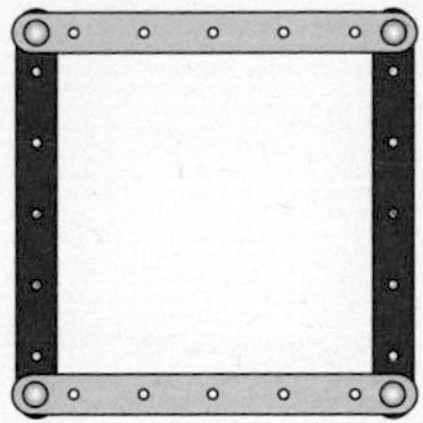

Card and paper fasteners

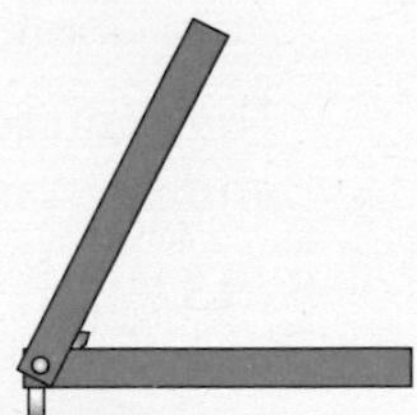

Pencils or rulers

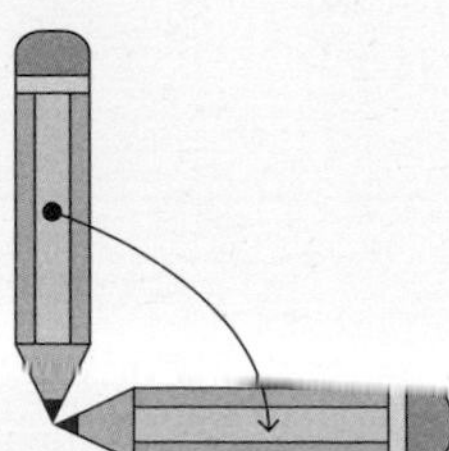

Parts of the body

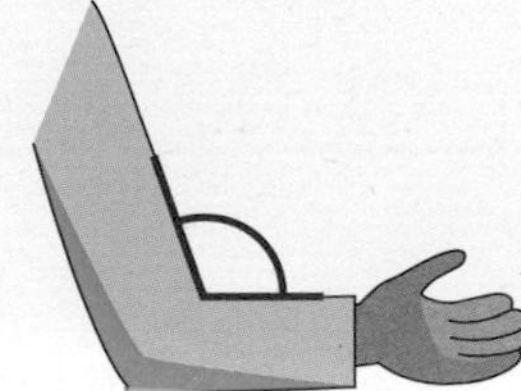

11.2 Working with angles

A full turn is 360 degrees. The symbol ° is used to stand for degrees.

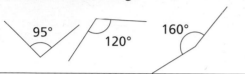

A quarter of a turn is called a **right angle**.
A right angle measures 360° ÷ 4 = 90°

90° 90° 90° 90°

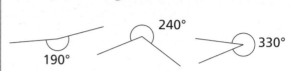

A turn less than 90° is called an **acute angle**

20° 50° 80°

A turn more than 90° but less than 180° is called an **obtuse angle**

95° 120° 160°

A turn more than 180° but less than 360° is called a **reflex angle**

190° 240° 330°

Knowing the type of angle helps you to draw and measure angles.

Example 1

For each angle

i state the type of angle

ii estimate the angle

iii measure the angle.

a **b**

a i The angle is obtuse

ii 135°

iii 132°

Drawing or imagining a dotted line can help you to see that the turn is more than a right angle.

The angle looks roughly half way between a right angle and a half turn.

The protractor shows 0° and 180° on one arm and 132° and 48° on the other arm.

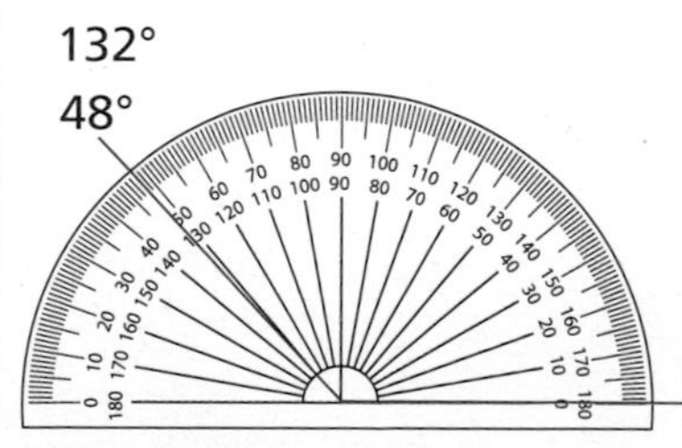

As the angle is obtuse you know it must be 132° not 48°

b i The angle is acute

ii 75°

iii 75°

Drawing or imagining a dotted line can help you to see that the turn is less than a right angle.

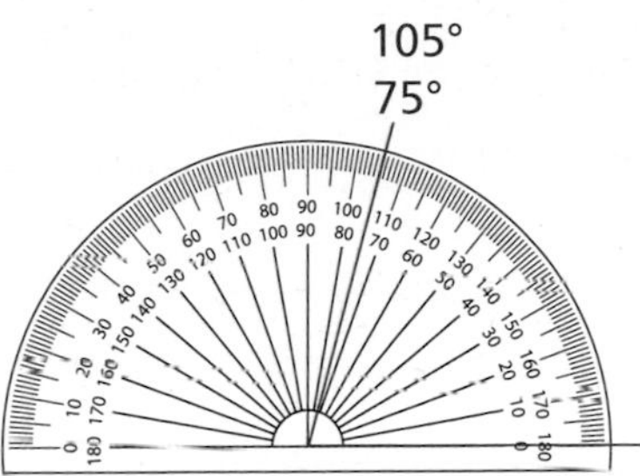

The angle looks fairly close to a right angle.

The protractor shows 0° and 180° on one arm and 105° and 75° on the other arm.

105°
75°

As the angle is acute you know it must be 75° not 105°

Example 2

Draw line segments AB and BC, both 6 cm long, so that angle ABC = 65°

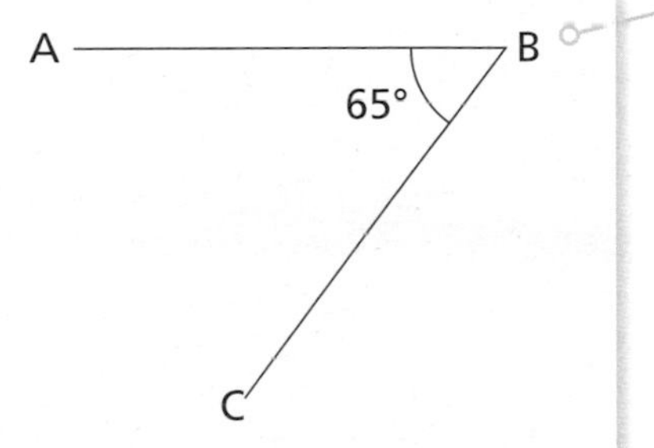
A
B
65°
C

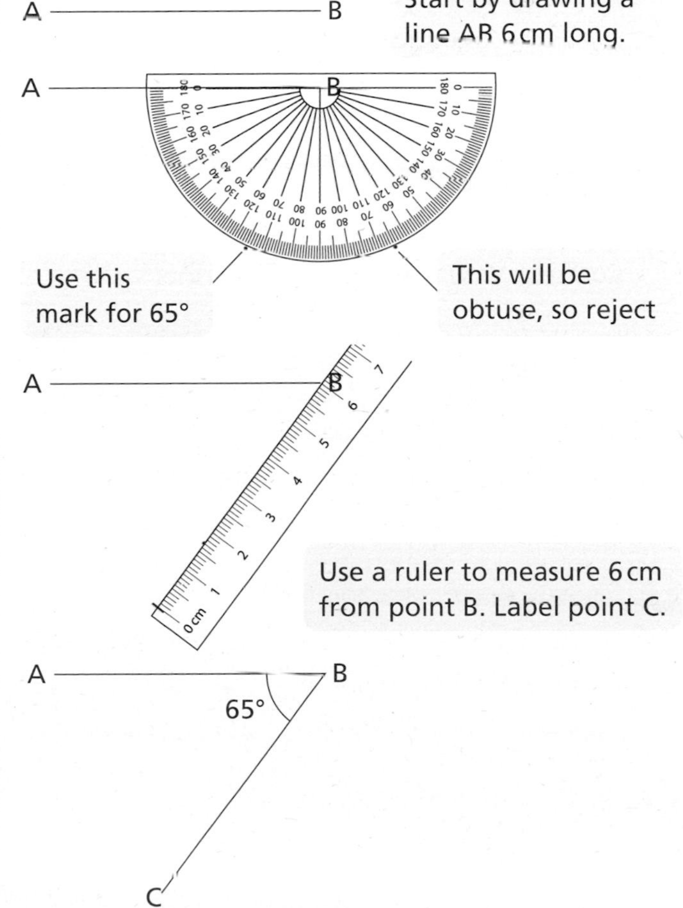
A
B
Start by drawing a line AB 6 cm long.
Use this mark for 65°
This will be obtuse, so reject
Use a ruler to measure 6 cm from point B. Label point C.
65°
C

Practice 11.2A

1. i Decide whether each angle is acute or obtuse.
   ii Estimate the size of each angle. iii Measure each angle using a protractor.

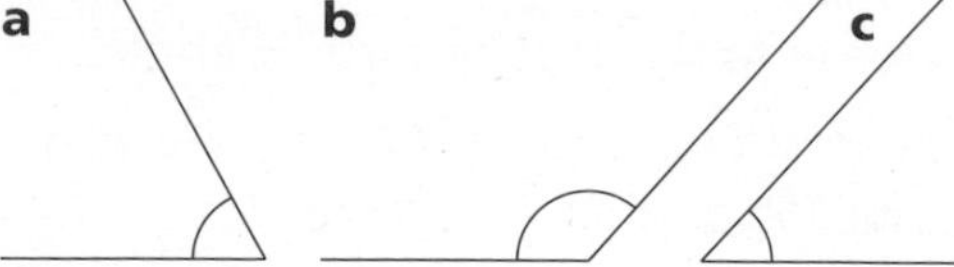

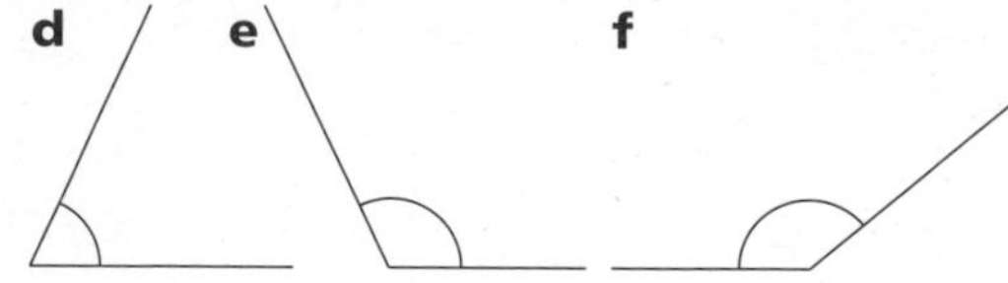

2. i Decide whether each angle is acute or obtuse.
   ii Use a ruler and protractor to draw each of these angles.
   a 110° b 80° c 70° d 147° e 111°

3. Draw line segments PQ and QR, both 6 cm long, so that angle PQR = 115°

4. Draw line segments AB and BC, both 6 cm long, so that angle ABC = 32°

5. a Draw any triangle and label it XYZ. Measure all three angles and all three sides. Write your answers using three letters to name the angles and two letters to name the sides.
   b What is the total of the three angles in your triangle?

   Compare your answers with a partner's. Check each other's measurements.

6. a Draw any quadrilateral and label it PQRS. Measure all four angles and all four sides. Write your answers using three letters to name the angles and two letters to name the sides.
   b What is the total of your four angles?

   Compare your answers with a partner's. Check each other's measurements.

What do you think?

1. Can you draw a triangle with
   a no acute angles b one acute angle c two acute angles d three acute angles?
   Draw an example of each possible type of triangle and measure the angles and sides.

2. Can you draw a quadrilateral with
   a no obtuse angles b one obtuse angle c two obtuse angles
   d three obtuse angles e four obtuse angles?
   Draw an example of each possible quadrilateral and measure the angles and sides.

3. Can you draw a quadrilateral with
   a no right angles b one right angle c two right angles
   d three right angles e four right angles?
   Draw an example of each possible quadrilateral and measure the angles and sides.

You can draw reflex angles using a 180° protractor.

Example 3

Draw an angle of 310°

Method A

Start with a line segment AB

A ——————— B

Half a turn is 180°, so you need
310° – 180° = 130° extra.

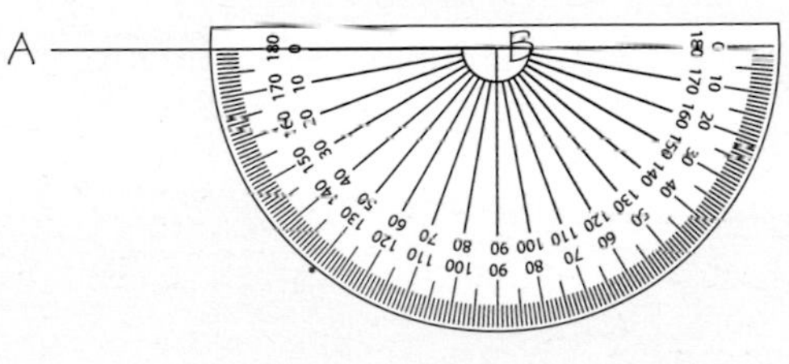

This is 130° after the half turn A to B.
Mark this point

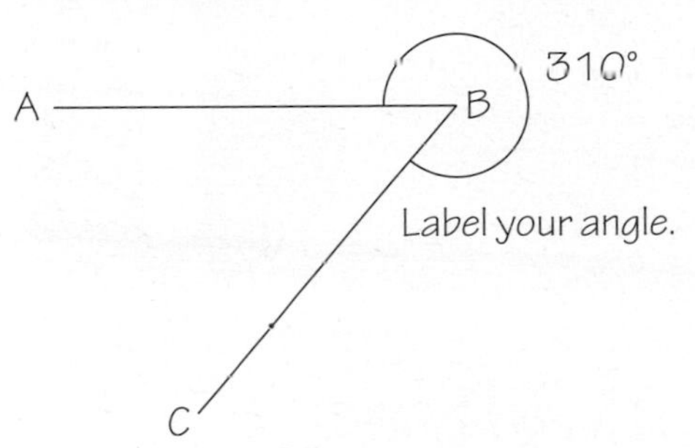

Draw a line from B through the point.

Method B

360° – 310° = 50°

Subtract the reflex angle from 360°.

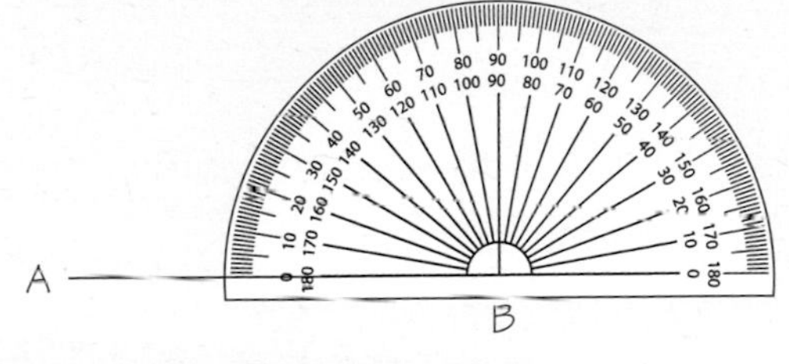

Draw an angle of 50°

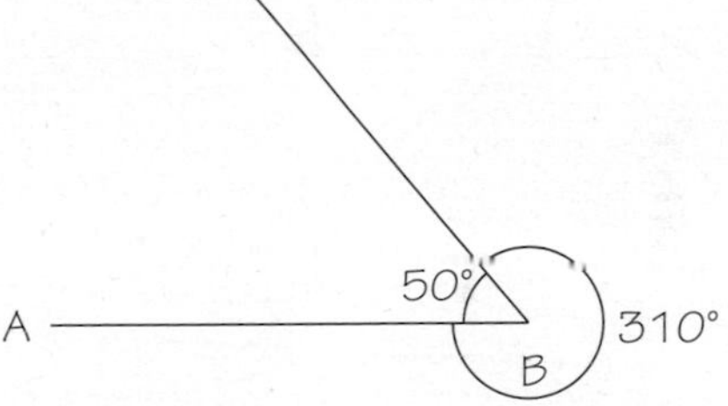

Label the angle outside the arms of the 50° angle.

Practice 11.2B

1. Look back at the methods used to draw a reflex angle in Example 3. How can you use a 180° protractor to measure a reflex angle?

2. Measure these reflex angles.

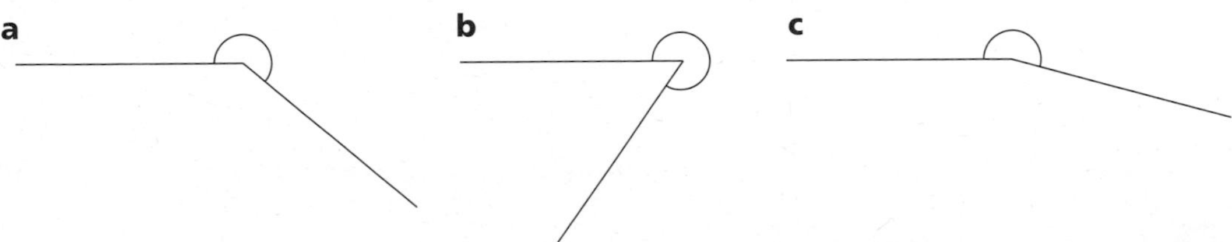

3. Use a ruler and protractor to draw each of these angles.

   a 300° **b** 205° **c** 185°

4 **a** Draw line segments AB and BC, both 6.5 cm long, so that angle ABC = 280°

b Draw line segments AB and AC both 6.5 cm long, so that angle BAC = 280°

5 Seb measures these angles and writes a = 82° and b = 284°

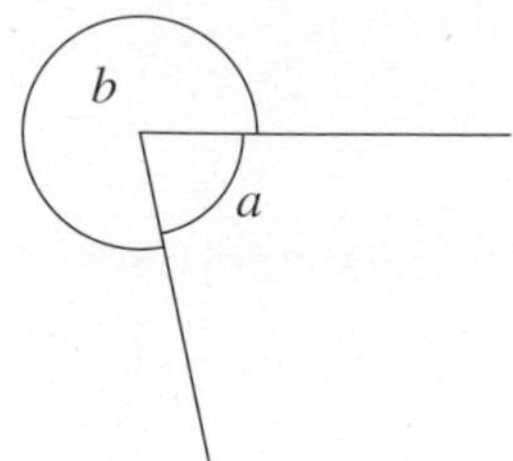

a Explain how you know without measuring that Seb must have measured incorrectly.

b Measure angles a and b.

What do you think?

1 What types of angles can you see on these road signs?

2 Draw a quadrilateral with a reflex angle and three acute angles.

Consolidate – do you need more?

1 **i** Decide whether each angle is acute, obtuse or reflex.

ii Estimate the size of each angle.

iii Measure each angle using a protractor.

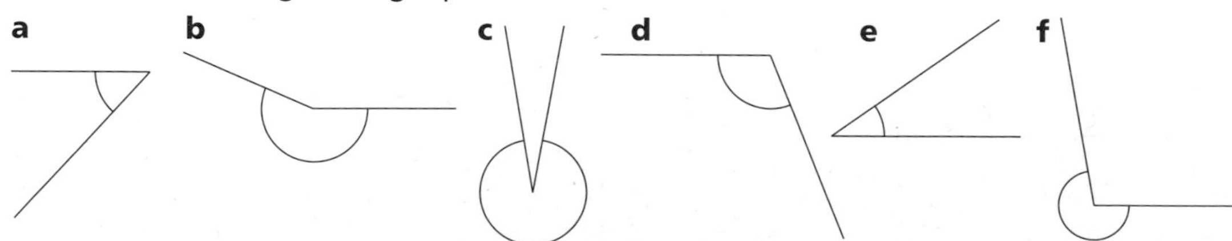

2 **i** Decide whether each angle is acute, obtuse or reflex.

ii Use a ruler and protractor to draw each of these angles.

a 85° **b** 95° **c** 195° **d** 155° **e** 205°

3 **a** Draw line segments PQ and QR, both 6 cm long, so that angle PQR = 115°

b Draw line segments PQ and QR, both 6 cm long, so that angle PQR = 35°

c Draw line segments PQ and QR, both 6 cm long, so that angle PQR = 305°

Stretch – can you deepen your learning?

1 Decide whether these statements are always true, sometimes true or never true. Give examples to justify your answers.

a The sum of two acute angles is an obtuse angle.

b The sum of a right angle and an acute angle is an obtuse angle.

c Two obtuse angles and an acute angle together make a full turn.

2 Investigate the maximum and minimum sums of three acute, three obtuse and three reflex angles. Show your answers using diagrams. What other angle facts like this can you investigate?

Reflect

1 Draw an example of

a an acute angle **b** a right angle **c** an obtuse angle **d** a reflex angle.

Explain how to identify each type of angle.

2 Explain how to draw and measure each type of angle using a protractor.

11.3 Lines and shapes

Small steps

- Identify perpendicular and parallel lines
- Recognise types of triangle
- Recognise types of quadrilateral
- Identify polygons up to a decagon

Key words

Polygon – a closed shape with straight sides

Regular polygon – a polygon whose sides are all equal in length and whose angles are equal in size

Parallel – always the same distance apart and never meeting

Perpendicular – at right angles to

Are you ready?

1 Measure the sides of this rectangle.

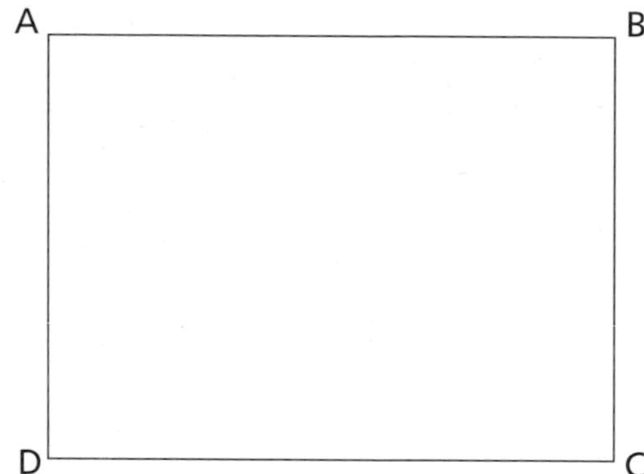

2 Measure each of the angles in this triangle.

3 Name the types of angles you can see in each shape.

a

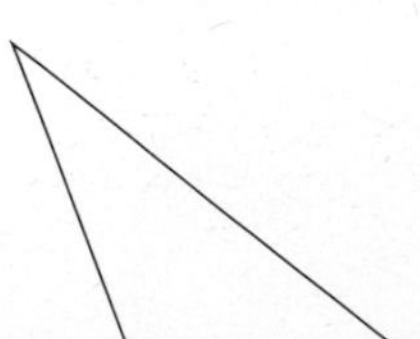

b

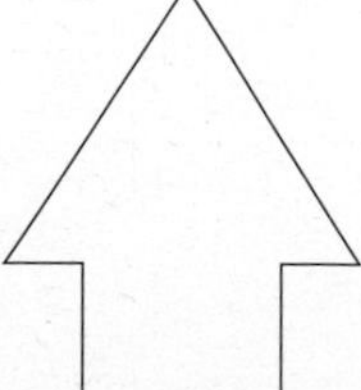

c

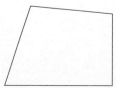

Models and representations

Geo strips

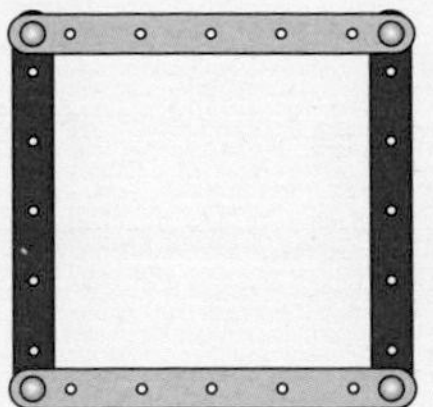

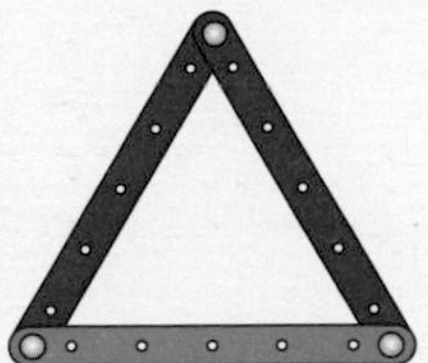

Card and paper fasteners

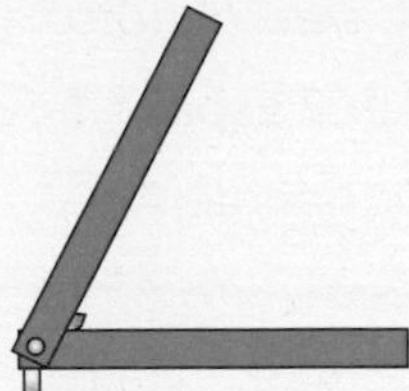

Rods and geoboards (or geoboard apps)

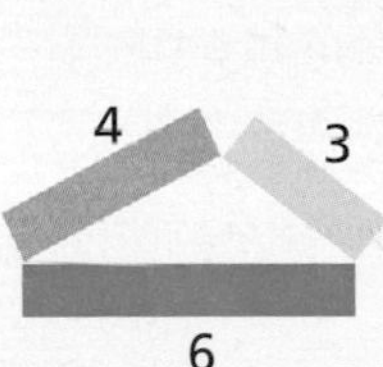

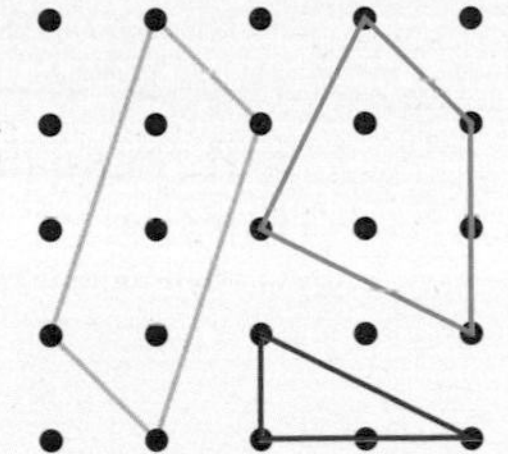

Set squares

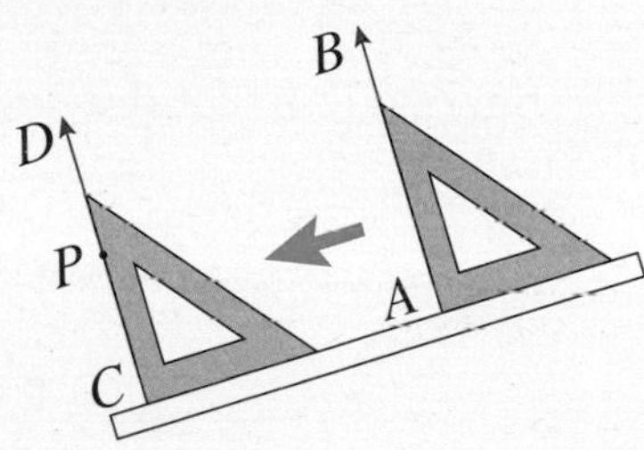

Parallel lines never meet. They are always the same distance apart. Arrows are used to show that lines are parallel.

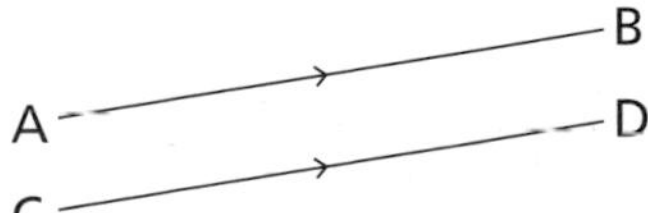

These arrows show that line segments AB and CD are parallel.

Two lines that meet at right angles are called **perpendicular**. This is shown using the right angle symbol.

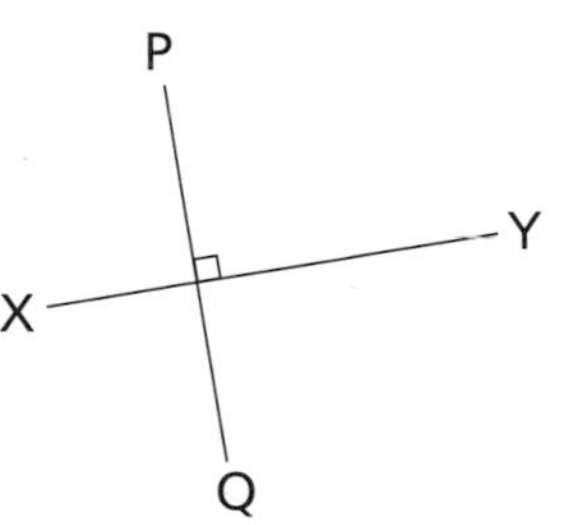

PQ is perpendicular to XY.

In this kite, AB and DA both have one hatch mark, so you know that AB = DA.

BC and CD both have two hatch marks, so you know that BC = CD, but these are different lengths to AB and DA.

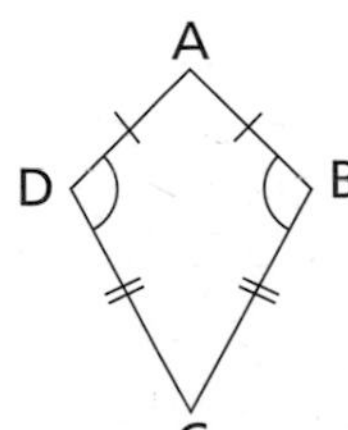

Hatch marks are used to show lines that are equal to each other.

Arcs are used to show angles that are equal to each other.

The arcs show that $\angle ADC = \angle ABC$.

You can use hatch marks to identify types of triangle.

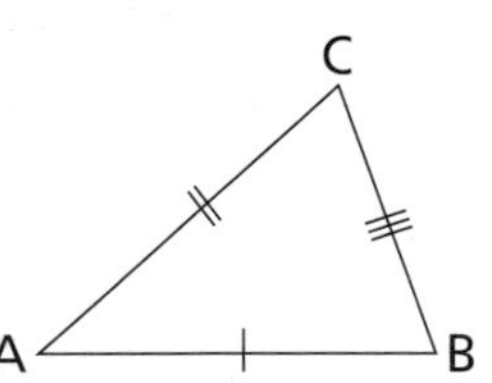

The three sides of this triangle have different lengths.

This is a **scalene** triangle.

The three angles will be different sizes too.

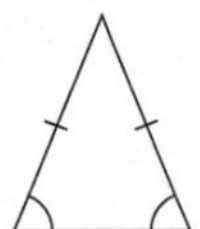

An **isosceles** triangle has one pair of equal sides and one pair of equal angles.

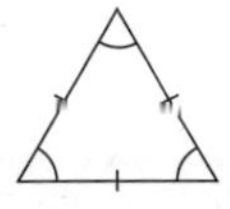

In an **equilateral** triangle, all three sides and all three angles are equal.

Example 1

a List all the pairs of parallel line segments in the diagram.

b Name a line segment that is perpendicular to BD.

c What type of triangle is ABC? Explain how you know.

d Name any equal angles in triangle ABC.

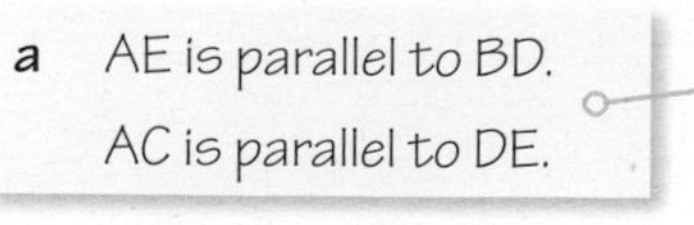

a AE is parallel to BD.
AC is parallel to DE.

The single arrows show that AE is parallel to BD.

The double arrows show that AC is parallel to DE, but these lines are not parallel to AE and BD.

b AB.

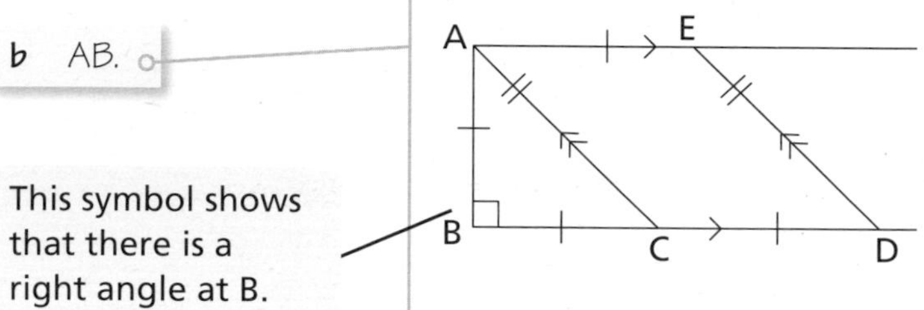

This symbol shows that there is a right angle at B.

c ABC is an isosceles triangle because AB and BC are equal, but AC is different.

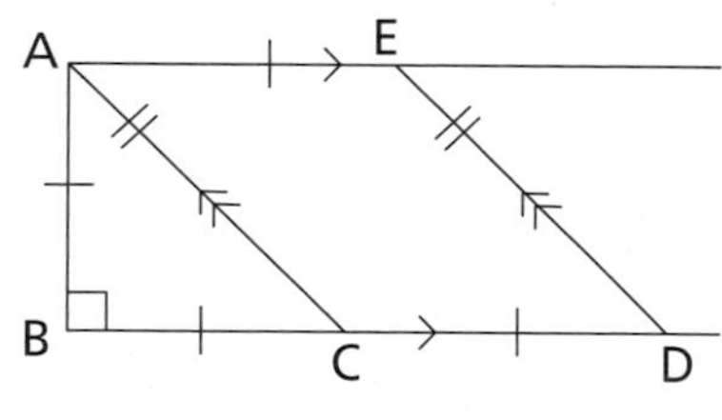

The double hatch mark shows that AC is different.

The single hatch marks show that AB = BC.

d BAC and ACB.

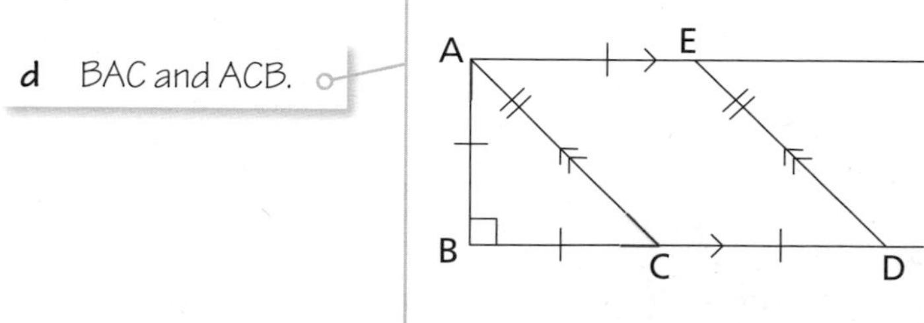

The equal angles in an isosceles triangle are found where the equal sides meet the third side (the base of the triangle).

Practice 11.3A

1 The diagram is drawn on a centimetre square grid.

Look at the diagram and decide whether the statements are true or false.

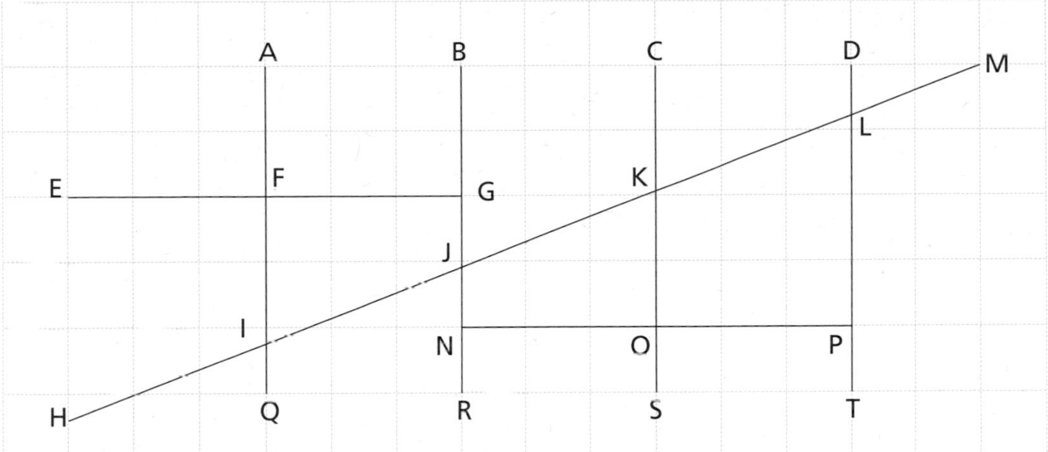

a AQ is parallel to BR.

b BR is perpendicular to HM.

c NP is perpendicular to CS.

d BR is parallel to DT.

e Copy the diagram and add arrows and right angle symbols to show parallel and perpendicular lines.

2 Huda says that AB is not parallel to PQ because XY is in the way.

Do you agree?

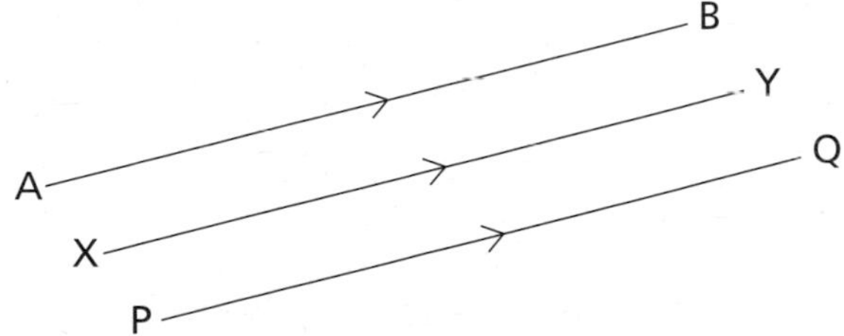

3 Copy and complete these statements.

a BD is parallel to ___

b ABC is an ________________ triangle

c ACD is a ________________ triangle

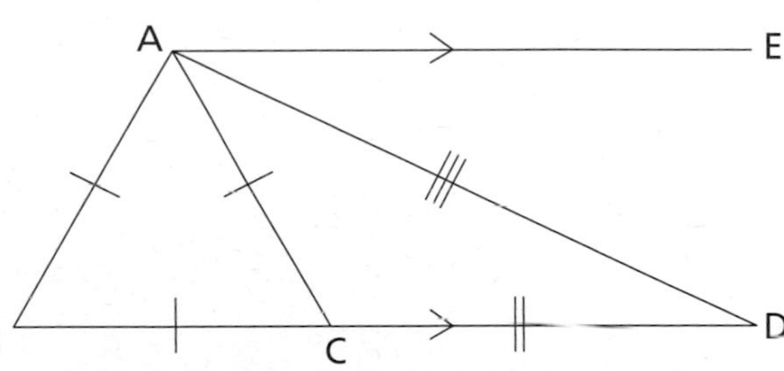

 **4**

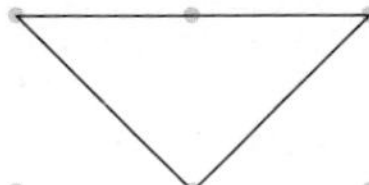

This shows a triangle on a 3 by 3 grid. The triangle is both isosceles and right-angled. What other types of triangle can you make on a 3 by 3 grid? Are there any types you cannot make?

5 Draw the triangles given by these descriptions.

a ABC, isosceles with AB = BC.

b ABC, isosceles with AB = AC.

c XYZ, isosceles, angle ZXY = angle ZYX.

d PQR, scalene and not right-angled.

e PQR, scalene with a right angle at RPQ.

6 Which line segments are parallel and which are perpendicular?

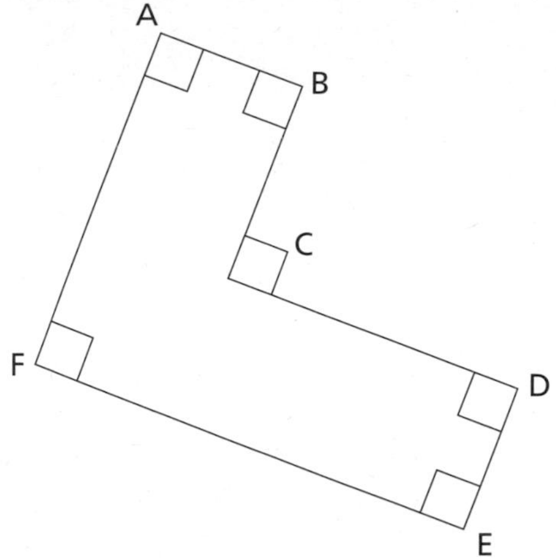

What do you think?

1 Are each of these statements true or false?

a If two sides of a triangle are equal in length, then two of the angles must also be equal in size.

b If two of the angles of a triangle are equal in size, then two of the sides must also be equal in length.

c If all three angles of a triangle are equal in size, then all three sides must also be equal in length.

d If all three sides of a triangle are equal in length, then all three angles must also be equal in size.

2 How many pairs of parallel and perpendicular lines can you see on your desk? How many can you see in the classroom?

3 If all three angles of a triangle are acute, the triangle is called an acute-angled triangle. Which of these triangles are possible?

A An acute-angled scalene triangle.

B An acute-angled isosceles triangle.

C An acute-angled equilateral triangle.

4 If any angle in a triangle is obtuse, the triangle is called an obtuse-angled triangle. Which of these triangles are possible?

A An obtuse-angled scalene triangle.

B An obtuse-angled isosceles triangle.

C An obtuse-angled equilateral triangle.

This table gives some important properties of some special quadrilaterals.

Name	Diagram	Some key properties
Trapezium		A quadrilateral with one pair of parallel sides. If the non-parallel sides are equal in length the trapezium is an isosceles trapezium.
Parallelogram		A quadrilateral with two pair of parallel sides and two pairs of equal angles. The opposite sides are also equal in length.
Kite		A quadrilateral with two pairs of adjacent sides equal in length. One pair of opposite angles is equal.
Rhombus		A quadrilateral with four equal sides and two pairs of equal angles. The opposite sides are also parallel.
Rectangle		A quadrilateral with four right angles and two pairs of opposite sides of equal length.
Square		A rectangle whose sides are all equal in length.

Example 2

What is the mathematical name for each of these quadrilaterals?

a

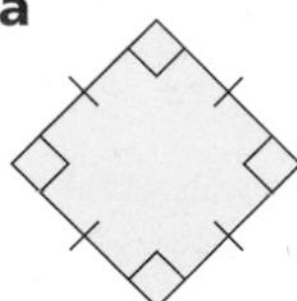

b

c

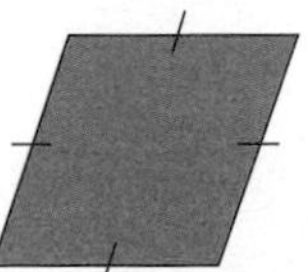

a Square.

You need to look at the marks on the shape to find out its properties.

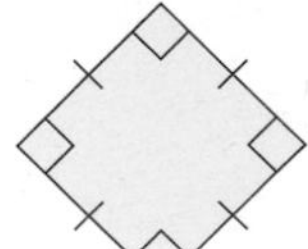

The shape has 4 right angles.

The shape has 4 equal sides.

Sometimes people do not recognise a square when it has been rotated like this one.

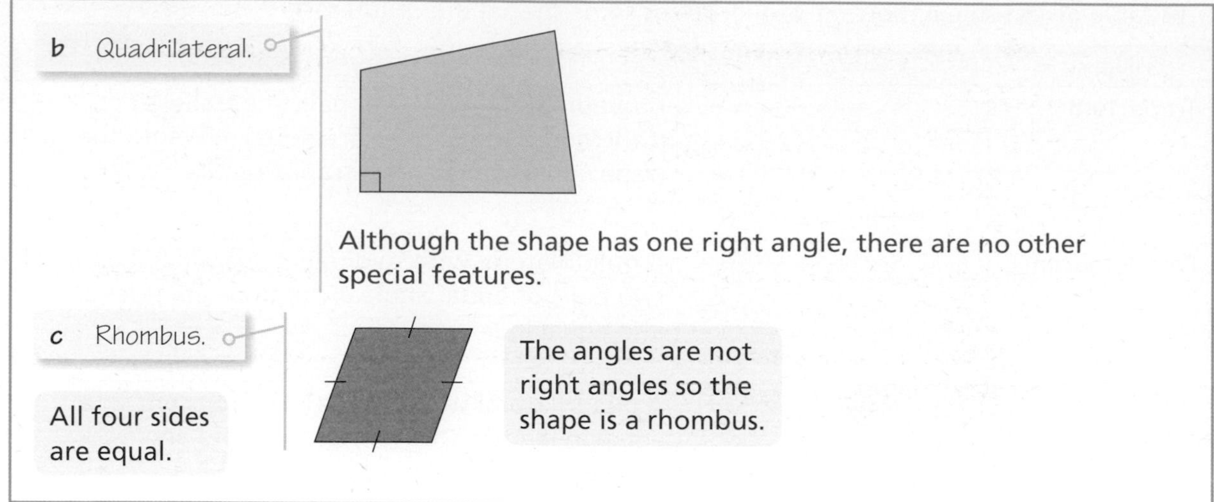

b Quadrilateral.
Although the shape has one right angle, there are no other special features.
c Rhombus.
All four sides are equal.
The angles are not right angles so the shape is a rhombus.

Practice 11.3B

1. Name these quadrilaterals. Discuss with a partner how you identify each one. Which of them have pairs of equal angles?

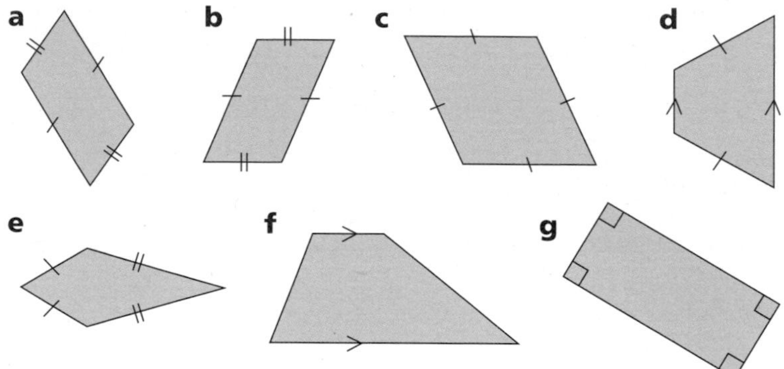

2. Copy and complete this Frayer model for a rectangle. You should give at least three more examples and non-examples.

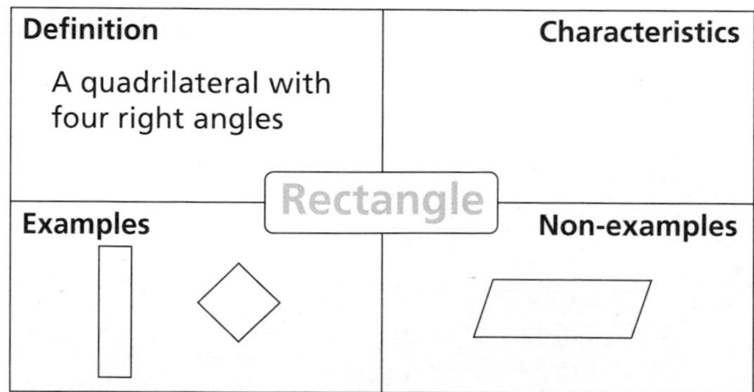

3 **a** What quadrilaterals can you make by putting two identical isosceles right-angled triangles next to each other edge-to-edge without overlapping? Sketch your answers and give their names.

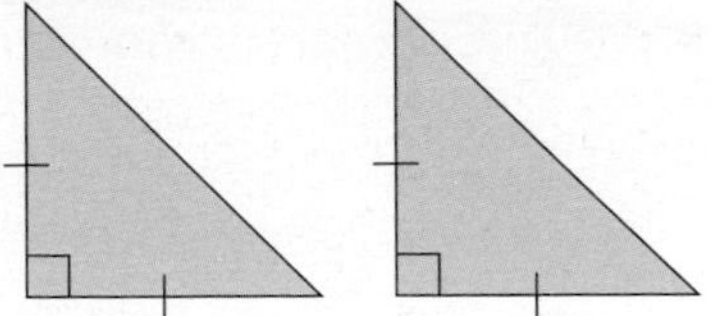

b What if the triangles were right-angled but not isosceles?

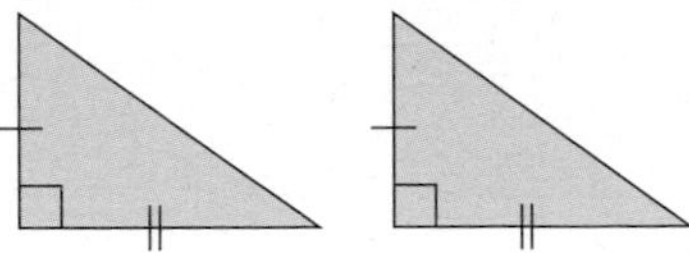

c What if the triangles were isosceles but not right-angled?

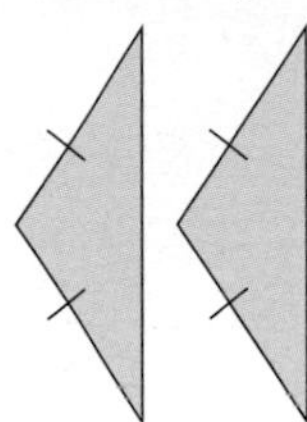

4 **a** Sketch a trapezium, ABCD, with AB parallel to CD.

b Sketch a trapezium, ABCD, with ABC acute.

c Sketch a kite, PQRS, with ∠PQR and ∠PSR both obtuse.

5 Which type or types of quadrilaterals could fit these descriptions?

a A quadrilateral with two pairs of different equal sides.

b A quadrilateral with two right angles.

c A quadrilateral with one right angle.

d A quadrilateral with two pairs of parallel sides.

6 Copy each diagram onto square dotty paper. Join dots on each grid to complete the shapes.

Square

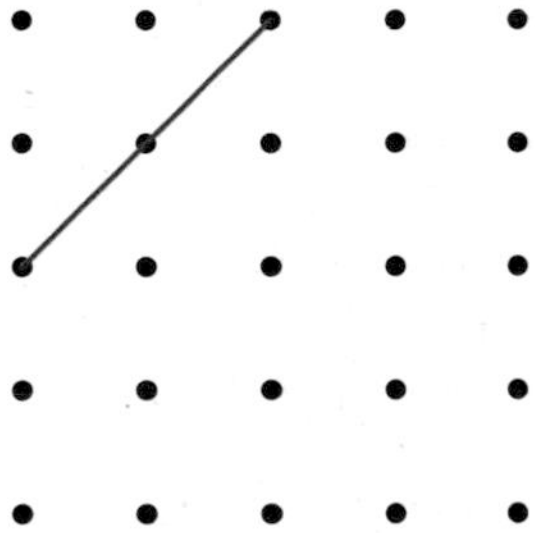

Trapezium

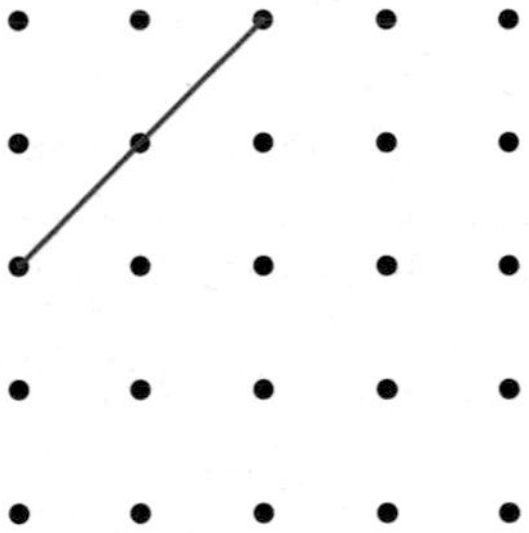

Rectangle

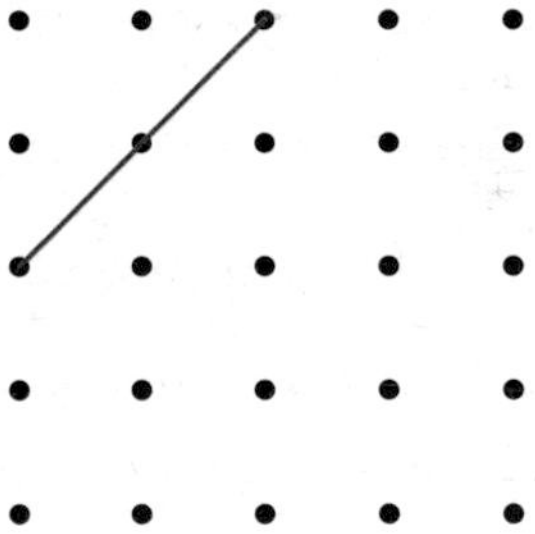

Rhombus

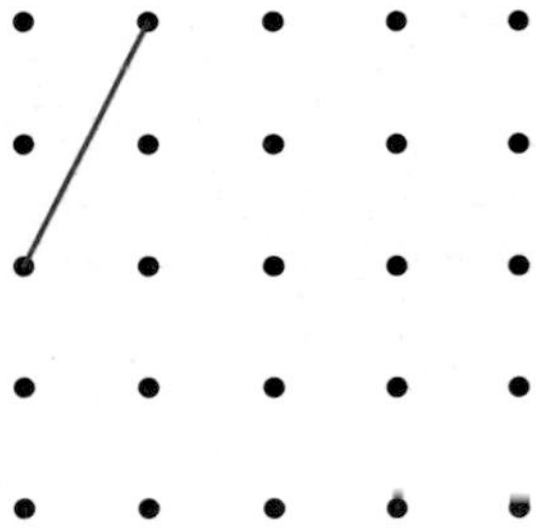

Kite

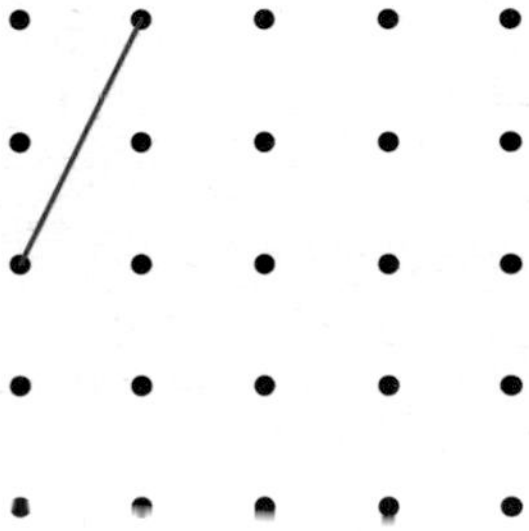

Parallelogram

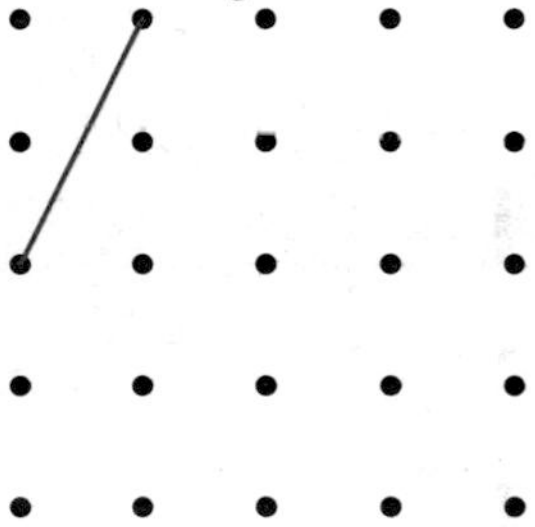

In how many different ways can you complete each shape?

7 Compare a rhombus and a parallelogram. What's the same and what's different?

What do you think?

1 The diagram shows a delta.

Describe the similarities and differences between this shape and a kite.

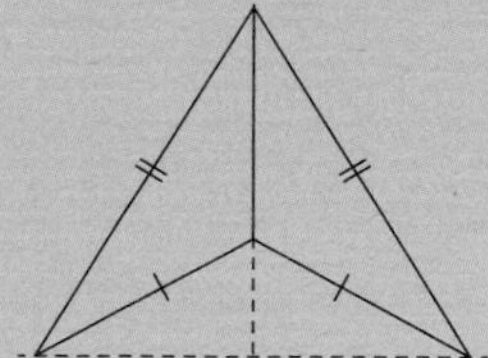

The shape is also sometimes known as an arrowhead.

2 Are each of these statements true or false?

a All squares are rectangles.

b All rectangles are squares.

c All rectangles are parallelograms.

d A square is a special type of kite.

3 Investigate how many different quadrilaterals you can make on a 3 by 3 geoboard.

How do you decide what you mean by 'different'?

A **polygon** is a closed shape with straight sides. Triangles and quadrilaterals are examples of polygons.

Number of sides	Name	Related words
3	Triangle	tripod, tricycle
4	Quadrilateral	quadrant, quadruplets
5	Pentagon	pentathlon, pentameter
6	Hexagon	hexapod
7	Heptagon	heptad
8	Octagon	octopus, octuplet
9	Nonagon	nonagenarian
10	Decagon	decimal, decathlon, decade

Many polygons have clues in their name which can help you to remember the number of sides.

In a **regular polygon** all its sides are equal and all its angles are equal. Otherwise it is **irregular**.

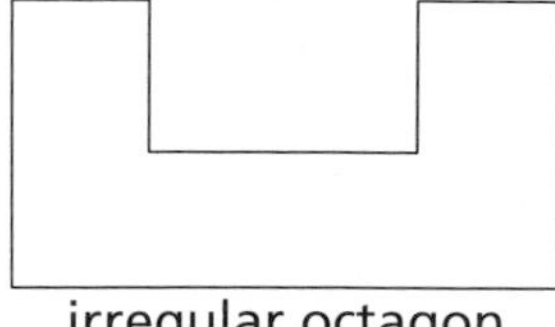

irregular octagon

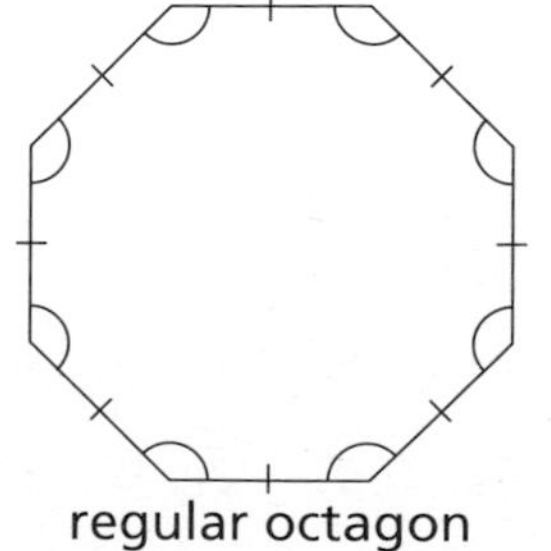

All the sides are equal
All the angles are equal

regular octagon

Example 3

Which of these shapes are polygons? Are any of them regular?

A **B** **C** **D**

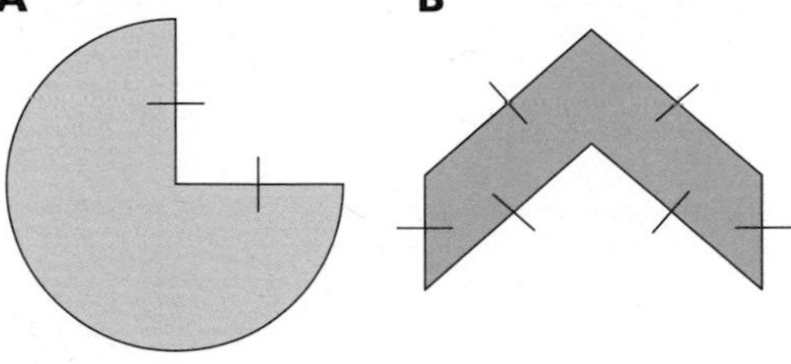

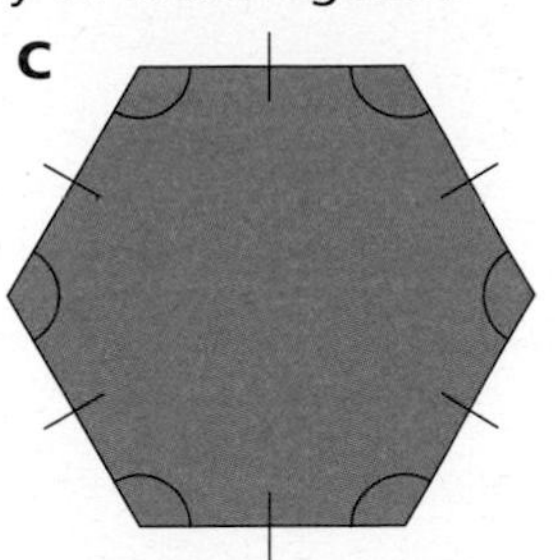

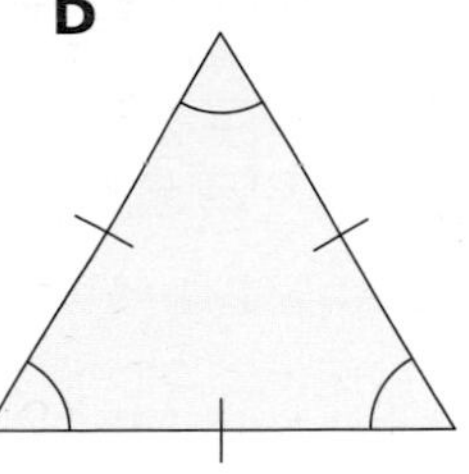

B, C and D are polygons

C and D are regular.

A is not a polygon because of the curved side.

B is not regular – although all the sides are equal, the angles are not.

Practice 11.3C

1 Which of the shapes are hexagons?

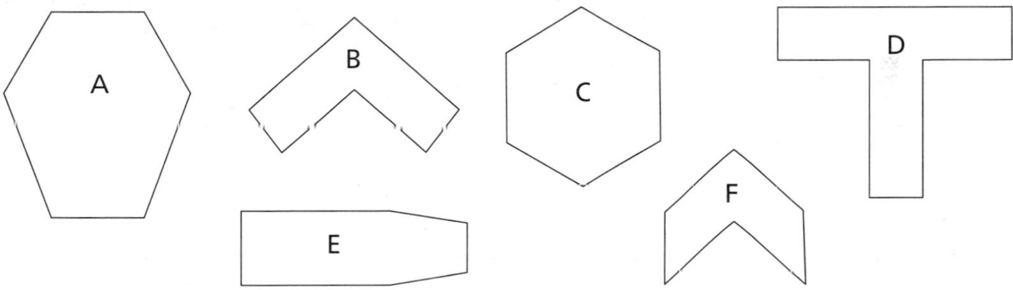

2 Draw a pentagon with

a no right angles **b** one right angle

c two right angles **d** three right angles.

 **3**

If you join two quadrilaterals, you get an octagon.

Do you agree with Flo? Why or why not?

4 **a**

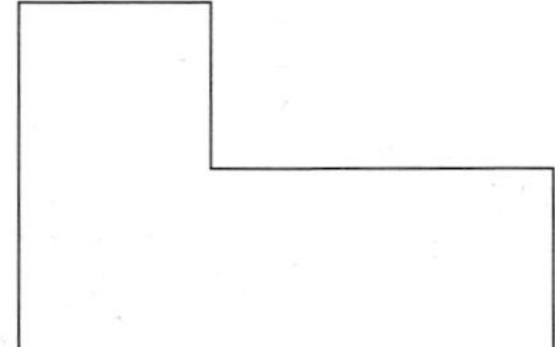

Draw a sketch to show how you can split this hexagon into

i 2 rectangles **ii** 2 trapezia.

b What other shapes can you make by joining pairs of quadrilaterals?

c What different shapes can you make by overlapping two squares?

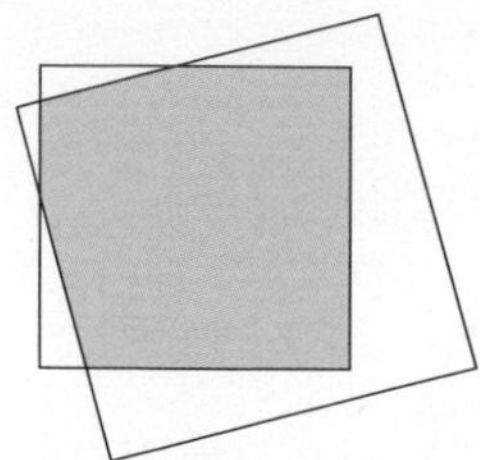

5 Which of the shapes are regular?

equilateral triangle	rectangle	rhombus	square
isosceles triangle	isosceles trapezium		

What do you think?

1 Are each of these statements always true, sometimes true or never true?

a A quadrilateral can be split into two triangles.

b A pentagon can be split into a quadrilateral and a triangle.

2 Draw a square and join its diagonals.

a What types of triangle are formed?

b If you remove one of triangles from the square, what shape is left?

c Investigate joining diagonals and removing triangles from other shapes.

Consolidate – do you need more?

1 The diagram is drawn on a centimetre square grid. Look at the diagram and decide whether each of the statements is true or false.

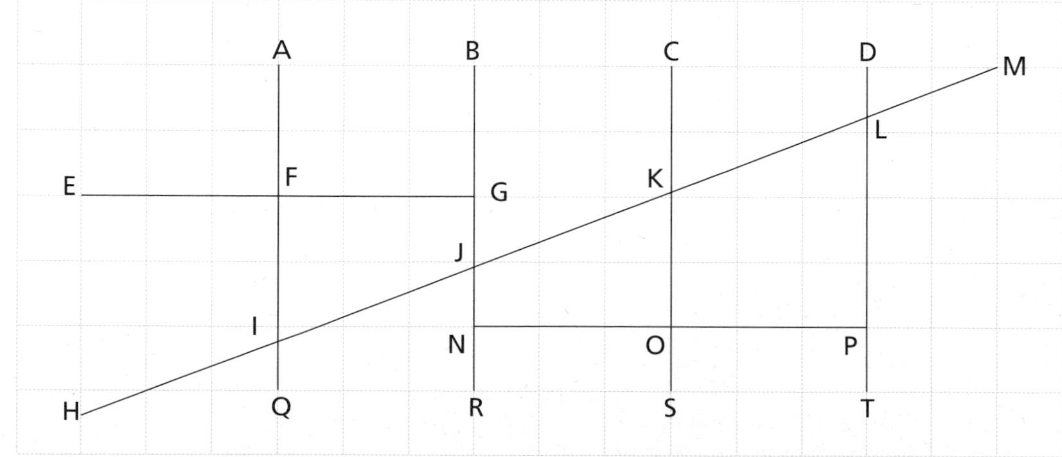

a DT is parallel to BR.

b BR is perpendicular to EG.

c Angle IJG is obtuse.

d EG is parallel to NP.

2 Copy the table and write the names of the special quadrilaterals in the most appropriate cell.

	One pair of equal angles	Two pairs of equal angles	All four angles equal
One pair of parallel sides			
Two pairs of parallel sides			
Opposite sides equal in length			
Adjacent sides equal in length			

3 Classify these triangles.

a 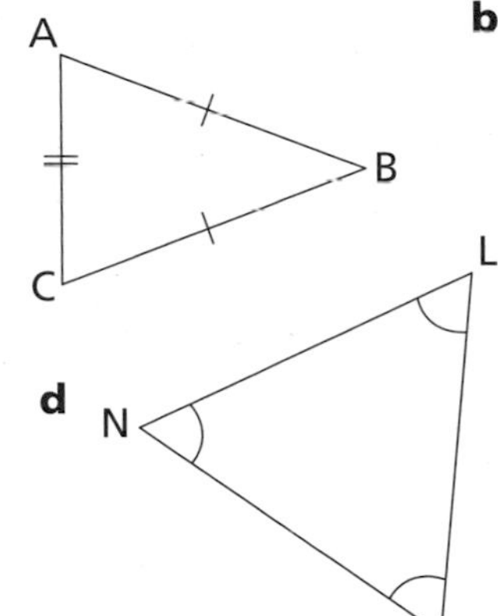

b 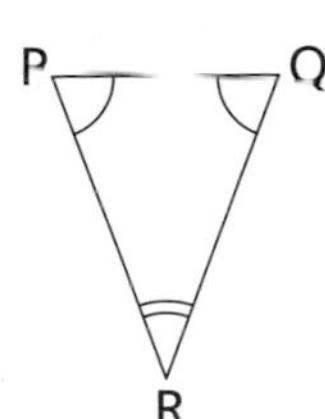

c 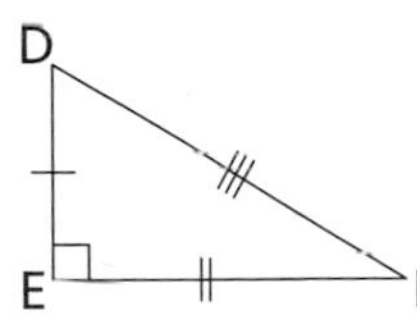

d

e 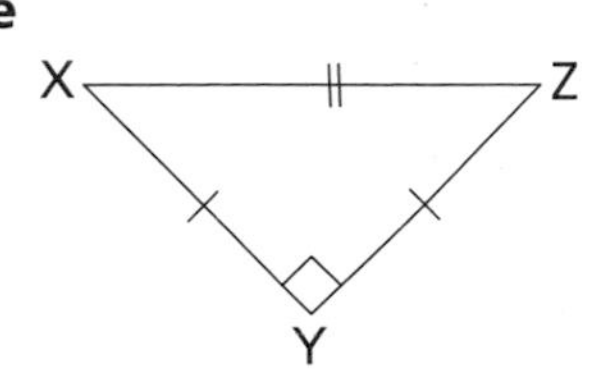

4 Draw each of the shapes described.

a An octagon with at least one right angle

b A hexagon with at least one acute angle

c An irregular nonagon

d A regular heptagon

e A regular quadrilateral

Stretch – can you deepen your learning?

1 

A triangle cannot have a pair of parallel sides.

Do you agree with Ed? Why or why not?

2 

The shortest side of a triangle is opposite the smallest angle.

Investigate Emily's claim.

3 'All quadrilaterals tessellate in the plane.' Find out what this statement means and investigate whether it is true. Use some different quadrilaterals to explain your answer.

4 A polygon is convex if all its angles are less than 180°. Otherwise a polygon is concave.

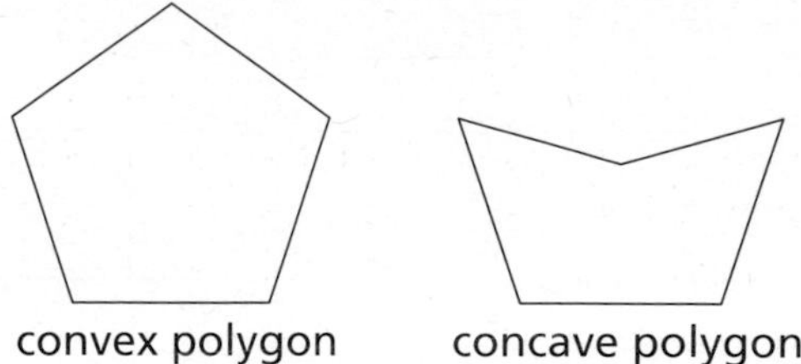

Investigate the maximum number of right angles that can be found in

a a convex polygon

b a concave polygon.

5 Investigate how to use a set square to draw

a parallel lines

b perpendicular lines

c angles of 30, 45 and 60 degrees.

Reflect

1 Explain the difference between parallel lines and perpendicular lines.

2 How would you describe the different types of triangles and quadrilaterals to someone who had never seen them before?

11.4 Constructing shapes

Small steps

- Construct triangles using SSS, SAS and ASA
- Construct more complex polygons

Key words

Construct – draw accurately using a ruler and compasses

Side – a line segment that joins two vertices in a 2-D shape

Are you ready?

1 What is the mathematical name of each of these triangles?

a

b

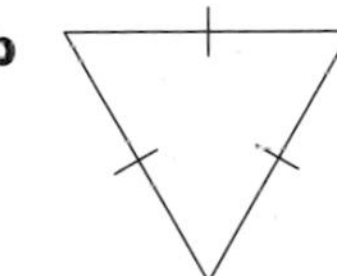

c

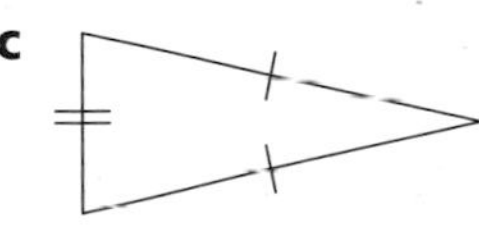

2 What is the mathematical name of each of these quadrilaterals?

a

b

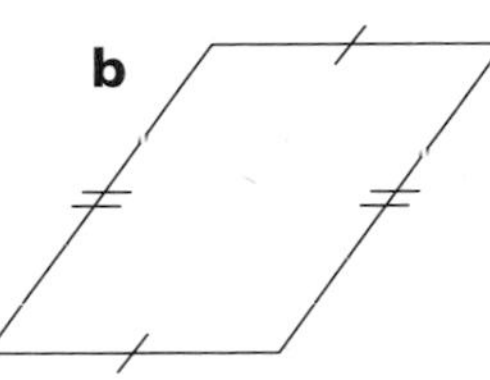

c

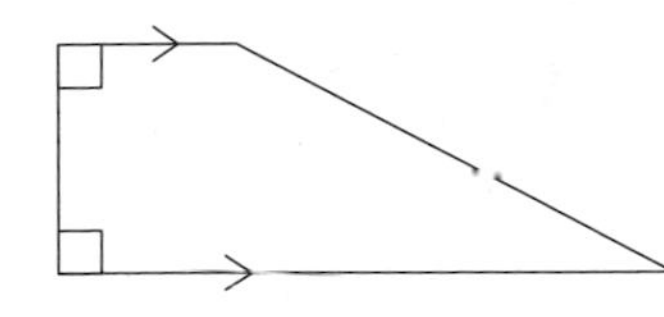

3 Use a protractor to draw each of these angles.

a 30° **b** 130° **c** 230°

4 **a** Name the shape.

b Measure its sides.

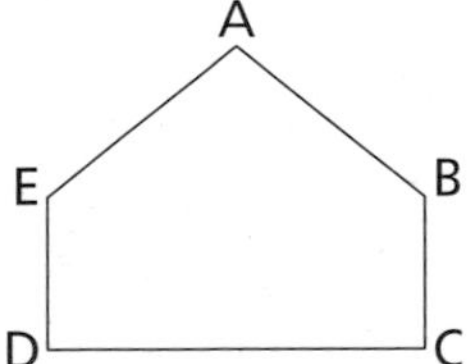

Models and representations

Dynamic geometry software

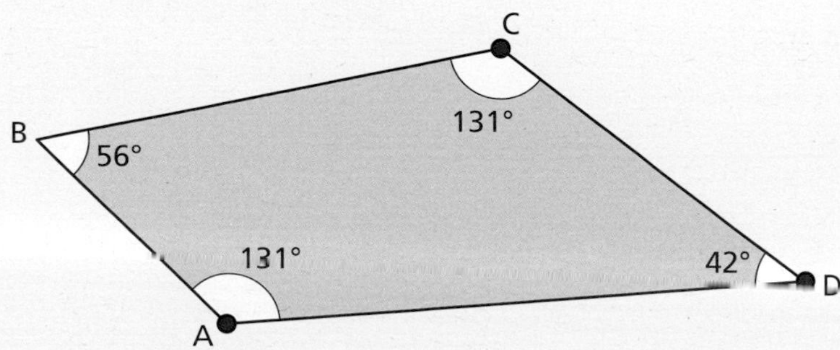

This is useful for checking constructed diagrams.

11.4 Constructing shapes

In this lesson, you will **construct** triangles and more complex shapes.

You will use your knowledge of polygons and how to draw angles and lines.

Example 1

Construct a triangle with **sides** 7 cm, 5 cm and 4 cm.

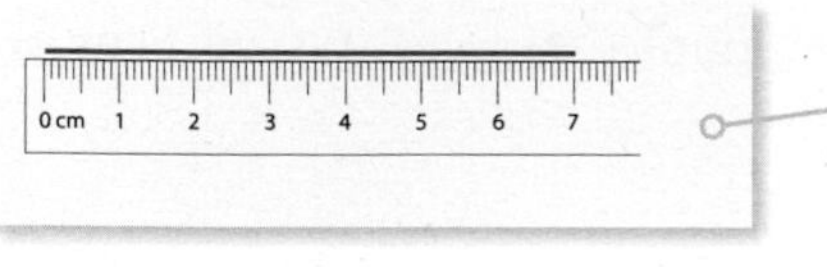

First draw a line segment 7 cm long.

Start by drawing the longest line.

It is a good idea to draw the triangle underneath this line so you don't need to work out how much room to leave above.

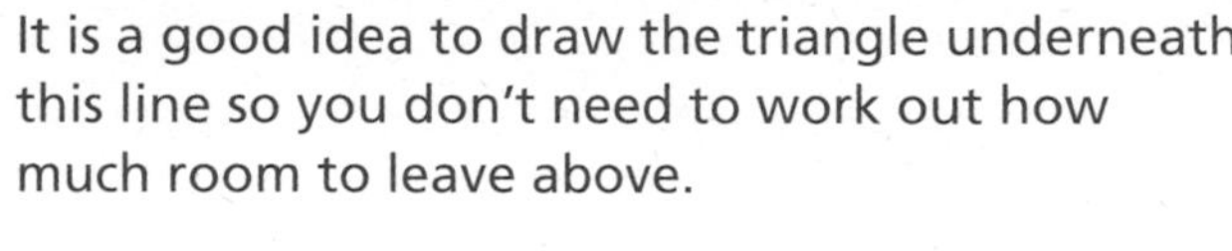

Set your compasses to 5 cm and draw an arc from one end of the line segment.

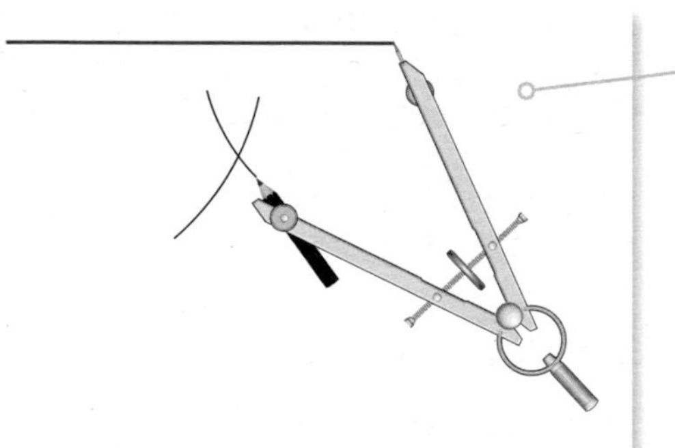

Set your compasses to 4 cm and draw an arc from the other end of the line segment.

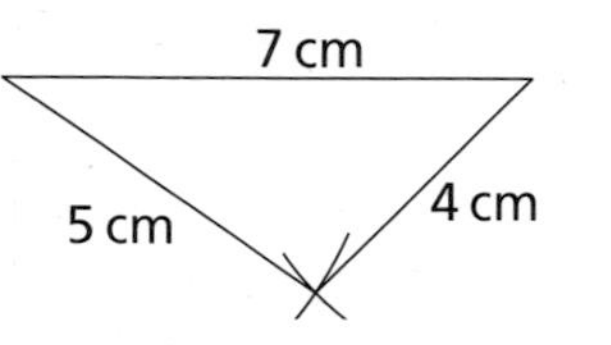

Do not rub out the construction arcs.

Join the point where the arcs meet to the ends of your first line to complete the triangle. Label the sides.

Example 2

Construct triangle ABC with AB = 5 cm, ∠ABC = 45° and BC = 4 cm

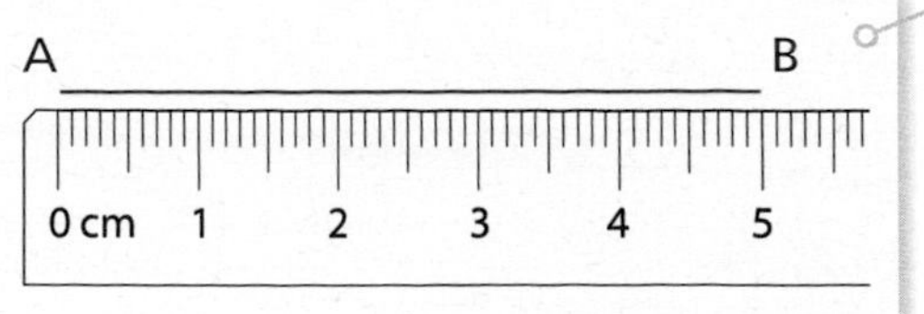

Start by drawing line segment AB 5 cm long.

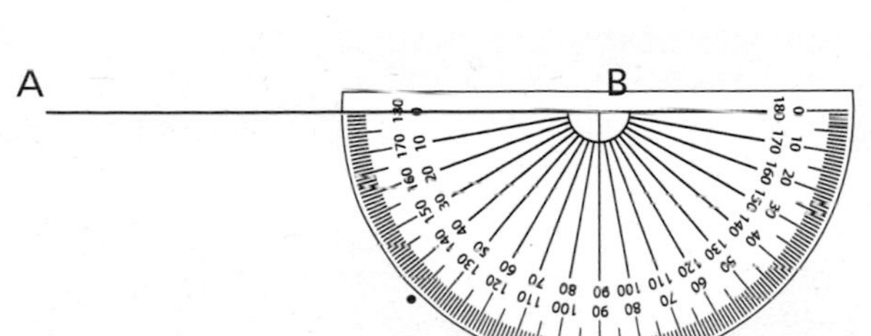

Mark a 45° angle from B.

45° is an acute angle so you need to mark this point.

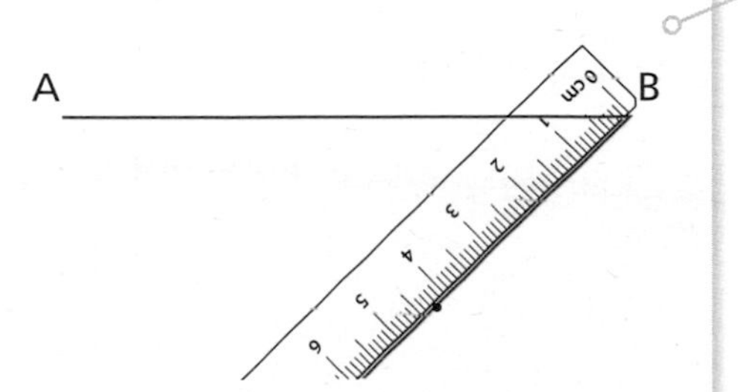

Line up your ruler with point B and this mark. Draw a line 4 cm long to find the position of C.

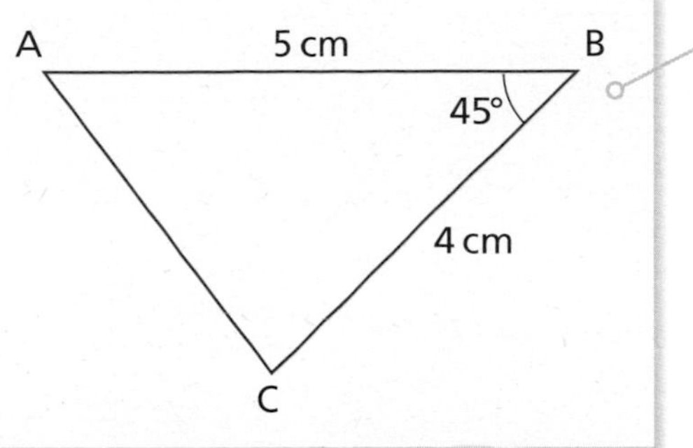

Draw line AC to complete the triangle. Label with the given information.

Example 1 shows how to construct a triangle given all three sides, which is written SSS.

Example 2 shows how to construct a triangle given two sides and the angle between them (called the 'included angle') which is written SAS.

You will look at a third way of constructing a triangle in question 3 of Practice 11.4A. This is called ASA. What do you think it means?

Practice 11.4A

1. Use a ruler and pair of compasses to construct these triangles.
   - **a** ABC with AB = 8 cm, AC = 5 cm and BC = 5 cm.
   - **b** PQR with PQ = 8 cm, PR = 5 cm and QR = 7 cm.
   - **c** Why can't you construct a triangle with sides 8 cm, 4 cm and 2 cm?

2. Use a ruler and protractor to construct these triangles.
   - **a** ABC with AB = 8 cm, ∠ABC = 65° and BC = 7 cm.
   - **b** PQR with PQ = 6 cm, ∠PQR = 105° and QR = 4 cm.

3. Follow these steps to construct triangle ABC with AB = 7 cm, ∠ABC = 55° and ∠BAC = 50°.
   - Draw line segment AB 8 cm long.
   - Mark a 55° angle from point B towards C underneath AB.
   - Draw a thin line from B through your mark.
   - Mark a 50° angle from point A towards C underneath AB.
   - Draw a thin line from A through your mark.
   - C is the point where the two lines meet.
   - Label point C. Draw lines AC and BC to complete the triangle.

4. **a** Construct an equilateral triangle with sides 6 cm long without using a protractor.
   b Construct an equilateral triangle with sides 6 cm long without using a pair of compasses.

5. Construct these triangles.

a

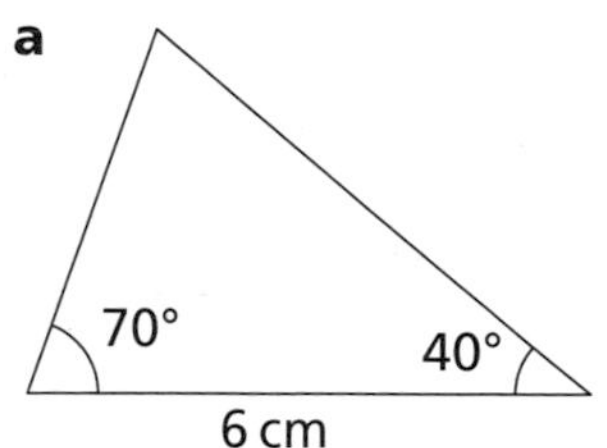

b

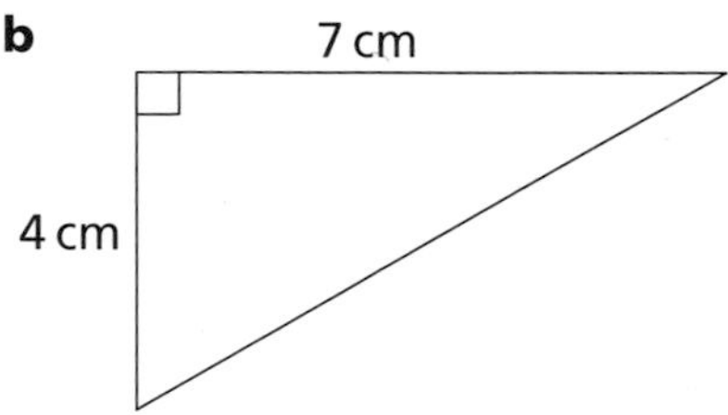

c

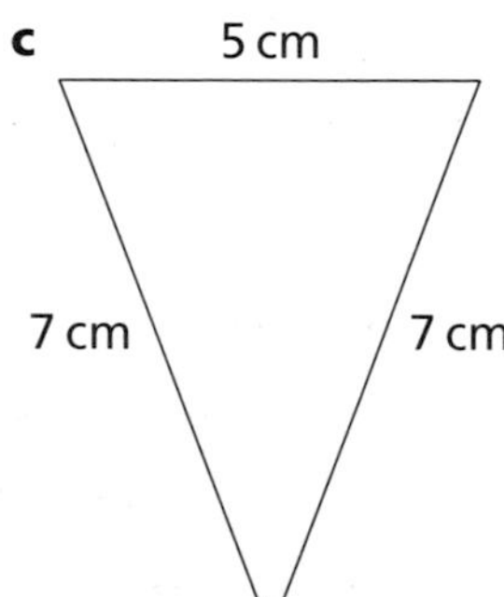

What do you think?

1 Which of these triangles are impossible to construct?

A Sides of length 10 cm, 5 cm, 5 cm.

B One side 10 cm, one angle 40° and one angle 70°.

C One side 10 cm, one angle 80° and one angle 100°.

2 **a** Discuss with a partner why the diagram shows that there are two possible triangles ABC with AB = 7 cm, ∠ABC = 40° and AC = 5 cm.

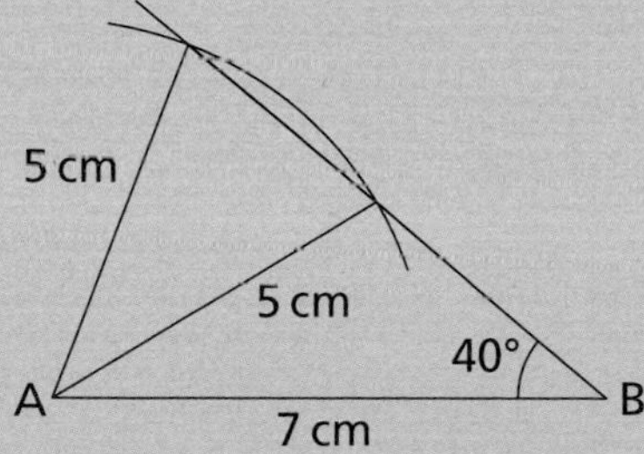

b How many possible triangles are there for each of these descriptions?

i DEF with DE = 6 cm, ∠DEF = 140° and EF = 5 cm.

ii PQR with PQ = 8 cm, QR = 5 cm and PR = 7.5 cm.

iii XYZ with XY = 4 cm, YZ = 3 cm and YXZ = 35°.

iv LMN with ∠LMN = 90°, ∠LNM = 40° and MLN = 50°.

v GHK with GH = 7 cm, HK = 12 cm and KG = 3 cm.

In this section, you will construct polygons.

You can construct polygons by joining triangles together or by working your way around a shape.

Example 3

Construct this quadrilateral.

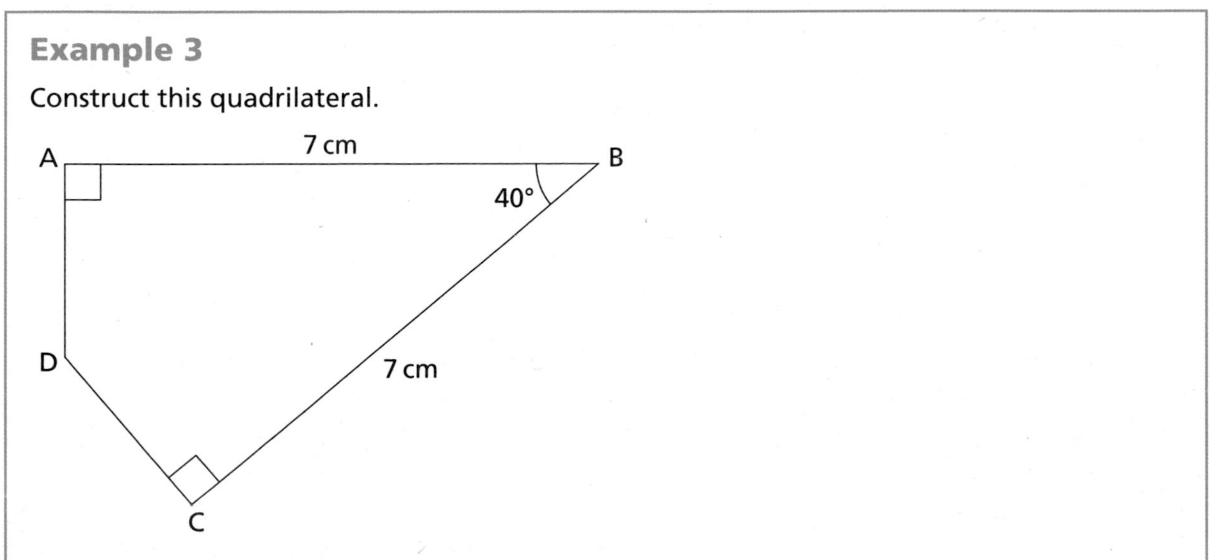

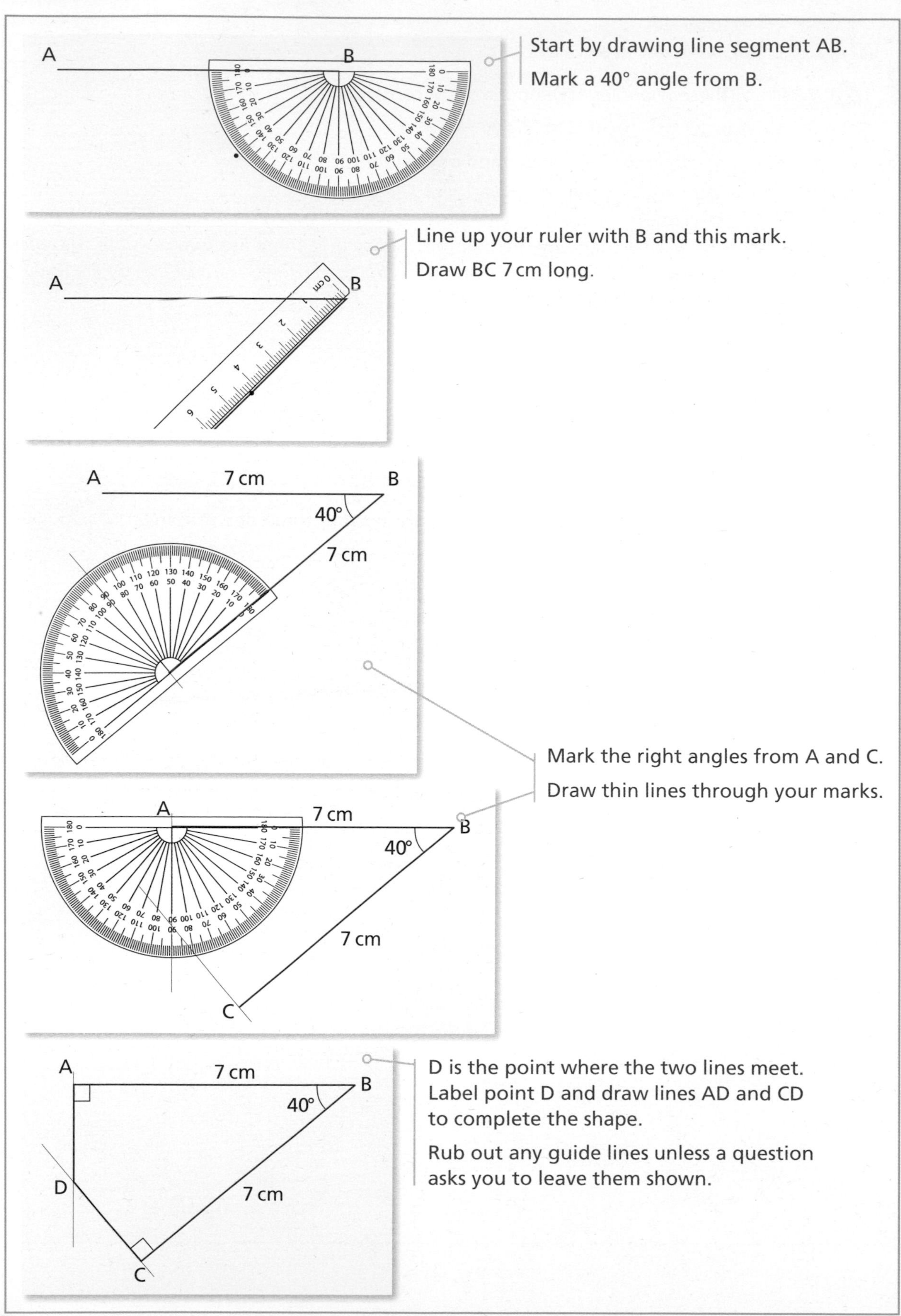
A
B
Start by drawing line segment AB.
Mark a 40° angle from B.
A
B
Line up your ruler with B and this mark.
Draw BC 7 cm long.
A
7 cm
B
40°
7 cm
Mark the right angles from A and C.
Draw thin lines through your marks.
A
7 cm
B
40°
7 cm
C
A
7 cm
B
40°
D
7 cm
C
D is the point where the two lines meet. Label point D and draw lines AD and CD to complete the shape.
Rub out any guide lines unless a question asks you to leave them shown.

Practice 11.4B

1 a Use a ruler and a pair of compasses to construct quadrilateral ABCD.

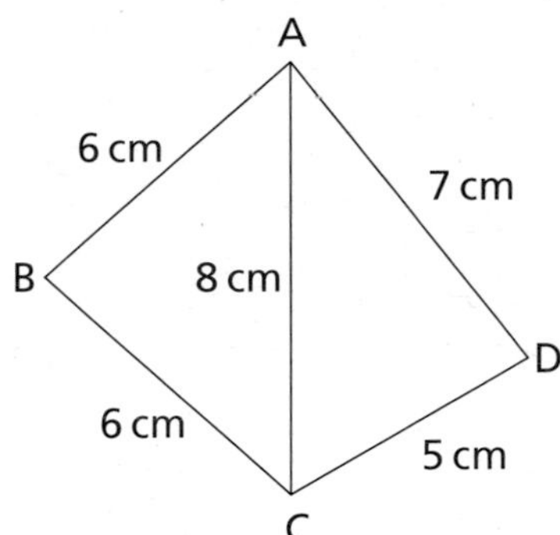

Start by drawing line AB and then triangle ABC.

b Measure angles ABC and ADC.

2 Construct a parallelogram, WXYZ, with WX = 6.5 cm, XY = 3.5 cm, angle WXY = 150° and angle XWZ = 30°.

3 By drawing two isosceles triangles, construct rhombus PQRS.

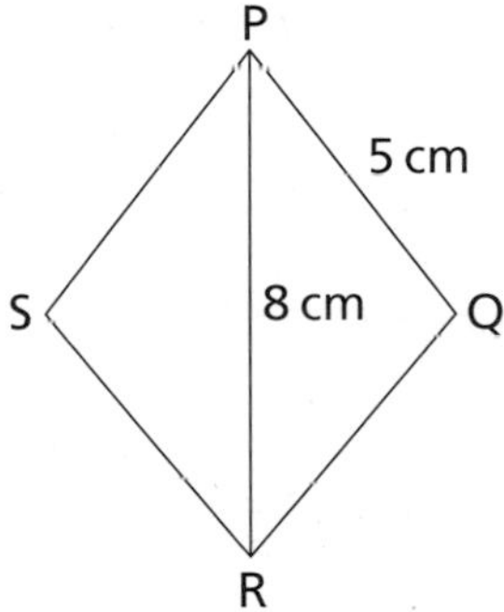

4 Construct quadrilateral WXYZ.

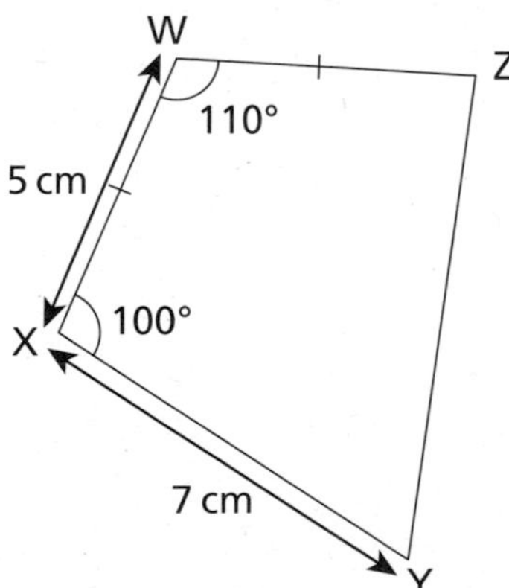

5 Construct delta ABCD with
AC = BC = DC = 5 cm

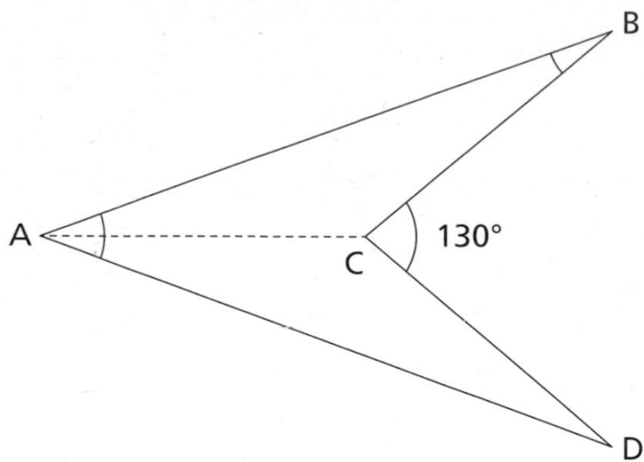

What do you think?

1 The angles in a regular pentagon are all 108°. Use this information to construct regular pentagon ABCDE with sides of length 5 cm.

2 How much information do you need to be given to construct a parallelogram?

3 What shapes do you get if you construct

a two identical equilateral triangles that meet edge to edge

b three identical equilateral triangles that meet edge to edge

c four identical equilateral triangles that meet edge to edge? (3 possible answers)

Check your answers by sketching or constructing the shapes.

Consolidate – do you need more?

1 **a** How long is each side of an equilateral triangle with perimeter 225 mm?
b Construct an equilateral triangle with perimeter 225 mm.

2 Construct accurate copies of these triangles

a

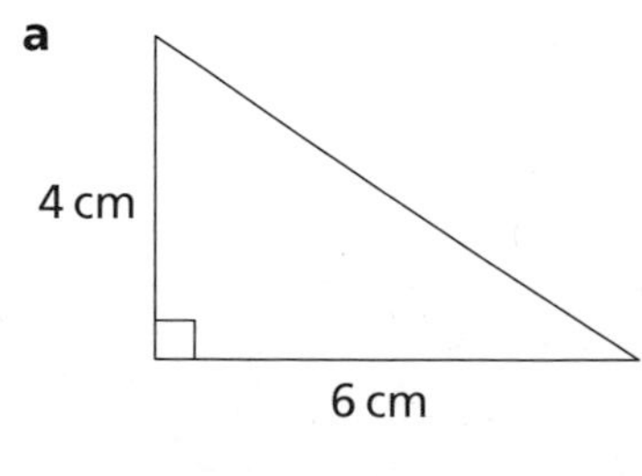

b

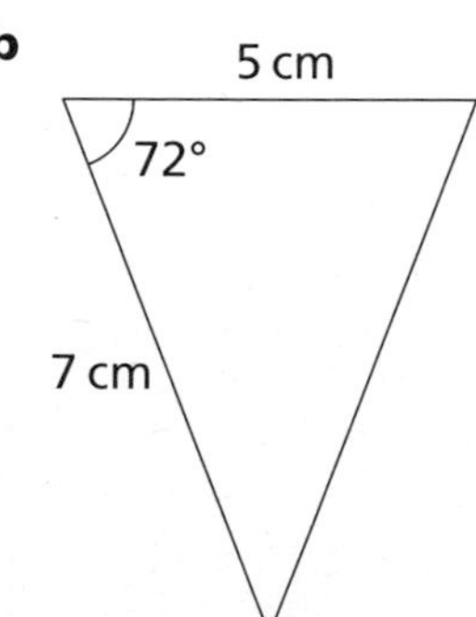

c

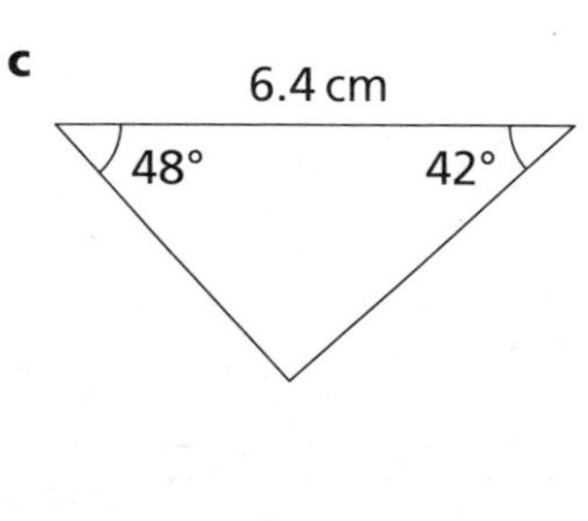

3 Construct these quadrilaterals

a

5 cm

4.2 cm

80°

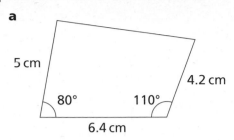

110°

6.4 cm

b

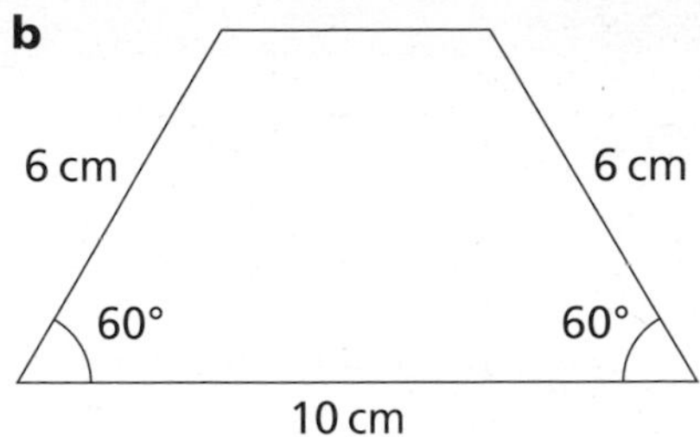

4 Construct a kite with longest diagonal 5 cm and two pairs of sides of length 3 cm and 4 cm.

Stretch – can you deepen your learning?

1 If you have four rods, with length 5 cm, 4 cm, 3 cm and 2 cm, in how many ways can you join the rods to produce triangles? What types of triangle can you make?

2 Investigate the triangles you can make if you have two rods of each of these lengths: 5 cm, 4 cm, 3 cm and 2 cm. What if you have three of each length?

Reflect

How much information do you need to be given to construct a triangle? Describe the steps you would take in each case.

11.5 Pie charts

Small steps

- Interpret simple pie charts using proportion
- Interpret pie charts using a protractor
- Draw pie charts

Key words

Pie chart – a circle divided into parts to show relative sizes of data

Proportion – a part or a share, usually given as a fraction

Sector – part of a circle between two radii and an arc

Frequency – the number of times something happens

Are you ready?

1 Draw and label an angle of

a 30° **b** 130° **c** 230°

2 How many degrees are there in

a a full turn **b** a half-turn **c** a quarter turn?

3 Work out

a $\frac{1}{5}$ of 360 **b** $\frac{1}{3}$ of 360 **c** $\frac{2}{9}$ of 360

4 Write these fractions in their simplest form.

a $\frac{60}{360}$ **b** $\frac{150}{360}$ **c** $\frac{270}{360}$ **d** $\frac{45}{360}$

Models and representations

36 counters placed around the edge of a circle

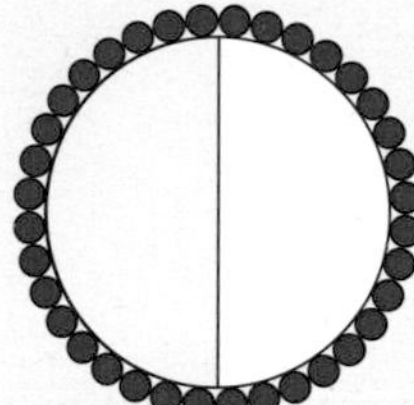
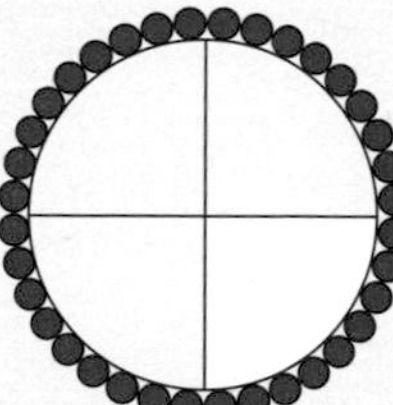
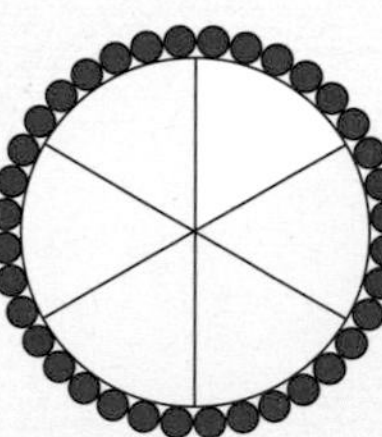

Sectors of equal size cut from a circle

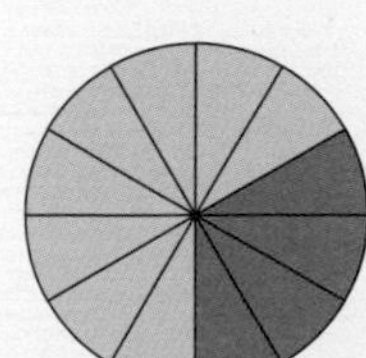
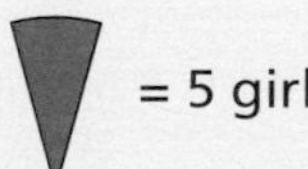

= 5 boys

= 5 girls

Pie charts are circular models used to represent information. They show the size of each item as a part of the whole. This shows the planned UK government spending for 2020.

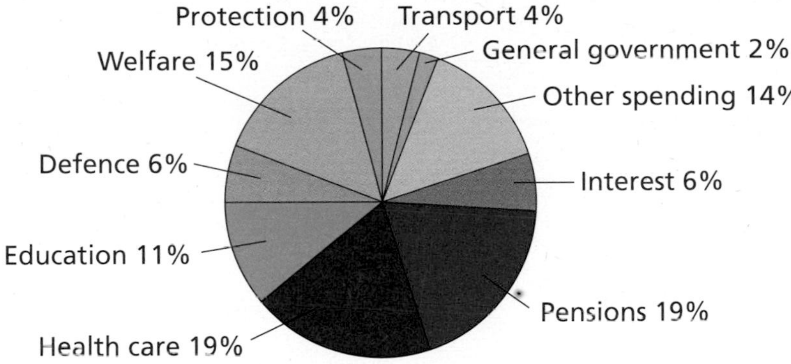

Example 1

A group of people were asked to name their favourite type of movie. The pie chart shows the results.

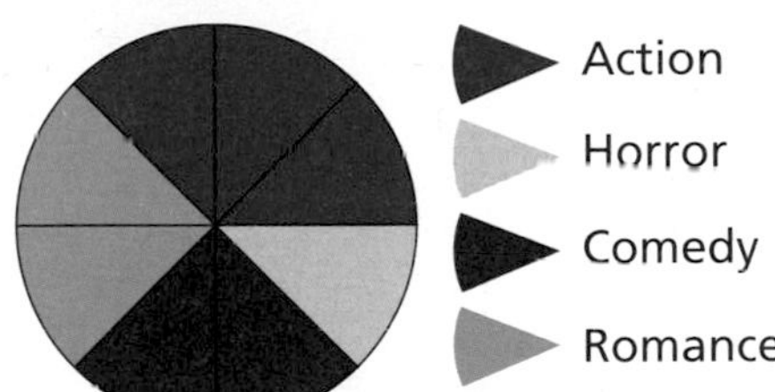

Some pie charts are split into equal sections to make it easier to compare the sizes of the each category.

a What fraction of the people chose action movies?

b 48 people chose action movies. How many people chose romance movies?

c How many people were asked altogether?

a $\frac{3}{8}$ — 3 out of 8 equal parts are shaded red.

b $48 \div 3 = 16$ people — 3 sections are 48 people, so each section is $48 \div 3 = 16$ people.

$2 \times 16 = 32$ people — Romance is 2 sections, so this is $2 \times 16 = 32$ people.

c $8 \times 16 = 128$ — There are 8 sections altogether.

This is similar to the way you worked with fractions and wholes in Chapter 8.1.

Practice 11.5A

1 The pie chart shows the **proportion** of games won, lost and drawn by a football team one season. They played 44 games altogether.

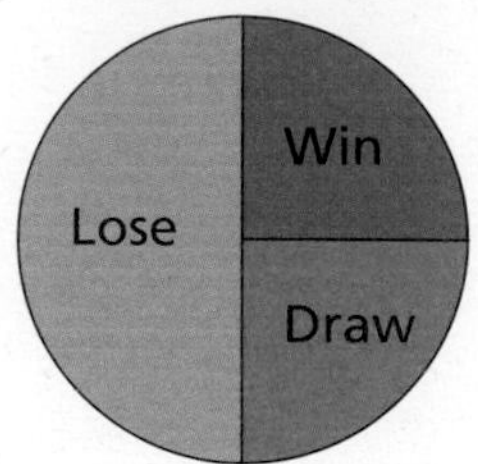

a How many games did they win?

b How many games did they not win?

2 The pie chart shows the favourite fruits of a group of students.

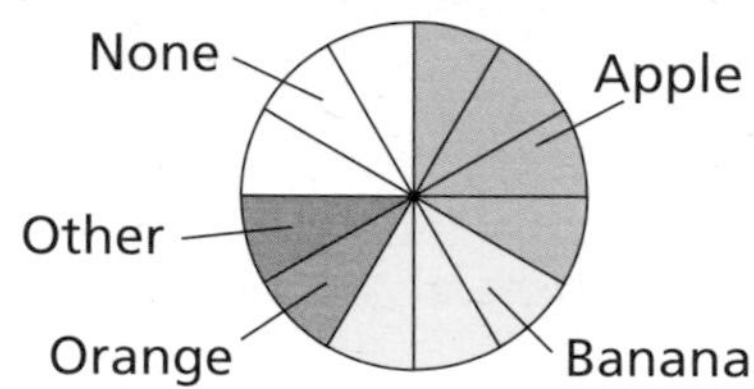

a Marta says that 42 people chose apples. Explain why Marta is probably wrong.

b 32 people chose apples as their favourite fruit. How many people chose

i oranges ii bananas iii none?

c How many people were asked altogether?

3 160 students were asked how they travel to school. The results are shown on the pie chart.

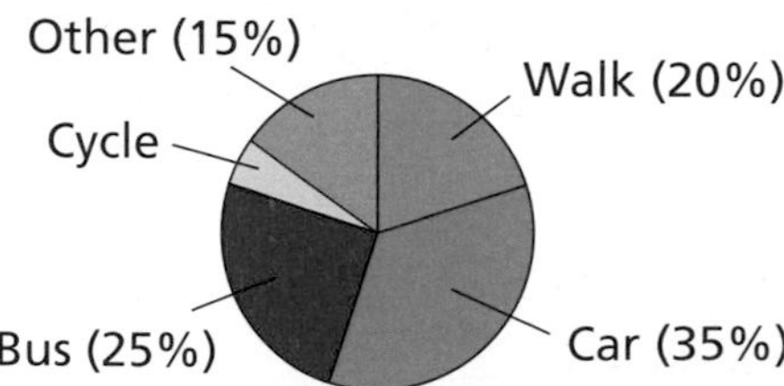

a How many students walk to school?

b What percentage of students cycle to school?

c How many more students travel to school by car than by bus?

4 The pie chart shows the different fund-raising events some students took part in to raise money for a charity.

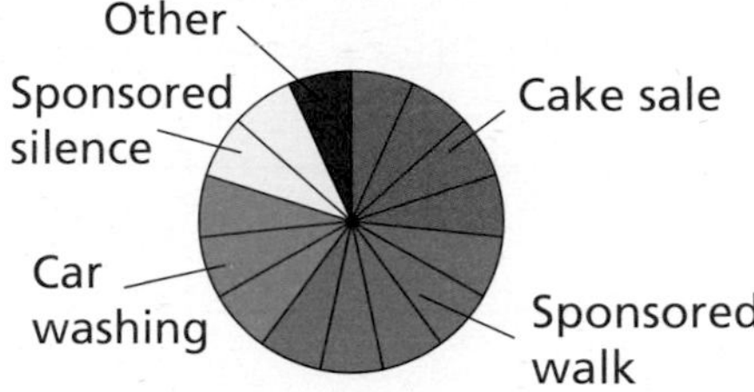

- **a** Write, as fractions in their simplest form, the proportion of students who took part in
  - **i** car washing
  - **ii** cake sale
  - **iii** sponsored walk
  - **iv** any sponsored event.
- **b** 24 students washed cars. How many students took part in fund-raising events altogether?

5 The pie chart shows information about the ages of visitors to a museum.

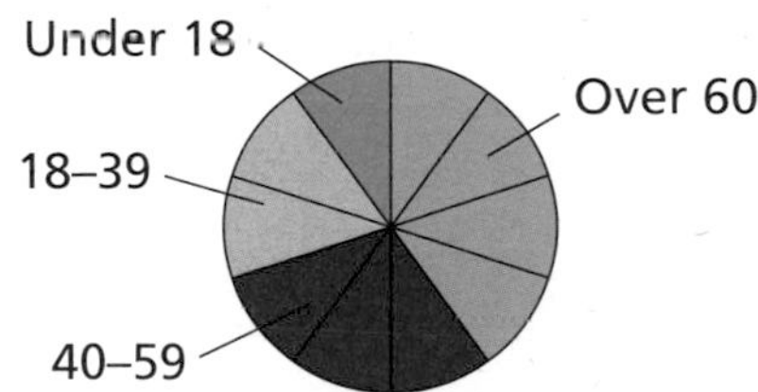

- **a** What percentage does each sector of the pie chart represent?
- **b** Decide whether each of these statements is definitely true or definitely false, or whether you cannot tell.
  - **i** Half of the visitors were aged between 18 and 59
  - **ii** Half of the visitors were 50 or over
  - **iii** 80% of the visitors were adults
  - **iv** There were twice as many over 60s as people aged 40–59
- **c** 87 visitors were aged 40–59. How many visitors were over 60?

What do you think?

1 A hockey team loses $\frac{1}{4}$ of the matches it plays one season. The team wins three times as many matches as they draw.

- **a** Show that this information cannot be represented on a pie chart with 8 equal sections.
- **b** What is the minimum number of equal sections needed?
- **c** Investigate the minimum number of equal sections needed for different fractions.

You can construct and interpret pie charts using angles instead of equal sections.

Remember that there are 360° in a full turn.

Example 2

The table shows information about the flavours of ice cream sold by a shop one day. Draw a pie chart to represent the information.

Ice cream flavour	Vanilla	Strawberry	Chocolate
Number sold	55	27	38

Method A

a $55 + 27 + 38 = 120$

First work out the angles to represent each section. Start by finding the total number of ice creams sold.

$360° \div 120 = 3°$ for each ice cream

Work out 360° ÷ total frequency

Then work the angle needed to represent 1 ice cream.

Vanilla $= 55 \times 3° = 165°$

Strawberry $= 27 \times 3° = 81°$

Chocolate $= 38 \times 3° = 114°$

Multiply by the number sold to find the angle for each flavour.

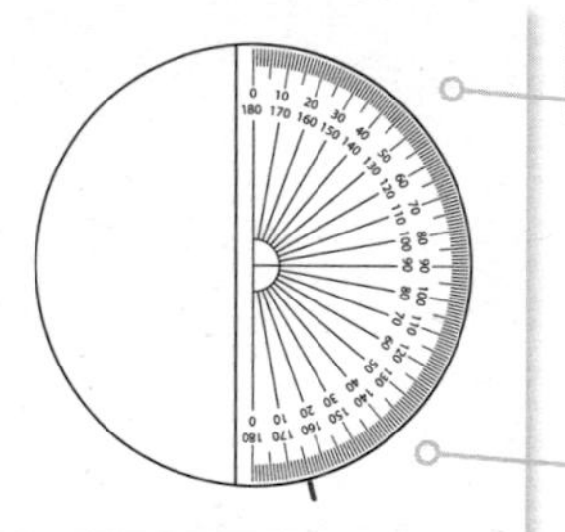

Draw a circle with the same radius as your protractor and mark on a radius.

This makes it easier to measure angles for the sectors.

Measure and mark the first angle.

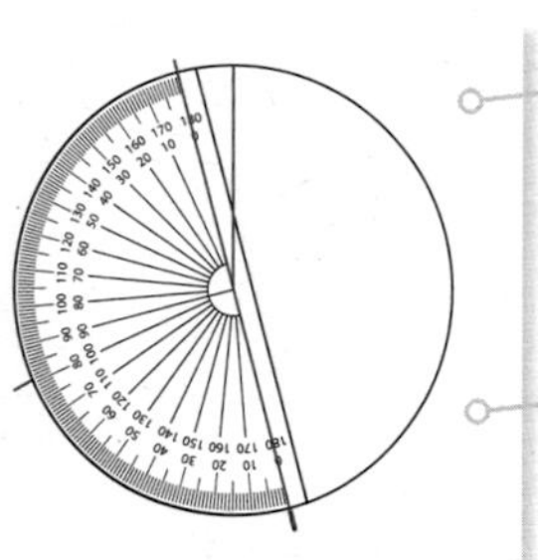

Line up the 0° on the protractor with the end of the first sector.

Complete the first sector and mark the next angle.

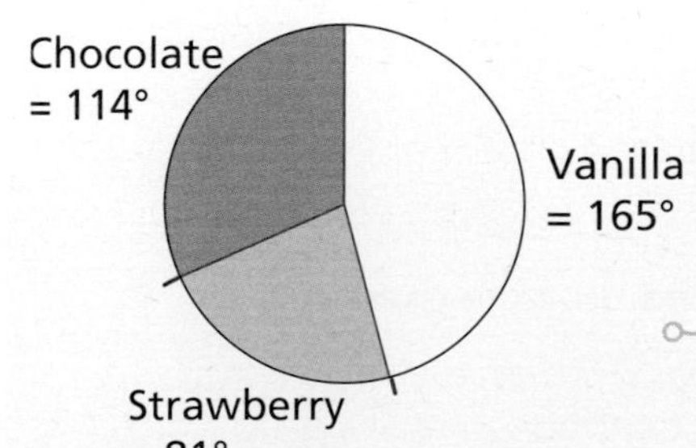

Finish drawing the pie chart.

Remember to label each sector or give a key.

Method B

a Vanilla $= \frac{55}{120} \times 360° = 165°$

Strawberry $= \frac{27}{120} \times 360° = 81°$

Vanilla $= \frac{38}{120} \times 360° = 114°$

You could work out the angles to represent each sector by working out these fractions of 360°.

Work out $\frac{\text{frequency of sector}}{\text{total frequency}} \times 360°$

Then draw the pie chart in the same way.

You can use method B for any numbers, but method A only works easily if the total is a factor or a multiple of 360°.

If you know the angles in a pie chart, how can you work out the numbers that each section represents?

Practice 11.5B

1 The table shows information about the eye colours of 60 students. Draw a pie chart to show the information.

Eye colour	Brown	Blue	Green	Other
Frequency	22	25	8	5

2 **a** 180 students voted for the Year 7 school council representative. Draw a pie chart to show the results.

Candidate	Rhys	Beth	Ali	Lydia
Number of votes	28	54	45	53

b Which candidate had exactly $\frac{1}{4}$ of the votes?

c Is it easy to tell the winner from your pie chart? Explain why or why not.

3 **a** What fraction, in simplest form, is a sector of a pie chart with each of these angles?

i 90°

ii 120°

iii 30°

iv 45°

b A pie chart represents 600 pieces of data. How many people are represented by an angle of

i 90°

ii 120°

iii 30°

iv 45°?

4 **Vegetables people dislike**

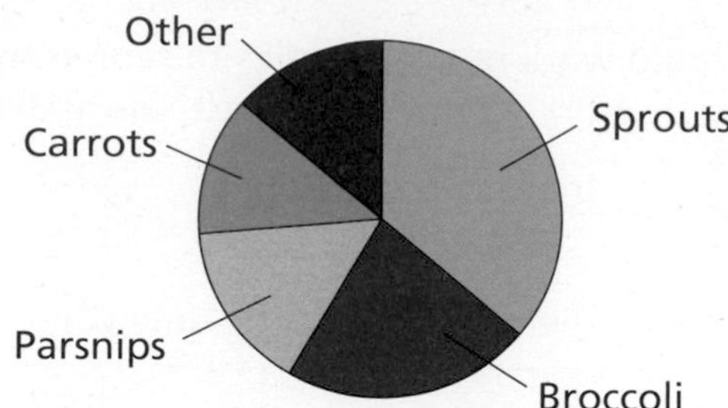

The pie chart shows the results of a survey about which vegetables people dislike the most.

a Which was the least popular vegetable?

b Measure the angle for broccoli.

32 people chose broccoli.

c How many people does 10 degrees represent?

d How many people were surveyed altogether?

5 300 people responded to a survey about the local library.

Opinion	Very satisfied	Quite satisfied	Disappointed	Don't know	Never go
Frequency	90	150	30	10	20

a What fraction of the respondents never go to the library?

b Why is it easy to work out the angle on a pie chart to represent the people who were "quite satisfied"?

c Draw a pie chart to show the results.

What do you think?

1 **a** Ed says, "To make a pie chart with 6 sectors, you only need to draw 5 angles." Explain why Ed is correct.

b How many angles do you need to draw to make a pie chart with n sectors?

2 Find the missing numbers.

a B is ______ % of the whole.

b The angle for C is ______.

c The fraction of the whole represented by A is ______.

d C as a fraction of A is ______.

e What other fraction–percentage connections can you find?

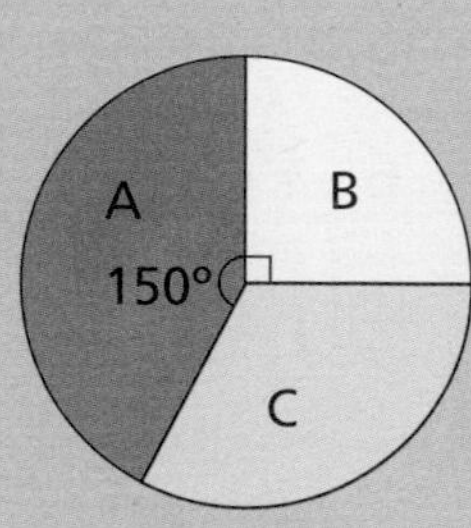

Consolidate – do you need more?

1 **Pet survey**

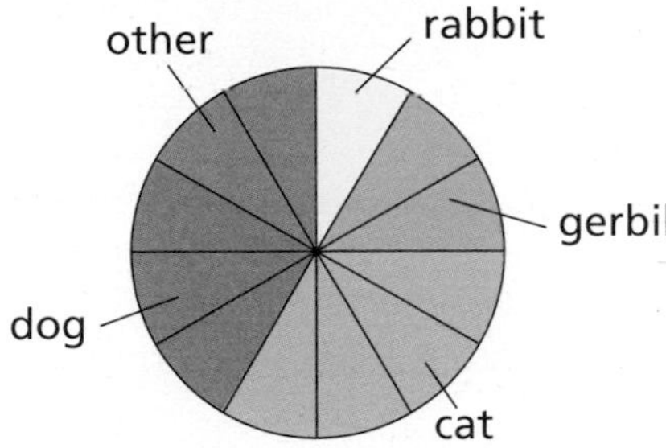

The pie chart shows the results of a survey about pets.

a What fraction of the people surveyed have a rabbit?

b 7 people own a rabbit. How many people

i own a cat

ii own a cat or a dog

iii were asked altogether?

2 The pie chart shows how Mr Patel spent his money one month.

a What did he spend the largest amount of money on?

b What fraction of his money did Mr Patel spend on transport?

c Mr Patel spent £150 on food. How much did he spend altogether?

d Work out how much he spent on

i rent **ii** transport.

Mr Patel's spending

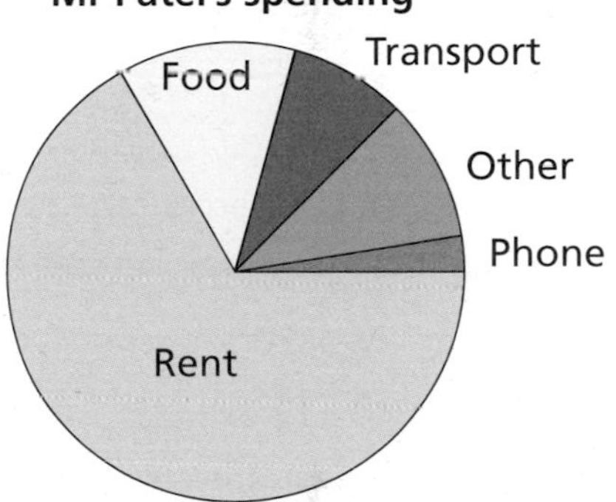

3 The table shows the instruments played by a group of 90 students. Draw a pie chart to show the information.

Instrument	Guitar	Piano	Drums	Other	None
Frequency	12	9	7	23	39

4 The table shows the number of sandwiches sold in a café one day. Draw a pie chart to show the information.

Filling	Cheese	Tomato	Egg	Tuna	Other
Number sold	200	150	75	25	50

Stretch – can you deepen your learning?

1

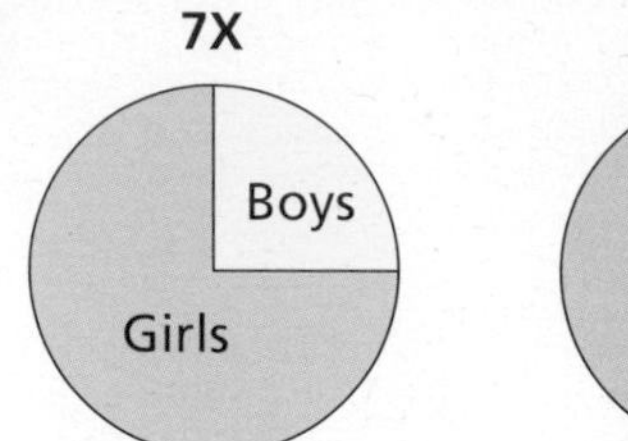

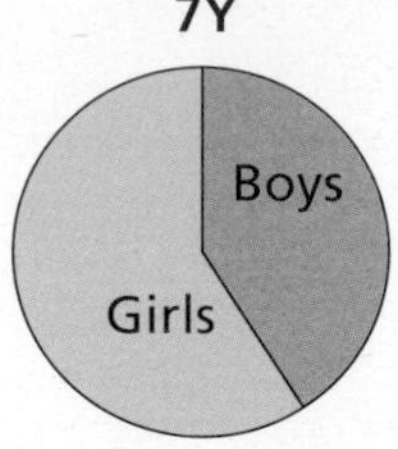

Zach says that there are more boys in 7Y than in 7X. Explain why Zach might be wrong.

2 The sectors in a pie chart have angles of 90°, 120°, x° and $2x$°.

- **a** Work out value of x.
- **b** Explain how you know that the pie chart does not represent 117 pieces of information.
- **c** What is the minimum number of pieces of information that the pie chart could represent?

3 Some pie charts are labelled in percentages.

- **a** What angle would represent
  - **i** 20%
  - **ii** 35%?
- **b** What percentage is represented by
  - **i** 108°
  - **ii** 270°
  - **iii** 306°?
- **c** Investigate the links between other angles and percentages.

4 When would it be more useful to use a pie chart rather than a bar chart to represent data? When might a bar chart be better?

Reflect

1 Describe how to construct a pie chart to represent information about 180 people.

2 How would your method be different if there were 200 people?

11 Construction and measuring

Chapters 11.1–11.5

I have become fluent in…

- Using letters to name angles and sides
- Using geometric notation
- Drawing and measuring angles
- Constructing shapes

I have developed my reasoning skills by…

- Using information to identify properties of shapes
- Classifying shapes using their properties
- Identifying parallel and perpendicular sides
- Working out how to construct a shape

I have been problem-solving through…

- Representing information on pie charts
- Use pie charts to find out information
- Finding fractions and percentages of amounts in different contexts

Check my understanding

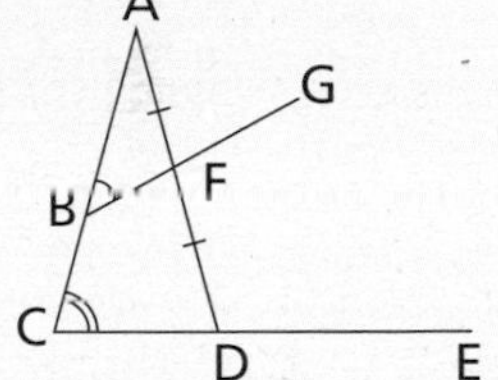

1 **a** Write the name of the angle shown by

  i a single arc **ii** a double arc

b Write the name of any obtuse angle in the diagram.

c Which two line segments are equal in length? How do you know?

d Explain why the diagram does not show that BG is perpendicular to AD.

e Measure angle ADC.

2 Draw an angle of

a 40° **b** 110° **c** 300°

3 Write the mathematical names of the shapes.

a
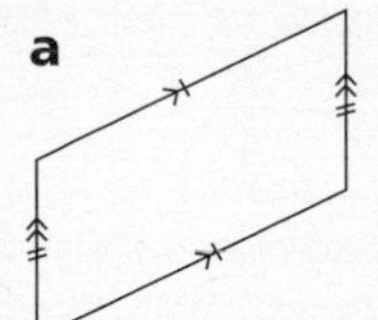

b
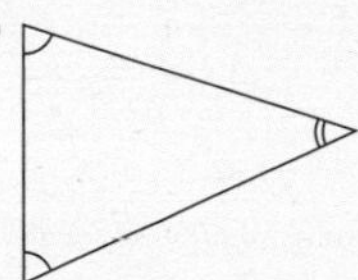

c
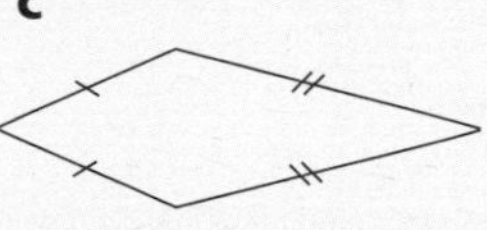

d

e
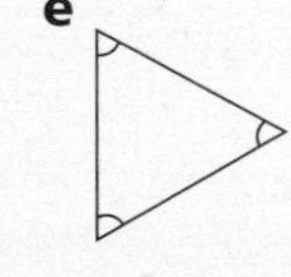

4 Construct the triangles.

a ABC with BC = 7 cm, ABC = 55° and ACB = 65°

b DEF with DE = 7 cm and EF = DF = 5.2 cm

5 **a** Construct a pie chart to show the information in the table.

Favourite colour	Red	Green	Blue	Black	Yellow
Frequency	75	60	10	30	25

b On another pie chart, 15 people are represented by an angle of 40°. How many people does the whole pie chart represent?

12 Geometric reasoning

In this block, I will learn...

the sum of the angles at a point and a straight line, and how to identify vertically opposite angles

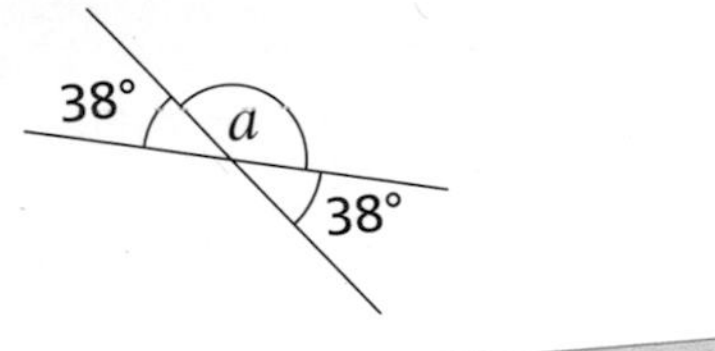

how to find missing angles in triangles

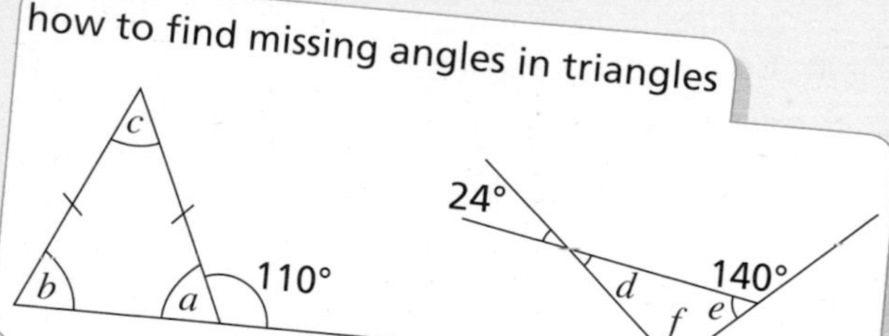

how to find missing angles in quadrilaterals

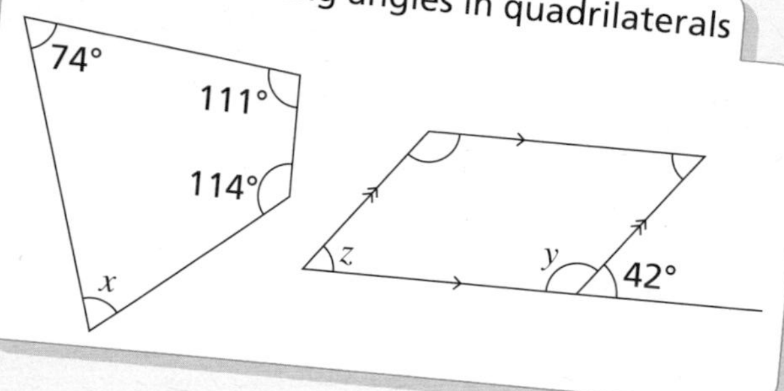

how to solve complex angle problems

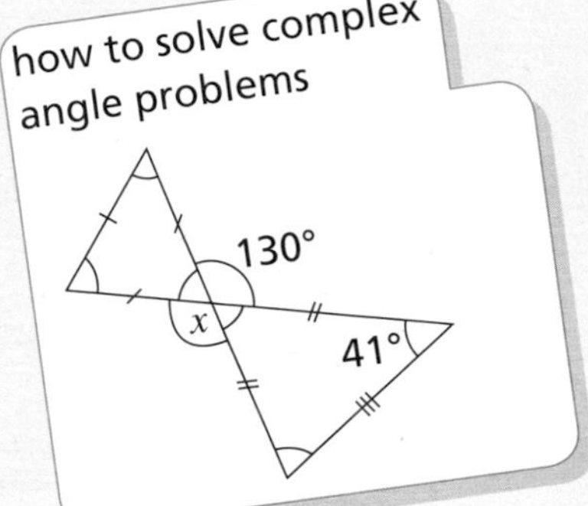

how to find missing angles in polygons (H)

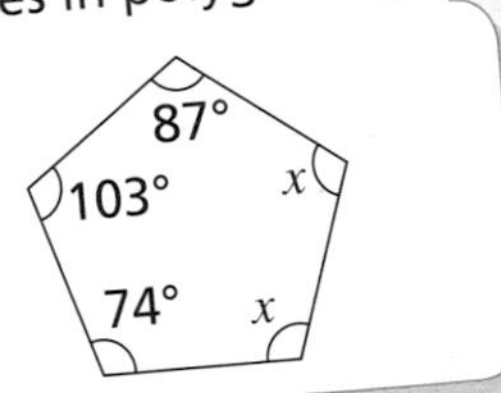

about angles found in parallel lines (H)

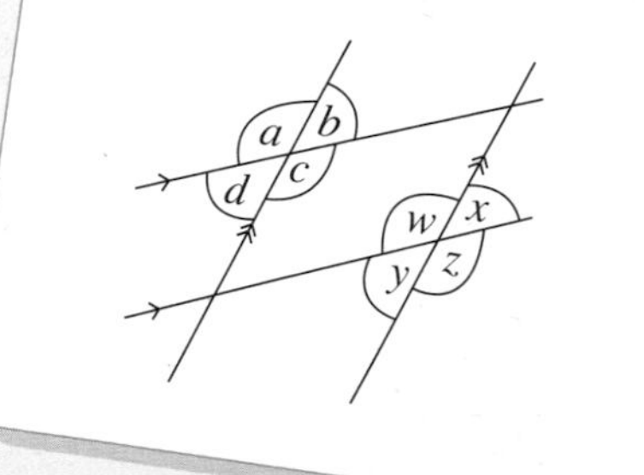

how to prove geometrical facts (H)

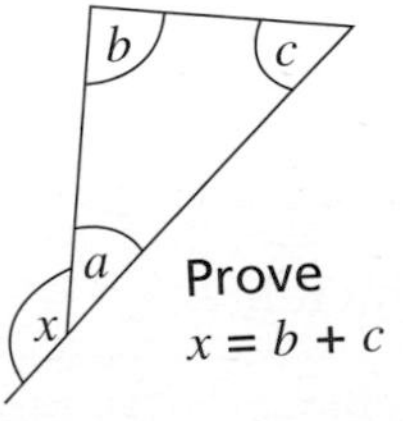

Prove
$x = b + c$

12.1 Points and lines

Small steps

- Understand and use the sum of angles at a point
- Understand and use the sum of angles on a straight line
- Understand and use the equality of vertically opposite angles

Key words

Adjacent – next to each other

Vertex – a point where two line segments meet

Vertically opposite angles – angles opposite each other when two lines cross

Are you ready?

1 Work out

a 360 – 140 **b** 360 – (80 + 60) **c** 360 – 149 **d** 360 – 189

2 Work out

a 180 – 100 **b** 180 – 130 **c** 180 – 137 **d** 180 – 64

3 Classify each of these angles as acute, obtuse, reflex or a right angle.

a

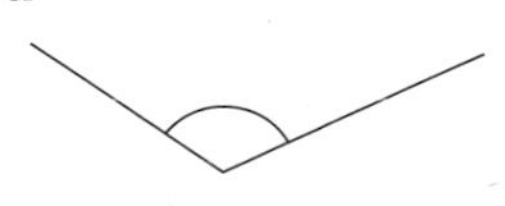

b

c

d

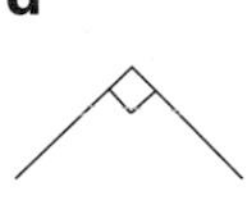

e

4 Solve these equations.

a $5x = 180$ **b** $5x + 30 = 180$ **c** $5x + 30 = 360$

Models and representations

Geoboards

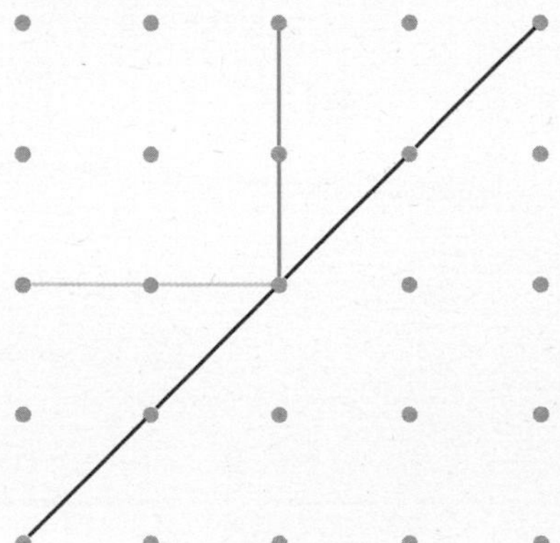

Straws

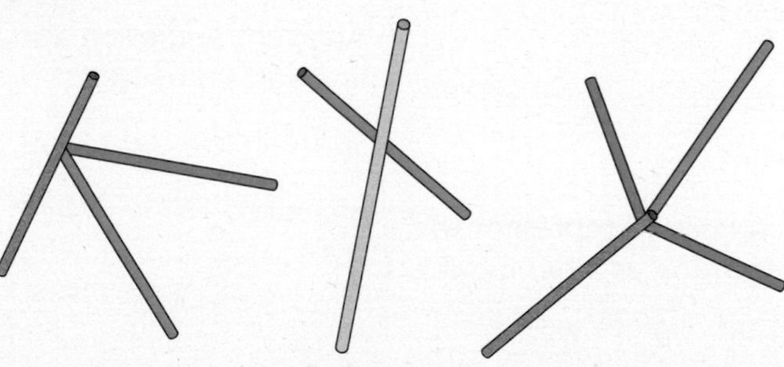

You can also use pencils, rulers and arms.

12.1 Points and lines

A full turn is 360°

360°

You learnt this in Block 11.

So half a turn is half of 360° which is 180°

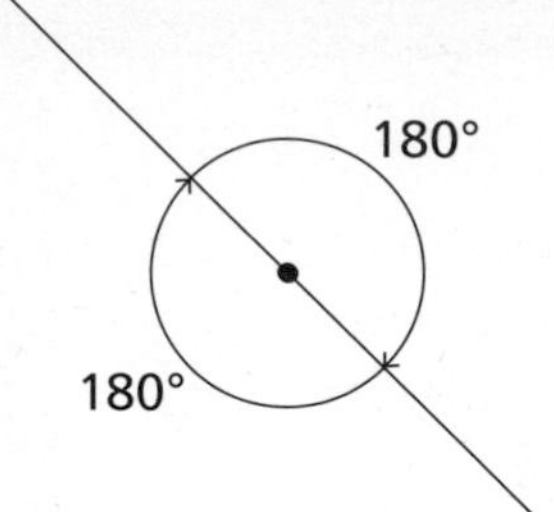

$360° \div 2 = 180°$

Half a turn is a straight line, so angles **adjacent** to each other on a straight line add up to 180°

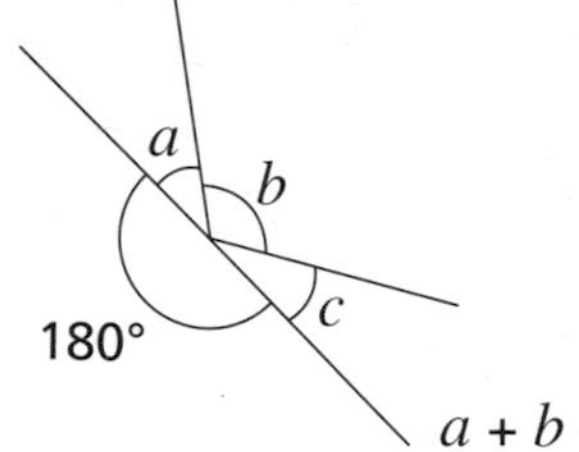

$a + b + c = 180°$

Angles at a point add to 360°

Angles on a straight line add to 180°

Example 1

Work out the unknown angles.

$a = 360° - 52° = 308°$

Angles in a full turn add up to 360°

So $a + 52° = 360°$

Using inverse operations, $a = 360° - 52°$

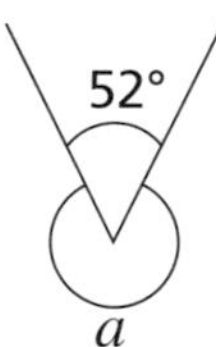

$38° + b + 75° = 180°$ — Add the two angles you know

$b + 113° = 180°$ — Adjacent angles on a straight line add up to 180°

$b = 180° - 113° = 67°$ — Subtract the total from 180°

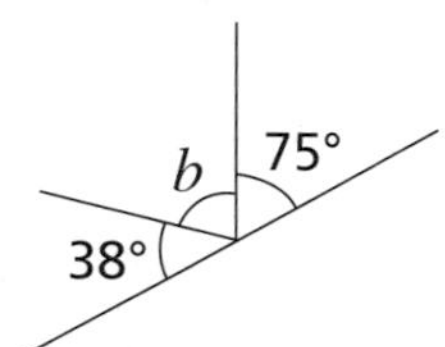

You could represent the solution as a bar model.

38°	b	75°
180°		

Practice 12.1A

1 Angles at a point add up to 360°. Use this fact to find the size of each angle labelled with a letter.

a

a 150°

b

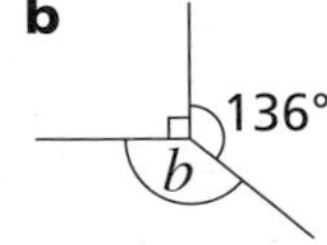

c

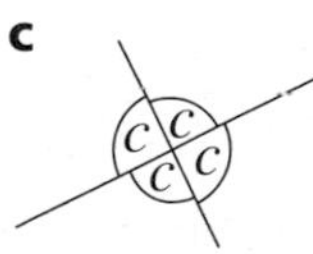

d

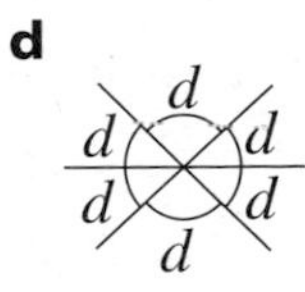

e

222° e

f

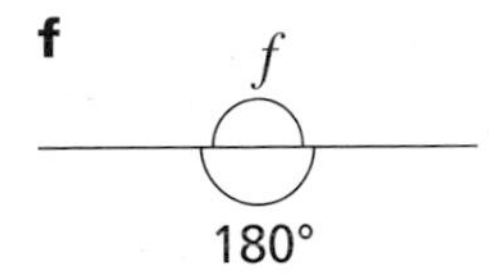

g

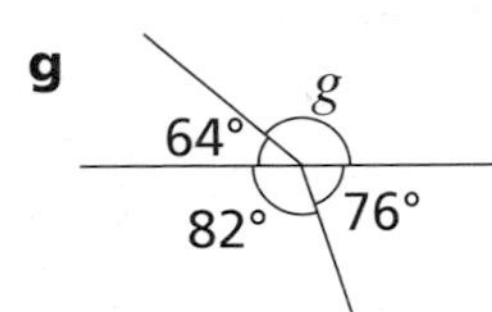

2 Adjacent angles on a straight line add up to 180°. Use this fact to find the size of each angle labelled with a letter.

a

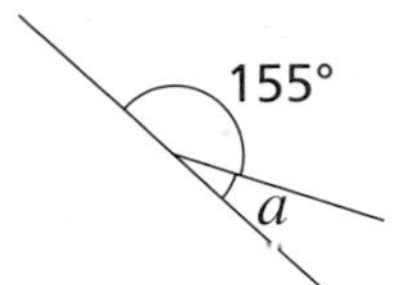

b

44° b 58°

c

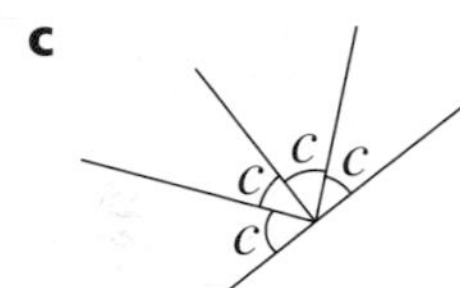

d

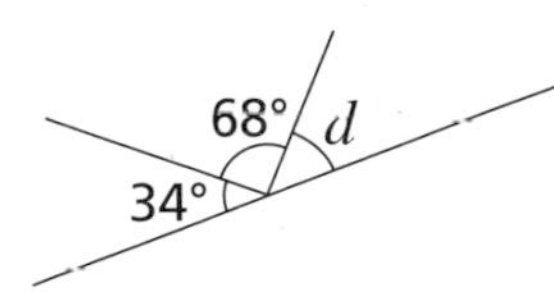

e

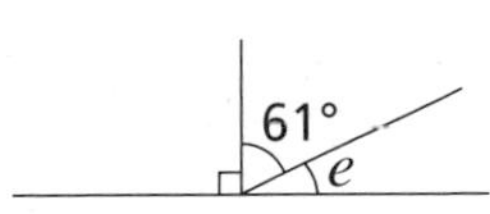

f

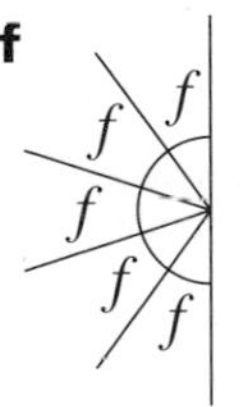

3 Work out the sizes of the unknown angles.

a

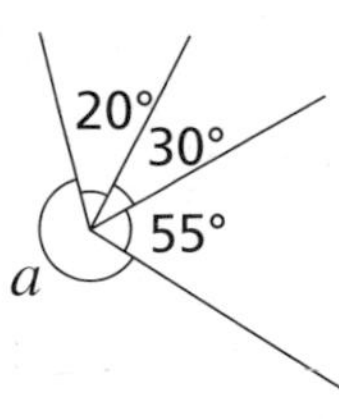

b

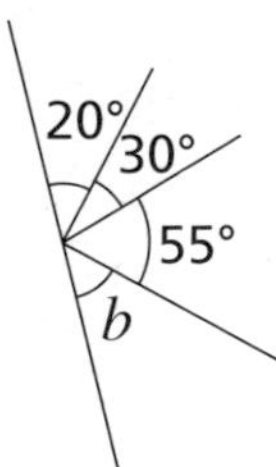

c

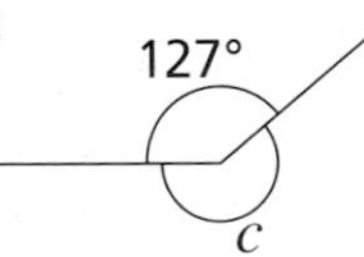

d

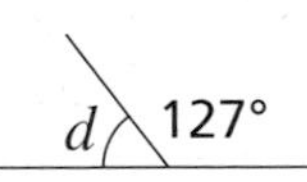

e

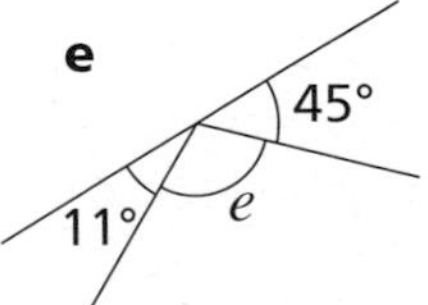

f

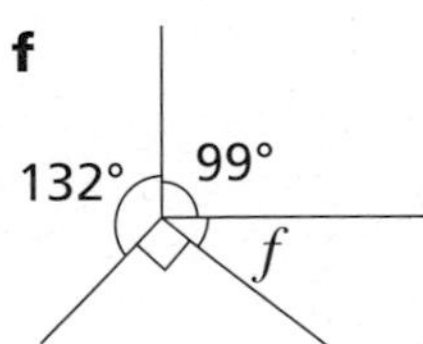

g

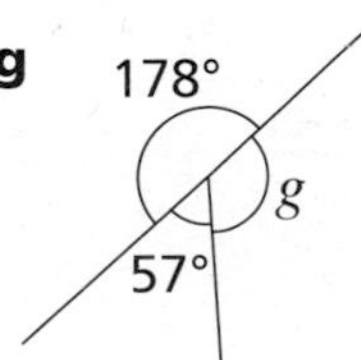

4 Use angle facts to form an equation for each diagram. Then solve your equations to work out the value of each letter.

a

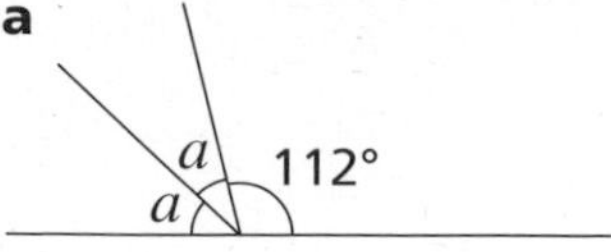

b

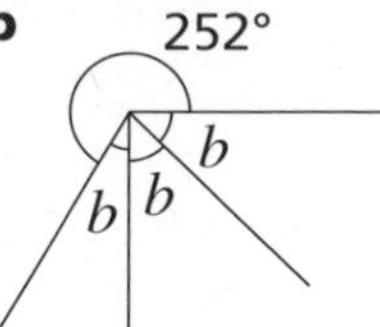

c

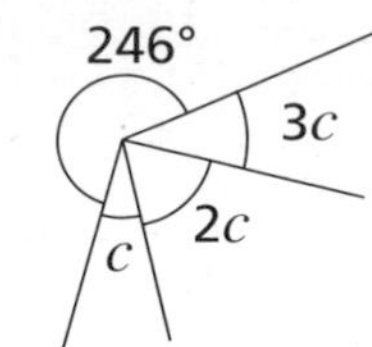

d

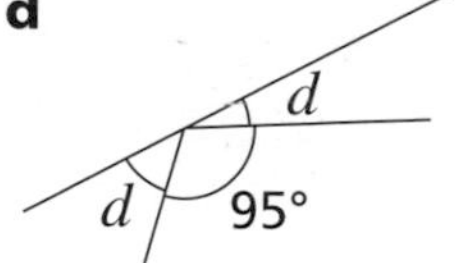

e

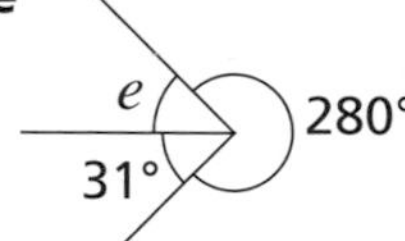

f

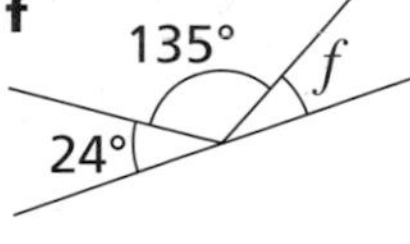

What do you think?

1 Marta thinks that $x = 180° - 41° = 139°$ because angles on a straight line add up to 180°. Do you agree with Marta? Why or why not?

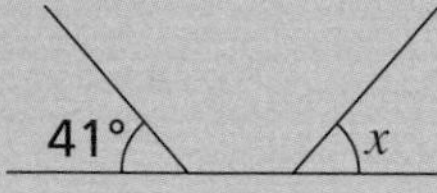

2 PQ and SR meet at O.

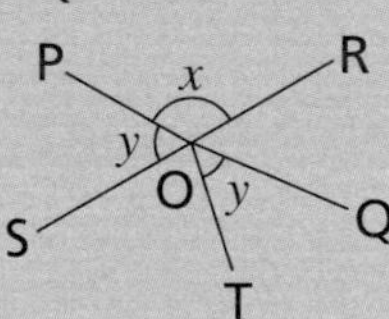

a Write expressions in terms of x and/or y for

i angle ROQ **ii** angle SOT

b

Angle TOS must be 60 degrees.

Do you agree with Jackson? Why or why not?

3 Faith has measured angles a, b and c.

Is she correct? How do you know?

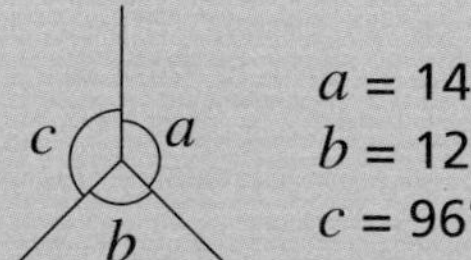

$a = 147°$
$b = 121°$
$c = 96°$

4 Decide whether each set of angles would form a straight line if they were placed next to each other.

a 88°, 92° **b** 25°, 165° **c** 55°, 65°, 70° **d** $x°, x°, (180 - 2x)°$

When two straight lines cross they meet at a **vertex**. The two pairs of angles formed are called **vertically opposite angles**.

This is because they are opposite each other at a vertex – not because they are vertical.

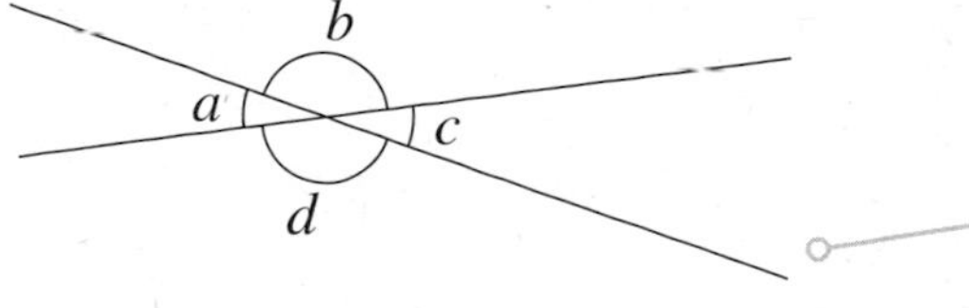

You will explore this in the next exercise and prove the result in Chapter 12.6H.

$a = c$ and $b = d$

Example 2

Work out sizes of the unknown angles.

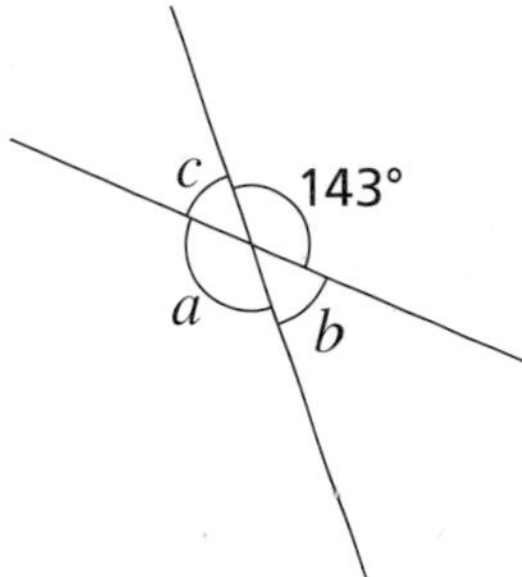

$a = 143°$ — a and the 143° angle are vertically opposite. Vertically opposite angles are equal.

$143° + b = 180°$ — Adjacent angles on a straight line add up to 180°

$b = 180° - 143° = 37°$

$c = 37°$ — c and b are vertically opposite angles. Vertically opposite angles are equal.

What other angle facts could you have used to work out the size of angle c?

Practice 12.1B

1. Draw two pairs of straight lines crossing each other. Measure the four angles formed in both of your diagrams. Check that each pair of vertically opposite angles are equal.

2. Which diagrams show pairs of vertically opposite angles?

A

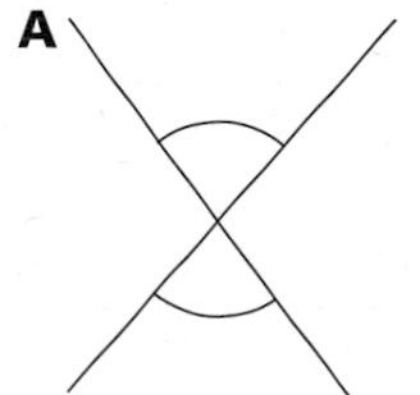

B

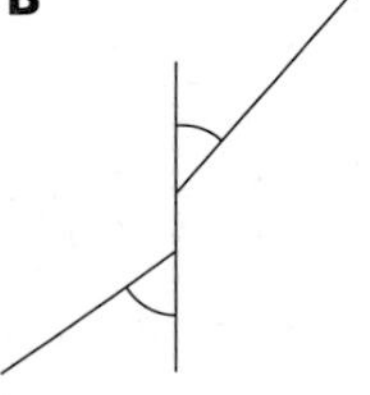

C

D

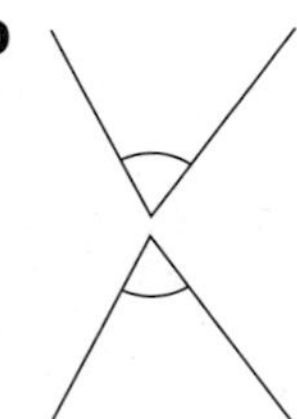

3 Work out the sizes of the unknown angles.

a

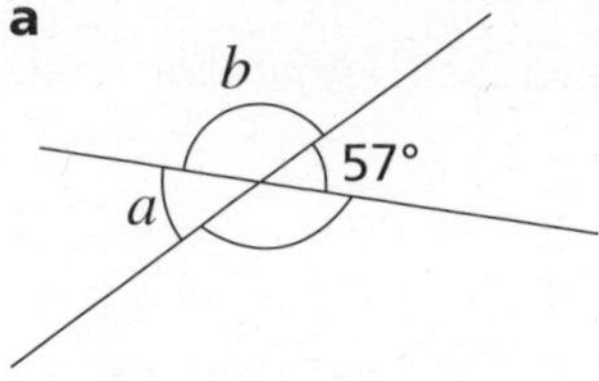

b

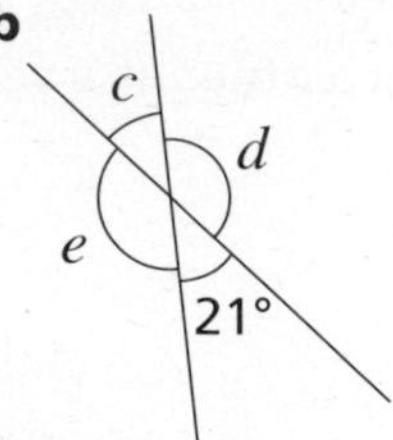

c

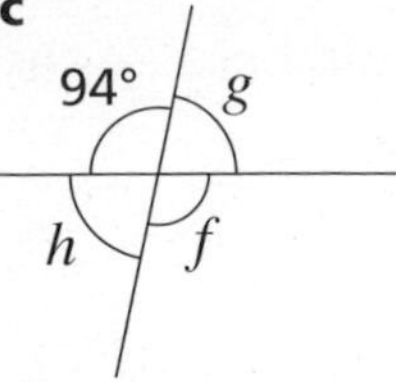

What do you think?

1 Explain why vertically opposite angles a and c are equal.

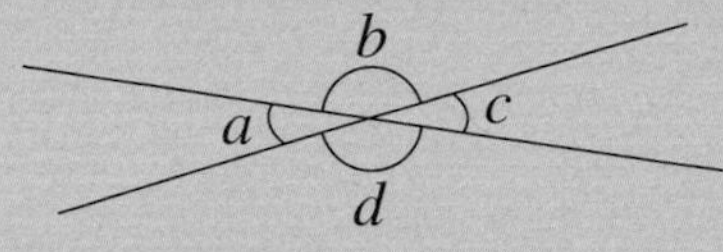

We will prove this to be always true in Chapter 12.6H.

2 In both diagrams, all the line segments are straight. ∠POQ = x and ∠BFC = y

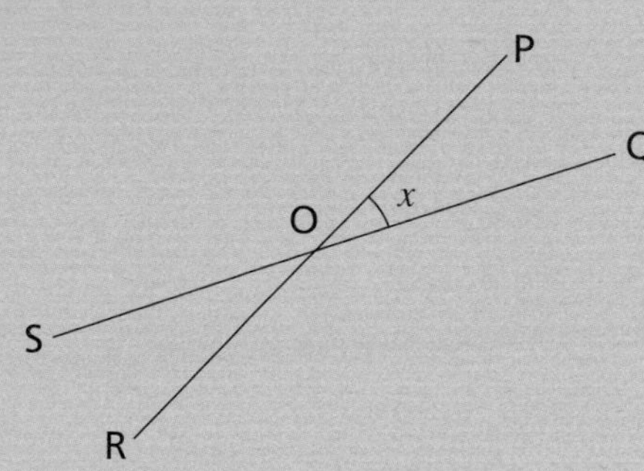

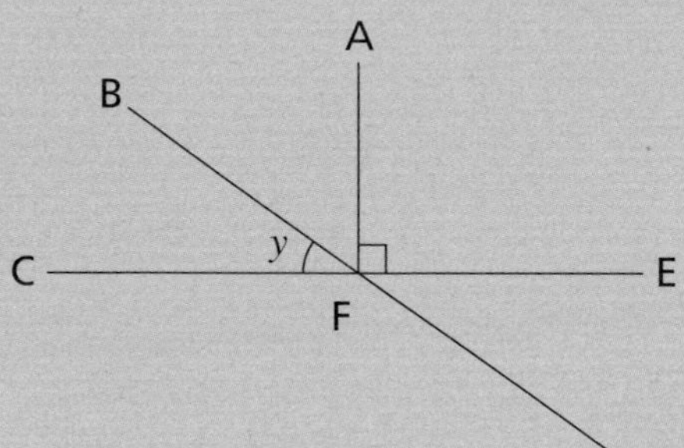

a Write an expression for

i ∠ROS **ii** ∠POS **iii** ∠ROQ **iv** ∠DFC **v** ∠EFD

b Explain why ∠AFB = 90 – y

Consolidate – do you need more?

1 Work out the size of each angle labelled with a letter.

a

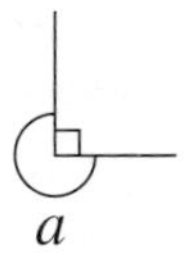

b

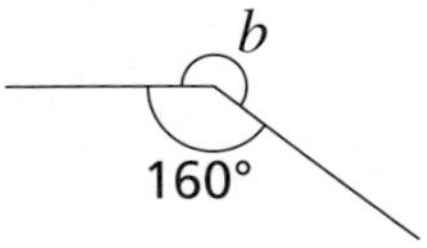

c

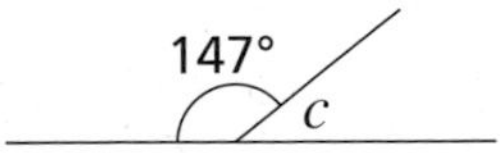

d

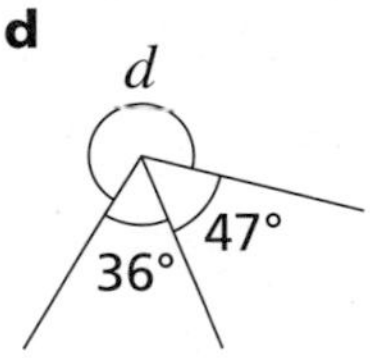

e

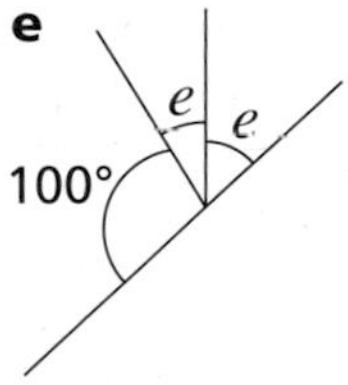

f

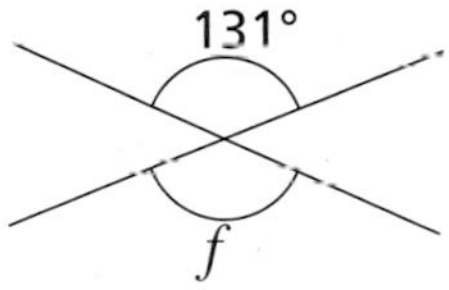

g

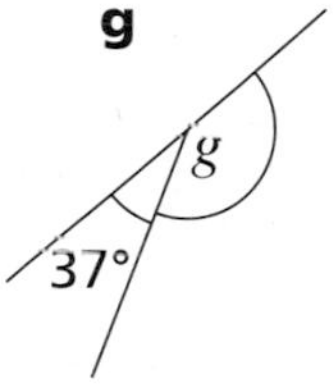

h

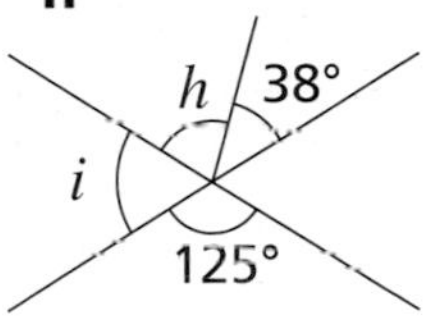

i

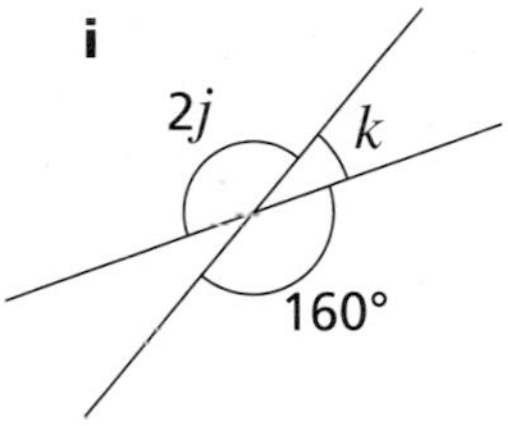

2 In this circle, the green sector is three times the size of the blue sector. The red sector is five times the size of the blue sector. Work out the sizes of the angles at the centres of the sectors.

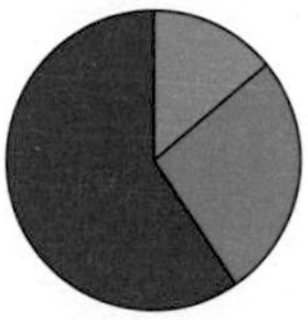

3 Use angle facts to form an equation for each diagram. Then solve your equations to find the value of each letter.

a

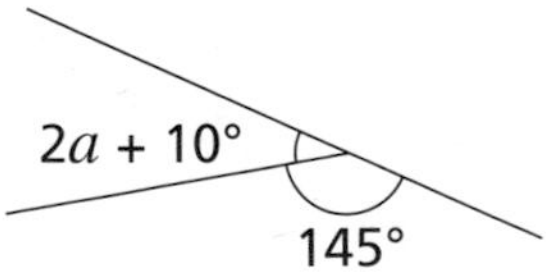

b

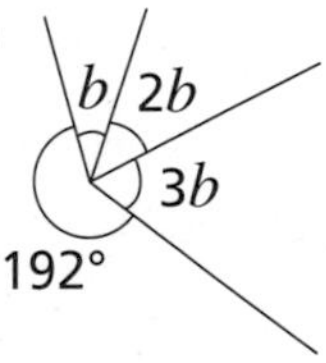

c

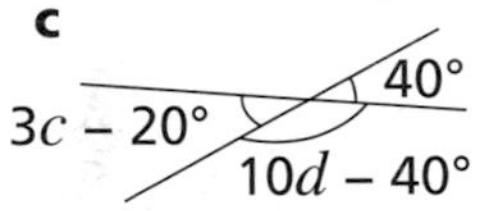

Stretch – can you deepen your learning?

1 Which is bigger, f or g? How do you know?

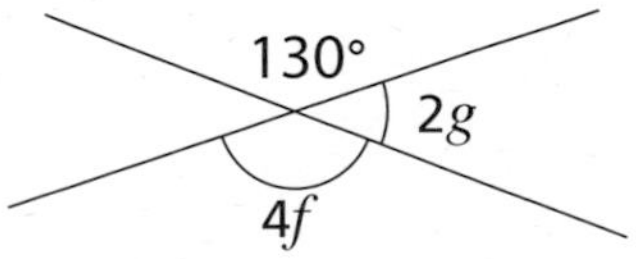

2 A company is designing a new logo, featuring a circle split into equal parts. Here are four possible designs.

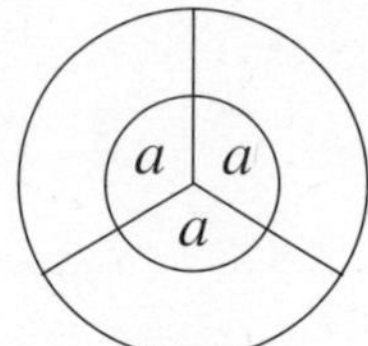

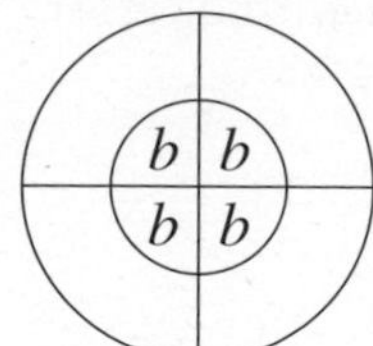

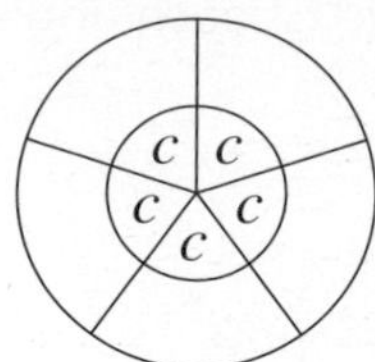

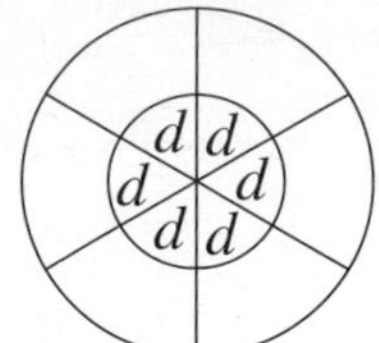

a Work out the sizes of angles a, b, c and d.

b What other numbers of parts can the circle be split into so that all the angles formed are integers?

c Why do you think 360° was chosen for a full turn instead of, for example, 100° or 400°?

3

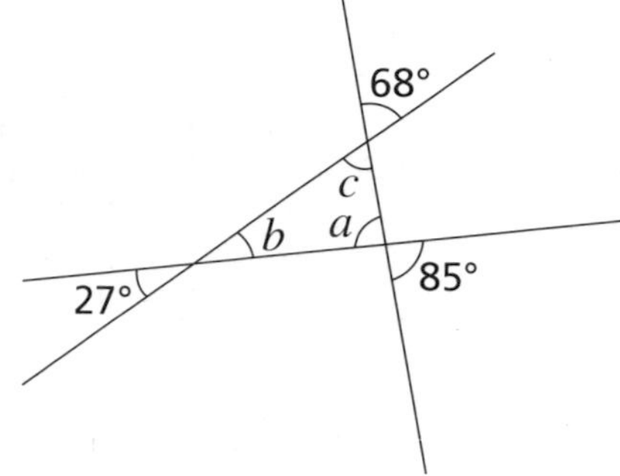

a Work out the sizes of angles a, b and c in the triangle.

b Find the sum of a, b and c. Why is the answer familiar?

Reflect

What rules do you know about angles? Draw labelled diagrams to illustrate them.

12.2 Angles in triangles

Small steps

- Know and apply the sum of angles in a triangle
- Solve complex angle problems

Key words

Find – work out the value of

Give a reason – state the mathematical rule(s) you have used, not just the calculations you have done

Are you ready?

1 Name the three different types of triangle based on their side lengths. What are their special properties?

2 Write the three-letter names of the angles labelled x and y.

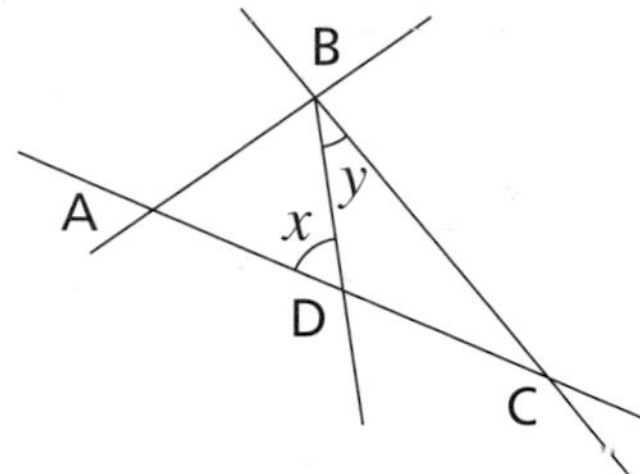

3 Use the angles rules you have already learnt to work out the values of a and b.

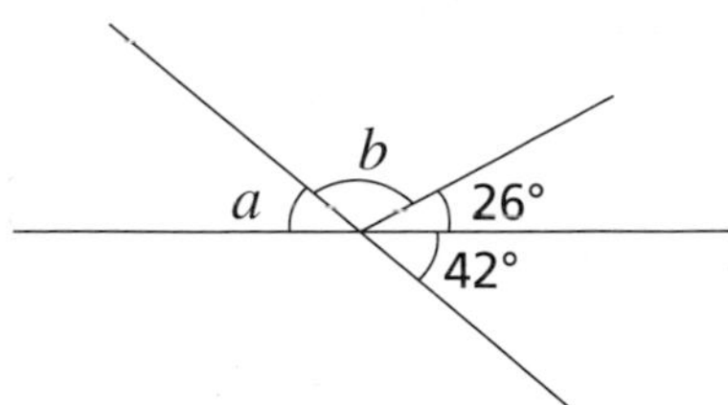

4 Simplify the expressions.

a $a + 2a + 3a$

b $b + 40 + 60$

c $26 + c + 109$

d $d + 30 + 2d + 45$

Models and representations

Remember you can use bar models to represent additions and subtractions.

180		
78	47	?

What additions and subtractions does this bar model represent?

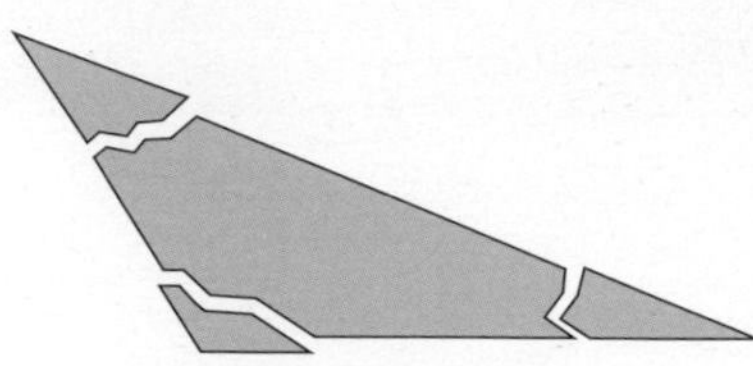

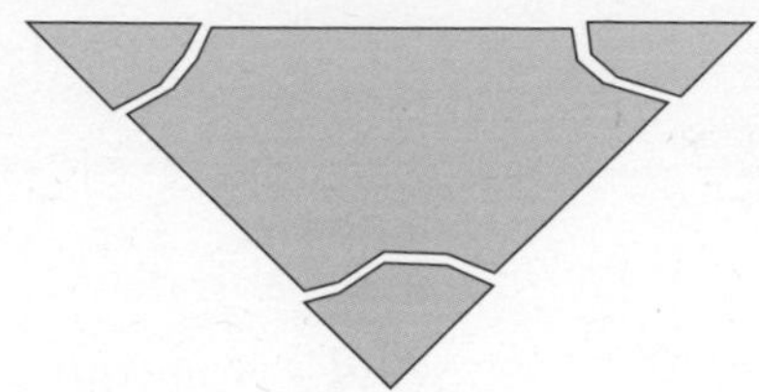

You can demonstrate that the angles in a triangle add up to 180° by tearing off the corners and arranging them to form a straight line. You already know the angles in a straight line add up to 180°.

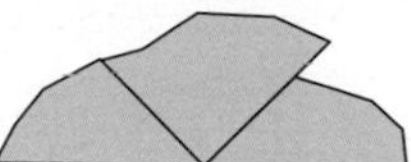

Later you will be able to prove that this result is true for all triangles.

Example 1

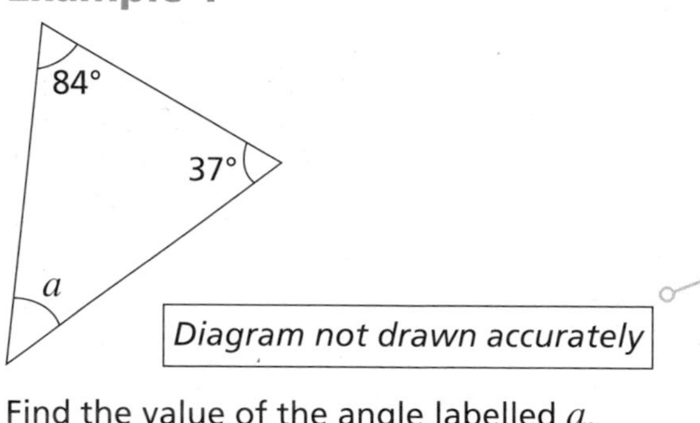

Remember this means you cannot measure to find the size of angle a, you need to use reasoning to work out its size.

Find the value of the angle labelled a.

Method A

$a + 37° + 84° = 180°$ — Set up an equation with the values adding up to 180°.

$a + 121° = 180°$ — Simplify the equation.

$a = 180° - 121° = 59°$ — Solve the equation to find the value of a.

Method B

$a = 180° - (37° + 84°)$ — We know the angles in a triangle sum to 180, so we can draw a bar model.

180		
37	84	a

$a = 180° - 121°$

$a = 59°$

Simplify and solve the equation to find the value of a.

Practice 12.2A

1 These triangles are not drawn accurately. Work out the sizes of the missing angles.

a

A 57° 24° B a C

b

D f 93° 48° E F

c

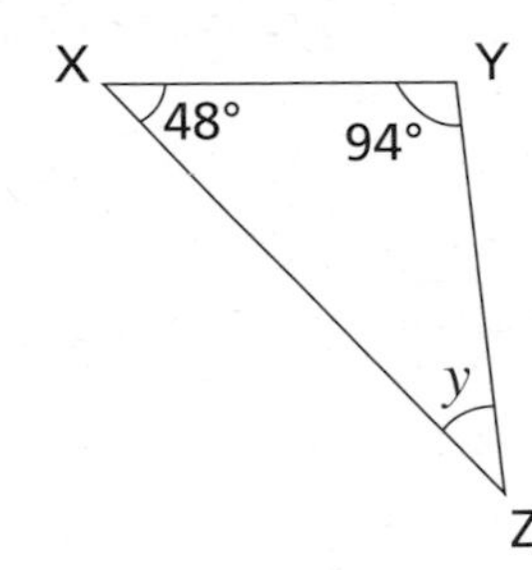

d Which of the three triangles below could be an isosceles triangle? How can you tell?

A

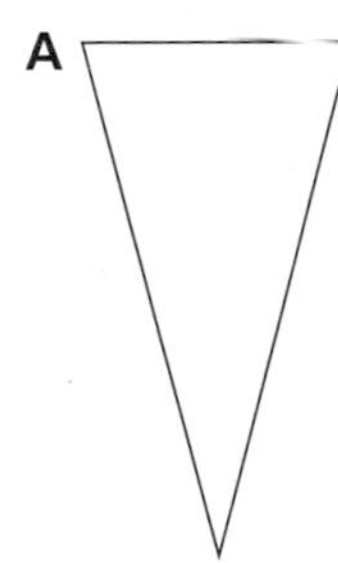

B

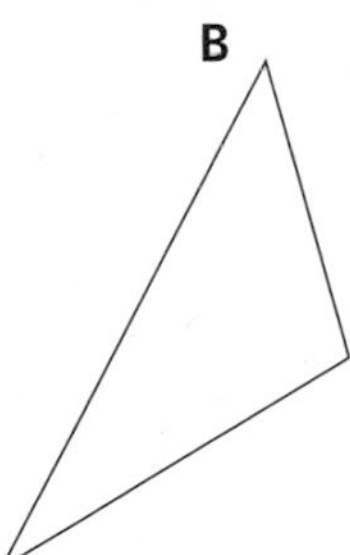

C

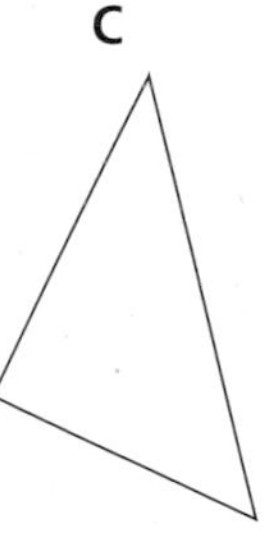

2 Two of the angles in a triangle are 74° and 104°. Work out the size of the third angle.

3 Jakub says the angles in an equilateral triangle must always be 60° each. Is Jakub right? Explain your thinking.

4 Both of the triangles shown are isosceles.

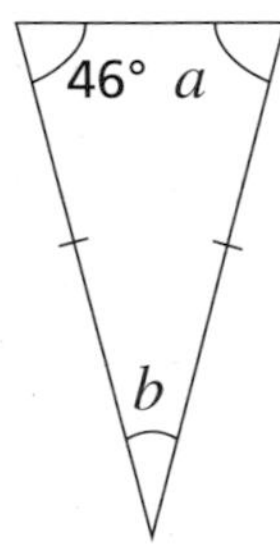

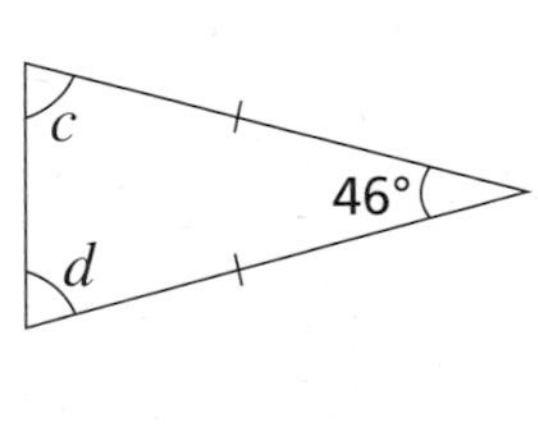

a How can you tell from the diagrams that they are isosceles triangles?

b Work out the missing angles a, b, c and d.

c The information in the triangles is the same. Why are angles c and d different to angles a and b?

5 **a** In triangle ABC, AB = AC and ∠BAC = 75°. Sketch triangle ABC and work out the size of ∠ACB.

b In triangle DEF, DF = EF and ∠EFD = 82°. Sketch triangle DEF and work out the size of ∠DEF.

6 **a** One of the angles in an isosceles triangle is 160°. Work out the sizes of the other two angles.

b One of the angles in an isosceles triangle is 80°. Work out the sizes of the other two angles (there are two possible sets of answers).

c Why are there two sets of possible answers to part **b** but only one possible set of answers to part **a**?

What do you think?

Abdullah and Beca both draw a triangle and measure the angles.

Abdullah's results: 74°, 39°, 69°

Beca's results: 50°, 60°, 110°

a Why do you think Abdullah's angles do not add up to 180°?

b Beca has measured one of her angles incorrectly. Which one? How can you tell?

Now we are going to look at some more complicated problems using our skills and knowledge of other angles rules, simplifying expressions and solving equations.

Example 2

Find the size of the largest angle in the triangle.

a $3a$ 64°

Method

180		
a	$3a$	64

$a + 3a + 64 = 180$

180	
$4a$	64

$4a + 64 = 180$

116
$4a$

$4a = 116$

29	29	29	29
a	a	a	a

$a = 29°$

So the angles are 64°, 29° and 3 × 29° = 87°. The largest angle is 87°.

Example 3

Find the size of angle PRQ, giving reasons for each step of your working.

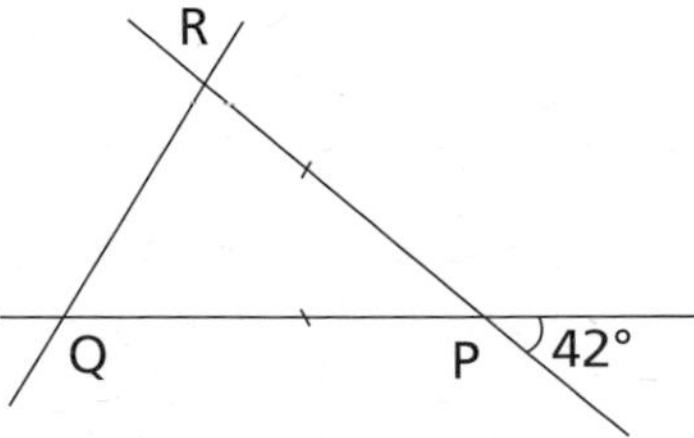

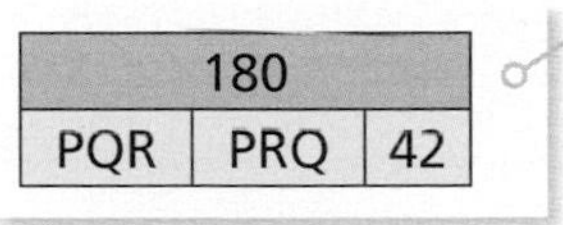

∠QPR = 42° (Vertically opposite angles are equal)

PQR + PRQ + 42 = 180

(Angles in a triangle add up to 180°)

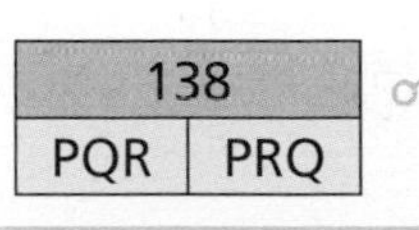

PQR + PRQ = 138°

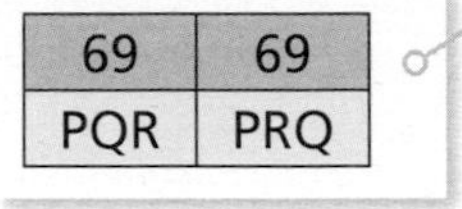

As PQR is an isosceles triangle, PQR and PRQ are equal.

So PRQ = 138 ÷ 2 = 69°

Practice 12.2B

1 **a** Work out

i the size of angle a **ii** the size of angle b

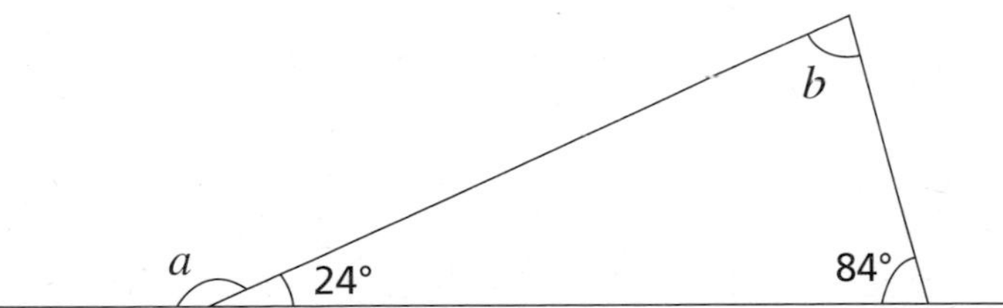

b Could you have worked out the size of angle b first and then angle a?

2 **a** Work out

i the size of angle a **ii** the size of angle b

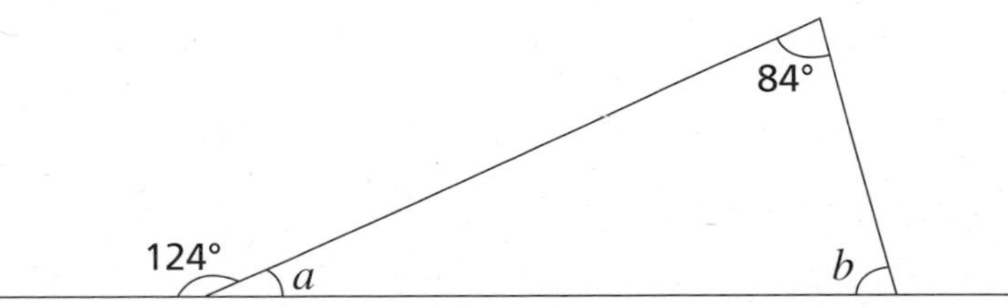

b Could you have worked out the size of angle b first and then angle a this time?

3 Work out the values of angles a, b and c, giving reasons for your answers.

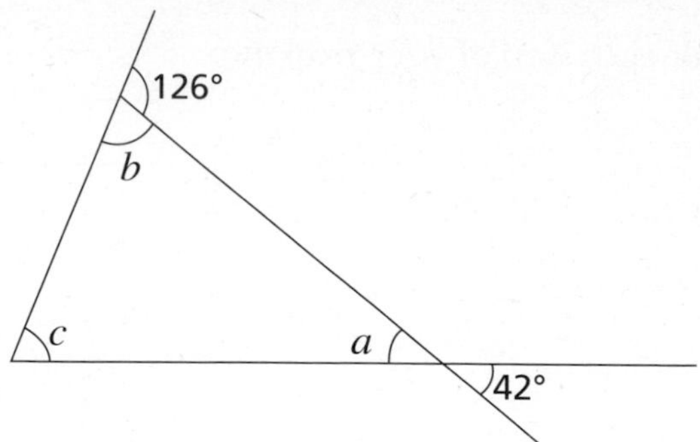

4

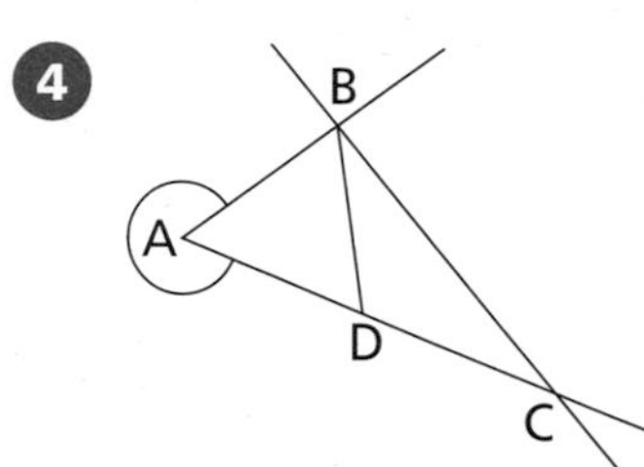

The reflex angle BAD is 295°, angle BDA is 70° and angle BCD is 31°.

What other angles can you work out? Give reasons for your answers.

What do you think?

What is the minimum amount of information needed to work out all the missing angles in these diagrams?

Consolidate – do you need more?

1 These triangles are not drawn accurately. Work out the sizes of the missing angles.

a

A, B, C; a, 120°, 34°

b

D, E, F; f, 18°, 13°

c

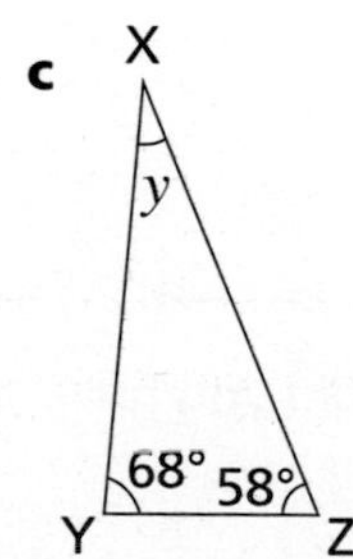

2 **a**

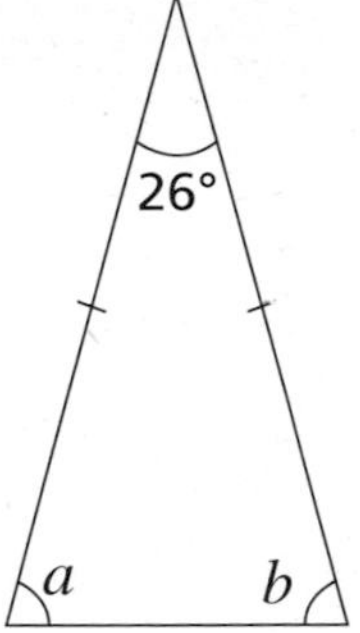

i Which two angles in this isosceles triangle are equal?

ii Work out the sizes of angles a and b.

b

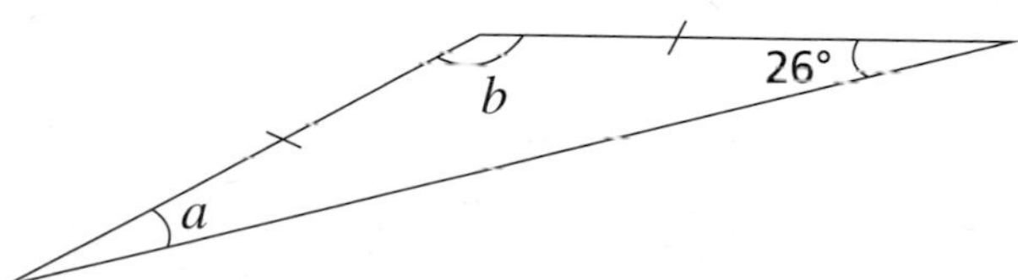

i Which two angles in this isosceles triangle are equal?

ii Work out the sizes of angles a and b.

3

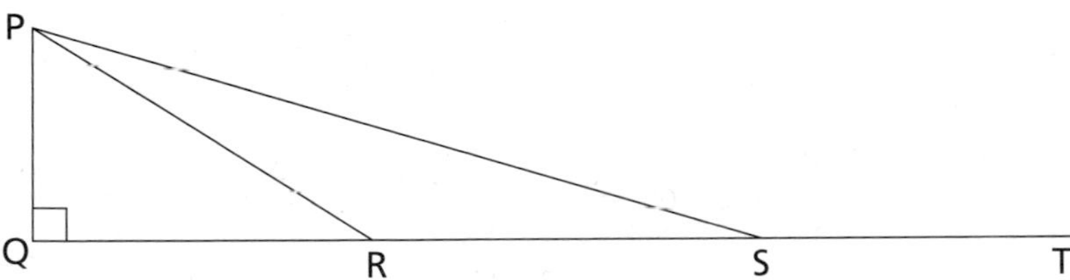

a Write down the size of ∠PQR

b If ∠PRQ = 48° and ∠TSP = 141°, work out the sizes of

i ∠RPQ **ii** ∠PSR **iii** ∠PRS **iv** ∠RPS

Stretch – can you deepen your learning?

1

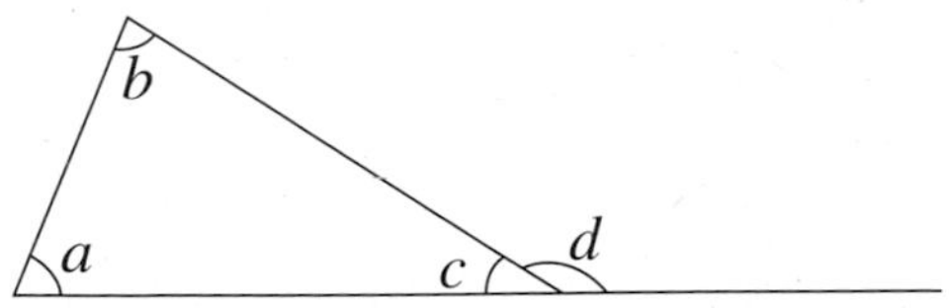

Work out these values for the given angles in parts **a** and **b**.

i c **ii** d **iii** $a + b$

a a = 75° and b = 85° **b** a = 81° and b = 57°

c Repeat for different values of a and b.

What do you notice?

2

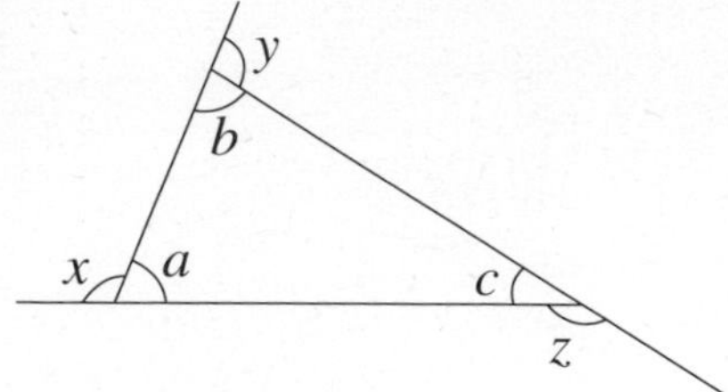

Work out these values for the given angles in parts **a** and **b**.

i c **ii** x **iii** y **iv** z **v** $x + y + z$

a $a = 75°$ and $b = 65°$

b $a = 61.3°$ and $b = 87.2°$

c Repeat for different values of a, b and c.

What do you notice?

Reflect

1 Construct an equilateral triangle with sides 8 cm. (You may need to refer back to chapter 11.4.) Measure the angles. How close are they to 60° each?

2 Construct a triangle with one side 7 cm and the other two sides both 5 cm. Are two of your angles equal in size?

3 Construct triangle ABC with AB = BC = 6 cm and angle ABC = 72°. Measure the other two angles. Are they the size you expect them to be?

12.3 Angles in quadrilaterals

Small steps

- Know and apply the sum of angles in a quadrilateral
- Recognise types of quadrilateral
- Solve angle problems using properties of triangles and quadrilaterals
- Solve complex angle problems

Key words

Properties – features of something that are always true

Parallelogram – a quadrilateral with two pairs of parallel sides that are equal in length

Kite – a quadrilateral with two pairs of adjacent sides that are equal in length

Are you ready?

1 Work out the size of each unknown angle.

a

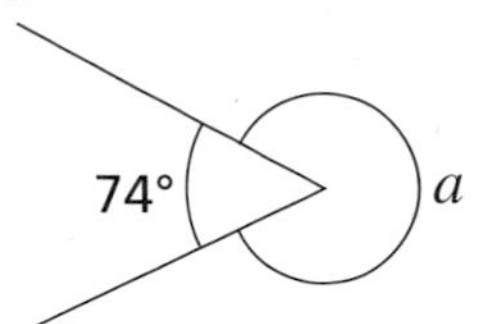

b

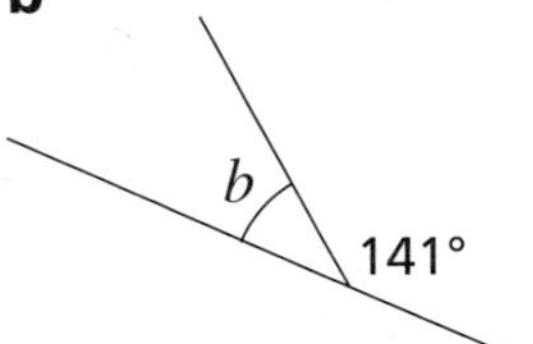

c

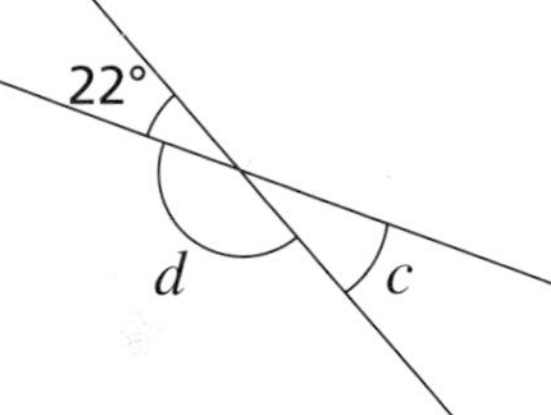

2 Find the size of each unknown angle in these triangles.

a

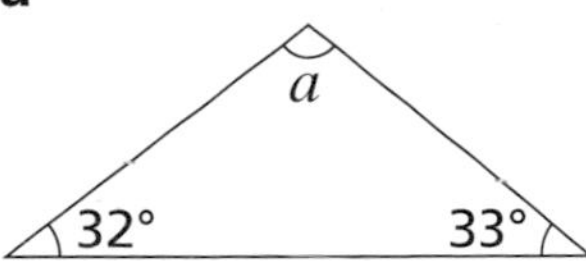

b

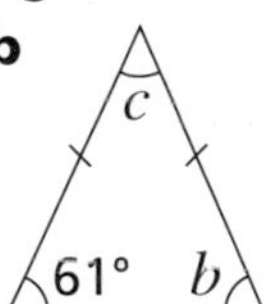

3 Name these quadrilaterals.

a

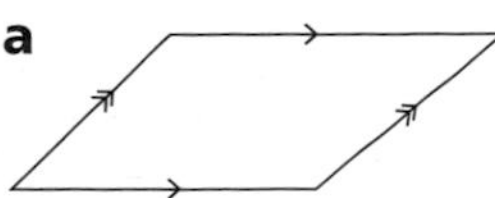

b

c

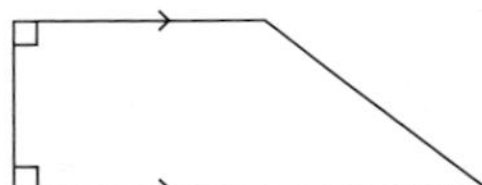

4 Solve these equations.

a $9x = 180$ **b** $9x + 90 = 360$ **c** $9x + 90 = 180$

Models and representations

Geoboards or geoboard apps

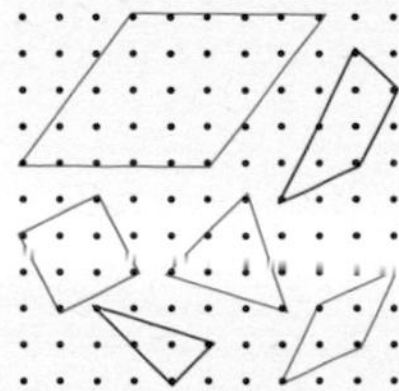

Paper strips

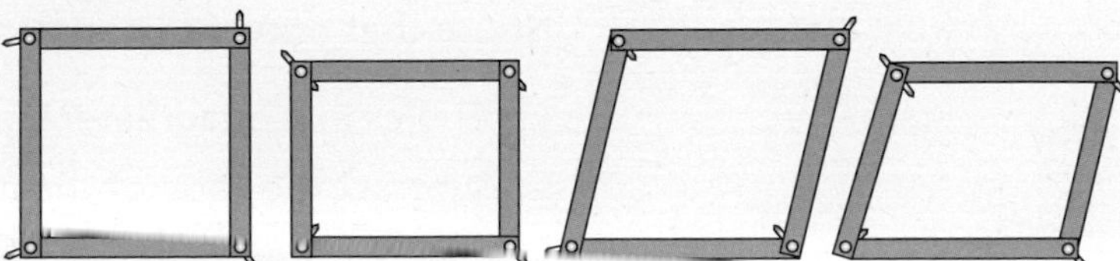

12.3 Angles in quadrilaterals

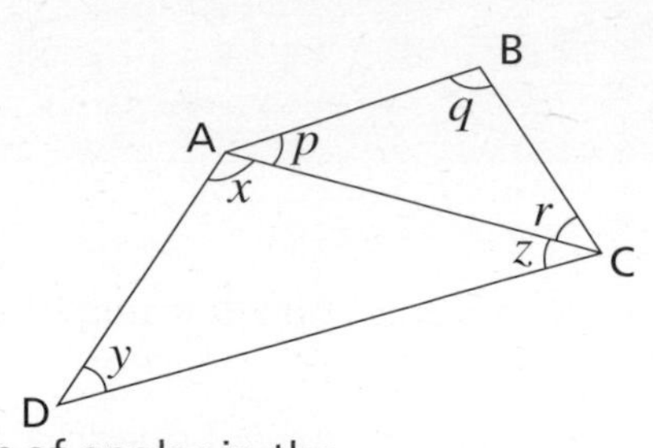

Any quadrilateral can be split into two triangles like this.

The angles of the quadrilateral are q, $(r + z)$, y and $(x + p)$.

The sum of the angles is $q + r + z + y + x + p$.

This is equivalent to $p + q + r + x + y + z$.

Angle sum of the quadrilateral $= (p + q + r) + (x + y + z)$

$= 180° + 180°$

$= 360°$

Both sets of angles in the triangles will add up to 180°.

Angles in a quadrilateral add to 360°.

You will use this fact to work out unknown angles in quadrilaterals. You will also explore the fact that many of the special quadrilaterals have pairs of equal angles.

Example 1

Work out the sizes of the unknown angles.

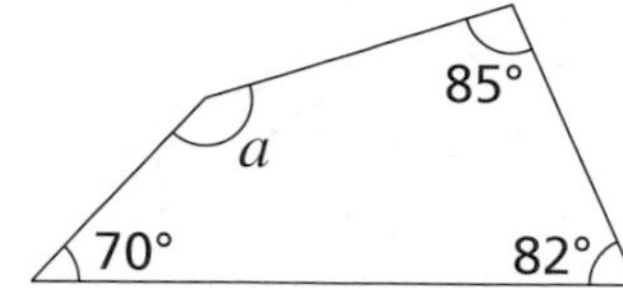

$85° + 70° + 82° = 237°$

Find the total of the angles you know.

So $a = 360° - 237° = 123°$

Angles in a quadrilateral add up to 360°.

Subtract the total of the angles you know from 360°.

Alternatively, you could have written an equation and solved it to find *a*.

$a + 85° + 70° + 82° = 360°$

$a + 237° = 360°$

$a = 123°$

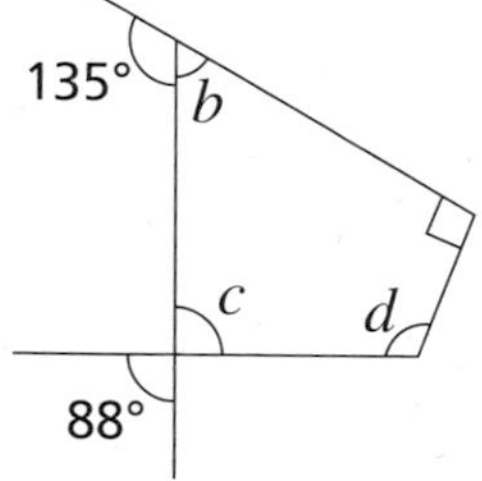

$b = 180° - 135° = 45°$

Adjacent angles on a straight line add up to 180°.

$c = 88°$

c is vertically opposite the 88° angle.

Vertically opposite angles are equal.

$45° + 88° + 90° = 223°$

Find the total of the angles you know.
Notice that one of the angles is a right angle.

So $d = 360° - 223° = 137°$

Angles in a quadrilateral add up to 360°.

Subtract the total of the angles you know from 360°.

Practice 12.3A

1 Draw four different quadrilaterals and show that they can all be split into two triangles.

2 Work out the size of each unknown angle in these quadrilaterals.

a

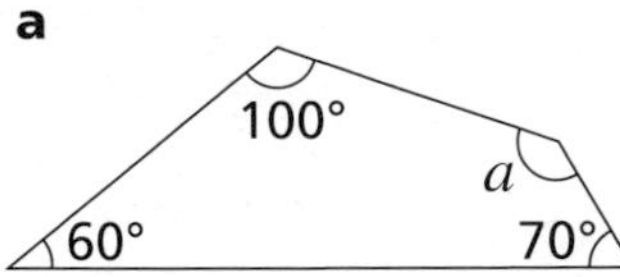

b

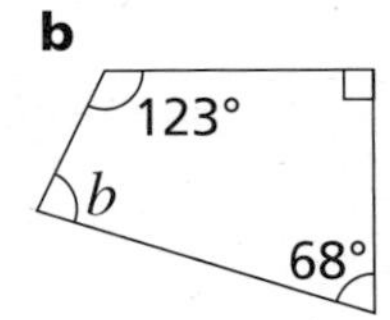

c

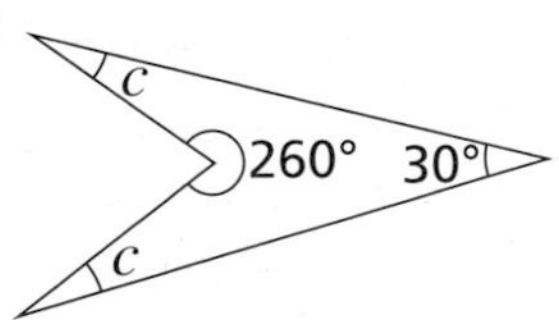

3 Use the angles facts you have studied to work out the size of each unknown angle.

a

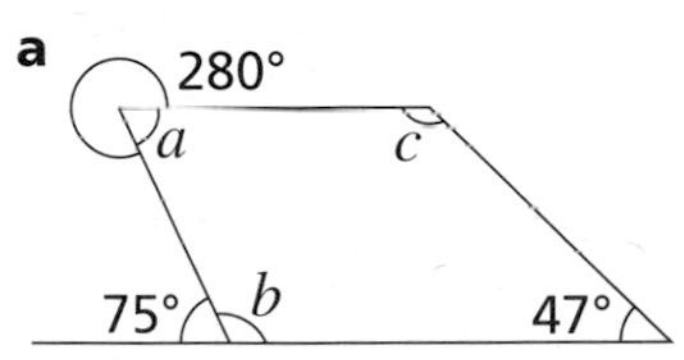

b

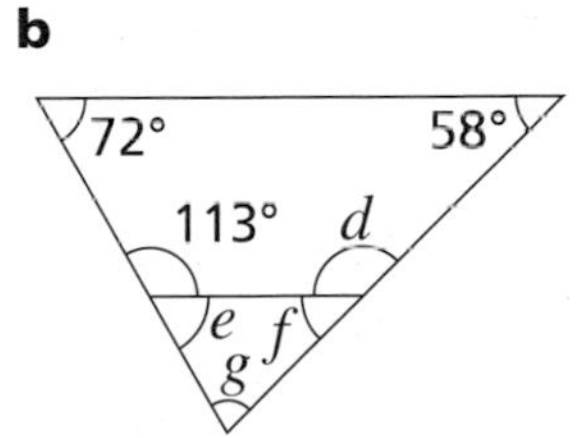

c

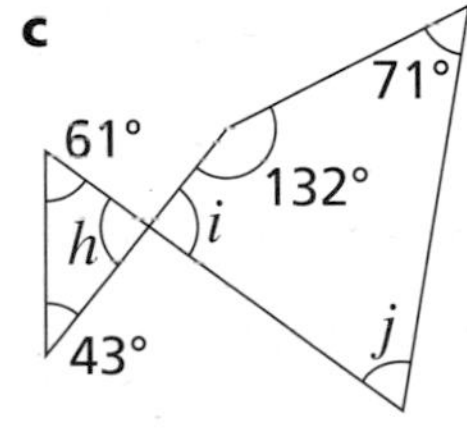

4 **a** Measure all the angles in these parallelograms. What do you notice about the opposite angles?

i

ii

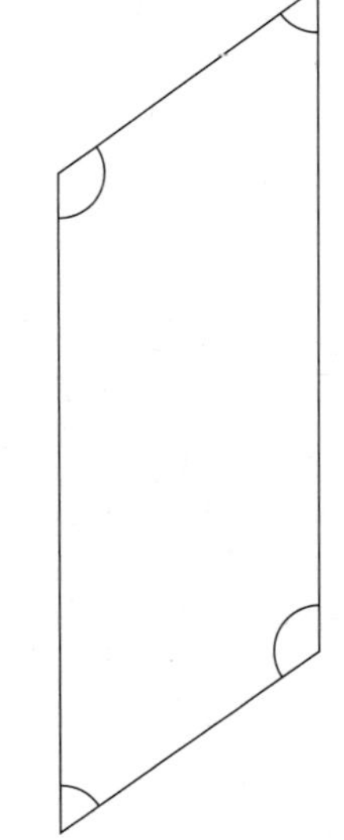

b Beca says, "Opposite angles of any rhombus are equal." Do you agree? Why or why not?

5 Work out the size of each unknown angle.

a

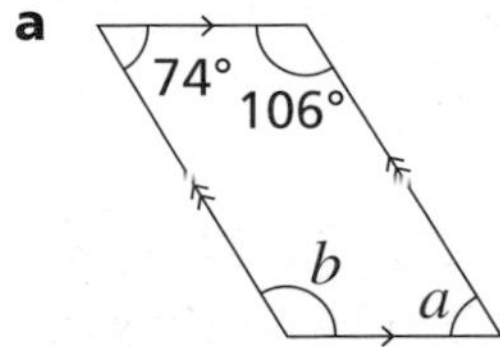

b

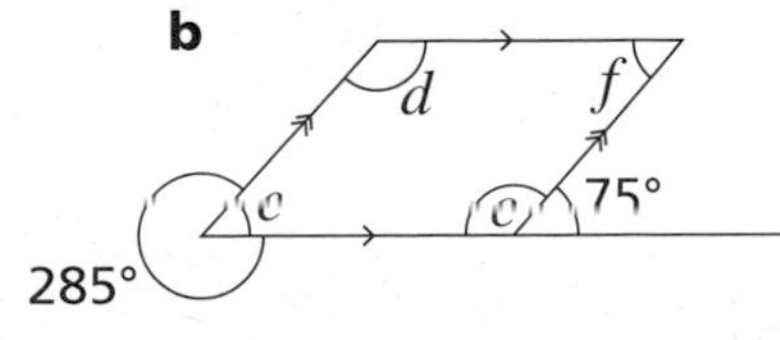

c

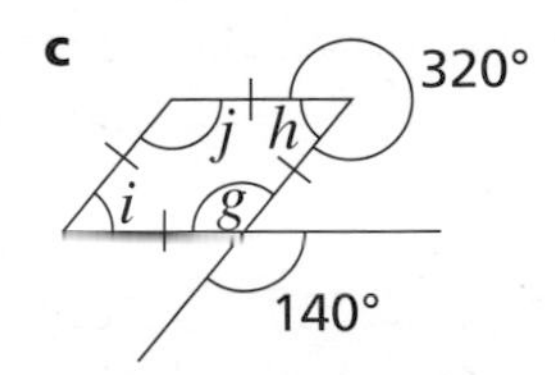

What do you think?

1 **a** Darius says that if you know one angle in a parallelogram or a rhombus, you can work out the other three angles. Is Darius correct? Explain how you know.

> Think about opposite angles in these shapes and the sum of the angles in a quadrilateral. You will meet this again in Chapter 12.6H.

b Work out

i ∠NOL **ii** ∠LMN **iii** ∠MLO **iv** ∠MNO

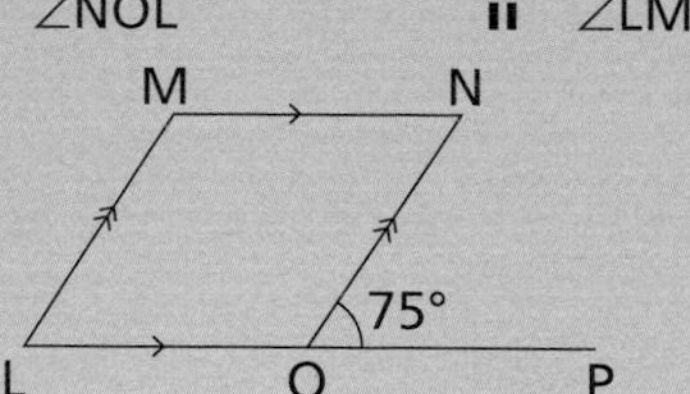

2 **a** PQRS is a rectangle. Work out the sizes of angles a and b.

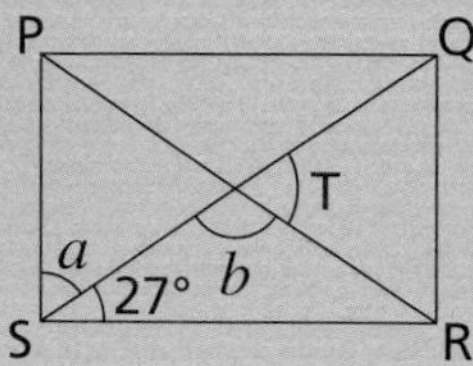

b ABCD is a rectangle. ∠BCE = 65°

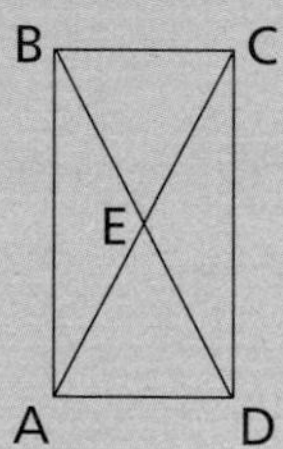

Work out

i ∠DCE **ii** ∠BEA

3 ABCD is a kite.

a What can you say about ∠DAB and ∠DCB?

b Work out ∠DAB and ∠DCB.

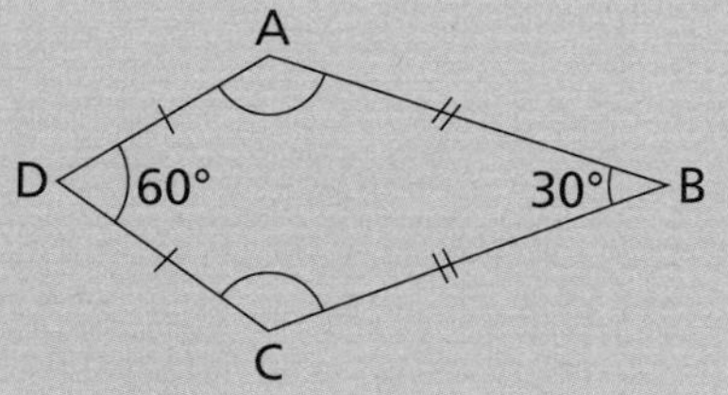

You are now going to use all the rules you know so far.

At each stage, you need to give a reason for the calculation or deduction you have made.

> Work step-by-step using the facts you are given to find the angles you are asked for.

Example 2

a PQRT is a rectangle. Work out ∠QST.

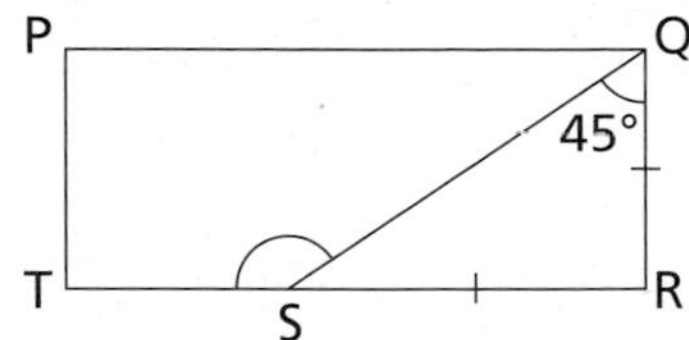

b Work out the reflex angle ∠CDE.

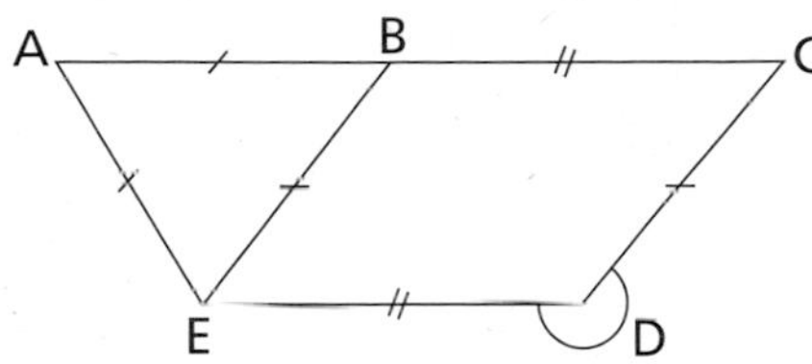

a ∠RSQ = 45° (angles at the base of an isosceles triangle are equal)

The hatch marks show that QRS is an isosceles triangle. So ∠RSQ = ∠SQR.

∠QST = 180° − 45° = 135° (angles on a straight line add to 180°)

∠QST and ∠RSQ are adjacent angles on a straight line.

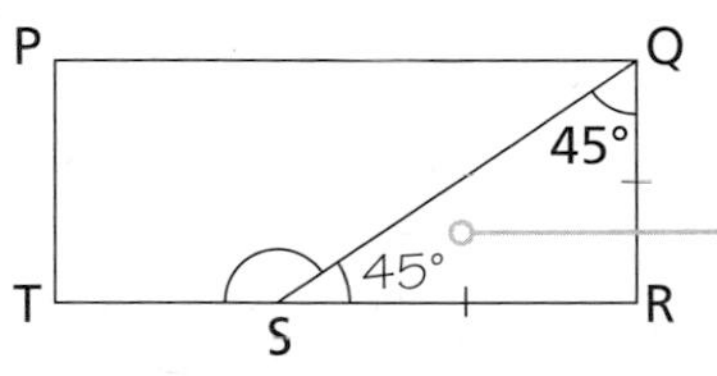

Add the angles you know to the diagram.

b ∠ABE = 180° ÷ 3 = 60° (ABE is an equilateral triangle)

The hatch marks show that AB = BE = EA so triangle ABE is equilateral.

The angles in an equilateral triangle are equal.

∠EBC = 180° − 60° = 120° (angles on a straight line add to 180°)

∠CDE = 120° (opposite angles of a parallelogram are equal)

The hatch marks show BCDE has two pairs of opposite equal sides, so it must be a parallelogram.

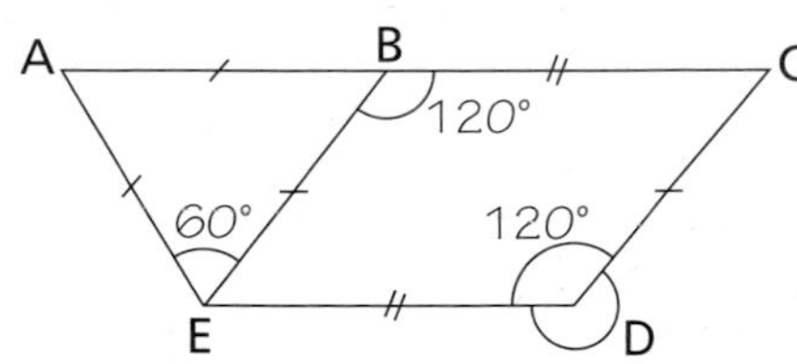

Reflex ∠CDE = 360° − 120° = 240° (angles at a point add to 360°)

Practice 12.3B

1 Work out the size of each unknown angle. Show all your working and give reasons for each step.

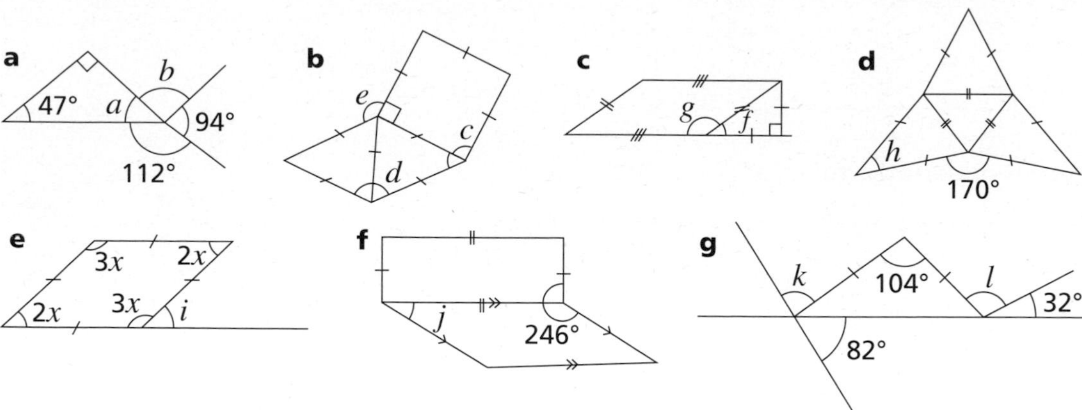

2 PQRS is a square and PQT is an equilateral triangle.

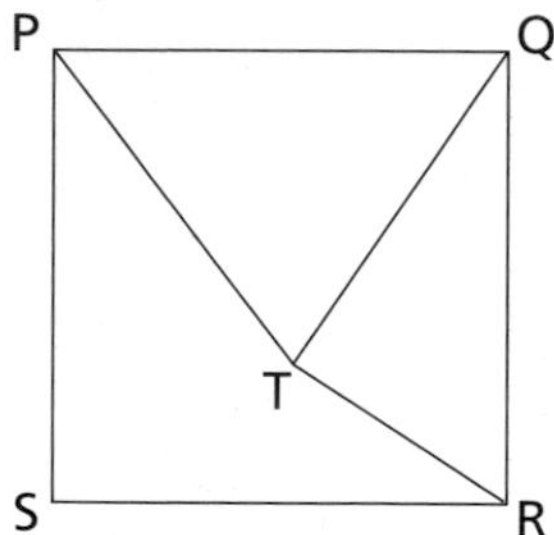

Work out the size of ∠QTR.

3 ABCD and CDEF are parallelograms.

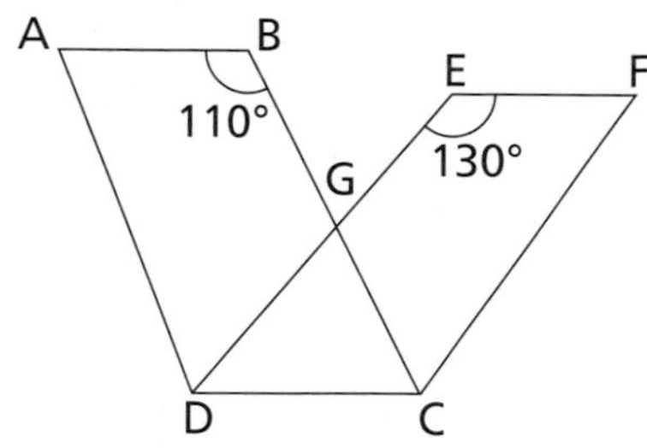

Work out the size of

a ∠EDC **b** ∠ADC **c** ∠DAB **d** ∠ADG **e** ∠BGD

What do you think? 

1

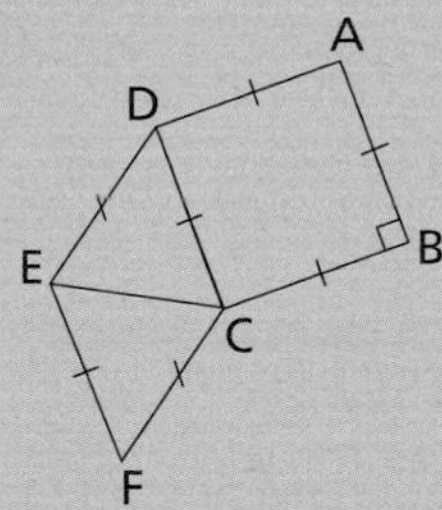

a ∠ADE = 126°. Work out the size of ∠BCF.

b What other angles can you find?

2 ABCD is a rectangle.

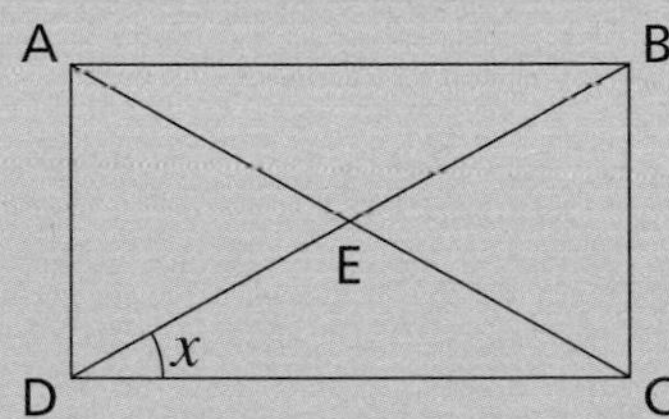

Write each of these angles in terms of x.

a ∠ADE

b ∠DEC

c ∠BEC

Consolidate – do you need more?

1 Work out the size of each unknown angle in these quadrilaterals.

a

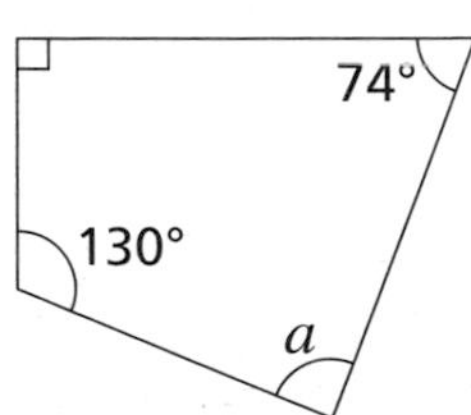

b

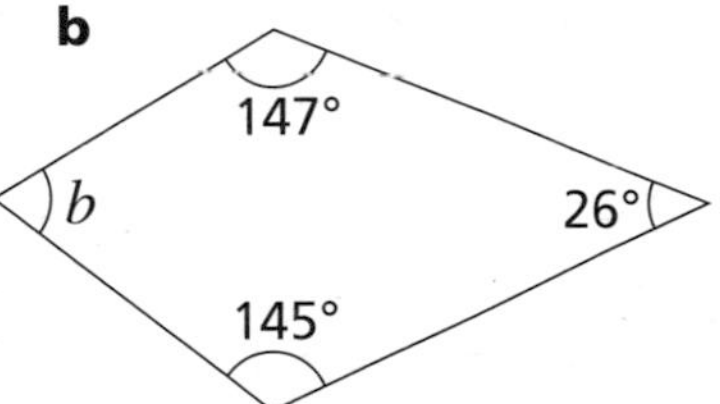

c

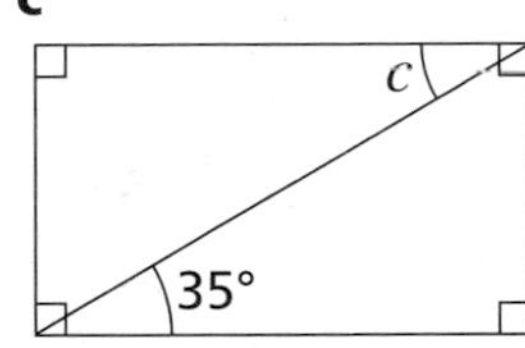

2 Use all the angle facts you know to work out the size of each unknown angle. Give reasons for each step of your working.

a

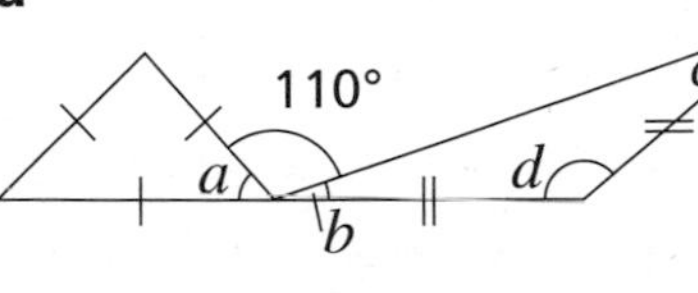

b

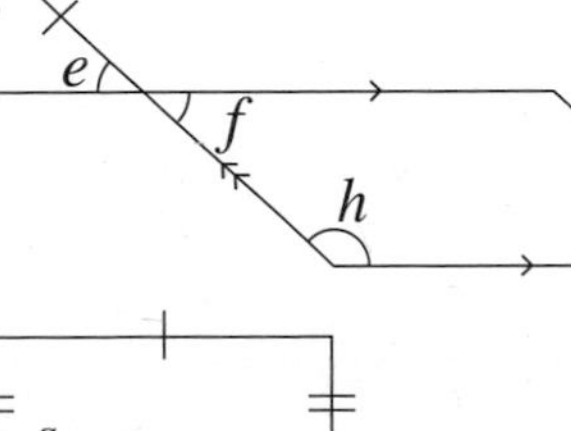

c

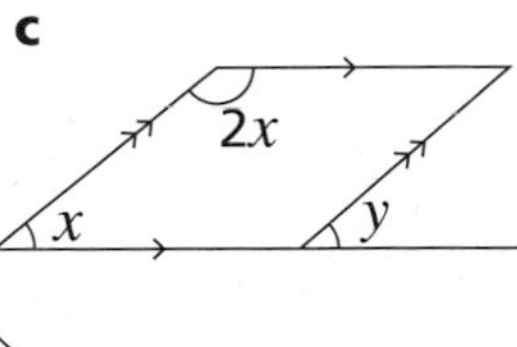

d

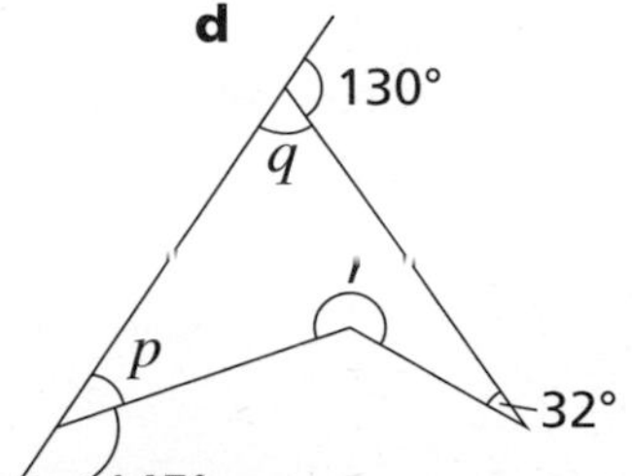

e

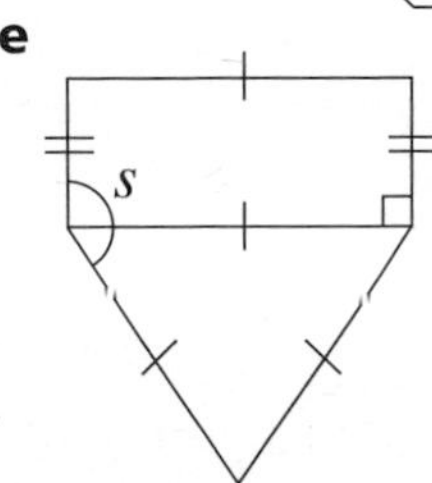

Stretch – can you deepen your learning?

1

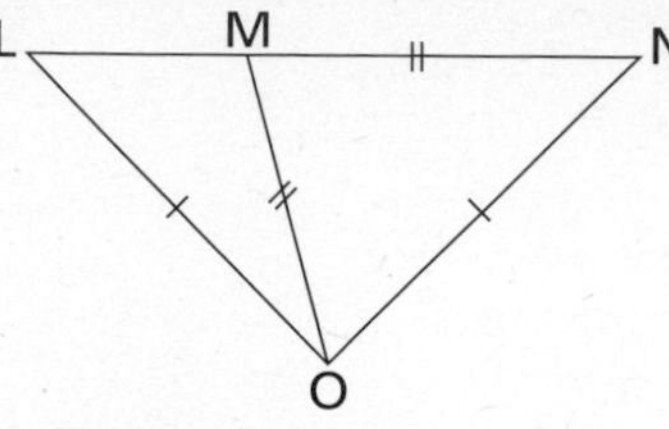

∠MNO = 52°

Sketch the diagram and mark on the sizes of all the other angles.

2

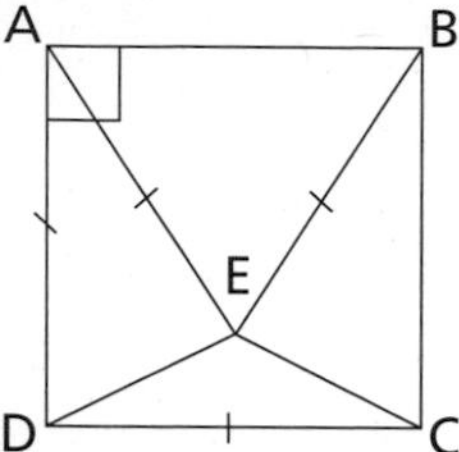

Work out the size of

a ∠DAE **b** ∠AED **c** ∠DEC

3

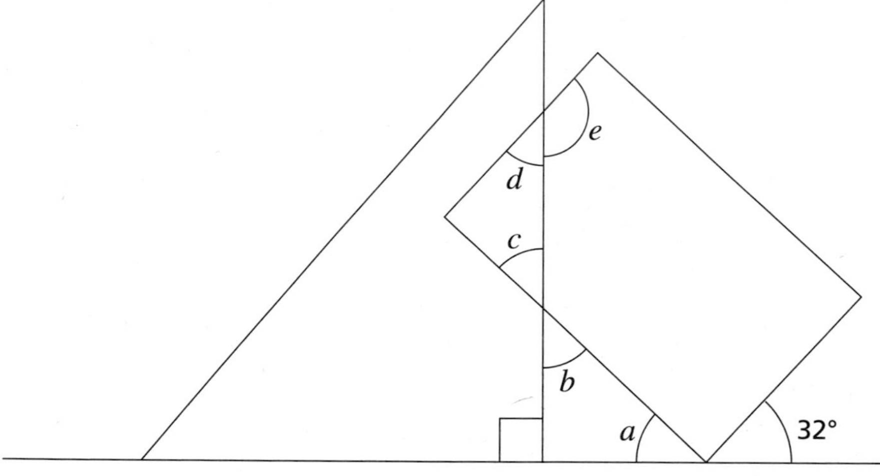

a The diagram shows a rectangle overlapping a right-angled triangle. Work out the size of each angle labelled with a letter.

b What happens to the sizes of angles a to e if the 32° angle is increased?

Reflect

Which quadrilaterals can you construct by joining two isosceles triangles? What can you deduce about their side and angle properties?

H 12.4 Angles in polygons

Small steps

- Find and use the angle sum of any polygon (H)
- Solve complex angle problems (H)

Key words

Interior angle – an angle on the inside of a shape

Polygon – a 2-D shape with straight sides

Regular polygon – a polygon whose sides are all equal in length and whose angles are all equal in size

Are you ready?

1 Write down the sum of the angles in

a a triangle

b a quadrilateral

2 Complete each of these statements using a number or a word.

a Angles on a straight line add up to ______.

b Angles in a full turn add up to ______.

c Opposite angles of a parallelogram are ______.

d Vertically opposite angles are ______.

3 Work out the missing angles.

a

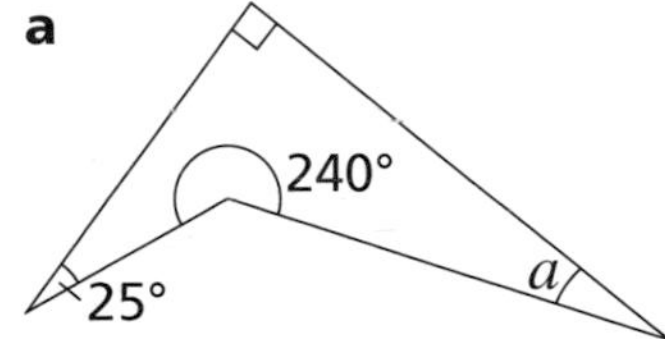

b

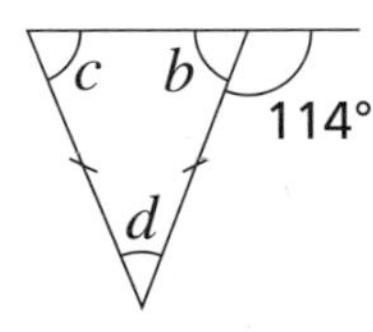

Models and representations

Geoboards or geoboard apps

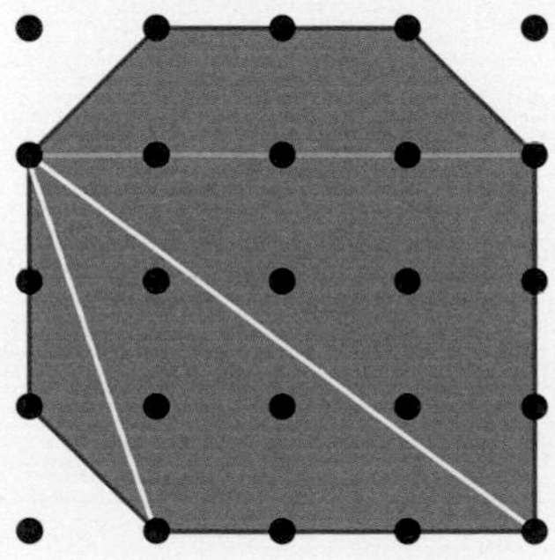

Paper strips

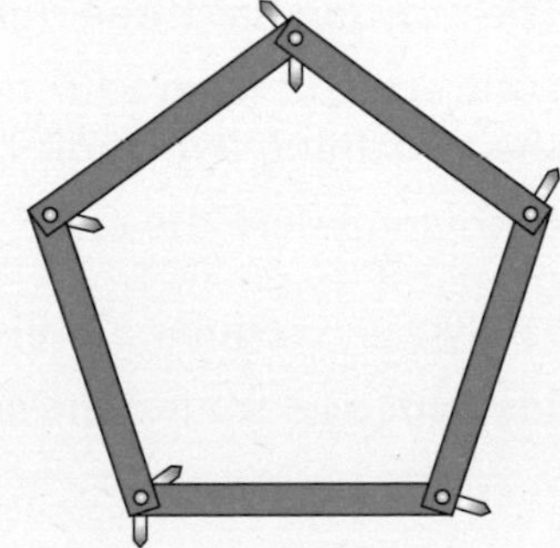

Dotty paper

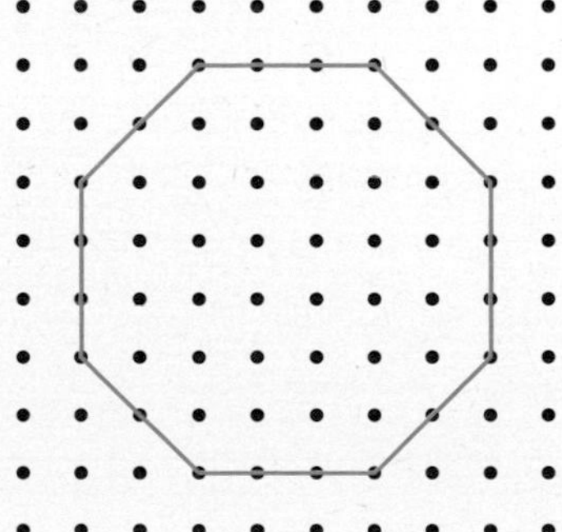

12.4 Angles in polygons H

In this lesson, you will learn how to find **interior angles** of **polygons**.

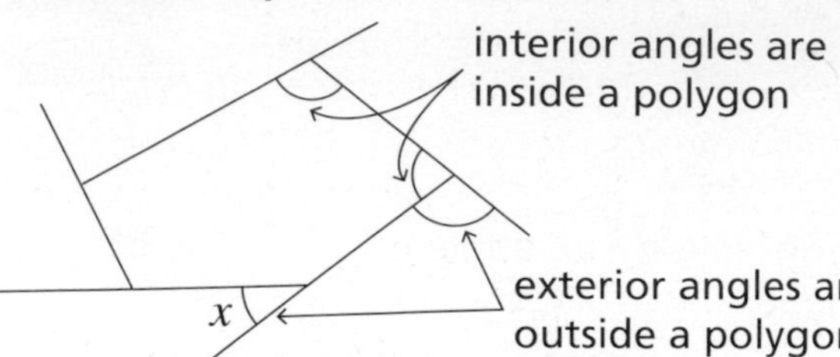

You will look at interior and exterior angles again in more detail in Book 2.

A pentagon can be split into three triangles.

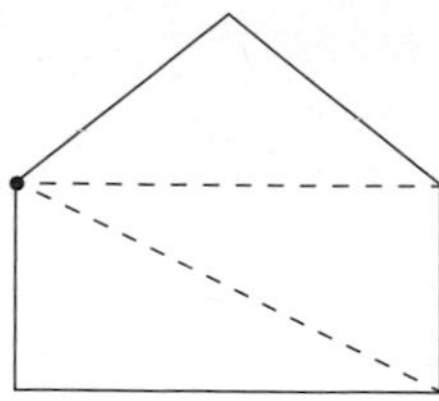

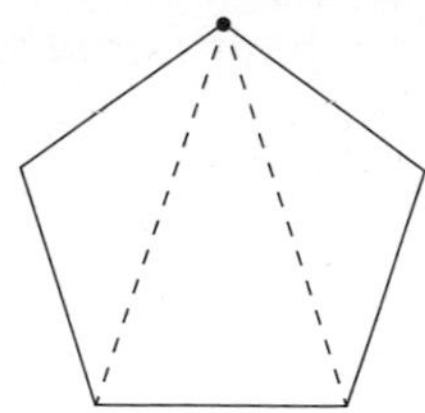

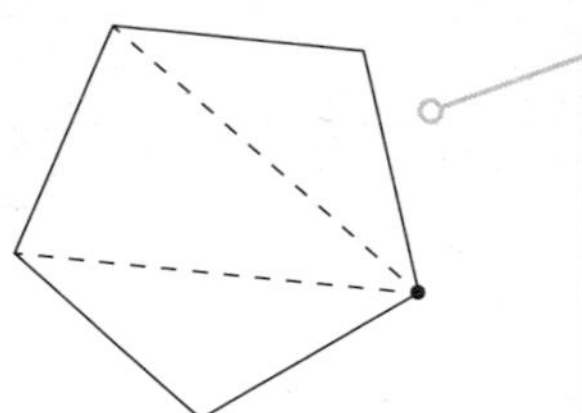

You know that the angles in a triangle add to 180° and that the angles in a quadrilateral add to 360° because it can be split into two triangles.

The sum of the interior angles of a pentagon is 3 × 180° = 540°

Example 1

Find the size of each of the unknown angles in these pentagons.

a

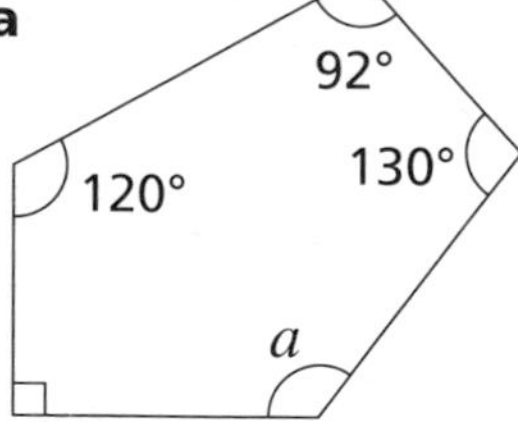

b

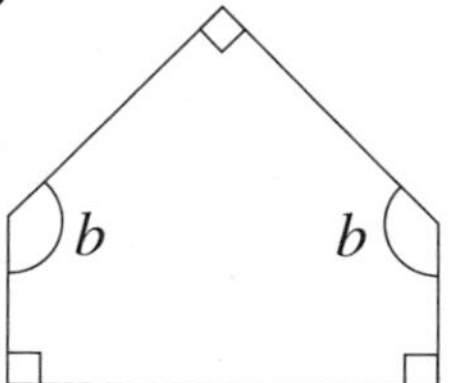

c

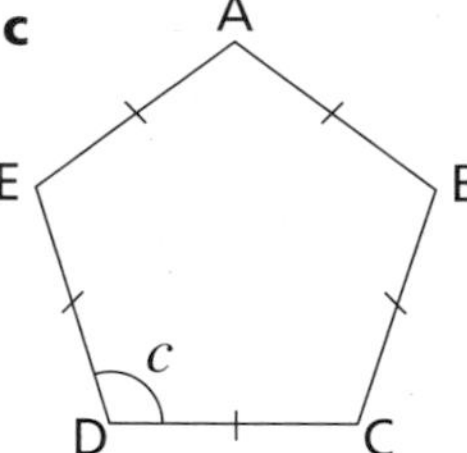

ABCDE is a regular pentagon

a $90° + 120° + 92° + 130° = 432°$ — Add up the angles you know.

So $a = 540° - 432° = 108°$ — Angles in a pentagon add to 540°. Subtract the sum of the angles you know from 540°.

b $3 \times 90° = 270°$ — The polygon has three right angles.

$540° - 270° = 270°$ — Subtract 270° from 540° to find the total of the remaining two angles.

So $b = 270° \div 2 = 135°$ — Both angles are the same size, so divide by 2

c $c = 540° \div 5 = 108°$ — The polygon is **regular**, so all the angles are equal. Divide 540° by 5 to find the size of each angle.

Practice 12.4A

1 The diagrams show a hexagon and a heptagon split into triangles from a vertex.

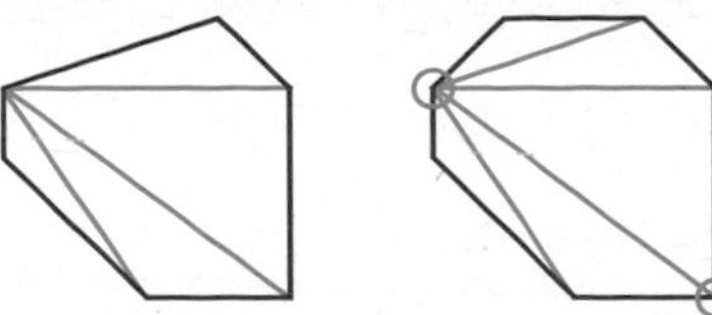

a Repeat this for an octagon and a nonagon.

b Copy and complete the table.

Shape	No. of sides	No. of triangles	Sum of interior angles
Triangle	3	1	1 × 180° = 180°
Quadrilateral	4	2	2 × 180° = 360°
Pentagon	5	3	3 × 180° = 540°
Hexagon	6	4	
Heptagon	7		
Octagon			
Nonagon			
Decagon			
n-sided polygon	n		

2 Work out the size of each unknown angle in these shapes.

a

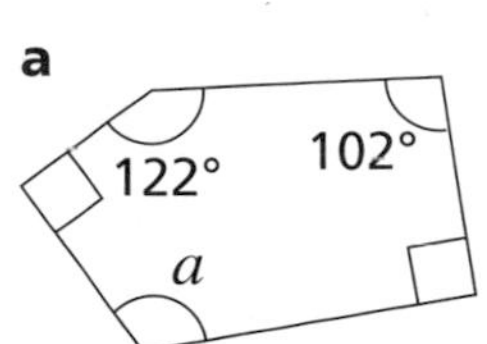

b

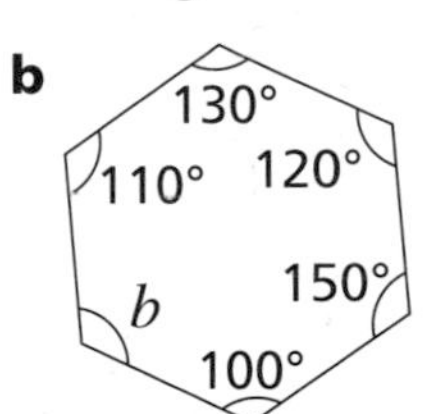

c

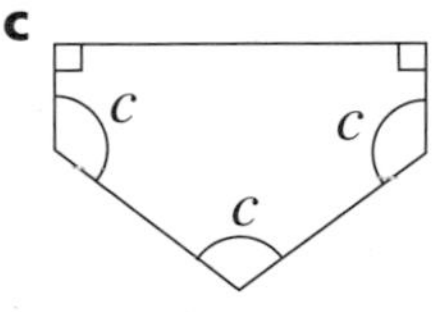

d

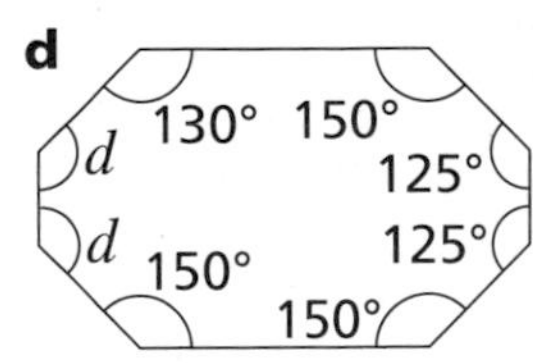

e

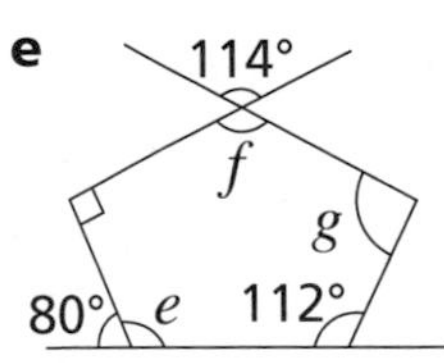

3 Find the size of each angle in

a a regular hexagon

b a regular nonagon

c a regular decagon

d a regular 12-sided shape (called a dodecagon)

4

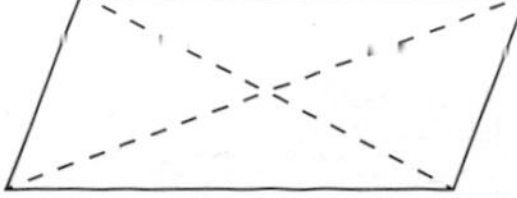

I have split the quadrilateral into four triangles so the angle sum of my quadrilateral is 4 × 180 = 720

What mistake has Chloe made?

What do you think?

1. The sum of the interior angles of a polygon with n sides is given by $(n - 2) \times 180°$.
   Find an expression for the size of an interior angle in a regular polygon with n sides.

2. **a** Both of these polygons are regular. Work out the size of each angle labelled with a letter.

   i

   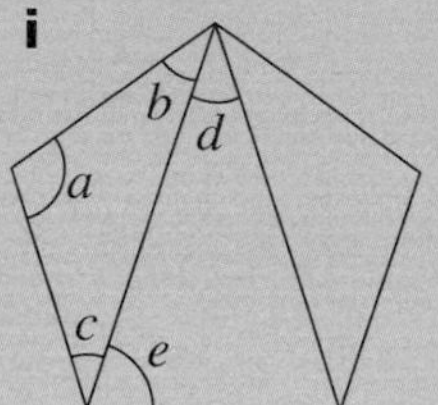

   ii

   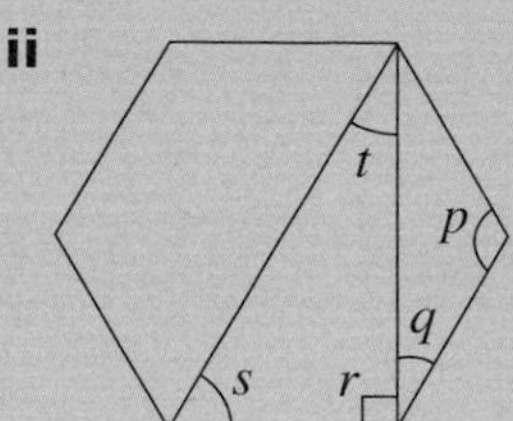

   b Investigate the sizes of the angles in the triangles formed by joining vertices of a regular octagon and a regular nonagon.

   c Why would it be more difficult to work with triangles in a regular heptagon?

Consolidate – do you need more?

1 Work out the size of each unknown angle in these shapes.

a **b** **c** **d**

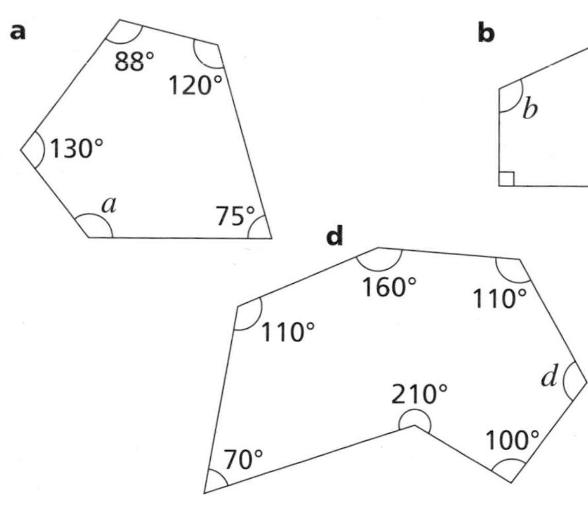

c **e**

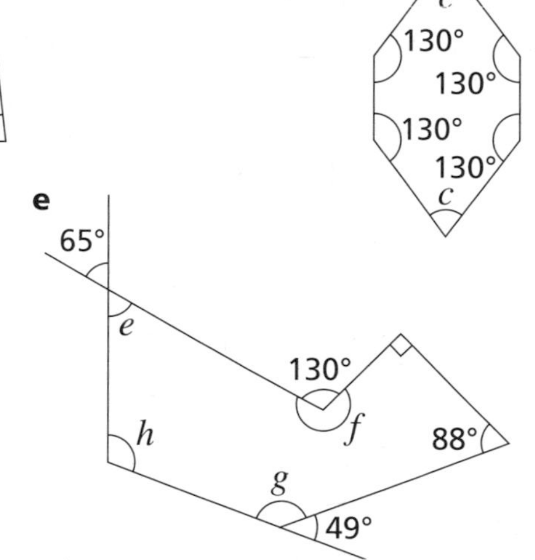

2 Find the size of each interior angle in

a a regular pentagon

b a regular octagon

c a regular 15-sided shape

d a regular 20-sided shape

3

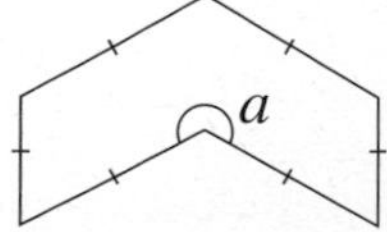

The sides of the hexagon are equal, so it is regular. $a = 720° \div 6 = 120°$

Explain why Abdullah is wrong.

Stretch – can you deepen your learning?

1 Each of the polygons below are regular.

a Work out the size of each angle labelled with a letter.

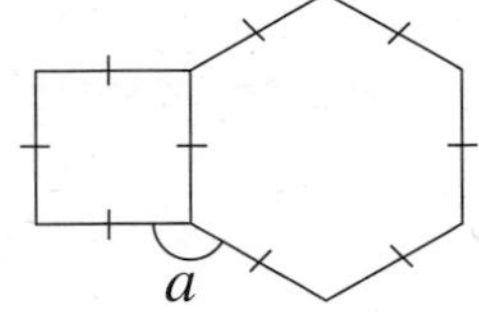

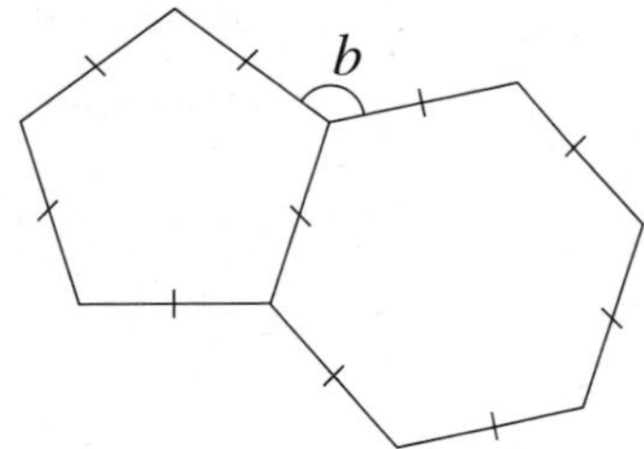

b Investigate the angles formed joining other regular polygons.

2 Four regular quadrilaterals can meet at a point.

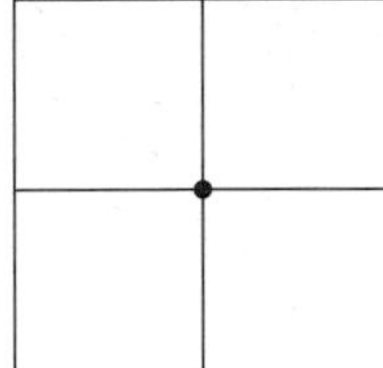

a Explain why a set of regular pentagons cannot meet at a point.

b Investigate which sets of regular polygons can meet at a point.

Reflect

Explain the rule for finding the sum of the angles in an n-sided polygon.

12.5 Angles and parallel lines

Small steps

- Investigate angles in parallel lines (H)
- Understand and use parallel line angle rules (H)

Key words

Transversal – a line that crosses at least two other lines

Corresponding angles – a pair of angles in matching positions compared with a transversal

Alternate angles – a pair of angles between a pair of lines on opposite sides of a transversal

Co-interior angles – a pair of angles between a pair of lines on the same side of a transversal

Are you ready?

1 What is the difference between parallel and perpendicular lines?

2 Work out size of each unknown angle.

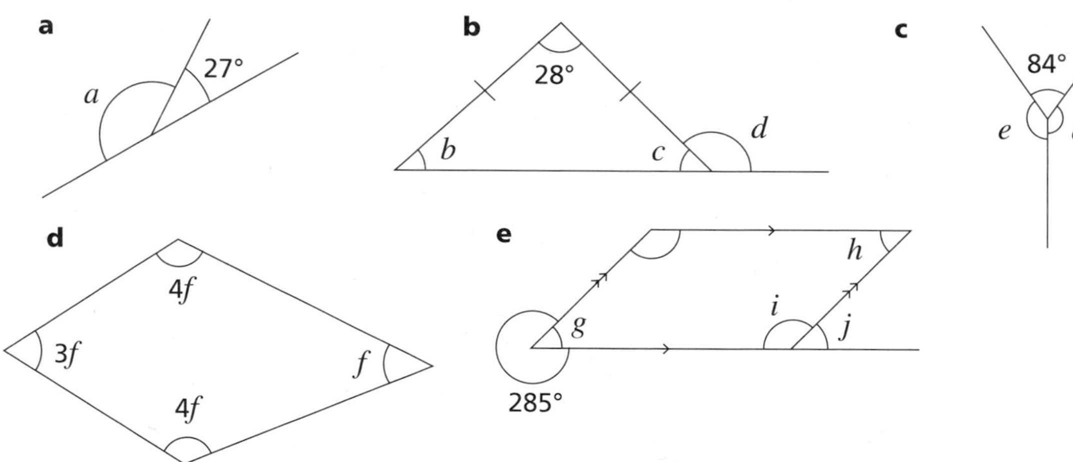

Models and representations

Geometry software

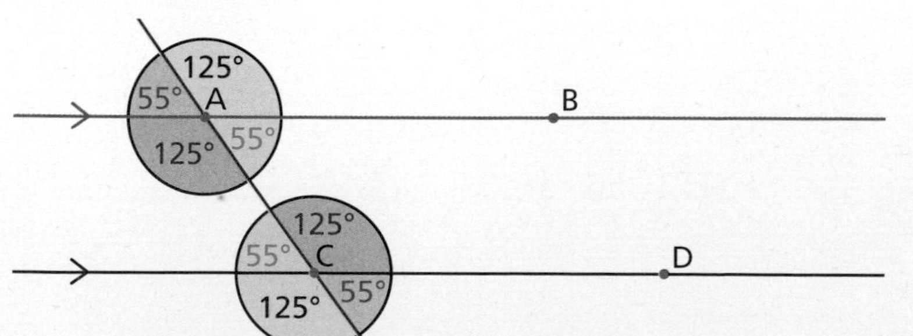

Straws

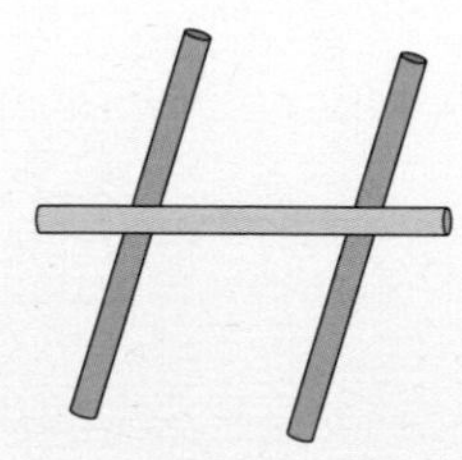

In this lesson, you will explore the angle properties of parallel lines and a line (called a **transversal**) going across them.

The rules you will discover are very important. You will look at them again in more detail in Books 2 and 3.

Practice 12.5A

1 **a** Draw a pair of parallel lines and a transversal line crossing them, as shown.

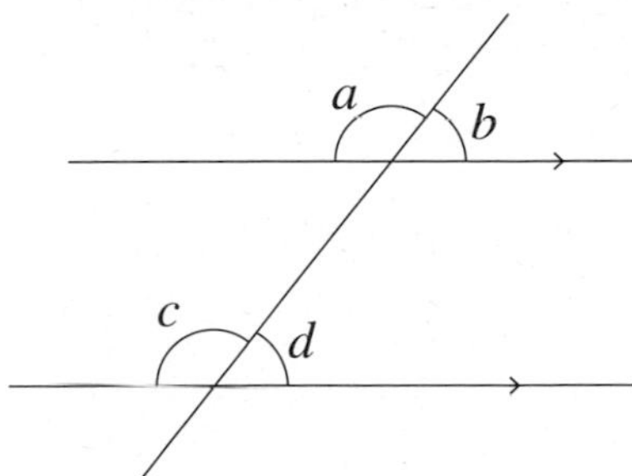

b Measure and write down the sizes of angles a, b, c and d.

c What do you notice? Compare your answers with a partner's.

2 **a** Draw a different pair of parallel lines and a transversal line crossing them, as shown.

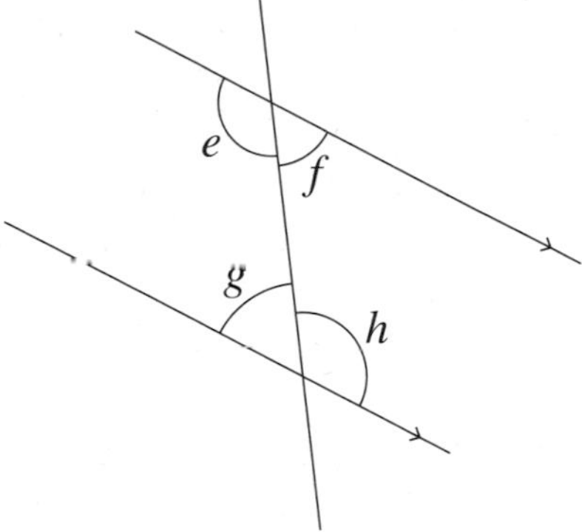

b Measure and record angles e, f, g and h.

c What do you notice? Compare your answers with a partner's.

3 **a** Now draw a pair of parallel lines and a transversal line crossing them, as shown.

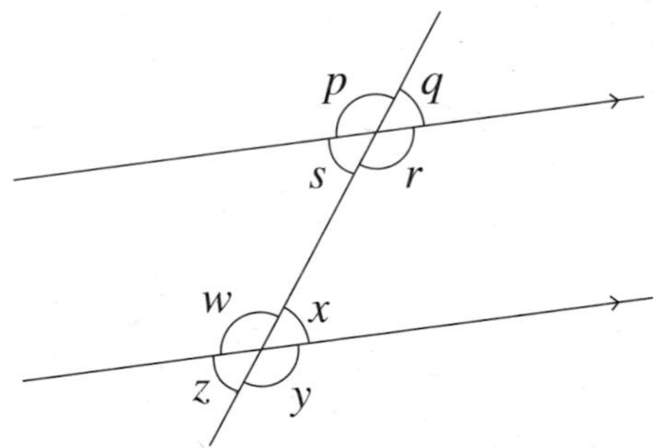

b Measure and record all eight angles.

c What do you notice? Deduce any possible rules that link the angles. Discuss your ideas with a partner.

What do you think?

1 Use your rules from Practice 12.5A question 3 part **c** to find the missing angles in parallelogram ABCD in terms of a.

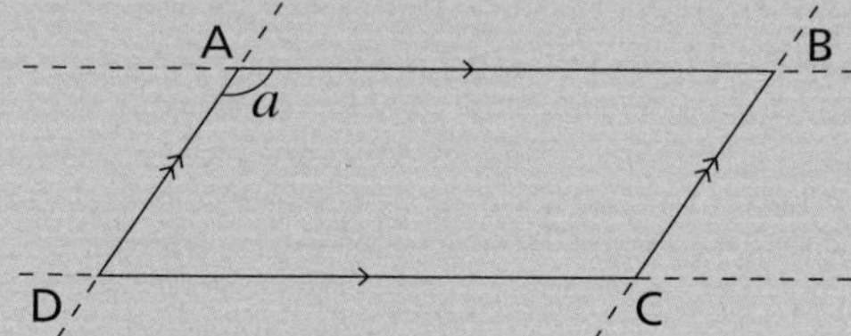

2 **a** What is the name of quadrilateral PQRS?

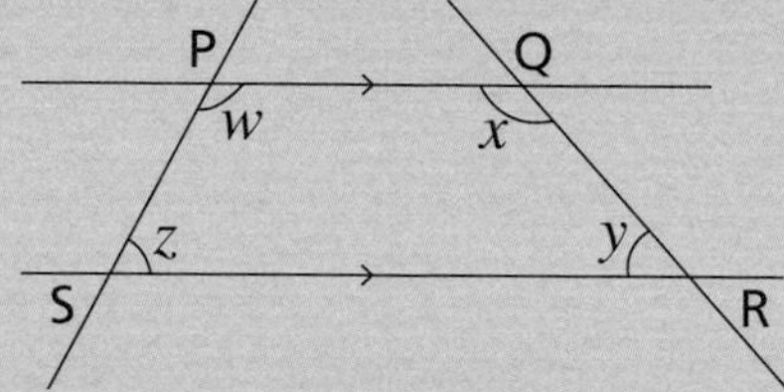

b What can you say about $w + z$ and $x + y$?

c Investigate other shapes formed by a pair of parallel lines and two non-parallel transversals. Is the result you found in part **b** always true?

In Figure 1, angles a and p are called **corresponding angles.**

They are to the right of the transversal and above the parallel lines – they are in corresponding positions.

Corresponding angles in parallel lines are equal.

Angles c and r are also corresponding angles so they are equal.

What other pairs of corresponding angles can you see?

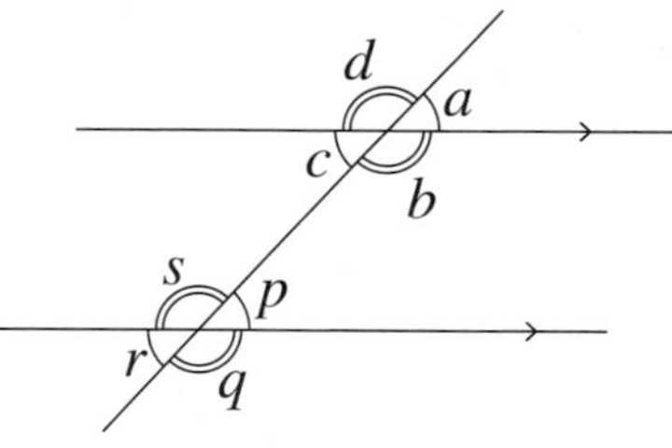

Figure 1

In Figure 2, angles w and z are called **alternate angles.**

They are between the parallel lines but on opposite (alternate) sides of the transversal.

Alternate angles in parallel lines are equal.

Angles x and y are also alternate angles so they are equal.

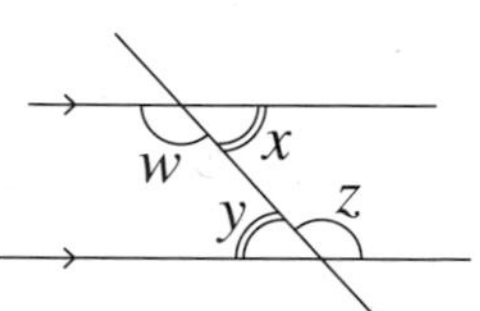

Figure 2

In Figure 3, angles w and y are called **co-interior angles.**

They are on the same side of the transversal and between the parallel lines.

Co-interior angles add up to 180°.

Angles x and z are also co-interior angles.

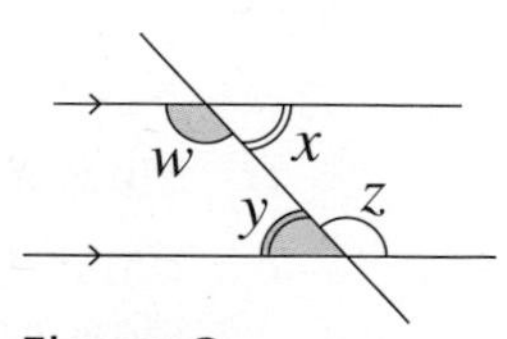

Figure 3

What pairs of angles in Figure 1 are co-interior?

Example 2

Find the value of each labelled angle. Give reasons for your answers.

a

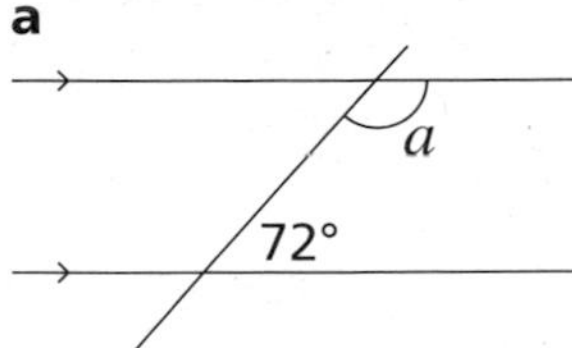

a $a = 180° - 72° = 108°$ (co-interior angles add up to 180°)

You need to recognise that a and the 72° angle are co-interior.

b

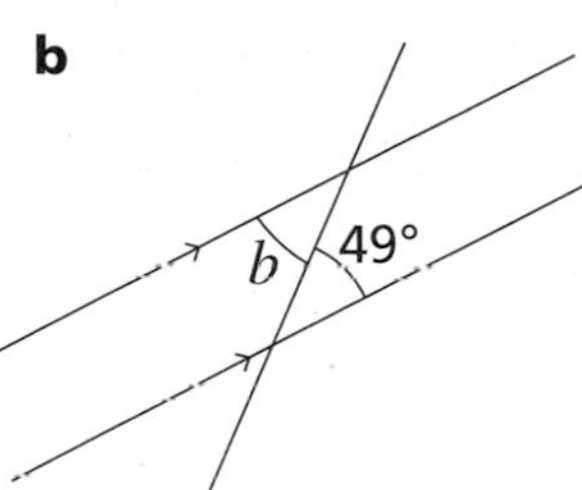

b $b = 49°$ (alternate angles are equal)

You need to recognise that b and the 49° angle are alternate.

c

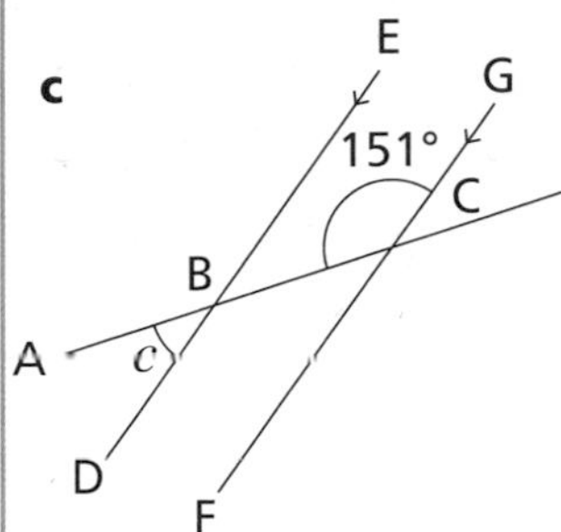

c $\angle BCF = 180° - 151° = 29°$ (angles on a straight line add up to 180°)

$c = 29°$ (corresponding angles are equal)

First find ∠BCF.

You cannot work out the angle you are asked to find in one step, but the angle adjacent to the 151° angle is corresponding to the angle you want.

Angle c and angle BCF are equal because they are corresponding angles.

Could you have worked out a different angle first?

Practice 12.5B

1 Write an angle that is

a corresponding to q

b alternate to r

c co-interior with z

d vertically opposite to n

e alternate to n

f corresponding to f

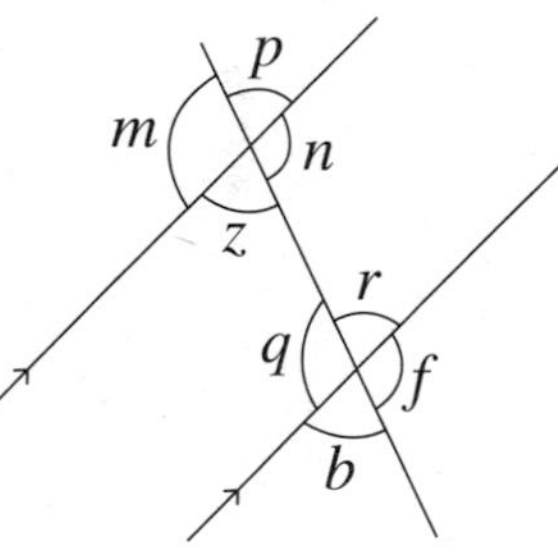

Is there more than one possible answer for any of the question parts **a** to **f**?

2 Work out the size of each unknown angle. Give reasons for your answers.

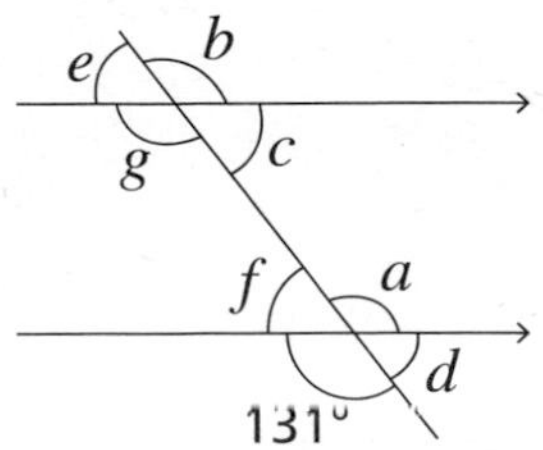

Compare your ways of finding the angles with a partner.

3 Work out the size of each unknown angle. Give reasons for your answers.

a a 128°

b 82° b

c c 82°

d 76° e d

e g f 31°

4 For each diagram, state whether the lines segments AB and CD are parallel or not.

a 112° B A 118° D C

b A 106° B C 74° D

5 Work out each of the unknown angles, giving reasons.

a 34° b c a

b

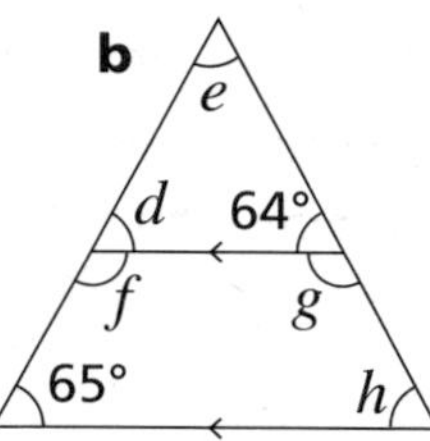

What do you think?

1 Which angles can you work out in this diagram?
Give reasons for your answers.

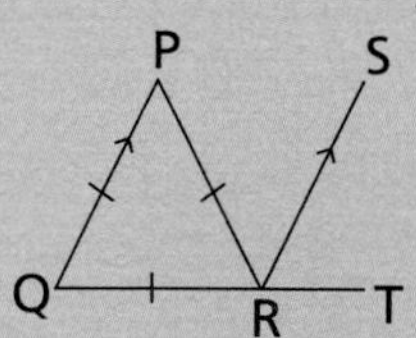

2 Find the values of x, y and z.

a $2x + 10$ 75° $3y$

b

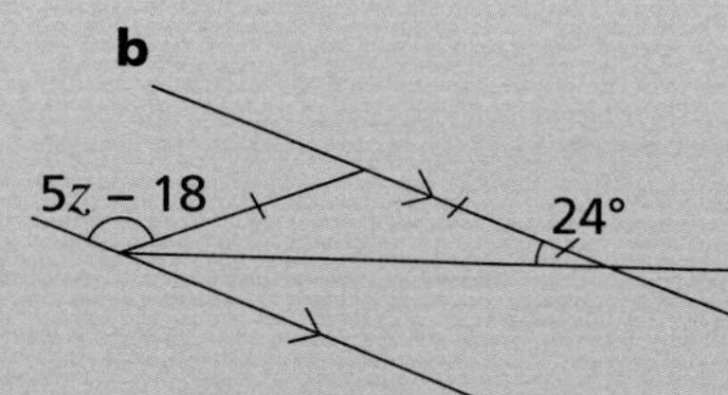

Consolidate – do you need more?

1 Write an angle that is

a corresponding to t

b alternate to t

c co-interior with t

d vertically opposite to t

e alternate to m

f corresponding to x

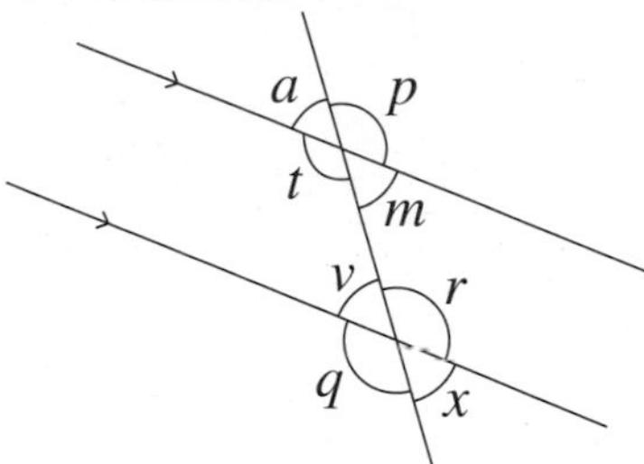

Is there more than one possible answer for any of questions in parts **a** to **f**?

2 Work out each of the unknown angles.

Give reasons for your answers.

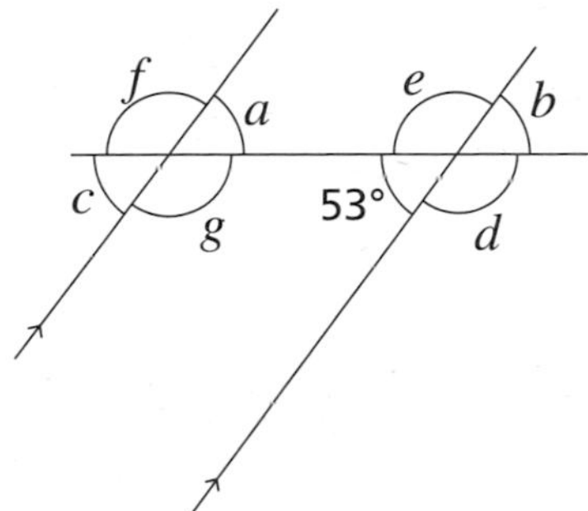

Compare your ways of finding the angles with a partner.

3 Work out each of the unknown angles, giving reasons.

a

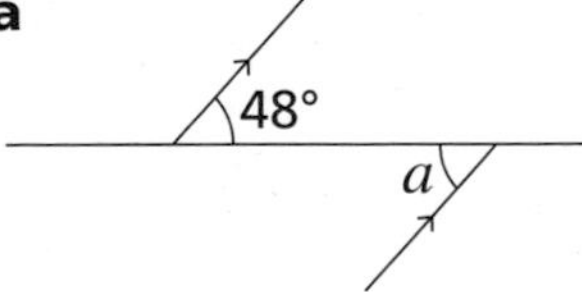

b

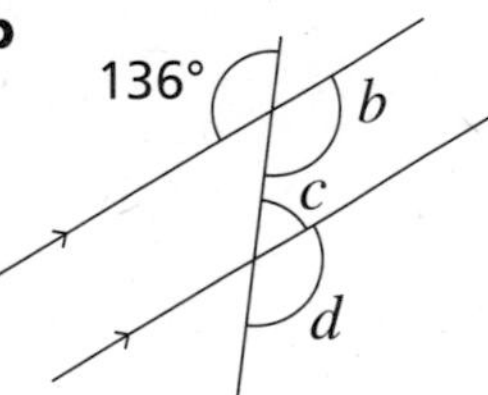

c

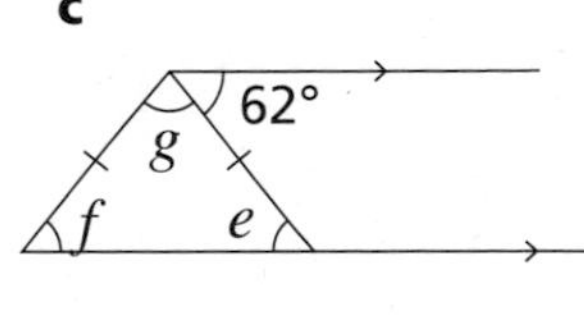

d

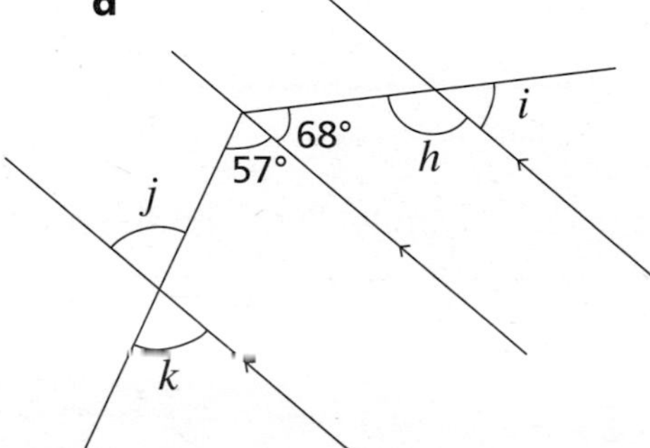

e

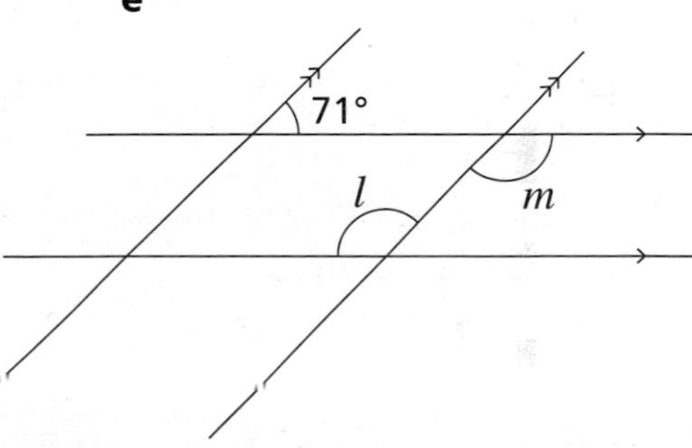

Stretch – can you deepen your learning?

1 **a** Find the size of ∠ABC.

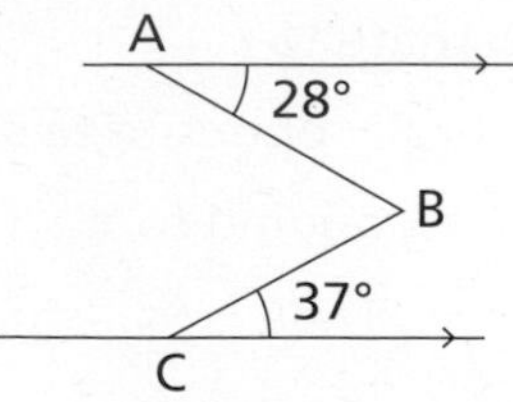

b Find the size of ∠BCD.

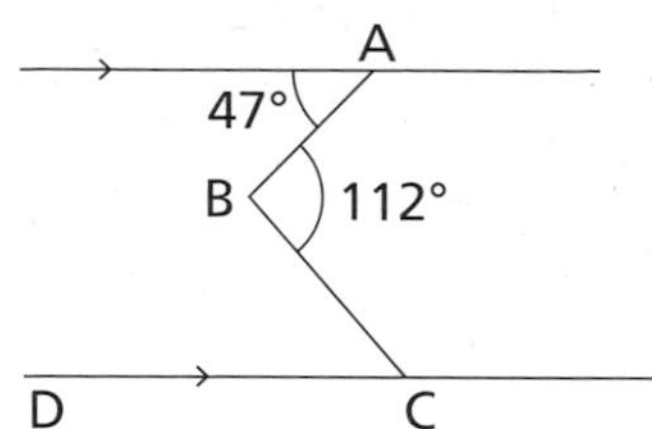

2 The angles labelled with arcs are called 'alternate exterior angles'.

What do you notice about them?

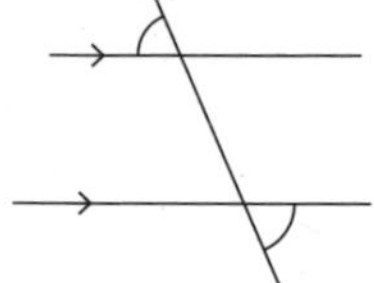

Can you find any other pairs of alternate exterior angles on the diagram? Copy the diagram and show them using coloured arcs.

3 Which angle is different from the other four? Why? Investigate other sets of angles where one is the 'odd one out'.

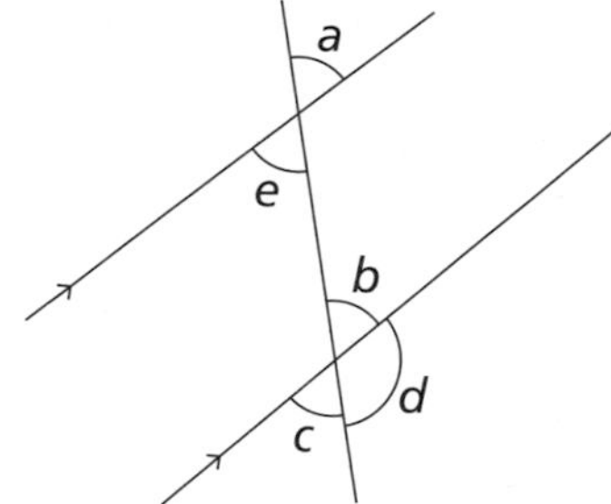

Reflect

Describe the connections between the rules for angles on parallel lines and the rules for equal angles in special quadrilaterals.

Ⓗ 12.6 Geometric proof

Small steps

- Use known facts to obtain simple proofs Ⓗ

Key word

Proof – an argument that shows that a statement is true

Are you ready?

1 Work out the value of the angle labelled x in each diagram.

a

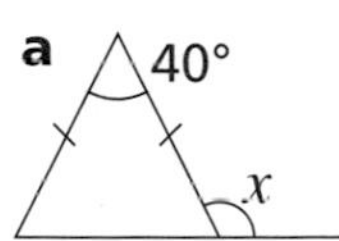

b

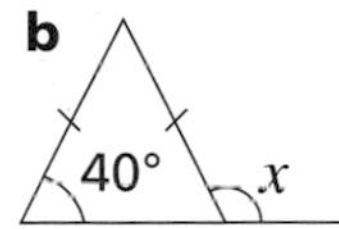

c

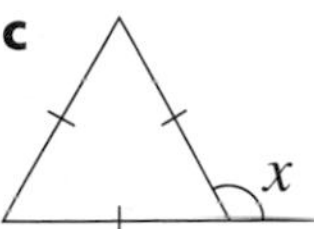

d

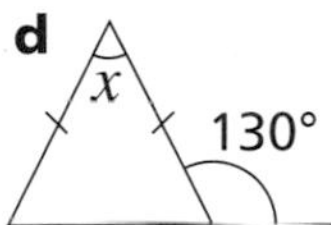

2 Work out the size of each unknown angle.

a

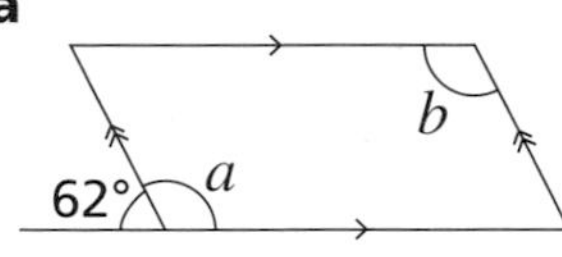

b

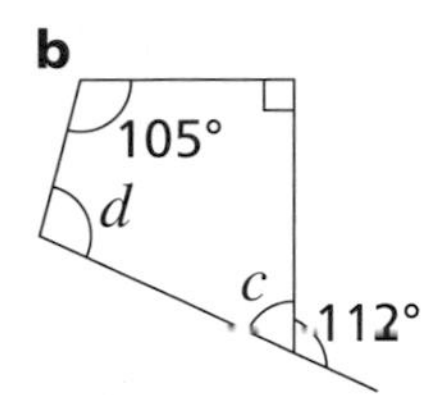

3 Work out the size of each unknown angle.

a

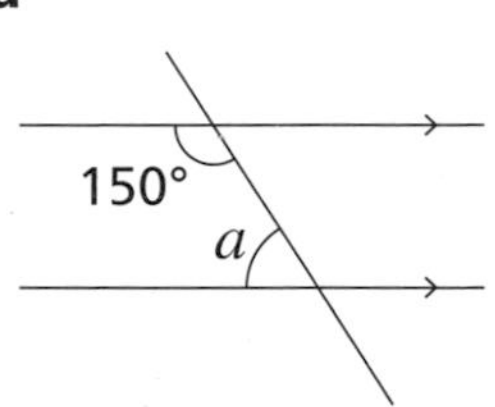

b

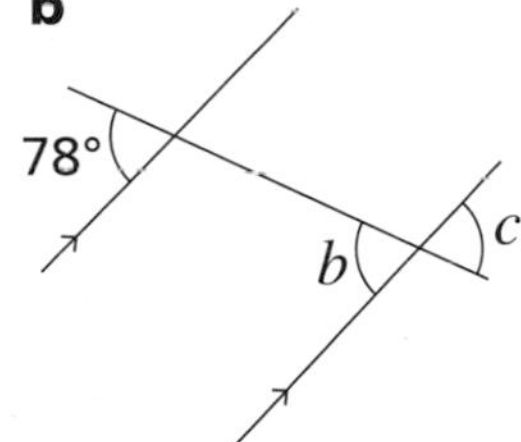

c

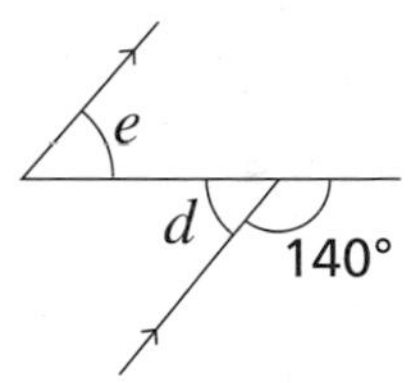

Models and representations

Bar models

a	b
180	

These can be used to show equality and help you to prove statements.

Proof is a formal way of 'explaining your answer'.

When writing a proof, each statement you make must follow on logically from the last statement or must be justified with a reason.

It shows that your claim is true for all cases and not just for some specific numerical examples.

Example 1

Use the diagram to show that $a = c$ and prove that vertically opposite angles are equal.

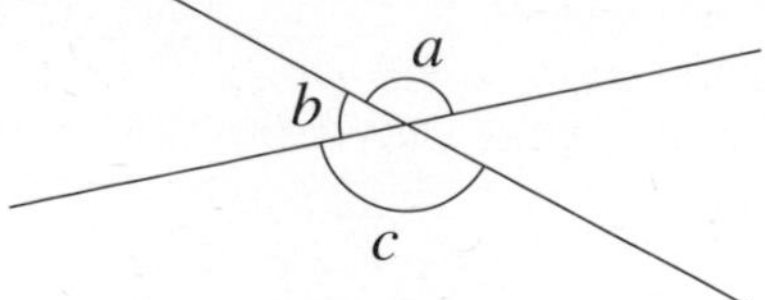

$a + b = 180°$ (angles on a straight line add up to 180°)

So $a = 180° - b$

$c + b = 180°$ (angles on a straight line add up to 180°)

So $c = 180° - b$

So $a = c$ as both are equal to $180° - b$.

In a proof, you should justify each of your statements with a reason.

This can be a rule you know already or something you have found out during your proof.

A useful way of showing that two things are equal to each other is showing there are both equal to something else.

You can use bar models.

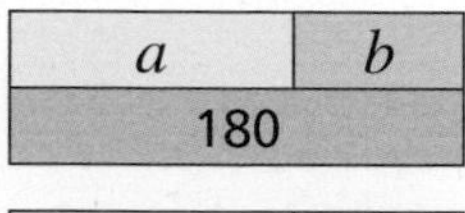

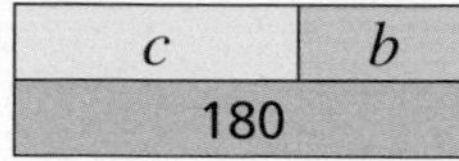

The two bar models show that both a and b and c and b add to 180°, so a and c must be equal.

Practice 12.6A

1 Prove that angle a is 45°.

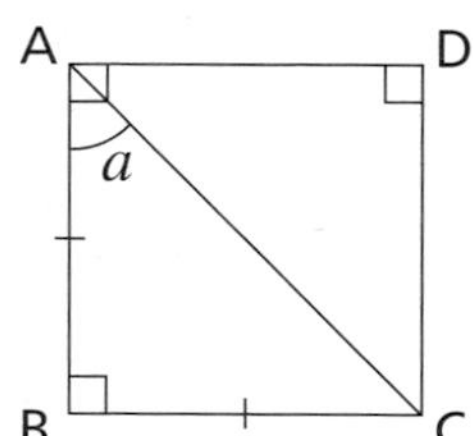

2 In both diagrams, ABCD is a rectangle.

i

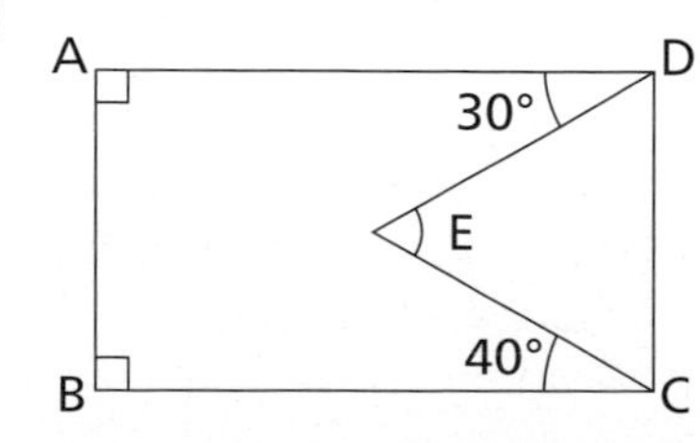

ii

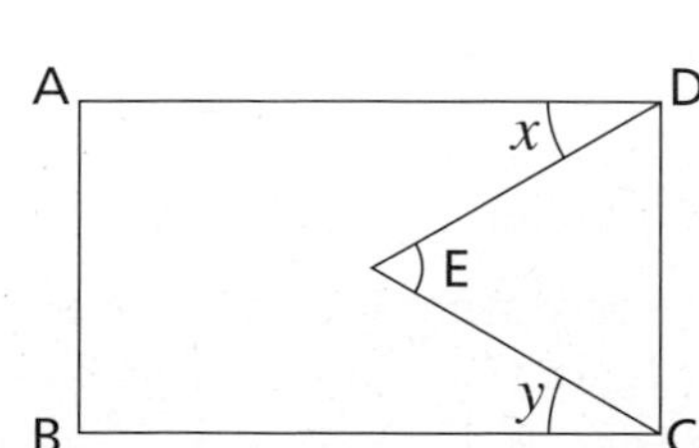

a Using diagram **i**, show that ∠DEC = 70°

b Using diagram **ii**, prove that ∠DEC = $(x + y)$

3 a Prove that $\angle CQY = 180° - a$

b Prove that $\angle DQY = a$

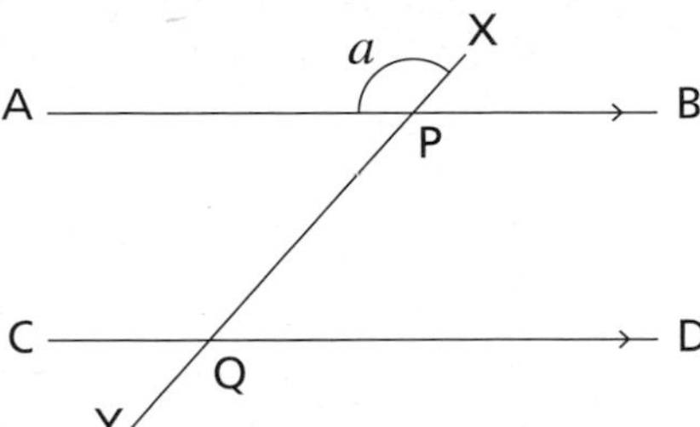

4 Prove that the opposite angles of a kite are equal in size.

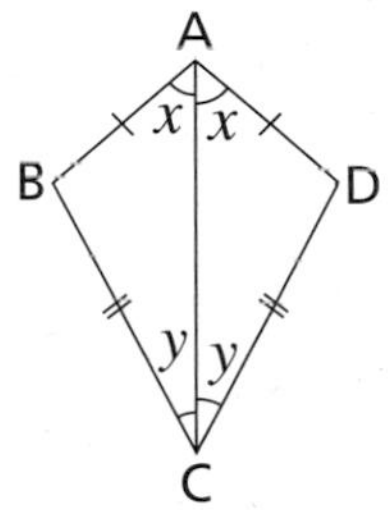

5 a Show that $a = 105°$

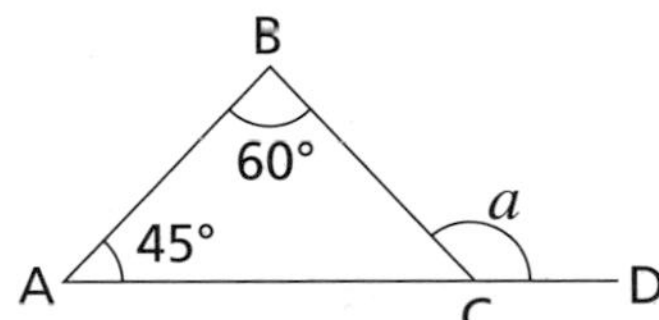

b Prove that $a = x + y$

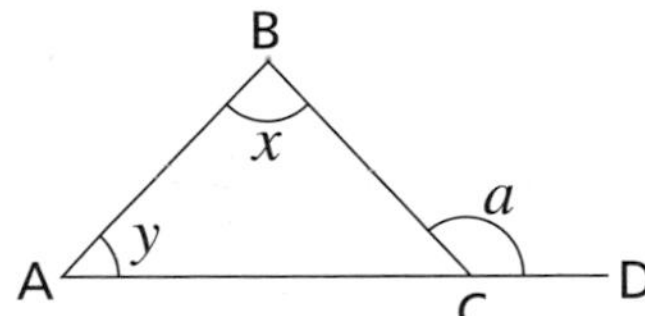

What do you think?

1 a Find expressions, in terms of x, for all the angles in this triangle.

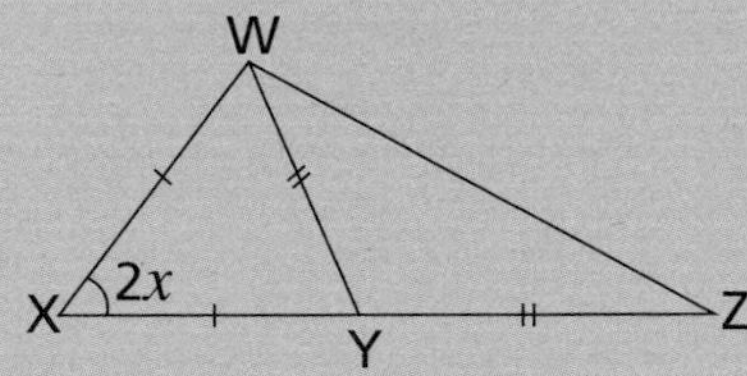

b Test your expressions by substituting a value for x.

2 Prove that $x + 4y = 180°$

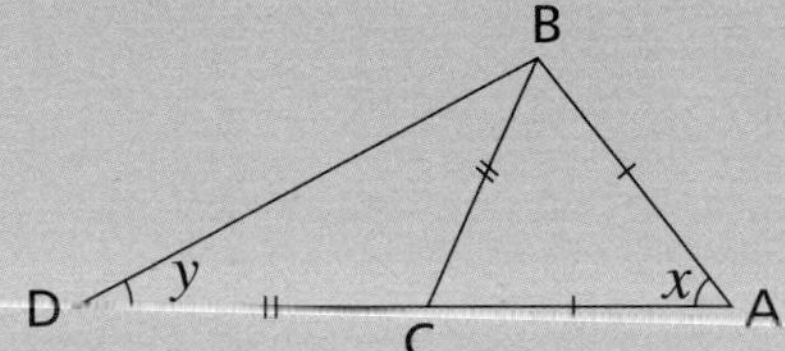

Consolidate – do you need more?

1 **a** Show that ∠AED = 150°

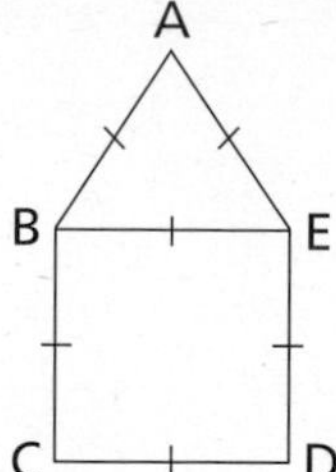

b Add BD to your diagram.

Find the size of ∠ABD. Give reasons for your answer.

2 Use the diagram to prove that the interior angles of a pentagon add to 540°.

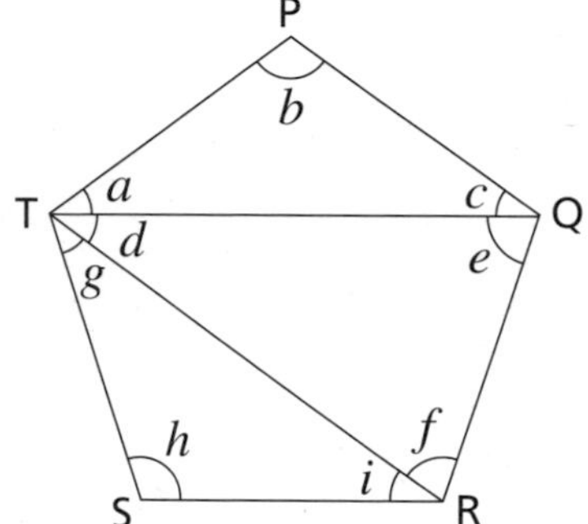

3 **a** Prove that $y = 180° - x$

b Prove that $∠BAC = 180° - 2x$

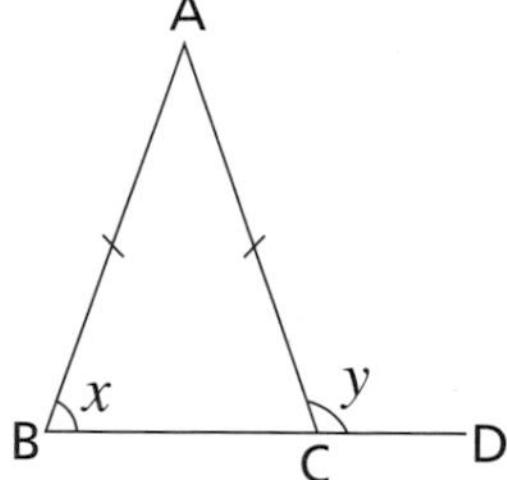

c Prove that $∠LNO = 90° + y$

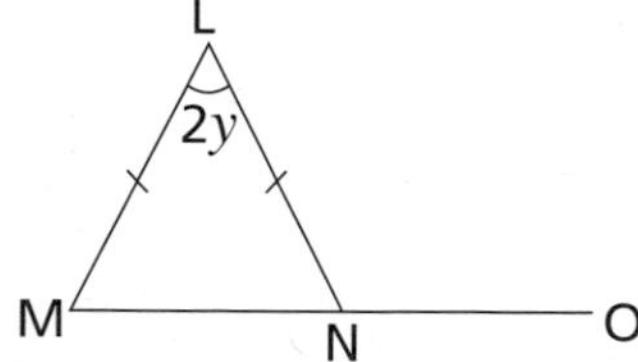

4 Prove that triangles ABC and ADE have the same angles.

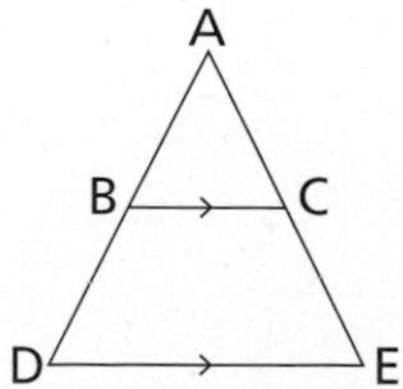

Stretch – can you deepen your learning?

1 **a** Use the diagram and your knowledge of alternate angles to prove that the angles of triangle BEF add to 180°.

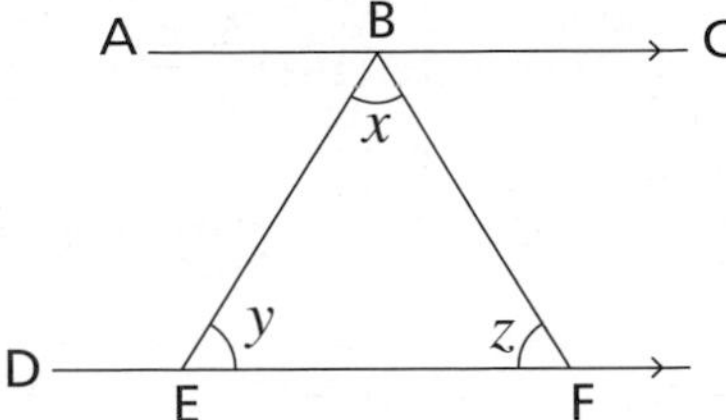

b Will your proof still hold if x is an obtuse angle?

2 **a** BD is parallel to EH and CF = FG. Work out the sizes of the angles labelled a, b, c, d and e.

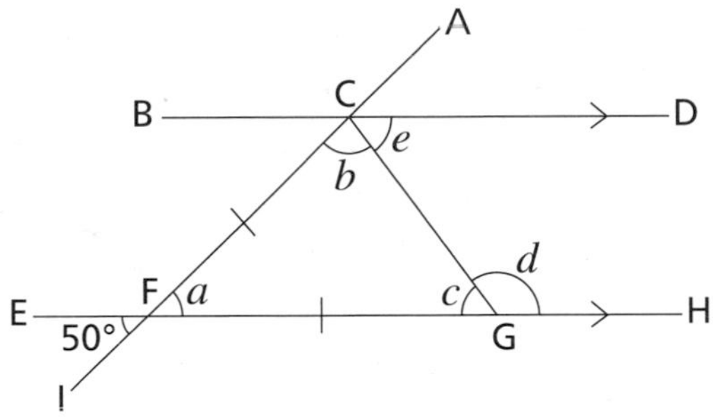

b Repeat for some different values of EFI. What do you notice? What results can you prove?

3 Use your knowledge of angle rules and the relationships between the sides of the special quadrilaterals to prove angles facts about them.

You learnt about angles in quadrilaterals in Chapter 12.3.

Reflect

What is the difference between a demonstration and a proof?

12 Geometric reasoning
Chapters 12.1–12.6

I have become fluent in…

- Finding missing angles on straight lines and at a point
- Identifying vertically opposite angles
- Finding missing angles in triangles
- Finding missing angles in quadrilaterals

I have developed my reasoning skills by…

- Justifying my calculations for missing angles
- Deciding which angle rules to use and when
- Forming chains of reasons
- Proving geometric facts (H)

I have been problem-solving through…

- Interpreting complex diagrams
- Solving multi-step angle problems
- Modelling angle problems given in words
- Finding missing angles in parallel line problems (H)

Check my understanding

1 Work out the angles marked with letters, giving reasons for your answers.

a **b**

2 One angle in an isosceles triangle is 42°. What might the other angles be? Find two possible sets of answers.

3 **a** Sketch a quadrilateral ABCD with ∠ABC = ∠BAD = 90° and ∠ADC = 112°

b Work out the size of ∠BCD.

4 Work out the angles marked with letters, giving reasons for your answers.

a **b**

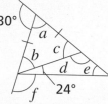

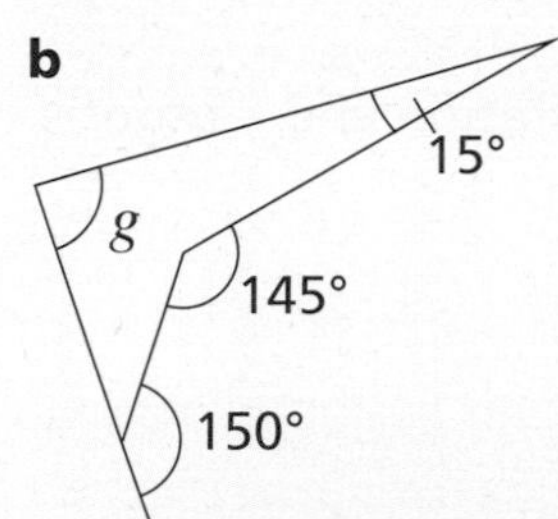

5 Explain why PQRS must be a parallelogram. (H)

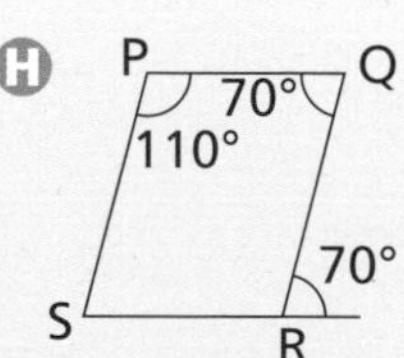

6 **a** Work out the size of each interior angle in a regular 15-sided polygon. (H)

b Prove that ∠DAC = $(180 - 4x)$°

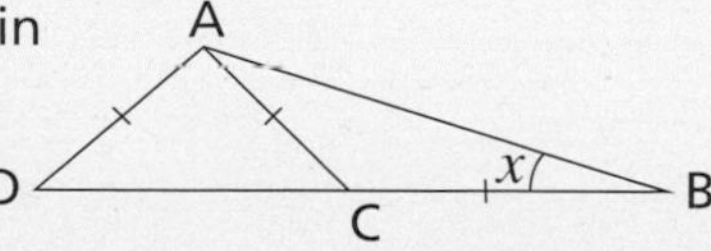

13 Developing number sense

In this block, I will learn...

how to use mental strategies for adding and subtracting

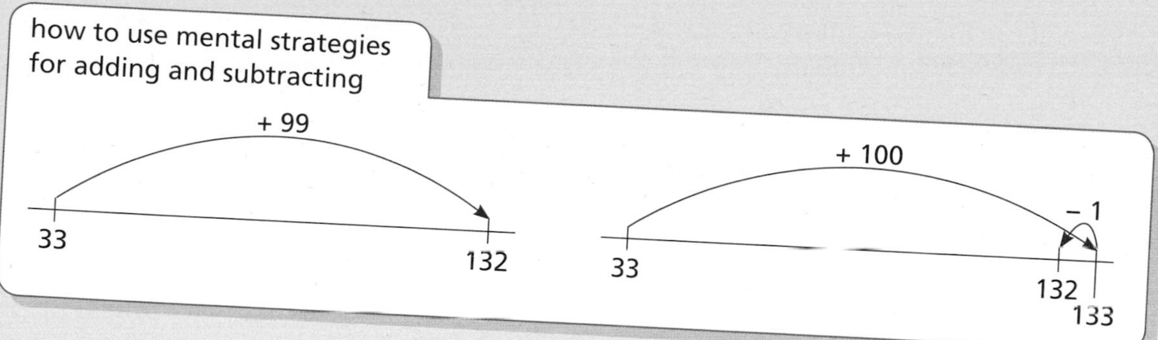

how to use mental strategies for multiplication and division

$3 \times 5 = 15$

$30 \times 5 = 150$

$30 \times 50 = 1500$

$0.3 \times 5 = 1.5$

$0.3 \times 0.5 = 0.15$

to make connections between decimals and fractions

$\frac{1}{10}$	$\frac{1}{10}$	$\frac{1}{10}$	$\frac{1}{10}$	$\frac{1}{10}$	$\frac{1}{10}$	$\frac{1}{10}$	$\frac{1}{10}$	$\frac{1}{10}$	$\frac{1}{10}$

0.1	0.1	0.1	0.1	0.1	0.1	0.1	0.1	0.1	0.1

to estimate answers to check a calculation makes sense

$3.9 \times 21.3 \approx 4 \times 20$ therefore the answer must be close to 80

$3.9 \times 21.3 = 83.07$

13.1 Making connections with integers

Small steps

- Know and use mental addition and subtraction strategies for integers
- Know and use mental multiplication and division strategies for integers

Key words

Mental strategy – a method which enables you to work out the answer in your head

Integer – a whole number

Are you ready?

1 Complete the additions.

a 47 + 53 **b** 765 + 119 **c** 5748 + 918 **d** 684 + 10 307

Did you use the same method for each part?

2 Complete the subtractions.

a 53 – 47 **b** 765 – 119 **c** 5748 – 918 **d** 10 307 – 684

Did you use the same method for each part?

3 Complete the calculations. Show each step in your working.

a 72 × 6 **b** 72 × 16 **c** 572 × 16 **d** 572 × 163

4 Complete the calculations. Show each step in your working.

a 72 ÷ 6 **b** 672 ÷ 6 **c** 12 672 ÷ 6 **d** 12 672 ÷ 36

Models and representations

Number line

A number line is useful when you are looking at mental methods for addition and subtraction. For example, adding 99 is the same as adding 100 and subtracting 1.

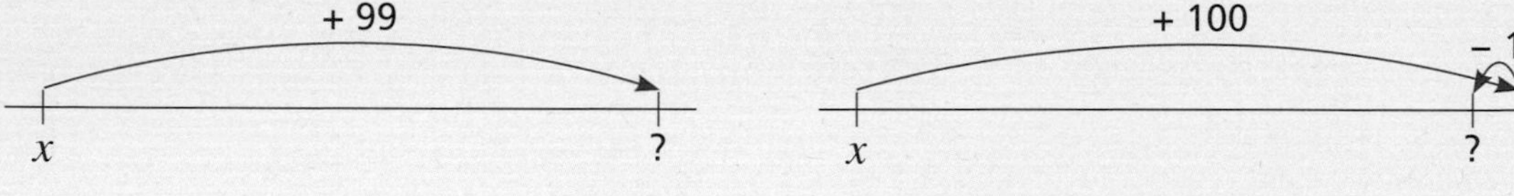

In this section you will look at **mental strategies** for addition and subtraction. Having a bank of mental methods will make you a more efficient mathematician.

Example 1

Complete the calculations.

a 1374 + 99 **b** 1374 – 99

Method A

a $1374 + 99 = 1374 + 100 - 1$
$= 1474 - 1$
$= 1473$

When adding 99 you are adding 1 less than 100, so you can rewrite the calculation.

Once you have added 100 you just need to subtract 1

b $1374 - 99 = 1374 - 100 + 1$
$= 1274 + 1$
$= 1275$

When subtracting 99 you are subtracting 1 less than 100, so you can rewrite the calculation.

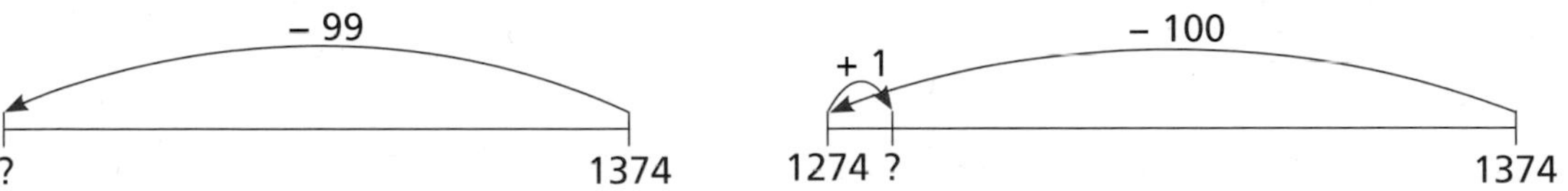

Once you have subtracted 100 you just need to add 1

Method B

a $1374 + 99 = 1373 + 100$
$= 1473$

If you add 1 to the 99 to make the calculation easier, you need to subtract 1 from the 1374 to keep the total the same.

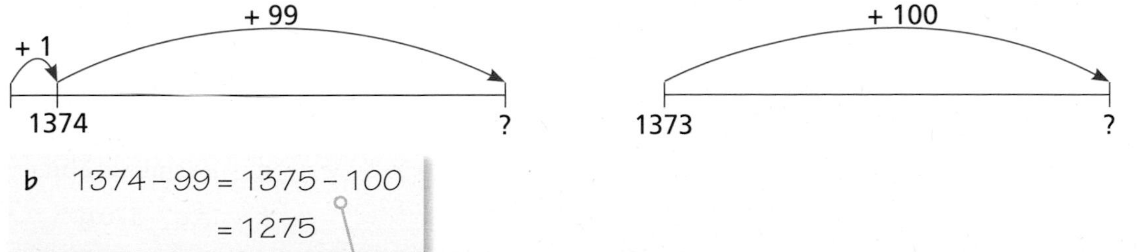

b $1374 - 99 = 1375 - 100$
$= 1275$

If you add 1 to the 99 to make the calculation easier, you need to add 1 to the 1374 to keep the difference the same.

Method C

a $1374 + 99 = 1374 + 90 + 9$
$= 1374 + 30 + 60 + 9$
$= 1404 + 60 + 9$
$= 1464 + 9$
$= 1473$

You can add 99 by partitioning it into 90 and 9

You might want to split the 90 into 30 and 60, to avoid the risk of numerical error.

b $1374 - 99 = 1374 - 90 - 9$
$= 1374 - 70 - 20 - 9$
$= 1304 - 20 - 9$
$= 1284 - 9$
$= 1275$

You can subtract 99 by partitioning it into 90 and 9. You could split the 90 into 70 and 20 to avoid the risk of numerical error.

Practice 13.1A

1 Work out the additions mentally.
Then write the steps in your thinking.

a 108 + 99 **b** 573 + 99 **c** 649 + 99 **d** 99 + 1876
e 13 825 + 99 **f** 482 + 199 **g** 612 + 299 **h** 399 + 1176

2 Work out the subtractions mentally.
Then write the steps in your thinking.

a 108 – 99 **b** 573 – 99 **c** 649 – 99 **d** 1876 – 99
e 13 825 – 99 **f** 482 – 199 **g** 612 – 299 **h** 1176 – 399

3 Work out the calculations mentally.
Then write the steps in your thinking.

a 57 + 21 **b** 39 + 42 **c** 128 + 43 **d** 189 + 76
e 57 – 21 **f** 42 – 39 **g** 128 – 43 **h** 189 – 76

4 Complete the calculations mentally.
Then write the steps in your thinking.

a 43 m + 99 m **b** 76 cm – 24 cm **c** 199 ml + 47 ml
d 517 g – 199 g **e** 83 km + 108 km **f** 1 kg – 570 g
g 3 km – 499 m **h** £764 + £999 **i** 3760 cm – 999 cm

What do you think?

1 Here are four calculation cards.

i	ii	iii
572 + 99	4762 − 199	999 + 5821

a By rounding to one significant figure, estimate the answer to each calculation.

b Work out each calculation mentally.

c How do your estimates support your mentally calculated answers?

2 Sven, Mario, Junaid and Kate are calculating 183 000 − 99 999

Here are their methods.

Sven

$$\begin{array}{r} \overset{0}{\cancel{1}}\,\overset{17}{\cancel{8}}\,\overset{12}{\cancel{3}}\,\overset{9}{\cancel{0}}\,\overset{9}{\cancel{0}}\,\overset{1}{0} \\ -\quad 9\,9\,9\,9\,9 \\ \hline 8\,3\,0\,0\,1 \\ \hline \end{array}$$

Mario

183 001 − 100 000

= 83 001

Junaid

183 000 − 90 000 = 93 000

93 000 − 9000 = 84 000

84 000 − 900 = 83 100

83 100 − 90 = 83 010

83 010 − 9 = 83 001

Kate

183 000 − 100 000 = 83 000

83 000 + 1 = 83 001

a Explain each person's method.

b Which methods can be completed mentally? Explain your choice.

c Which method is most efficient? Compare your answer with a partner.

3 Here is Benji's method for calculating 14 971 + 99

14 971 + 99 = 14 971 + 100 − 1

= 15 071 − 1

= 15 070

Explain how Benji's method can be adapted to calculate

a 14 971 + 98 **b** 14 971 + 97 **c** 14 971 + 199 **d** 14 971 + 396

4 Complete the calculations. Give your answers as improper fractions.

a $\frac{514}{4} + \frac{99}{4}$ **b** $\frac{514}{4} - \frac{99}{4}$ **c** $\frac{612}{7} + \frac{198}{7}$ **d** $\frac{499}{9} + \frac{523}{9}$

13.1 Making connections with integers

In this section you will look at mental strategies for multiplication and division.

Example 2

Complete the calculations.

a 9 × 47 **b** 700 ÷ 25

Method A

a

		4	7
×			9
	4	2_6	3

You can calculate 9 multiplied by 47 using a formal written method to get an answer of 423

b

$$25\overline{)7\ {}^{7}0\ {}^{20}0}\quad = 0\ 2\ 8$$

You could use a formal written method to calculate 700 divided by 25 to get an answer of 28

Alternatively, you could divide by 5 and then divide by 5 again. This is the same as dividing by 25

Method B

a $9 \times 47 = 9 \times 40 + 9 \times 7$

$9 \times 40 = 360$

$9 \times 7 = 63$

$9 \times 47 = 360 + 63 = 423$

You could partition 47 into 40 and 7

4 multiplied by 9 is 36, so 40 multiplied by 9 is 360 (because 40 × 9 = 4 × 10 × 9 = 4 × 9 × 10)

7 multiplied by 9 is 63

Adding them together gives 423, so 9 × 47 is equal to 423

b $25 \times 10 = 250$

$25 \times 10 = 250$

$25 \times 4 = 100$

$25 \times 4 = 100$

$250 + 250 + 100 + 100 = 700$

So $700 \div 25 = 10 + 10 + 4 + 4 = 28$

You can calculate 700 divided by 25 by 'chunking'. Use the facts that you know, such as

- 25 multiplied by 10 is 250
- 25 multiplied by 4 is 100

Then use these facts to 'build' 700

Method C

a $10 \times 47 = 470$

$9 \times 47 = 470 - 47 = 423$

Multiply 47 by 10, then subtract 1 lot of 47

This works because: 9 × 47 = (10 – 1) × 47 = 10 × 47 – 1 × 47

b $4 \times 25 = 100$

$7 \times 4 \times 25 = 700$

$28 \times 25 = 700$

$700 \div 25 = 28$

There are 4 lots of 25 in 100

There are 7 lots of 100 in 700

Therefore there are 7 lots of 4 lots of 25 in 700

so 700 divided by 25 must be equal to 28 (7 × 4)

Practice 13.1B

1 Here are four methods of calculating 24 × 5

i	ii	iii	iv
20 × 5 = 100 4 × 5 = 20 24 × 5 = 120	24 × 10 = 240 24 × 5 = 120	12 × 10 = 120	6 × 4 × 5 = 6 × 20 = 120

a Explain why each method works.

b Use your preferred method to complete the calculations.

i 36 × 5 **ii** 5 × 72 **iii** 5 × 19

iv 484 × 5 **v** 1096 × 5 **vi** 5 × 8762

Did you use the same method for each part?

2 Complete the multiplications mentally.

Then write down each stage in your thinking.

a 9 × 17 **b** 99 × 8 **c** 98 × 9 **d** 76 × 5

e 90 × 11 **f** 15 × 34 **g** 19 × 13 **h** 999 × 6

i 5 × 4026 **j** 11 × 624

3 Here are four methods of calculating 750 ÷ 5

i	ii	iii	iv
75 ÷ 5 = 15 750 ÷ 5 = 150	500 ÷ 5 = 100 250 ÷ 5 = 50 750 ÷ 5 = 150	750 ÷ 10 = 75 750 ÷ 5 = 2 × 75 = 150	1500 ÷ 10 = 150

Explain why each method works.

4 Complete the divisions mentally.

Then write down each stage in your thinking.

a 640 ÷ 8 **b** 800 ÷ 16 **c** 540 ÷ 5 **d** 864 ÷ 4

e 1080 ÷ 5 **f** 240 ÷ 6 **g** 960 ÷ 3 **h** 490 ÷ 7

i 4500 ÷ 9 **j** 6900 ÷ 5

5 Work out the calculations mentally.

Then write down each stage in your thinking.

a £750 ÷ 15 **b** 6000 g ÷ 25 **c** 425 kg ÷ 5 **d** 5 × 125 ml

e £79 × 9 **f** 2000 kg ÷ 50 **g** £50 × 99 **h** 98 × 70 mm

i 964 g ÷ 4 **j** 30 m × 999

What do you think?

1

200 cm

99 cm

a Estimate the area of the rectangle, giving your answer in square centimetres.

b Mentally calculate the area of the rectangle.

c How does your estimate support your mental calculation?

2 A pack of 24 tiles costs £96

A pack of two of the same tiles costs £9

Calculate mentally which pack of tiles is better value.

Write down each step in your thinking.

3 Use the fact that $5 \times 30 = 150$ to complete the calculations mentally.

a 50×30 b 5×300 c 500×30 d $15\,000 \div 30$ e $15\,000 \div 500$

How does the original calculation help you work out the others mentally?

Consolidate – do you need more?

1 Complete the additions mentally.

Write down each step in your thinking.

a 583 + 99 b 743 + 198 c £261 + £999 d 999 ml + 1506 ml

2 Complete the subtractions mentally.

Write down each step in your thinking.

a 742 – 99 b 1000 – 399 c 2507 g – 199 g d 100 000 kg – 9999 kg

3 Complete the multiplications mentally.

Write down each step in your thinking.

a 582×5 b 40×99 c £782 × 9 d 4 × 240 cm

4 Complete the divisions mentally.

Write down each step in your thinking.

a $630 \div 7$ b $480 \div 60$ c 980 ml ÷ 5 d 3600 g ÷ 24

Stretch – can you deepen your learning?

1 Solve the equations mentally.

a $99 + x = 714$

b $5y = 2400$

c $\frac{z}{7} = 99$

d $f - 299 = 1067$

2 Use a mental method to write each of the improper fractions as integers.

a $\frac{460}{5}$

b $\frac{980}{4}$

c $\frac{7200}{18}$

d $\frac{740p}{5p}$

3 **a** Complete this calculation mentally.

5800×5

b Without further calculation, which multiplication gives the greater answer?

5801×5 or 5800×6

Explain your choice.

Reflect

1 Explain why dividing by 5 is the same as dividing by 10 and then multiplying by 2

2 Describe four different methods which can be used to calculate 749 + 99

Which of the methods can be done mentally?

Which of the methods is most efficient?

13.2 Making connections with fractions and decimals

Small steps

- Know and use mental arithmetic strategies for decimals
- Know and use mental arithmetic strategies for fractions

Key words

Mental strategy – a method that enables you to work out the answer in your head

Decimal – a number with digits to the right of the decimal point

Fraction – a part of a whole

Are you ready?

1 Complete the additions and subtractions.

a 35 + 12 **b** 35 + 17 **c** 35 + 87 **d** 735 + 9870

e 96 – 23 **f** 96 – 27 **g** 96 – 87 **h** 9600 – 870

2 Complete the multiplications and divisions.

a 5 × 5 **b** 16 × 2 **c** 70 × 5 **d** 51 × 7

e 30 ÷ 6 **f** 48 ÷ 3 **g** 490 ÷ 7 **h** 490 ÷ 70

3 Complete the calculations.

a 27 × 10 **b** 2.7 × 10 **c** 2.7 × 100 **d** 0.27 × 100

e 5700 ÷ 10 **f** 570 ÷ 10 **g** 570 ÷ 100 **h** 57 ÷ 100

4 Work out the fractions of the amounts.

a $\frac{5}{6}$ of 30 **b** $\frac{2}{3}$ of 48 **c** $\frac{5}{7}$ of 490 **d** $\frac{51}{70}$ of 490

Models and representations

Bar models

Bar models are useful to demonstrate the method behind a mental calculation.

0.3 + 0.7 = 3 tenths + 7 tenths
= 1 whole (not 0.10)

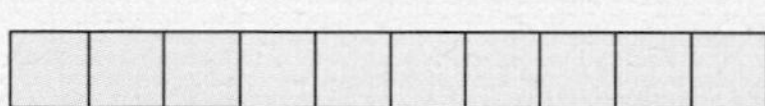

When finding $\frac{3}{5}$ of a whole, you divide by 5 then multiply by 3.

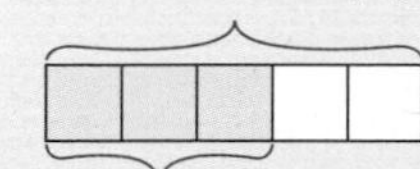

In this section you will look at **mental strategies** for calculating with **decimals**. You have already studied formal written methods, and now it is time to make your calculations more efficient.

Example 1

Calculate

a 32.4 + 21.5 **b** 0.3×0.4

a $32 + 21 = 53$

$0.4 + 0.5 = 0.9$

$53 + 0.9 = 53.9$

20 more than 32 is 52, so 21 more is 53

4 tenths plus 5 tenths is 9 tenths

Then if you combine these totals you get 53.9

b $0.3 \times 0.4 = 0.12$

$\times 10 \downarrow \quad \times 10 \downarrow \quad \uparrow \div 100$

$3 \times 4 = 12$

To calculate this mentally, adjust the numbers to make them easier to work with.

If you multiply both 0.3 and 0.4 by 10, then your calculation becomes $3 \times 4 = 12$

However, now your answer is 100 times greater, so you need to divide 12 by 100 to give a final answer of 0.12

Practice 13.2A

1 Complete the additions.

a 3 + 5 **b** 3 trees + 5 trees **c** 3 tenths + 5 tenths

d 0.3 + 0.5 **e** 0.03 + 0.05

What do you notice?

2 Complete the additions mentally. Then write down each step in your thinking.

a 0.6 + 0.1 **b** 0.6 + 0.3 **c** 0.6 + 0.5

d 0.06 + 0.05 **e** 0.63 + 0.55

3 Complete the subtractions.

a 7 – 6 **b** 7 eggs – 6 eggs **c** 7 tenths – 6 tenths

d 0.7 – 0.6 **e** 0.07 – 0.06

What do you notice?

4 Complete the subtractions mentally. Then write down each step in your thinking.

a 0.9 – 0.4 **b** 0.09 – 0.04 **c** 0.97 – 0.43

d 0.97 – 0.1 **e** 0.86 – 0.07

5 Complete the multiplications mentally. Then write down each step in your thinking.

a 0.4×6 **b** 7×0.2 **c** 0.6×8 **d** 9×0.9

e 5×0.8 **f** 0.4×0.6 **g** 0.7×0.2 **h** 0.6×0.8

i 0.9^2 **j** 0.5×0.8

6 Complete the divisions mentally. Then write down each step in your thinking.

a 48 ÷ 6 **b** 4.8 ÷ 6 **c** 0.48 ÷ 6 **d** 7.2 ÷ 12

e 9.5 ÷ 5 **f** 0.42 ÷ 7 **g** 3.3 ÷ 11 **h** 0.63 ÷ 9

i 9.63 ÷ 9 **j** 909.63 ÷ 9

7 Here is the cost of four items.

£1.49

£3.99

£9.99

£0.49

Work out each of these calculations mentally.

a How much would four mugs cost?

b How much would it cost to buy one of each item?

c If you paid for a newspaper with a £10 note, how much change would you get?

d How much more does a hat cost than a mug?

What do you think?

1 Use the fact that 1 whole = 10 tenths to complete the calculations mentally.

a 1 – 0.7 **b** 1 – 0.6 **c** 1 – 0.2 **d** 1 – 0.1 – 0.4 **e** 1 – 0.9 + 0.1

f 5 – 0.7 **g** 12 – 0.6 **h** 100 – 0.2 **i** 7 – 0.1 – 0.4 **j** 99 – 0.9 + 0.1

Compare your method with a partner.

2 Use the fact that 24 × 4 = 96 to work out the calculations mentally.

a 2.4 × 4 **b** 24 × 0.4 **c** 2.4 × 0.4 **d** 9.6 ÷ 4

3 Seb says "0.8 plus 0.9 is equal to 0.17"

a Draw a diagram to show that Seb is wrong.

b How would you help Seb to not make this mistake again?

c What is the correct answer?

Example 2

Calculate mentally.

a $\frac{1}{2}+\frac{1}{8}$ **b** $\frac{2}{3}$ of 45

a $\frac{1}{2}+\frac{1}{8}=\frac{4}{8}+\frac{1}{8}$
$=\frac{5}{8}$

$\frac{1}{2}$ is equivalent to $\frac{4}{8}$ so think of the question as $\frac{4}{8}+\frac{1}{8}$
The denominator is the same so just add the numerators.

b $\frac{1}{3}$ of 45 = 15
$\frac{2}{3}$ of 45 = 30

Dividing 45 into 3 equal parts gives 15 in each part.
Multiplying this by 2 tells you that $\frac{2}{3}$ of 45 is 30

Practice 13.2B

1 Complete the calculations mentally. You may use the fraction wall to help you.

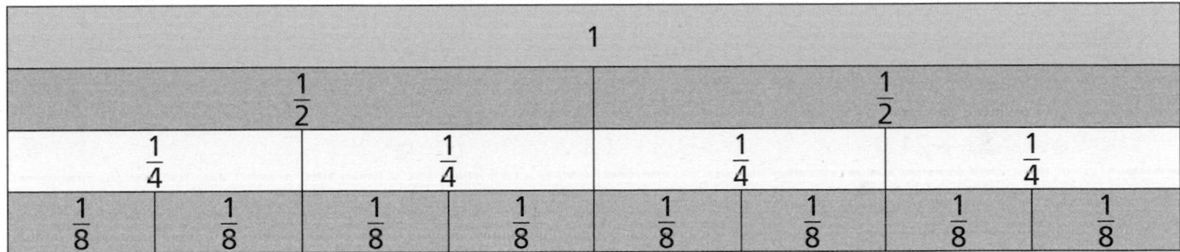

1							
$\frac{1}{2}$				$\frac{1}{2}$			
$\frac{1}{4}$		$\frac{1}{4}$		$\frac{1}{4}$		$\frac{1}{4}$	
$\frac{1}{8}$	$\frac{1}{8}$	$\frac{1}{8}$	$\frac{1}{8}$	$\frac{1}{8}$	$\frac{1}{8}$	$\frac{1}{8}$	$\frac{1}{8}$

a $\frac{1}{2}+\frac{1}{4}$ **b** $\frac{3}{8}+\frac{1}{4}$ **c** $\frac{1}{4}-\frac{1}{8}$ **d** $1-\frac{1}{2}$

e $1-\frac{3}{8}$ **f** $\frac{7}{8}-\frac{3}{4}$ **g** $\frac{1}{4}+\frac{1}{4}-\frac{1}{2}$ **h** $\frac{7}{8}+\frac{1}{8}-\frac{3}{4}$

2 Work out the fractions of amounts mentally.

Then write down each step in your thinking.

a $\frac{1}{8}$ of 80 **b** $\frac{1}{5}$ of 60 **c** $\frac{1}{3}$ of 90 **d** $\frac{1}{4}$ of 36

e $\frac{1}{9}$ of 72 **f** $\frac{1}{10}$ of 70 **g** $\frac{1}{6}$ of 66 **h** $\frac{1}{25}$ of 100

3 Work out the fractions of amounts mentally.

Then write down each step in your thinking.

a $\frac{5}{8}$ of 80 **b** $\frac{3}{5}$ of 60 **c** $\frac{2}{3}$ of 90 **d** $\frac{3}{4}$ of 36

e $\frac{8}{9}$ of 72 **f** $\frac{7}{10}$ of 70 **g** $\frac{5}{6}$ of 66 **h** $\frac{21}{25}$ of 100

4 A shop has a sale on with $\frac{1}{5}$ off all of the prices.

The normal price of each item is shown. Work out the sale price mentally.

a 

b 

c 

d 

Compare your method with a partner.

What do you think?

1. $\frac{1}{5}$ of a number is equal to 36

   Complete the questions mentally.

   a What is $\frac{1}{10}$ of the number? **b** What is the number?

2. Work out these fractions of amounts mentally.

   a $\frac{1}{2}$ of £200 **b** $\frac{1}{4}$ of £200 **c** $\frac{1}{8}$ of £200 **d** $\frac{1}{16}$ of £200

   What do you notice?

3. Remembering that there are 8 eighths in one whole, work out the calculations mentally.

   a $1 - \frac{5}{8}$ **b** $6 - \frac{3}{8}$ **c** $17 - \frac{7}{8}$ **d** $35 - \frac{1}{8}$

Consolidate – do you need more?

1. Complete the calculations mentally. Then write down each step in your thinking.

   a 3.5 + 2.3 **b** 8.1 + 7.7 **c** 2.5 + 7.5 **d** 3.6 + 6.5
   e 43.7 + 21.9 **f** 3.5 – 2.3 **g** 8.1 – 7.7 **h** 7.5 – 2.5
   i 6.5 – 3.6 **j** 43.7 – 21.5

2. Complete the calculations mentally. Then write down each step in your thinking.

   a 0.4 × 5 **b** 0.6 × 3 **c** 4 × 0.7 **d** 0.5 × 0.6
   e 0.2 × 0.9 **f** 1.2 ÷ 6 **g** 0.6 ÷ 3 **h** 8.1 ÷ 9
   i 0.49 ÷ 7 **j** 0.016 ÷ 8

3. Complete the calculations mentally. Then write down each step in your thinking.

   a $\frac{1}{2} + \frac{5}{12}$ **b** $\frac{1}{8} + \frac{3}{4}$ **c** $\frac{3}{10} + \frac{7}{20}$ **d** $\frac{31}{100} + \frac{1}{10}$
   e $\frac{1}{2} - \frac{5}{12}$ **f** $\frac{3}{4} - \frac{1}{8}$ **g** $\frac{7}{20} - \frac{3}{10}$ **h** $\frac{31}{100} - \frac{1}{10}$

Stretch – can you deepen your learning?

1. Here are three digit cards.

   B

   C

   You know the following facts

   A = 300 B = $\frac{5}{6}$ of A C = $\frac{7}{10}$ of B

   Work out the values of B and C mentally.

2 The greatest value in a data set is $\frac{17}{20}$

The smallest value in the same data set is $\frac{1}{4}$

Work out the range of the data set mentally.

Then write down each step in your thinking.

3 Solve the equations mentally.

a $5x = 6.5$

b $7x = 4.9$

c $x + 19.2 = 20$

d $36.5 - x = 10.2$

e $\frac{x}{6} = 0.4$

f $\frac{x}{8} = 0.07$

g $4x - 0.6 = 1.8$

h $15 - 3x = 11.4$

Reflect

1 Show two different methods which could be used to calculate $\frac{6}{7}$ of 63 mentally.

2 Explain why $1.7 + 3.5 \neq 4.12$

13.3 Making connections and estimating

Small steps

- Use factors to simplify calculations
- Use estimation as a method for checking mental calculations

Key words

Factor – a number or expression that divides into another number or expression exactly with no remainder

Estimate – work out something close to the correct answer

Are you ready?

1 List all of the factors of each number.

a 6 **b** 9 **c** 12 **d** 24

e 36 **f** 51 **g** 16 **h** 48

2 Round each number to one significant figure.

a 310 **b** 476 **c** 15 **d** 7.41

e 1682 **f** 0.605 **g** 9.89 **h** 999 999

3 Complete the calculations.

a $300 + 500$ **b** 20×8 **c** 2000×10

d $300 \div 20$ **e** $1\,000\,000 - 300$

4 Work out the value of each.

a $\sqrt{49}$ **b** 6^2 **c** $\sqrt{121}$

d 12^2 **e** 3^3 **f** $\sqrt[3]{8}$

Models and representations

Bar models and counters

A bar model can demonstrate why calculations can be simplified using factors.

900					
450			450		
150	150	150	150	150	150

$150 \times 3 \times 2 = 150 \times 6$

Also $900 \div 2 \div 3 = 900 \div 6$

You can use counters to find **factors** of a number.

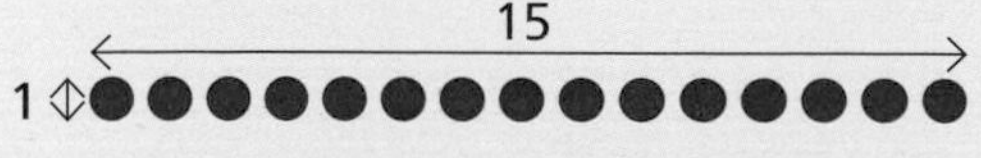

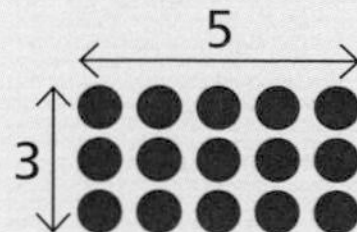

The factors of 15 are 1, 3, 5 and 15

Using the factors of a 2-digit number, you can simplify calculations to make them easier to work out both written and mentally.

Example 1

Use factors to complete the calculations.

a 25×12 **b** $204 \div 12$

Method A

a $12 = 4 \times 3$

$25 \times 12 = 25 \times 4 \times 3$

$25 \times 4 = 100$

$100 \times 3 = 300$

So $25 \times 12 = 300$

If you are multiplying by 12, you can multiply by 4 and then multiply by 3 (or the other way around).

25 multiplied by 4 is equal to 100

and then 100 multiplied by 3 is 300

b $12 = 4 \times 3$

$204 \div 2 = 102$

$102 \div 2 = 51$

$51 \div 3 = 17$

So $204 \div 12 = 17$

You can use the same pair of factors to simplify divisions. Calculate 204 divided by 4 by halving and halving again, then divide this answer by 3

Method B

a $12 = 2 \times 6$

$25 \times 12 = 25 \times 2 \times 6$

$25 \times 2 = 50$

$50 \times 6 = 300$

25 multiplied by 2 is 50

then multiplying this by 6 gives you the same answer as above of 300

b $12 = 2 \times 6$

$204 \div 2 = 102$

$102 \div 6 = 17$

Start by halving 204 (dividing by 2) then divide that answer by 6

You could break 102 into 60 + 42 to help with the final division.

Practice 13.3A

1. Complete the calculations.

   a $480 \div 2 \div 8$ **b** $480 \div 4 \div 4$ **c** $480 \div 8 \div 2$ **d** $480 \div 16$

   What do you notice? Why does this happen?

2. Complete the calculations.

   a $21 \times 2 \times 9$ **b** $21 \times 3 \times 6$ **c** $21 \times 9 \times 2$

   d $21 \times 2 \times 3 \times 3$ **e** 21×18

   What do you notice? Why does this happen?

3. Draw a diagram to show that $8 \times 20 = 8 \times 2 \times 10$

4. **a** Complete the divisions.

   i $640 \div 8$ **ii** $80 \div 5$

   b Without further calculation, work out the answer to $640 \div 40$
   Explain your reasoning.

5. Complete the multiplications using factors. Write down the factors that you use.

   a 27×4 **b** 60×18 **c** 40×27

   d 20×49 **e** 35×21

6. Complete the divisions using factors. Write down the factors that you use.

   a $108 \div 4$ **b** $160 \div 32$ **c** $240 \div 16$

   d $270 \div 15$ **e** $96 \div 6$

What do you think?

1. Decide whether each statement is true or false. Explain your answer.

   a $164 \times ab = 164 \times a \times b$ **b** $164 \div ab = 164 \div a \div b$

2. Martin is dividing 357 by 7

   He says, "I am going to divide using factors." Suggest why using factors will not be any more efficient for this example.

3. Work out 116×16 in three different ways using factors.

   Write down your method for each. Which method did you prefer? Why?

In this section you will use **estimation** as a method for checking mental calculations. It is easy to make mistakes when working out calculations mentally, and estimation is a great way to sense check your answers.

Example 2

Estimate the answer to each calculation. Then work out the actual answer mentally.

a 5320 + 998 **b** 620 × 6

a Estimate: 5000 + 1000 = 6000

5320 rounded to one significant figure is 5000

998 rounded to one significant figure is 1000

The total of 5000 and 1000 is 6000, therefore 5320 + 998 ≈ 6000

Actual answer

5320 + 998 = 5320 + 1000 − 2

= 6320 − 2

= 6318

The actual answer should be greater than 6000 because 5320 > 5000, and even though 998 < 1000 it is only 2 less.

First add 1000 and then subtract 2 from your answer.

6318 is close to 6000 but greater, therefore the estimate supports your answer.

b Estimate: 600 × 6 = 3600

620 rounded to one significant figure is 600 so 620 × 6 ≈ 3600

Actual answer

620 × 3 × 2 = 1860 × 2

= 3720

Your actual answer should be greater than 3600 as 620 > 600

6 = 2 × 3 so use factors to break the calculation down.

The final answer is 3720, and the estimate supports this.

Practice 13.3B

1 **a** Round 791 to one significant figure.

b Round 995 to one significant figure.

c Estimate the answer to 791 + 995

d Is the actual answer going to be more or less than your estimate?

e Work out the actual answer mentally. How does your estimate support your answer?

2 For each calculation

i Estimate the answer by rounding to one significant figure.

ii Decide whether the actual answer will be greater or less than your estimate.

iii Work out the answer to the calculation mentally.

a 9865 + 6993 **b** 709 × 9

c 294 ÷ 6 **d** 17 564 − 4990

3 Use estimations to explain why each statement is true.

a 479 + 977 < 1500 **b** 47 × 9 < 450

c $\sqrt{107} > 10$ **d** 21 572 + 34 916 > 50 000

4 Zach and Emily have each worked out the answer to a calculation mentally.

a Zach

596 + 193 = 803

b Emily

79 × 8 = 648

Use estimation to show that they are both wrong.

What mistake do you think each person made?

5 **a** Estimate the area of each shape.

i

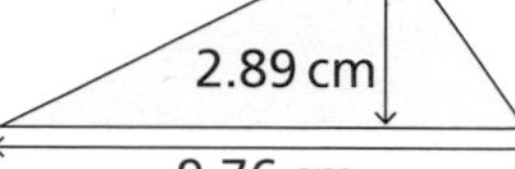

ii 76.34 mm

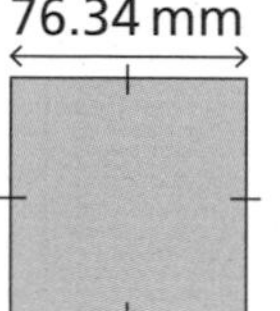

iii

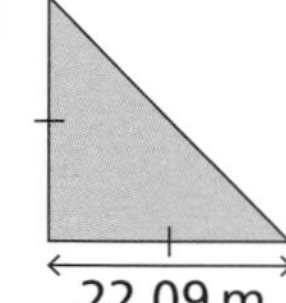

b Will the actual area of each shape be greater than or less than your estimate? Explain your answers.

6 Estimate the answer to each calculation.

a $\sqrt{403}$

b 7.81^2

c $\frac{2}{3}$ of £86

d 4.9 kg + 7.3 kg + 12.814 kg

What do you think?

1

a Estimate the total cost of buying one of each item.

b Beca wants to buy one of each item. She has £10 exactly. Does she have enough money? Did your estimate from part **a** support your answer?

2 Suggest why rounding to 1 significant figure is not the most appropriate method for estimating the value of each calculation.

a $\sqrt{47}$

b $\sqrt{65}$

c $\sqrt{27}$

d $\sqrt{8400}$

Estimate the answer to each calculation. Decide whether the actual answer will be greater or less than your estimate.

Consolidate – do you need more?

1 Complete the multiplications using factors.

a 412×8 **b** 307×12 **c** 24×25 **d** 35×20

2 Complete the divisions using factors.

a $416 \div 8$ **b** $300 \div 12$ **c** $585 \div 15$ **d** $1080 \div 18$

3 Estimate the answer to each calculation.

a $4782 + 997$ **b** $3649 - 1001$ **c** 712×5 **d** $618 \div 6$

4 Work out the answer to each calculation mentally.

a $4782 + 997$ **b** $3649 - 1001$ **c** 712×5 **d** $618 \div 6$

5 Compare your answers to questions 3 and 4. How do your answers to question 3 support your answers to question 4?

Stretch – can you deepen your learning?

1 Use division by factors to show that 78 divided by 18 is $4\frac{1}{3}$

2 x and y are rounded to 1 significant figure to give $x + y \approx 12\,000$

a List three possible pairs of values for x and y.

b What is the least possible value of $x + y$?

c $c \leq x + y < d$

Work out the values of c and d.

3 Estimate the solution to each equation.

a $3x = 297$ **b** $8x = 11.065$

c $\frac{y}{9} = 748.9$ **d** $6b + 10\,007 = 68\,971$

e $c^2 = 6413$ **f** $\sqrt{h} = 19.764$

g $\frac{p^2}{11} = 481$ **h** $12f = 29\,765$

4 A number, x, is the product of y and z.

Faith is multiplying and dividing different numbers by x.

She says "Using factors doesn't help to simplify any of my calculations."

a What must be true about x? **b** What must be true about y or z?

Reflect

1 Explain how using factors can help to simplify multiplications and divisions.

Use examples to support your answer.

2 Give an example of an addition where the estimated answer will be an underestimate.

13.4 Using known facts

Small steps

- Use known number facts to derive other facts
- Use known algebraic facts to derive other facts
- Know when to use a mental strategy, formal written method or a calculator

Key word

Derive – to find or discover something from existing knowledge

Are you ready?

1 Write the fact family for each bar model.

a

12	
5	7

b

1000	
300	700

c

96	
68	28

2 Write the fact family for each bar model.

a

12	
x	7

b

1000	
300	$5x$

c

96	
$3x$	y

3 Substitute $x = 7$ into the expressions.

a $5x$ **b** $3x$ **c** $19x$

d $x + 27$ **e** $x - 5$ **f** $3x - 21$

Models and representations

Bar models

You can use bar models to **derive** related facts.

163	
55	108

$55 + 108 = 163$

Also $163 - 55 = 108$ and $163 - 108 = 55$

42					
7	7	7	7	7	7

$6 \times 7 = 42$ $7 \times 6 = 42$

$42 \div 6 = 7$ $42 \div 7 = 6$

In this section you will use known facts to derive other facts. You can apply this logic when working with both numbers and algebra.

Example 1

Use the fact that $4x + 20 = 163$ to work out the value of

a $163 - 4x$ **b** $8x + 40$

a 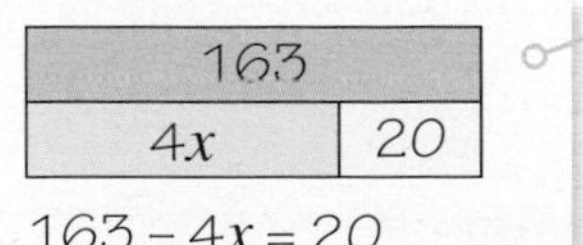

$163 - 4x = 20$

The bar model represents $4x + 20 = 163$

You can also see that if you subtract $4x$ from 163 you will be left with 20

b

163		163	
$4x$	20	$4x$	20

326	
$8x$	40

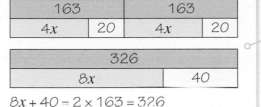

$8x + 40 = 2 \times 163 = 326$

$8x + 40$ is double $4x + 20$

If you place two of the bar models from part a side by side, in total there is $8x$ and 40

This means that the value of $8x + 40$ must be double 163

So $8x + 40 = 326$

Practice 13.4A

1 Here is a bar model.

170	
30	140

Use the bar model to write down the value of

a 30 + 140 **b** 140 + 30 **c** 170 – 30 **d** 170 – 140

2 Here is a bar model.

91						
13	13	13	13	13	13	13

Use the bar model to write down the value of

a 7×13 **b** 13×7 **c** $91 \div 7$ **d** $91 \div 13$

3 Here is a bar model.

84	
63	21

Use the bar model to work out the value of

a **i** 63 + 21 **ii** 21 + 63 **iii** 84 – 63 **iv** 84 – 21

b **i** 630 + 210 **ii** 210 + 630 **iii** 840 – 630 **iv** 840 – 210

c **i** 6300 + 2100 **ii** 2100 + 6300 **iii** 8400 – 6300 **iv** 8400 – 2100

d **i** 6.3 + 2.1 **ii** 2.1 + 6.3 **iii** 8.4 – 6.3 **iv** 8.4 – 2.1

e **i** $63x + 21x$ **ii** $21x + 63x$ **iii** $84x - 63x$ **iv** $84x - 21x$

What do you notice?

4 Use the fact that $24 \times 15 = 360$ to work out each calculation.

a 240×15
b 24×150
c 240×150
d 2.4×15
e 1.5×24
f 24000×15000
g $24x \times 15$
h $36000 \div 24$

5 Use the fact that $1730 + 687 = 2417$ to work out each calculation.

a $1730 + 688$
b $1729 + 687$
c $1720 + 687$
d $3420 - 687$

6 The bar model shows that $x + y = 18$

18	
x	y

Use this to work out the value of

a $2x + 2y$
b $3x + 3y$
c $5(x + y)$
d $18 - x$
e $\frac{1}{2}(x + y)$

7 **a** Draw a bar model to represent $5x + 25 = 210$

b Work out the value of

i $10x + 50$
ii $x + 5$
iii $5x$
iv $5x + 32$
v $500x + 2500$

8 If $3n = 17$, what is the value of these expressions?

a $6n$
b $6n - 2$
c $\frac{3n}{2}$

What do you think?

1 $32 \times 27 = 864$. Explain why

a $320 \times 270 \neq 8640$
b $3200 \times 2700 = 8640000$
c $3200 \times 27 = 27000 \times 3.2$
d $864 \div 2.7 \neq 3.2$

2 Write down three calculations which will have the same answer as 420×16
Compare your answers with a partner.

3 $a - b = 17$

What is the value of $b - a$? How do you know?

In this section you will identify when it is best to use a mental, written or calculator method. The focus is not on the answer, but instead on the most efficient and reliable way of getting to it.

Example 2

Decide whether a mental or written method is most appropriate for each calculation.

a 4763 – 2189 **b** 4763 – 2001

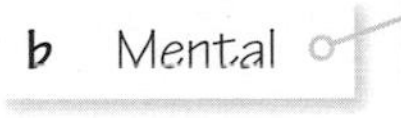

Neither of the numbers are close the nearest thousand, so you cannot use an efficient mental method. Calculating this mentally would take a little longer and also carry the risk of error, so a formal written method is more appropriate.

b Mental

It would take longer for you to write out the formal written method than it would to calculate it mentally. 2001 is only 1 more than 2000 so you can subtract 2000 from 4763, then subtract another 1

Practice 13.4B

1 Marta is calculating 579 + 405 using the column method.

- **a** Explain a more efficient method Marta could use.
- **b** Work out the answer to the calculation.

2 For each calculation, decide whether it is more efficient to work it out mentally or using a formal written method. Then work out the answer.

a 764 + 1005 **b** 86509 + 4719 **c** 576 × 27 **d** 799 × 5

Compare your answers with a partner.

3 A mug costs £4.99, a sports bottle costs £2.65 and a bag costs £12.01

For each question, first decide whether a mental or written method is more efficient, then work out the answer. Write down each stage in your thinking.

- **a** What is the total cost of seven mugs?
- **b** What is the total cost of 18 sports bottles?
- **c** What is the cost of buying one of each item?

Compare your answers with a partner. Did you use the same method?

4 Mr Smith has £200 to buy new football shirts for his team.

There are 19 players on the team.

Each shirt costs £9.95

Mr Smith thinks he needs to use a calculator to see if he has enough money.

Show using a mental estimation that Mr Smith's method is not the most efficient.

Write down each stage in your thinking.

5 Benji is selling some of his books for a total of £3.97

The customer pays for them with a £20 note.

Benji says "I need to use the column method to work out the change. It's going to need a lot of exchanges."

Suggest a more efficient method Benji could use.

What do you think?

1 Seb and Flo are working out 599 × 5

I'm going to use a written method

I'm going to calculate it mentally

a Whose method is more efficient?

b Why does it not matter that they are using different methods?

2 Write down five calculations that are quicker to work out mentally than using a written method.

3 Write down five calculations that are more suited to a written method than a mental one.

4 Emily is working out 586 + 19 998

I know that a mental method is appropriate here, but I'm not very confident

What should Emily do? Explain your answer.

Consolidate – do you need more?

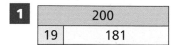

1 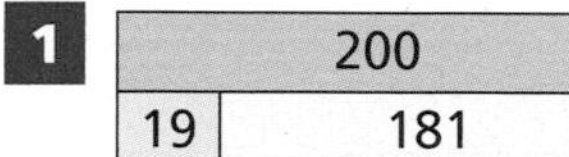

Use the bar model to work out the value of

a 19 + 181 b 181 + 19 c 200 – 181 d 200 – 19

2 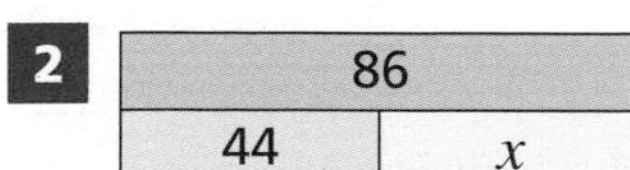

Use the bar model to find the value of

a x + 44 b 44 + x c 86 – 44 d 86 – x

3 Use the fact that 7 × 5 = 35 to find the value of

a 7 × 50 b 70 × 5 c 70 × 50 d 3500 ÷ 5 e 0.7 × 0.5

4 Copy the diagram into your book and fill in the missing values.

Write down each stage in your thinking.

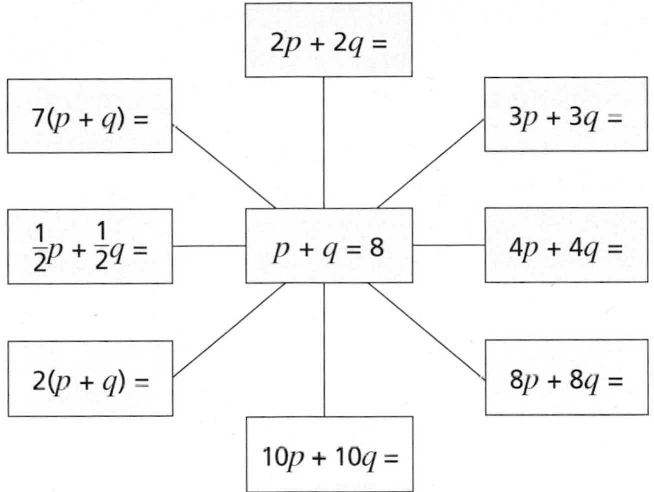

Stretch – can you deepen your learning?

1 Given that $r + s = 20$, write down expressions that will have the given values.

a 40 **b** 10 **c** 100 **d** 200

e 30 **f** 0.2 **g** 1 **h** 60 000

Compare your answers with a partner.

2 Given that $\frac{p}{9} = 39$, work out the value of

a $\frac{p}{9} + 1$ **b** p **c** $127 - \frac{p}{9}$ **d** $3p$

e $\frac{p}{18}$

3 $a + b = 17$, $a > b > 0$

Decide whether each statement is true or false. Explain your answers.

a $2a + b = 34$ **b** $2a + 2b = 34$

c $170 - b = 10a$ **d** $a + b - a + b = 0$

4 d, e, f and g are all greater than 0

$d + e = f - g$

Write an equivalent expression for each in terms of f and g.

a $2d + 2e$ **b** $\frac{1}{2}e + \frac{1}{2}d$ **c** $\frac{(d + e)}{5}$ **d** $(d + e)^2$

Reflect

1 Give an example of a calculation where a mental method is the most efficient way of calculating the answer.

2 If $m \times n = 170$, explain why $10m \times 10n = 17000$

What other expressions involving m and n can you work out easily?

13 Developing number sense

Chapters 13.1–13.4

I have become fluent in...

- Using mental methods to perform calculations involving integers, decimals and fractions
- Using factors to simplify calculations
- Estimating the answer to a calculation

I have developed my reasoning skills by...

- Choosing an appropriate mental method to perform a calculation
- Deciding when to use a mental, written or calculator method
- Using estimation to check accuracy of answers

I have been problem-solving through...

- Using known number facts to derive other facts
- Using known algebraic facts to derive other facts
- Using mental methods of calculating in different contexts

Check my understanding

1 Work out the calculations mentally.

Then write down each step in your thinking.

a 572 + 99 **b** 343 − 99 **c** 1265 + 2999 **d** 57 640 − 197

2 Work out the multiplications mentally.

a 28 × 5 **b** 9 × 35 **c** 19 × 40 **d** 8 × 99 **e** 5 × 362

3 Complete the divisions mentally.

a 540 ÷ 6 **b** 420 ÷ 7 **c** 6400 ÷ 80 **d** 81 000 ÷ 9 **e** 270 000 ÷ 3000

4 Complete the calculations mentally.

a 3.68 + 2.99 **b** 5.97 − 3.01 **c** 0.99 + 27.08

d 16.83 − 5.99 **e** $\frac{1}{2}+\frac{1}{4}$ **f** $\frac{2}{5}+\frac{1}{10}$

g $\frac{7}{8}-\frac{1}{2}$ **h** $\frac{97}{100}-\frac{3}{10}$

5 Estimate the answers to the calculations.

a 326 × 12 **b** 589 ÷ 6.3

c $\frac{417 + 791}{1.88}$ **d** $\frac{12372 - 5899}{0.512}$

6 Use the fact that 152 + 83 = 235 to write down the answer to each calculation.

a 235 − 83 **b** 235 − 152 **c** 15.2 + 8.3

d $152x + 83x$ **e** 15 200 + 8300 **f** 2350 − 830

g 83 − 235 **h** 0.83 + 1.52

7 Use the fact that 21 × 14 = 294 to write down the answer to each calculation.

a 294 ÷ 21 **b** 210 × 14 **c** 1400 × 210 **d** 2940 ÷ 14

14 Sets and probability

In this block, I will learn...

how to identify and represent sets

Curly brackets are used to indicate a set.

{1, 3, 5, 7, 9}

This is the set of odd numbers below 10

to interpret and create Venn diagrams

ξ

A

B

Objects here are in set A but not set B

Objects here are in set B but not set A

Objects here are neither in set A nor set B

Objects here are in both set A and set B

the vocabulary of probability

impossible unlikely

even chance likely

certain

how to generate sample spaces

outcomes: 0, 1, 2, 2, 0, 3, 1

sample space: 0, 1, 2, 3

A sample space does not include duplicates.

how to calculate probabilities

P(blue) = 0.25

$P(5) = \frac{1}{6}$

0 ———————— 1

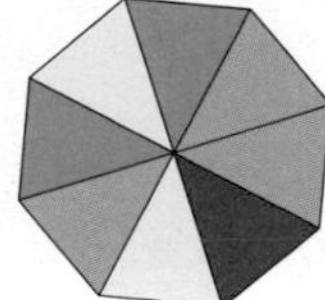

14.1 Representing sets

White
Rose
Maths

Small steps

- Identify and represent sets
- Interpret and create Venn diagrams

Key words

Set – a collection of objects or numbers

Element – a member of a set

Universal set – the set containing all relevant elements

Inclusive – including the end points of a list

Are you ready?

1 Write down the first five

a odd numbers **b** even numbers **c** square numbers **d** multiples of 3

2 List the factors of each of these numbers.

a 12 **b** 20 **c** 21

3 Here are the first six prime numbers and the first seven odd numbers.

Prime numbers: 2, 3, 5, 7, 11, 13

Odd numbers: 1, 3, 5, 7, 9, 11, 13

a Write down the numbers that are in both lists.

b Write down the next number that will appear in both lists.

c Write down the next odd number that is not prime.

Models and representations

Venn diagram

Objects here are in set A but not set B
ξ
A
B
Objects here are neither in set A nor set B
Objects here are in set B but not set A
Objects here are in both set A and set B

This symbol ξ should always be included. It stands for the **universal set**. The universal set includes everything we are interested in.

In this section you will systematically organise information into **sets**, which makes it easier to spot patterns.

Curly brackets { and } are used to denote a set. For example, {1, 3, 5, 7, 9} represents the set of positive odd integers less than 10

You need to be able to

- use set notation
- identify members of a set given a description
- describe simple sets given the **elements**.

Example 1

a Use set notation to list the elements of each of these sets.

i A = {square numbers between 20 and 50}

ii B = {the days of the week}

iii C = {letters in the word "integer"}

b Describe the elements of each set in words.

i {2, 3, 5, 7, 11, 13, 17, 19}

ii {12, 16, 20, 24, 28}

a i A = {25, 36, 49}

ii B = {Monday, Tuesday, Wednesday, Thursday, Friday, Saturday, Sunday}

Write the elements of each set in curly brackets.

Sets can include numbers, letters or words.

iii C = {i, n, t, e, g, r}

There should not be repeats within a set. Although the letter e appears twice in the word integer, you only include one e in the set.

b i Prime numbers less than 20

It is important to be exact when describing a set. You cannot just say 'prime numbers'; you need to make it clear that they are prime numbers less than 20

ii Multiples of 4 between 12 and 28 inclusive

The word **inclusive** needs to be used here to show that you are including 12 and 28 within the set.

Practice 14.1A

1 State whether or not the pairs of sets are equivalent. Explain your reasoning.

a	A = {C, D, E, F}	B = {F, E, D, C}
b	A = {1, 2, 3, 4}	B = {1, 2, 3}
c	A = {odd numbers}	B = {1, 3, 5, 7, 9, 11}
d	A = {types of sport}	B = {football teams}
e	A = {0.5, 1, 1.5, 2, 2.5}	B = {$\frac{1}{2}$, 1, $\frac{3}{2}$, 2, 2.5}

2 Geri asks 10 students in her class to name their favourite colour.

Geri writes the results as set A.

A = {green, purple, pink, red, blue, red, red, green, orange, red}

Explain the mistake that Geri has made.

3 List the elements of each of these sets. Use correct set notation.

a X = {multiples of 7 between 20 and 40}

b Y = {seasons in a year}

c Z = {letters in the word "science"}

4 Write down a possible description for the each of these sets.

a A = {1, 3, 5, 7, 9, 11}

b B = {10, 100, 1000, 10 000, 100 000, 1 000 000}

c C = {A, C, E, R}

d D = {1, 2, 3, 4, 6, 12}

5 ξ = {numbers from 1 to 50}

A = {odd numbers} B = {even numbers}

C = {multiples of 10} D = {square numbers}

a List all the elements of set D.

b List all the elements that are in both set B and set D.

c Explain why there are no elements that are listed in both set A and set C.

d List the elements that are in set C or D or both, for both questions.

What do you think?

1 ξ = {1, 2, 3, 4, 5, 6, 7, 8, 9, 10}

A = {numbers between 3 and 5} B = {7, 8, 9, 10, 11}

a Explain why 2.7 is not a member of set A.

b Explain why 5 is not a member of set A.

c Explain why set B is incorrect.

d List the elements that are not in set A. Discuss your answer with a partner.

2 ξ = {letters of the alphabet}

X = {letters in the word "beehive"} Y = {letters in the word "gatekeeper"}

a List the elements of set X and the elements of set Y.

b List the elements that are in both set X and set Y.

c List the elements of set X or set Y.

3 a What do you think is meant by the term "empty set"?

b A is the set of even numbers. B is the set of odd numbers. Explain why the set of elements of both A and B is empty.

You can use a Venn diagram to sort information efficiently and see whether or not sets intersect.

If two sets intersect, then they have at least one element in common.

Example 2

ξ = {1, 2, 3, 4, 5, 6, 7, 8, 9, 10} A = {even numbers} B = {factors of 20}

a List the elements of set A.

b List the elements of set B.

c List the elements that are in both set A and set B.

d Draw a Venn diagram to show this information.

a A = {2, 4, 6, 8, 10}

b B = {1, 2, 4, 5, 10}

c 2, 4, 10

d

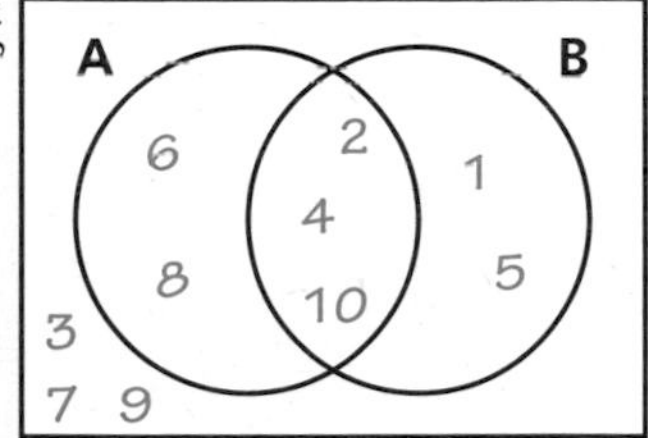

When drawing your Venn diagram ensure that you include the universal set symbol and write any numbers that do not fit into set A or set B outside the circles but inside the box.

Practice 14.1B

1 Make two copies of this diagram in your book.

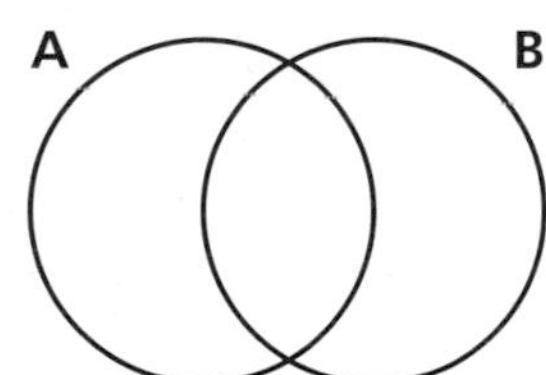

a Shade the region that contains elements of both set A and set B.

b In a different colour, shade the region that contains elements that are in set A only.

2 ξ = {1, 2, 3, 4, 5, 6, 7, 8, 9, 10}

A = {even numbers} B = {factors of 12}

a List the elements of set A.

b List the elements of set B.

c List the elements that are in both sets A and B.

d Draw a Venn diagram to represent this information.

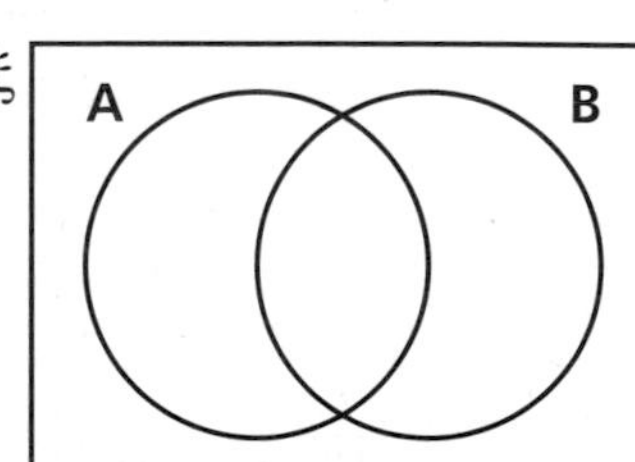

3 ξ = {numbers from –5 to 5 inclusive}

A = {–2, 0, 2, 4}　　　　B = {–3, –2, –1, 0, 1, 2}

Draw a Venn diagram to represent this information.

4 Here is a Venn diagram

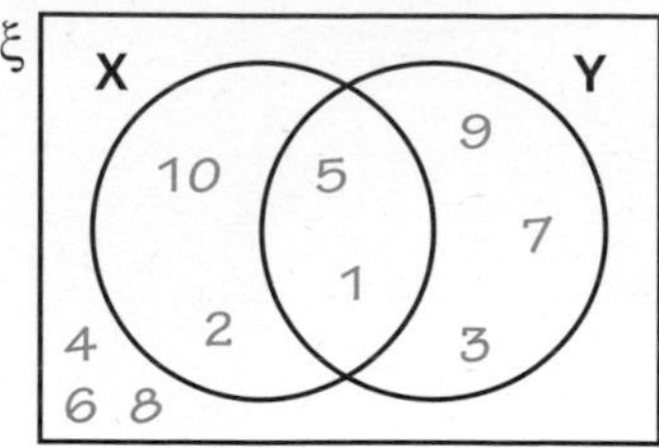

a List the elements of set X.

b List the elements of set Y.

c Write down the elements of both set X and set Y.

d Write down the elements of set X or set Y.

e Write down the elements that are not in set X.

f Write down the elements that are not in set Y.

g Write down a possible description of set X and of set Y.

5 The Venn diagram shows the number of students who study history and geography in Year 10 at a school.

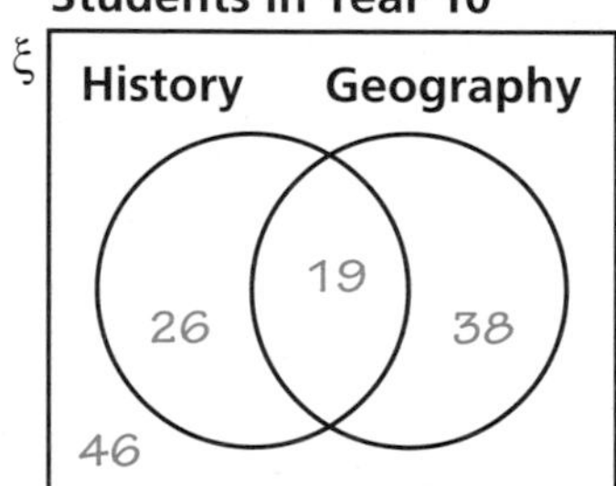

a Is it possible to tell at a glance whether more students study history or geography? Explain why.

b How many students study history?

c How many students study geography?

d How many students study history and geography?

e How many students study history or geography?

f How many students are there in Year 10?

g What percentage of the students in Year 10 study history?

Kate is a new Year 10 student at the school.

She chooses to study history and geography.

h Which number(s) change? Explain why.

6 **Numbers from 11 to 19**

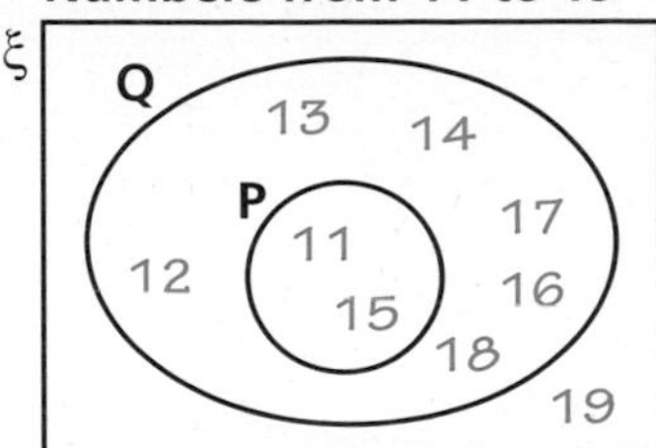

a List the elements of set P.

b Sven says that 11 is not an element of set Q. Is Sven correct?

c Write down the elements of set Q.

d How many elements are not in set Q?

e How many elements are not in set P?

f Explain why the elements of both set P and set Q are the same as those of set P.

What do you think? 

1 ξ = {1, 2, 3, 4, 5, 6, 7, 8}   A = {1, 2, 3, 4, 5}   B = {2, 3, 4}

Which of the Venn diagrams would you use to represent these sets?

1

ξ A B

2

ξ A B

3

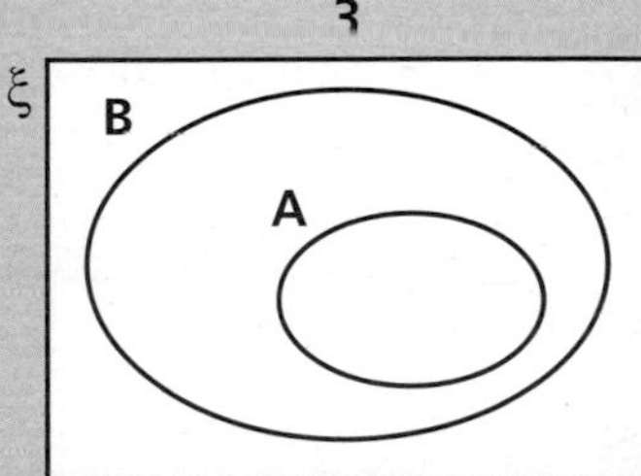

Explain your answer.

2 ξ = {20, 21, 22, 23, 24, 25, 26, 27, 28, 29}

P = {20, 22, 24, 26, 28}   Q = {21, 24, 27}

a How do you know that the two sets in the Venn diagram overlap?

b How do you know that there are some numbers outside the two sets P and Q?

c Is it possible to fill in the numbers in this diagram?

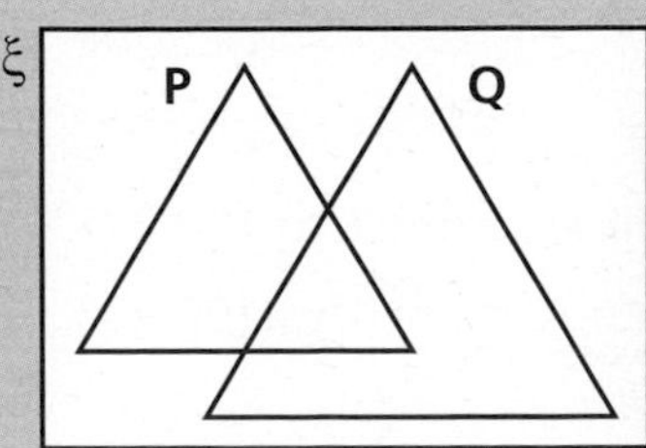

3 The Venn diagram represents two sets, A and B.

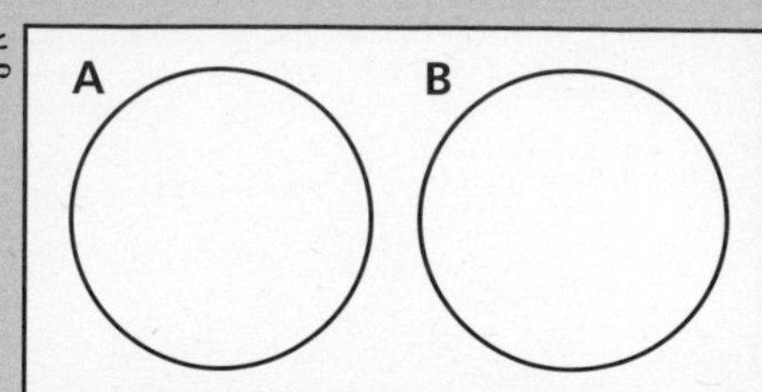

a How do you know that there are no elements in both set A and set B?

b How could you describe sets A and B? Compare your answer with a partner's.

Consolidate – do you need more?

1 List the elements of these sets using correct set notation.

a Set A: months of the year

b Set B: letters in "parallelogram"

c Set C: even numbers below 12

d Set D: factors of 20

2 A = {factors of 12} B = {factors of 30}

a List the elements of set A.

b List the elements of set B.

c Which numbers are in both set A and set B?

3 ξ = {integers between 1 and 20 inclusive}

List the elements of these sets.

a A = {odd numbers}

b B = {multiples of 2}

c Draw a Venn diagram to represent the information.

d Why are there no elements in both set A and set B?

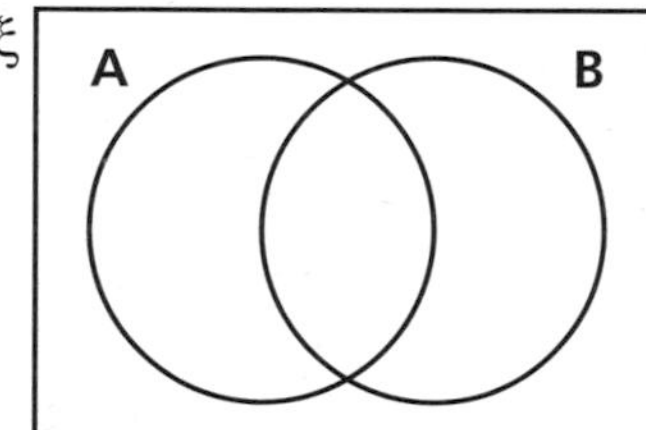

Stretch – can you deepen your learning?

1 A = {multiples of 9}

a How many elements of set A are there if ξ = {numbers from 1 to 100}?

b How many elements of set A are there if ξ = {numbers from 1 to 1000}?

Explain your method.

2 ξ = {integers from 1 to 50 inclusive}

Write down a possible description of each of these sets.

P = {13, 26, 39} Q = {4, 16, 36} R = {19, 28, 37, 46}

S = {23, 29, 31, 37} T = {44}

Compare your answers with a partner's.

3 State whether or not this set has an infinite number of elements. Explain your answer.

A = {fractions between $\frac{1}{2}$ and $\frac{3}{4}$}

4 Write down the possible members of the universal set, elements of set A and elements of set B if

- the elements of set A and B are {14, 21, 28}
- the elements not in set A are {7}
- the elements of set A or set B are {10, 14, 15, 20, 21, 25, 28}.

5 Draw Venn diagrams to represent these sets. ξ = {numbers from 10 to 25 inclusive}

a A = {square numbers} B = {even numbers]

b A = {odd numbers} B = {prime numbers}

c A = {multiples of 5} B = {factors of 24}

d What's the same and what's different about your answers to parts **a** to **c**?

6 A class contains 30 students.

14 of the students own a dog.

18 of the students own a cat.

10 students own a cat and a dog.

How many students do not own a cat or a dog?

Reflect

What could each diagram represent?

a 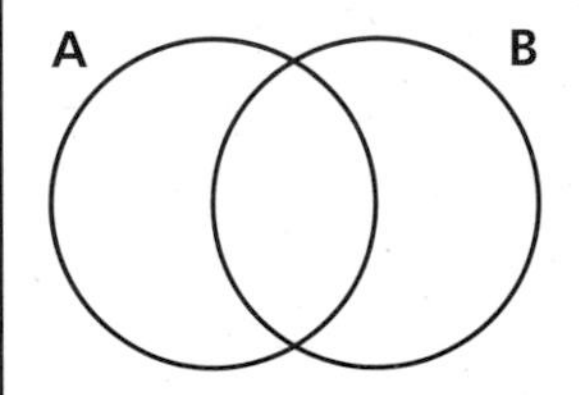

b 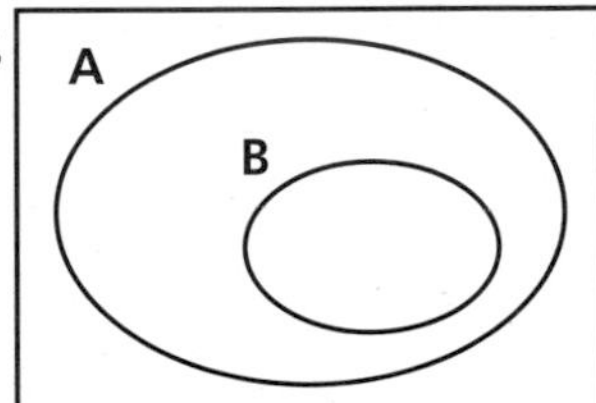

c 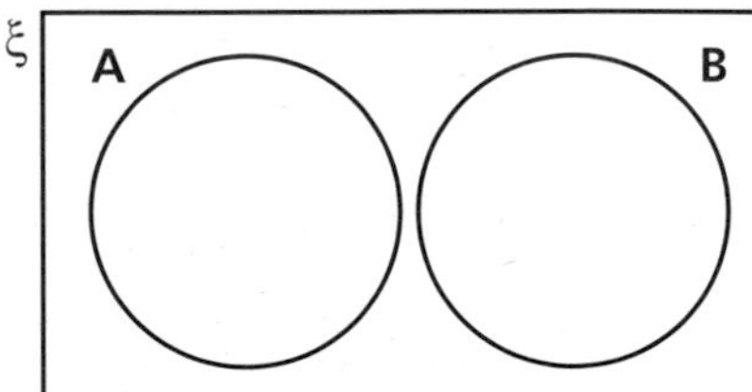

14.2 Intersections and unions

Small steps

- Understand and use the intersection of sets
- Understand and use the union of sets
- Understand and use the complement of a set (H)

Key words

Intersection – the set containing all the elements of A that also belong to B

Union – the set containing all the elements of A or B or both A and B

Complement – the elements of the universal set that do not belong to a set

Are you ready?

1 List the elements of each set.

a A = {days of the week}

b B = {months of the year}

c C = {factors of 18}

d D = {letters in the word "mathematics"}

2 ξ = {numbers between 1 and 20 inclusive}

List the elements of each set.

a) A = {odd numbers}

b B = {even numbers}

Models and representations

Venn diagrams

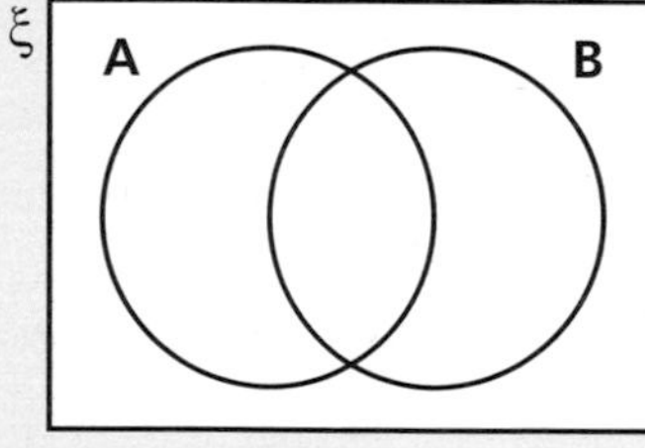

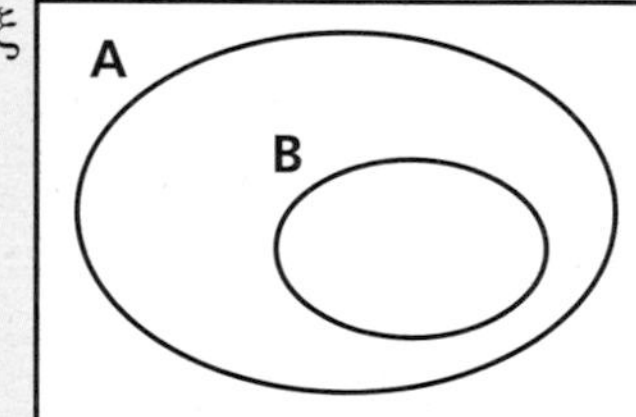

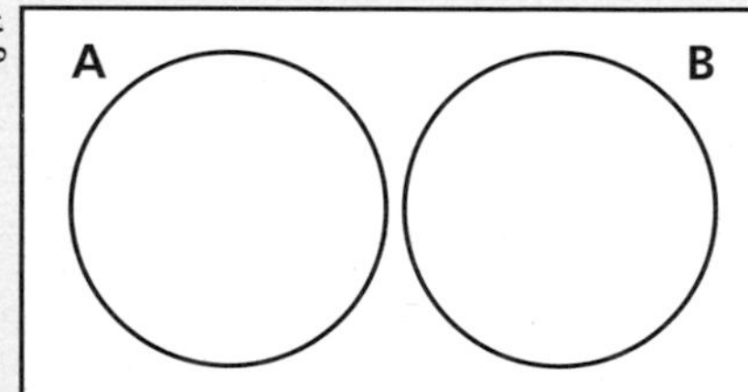

The **intersection** represents the elements that belong to two or more sets.

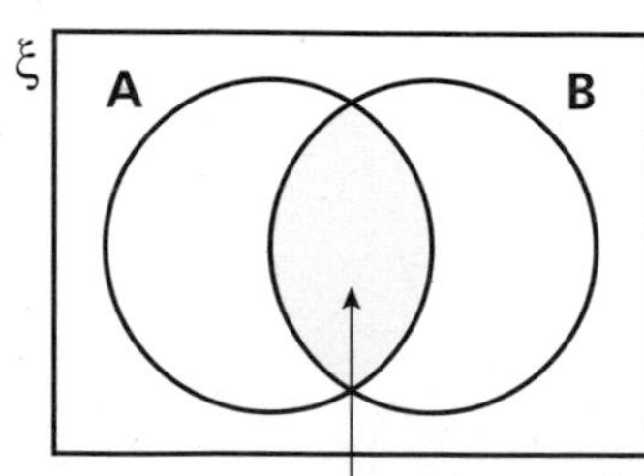

This region represents the intersection of sets A and B. This is denoted by $A \cap B$

For example, if A = {multiples of 5} and B = {factors of 5}, then 5 would be in the intersection because it belongs to both sets.

Example 1

A = {1, 2, 3, 4, 5, 6, 7, 8, 9, 10} B = {2, 4, 6, 8, 10, 12, 14, 16, 18, 20}

List the elements of A ∩ B.

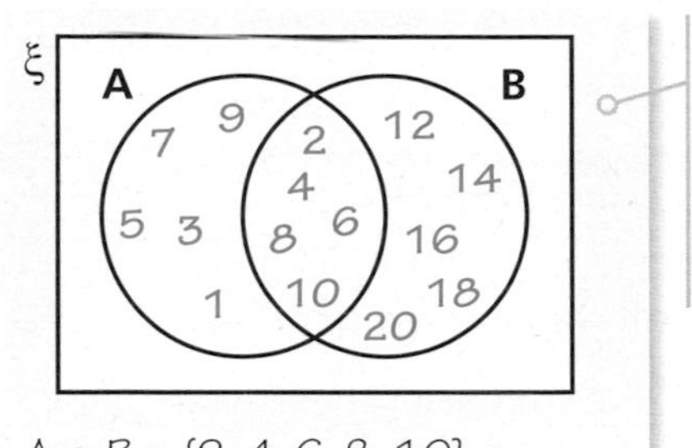

A ∩ B = {2, 4, 6, 8, 10}

You can sort the sets into a Venn diagram.

2, 4, 6, 8 and 10 go into the intersection of the Venn diagram because they are elements of both sets A and B.

This means that 2, 4, 6, 8 and 10 are all elements of A ∩ B.

Practice 14.2A

1. A = {10, 11, 12, 13, 14, 15} B = {12, 13, 14}

   List the elements of A ∩ B.

2. A = {red, white, blue} B = {green, red, yellow, black}

   List the elements of A ∩ B.

3. Copy the Venn diagram into your book and shade the region A ∩ B.

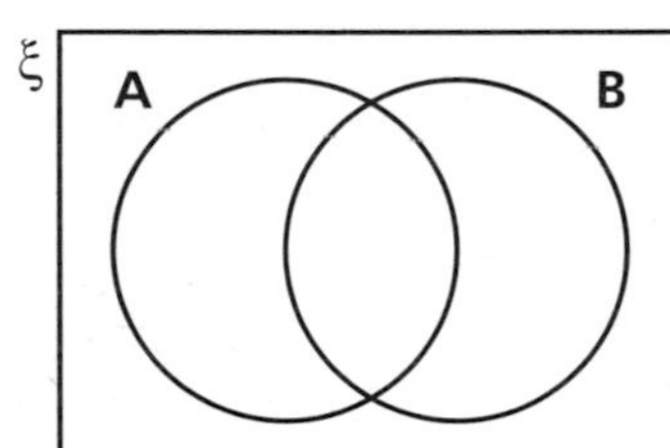

4. X = {multiples of 5 between 12 and 33}

   Y = {even numbers}

   List the elements of X ∩ Y.

5. The Venn diagram shows two sets, A and B.

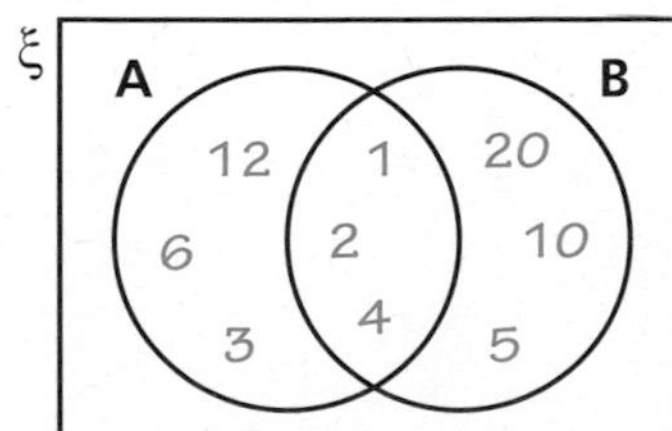

   a Write down the elements of both A and B.

   b Explain what you think the elements of A and B describe

   c What do the elements of A ∩ B represent?

6 ξ = {numbers from 1 to 100}

A = {5, 9, 11, 17, 29, 33}

A ∩ B = {9, 17}

Set B has seven elements. Write down three possible sets for B.

What do you think?

1 The Venn diagram shows two sets, A and B.

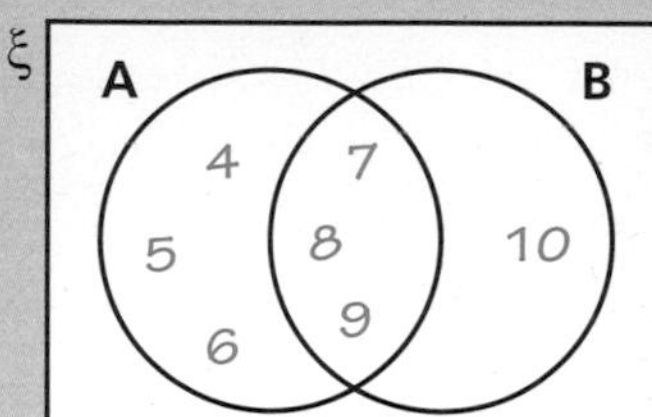

a Write down the elements that are in set A and set B.

b List the elements of A ∩ B.

What do you notice?

2 Here are two sets represented on a Venn diagram.

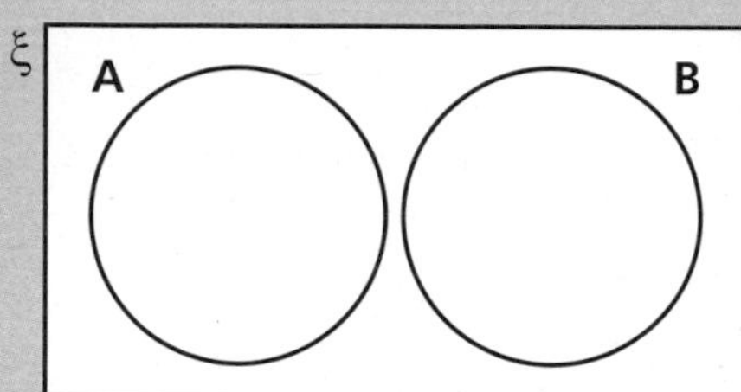

a Explain why A ∩ B does not contain any elements.

b Give an example of two possible sets A and B.

3 E = {English, maths, science, PE, history, French}

F = {maths, geography, art, science, computing}

Ali says that E ∩ F is {maths, maths, science, science}. What mistake has Ali made?

4 Here are two sets represented on a Venn diagram.

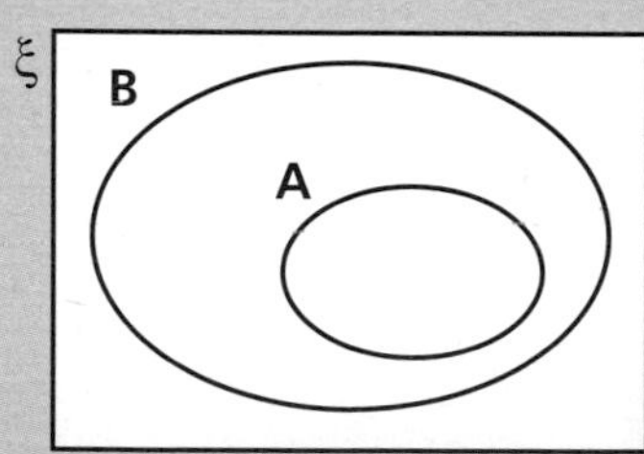

a Explain why A ∩ B is equal to the elements of set A.

b Give an example of two possible sets A and B.

5 True or false? A ∩ B = B ∩ A

Explain your answer.

The **union** is the set that contains all the elements of two or more sets.

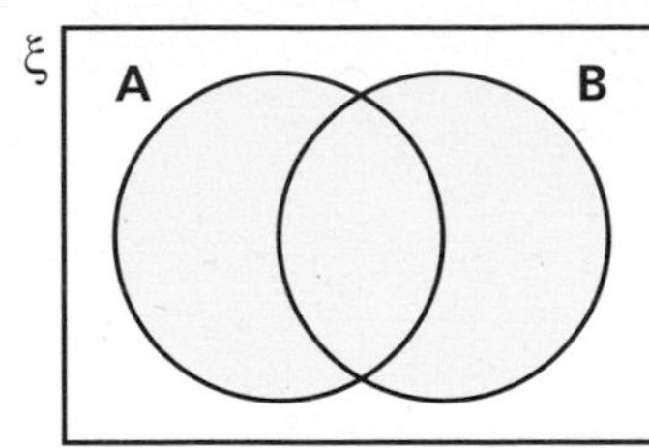

The region shaded represents the union of sets A and B. This is denoted by A ∪ B

For example, if A = {days of the week} and B = {months of the year}, then the union would contain all the days of the week and all the months of the year.

Example 2

A = {1, 2, 3, 4, 5, 6, 7, 8, 9, 10} B = {2, 4, 6, 8, 10, 12, 14, 16, 18, 20}

List the elements of A ∪ B.

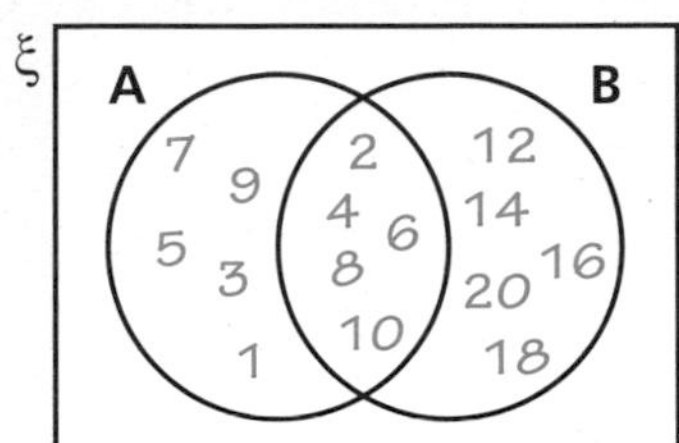

A ∪ B represents all the elements of both set A and set B. This is all the numbers that are written in the Venn diagram.

Even though 2 is in both set A and set B, you only write it once in A ∪ B.

A ∪ B = {1, 2, 3, 4, 5, 6, 7, 8, 9, 10, 12, 14, 16, 18, 20}

Practice 14.2B

1 Mateusz and Hannah are writing down as many three-letter words beginning with 't' as possible in 30 seconds.

M = {tab, tan, tam, the, tea, ten}

H = {tap, tax, tan, the, ten}

a List the elements of M ∪ H.

b Describe what M ∪ H tells you.

2 A = {red, white, blue} B = {green, red, yellow, black}

Explain why A ∪ B is not {red, white, blue, green, red, yellow, black}.

3 The Venn diagram shows two sets, A and B.

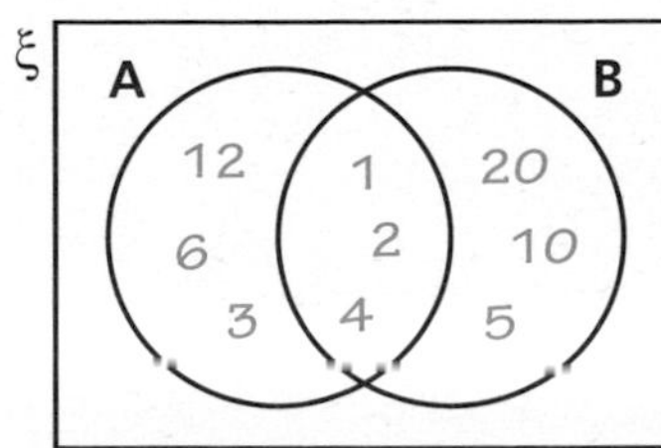

Write down the elements of A ∪ B.

4 ξ = {multiples of 5 between 1 and 49}

A = {multiples of 3} B = {multiples of 2}

a Write down the elements of A ∪ B.

b Write down the elements of A ∩ B.

5 ξ = {1, 2, 3, 4, 5, 6, 7, 8, 9, 10}

A = {2, 4, 6, 8, 10} B = {5, 10} C = {2, 3, 5, 7}

Write down the elements of each of these sets.

a A ∪ B **b** A ∪ C **c** B ∪ C **d** (A ∪ B) ∪ C

e Is (A ∪ B) ∪ C the same as (A ∪ C) ∪ B?

6 The Venn diagram represents the children in a class. B is the set of children who own a bike. S is the set of children who own a skateboard.

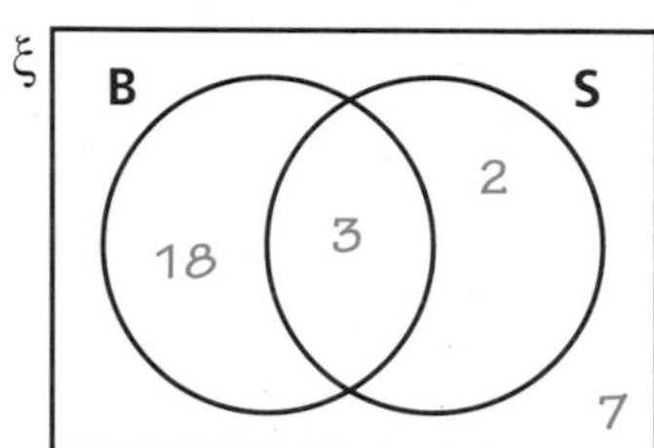

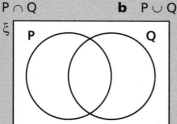

a How many children own a bike and a skateboard?

b How many children own a bike or a skateboard?

c How many children own just a skateboard?

What do you think?

1 Copy the Venn diagram into your book and shade each of these regions.

a P ∩ Q **b** P ∪ Q

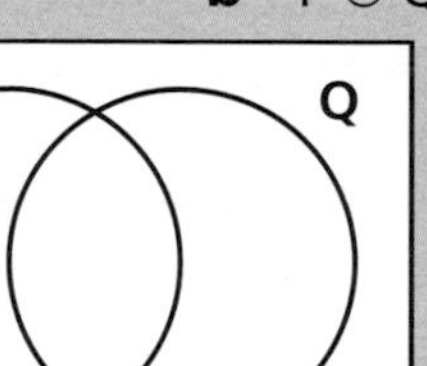

2 ξ = {countries of the world}

A = {France, South Africa, Brazil, Canada, Russia}

B = {Spain, Cuba, China}

a Draw a Venn diagram to represent sets A and B.

b Explain why A ∪ B is just all the countries in sets A and B put together.

C = {Japan, China, Australia, Canada}

c Draw a Venn diagram to represent sets A, B and C.

3 A ∪ B = {2, 5, 6, 8, 9, 10}

Write down possible sets for A and B that match each of these diagrams.

a

ξ A B

b

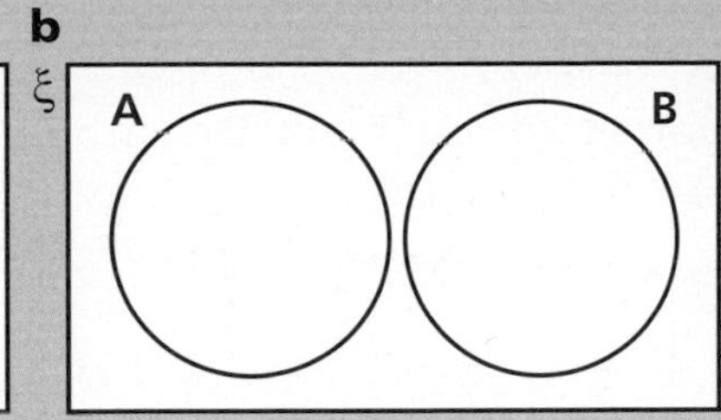

c

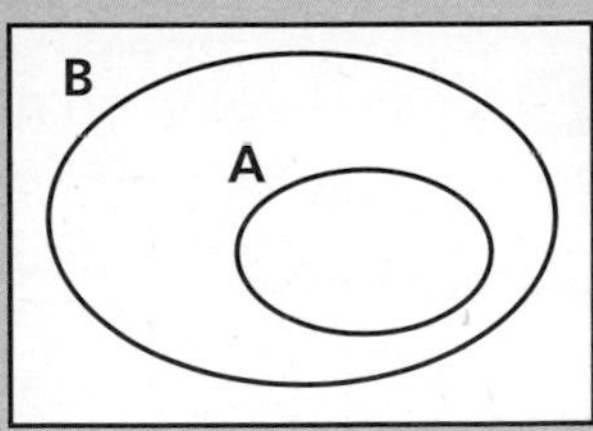

The **complement** of a set is the elements of the universal set that are not in a given set. H

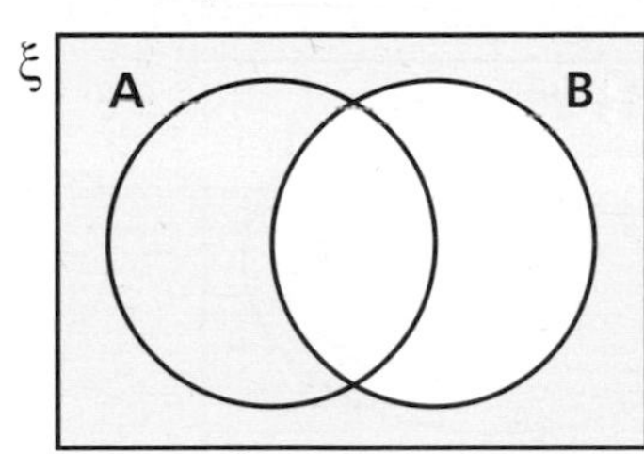

The shaded region represents all the elements that are not in B. This is denoted by B'

For example, if ξ = {numbers between 1 and 20 inclusive} and A = {even numbers}, then the complement of set A (A') would be odd numbers between 1 and 20

Example 3

ξ = {integers between 1 and 20 inclusive}

A = {1, 2, 3, 4, 5, 6, 7, 8, 9, 10}

B = {2, 4, 6, 8, 10, 12, 14, 16, 18, 20}

List the elements of A'.

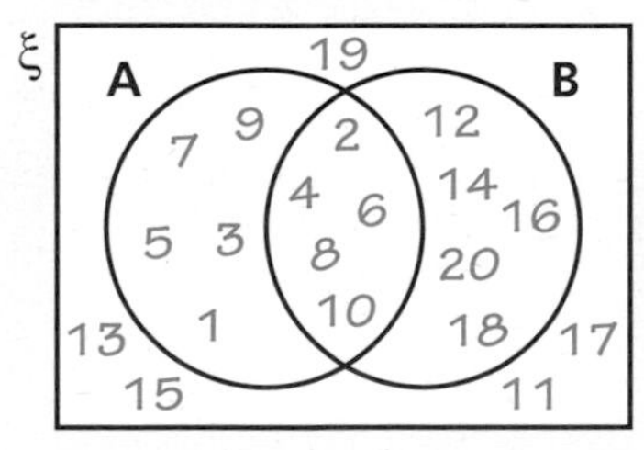

You need to include the elements in set B that are not in the intersection.

A' = {11, 12, 13, 14, 15, 16, 17, 18, 19, 20}

The complement of set A is the set of elements that are not in A.

Practice 14.2C Ⓗ

1 ξ = {2, 4, 6, 8, 10, 12, 14} A = {2, 4, 8, 12}

Write down the elements of A′.

2 Copy these Venn diagrams into your book. On each diagram, shade the region A′.

a 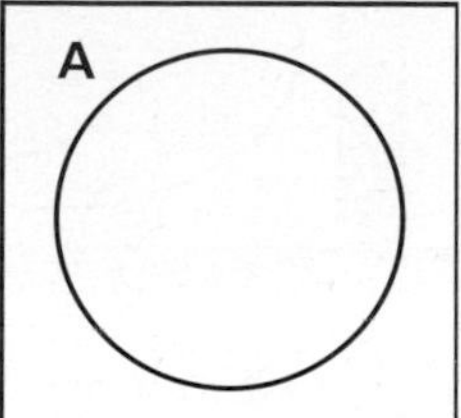

b 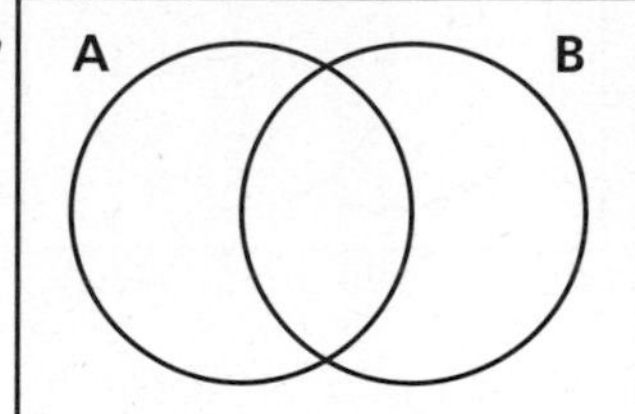

c 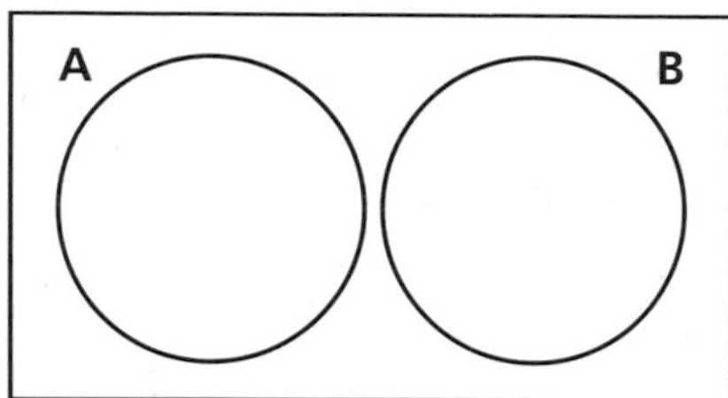

d 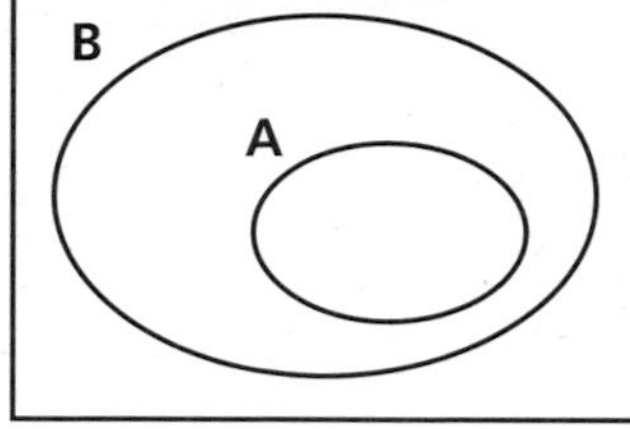

3 Here are two sets, A and B.

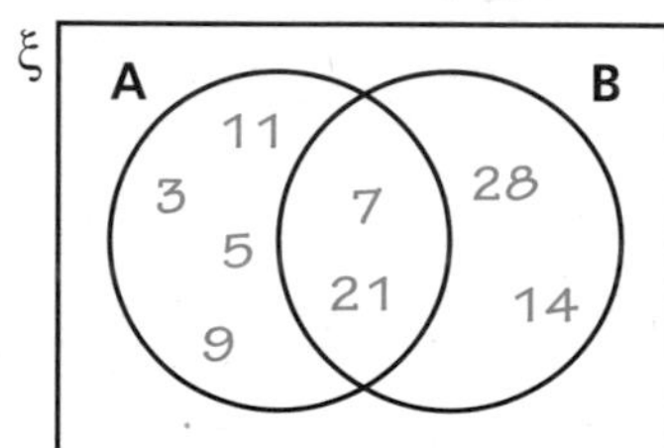

a Write down the elements of A′.

b Write down the elements of B′.

4 ξ = {e, g, h, m, n, p, s, t}

A = {h, p, s, t} B = {e, g, h, m, p, t}

Write down the elements of each of these sets.

a A′

b B′

c (A ∩ B)′

d (A ∪ B)′

What do you think?

1 Copy each Venn diagram into your book.

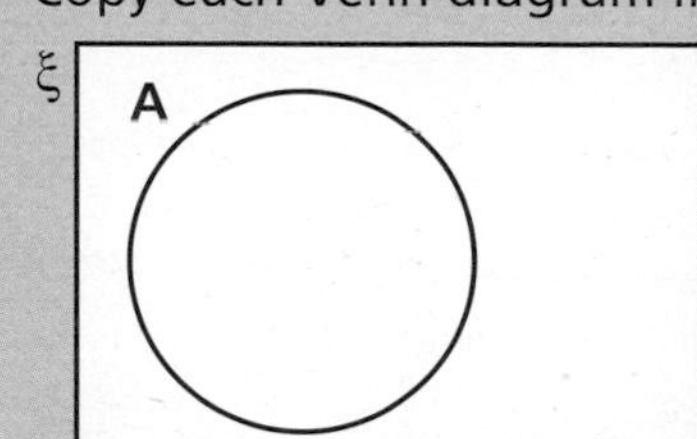

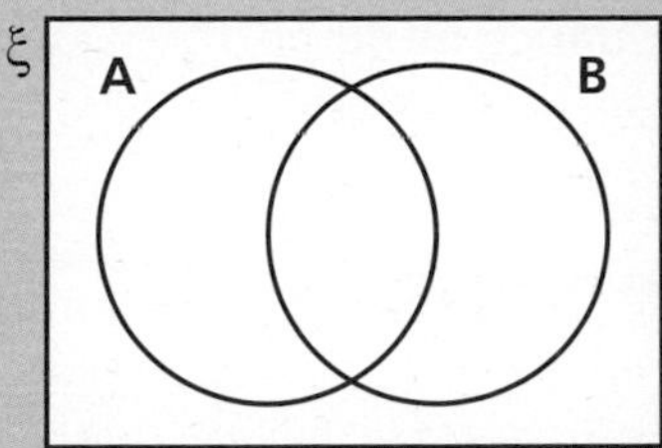

- **a** Shade the region which represents A.
- **b** Shade the region which represents A′.
- **c** What do you notice about A and A′? Why do you think A′ is called the complement?

2 ξ = {integers from 1 to 20 inclusive}

- **a** List the elements of the complement of A.
- **b** List the elements of the complement of B.

Consolidate – do you need more?

1 The elements of two sets, P and Q, are shown in the Venn diagrams.

A **B** **C** **D**

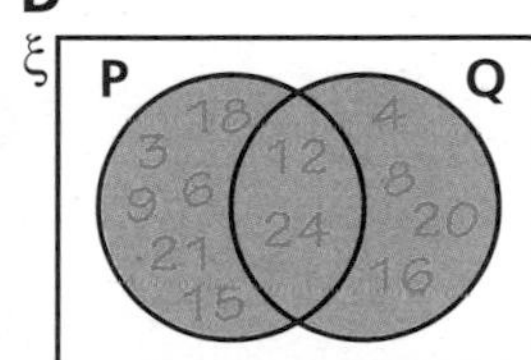

- **a** Which Venn diagram shows the intersection of P and Q? List the element of P ∩ Q.
- **b** Which Venn diagram shows the union of sets P and Q? List the elements of P ∪ Q.
- **c** Which Venn diagram shows the complement of set P? List the elements of P′.
- **d** Which Venn diagram shows the complement of set Q? List the elements of Q′.

2 The elements of two sets, A and B, are shown in the Venn diagram.

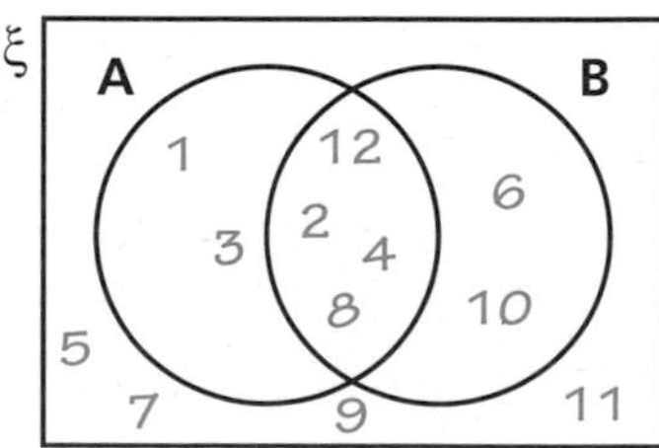

- **a** List the elements of A ∩ B.
- **b** List the elements of A ∪ B.
- **c** List the elements of A′.
- **d** List the elements of B′.

Stretch – can you deepen your learning?

1 40 children take part in a school show.

Children either sing, dance or do both.

35 children dance.

28 children sing.

How many children sing and dance? Use a Venn diagram to help you.

2 Here are three sets X, Y and Z.

Write down the elements of

a $X \cap Y$

b $Y \cap Z$

c $(X \cap Y) \cap Z$

d Y'

e $X' \cup Y'$

f $X \cup Y'$

g $(X \cup Y)'$

h $(X \cup Y) \cap Z'$

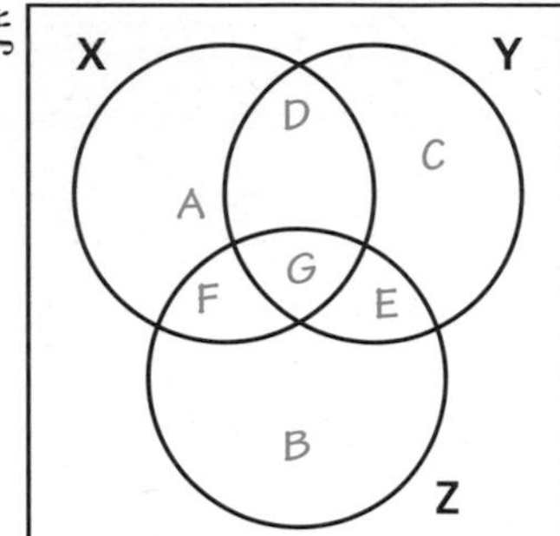

3 30 students are doing PE.

18 of the children are wearing a bib.

10 children are holding a cone.

80% the children are either wearing a bib or holding a cone.

How many children are just wearing a bib?

Reflect

1 In your own words, explain what is meant by the intersection of two or more sets.

Draw a diagram to support your answer.

2 In your own words, explain what is meant by the union of two or more sets.

Draw a diagram to support your answer.

14.3 Introducing probability

Small steps

- Know and use the vocabulary of probability
- Generate sample spaces for single events

Key words

Probability – how likely an event is to occur

Random – happening without method or conscious decision; each outcome is equally likely to occur

Sample space – the set of all possible outcomes or results of an experiment

Are you ready?

1 List the numbers that are on a standard six-sided dice.

2 What are the two possible outcomes of flipping a coin?

3 What is the difference between a fair coin and a biased coin?

Models and representations

You can describe the **probability** of an event happening using the words "impossible", "unlikely", "even chance", "likely" or "certain".

Probability scale

Impossible ——————— Certain

A **sample space** is a list of possible outcomes.

S = {1, 2, 3, 4, 5, 6}

Example 1

impossible	unlikely	even chance	likely	certain

Marta has a fair six-sided dice. Use a word from the box to describe the probability that

a she rolls a number below 7

b she rolls a 7

c she rolls a number above 3

Method A

a Certain — All the numbers on a dice are below 7 so she will definitely roll a number below 7

b Impossible — The highest number on the dice is 6, so she cannot roll a 7

c Even chance — There are three numbers on a dice that are above 3 (4, 5 and 6). There are six numbers on the dice in total so there is the same chance of rolling a 4, 5 or 6 as of rolling a 1, 2 or 3

14.3 Introducing probability

Practice 14.3A

1 Match each statement to the correct word to describe the probability of that event happening.

	Statement		Word
a	You will blink today		Likely
b	You will get a head when you a flip a fair coin		Impossible
c	You will go to bed by midnight		Unlikely
d	It will snow tomorrow		Certain
e	You will get a 7 if you roll a fair dice.		Even chance

2 Jakub puts these numbers cards into a bag.

1	4	5	5	5
7	9	10	11	12

Jakub removes a card without looking. Use a word from the list to describe the likelihood of Jakub choosing each type of number.

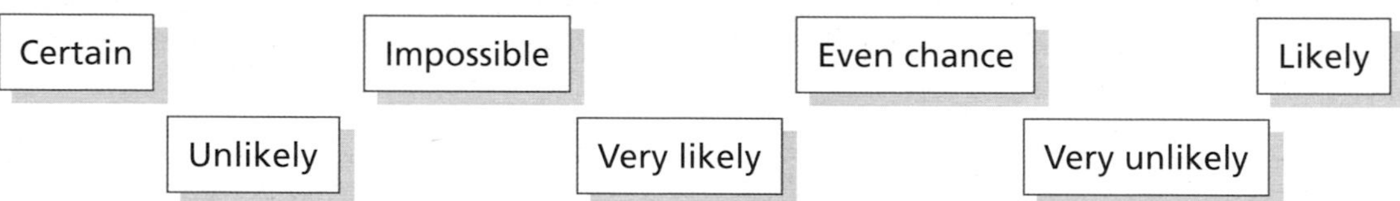

a a number greater than 6

b the number 1

c a number less than 20

d the number 30

e a number above 2

f a number below 8

3 Here is a fair, numbered spinner.

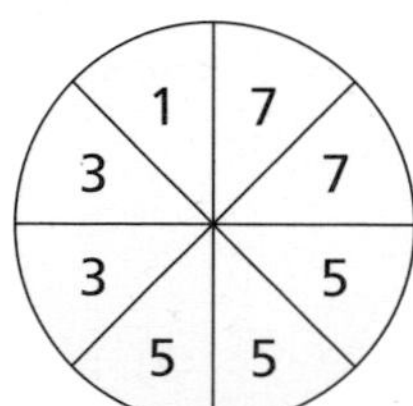

a Which is the most likely number the spinner will land on?

b Which is the least likely number the spinner will land on?

c Do any numbers have the same chance of occurring?

d Is it more likely that the spinner will land on a number greater than 4 or less than 4?

Give a reason for each of your answers.

4 Here are two fair spinners.

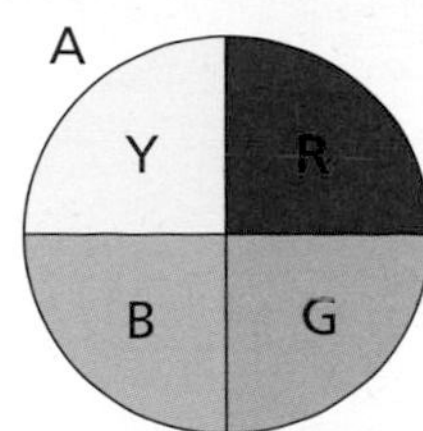

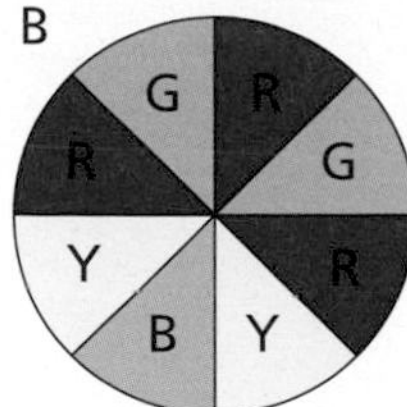

a Jackson says, "The probability of spinning blue on each spinner is the same because there is one blue section on each spinner." Explain why Jackson is incorrect.

b The spinners are used in a game. To win the game, you must choose a spinner and spin red.
Is it more likely that you will win using spinner A or spinner B? Explain your answer.

5 Jakub buys a raffle ticket.

There are 500 tickets in the raffle and there is only one winning ticket.

Jakub says that he has an even chance of winning the prize because he can win or lose.

Is Jakub correct?

What do you think?

1 **a** Write down an event that is certain to happen today.

b Write down an event that is very unlikely to happen today.

c Write down an event that has an even chance of happening today.

Compare your events with your partner's. Do you agree?

2 Flo and Faith are playing a game. One of them will win. Flo says, "There is an even chance that I will win, because there are two people and one of us will win." Explain why Flo might be wrong.

3 Discuss each of these statements.

> If I flip a fair coin 10 times, I will get 5 heads and 5 tails.

> If I flip a fair coin 50 times, it is impossible that it will land on heads every time.

You can use a **sample space** to systematically list all possible outcomes of an experiment.

This follows from the way you systematically listed information using a Venn diagram.

Example 2

a Here is a spinner.

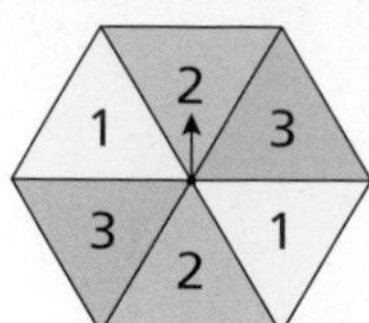

Beca says that the sample space is S = {1, 1, 2, 2, 3, 3}. Is Beca correct? Explain your answer.

b One of these cards is chosen at **random**.

B E E K E E P E R

Write the sample space for the possible outcomes.

a *No, Beca is wrong. You only write each number once. The sample space is S = {1, 2, 3}.*

Sample space means all the possible outcomes. The only possible outcomes could be to land on a 1, 2 or a 3

b *S = {B, E, K, P, R}*

To show a sample space always use S = {…}. Use commas between each outcome.

Practice 14.3B

1 A fair six-sided dice is rolled. Write down the sample space for the event using set notation; S = {…}.

2 Here is a spinner.

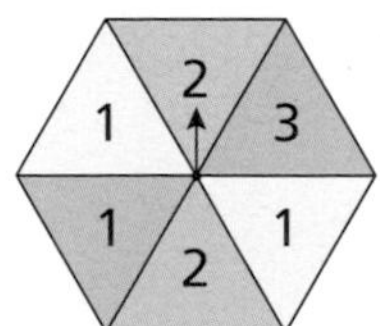

Zach says that the sample space for the event is S = {1, 1, 1, 2, 2, 3}. Explain why Zach is incorrect.

3 Write down the sample space for each of these events.

a Flipping a coin.

b The outcomes from spinning this spinner.

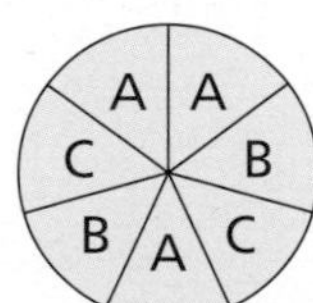

4 These letter cards are placed in a bag.

M A T H E M A T I C A L

A letter is picked out at random. Write down the sample space for the event.

5 A spinner has 12 sections.

The sample space for the event is S = {red, green, yellow, blue}.

a Draw a possible spinner if the outcomes in the sample space are equally likely.

b Draw a possible spinner if all the outcomes have a different probability of happening.

6 Here are 5 number cards.

The cards are put into a bag and a card is chosen at random. The sample space of the outcomes is S = {10, 11, 12, 13, 14}.

a What are the numbers on the cards?

Here are 9 number cards.

The cards are put into a bag and a card is chosen at random. The sample space for the outcomes remains the same as before.

b What could the numbers on the cards be?

c Why is it not possible to determine which number is most likely to be removed when you have 9 cards?

What do you think?

1 Here is a bowl of fruit.

A piece of fruit is chosen at random. What is a possible sample space?

2 Here are the contents of two boxes of balls.

Box A	Box B
1 pink ball	5 pink balls
1 orange ball	4 orange balls
1 yellow ball	3 yellow balls

a A ball is removed out of box A at random and the colour is noted.
What is the sample space for the colour of the ball?

b A ball is removed out of box B at random and the colour is noted.
What is the sample space for the colour of the ball?

c What's the same and what's different about your answers to parts **a** and **b**?

3 Discuss the following statement.

"You can't use the sample space for an event to work out the probability of an outcome."

Consolidate – do you need more?

1 Unscramble the letters to find the words related to probability.

a KELLIY **b** CRINTAE **c** LUEKILNY **d** ELSPOSBIMI **e** EEVN HANCCE

2 Describe the likelihood of each event happening.

a Coming to school tomorrow.

b It snows next week.

c Getting a head or a tail when you flip a fair coin.

d Bumping into a giraffe on your way home.

Compare your answers with a partner's. Do you agree? Would your answers be different if it were a different day or time of year?

3 Write down the sample space for each spinner.

a

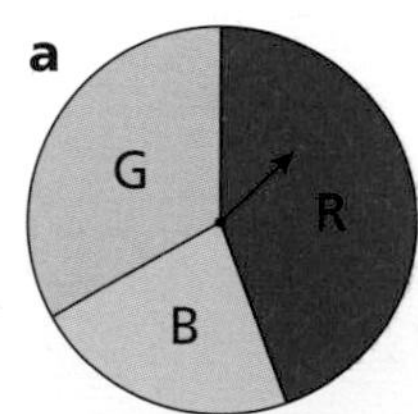

b

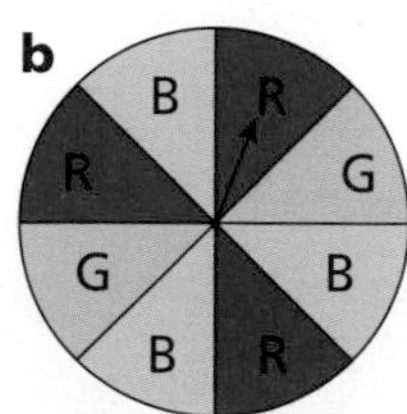

What do you notice? Why does this happen?

Stretch – can you deepen your learning?

1 Abdullah has five number cards.

He chooses a card at random.

- The probability that he chooses a number less than 10 is certain.
- The probability that he chooses a number greater than 5 is the same as the probability that he chooses a number less than 5.
- The most likely number that he chooses is 4.

What could the numbers on the cards be?

2 Some counters are labelled with numbers. A counter is chosen at random. The sample space of this event is S = {4, 6, 8, 10}.

a Write down the probability of getting an odd number.

b Write down the probability of getting an even number.

c Explain why you cannot work out the probability of getting a number greater than 6 with this information.

3 Here are 5 counters.

A counter is chosen at random. Write down the sample space for this event.

4 Here are 16 cards. Each card has a number on it.

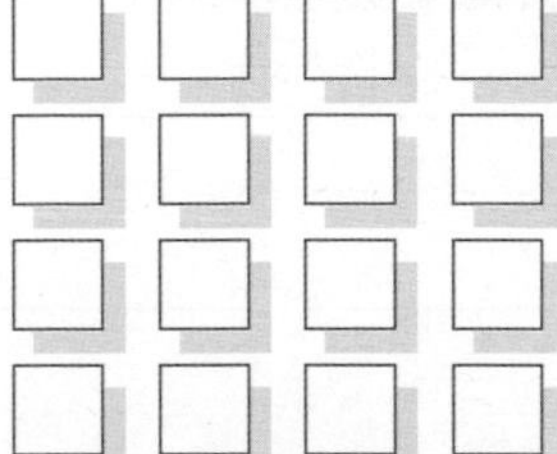

A card is chosen at random. The sample space for the number chosen is S = {1, 3, 6, 12, 24, 48}.

The probability of choosing a card numbered 48 is greater than the probability of choosing a card numbered 12.

It is equally likely that a card numbered 3, 6, 12 or 24 will be chosen.

How many cards of each number could there be?

Reflect

1 Make up your own probability questions with these answers: impossible, certain, unlikely, likely, even chance.

Which did you find the most difficult to write?

2 Explain, in your own words, what a sample space is.

14.4 Calculating probability (1)

Small steps

- Calculate the probability of a single event
- Understand and use a probability scale

Key words

Random – happening without method or conscious decision; each outcome is equally likely to occur

Biased – all possible outcomes are not equally likely

Equally likely – having the same chance of happening

Are you ready?

1 Simplify these fractions.

a $\frac{50}{100}$ **b** $\frac{4}{12}$ **c** $\frac{16}{64}$ **d** $\frac{25}{30}$ **e** $\frac{21}{70}$

2 Write the sample space of the outcomes when this spinner is spun.

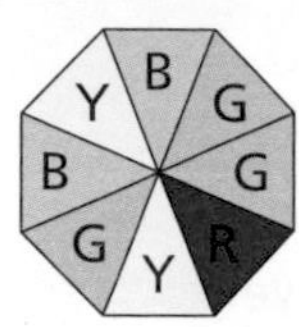

3 Work out the intervals on these number lines. Copy the number lines and complete them using fractions.

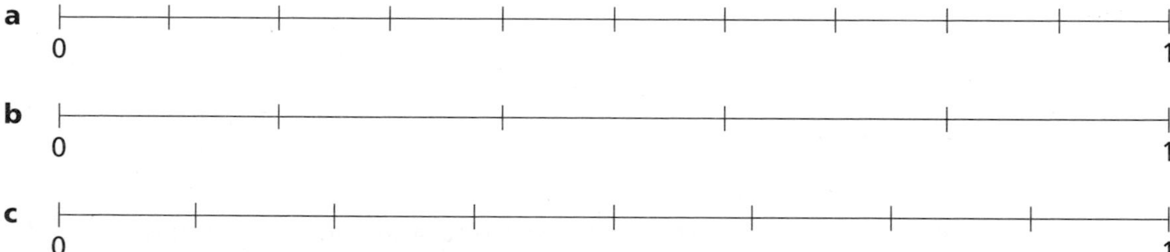

Models and representations

Probability scale

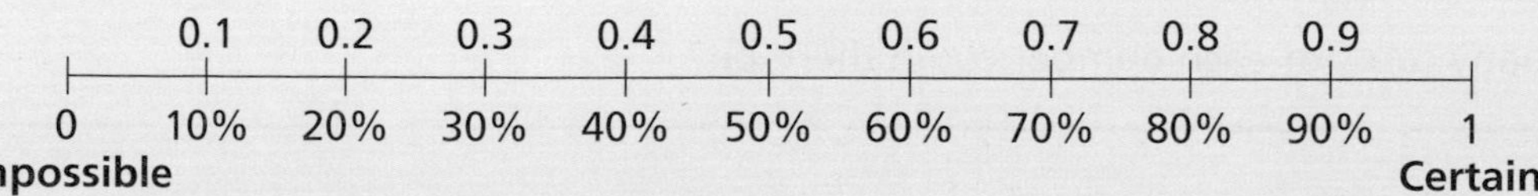

P(Event) is a short way of writing "the probability of an event happening".

In this chapter you will calculate the probability of single events. You can give probabilities as fractions, decimals or percentages.

Probabilities should always be between 0 and 1, so, for example, 60% is a possible probability, but 60 is not.

Note: ratio notation is not an acceptable way of writing probabilities.

Example 1

This fair spinner is spun.

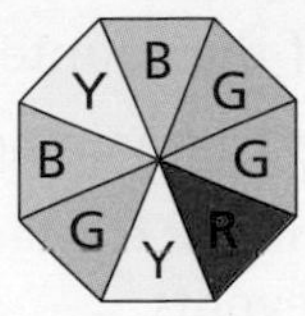

a What is the probability that it lands on red?

b What is the probability that it lands on green or yellow?

c What is the probability that it does not land on blue?

It is important to know that the spinner is fair because that means that it is **equally likely** to land on any of the sections, i.e. it is not **biased**.

a $P(\text{red}) = \frac{1}{8}$

The notation P(red) is a short way of writing 'the probability of getting red'.

There are 8 sections in total, and one of them is red, so the probability that it lands on red is 1 out of 8 or $\frac{1}{8}$

b $P(\text{green or yellow}) = \frac{5}{8}$

There are 3 green sections and 2 yellow sections.

This means that there are 5 green or yellow sections, so the probability is $\frac{5}{8}$

c $P(\text{not blue}) = \frac{6}{8} = \frac{3}{4}$

The spinner has 2 blue sections. So there are 6 sections that are not blue (because there are 8 sections in total). Therefore the probability of the spinner not landing on blue is $\frac{6}{8}$, which simplifies to $\frac{3}{4}$

Practice 14.4A

1 A fair six-sided dice is rolled.

What is the probability of each of these events?

a Rolling a 4

b Rolling a 7

c Rolling an odd number

d Not rolling a 4

e Rolling a multiple of 3

f Rolling a square number

g Rolling a number greater than 1

h Rolling a number less than 10

2 A drawer contains some socks.

There are 4 grey socks in the drawer.

There are 6 blue socks in the drawer.

A sock is taken out of the drawer at **random**.

a What is the probability that the sock is grey?

b What is the probability that the sock is blue?

c What is the probability that the sock is black?

3 A bag contains some coloured counters.

- 4 of the counters are red.
- 3 of the counters are yellow.
- The 1 remaining counter is blue.

A counter is chosen at random from the bag. Work out the probability of each of these events.

a The counter chosen is yellow.

b The counter chosen is either red or blue.

c The counter chosen is not yellow.

d What do you notice about your answers to parts **b** and **c**? Why do you think this is the case?

4 Here is a fair spinner.

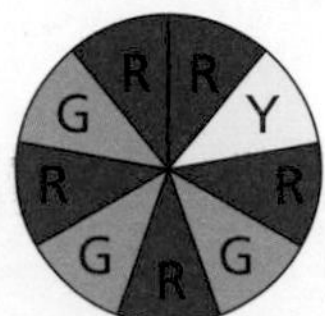

The spinner is spun. Work out

a P(spinner lands on red)

b P(spinner lands on green)

c P(spinner lands on pink)

d P(spinner lands on red or green)

5 A jar contains some sweets.

There are 60 mints, 35 chocolates and 25 jellies.

A sweet is chosen at random from the jar.

What is the probability of choosing a jelly sweet?

6 A bug box contains some creatures.

The table below shows the number of each species of creature.

Creature	Beetle	Caterpillar	Spider	Worm
Number	18	30	47	5

A creature is chosen at random from the box.

Work out

a P(choosing a beetle)

b P(choosing a spider)

c P(choosing a spider or worm)

d P(not choosing a beetle)

7 There are 12 counters in a box.

P(orange counter) = $\frac{5}{12}$ P(yellow counter) = $\frac{1}{4}$ P(white counter) = $\frac{1}{3}$

a How many of each colour counter are in the box?

b Are there any other coloured counters in the box? How do you know?

c Samira says that she is more likely to get an orange counter than not get an orange counter. Explain why she is incorrect.

d What is the smallest number of counters that Samira needs to select from the box so that she is guaranteed to remove an orange counter? Explain your reasoning.

8 Some cards have some letters and symbols on them.

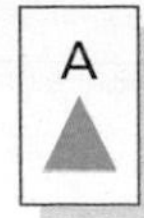

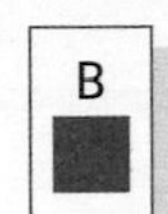

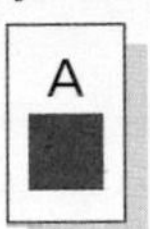

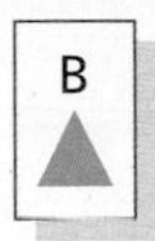

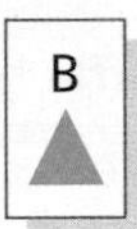

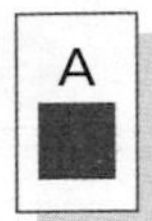

A card is chosen at random. What is the probability that the card

a has the letter A on it

b has a square on it

c has an A and a square on it

9 A spinner has 10 sections.

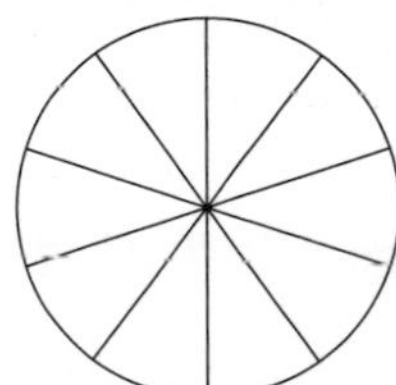

Copy the spinner and colour it in so that P(red) = $\frac{3}{10}$, P(blue) = $\frac{1}{5}$ and P(green) = 0.5

What do you think? 

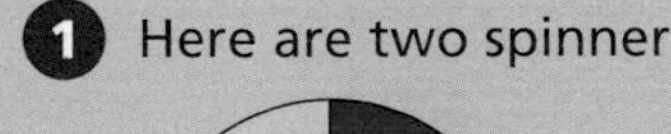

1 Here are two spinners.

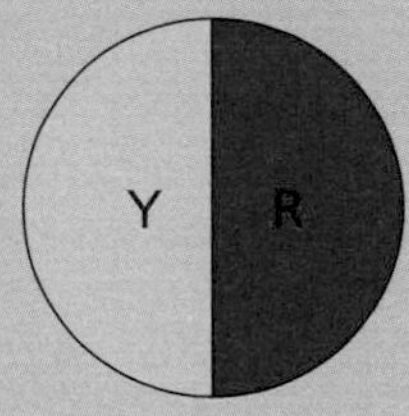

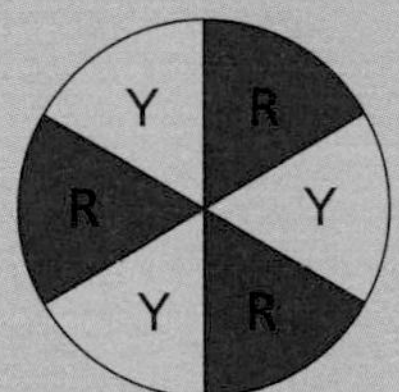

It is more likely that you will spin red on the second spinner as there are more red sections

Do you agree with Beca? Explain your reasons.

2 A fair six-sided dice has an equal chance of landing on an odd number or an even number.

Write this probability as a fraction, a decimal and a percentage.

3 Benji and Faith are playing a game with a fair eight-sided dice, numbered 1 to 8.

Benji wins if he rolls an even number.

Faith wins if she rolls a number greater than 5.

a Decide which player is more likely to win.

b If they used a 12-sided dice (numbered 1 to 12) would your answer change?

c What about a 10-sided dice?

4 A dice is rolled. The probability of rolling a 3 on this dice is $\frac{1}{4}$. Give two possible reasons why this might be the case.

5 A raffle has a prize.

There are 200 raffle tickets.

Jackson buys 1 ticket and Tamsin buys 2 tickets.

Tamsin says that she twice as likely to win a prize as Jackson.

Is Tamsin correct? Justify your answer.

Probabilities are represented by numbers on a scale between 0 and 1, where 0 is impossible and 1 is certain.

Example 2

This spinner is fair.

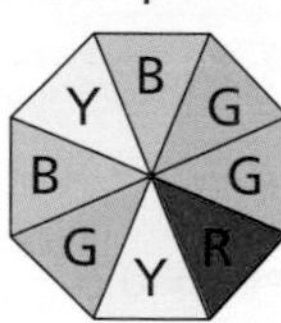

Copy the number line shown and draw an arrow to show each probability.

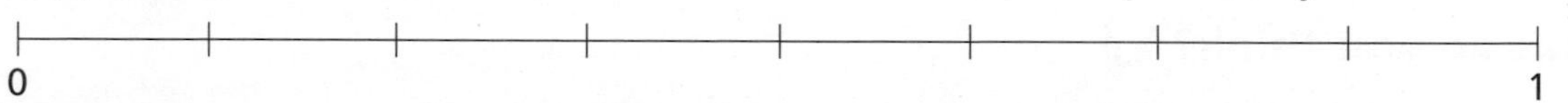

a P(red) **b** P(green or blue) **c** P(red or green or yellow or blue)

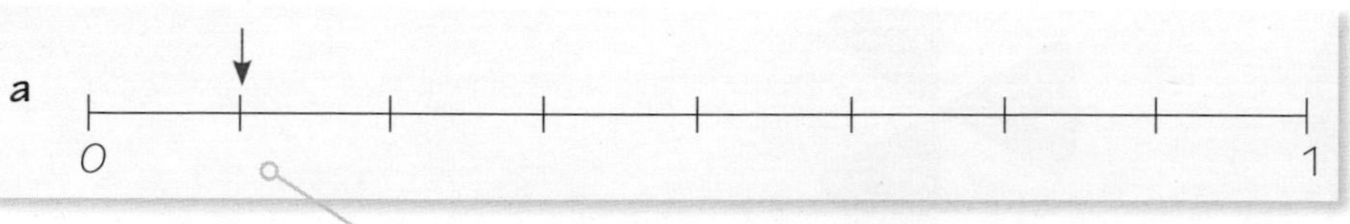

The probability of the spinner landing on red is $\frac{1}{8}$

1 out of the 8 sections is red.

The number line is going up in eighths so the arrow is pointing to $\frac{1}{8}$

The number line is divided into 8 equal parts.

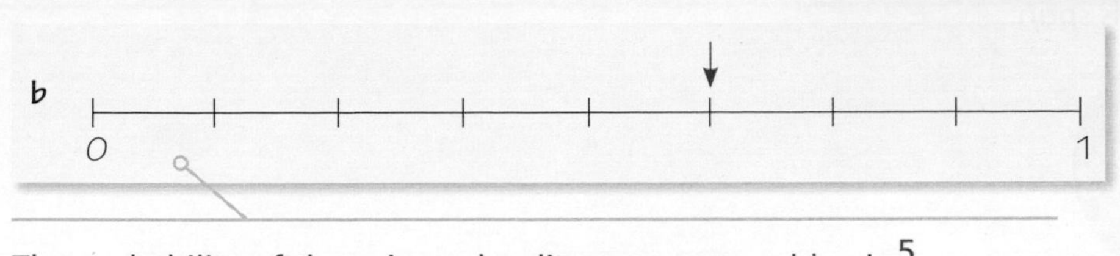

The probability of the spinner landing on green or blue is $\frac{5}{8}$

There are 2 blue sections and 3 green sections: 2 + 3 = 5

The number line is going up in eighths so the arrow is pointing to $\frac{5}{8}$

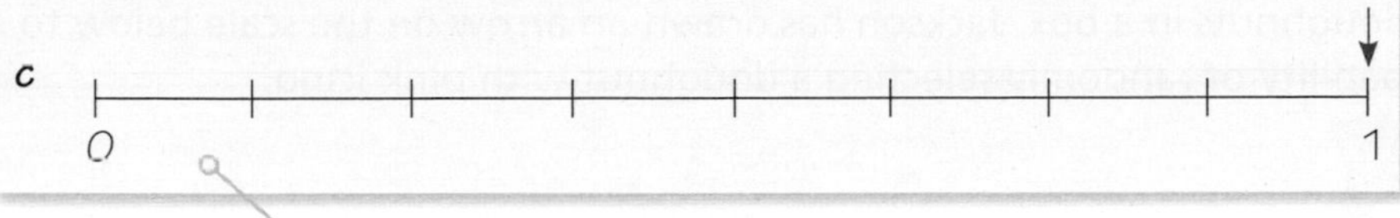

It is certain that the spinner will land on either red, green, yellow or blue so the arrow is pointing to $\frac{8}{8}$ or 1 whole.

Practice 14.4B

1 Here is a probability scale. Three probabilities are labelled A, B and C.

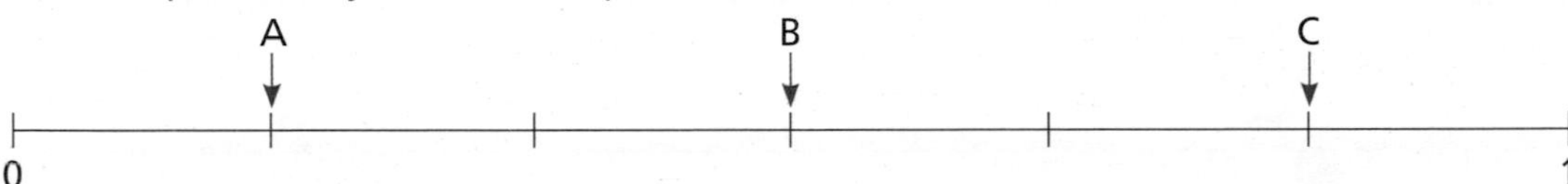

A fair six-sided dice numbered 1 to 6 is rolled. Write down the letter that matches each of these probabilities.

a The probability of rolling an even number.

b The probability of rolling a 3.

c The probability of rolling a number greater than 1.

2 Here is a probability scale.

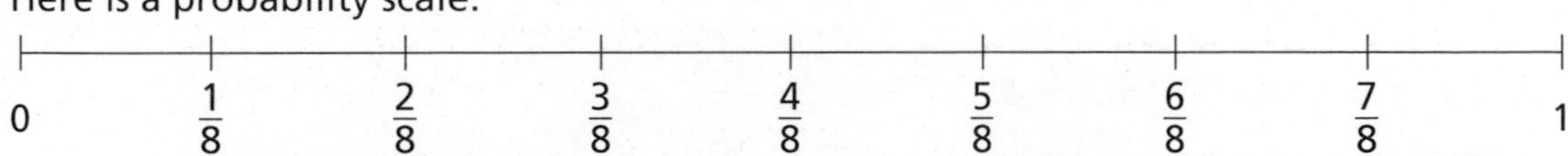

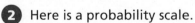

A bag contains 8 counters. 5 of the counters are red and 3 of the counters are blue. On a copy of the probability scale, mark

a the probability of getting a red counter

b the probability of getting a blue counter

c the probability of getting a green counter

3 This spinner is spun.

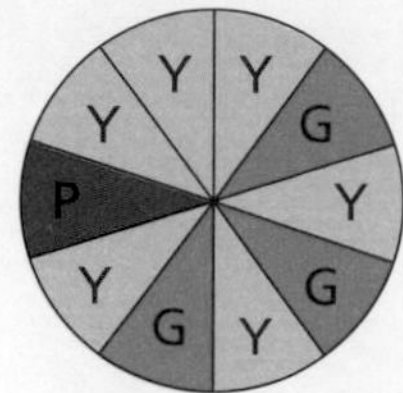

Draw a probability scale from 0 to 1.

Mark on the probability scale, the probability of spinning

a yellow **b** green **c** purple

4 There are 12 doughnuts in a box. Jackson has drawn an arrow on the scale below to show the probability of randomly selecting a doughnut with pink icing.

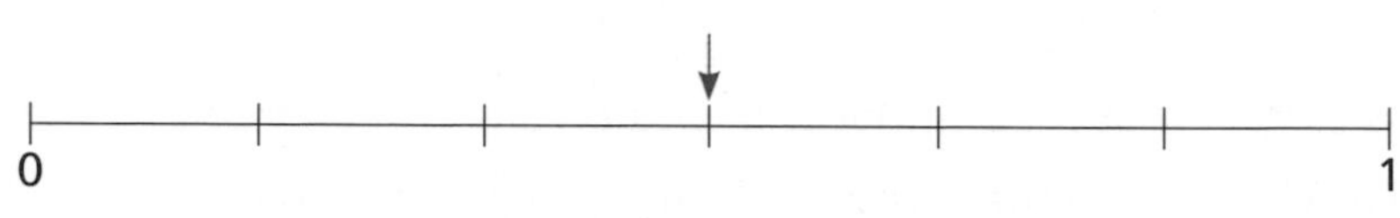

Is Jackson correct? Explain your answer.

5 Some coloured balls are placed in a bag. A ball is picked at random from the bag.

Copy the scale and mark the probability of getting a ball of each colour.

0 1

What do you think?

1 Here is a probability scale.

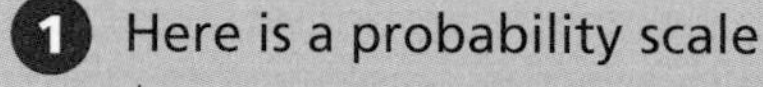

Copy the scale and draw arrows to show the probabilities of these events.

- Event A which is certain.
- Event B which is unlikely.
- Event C which is very unlikely.
- Event D which has equal chance.

Discuss your answers with your partner.

2 The scale shows the probabilities of two events.

X ↓ Y ↓

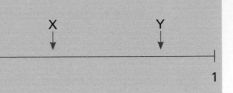

0 ———— $\frac{1}{2}$ ———— 1

a Which event is more likely, X or Y? How do you know?

b Estimate the probability of event X happening.

c Estimate the probability of event Y happening.

Consolidate – do you need more?

1 A box contains some counters.

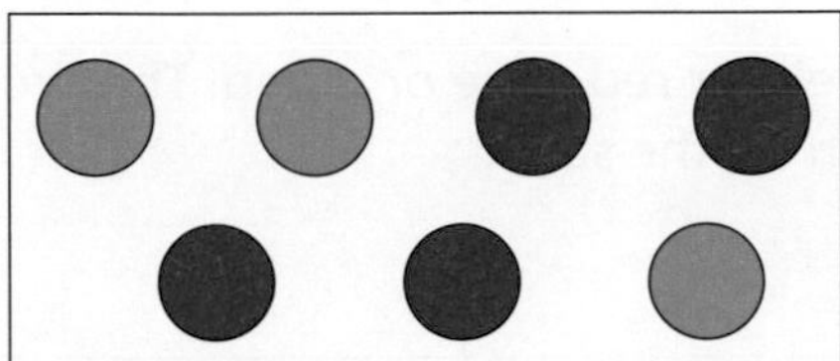

a Copy and complete these sentences.

i ___ out of ___ counters are red.

ii ___ out of ___ counters are blue.

b A counter is chosen at random.

i Work out the probability that the counter is red.

ii Work out the probability that the counter is blue.

c Copy and complete the number line.

Mark on the number line the probability of getting a counter of each colour.

0 ———— 1

2 This spinner is fair.

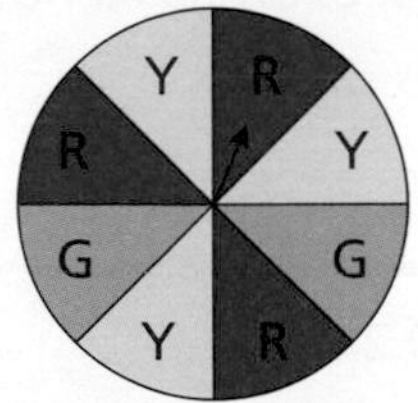

a Copy and complete these sentences.

i ___ out of ___ sections are red.

ii ___ out of ___ sections are green.

iii ___ out of ___ sections are yellow.

b The spinner is spun. Work out the probability that it lands on each colour.

i P(red) **ii** P(green) **iii** P(yellow)

c Copy and complete the number line.

Mark on the number line the probability of the spinner landing on each colour.

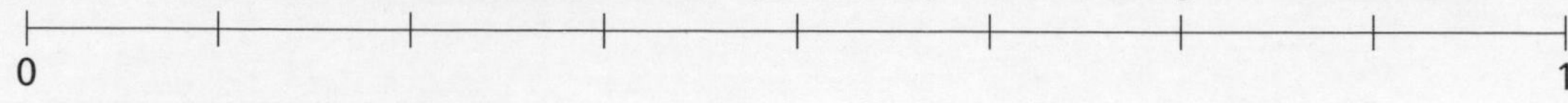

3 In a school there are 100 students.

47 of the students are girls, the rest are boys.

A student is selected at random.

a What is the probability that the student is a girl? Give your answer as a fraction, a decimal and a percentage.

b Copy the number line and draw an arrow on the scale to estimate the probability of selecting a boy.

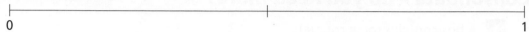

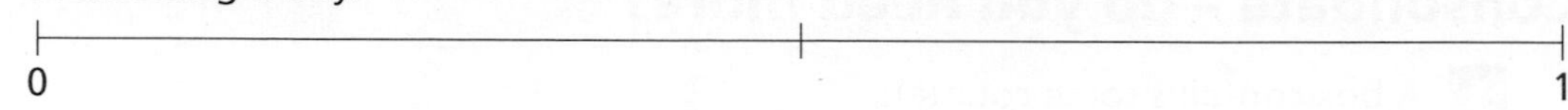

4 A fair spinner is split into 10 sections that are either red, blue or green. The probability of the spinner landing on each colour is shown on the scale.

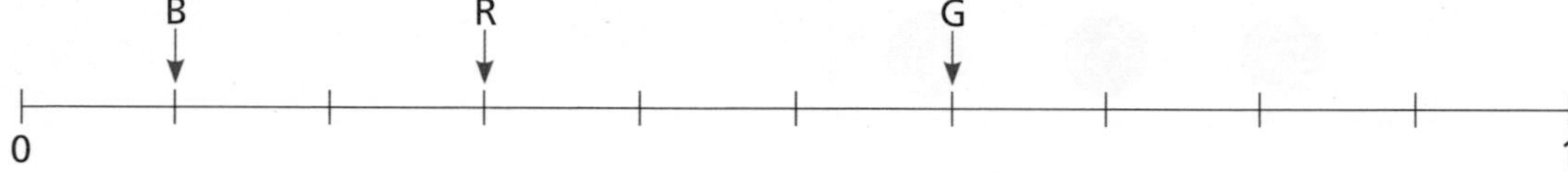

a Write the probability of the spinner landing on each colour.

b Draw a diagram to show what the spinner might look like.

Stretch – can you deepen your learning?

1 20 000 people enter a competition.

There are going to be 10 winners.

Jakub says that there is less than a 1% chance of winning the competition.

Is Jakub correct?

2 Marta has a spinner that is divided into 10 equal sections. Each section has a number written on it. Marta spins the spinner.

- The probability that she gets an even number is 0.3
- The probability that she gets a number less than 7 is 0.6
- The probability that she gets an even number less than 7 is 0.2

Draw a diagram to show what Marta's spinner might look like.

3 Abdullah has 20 numbered cards. The probability of choosing a card less than 3 is shown on the scale.

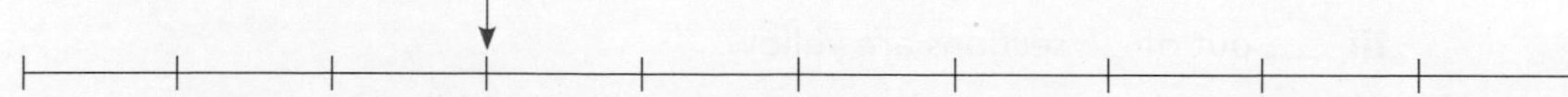

How many more cards less than 3 do you need to add so that the probability of getting a number less than 3 is 0.5? Show all your working.

Reflect

1 Max says, "The probability that the spinner lands on yellow is $\frac{1}{2}$ because there are two sections and one of them is yellow."

Explain the mistake Max has made.

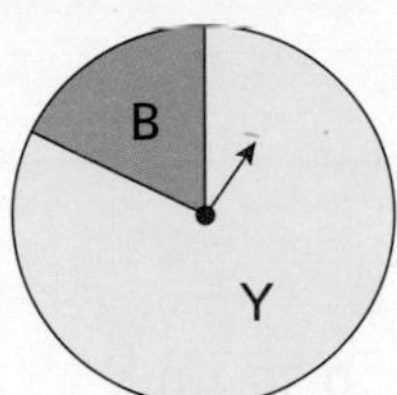

2 Here is a probability scale, with three probabilities marked.

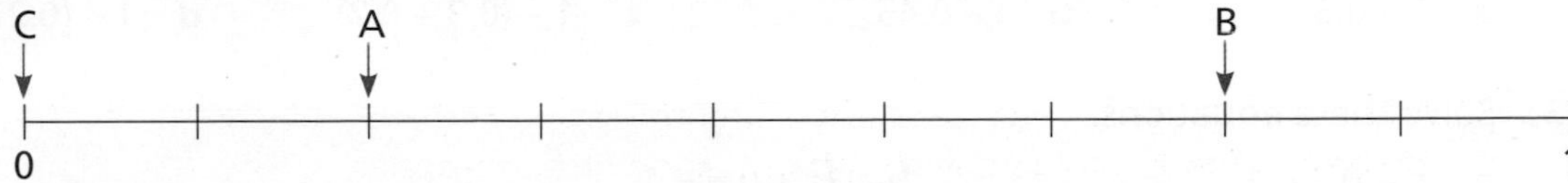

What could this represent? Compare your answer with a partner's.

14.5 Calculating probability (2)

Small steps

- Know that the sum of probabilities of all possible outcomes is 1

Key words

Outcome – the possible result of an experiment

Are you ready?

1 Work out

a 0.2 + 0.5 **b** 0.2 + 0.05 **c** 0.36 + 0.4

2 Work out

a 1 – 0.6 **b** 1 – 0.45 **c** 1 – (0.3 + 0.2) **d** 1 – (0.27 + 0.3)

3 Solve these equations.

a $x + 0.2 = 1$ **b** $2x + 0.6 = 1$

Models and representations

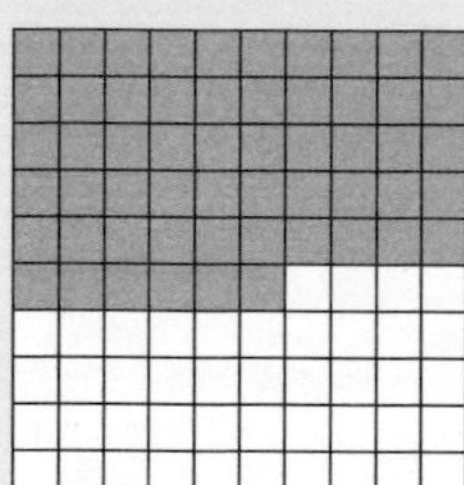

P(orange) = 56%, P(Not orange) = 100% – 56% = 44%

The sum of the probabilities for all the possible **outcomes** of an event is 1.

You can calculate unknown probabilities using this fact. You can also use it to calculate the probability of an event not happening.

Example 1

The probability of it raining is 0.32

What is the probability of it not raining?

1 – 0.32 = 0.68

If the chance of it raining is 0.32, then to find out the probability of it not raining, subtract this from 1

It is certain that it will either rain or not rain so the sum of the probabilities of the two events is 1

Example 2

The table shows the probability of a spinner landing on a particular colour.

Red	Yellow	Blue	Green
$\frac{1}{10}$	$\frac{5}{10}$		

The probability of landing on blue is the same as the probability of landing on green.

Work out the probability of the spinner landing on green.

$\frac{1}{10} + \frac{5}{10} = \frac{6}{10}$ — Add up the probabilities you are given.

$1 = \frac{10}{10}$ — The total of all the probabilities must be 1

$\frac{10}{10} - \frac{6}{10} = \frac{4}{10}$

The spinner can only land on red, yellow, blue or green.

The probabilities are given as fractions.

$1 = \frac{10}{10}$ so to work out the missing values you need to subtract $\frac{6}{10}$ from $\frac{10}{10}$

$\frac{4}{10} \div 2 = \frac{2}{10}$ — You know that P(green) = P(blue), so the final step is to divide $\frac{4}{10}$ by 2

The probability of landing on green is $\frac{2}{10}$ or $\frac{1}{5}$

Practice 14.5A

1. The probability of flipping a head on a biased coin is 0.4.

   What is the probability of flipping a tail?

2. The probability that the Spencer family go on holiday to Greece is $\frac{7}{10}$

   What is the probability the Spencer family do not go to Greece for their holiday?

3. There are 3 types of chocolates in a box of chocolates. The box contains 15 chocolates in total.

   - There are 7 milk chocolates.
   - There are 3 dark chocolates.
   - The rest are white chocolates.

   A chocolate is chosen at random from the box.

   a What is the probability of choosing a milk chocolate?

   b What is the probability of choosing a white chocolate?

4 A spinner has different 3 different coloured sections.

Colour	Red	Green	Blue
Probability	0.2	0.3	

The spinner is spun. What is the probability of the spinner landing on blue?

5 Two teams play a game of football.

The probability that team A wins is 0.3

The probability that team B wins is 0.23

a Which team is more likely to win?

b Chloe is working out the probability that the result is a draw.

1 – (0.3 + 0.23) = 1 – 0.26 = 0.74

What mistake has Chloe made?

What is the probability that the result is a draw?

6 A spinner has 20 sections. Each section is labelled A, B, C or D. The table shows the probability of spinning each of the letters.

Letter	A	B	C	D
Probability	$\frac{1}{5}$	$\frac{1}{10}$	$\frac{2}{5}$	

a What is the probability of spinning the letter A or B?

b What is the probability of spinning the letter D?

c How many sections are labelled C?

7 The table shows the probabilities of students being picked at random from each year group.

Year 7	Year 8	Year 9	Year 10	Year 11
0.3	0.15	0.1	0.05	x

A student is selected at random. Find these probabilities.

a P(Year 11) **b** P(Year 7 or Year 8) **c** P(not Year 9)

8 The probability of each outcome, A to D, is shown in the table.

Outcome	A	B	C	D
Probability	0.3	0.24		

The probability of outcome C is the same as outcome D. What is the probability of getting outcome B or C?

What do you think?

1. The probability that Ed gets maths homework is x. What is the probability that Ed does not get maths homework?

2. A spinner has 3 sections, labelled A, B and C. The table shows the probability of getting each outcome.

Section	A	B	C
Probability	x	y	z

What do you know about the sum of x, y and z?

3. A game has two outcomes, A and B. The probability scale shows the probability of outcome A.

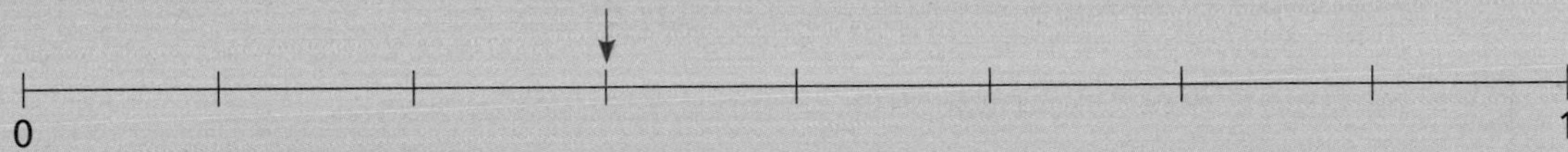

Copy the probability scale and mark the probability of outcome B. Discuss your strategy.

4. Design your own spinner with the colours red, blue, green and yellow that fits the following criteria

P(red) = $\frac{1}{10}$ P(blue) = 0.5 P(green) = P(yellow)

Consolidate – do you need more?

1. The probability of it snowing on New Year's Day is 0.1. What is the probability of it not snowing on New Year's Day?

2. Benji plants some flower seeds. They will grow into either pink or orange flowers. The probability of them growing into pink flowers is $\frac{2}{5}$. What is the probability of them growing into orange flowers?

3. A bowl contains some pieces of fruit. The table shows the types of fruit in the bowl and the probability of choosing each type of fruit. A piece of fruit is chosen at random from the bowl.

Apple	Banana	Mango	Kiwi
0.32	0.27	0.1	

a What is the probability of choosing a banana?

b What is the probability of choosing an apple or a mango?

c Work out the probability that the fruit chosen is a kiwi.

Stretch – can you deepen your learning?

1 Two boxes contain some blue and yellow counters.

There are 200 counters in total in each box.

The probability of choosing a yellow counter from box A is 0.27

You are twice as likely to choose a yellow counter from box B than from box A.

What is the probability of choosing a blue counter from box B?

2 A bag of sweets contains three types of sweet: strawberry, lime and orange. A sweet is chosen at random from the bag.

It is twice as likely that a strawberry sweet is chosen than a lime sweet.

It is three times as likely that an orange sweet is chosen than a lime sweet.

What is the probability of choosing a lime sweet?

3 The table shows the probabilities of three different outcomes.

Outcome	A	B	C
Probability	1.3	–0.2	–0.1

Explain why this table of outcomes must be incorrect.

4 There are 60 socks in a drawer.

The socks are either black, navy or grey.

Mario chooses a sock at random from the drawer.

The probability of choosing a black sock is $\frac{1}{3}$ and the probability of choosing a navy sock is $\frac{5}{12}$.

How many grey socks are in the drawer?

Reflect

1 What do probabilities always sum to?

2 An event has three outcomes. What can you say about the probability of the outcomes? Give a possible value of each outcome.

14 Sets and probability

Chapters 14.1–14.5

I have become **fluent** in...

- Identifying and representing sets
- Creating and interpreting Venn diagrams
- Calculating the probability of single events
- Using the probability scale

I have developed my **reasoning** skills by...

- Explaining why the sum of probabilities is 1
- Generating sample spaces for events
- Describing the probabilities of different things in words

I have been **problem-solving** through...

- Listing the elements of a set given a Venn diagram
- Applying knowledge of probability in different contexts
- Identifying the possible outcomes of an event given information about probabilities

Check my understanding

1 Use set notation to list the elements of each set.

a A = {letters in the word "probability"} **b** B = {odd numbers between 0 and 10}

c C = {factors of 48} **d** D = {multiples of 3 between 40 and 70}

2 ξ = {integers from 0 to 20 inclusive} A = {even numbers} B = {square numbers}

a List the elements of set A.

b List the elements of set B.

c List the elements that are in both sets.

d Draw a Venn diagram to represent the information.

3 Here is a Venn diagram.

List the elements of each set.

a A **b** B **c** $A \cap B$ **d** $A \cup B$ **e** A′ (H)

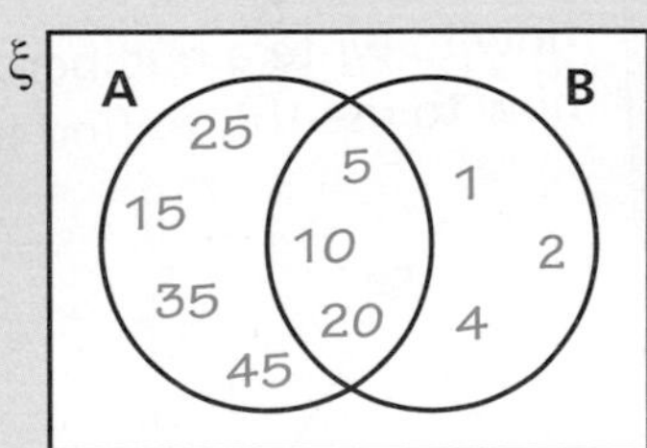

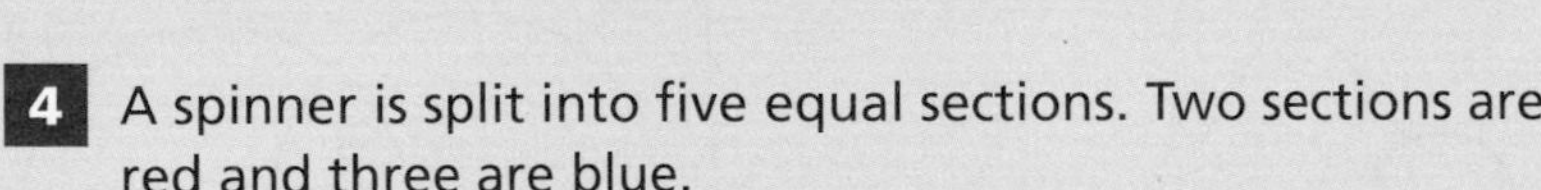

4 A spinner is split into five equal sections. Two sections are red and three are blue.

a What is the probability that the spinner lands on red?

b What is the probability that the spinner lands on blue?

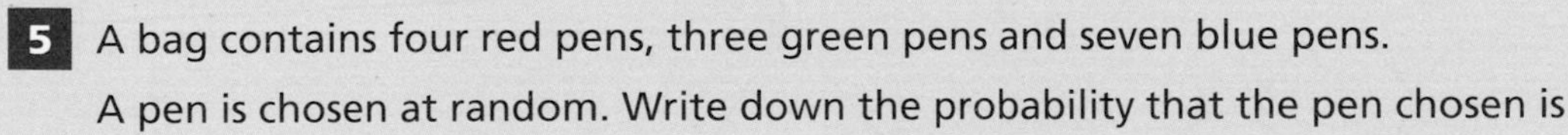

5 A bag contains four red pens, three green pens and seven blue pens.

A pen is chosen at random. Write down the probability that the pen chosen is

a red **b** green **c** blue **d** not red **e** not blue **f** orange

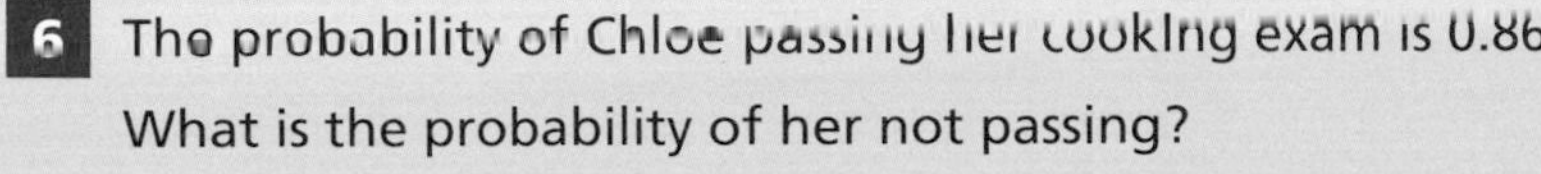

6 The probability of Chloe passing her cooking exam is 0.86

What is the probability of her not passing?

15 Primes and proof

In this block, I will learn...

to identify and use factors and multiples

$1 \times 4 = 4$ $2 \times 4 = 8$ $3 \times 4 = 12$

$4 \times 4 = 16$ $5 \times 4 = 20$

to recognise prime, square and triangular numbers

Prime numbers  Square numbers  Triangular numbers

how to write a number as a product of its prime factors; how to use this to find the LCM and HCF (H)

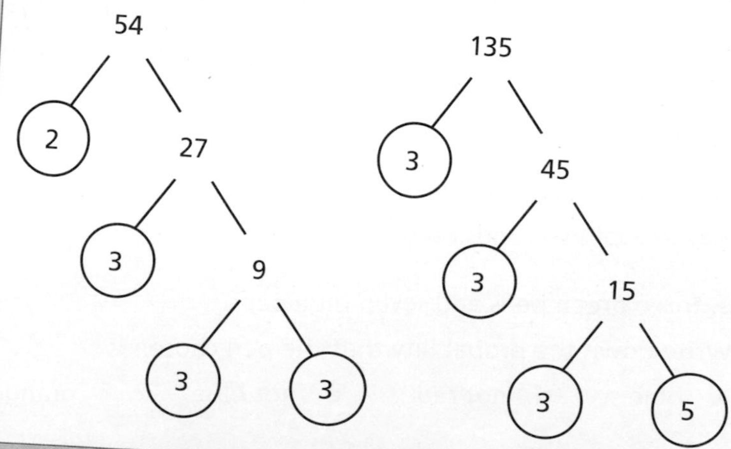

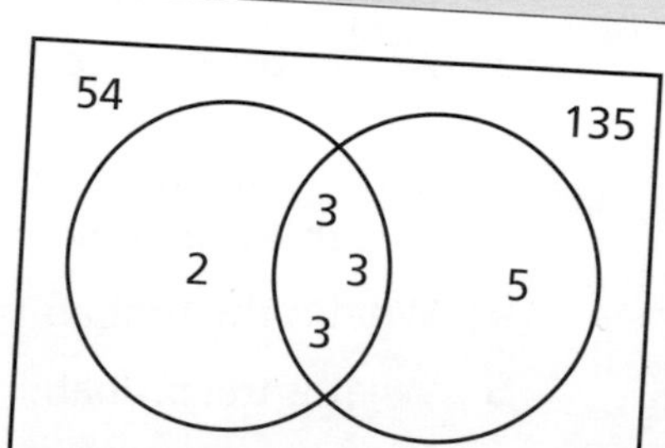

15.1 Factors and multiples

Small steps

- Find and use multiples
- Identify factors of numbers and expressions

Key words

Multiple – the result of multiplying a number by a positive integer

Factor – a number that divides into another number exactly without leaving a remainder

Array – an arrangement of objects, pictures or numbers in columns and rows

Remainder – the amount left over after dividing one integer by another

Are you ready?

1 What calculations can be seen from the array?

2 Work out the multiplications.

a 1×72 b 36×2 c 6×12

d 18×4 e 3×24 f 8×9

What do you notice?

3 Work out the divisions.

a $42 \div 7$ b $28 \div 4$ c $63 \div 9$

d $108 \div 12$ e $120 \div 3$ f $555 \div 5$

4 Work out the divisions, showing the remainder each time. The first one has been done for you.

a $43 \div 7 = 6r1$ b $31 \div 4$ c $70 \div 9$

d $119 \div 12$ e $119 \div 3$ f $551 \div 5$

Models and representations

Arrays

Arrays can be used to help identify **factors** of a number.

These arrays show that the factors of 14 are 1, 14, 2 and 7. These are the only ways the 14 counters can be arranged to make a complete rectangle.

15.1 Factors and multiples

Algebra tiles

Algebra tiles can be used when finding **multiples** and factors of algebraic expressions.

These algebra tiles show that $4x + 4$ is a multiple of $2x + 2$, because $2 \times (2x + 2) = 4x + 4$

x	1	x	1
x	1	x	1

These algebra tiles show that 2 and $x + 1$ are factors of $2x + 2$, because $2 \times (x + 1) = 2x + 2$

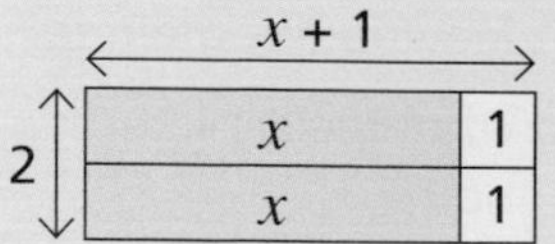

You could also draw bar models if you do not have access to algebra tiles.

In this section you will recap learning from earlier in this book by identifying multiples of integers.

Example 1

Write down the first five multiples of 4

4, 8, 12, 16, 20

$1 \times 4 = 4$ $2 \times 4 = 8$ $3 \times 4 = 12$

$4 \times 4 = 16$ $5 \times 4 = 20$

You can use counters to build the multiples of a number.
Or you can just use your knowledge of multiplication facts.

Practice 15.1A

1 Write down the first five multiples of each number.

a 4 **b** 7 **c** 15 **d** 21

2 Write down the first five multiples of each number.

a 8 **b** 14 **c** 30 **d** 42

What do you notice about your answers to questions 1 and 2?

3 Jakub is trying to find three multiples of 6 between 3000 and 4000. He says "It is going to take me a long time to write out all those multiples of 6."

a Explain a more efficient way that Jakub could find three multiples of 6 between 3000 and 4000

b Write down three multiples of 6 between 3000 and 4000

4 Write down a multiple of 5 greater than

a 100　　b 1000　　c 10 000

5 At an airport a plane takes off every 12 minutes between 7 a.m. and 6 p.m.
How many planes take off between 9 a.m. and 5 p.m.?

6 a List all the multiples of 2 up to 20
b List all the multiples of 4 up to 20
c What do you notice about the multiples of 2 and 4?
d Is the number 47 a multiple of 4? Explain how you know.

7 Here are some number cards.

17	40	22	15	35	20	27	32	25	30	12

a Write down the number cards that are multiples of

i 2　　ii 3　　iii 4　　iv 7　　v 10　　vi 5

b Which card was not used? Explain why.
c What do you notice about your answers to **v** and **vi**? Why does this happen?

What do you think?

1 The table shows some multiples of 3

Multiples of 3	75	126	432	9735
Sum of the digits	12			

a Copy and complete the table.
b What do you notice about the sum of the digits?
c Write down another multiple of 3 that is greater than 5000

2 Abdullah's age is a multiple of 8 and 12. His age is one year away from a multiple of 7
He is younger than 50 years old. How old is Abdullah?

3 x is a positive integer.
a What value of x would make the expression $4x + 3$

i a multiple of 7　　ii a multiple of both 3 and 5?

b Is there more than one solution for each part?

4 The algebra tiles show the first multiple of $x + 2$

x	1	1

a How do the algebra tiles show $1 \times (x + 2)$?

b Use the algebra tiles to write down the second multiple of $x + 2$

x	1	1
x	1	1

c List the first five multiples of $x + 2$

In this section you will recap learning from earlier in this book by identifying factors of integers. It is crucial you understand the difference between factors and multiples as this often causes confusion.

For any given number, there is a finite number of factors, but there is an infinite number of multiples.

Example 2

List the factors of 12

a 1, 2, 3, 4, 6 and 12

To find the factors of 12, you can use 12 counters to form as many different arrays as possible.

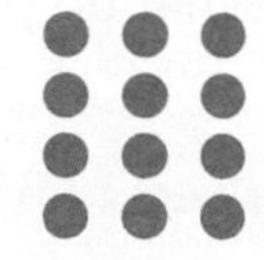

These are the different arrays you can make using 12 counters. You can make 1 row of 12, 2 rows of 6, or 4 rows of 3. Therefore the factors are 1, 2, 3, 4, 6 and 12

Practice 15.1B

1 Find all the factors of these numbers.

a 6 **b** 8 **c** 20 **d** 22

2 Benji says, "The greater the number, the more factors it has."

Do you agree with Benji? Explain your answer.

3 **a** Find all the factors of these numbers.

i 7 **ii** 11 **iii** 13 **iv** 17

b What is the same about all of your answers?

c Marta says, "All odd numbers only have two factors, 1 and itself." Use an example to show that Marta is wrong.

4 **a** Draw arrays to show that

i 6 is a factor of 18 **ii** 4 is not a factor of 15

b In part **a**, did you draw the same arrays as a partner?

5 Here are some number cards.

30	3	12	36	4	15	24	20

a Write down all the cards that are

i factors of 12 ii multiples of 12

b What's the same and what's different about your answers to part **i** and part **ii**?

c Flo says, "Four of the cards are factors of 6." What mistake has Flo made?

6 a List the factors of 24

b List the factors of 48

c True or false?

48 has double the amount of factors as 24, because 48 is double 24

Explain your answer.

What do you think?

1 Is this statement always, sometimes or never true?

The greater a number, the more factors it has.

Explain your answer.

2 Seb says, "Every number has exactly two factors."

a Give three examples that would support Seb's statement.

b Give three examples that disprove Seb's statement.

3 Beca has a brother. She says, "My brother's age is a factor of every integer." How old is Beca's brother?

4 Use the algebra tiles to help you list the factors of $4x + 8$. The first block has been completed for you.

$4x + 8$

1	x	x	x	x	1	1	1	1	1	1	1	1

x	x	1	1	1	1
x	x	1	1	1	1

x	1	1
x	1	1
x	1	1
x	1	1

Consolidate – do you need more?

1 List the first five multiples of each number.

a 3 b 7 c 11 d 15

e 20 f 100 g 200

2 List all the factors of each number.

a 4 **b** 12 **c** 16 **d** 28
e 32 **f** 60 **g** 100

3 **a** Write down the seventh multiple of 5
b Write down the twelfth multiple of 6
c Write down the hundredth multiple of 9

4 Write down a multiple of each number that is greater than 2000 but less than 3000

a 5 **b** 2 **c** 4 **d** 25
e 400 **f** 50 **g** 6

Stretch – can you deepen your learning?

1 Write down the first five multiples of each expression.

a x **b** $2x$ **c** $3y$ **d** $6ab$ **e** $7x^2$

2 Write down the first five multiples of each expression.

a $x + 5$ **b** $2x - 1$ **c** $4 + 3y$ **d** $6ab + 11$ **e** $7x^2 - 9$

3 Explain how the diagrams show that 1, 2, 4, $4x$, $2x$ and x are all factors of $4x$.

x	x	x	x

x
x
x
x

x	x
x	x

4 List the factors of each expression.

a $3x$ **b** $8x$ **c** $12y$ **d** $18z$ **e** $4ab$

5 List the factors of each expression.

a $3x + 1$ **b** $8x + 4$ **c** $12y - 6$ **d** $18z + 27$ **e** $4ab + 2$

Reflect

Use your own words to explain the following.

1 How to find the multiples of a given number.

2 How to find the factors of a given number.

3 The difference between a factor and a multiple.

4 Why it is impossible to list all the multiples of a given number.

15.2 Special numbers

Small steps

- Recognise and identify prime numbers
- Recognise square and triangular numbers

Key words

Prime number – a positive integer with exactly two factors; 1 and itself

Square number – a positive integer that is the result of an integer multiplied by itself

Triangular number – a positive integer that is the sum of consecutive positive integers starting from 1

Are you ready?

1 List the factors of each number.

a 2 **b** 4 **c** 18 **d** 25 **e** 29

2 Complete the calculations.

a 1×1 **b** 2×2 **c** 5×5 **d** 9×9 **e** 15×15

3 Complete the calculations.

a $1 + 2$ **b** $1 + 2 + 3$ **c** $1 + 2 + 3 + 4$ **d** $1 + 2 + 3 + 4 + 5$

Models and representations

Counters

Counters can be used to support understanding of prime, square and triangular numbers.

7 is a **prime number** because the only array that can be made using 7 counters is 1 row of 7 (or 7 rows of 1).

4 is a **square number** because 4 counters can be arranged to form a completed square.

6 is a **triangular number** because 6 counters can be arranged to form a triangle.

In this section you will focus on prime numbers. A prime number is a positive integer that has exactly two factors: 1 and itself.

15.2 Special numbers

Example 1

a Is 5 a prime number? Explain your answer.

b Is 9 a prime number? Explain your answer.

a *Yes because 5 has exactly two factors: 1 and 5*

The only array that can be made using exactly five counters is a 1 by 5 array.

This means that the only factors of 5 are 1 and 5

So 5 is a prime number.

b *No because 9 has three factors: 1, 3 and 9*

You can make two different arrays using exactly 9 counters. A 1 by 9 array or a 3 by 3 array.

This means that 9 is not a prime number, because it does not have exactly two factors.

Practice 15.2A

1 Copy the number grid into your book.

1	2	3	4	5	6	7	8	9	10
11	12	13	14	15	16	17	18	19	20
21	22	23	24	25	26	27	28	29	30
31	32	33	34	35	36	37	38	39	40
41	42	43	44	45	46	47	48	49	50

a Cross out

- 1
- multiples of 2, apart from 2
- multiples of 3, apart from 3
- multiples of 5, apart from 5
- multiples of 7, apart from 7

b List the prime numbers between 1 and 50.

c Why did you not need to cross out multiples of 4, 6 or 8?

2 **a** List the factors of each number.

i 1 **ii** 2 **iii** 16 **iv** 19 **v** 22 **vi** 3 **vii** 31

b Which of the numbers from those in part **a** are prime numbers? How do you know?

3 **a**

21 is a prime number because it is odd.

Draw an array to show that Beca is incorrect.

b

2 is not a prime number because it is even.

Explain why Jackson is incorrect.

4 Which of these numbers are prime?

11 25 37 39 53 63 99 121 963

How did you decide?

5 Decide whether each statement is true or false. Explain your answer for each.

a 1 is the smallest prime number.

b All prime numbers are odd.

c 43 is a prime number.

6 Copy the Venn diagram into your book.

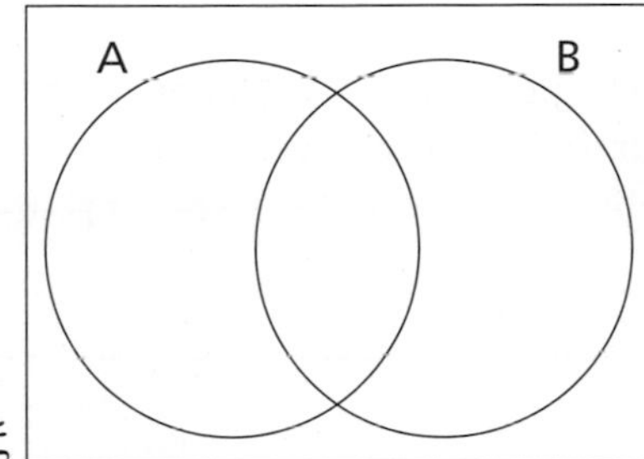

a ξ = {integers from 1 to 20} A = {prime numbers} B = {odd numbers}

Represent this information in the Venn diagram.

b List the elements of sets A and B using correct set notation.

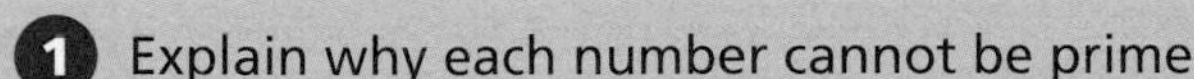

What do you think?

1 Explain why each number cannot be prime.

a 12 476 582 **b** 708 611 935 **c** 37 907 695 836

2 **a** Darius rolls a fair six-sided dice. What is the probability that it lands on a prime number?

b Beth rolls two fair six-sided dice and adds the scores together. What is the probability that her total is a prime number?

c Sven rolls two fair six-sided dice and finds the product of his scores. What is the probability that his product is not prime?

3 a and b are prime numbers.

a is greater than b.

$a + b = 73$

a Find the values of a and b.

b Explain why there is only one possible value each for a and b.

In this section you will look at square and triangular numbers.

- A square number is the result of multiplying an integer by itself.
- A triangular number is the result of adding a series of consecutive positive integers starting from 1

Square and triangular numbers can be represented using counters.

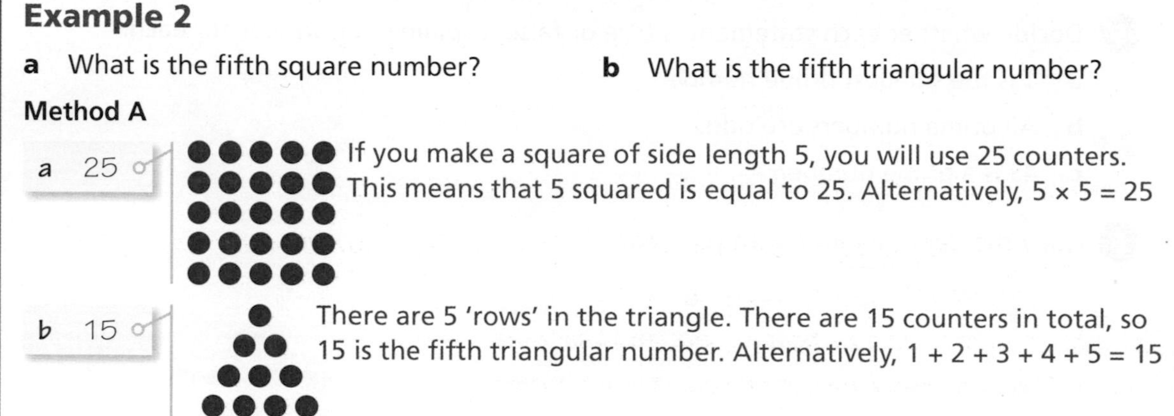

Example 2

a What is the fifth square number? **b** What is the fifth triangular number?

Method A

a 25 — If you make a square of side length 5, you will use 25 counters. This means that 5 squared is equal to 25. Alternatively, $5 \times 5 = 25$

b 15 — There are 5 'rows' in the triangle. There are 15 counters in total, so 15 is the fifth triangular number. Alternatively, $1 + 2 + 3 + 4 + 5 = 15$

Practice 15.2B

1 Draw a diagram to show that the fourth square number is 16

2 Draw a diagram to show that the fourth triangular number is 10

3 List the first 10 square numbers.

4 List the first 10 triangular numbers.

5 **a** Flo says "17 squared is 34 because 17 + 17 is equal to 34"

i Explain why Flo is incorrect. **ii** Work out 17 squared.

b Ed says "8 is a square number because I have made a square using 8 counters."

Here is Ed's model.

Do you agree with Ed? Explain your answer.

6 Work out the value of each.

a 5^2 b 11^2 c 15^2 d 20^2

7 a What is the twelfth triangular number?

b What is the thirteenth triangular number?

c What is the twentieth triangular number?

What do you think?

1 3 and 6 are consecutive triangular numbers.

3 plus 6 is equal to 9 which is a square number.

Add other pairs of consecutive triangular numbers.

What do you notice?

2 If you cut a square in half from corner to corner you get two triangles. Therefore if you halve a square number you get a triangular number.

Do you agree with Benji? Explain your answer.

3 Decide whether each statement is true or false. Explain your answers.

a A number can be both square and prime.

b A number can be both square and triangular.

c A number can be both triangular and prime.

Consolidate – do you need more?

1 Here are some numbers.

3 15 100 25 11 16 2 21

a List the factors of each number.

b Which of the numbers have exactly two factors?

c Which of the numbers have an odd number of factors?

d Copy the table and sort your numbers into the table.

Prime	Square	Neither

2 The sequence of triangular numbers is shown using counters.

a Draw the next four terms in the sequence.

b How many counters would be in the eighth term of the sequence?

c List the first ten triangular numbers.

Stretch – can you deepen your learning?

1 Sven says, "Eleven is prime, so negative eleven must also be prime." Explain why Sven is wrong.

2 Is each statement always, sometimes or never true?

a Square numbers have an odd number of factors.

b Prime numbers have an even number of factors.

3 a, b and c are integers.

$a^2 + b^2 = c^2$

Find two sets of possible values for a, b and c.

4 Marta is investigating square numbers.

She says, "When you square a number the answer is greater than the number you started with."

a Give an example of when Marta is correct.

b Give an example of when Marta is incorrect.

c Generalise your findings to explain when $x^2 > x$.

5 n is an integer greater than 1

a Explain why $2n$ cannot be prime.

b Find a value of n such that $2n + 1$ is prime.

c Find a value of n such that $2n + 1$ is not prime.

d a is an integer such that $2n + a$ can never be prime.

What must be true about a?

Reflect

1 Explain why 1 is not a prime number.

2 Explain why 2 is the only even prime number.

3 What calculation would you need to do to work out the seventh triangular number?

4 What calculation would you need to do to work out the thirty-third square number?

15.3 HCF and LCM

Small steps

- Find common factors of a set of numbers including the HCF
- Find common multiples of a set of numbers including the LCM

Key words

Factor – a positive integer that divides exactly into another positive integer

Multiple – the result of multiplying a number by a positive integer

Highest common factor (HCF) – the highest factor shared by two or more numbers

Lowest common multiple (LCM) – the lowest multiple shared by two or more numbers

Are you ready?

1 Identify the greatest number in each set.

a 1, 8, 2, 11, 7, 16, 5 **b** 7.1, 8.2, 0.9, 6.4, 12.7 **c** 1706, 307, 92, 17.06, 29.058

2 Identify the smallest number in each set.

a 1, 8, 2, 11, 7, 16, 5 **b** 7.1, 8.2, 0.9, 6.4, 12.7 **c** 1706, 307, 92, 17.06, 29.058

3 List the **factors** of each number.

a 27 **b** 16 **c** 18 **d** 28 **e** 49

4 List the first five **multiples** of each number.

a 7 **b** 10 **c** 12 **d** 15 **e** 40

Models and representations

Arrays

Using counters to make arrays can be a useful way to explore factors. For example, these arrays show the factors of 6 are 1, 2, 3 and 6

●●●●●●

1 row of 6

1 × 6

●●●
●●●

2 rows of 3

2 × 3

In this section you will extend your knowledge of factors to find the **highest common factor** of two or more numbers. A common factor is a factor that is shared by two or more numbers. For example, 2 is a factor of both 6 and 8, so it is a common factor.

Example 1

Work out the highest common factor of 12 and 16

4

The factors of 12 are **1**, **2**, 3, **4**, 6 and 12

The factors of 16 are **1**, **2**, **4**, 8 and 16

The common factors of 12 and 16 are **1**, **2** and **4**

The highest common factor is therefore 4

Practice 15.3A

1 **a** List the factors of 18

b List the factors of 27

c List the common factors of 18 and 27

d What is the highest common factor of 18 and 27?

2 **a** List the factors of 16

b List the factors of 32

c List the common factors of 16 and 32

d What is the highest common factor of 16 and 32?

e What do you notice? Why does this happen?

3 **a** List the factors of 7

b List the factors of 9

c What is the highest common factor of 7 and 9?

d What do you notice? Why does this happen?

4 Work out the highest common factor of each pair of numbers.

a 12 and 15 **b** 20 and 25 **c** 18 and 24 **d** 11 and 21

e 8 and 40 **f** 12 and 6 **g** 36 and 48 **h** 37 and 32

5 a List the factors of 15

b List the factors of 20

c List the factors of 25

d What is the highest common factor of 15, 20 and 25?

6 Work out the highest common factor of each set of numbers.

a 15, 9 and 12 **b** 20, 18 and 10 **c** 7, 11 and 13

d 45, 15 and 30 **e** 28, 36 and 42 **f** 96, 100 and 80

What do you think?

1. Here are some fractions.

   i $\frac{14}{28}$ ii $\frac{25}{30}$ iii $\frac{72}{108}$ iv $\frac{96}{100}$

   a For each fraction, work out the highest common factor of the numerator and denominator.

   b Use the highest common factor to simplify each fraction.

   c What advantage does knowing the highest common factor give you when simplifying fractions?

2. a Work out the highest common factor of 72 and 100

   b Use your answer to part **a** to work out the highest common factor of each pair of numbers. Explain each step in your thinking.

   i 720 and 1000 ii 144 and 200 iii 36 and 50

3. The areas of two rectangles are shown.

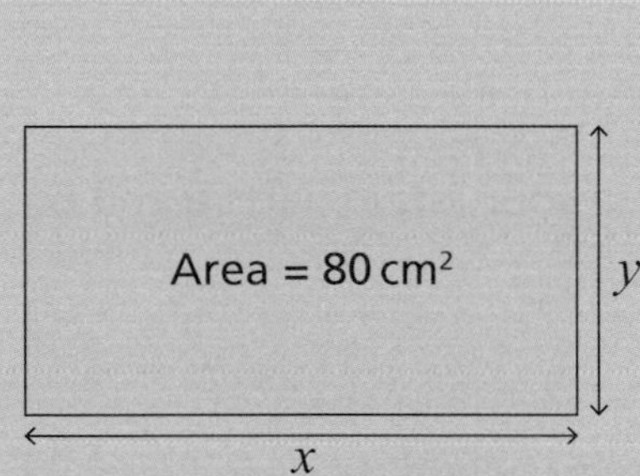

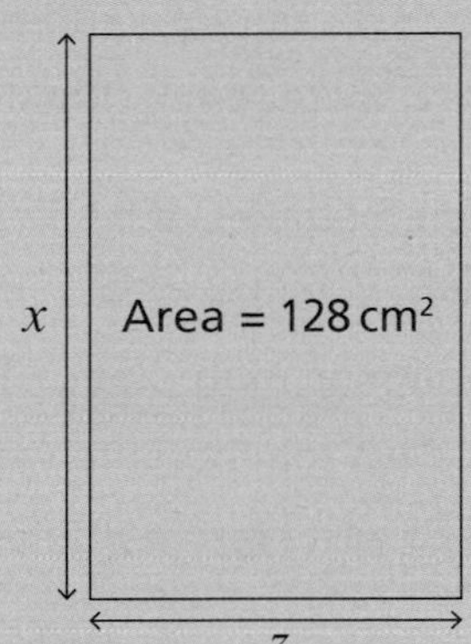

   x, y and z are integers.

   a What are the possible values of x?

   b For each value of x, work out the corresponding values of y and z

4. a, b and c are integers.

   a The highest common factor of a and b is 6. Give three possible values of a and b.

   b The highest common factor of a, b and c is 1. Suggest three possible values of c.

   c What is the highest common factor of a and c? What is the highest common factor of b and c? How do you know?

In this section you will focus on finding the **lowest common multiple**.

- A common multiple is a shared multiple of two or more numbers.

  For example, 100 is a multiple of both 10 and 20, so it is a common multiple.

- The lowest common multiple is the lowest shared multiple of two or more numbers.

Example 2

Work out the lowest common multiple of 15 and 6

Multiples of 15:

15, 30, 45, 60, 75, 90, 105, 120, 135, 150 ...

Multiples of 6:

6, 12, 18, 24, 30, 36, 42, 48, 54, 60 ...

The common multiples of 15 and 6 shown in the lists are 30 and 60

The lowest common multiple is 30

The lowest common multiple is therefore 30

You could have stopped writing the multiples of 15 at 30, because you know that 30 is also a multiple of 6

Practice 15.3B

1 **a** List the first ten multiples of 8

b List the first ten multiples of 10

c Identify the common multiples of 8 and 10 from your lists in parts **a** and **b**.

d What is the lowest common multiple of 8 and 10?

e Did you need to write ten multiples for each?

2 **a** List the first ten multiples of 12

b List the first ten multiples of 6

c Identify the common multiples of 12 and 6 from your lists in parts **a** and **b**.

d What is the lowest common multiple of 12 and 6?

e Did you need to write ten multiples for each?

3 Work out the lowest common multiple for each set of numbers.

a 5 and 8 **b** 6 and 7 **c** 4 and 3

d 2 and 11 **e** 13 and 4 **f** 9 and 14

What do you notice? Why does this happen?

4 Work out the lowest common multiple for each set of numbers.

a 6 and 12 **b** 27 and 3 **c** 15 and 5

d 10 and 30 **e** 24 and 8 **f** 51 and 17

What do you notice? Why does this happen?

5 Work out the lowest common multiple for each set of numbers.

a 4 and 18 **b** 9 and 12 **c** 16 and 36

d 14 and 6 **e** 32 and 24 **f** 15 and 18

What do you notice? Why does this happen?

6 Work out the lowest common multiple of each set of numbers.

a 3, 4 and 5
b 12, 15 and 10
c 6, 12 and 36

What do you think?

1 Is the statement always, sometimes or never true? Give examples to support your answer.

"To find the lowest common multiple of a set of numbers you multiply the numbers together."

2 A number 17 bus leaves the station every 12 minutes.

A number 31 bus leaves the station every 15 minutes.

Both buses leave the station at 9 a.m.

a What time will they next both leave the station at the same time?

b How many times will a number 17 bus and number 31 bus leave the station at the same time between 8:55 a.m. and 6:05 p.m.?

3 Jackson, Emily and Faith each went for a run on 1st February.

Jackson ran every 2 days after that.

Emily ran every 3 days after that.

Faith ran every 4 days after that.

When they run on the same day, they run together.

a On what date will they next run together?

b How many times will they run together in February?

c Explain why every time Faith runs, she runs with Jackson.

Consolidate – do you need more?

1 List the factors of each number.

a 15
b 18
c 25
d 29
e 45
f 12

2 Identify the highest common factor of each pair of numbers.

a 15 and 18
b 15 and 25
c 15 and 29
d 15 and 45
e 15 and 12
f 18 and 25
g 18 and 29
h 18 and 45
i 18 and 12
j 25 and 29
k 25 and 45
l 25 and 12
m 29 and 45
n 29 and 12
o 45 and 12

3 List the first 10 multiples of each number.

a 3
b 6
c 8
d 12
e 15
f 7

4 Identify the lowest common multiple of each pair of numbers.

a 3 and 6 **b** 3 and 8 **c** 3 and 12 **d** 3 and 15
e 3 and 7 **f** 6 and 8 **g** 6 and 12 **h** 6 and 15
i 6 and 7 **j** 8 and 12 **k** 8 and 15 **l** 8 and 7
m 12 and 15 **n** 12 and 7 **o** 15 and 7

Stretch – can you deepen your learning?

1 The highest common factor of two numbers is 1

What must be true about the two numbers?

2 The highest common factor of two numbers is 5

The lowest common multiple is 150

What could the two numbers be? Is there more than one answer?

3 The highest common factor of a and b is 1

Write an expression for the lowest common multiple of a and b.

4 x and y are positive integers greater than 1

Work out the highest common factors of the expressions.

a $8x$ and $2x$ **b** $8x$ and $2y$
c $8x^2$ and $2x$ **d** $8x^3y$ and $2x^2y$

5 **a** Work out the lowest common multiple of $3x$ and x.

b How does your answer to part **a** help you to calculate $\frac{1}{3x} + \frac{1}{x}$?

c Complete the calculations.

i $\frac{1}{x} - \frac{1}{3x}$ **ii** $\frac{1}{3x} + \frac{2}{x}$ **iii** $\frac{7}{6x} + \frac{2}{3x}$

Reflect

1 Decide whether each statement is true or false. Explain your answer.

a Two or more numbers always share at least one common factor.

b Two or more numbers always share at least one common multiple.

c The HCF and LCM of a pair of numbers can be equal.

15.4 Using prime numbers

Small steps

- Write a number as a product of its prime factors
- Use a Venn diagram to calculate the HCF and LCM (H)

Key words

Product – the result of multiplying

Prime number – a positive integer with exactly two factors, 1 and itself

Highest common factor (HCF) – the greatest number that is a factor of every one of a set of numbers

Lowest common multiple (LCM) – the smallest number that is a multiple of every one of a set of numbers

Prime factor decomposition – writing numbers as a product of their prime factors

Are you ready?

1 Decide whether each statement is true or false.

a $3 \times 2 \times 1 = 6$ **b** $2 \times 2 \times 2 \times 2 = 8$ **c** $3 \times 5 \times 7 = 105$

2 Which of these numbers are prime?

1 2 3 4 5 6 7 8 9 10

How do you know?

3 Copy the Venn diagram into your book. Sort these numbers into the Venn diagram.

5 12 20 23 25 30 37 40

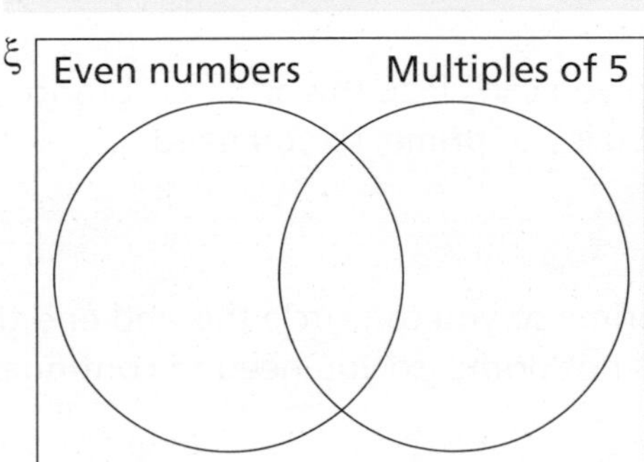

4 Work out the HCF and LCM of each pair of numbers.

a 3 and 5 **b** 6 and 9 **c** 12 and 24 **d** 20 and 30

Models and representations

Factor tree

A factor tree is useful when finding prime factors of numbers. The prime factors are circled, leaving you with all the prime factors of a number.

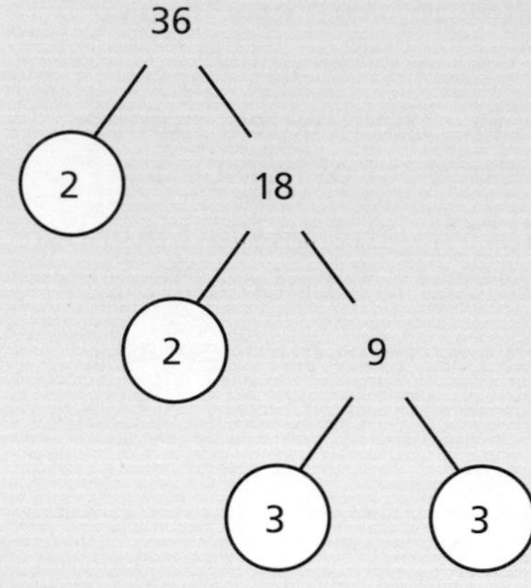

Any number can be written as a **product** of its factors, for example 8 = 2 × 4. However, as 4 is not **prime**, this is not the product of prime factors. When writing a number as the product of its prime factors, all numbers in the product are prime.

Example 1

Write 180 as the product of its prime factors.

Method A

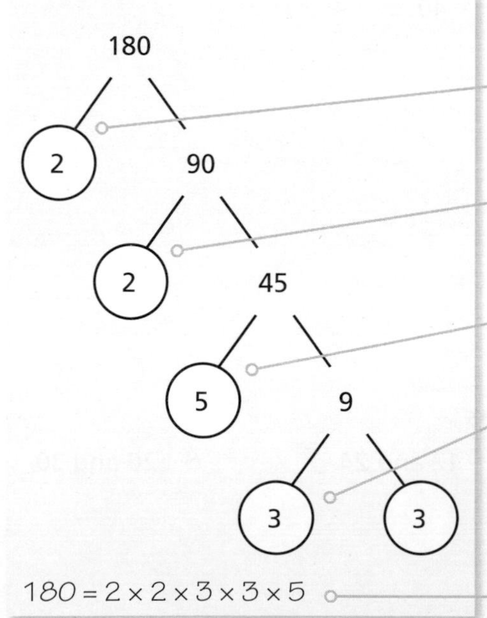

To start a prime factor tree, first identify a pair of factors.

180 = 2 × 90

2 is prime, so you can circle this and that branch ends there. 90 is not prime, so you need to continue.

90 = 2 × 45

Again, 2 is prime so you can circle this and end that branch. 45 is not prime, so you need to continue.

45 = 5 × 9

5 is prime, but 9 is not.

9 = 3 × 3

Now that all branches are ended, you can write 180 as the produce of its prime factors.

Write the prime factors in ascending order, so 180 = 2 × 2 × 3 × 3 × 5

You can check your answer by working out 2 × 2 × 3 × 3 × 5. It should give you an answer of 180

This is the prime factor decomposition of 180

Method B

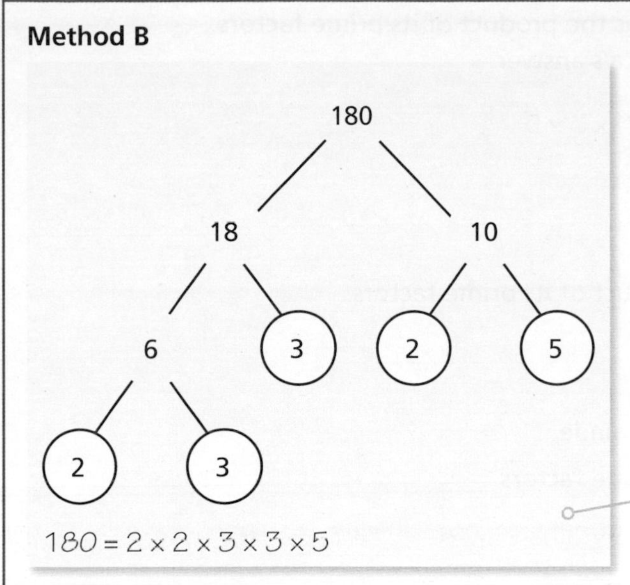

You could have started with a different pair of factors and you would still end up with the same final answer. Try it.

Practice 15.4A

1. A factor tree has been drawn for each number.
   Write each number as the product of its prime factors.

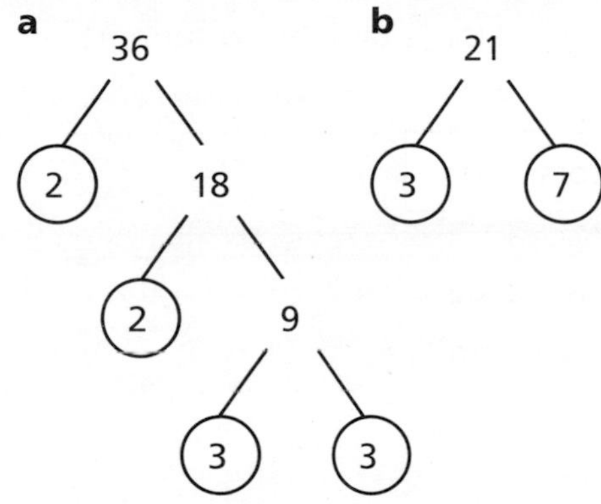

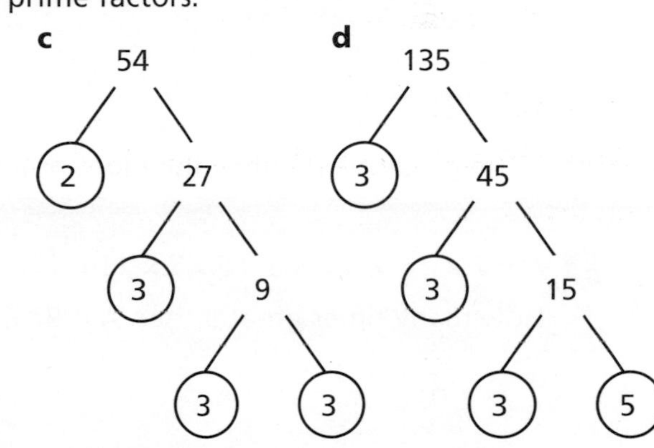

2. Write each number as the product of its prime factors.

   a 72 **b** 42 **c** 108 **d** 270

   Compare your answers to questions 1 and 2. What do you notice? Why does this happen?

3. Write each number as the product of its prime factors.

   a 48 **b** 95 **c** 63 **d** 72 **e** 242
   f 250 **g** 300 **h** 460 **i** 207 **j** 513

4. 1000 = 2 × 2 × 2 × 5 × 5 × 5

   Use this to write each number as the product of its prime factors.

   a 2000 **b** 3000 **c** 27 000 **d** 63 000

5 Seb and Beca have each written 120 as the product of its prime factors.

Seb's answer

$2 \times 2 \times 2 \times 3 \times 5$

Beca's answer

$2^3 \times 3 \times 5$

Who is correct? Explain your answer.

6 Abdullah has written 126 as the product of its prime factors.

$126 = 2 \times 7 \times 9$

a Explain the mistake Abdullah has made.

b Write 126 as the product of its prime factors.

What do you think?

1 $684 = 2 \times 2 \times 3 \times 3 \times 19$

Decide whether each number is a factor of 684. Explain your answer for each.

a 4 **b** 8 **c** 38 **d** 9 **e** 36

2 Is this statement always, sometimes or never true? Use examples to support your reasoning.

The greater a number, the more prime factors it has.

3 $A = 2 \times 2 \times 2 \times 2 \times 3 \times 5 \times 7 \times 11 \times 17$ $B = 2 \times 2 \times 2 \times 2 \times 3 \times 5 \times 5 \times 7 \times 11 \times 17$

How many times greater than A is B? Explain your answer.

In this section you use **prime factor decomposition** and Venn diagrams to calculate the **highest common factor** and **lowest common multiple** of a set of numbers. (H)

Example 2

a Work out the highest common factor of 72 and 60

b Work out the lowest common multiple of 72 and 60

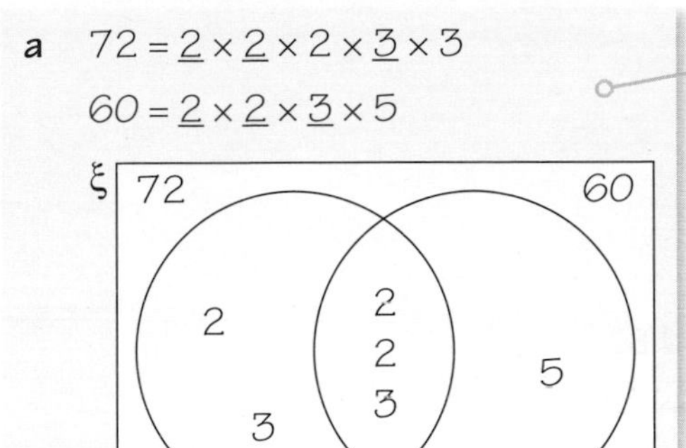

First, you need to write each number as the product of its prime factors.

Looking at the prime factor decomposition of each number, 72 and 60 have three prime factors in common: two 2s and a 3

These go in the intersection of the Venn diagram.

The highest common factor is the product of these three numbers.

You can check that you have correctly organised your prime factors by multiplying.

The circle for 72 contains three 2s and two 3s

$2 \times 2 \times 2 \times 3 \times 3 = 72$

The circle for 60 contains two 2s, a 3 and a 5

$2 \times 2 \times 3 \times 5 = 60$

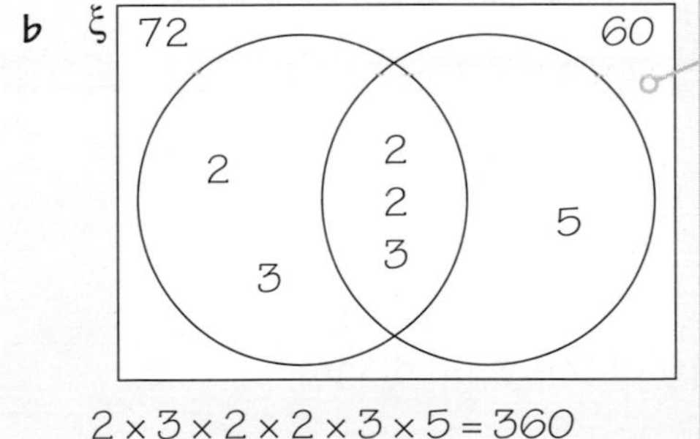

To calculate the lowest common multiple, you need to find the product of all the numbers in the Venn diagram.

Practice 15.4B Ⓗ

1 The prime factors of two numbers, A and B, are shown in the Venn diagram.

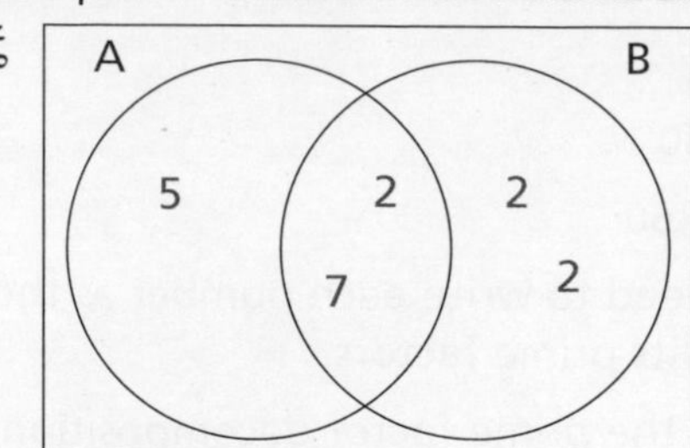

a What is the value of A?

b What is the value of B?

c What is the highest common factor of A and B?

d What is the lowest common multiple of A and B?

2 The prime factors of two numbers, X and Y, are shown in the Venn diagram.

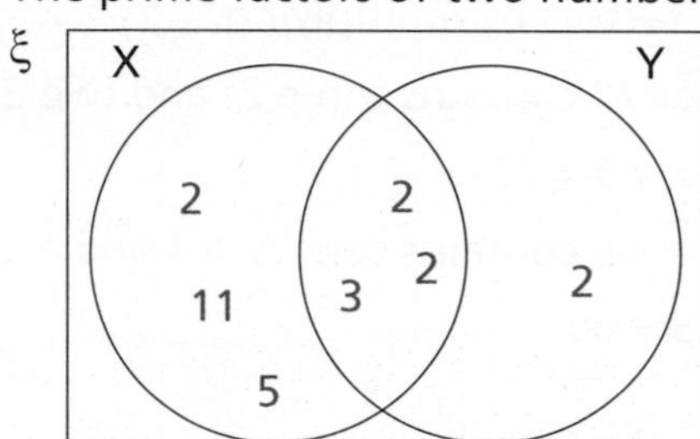

a What is the HCF of X and Y?

b What is the LCM of X and Y?

3 **a** Write 48 as the product of its prime factors.

b Write 64 as the product of its prime factors.

c Write the prime factors of 48 and 64 in a copy of the Venn diagram.

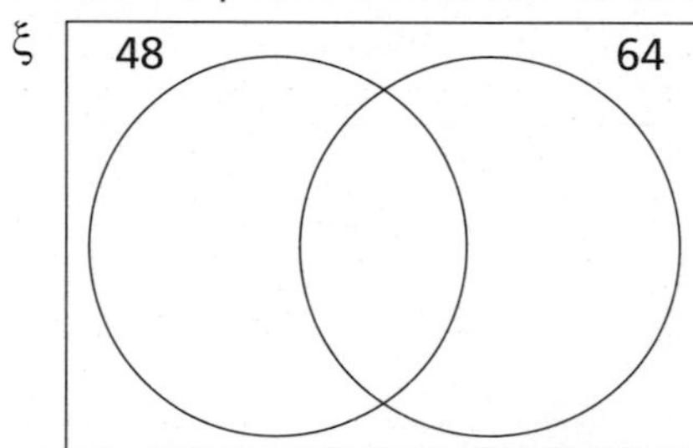

d What is the HCF of 48 and 64?

e What is the LCM of 48 and 64?

4 For each pair of numbers.

- Write each number as the product of its prime factors.
- Work out the HCF and LCM of the numbers using a Venn diagram.

a 48 and 28 **b** 110 and 125 **c** 315 and 45

d 560 and 140 **e** 1080 and 180 **f** 63 and 721

What do you think?

1 Marta has written the prime factors of 28 and 51 in a Venn diagram.

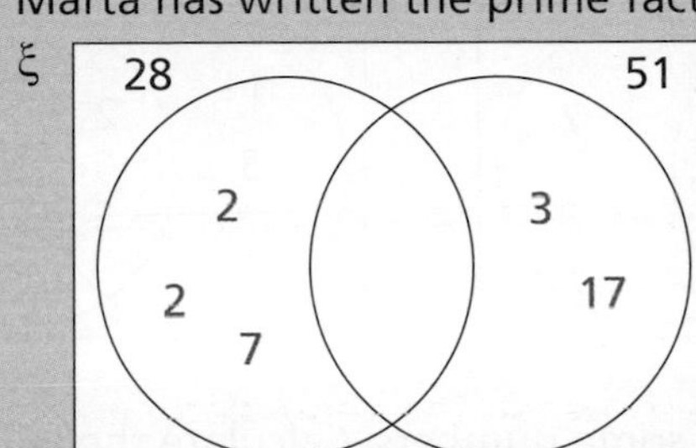

She says "There is nothing in the intersection, so the highest common factor of 28 and 51 is 0"

Do you agree? Explain your answer.

2 Abdullah has written the prime factors of two numbers in a Venn diagram.

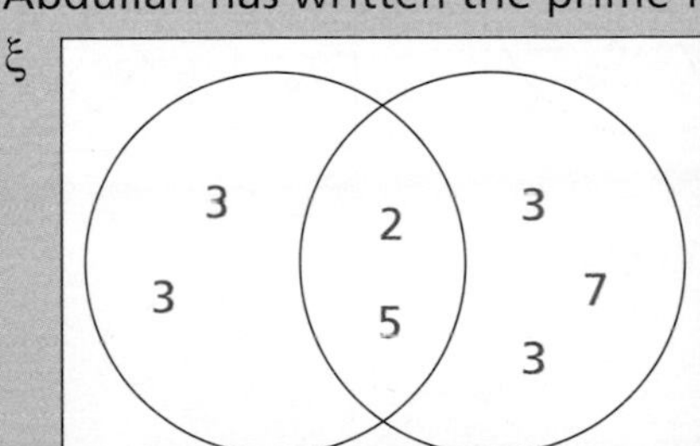

How can you tell that he has made a mistake?

3 The Venn diagram shows the prime factors of two numbers, P and Q.

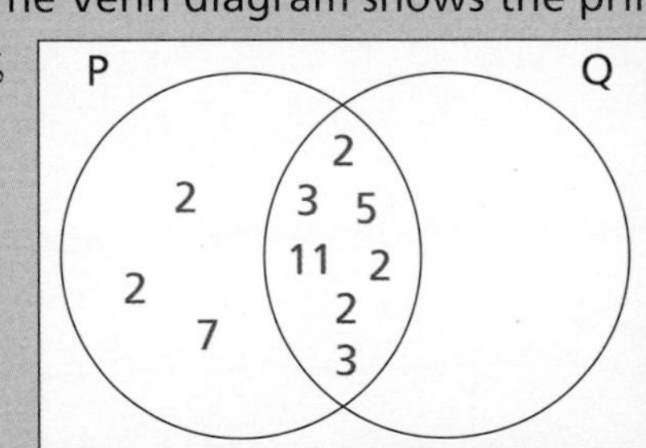

Decide whether each statement is true or false. Explain your answer.

a P and Q are both even. **b** 7 is a factor of both numbers.

c 66 is a factor of both P and Q. **d** Q is a factor of P.

Consolidate – do you need more?

1 Write each number as the product of its prime factors.

a 24 **b** 72 **c** 100 **d** 125 **e** 840

f 48 **g** 216 **h** 700 **i** 750 **j** 84 000

2 Each Venn diagram shows the prime factors of some numbers. Calculate the HCF of each pair of numbers.

a

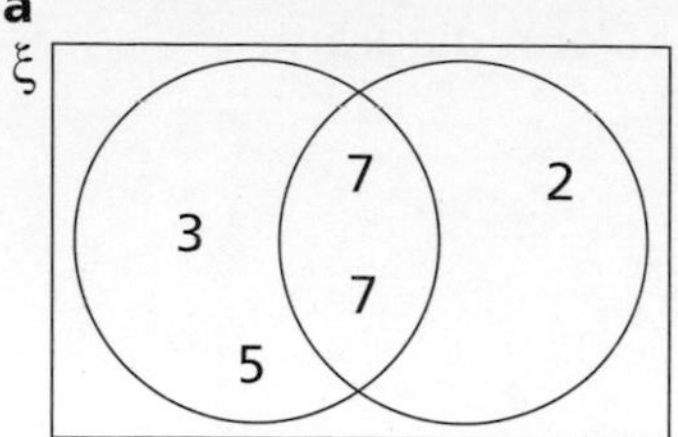

b

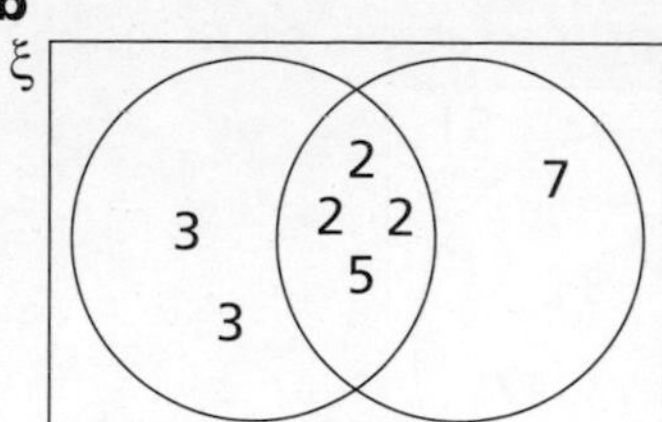

c

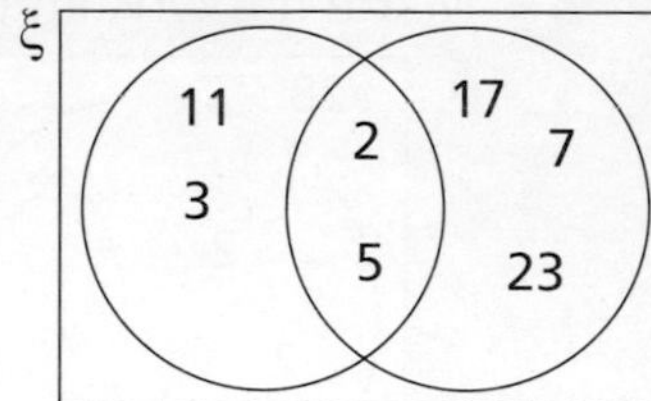

3 Each Venn diagram shows the prime factors of some numbers. Calculate the LCM of each pair of numbers.

a

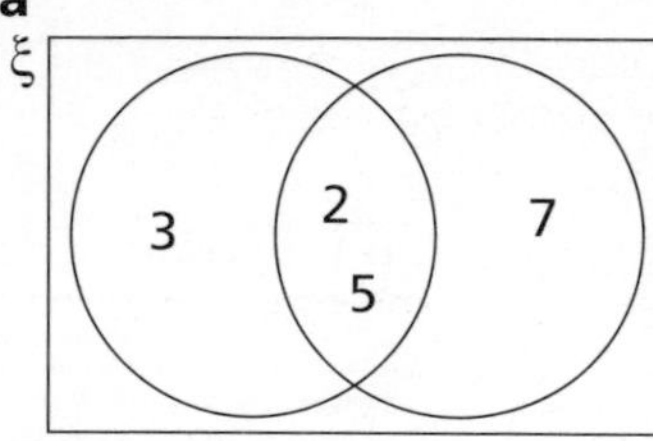

b

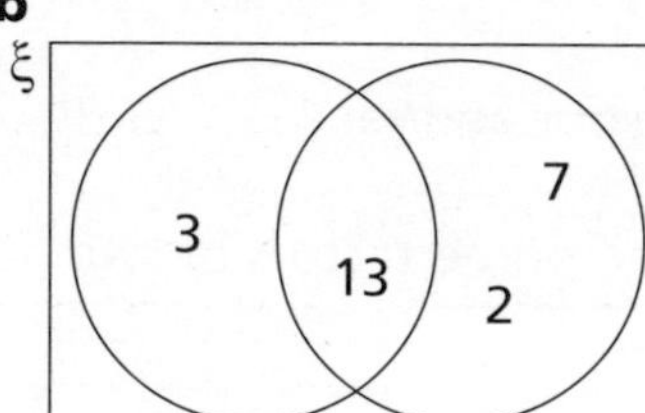

c

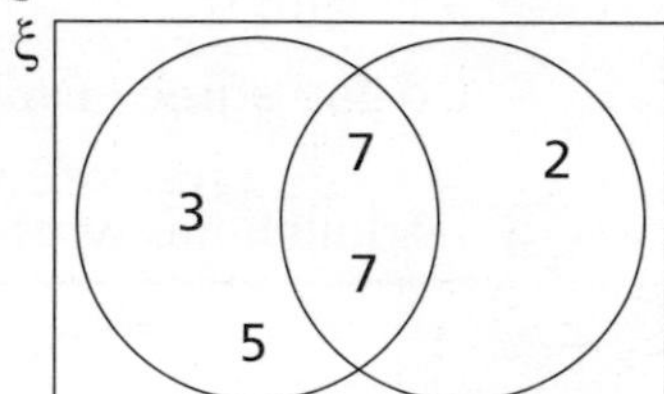

4 Calculate the HCF and LCM of each pair of numbers.

a 24 and 72

b 100 and 125

c 840 and 48

Stretch – can you deepen your learning?

1 The highest common factor of two numbers is 12. The lowest common multiple of the numbers is 180. Use a Venn diagram to work out possible values of the two numbers.

2 X = $27pq$ Y = $36p^2q$

p and q are **prime numbers**. Use a Venn diagram to work out the HCF and LCM of X and Y.

3 The HCF of two numbers can be calculated using $2^4 \times 3^2 \times 7$. The LCM of the numbers can be calculated using $2^4 \times 3^2 \times 7 \times 9 \times 11$. What could the numbers be?

4 **a** $524 = a^2b$ where a and b are prime. Work out the values of a and b.

b $728 = c^3de$. c, d and e are prime numbers such that $c < d < e$. Work out the values of c, d and e.

5 **a** Write 1024 as the product of its prime factors.

b How does the prime factor decomposition tell you that 1024 is a square number?

c $1024 = p^2$. Work out the value of p.

6 **a** Use prime factor decomposition to show that 5832 is a cube number.

b Calculate the cube root of $5832x^3y^6$.

Reflect

1 What is the difference between finding the factors of a number and finding the prime factors of a number?

2 **a** Give an example of when a Venn diagram is the most appropriate method for calculating the HCF and LCM of two or more numbers.

b Give an example of when a Venn diagram is *not* the most efficient method for calculating the HCF and LCM of two or more numbers.

15.5 Conjectures and proof

Small steps

- Make and test conjectures
- Use counterexamples to disprove a conjecture

Key words

Conjecture – a statement that might be true that has not yet been proved

Counterexample – an example that disproves a statement

Are you ready?

1 Here are some numbers.

16	5	2	1	24	36	100	51
47	28	11	9	49	17	18	21

Which of the numbers are

a odd **b** even **c** prime **d** square?

2 Complete the additions.

a $21 + 43$ **b** $57 + 81$ **c** $7 + 95$ **d** $101 + 99$

e $42 + 50$ **f** $26 + 32$ **g** $198 + 8$ **h** $306 + 222$

3 Complete the multiplications.

a 3×5 **b** 17×9 **c** 11×7 **d** 21×19

e 4×6 **f** 22×10 **g** 18×4 **h** 8×30

Models and representations

In this chapter you will look at forming **conjectures**, as well as proving or disproving them. You could use algebra titles to support your arguments and proofs.

Example 1

Seb conjectures that the product of two prime numbers is always odd.

a Find an example that supports Seb's conjecture.

b Use a **counterexample** to disprove Seb's conjecture.

a $7 \times 5 = 35$

Choose two prime numbers that support Seb's conjecture.

7 and 5 are both prime, and their product 35 is odd. Therefore this example supports Seb's conjecture.

b $2 \times 11 = 22$

Now find an example that does not support it.

You should know that 2 is the only even prime number. Multiplying 2 by another prime, in this case 11, gives an even product.

Therefore Seb's conjecture is incorrect.

Practice 15.5A

1 Marta conjectures that the sum of two prime numbers is always even.

a Find an example that supports Marta's conjecture.

b Use a counterexample to disprove Marta's conjecture.

2 Beca conjectures that the lowest common multiple of two numbers is equal to the product of the two numbers.

a Find an example that supports Beca's conjecture.

b Use a counterexample to disprove Beca's conjecture.

3 **a** Investigate each of these conjectures.

- The sum of two odd numbers is always even.
- The sum of two even numbers is always even.

b Explain your findings.

c Is it possible to find a counterexample for either of these conjectures?

4 Decide whether each conjecture is always, sometimes or never true.

Use examples to support your reasoning.

a $x + y = y + x$ **b** $x - y = y - x$ **c** $xy = yx$ **d** $\frac{x}{y} = \frac{y}{x}$

5

The sum of two proper fractions is always a proper fraction.

a Give an example of when Jackson's conjecture is correct.

b Use a counterexample to disprove Jackson's conjecture.

6

Every number has at least two factors.

a Give three examples that support Flo's conjecture.

b Use a counterexample to disprove Flo's conjecture.

7

When you multiply two numbers, the answer is always greater than both the numbers you started with.

a Give three examples that support Benji conjecture.

b Use a counterexample to disprove Benji's conjecture.

8 A sequence has first term 2 and second term 4

The third term in the sequence must be 6

a What sequence is Abdullah thinking of? What is the term-to-term rule for this sequence?

b Give an example of a sequence with first term 2 and second term 4 that disproves Abdullah's conjecture.

What do you think?

1 If the side length of a square is an integer, then its area is a square number.

a Investigate this conjecture. b Summarise your findings.

2 If a and b are prime numbers, then the HCF of a and b is 1

a Investigate this conjecture. b Summarise your findings.

3 a Predict the next two terms in each sequence. Explain your prediction.

i 1, 4 … ii 1, 1 … iii 16, 8 … iv 5, 9 …

b The next two terms in each sequence are shown. For each one, explain if your prediction was correct and why.

i 1, 4, 9, 16 ii 1, 1, 2, 3 iii 16, 8, 4, 2 iv 5, 9, 13, 17

c Now predict the fifth and sixth terms in each sequence. Do you think your predictions are more or less accurate than in part **a**?

Consolidate – do you need more?

1 Two fair six-sided dice are rolled, and the totals are added together.

a Copy and complete the sample space for the possible outcomes.

+	1	2	3	4	5	6
1	2	3	4			
2						
3						
4						
5						
6						

b Ed says, "When rolling two dice, the total of the scores is even."

i From your sample space, give three examples of when Ed is correct.

ii From your sample space, give a counterexample to disprove Ed's conjecture.

2 **a** Work out the value of $x + y$ and xy in each case.

i $x = 3, y = 7$ **ii** $x = -3, y = 7$ **iii** $x = 0.5, y = 20$ **iv** $x = -5, y = -9$

b Faith conjectures that for all values of x and y, xy is greater than $x + y$.

i Which of your answers from part **a** support Faith's conjecture?

ii Which of your answers from part **a** disprove Faith's conjecture?

3 **a** Work out the HCF of each pair of numbers.

i 4 and 6 **ii** 12 and 18 **iii** 7 and 9 **iv** 9 and 27

b Emily conjectures "The HCF of two numbers is always less than either of the numbers."

i Which of your answers from part **a** support Emily's conjecture?

ii Which of your answers from part **a** disprove Emily's conjecture?

Stretch – can you deepen your learning?

1 n is an integer.

a Explain why $2n$ is always even and $2n + 1$ is always odd.

b Prove that the sum of two consecutive numbers is always odd.

c Why did you need the information that n is an integer?

2 Rob conjectures that for two positive integers x and y, x subtract y is always positive.

a Use a counterexample to disprove Rob's conjecture.

b What would be needed for Rob's conjecture to always be true?

3 Two numbers, A and B, have been written as the product of their prime factors.

A = $f^2 \times g \times h \times i$ B = $f^2 \times g \times j$

Jakub says, "A is greater than B because it has more prime factors."

a Give values of f, g, h, i and j that support Jakub's conjecture.

b Give values of f, g, h, i and j that disprove Jakub's conjecture.

4 Seb says that for all values of x, $\frac{1}{x} + \frac{5}{7x}$ is a proper fraction.

a Choose a value of x that supports Seb's conjecture.

b Choose a value of x that disproves Seb's conjecture.

5 The sum of two consecutive triangular numbers is a square number.

Investigate this conjecture and explain your findings.

Reflect

1 Explain why only one counterexample is needed to disprove a conjecture.

2 Explain why you cannot prove a conjecture to be true using just numerical calculations.

15 Primes and proof
Chapters 15.1–15.5

I have become **fluent** in…

- Finding factors and multiples
- Identifying the HCF and LCM of two or more numbers
- Recognising prime, square and triangular numbers
- Writing a number as the product of its prime factors

I have developed my **reasoning** skills by…

- Using diagrams to explain why a number is a factor/multiple of another
- Explaining why a number cannot be prime
- Making and testing conjectures
- Using counterexamples to disprove a conjecture

I have been **problem-solving** through…

- Identifying a pair of numbers which have a given HCF/LCM
- Using the prime factor decomposition of a number to write another number as the product of its prime factors

Check my understanding

1 Write down the first five multiples of each number.

a 8 **b** 20 **c** 15 **d** 12 **e** 16

2 List the factors of each number.

a 8 **b** 20 **c** 15 **d** 12 **e** 16

3 Identify a prime number between 10 and 20 and explain why it is prime.

4 Draw a diagram to show that 25 is a square number.

5 Calculate the fifteenth square number.

6 Find the HCF and LCM of each pair of numbers.

a 5 and 8 **b** 12 and 6 **c** 16 and 12 **d** 20 and 50 **e** 45 and 15

7 Write each number as the product of its prime factors.

a 27 **b** 50 **c** 120 **d** 144 **e** 960

8 **a** Write 720 as the product of its prime factors.

b Write 450 as the product of its prime factors.

c Represent the prime factors of 720 and 450 in a Venn diagram. H

d Calculate the HCF of 450 and 720

e Calculate the LCM of 450 and 720

Glossary

Acute angle – an angle less than 90°

Adjacent – next to each other

Alternate angles – a pair of angles between a pair of lines on opposite sides of a transversal

Angle – a measure of turn between two lines around their common point

Anticlockwise – in the opposite direction to the way the hands of an analogue clock move

Area – the space inside a two-dimensional shape

Array – an arrangement of objects, pictures or numbers in columns and rows

Ascending – increasing in size

Associative law for addition – when you add numbers it does not matter how they are grouped

Axis – a line on a graph that you can read values from

Balance – an amount of money in an account

Bar chart – this uses horizontal or vertical rectangles to show frequencies

Base – a side of a shape that is used as the foundation

Biased – all possible outcomes are not equally likely

Centi – one hundredth

Check – find out if you are correct

Clockwise – in the same direction as the hands of an analogue clock move

Coefficient – a number in front of a variable, for example for $4x$ the coefficient of x is 4

Co-interior angles – a pair of angles between a pair of lines on the same side of a transversal

Collect like terms – put like terms in an expression together as a single term

Common denominator – two or more fractions have a common denominator when their denominators are the same

Commutative – when an operation can be in any order

Complement – the elements of the universal set that do not belong to a set

Conjecture – a statement that might be true that has not yet been proved

Constant – not changing

Construct – draw accurately using a ruler and compasses

Convert – change from one form to another, for example a percentage to a decimal

Corresponding angles – a pair of angles in matching positions compared with a transversal

Counterexample – an example that disproves a statement

Credit – an amount of money paid into an account

Debit – an amount of money taken out of an account

Decimal – a number with digits to the right of the decimal point

Decreasing (or descending) sequence – a sequence where every term is smaller than the previous term

Demonstrate – show how to do something

Denominator – the bottom number in a fraction; it shows how many equal parts one whole has been divided into

Derive – find or discover something from existing knowledge

Descending – decreasing in size

Describe – say what you see, or what is happening

Diagram – a simplified drawing showing the appearance, structure, or workings of something

Difference – in arithmetic, the result of subtracting a smaller number from a larger number; in sequences, the gap between numbers in a sequence

Digits – the numerals used to form a number

Directed numbers – numbers that can be negative or positive

Dividend – the amount you are dividing

Division – the process of splitting a number into equal parts

Divisor – the number you are dividing by

Element – a member of a set

Equal – having the same value. We use the sign = between numbers and calculations that are equal in value, and the sign ≠ when they are not

Equally likely – having the same chance of happening

Equation – a statement with an equals sign, which states that two expressions are equal in value

Equivalent – numbers or expressions that are written differently but are always equal in value

Glossary

Estimate – give an approximate answer

Evaluate – work out the numerical value of

Exchange – when we change numbers for others with equal value, for example replacing 1 ten with 10 ones or replacing 10 tens with 1 hundred

Expression – a collection of terms involving mathematical operations

Fact family – a list of related facts from one calculation

Factor – a positive integer that divides exactly into another positive integer

Factor pair – a pair of numbers that multiply together to give a number

Fibonacci sequence – the next term in a Fibonacci sequence is found by adding the previous two terms together

Find – work out the value of

Fraction – a number that compares equal parts of a whole

Frequency – the number of times something happens

Frequency tree – a diagram showing a number of people/objects grouped into categories

Function – a relationship with an input and an output

Geometric sequence – a sequence is geometric if the value of each successive term is found by multiplying or dividing the previous term by the same number

Give a reason – state the mathematical rule(s) you have used, not just the calculations you have done

Graph – a diagram showing how values change

Highest common factor (HCF) – the greatest number that is a factor of every one of a set of numbers

Hundredth – one of the parts when a whole is split into 100 equal parts

Improper fraction – a fraction in which the numerator is greater than the denominator

Inclusive – including the end points of a list

Increasing (or ascending) sequence – a sequence where every term is greater than the previous term

Indices – an index number (or power) tells you how many times to multiply a number by itself

Inequality symbol – a symbol comparing values showing which is greater and which is smaller

Integer – a whole number

Interior angle – an angle on the inside of a shape

Intersection – the set containing all the elements of A that also belong to B

Inverse – the opposite of a mathematical operation; it reverses the process

Is equivalent to – is shown as $\equiv$ (the same way as = means "is equal to")

Kilo – one thousand

Kite – a quadrilateral with two pairs of adjacent sides that are equal in length

Like terms – terms whose variables are the same, for example $7x$ and $12x$

Line graph – this has connected points and shows how a value changes over time

Line segment – a part of a line that connects two points

Linear equation – an equation with a simple unknown like a, b, or x

Linear sequence – a sequence whose terms are increasing or decreasing by a constant difference

Loss – if you buy something and then sell it for a smaller amount, loss = amount paid – amount received

Lowest common multiple (LCM) – the smallest number that is a multiple of every one of a set of numbers

Mean – the result of sharing the total of a set of data equally between them

Measure of spread – shows how similar or different a set of values are

Median – the middle number in an ordered list

Mental strategy – a method that enables you to work out the answer in your head

Milli – one thousandth

Mixed number – a number presented as an integer and a proper fraction

Multiple – the result of multiplying a number by a positive integer

Negative numbers – numbers less than zero

Non-linear sequence – a sequence whose terms are not increasing or decreasing by a constant difference

Non-unit fraction – a fraction with a numerator that is not 1

Number line – a line on which numbers are marked at intervals

Numerator – the top number in a fraction that shows the number of parts

Obtuse angle – an angle more than 90° but less than 180°

Operation – a mathematical process such as addition, subtraction, multiplication or division

Order of magnitude – size of a number in powers of 10

Original – referring to a number, the number that you started with

Outcome – the possible result of an experiment

Parallel – always the same distance apart and never meeting

Parallelogram – a quadrilateral with two pairs of parallel sides

Partition – break up a number into smaller parts

Per cent – parts per hundred

Percentage – the number of parts per hundred

Perimeter – the total distance around a two-dimensional shape

Perpendicular – at right angles to

Pie chart – a graph in which a circle is divided into sectors that each represent a proportion of the whole

Polygon – a closed 2-D shape with straight sides

Power (or exponent) – this is written as a small number to the right and above the base number, indicating how many times to use the number in a multiplication. For example, the 5 in 2^5

Powers of 10 – the result of multiplying 10 by itself a number of times to give a value such as 10, 100, 1000, 10 000 and so on

Predict – use given information to say what will come next

Prime factor decomposition – writing numbers as a product of their prime factors

Prime number – a positive integer with exactly two factors, 1 and itself

Priority – a measure of the importance of something

Probability – how likely an event is to occur

Product – the result of a multiplication

Profit – if you buy something and then sell it for a higher amount, profit = amount received – amount paid

Proof – an argument that shows that a statement is true

Properties – features of something that are always true

Proportion – a part, share, or number considered in relation to a whole

Quotient – the result of a division

Random – happening without method or conscious decision; each outcome is equally likely to occur

Range – the difference between the greatest value and the smallest value in a set of data

Reflex angle – an angle more than 180° but less than 360°

Regular polygon – a polygon whose sides are all equal in length and whose angles are all equal in size

Remainder – the amount left over after dividing one integer by another

Represent – draw or show

Right angle – an angle of exactly 90°

Root – the nth root of a number x is the number that is equal to x when multiplied by itself n times

Round – give an approximate value of a number that is easier to use

Sample space – the set of all possible outcomes or results of an experiment

Sector – a part of a circle formed by two radii and a fraction of the circumference

Sequence – a list of items in a given order, usually following a rule

Set – a collection of objects or numbers

Side – a line segment that joins two vertices in a 2-D shape

Significant figures – the most important digits in a number that give you an idea of its size

Simplify – rewrite in a simpler form, for example rewrite $8 \times h$ as $8h$

Solution – a value you can substitute in place of the unknown in an equation to make it true

Solve – find a value that makes an equation true

Square number – a positive integer that is the result of an integer multiplied by itself

Standard form – a number written in the form $A \times 10^n$ where A is at least 1 and less than 10, and n is an integer

Substitute – to replace letters with numerical values

Successive – coming after another term in a sequence

Glossary

Tabular – organised into a table

Tenth – one of the parts when a whole is split into 10 equal parts

Term – in algebra, a single number or variable, or a number and variable combined by multiplication or division; in sequences, one of the members of a sequence

Term-to-term rule – a rule that describes how you get from one term of a sequence to the next

Thousandth – one of the parts when a whole is split into 1000 equal parts

Timetable – a table showing times

Total – the result of adding two or more numbers

Transversal – a line that crosses at least two other lines

Trapezium – a quadrilateral with one pair of parallel sides

Trial and improvement – a method of finding a solution to a mathematical problem where you make a guess (a trial), see if it works in the problem, and then refine it to get closer to the actual answer (improvement)

Triangular number – a positive integer that is the sum of consecutive positive integers starting from 1

Two-step – when a calculation involves two processes rather than one

Two-way table – this displays two sets of data in rows and columns

Union – the set containing all the elements of A or B or both A and B

Unit fraction – a fraction with a numerator of 1

Universal set – the set containing all relevant elements

Unknown – a variable (letter), whose value is not yet known

Unlike terms – terms whose variables are not exactly the same, for example $7x$ and 12 or $5a$ and $5a^2$

Variable – a numerical quantity that might change, often denoted by a letter, for example x or t

Vertex – a point where two line segments meet

Vertically opposite angles – angles opposite each other when two lines cross

Zero pairs – for example, +1 and –1 make zero

Answers

Block 1 Sequences

Chapter 1.1

Are you ready?

1 a Red **b** B **c** 1 **d** A
2 Charlie Rob Huda
3 a 666 666 **b** 100 000
4 5th, 6th, 7th

Practice 1.1A

1 a There are two rows of squares. The top row always has one square. The bottom row starts with zero, then a single square under the original. The second row then increases by one with each term.
b 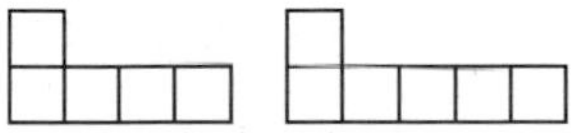 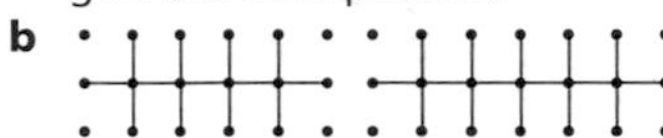
2 a Two vertical and one horizontal lines are added to the end of each pattern to give the next pattern.
b
3 a Start with one circle, then add on two below the first circle, then add on three. Add on a new row with one more circle every time.
b 
Filipo is incorrect as there are 15 circles in the fifth term of the sequence.
4 

What do you think?

a Faith has doubled the number of counters each time. Benji has made a repeating sequence of one counter then two counters.
b Compare answers as a class – there are an infinite number of possibilities. Examples are 1, 2, 3, 6, 10 … (add one more each time) or 1, 2, 5, 14, 41 … (multiply by 3 and subtract 1) or 1, 2, 3, 5, 8 (add the last two terms together), etc.
c Compare answers as a class – it depends on the structure of the sequence.

Practice 1.1B

1 a 1, 2, 3 **b** 4, 7, 10
c 13, 16, 19

2 a 1, 4, 9 **b** 4, 12, 24
c 40, 60, 84
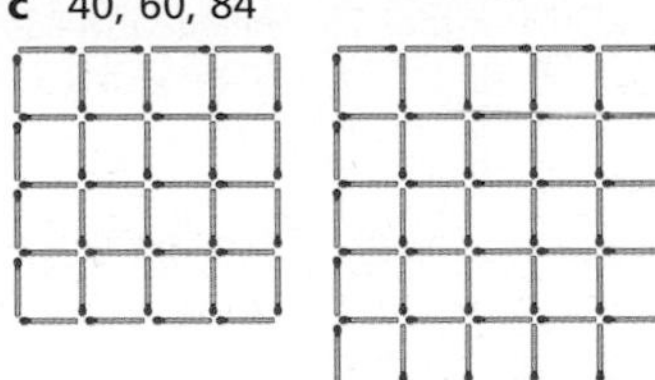
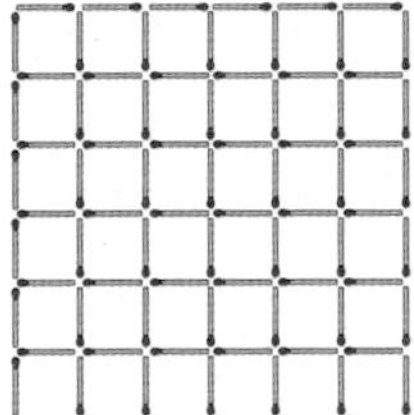
3 a With each term, the number of tables increases by one, the number of chairs increases by four, and the total number of pieces of furniture increases by five.
b 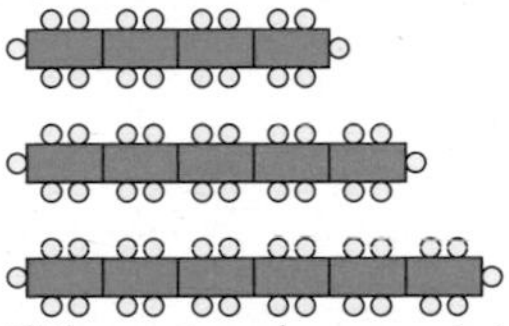 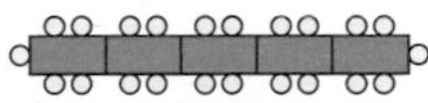
c If the rectangles were joined long edge to long edge, the sequence would look like this.
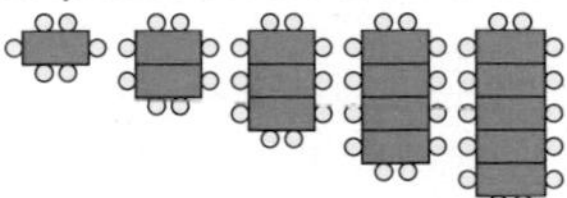
So with each term, the number of tables increases by one, the number of chairs increases by two, and the total number of pieces of furniture increases by three.
4 a 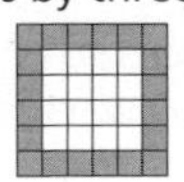 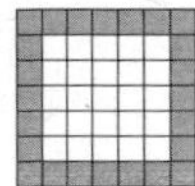
b Grey: 1, 4, 8, 12, 16, 20, 24. Increases by 3 first then by 4 every time. 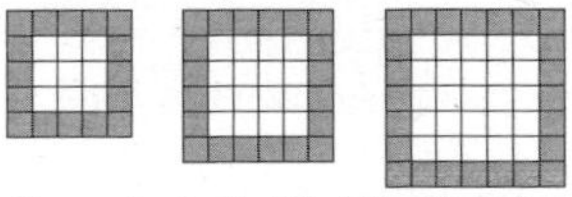
White: 0, 0, 1, 4, 9, 16, 25. Increases by 0 first then by 1, 3, 5, etc. The difference increases by 2 every time.

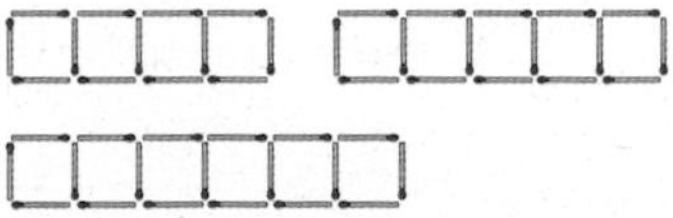

What do you think?

1 The next term would require removing three counters. However, it is impossible to have a negative number of counters, so Marta is correct.
2 Compare answers as a class. The sequence could be increasing or decreasing by a constant, or a repeating pattern, etc.

Consolidate

1 a 13, 16
b 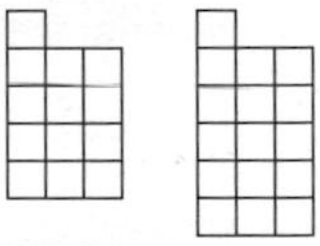
2 a 15, 21
b 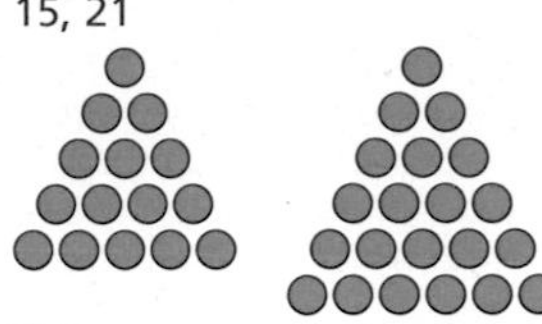
3 a 7, 9
b 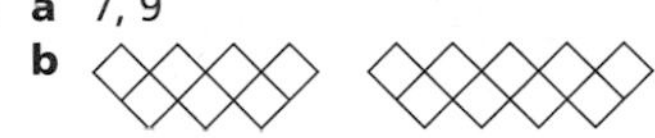

Stretch

1 Both sequences contain polygons, each one having one more side than the last.
Sequence A contains regular polygons, but sequence B contains irregular polygons.
2 This is one possible explanation. The 'dog' contains four parts. The 'head and neck' is the three blocks at the top left, and this remains constant. The 'body' is a row of 3 blocks in the first 'dog' and this increases by 1 block with each term. The 'legs' are 1 block each in the first 'dog' and increase by 1 block each time for the next term. The total number of blocks starts at 8 and increases by 3 every time.

Chapter 1.2

Are you ready?

1 a 5 **b** 6 **c** 1.4 **d** 3.6
2 a Every number is 10 more than the last number
b Every number is double the last number
3 a 8 **b** 10
4 a 3 **b** 0

Practice 1.2A

1 a The next term is 7 more every time
b The next term is 0.41 more every time
c The next term is 16 smaller every time
d The differences are 0.34, then 0.35, then 0.36, then 0.37
e The next term is 1000 more every time
f The differences are 1 less, then 4 less, then 9 less, then 16 less
g The next term is $\frac{8}{100}$ more every time
Sequences **a**, **b**, **c**, **d** and **g** are linear
2 a i Add 14 every time

ii 122
b i Subtract 25 every time
ii 28
c i Add 2.49 every time
ii 6.46
d i Subtract 20.8 every time
ii 8.4
e i Add 900 every time
ii 1900
3 a 64, 68, 72 … is the only linear sequence
b 64, 68, 72 … and 64, 60, 56 …
4 20.76, 22.82, 24.88
5 16.64, 14.58, 12.52
6 a Yes, the number of squares increases by 1 every time
b No, the number of squares increases by 1, then by 2, then by 4
c Yes, the number of squares increases by 3 every time
7 111

What do you think?

Yes. You can use the first two terms to find the common difference and then you can continue the sequence in either direction. The first two terms could also tell you if the sequence is increasing or decreasing.

Practice 1.2B

1 a 4 and 43 b 10.3 and 7
c 32 and 91 d $6\frac{1}{7}$ and 7
2 a (200), 248, 296, 344
b 152, (200), 248, 296
c 104, 152, (200), 248
d 56, 104, 152, (200)
3 a (200), 152, 104, 56
b 248, (200), 152, 104
c 296, 248, (200), 152
d 344, 296, 248, (200)
4 11 450
5 14

What do you think?

1 Flo is correct. There is no information about whether the sequence is increasing or decreasing. The first term will be 200 − 30 − 30 − 30 = 110
2 Although the difference is always 5, the sequence is not linear as the terms alternate between being 5 more and 5 less than the previous term.

Consolidate

1 a Yes
b No
c Yes
2 a linear because the terms decrease by 10 every time
b not linear
c linear as the terms decrease by 0.02 every time
d not linear
3 a 126 b 45 c 12.6
d 4.5 e 1260 f 450

Stretch

1 a Examples for the first rule:
■ 2, 12, 22, 32 …
■ 2, 102, 202, 302 …
■ 62, 82, 102 …
■ 1002, 902, 802 …
Any sequence where the first term ends in 2 and the common difference is a multiple of 10
b Examples for the second rule:
■ 8, 13, 18, 23 …
■ 8, 113, 218 …
■ 48, 73, 98 …
■ 1468, 1443, 1418 …
Any sequence where the first term ends in 8 and the common difference ends in 5
2 Add 18 × 98 to the second term

Chapter 1.3

Are you ready?

1 a Not linear; the differences between the terms change
b Linear; the differences between terms are the same
2 5, 10, 15, 20
3 19, 17, 15, 13
4 Various answers, e.g. 10, 12, 14, 16 … or 11, 12, 13, 14 … or 15, 12, 9, 6 …

Practice 1.3A

1 a 10, 20, 40, 80
b 10, 5, 2.5, 1.25
c 8, 9, 11, 14 d 67, 78, 67, 78
2 a linear b non-linear
c linear d non-linear
e non-linear f non-linear
g linear
3 a (5, 20), 35, 50, 65
b (5, 20), 80, 320, 1280
c (5, 20), 25, 45, 75
4 a ×5, 3125
b ÷2, 200
c −19 then −17 then −15 etc., 25
d Cube numbers, 216
e ÷10, $\frac{1}{100\,000}$
f Add the two previous terms, 71
5 Various answers are possible, e.g. 100, 50, 25, 12.5 … or 100, 50, 150, 200 … or 100, 50, 0, −50, etc.

What do you think?

1 Compare answers as a class
2 Compare answers as a class. Possibilities include:
■ linear 2, 12, 22, 32 …
■ geometric 2, 12, 72, 432 …
■ multiply previous two terms together 2, 12, 24, 288 …
■ Fibonacci 2, 12, 14, 26 …

Practice 1.3B

1 a 20 and 540 b 60 and 7.5
c 0.08 and 0.8 d 20 and 67.5
2 a i (3), 5, 9, 17, 33
ii (8), 15, 29, 57, 113
iii (8.5), 16, 31, 61, 121
iv (1), 1, 1, 1, 1
b After the first term, all the terms in **i** and **ii** have the same final digit
3 a i (3), 7, 15, 31, 63
ii (8), 17, 35, 71, 143
iii (8.5), 18, 37, 75, 151
iv (1), 3, 7, 15, 31
b After the first term, all the terms in **i** and **ii** have the same final digit
After the first term, **iv** becomes the same as **i**
4 a 144 b 400
5 4 095 875
6 a i (1, 1), 2, 3, 5, 8, 13
ii (1, 2), 3, 5, 8, 13, 21
iii (2, 2), 4, 6, 10, 16, 26
b Various answers; e.g. **ii** is the same as **i** moved along one term, every term in **iii** is double the terms in **i**, etc.

What do you think?

1 After the first term, the last digits are all 5. Compare investigation answers with a class. Possible findings include:
■ If first term ends in 5 and rule is ×5, ×15, etc. then all terms end in 5
■ If first term ends in 6 and rule is ×6, ×16, etc. then all terms end in 6
■ If rule is ×10 or any multiple of 10, all terms after the first end in 0, etc.
2 Four; some Fibonacci sequences look linear for the first three terms e.g. 5, 10, 15, 25 ...

Consolidate

1 a non-linear b non-linear
c non-linear d linear
2 a linear; difference +100 every time
b linear; difference +1000 every time
c non-linear; differences change
d non-linear; differences change
e non-linear; differences change
f non-linear; differences change
g linear; difference −100 every time
h non-linear; differences change
3 a i ×3 ii 4860
b i ÷10 ii 0.008
c i ×5 ii 125 000
d i ÷4 ii 256
e i ÷12 ii 12
f i ×100 ii 600 000
4 a 60 b 67.5 c 75
5 a 15 b 20 c 15

Stretch

The terms in the sequence get closer and closer to 6, whatever the starting number.
If instead you add 4, the terms in the sequence get closer and closer to 8.
If instead you add 3.5, the terms in the sequence get closer and closer to 7.
Generally, the terms get closer and closer to double the number you are adding on.

Chapter 1.4

Are you ready?

1 a 11 **b** 17 **c** 2nd

2 A is linear and B is non-linear

3 17, 22 and 27

4 Various answers, e.g. 12, 11, 10, 9 … or 12, 10, 8, 6 … or 12, 9, 6, 3 …

Practice 1.4A

1 a A

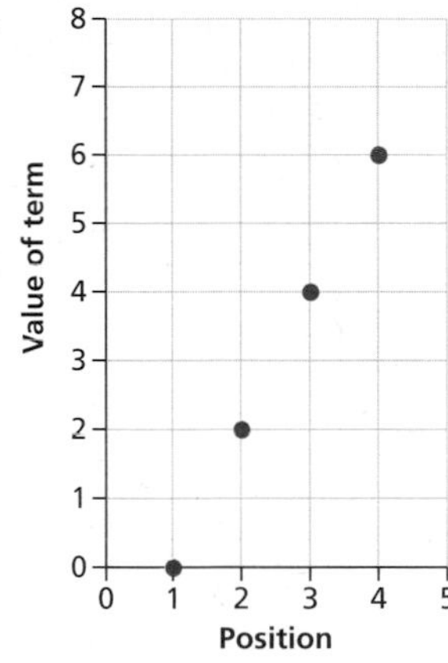

B

C

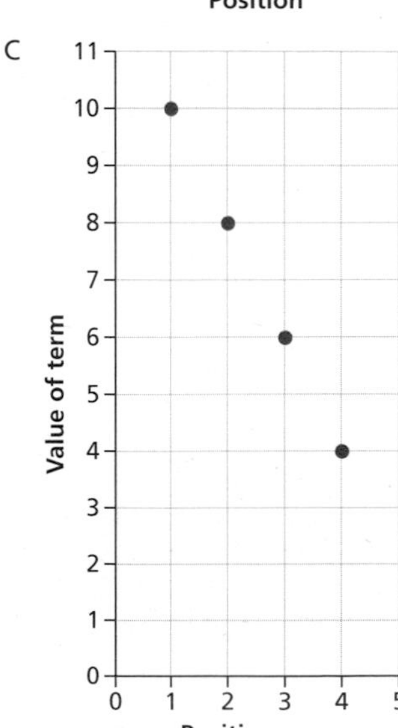

b A **i** Add 2 every time

ii 10

B **i** Add 1 every time

ii 8

C **i** Subtract 2 every time

ii 0

2 a A

Position	1	2	3	4
Term	2	3	4	5

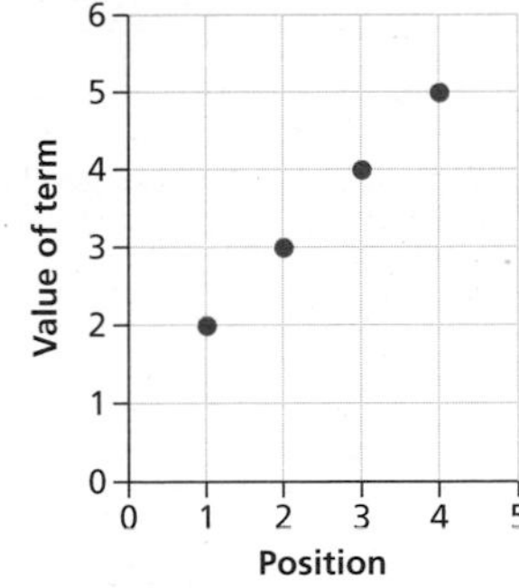

B

Position	1	2	3	4
Term	8	7	6	5

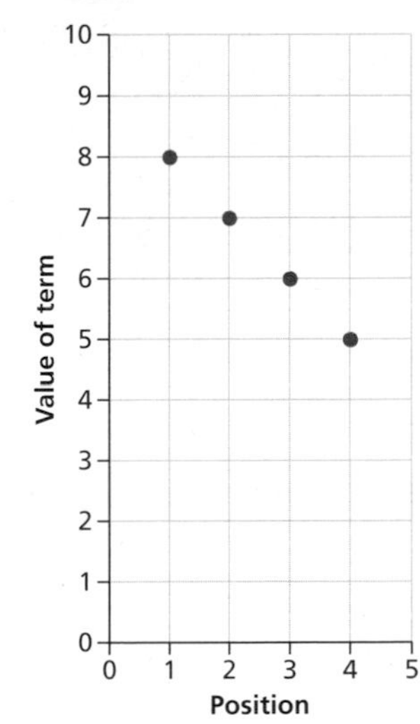

b A **i** Add 1 every time

ii

B **i** Subtract 1 every time

ii

3 a

Position	1	2	3	4
Term	4	6	8	10

b Start at 4, increase by 2 every time

c 16

d No; it starts with an even number and adds on 2 every time. This will miss all the odd numbers.

4 a

Position	1	2	3	4
Term	11	9	7	5

b Start at 11, decrease by 2 every time

c 1

d No; it starts with an odd number and adds on 2 every time. This will miss all the even numbers.

5 a You don't know the first term

b

Position	1	2	3	4
Term	5	8	11	14

6 a

Position	1	2	3	4
Term	1	$\frac{6}{7}$	$\frac{5}{7}$	$\frac{4}{7}$

b Start at 1, decrease by $\frac{1}{7}$ every time

c $\frac{3}{7}$

d 8th

What do you think?

1 a There isn't a 2.5th term in the sequence

b It might help you to 'see' that the points lie on the same line, but the points between the integer positions on the x-axis don't exist.

2 Compare answers as a class.

Practice 1.4B

1 a

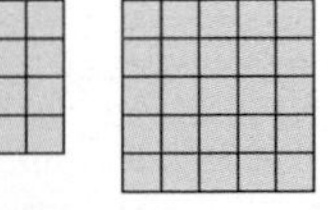

b

Position	1	2	3	4	5
Term	1	4	9	16	25

c

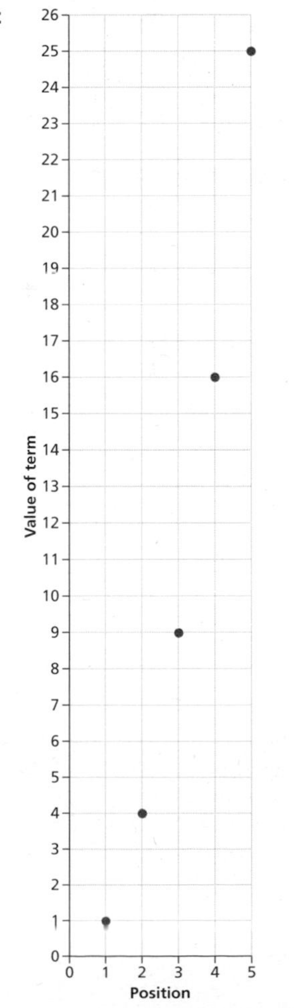

2 a

Position	1	2	3	4
Term	15	13	10	6

b The numbers decrease by one more each time

c 1

3 a

Position	1	2	3	4
Term	1	3	6	10

b The numbers increase by one more each time

c 15

4 a Yes **b** Yes **c** No **d** No
e No **f** No **g** Yes **h** No

5 a i (1, 1), 2, 3, 5, 8
ii (1, 2), 3, 5, 8, 13
iii (1, 3), 4, 7, 11, 18

b i

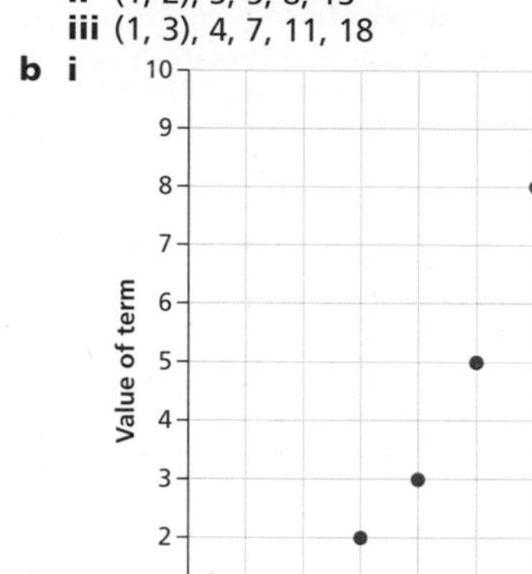

ii

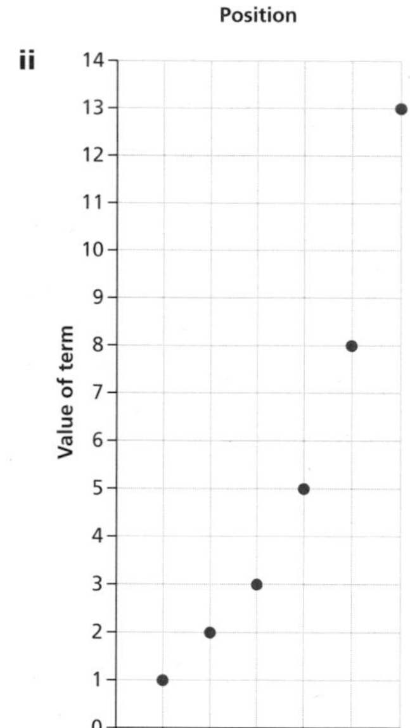

iii

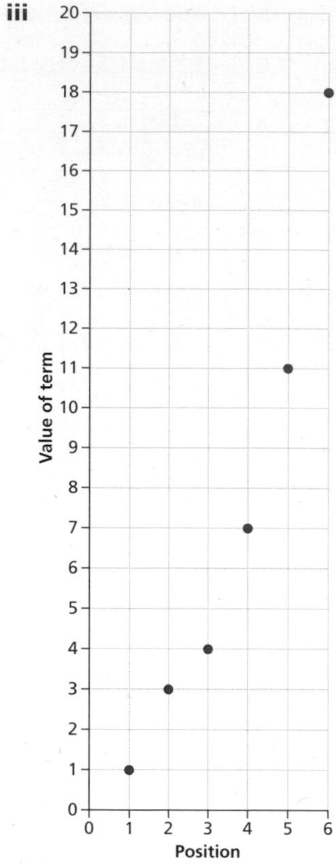

What do you think?

1 a 1, 2, 3, 4, 5
b 1, 2, 4, 8, 16
c 1, 2, 3, 5, 8

2 a

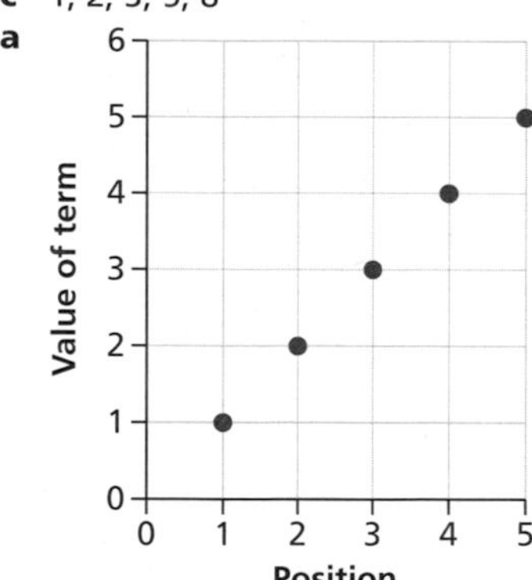

b

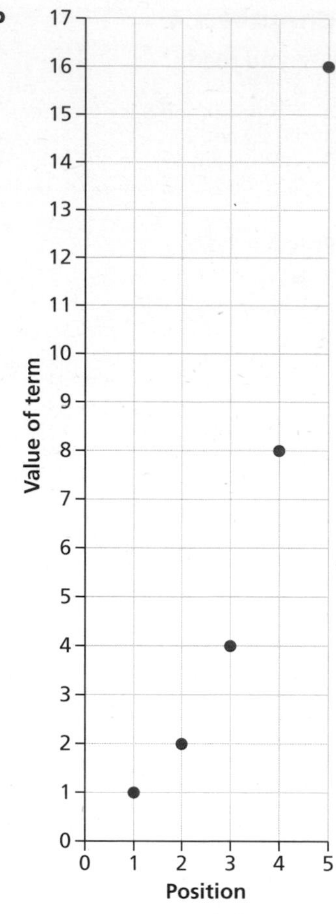

c

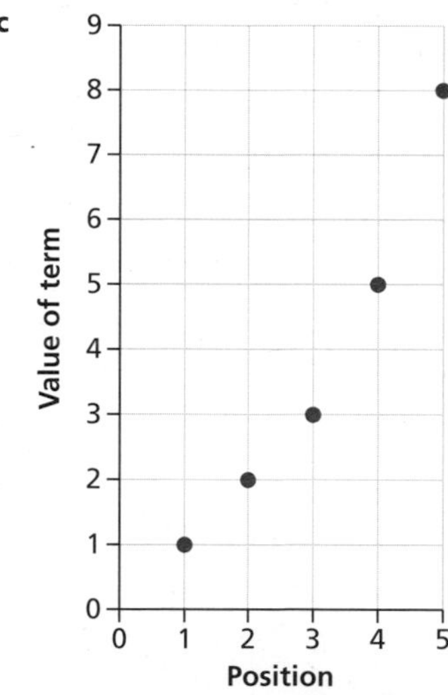

Various answers, e.g. points lie in a straight line for the first graph only, second graph grows most quickly, etc.

Consolidate

1 a A

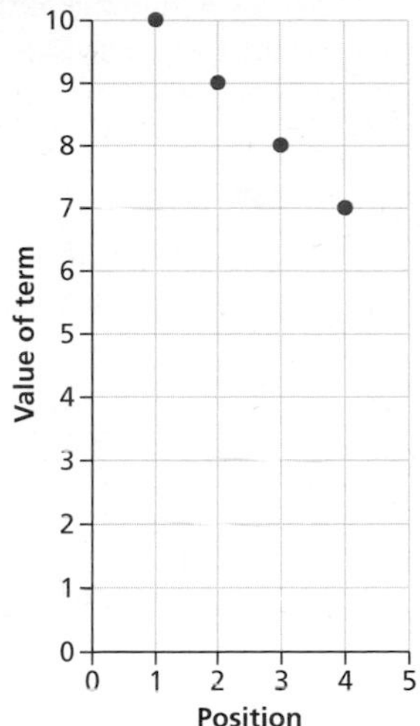

B

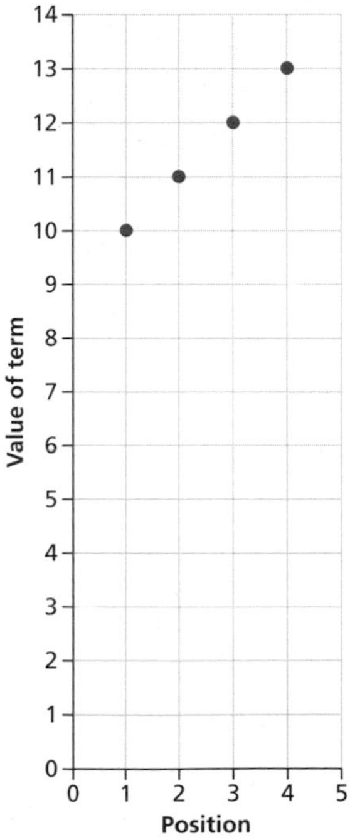

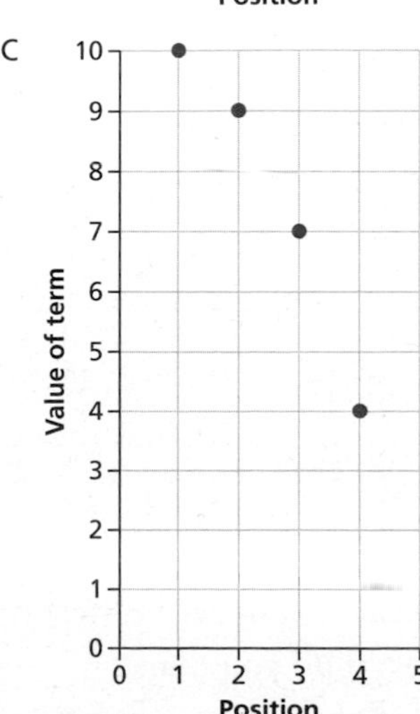

C

Value of term

Position

b A **i** Decrease by 1 every time
ii 6
B **i** Increase by 1 every time
ii 14
C **i** Decrease by 1 then 2 then 3, etc.
ii 0

2 a

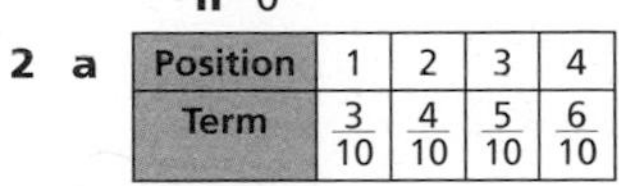

Position	1	2	3	4
Term	$\frac{3}{10}$	$\frac{4}{10}$	$\frac{5}{10}$	$\frac{6}{10}$

b Start at $\frac{3}{10}$ and go up in steps of $\frac{1}{10}$
c $\frac{7}{10}$ and $\frac{8}{10}$
d 8th

3

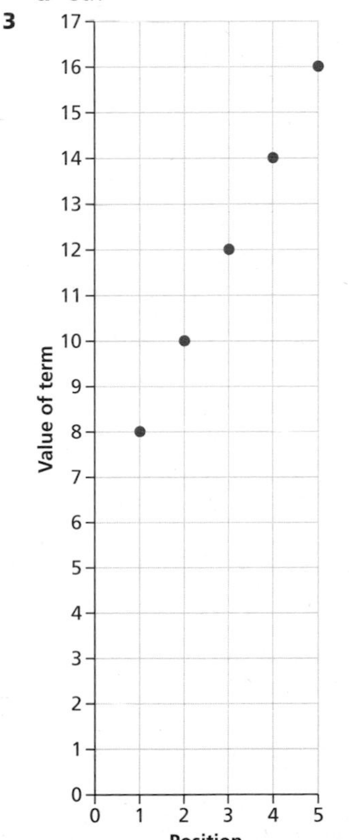

4 a Yes **b** Yes **c** Yes **d** No
e Yes **f** No **g** Yes **h** No

Stretch

1 a You go up 9 squares on the vertical axis
b Compare answers as a class

Chapter 1.5

Are you ready?

1 a 15 **b** 0
c The difference between the terms is the same (5) but for **a** you add and for **b** you subtract
2 a 20 **b** 2.5
c For **a** you multiply by 2 but for **b** you divide by 2
3 a 15 **b** 5
c For **a** you find the sum of the second and third terms but for **b** you find the difference between them
4 a 3 **b** 4

Practice 1.5A

1 a 70 **b** 25 **c** 20, 30
d 17.5, 25, 32.5
e 16, 22, 28, 34
f 15, 20, 25, 30, 35
g 35, 30, 25, 20, 15 **h** 30, 20
2 a There are only 9 gaps, not 10
b 58
3 a The 20th term is not 10 times as big as the 2nd term
b 96
4 a 97 **b** 997 **c** 9997
d 10 996 **e** 11 995
5 84
6 0

What do you think?

1 a The 5th term
b Every 5th term
c No; the integer terms form the sequence 2, 4, 6, 8 …, which are all even
2 Compare answers as a class.
3 Compare answers as a class.

Practice 1.5B

1 a 160 **b** 20 **c** 2.5 **d** 20
e 6 **f** 15 **g** 12, 48
h 100 **i** 20 **j** 200, 50
2 a 8 **b** 4, 16
c 2, 4, 8, 16, 32
3 a 1 and 3 **b** 3 and 4 **c** 2 and 5
d 7 and 15 **e** 0 and 1
f 2, 6, 8 and 36
4 a 50 **b** 30 **c** 80
5 a 6 **b** 8 **c** 6

What do you think?

1 a 5, 6 **b** 2, 10 **c** 7, 10, 17
2 Various answers, e.g. 2, $6\frac{2}{3}$, $11\frac{1}{3}$, 16 …, 2, 4, 8, 16 …, 2, 7, 9, 16

Consolidate

1 a 5 and 20 **b** 15 and 0
c 1.5 and 0
2 a 7 and 13 **b** 10 and 14
c 13 and 15 **d** 14.5 and 15.5
e 6 and 14 **f** 7.5 and 14.5
g 7 and 15
3 a 50 **b** 100 **c** 235
4 a 54 **b** 18 **c** 14, 22
d 12, 18, 24
e 10, 14, 18, 22, 26
f 9, 12, 15, 18, 21, 24, 27
5 a 4 **b** 2, 4, 8 **c** 9
d 3, 9, 27 **e** 24 **f** 20, 50
g 12, 18
6 Compare answers as a class.

Stretch

1 a $\frac{2}{10}$, $\frac{3}{10}$ **b** $\frac{1}{5}$, $\frac{1}{20}$
c 0.01, 0.01
d 1.55, 1.3, 1.05, 0.55

2 **a** and **c** are always possible. **b** is not always possible, e.g. if one of the given terms was zero and the other was non-zero.

Check my understanding

1 **a** 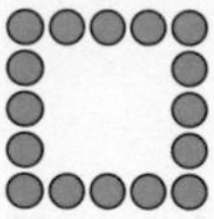

b 20

2 **a** 5, 12, 19, 26 **b** 5, 30, 180, 1080

c 5, 9, 21, 57

3 **a** **i** Start at 80, subtract 10 from the previous term

ii Start at 80, divide the previous term by 2

iii Start at 80, add 5 then 10 then 15, etc. increasing the gap by 5 every time

b Sequence **i** is linear as the difference between the terms is constant, the others have changing differences so are not linear.

4 **a** (4, 8), 12, 16 **b** (4, 8), 16, 32

c (4, 8), 12, 20

5

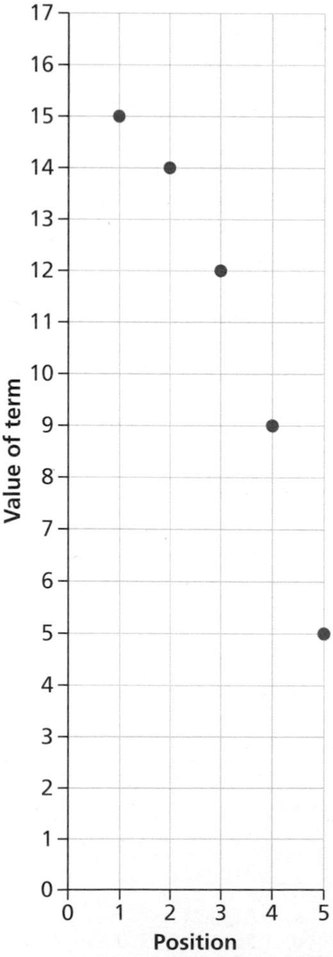

6 **a** **i** (6), 9, (12) **ii** (6), 8, 10, (12)

iii (6), 7.5, 9, 10.5, (12)

iv (12), 10.5, 9, 7.5, (6)

b Yes it is possible. The answers given are for linear sequences, but **i** could be Fibonacci, giving 6, 6, 12, etc.

Block 2 Algebraic notation

Chapter 2.1

Are you ready?

1 +, ×, −, ÷

2 They all have the same answer.

3 **a** 4 × 6 **b** 8 × 7 **c** 7 × 5

4 **a** 7 × 7

b Compare answers as a class.

c **i** 100 **ii** 400

iii 31.36 **iv** 41 209

Practice 2.1A

1 **a** 70 **b** 30 **c** 250 **d** 2.5

e 50.8 **f** 49.2 **g** 40 **h** 62.5

2 **a** 5, 6, 7, 8, 9, 10

b −5, −4, −3, −2, −1, 0

c 0, 5, 10, 15, 20, 25

d 5, 4, 3, 2, 1, 0

3 **a** 78.3, 463.5, 1611

b 29.3, 114.9, 369.9

c −6.3, 79.3, 334.3

d 8517.4, 8603, 8858

4 **a** 4 cubes

b 2 cubes, 2 counters

c 1 cube

5 **a** $x + 4$ **b** $x - 4$ **c** $4x$ **d** $4 - x$

6 **a** $5y$ **b** $y + 5$ **c** $y - 5$

What do you think?

1 **a** $6x$, agree with Seb

b $15p$ **c** $30t$ **d** h

2 Compare answers as a class

3 Yes, Rob is correct; '+7' will always be seven more than the input.

Practice 2.1B

1 **a** **i** 120 **ii** $60b$ **iii** 30

iv $\frac{60}{b}$ **v** $60 + b$ **vi** $60 - b$

b **i** $2x$ **ii** xb **iii** $\frac{x}{2}$

iv $\frac{x}{b}$ **v** $x + b$ **vi** $x - b$

c **i** $2b$ **ii** b^2 **iii** $\frac{b}{2}$

iv 1 **v** $2b$ **vi** 0

2 **a** 2, 8, 49, $t + 2$ **b** −3, 3, 44, $t - 3$

c 0, 30, 235, $5t$ **d** 0, 36, 2209, t^2

e 0, $6m$, $47m$, tm

f 0, $1\frac{1}{2}$, 11.75, $\frac{t}{4}$

g 0, $\frac{6}{p}$, $\frac{47}{p}$, $\frac{t}{p}$

3 **a** **i** $x + 5$ **ii** $3x$ **iii** x^2

iv xy **v** $\frac{x}{2}$ **vi** 1

b **i** $4x + 5$ **ii** $12x$ **iii** $4x^2$

iv $4xy$ **v** $\frac{4x}{2} = 2x$ **vi** 4

4 Filipo is correct: $\frac{1}{2}x$ is the same as $\frac{x}{2}$.

What do you think?

1 **a** Rhys is correct as ab shows the operator × b OR Samira is correct as multiplication is commutative.

b If the function machine had been ÷ b then the output would be $\frac{a}{b}$ as division is not commutative.

Consolidate

1 **a** **i**  **ii** 

iii  + 2 **iv** 0

b **i**  **ii** 

iii  + 2 **iv** 

c **i** $4a$ **ii** a

iii $2a + 2$ **iv** a

d **i** 216 **ii** 54

iii 110 **iv** $108 - a$

e **i** 0.32 **ii** 0.08

iii 2.16 **iv** $0.16 - a$

2 **a** $3a$ **b** $7b$ **c** $\frac{c}{5}$ **d** $\frac{4}{d}$

e $5e$ **f** f^2 **g** gh

3 **a** The product of 4 and p

b f divided by 5

c The square of h

d t subtract 5

e The total of 5 and d

f The total of k and 4

g 6 subtract n

4 **a** 6 **b** 8 **c** 12 **d** $\frac{1}{2}$

5 **a** 130, 30.06, 62, $p + 30$

b 70, −29.94, 2, $p - 30$

c 200, 0.12, 64, $2p$

d 10 000, 0.0036, 1024, p^2

e $100m$, $0.06m$, $32m$, mp

f 25, 0.015, 8, $\frac{p}{4}$

g $\frac{100}{p}$, $\frac{0.06}{p}$, $\frac{32}{p}$, 1

Stretch

1 **a** **i** 10

ii Compare answers as a class

b **i** 2

ii Compare answers as a class

2 **a** To multiply the number by itself three times

b 1, 0.008, 1000, x^3

Chapter 2.2

Are you ready?

1 **a** Subtract 10 **b** Add 10

c Divide by 10 **d** Multiply by 10

2 **a** $8a$ **b** $5b$ **c** $\frac{5}{c}$ **d** $\frac{d}{5}$

e $3e$ **f** f^2 **g** hg

3 **a** $2x$ **b** $\frac{x}{3}$ **c** $x + 4$ **d** $x - 5$

Practice 2.2A

1 **a** − 5 **b** + 3 **c** ÷ 8 **d** × 6

2 **a** 45 **b** 55 **c** 10 **d** 250

3 **a** 48 **b** 58.75 **c** 75 **d** 61.25

4 **a** 162 **b** 16.39 **c** 0.218

d 12.604 **e** 5476 **f** 0

g 1 **h** 148 **i** 74

j 0

5 **a** 20.55 **b** 11.85 **c** 7598

d 7598 **e** 200 **f** 2000

g 2000 **h** 180 **i** 0.3

j 0

6 **a** $t + 7$ **b** $\frac{t}{2}$ **c** $7t$ **d** $t - 7$

7 **a** $6t$ **b** $24t$ **c** $14t$ **d** $10t$

What do you think?

1 **a** $c + 2$ **b** $c + 6$

c $\frac{1}{2}c + 2$ d $2c + 8$

2 a $c - 6$ b $c - 2$

c $\frac{1}{2}c - 2$ d $2c - 8$

3 a $2c + 2$ b $2c + 6$

c $c + 2$ d $4c + 8$

Practice 2.2B

1 a 5 b 9 c 12 d 30

e 630 f 0.5 g $\frac{1}{3}$

2 Ed is incorrect as the numbers between 16 and 25 do have a square root (though it would not be a whole number).

3 a 32 b 29 c 71

4 a

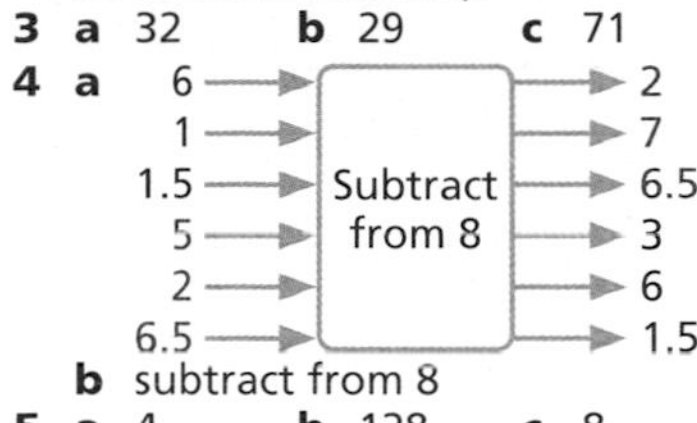

b subtract from 8

5 a 4 b 128 c 8

d 468 e 52.2

What do you think?

1 0 and 1

2 a square b cube root

c i 4 ii 256 iii 4

iv 262 144 v 1000 vi 100

Consolidate

1 a ÷ 4 b + 9 c × 0.5

d – 3.5 e square root $\sqrt{\ }$

2 a 37 b 231

c 13.97 d 14.62

3 a 0 b 2000 c 1

d 1 000 000

4 a 9 b 27 c 162

d 729 e 45 f 6561

5 a b b $x + 9$ c 5 d $g + 5$

e $g + 7$ f b g k h 16

i $24f$ j 30 k 16 l 30

Stretch

1 – 0, × 1, ÷ 1

2 Compare answers as a class

Chapter 2.3

Are you ready?

1 a The product of 6 and x

b The sum of 6 and x

c x divided by 6

d x subtract 6

e 6 subtract x

2 a Each term is increasing by + 5

b Each term is increasing by × 10

3 $y^2 = y$ multiplied by y, $2y = y$ multiplied by 2

4 a 7

b 30 and 30, or any other pair of numbers that add to 60

Practice 2.3A

1 a $5n$ b $3q$ c $\frac{x}{3}$

d $t + 7$ e $t - 12$ f $30 - y$

2 a

$b + 4$	
b	4

b

b	
$b - 4$	4

c

b			
$\frac{b}{4}$	$\frac{b}{4}$	$\frac{b}{4}$	$\frac{b}{4}$

d

$4b$			
b	b	b	b

3 a 16.5 b 8.5

c 3.125 d 50

4 a 40 b 0.025 c 4

d 0.25 e 20 f 0.05

g 2 h 0.5 i 1

j 1 k 0.4 l 2.5

5 a 10.4 b 1 c 0 d 27.04

e 0 f 1 g 10.4

6 a $\frac{x}{10}$ and $\frac{10}{x}$, $x + 10$ and $10 + x$, $10 - x$ and $x - 10$

b $x + 10$ and $10 + x$ will be equal for all values of x

7 a $x \times x$

b i 64 ii 6400 iii 0.64

iv 640 000

8 a The square root of t

b i 18 ii 1.8

iii 0.063 245 553 iv 20

9 a Always equal

b Sometimes equal

c Sometimes equal

What do you think?

1 $n = 0$

2 a $n = 0$ and $n = 1$

b $n = 0$ and $n = 2$

c $n = 3$ d $n = 1\frac{1}{2}$

3 a Chloe is wrong because p cannot be divided by 0

b $\frac{p}{5} = \frac{5}{p}$ if $p = 5$

Practice 2.3B

1 a i 5, 6, 7, 8, 9

ii 1.4, 2.4, 3.4, 4.4, 5.4

iii 41, 42, 43, 44, 45

iv 39, 38, 37, 36, 35

b In each of the sequences the difference between the terms is 1. The first three sequences are ascending; the final sequence is descending.

c i 10th term = 14, 100th term = 104

ii 10th term = 10.4, 100th term = 100.4

iii 10th term = 50, 100th term = 140

iv 10th term = 30, 100th term = –60

2 a i 6, 12, 18, 24, 30

ii 7, 8, 9, 10, 11

iii 6, 3, 2, 1.5, 1.2

iv 0.166, 0.33, 0.5, 0.66, 0.833

b i Multiples of 6 starting from 6

ii Whole numbers starting from 7

iii Start from 6 and divide by 2

iv Start from $\frac{1}{6}$ and add $\frac{1}{6}$

Compare answers as a class

c i 10th term = 60, 100th term = 600

ii 10th term = 16, 100th term = 106

iii 10th term = 0.6, 100th term = 0.06

iv 10th term = 1.66, 100th term = 16.66

3 a 1, 4, 9, 16, 25

b Square numbers

c 10th term = 100, 100th term = 10 000

d 1, 8, 27, 64, 125

e Cube numbers

f 10th term = 1000, 100th term = 1 000 000

g Compare answers as a class

4 a i 20 ii 0 iii 100

iv 1 v 1 vi 20

b Students' own predictions

c i 110 ii 90 iii 1000

iv 10 v 0.1 vi 110

5 a 2, 4, 8, 16, 32, 64, 128, 256, 512, 1024

b Sequence is 2, 4, 8, 6, 2, 4, 8, 6 …

c 20th term has a 6 as the final digit

d 20th term of 10^n would have 0 as the final digit.

What do you think?

1 a Abdullah is correct for all examples except $n = 3$

b $n = 3$, where the values are the same

c 3^n is greater more often

2 2^n = 2 4 8 16 32 64 128 512 1024

n^2 = 1 4 9 16 25 36 49 64 100

Between $n = 1$ and $n = 4$, 2^n starts off larger, then n^2 is larger. When $n = 4$, both are equal, then for $n > 4$, 2^n is always larger.

Consolidate

1 a

$x + 5$	
x	5

b

x	
3	$x - 3$

c

$5x$				
x	x	x	x	x

d

50	
x	$50 - x$

2 a 130 b 32 c 343

d 7 e 2401 f 7

3 a 0.1 b 0 c 10

d 0.01 e 1

4 a i 9, 10, 11, 12

ii 0.125, 0.25, 0.375, 0.5

iii 8, 16, 24, 32

iv 7, 6, 5, 4

b All of the sequences are linear.

5 a 855 b 0 c 195

d 1.5 e 30

Answers

Stretch

1 If n is a whole number then $2n$ must be an even number, so will not end in the number 5

2 Yes, for example if $n = 5$ then $n^2 = 25$

3 No, Seb is not correct. $10^2 = 100$, so 0 can also be a final digit. The sequence for $n = 2, 4, 6, 8, 10 \ldots$ is 4, 6, 6, 4, 0 …
Other patterns could include even and odd terms ending in even and odd numbers or 10th, 20th, 30th etc. … ending in zero.

Chapter 2.4

Are you ready?

1 **a** 11 **b** 106 **c** □ + 6 **d** $a + 6$

2 **a** 8 **b** □ + 2 **c** 4 **d** $b + 2$

3 **a** ÷ 4 **b** × 3

4 **a** 31 **b** 211 **c** 11

5 **a** $4a$ **b** $\frac{b}{2}$ **c** $4c$ **d** $\frac{20}{d}$ **e** e^2

6 **a** $x + 2$ **b** $2x + 2$ **c** $5x - 3$

Practice 2.4A

1 **a** 30 **b** 35 **c** 5.2 **d** 7

2 **a** 15, 18, 21, 24, 27, 30
b This forms an ascending linear sequence, the multiples of 3

3 **a** Input → (× 4) → (− 2) → Output
b **i** 46 **ii** 49.2 **iii** 478 **iv** $4m - 2$

4 **a** Input → (− 2) → (× 4) → Output
b **i** 40 **ii** 43.2 **iii** 472 **iv** $4(m - 2)$

5 **a** Input → (÷ 4) → (− 2) → Output
b **i** 1 **ii** 1.2 **iii** 28 **iv** $\frac{m}{4} - 2$

6 **a** Input → (− 2) → (÷ 4) → Output
b **i** 2.5 **ii** 2.7 **iii** 29.5 **iv** $\frac{m-2}{4}$

What do you think?

1 **a** Inputs that show Darius is wrong could include inputs of 2, 3, 4 or 5
b An input that would give the same number would be 1

2 **a** Answers vary though 0, 1 or 2 would be useful illustrators
b 5 × ■ + 1 ≠ 5(■ + 1)

3 **a** could be replaced with × 6
b could be replaced with + 5
c cannot be replaced
d cannot be replaced
e could be replaced with × 1.5
f could be replaced with ÷ 6

Practice 2.4B

1 **a** Input → (+ 5) → (× 2) → Output
b 10 **c** 50 **d** 0.5

2 **a** 12 **b** 27 **c** 69 **d** $2x$ **e** x **f** 0

3 **a** 70 **b** 60 **c** 398 **d** 380

4 **a** one cube
b 30 cubes and 80 counters

5 one cube and 3 counters

6 **a** $a, 2a, 5$ **b** $2a, 4a$ **c** a, a^2

7 **a** $6a + 4$ **b** $6a$ **c** $3a + 1$ **d** $a + 2$

What do you think?

1 **a** The input and output values will always be the same as the two function machines in each pair have inverse operations.
b You could add a function machine of × 1 (or + 0) to maintain this result over three function machines.

Consolidate

1 **a** Input → (− 10) → (× 3) → Output
b 60 **c** 20

2 **a** Input → (× 3) → (− 10) → Output
b 50 **c** 10

3 **a** $2(a + 5)$ **b** $2a + 5$ **c** $5(a + 2)$ **d** $5a + 2$ **e** $\frac{a+5}{2}$ **f** $\frac{a}{2} + 5$ **g** $\frac{a+2}{5}$ **h** $\frac{a}{5} + 2$

4 **a** $6b$ **b** $6b$ **c** $b + 5$ **d** $b + 5$
Machines **a** and **b** give the same result. Machines **c** and **d** also give the same result.

5 **a** 4, 10, 80 **b** $x \div 6$

Stretch

1 **a** To get the same output, the input would need to be zero for both pairs of machines
b The output for the second pair of machines is double that of the first

2 **a** Input needs to be 0
b The output for the second pair of machines is 4 times that of the first

Chapter 2.5

Are you ready?

1 **a** 9 **b** 6 **c** 1 **d** 0 **e** 0 **f** 1 **g** 9

2 **a** 40 **b** 24 **c** 10 **d** 1 **e** 180

3 50

4 **a** $2x + 5$ **b** $2x - 5$ **c** $5 - 2x$

Practice 2.5A

1 **a** $4a + 5$ **b** $2a + 12$ **c** $3a - 2$ **d** $4a - 6$ **e** $10 - 2a$

2 **a**

b	b	b	4
$3b + 4$			

b

b	b	b
$3b - 4$	4	

c

b	
$\frac{b}{4}$	
$\frac{b}{4}$	7

d

10	b	b	b	b
$10 + 4b$				

e

10	
$10 - 4b$	$4b$

3 **a** 28.51 **b** 20.51 **c** 9.04 **d** 42.68 **e** −22.68

4 **a** 13 **b** 3 **c** 805 **d** 13 **e** 795 **f** 6 **g** 9.75 **h** 10.5 **i** 40.125 **j** 5000

5 **a** 10.4 **b** 15.6 **c** 0 **d** 5.2 **e** 20.8 **f** 14.8 **g** 9.6 **h** 7.04 **i** 47.04 **j** 22.6

6 **a** Sven is correct. $3a^2$ is equal to $3 \times a^2$
b $2a^3 = 2000$, $5a^2 = 500$, so the difference is 1500

7 **a** 5, 9, 13, 17, 21
b 7, 11, 15, 19, 23
c 1, 5, 9, 13, 17
d 5, 11, 17, 23, 29
e 8, 14, 20, 26, 32
f 2, 8, 14, 20, 26
g $4\frac{1}{2}, 5, 5\frac{1}{2}, 6, 6\frac{1}{2}$
h $-2\frac{1}{2}, -2, -1\frac{1}{2}, -1, -\frac{1}{2}$
i $4\frac{1}{2}, 4, 3\frac{1}{2}, 3, 2\frac{1}{2}$

8 **a** All of the sequences are linear
b All of the sequences except **i** are increasing. In **a**, **b** and **c** the differences are + 4 each time; in **d**, **e** and **f** the differences are + 6 each time; in **g** and **h** the differences are $+\frac{1}{2}$ each time, whereas in **i** the differences are $-\frac{1}{2}$

What do you think?

1 Darius is correct. Although both sequences are linear and begin with 5, the first increases by 3 each term whereas the second increases by 2 each term.

2 **a** Same **b** Different **c** Different **d** Different

3 Compare answers as a class.

Practice 2.5B

1 **a** 65 **b** 15 **c** 1000 **d** 1.6 **e** 0.625 **f** 1.25 **g** 3.2

2 **a** 8.2 **b** 27.4 **c** 14.4 **d** −14.4 **e** 30.95 **f** 121.19 **g** 70.45 **h** 70.45 **i** 64.85 **j** 19.65

3 **a** 21 **b** 20.2 **c** 26 **d** 25.25 **e** 11 **f** 10.1 **g** 22 **h** 30 **i** 2002 **j** 2010 **k** 1998 **l** 1990 **m** 10 **n** 100 **o** 2 **p** 15 **q** 15 **r** 51

4 **a** 9 **b** 9 **c** 9 **d** 9
e The difference is 9 each time. The difference will be 9 for any value of a.
f 9

5 **a** $m + n = 4.2$, $\frac{m}{n} = 20$, $\frac{n}{m} = 0.05$, $10m - n = 39.8$, $\frac{m^2}{n} = 80$, $\frac{m}{n^2} = 100$

Greatest value is $\frac{m}{n^2} = 100$,

least value is $\frac{n}{m} = 0.05$

b **i** Yes. Greatest value is $\frac{m}{n^2} = 10$,
least value is $\frac{n}{m} = 0.5$

ii No. Greatest value is $\frac{m^2}{n} = 8000$,
least value is $\frac{n}{m} = 0.005$

iii Yes. Greatest value is $\frac{m}{n^2} = 10\,000$,
least value is $\frac{n}{m} = 0.005$

What do you think?

1 $3(p + 2)$ will give a greater value than $3p + 2$

p	2	p	2	p	2

p	p	p	2

2 Kate is correct. Mario has worked out $\frac{a}{2}$ and then multiplied the answer by b (instead of dividing by b).

3 **a** $3k^2 = 300$, $(3k)^2 = 900$
b $3k^2 = 0.03$, $(3k)^2 = 0.09$
c $3k^2 = 30\,000$, $(3k)^2 = 90\,000$
d The value of k where $3k^2$ and $(3k)^2$ are equal is $k = 0$

Consolidate

1 **a** **i** 10, 13, 16, 19, 22
ii 10, 17, 24, 31, 38
iii 10, 17, 24, 31, 38
iv 8, 11, 16, 23, 32
v 0, 3, 8, 15, 24
vi 1.5, 2, 2.5, 3, 3.5
vii −0.5, 0, 0.5, 1, 1.5
viii 9.5, 9, 8.5, 8, 7.5
ix 99, 96, 91, 86, 75
b **iv**, **v** and **ix** are not linear
c The sequences that are not linear all include a squared value of n.

2 **a** 53 **b** 5 **c** 47 **d** 3.5
e 80 **f** 1.25 **g** 20 **h** 3.5
i 1.66 **j** 9.4 **k** 5 **l** 1

3 **a** 12.6 **b** 24.3 **c** 24.6
d 23.4 **e** 27.6 **f** 3
g 42 **h** 38 **i** 5.22
j 46.66 **k** 6.15 **l** 2.025
m 1.975 **n** 33.33 **o** 11.4
p 5.6 **q** 1.6 **r** 42
s 480 **t** 0.0075

4 **a** 3 **b** 3 **c** 3 **d** 3
e 0 **f** 0 **g** 1.5 **h** 1.5
i 9 **j** 9 **k** 3 **l** 6

5 **a** 7 **b** 5 **c** 36
d 48 **e** 12

Stretch

1 **a** **i** 27 **ii** 729 **iii** 19 683
iv 27 **v** 27
b If $a = 4$ and $b = 2$ then $a^b = b^a$ and $b^{\sqrt{a}} = \sqrt{a^b}$

2 **a** 1856 **b** 8400 **c** 5568
d 2368 **e** 416 **f** 384
g 24 **h** −3360 **i** 3232
j 29 **k** 1.04 **l** 10.4

3 Compare answers as a class

Chapter 2.6

There are many other possible correct answers throughout this chapter. For most questions only some of the possibilities are shown. Check your answers with a partner.

Are you ready?

1 **a** 0 **b** 0 **c** 1 **d** 1
e 25 **f** 10 **g** 10

2 **a** **i** 50 **ii** 2 **iii** 15 **iv** 5
b **i** $5n$ **ii** $\frac{n}{5}$ **iii** $n + 5$
iv $n - 5$

3 **a** **i** 14 **ii** 5 **iii** 14 **iv** $\frac{1}{2}$
b **i** $3m + 2$ **ii** $\frac{m}{2} + 3$
iii $2(m + 3)$ or $2m + 6$
iv $\frac{m-3}{2}$

4 $3m + 2$ means $3 \times m$ then add 2, but $3(m + 2)$ means $(m + 2)$ three times, which is $3m + 6$.

Practice 2.6A

1 **a** + 3 **b** × 10
c ÷ 7 (or − 6) **d** × 3
e ÷ 10 **f** × 1000
g × 1

2 **a** × 2 (or + t) **b** − 4
c ÷ 5 **d** + 3
e + 3 **f** square
g square root **h** × a

3 **a** × 0.04 (or ÷ 25)
b × 8000 **c** ÷ 1000
d × 0.7 **e** × 0.003
f × 1001 **g** × 10 010

4 **a** × 10 **b** + 4 **c** ÷ 3 **d** ÷ 3
e + 4 **f** + 8 **g** × 2

5 − 1 or × 0

What do you think?

1 square, + 2, × 2

2 **a** × 1, + 0
b × 1, × 2, × (any number), ÷ (any number), + 0
c + 0, × 1, square, square root, cube, cube root, etc.

Practice 2.6B

1 **a** × 2, + 4 **b** × 6, + 2
c × 3, − 6 **d** × 1, − 1
e × 2, − 6 **f** × 30, + 10
g × 1000, + 1

2 **a** + 2, × 2 **b** + 2, × 4
c − 2, × 3 **d** − 1, × 1
e + 10, × 0 **f** + 7, × 10
g + 2998, × 1

3 Both are correct. This is because $3n + 9 = 3(n + 3)$

4 **a** × 2, + 10 **b** × 2, + 10
c + 4, × 2 **d** ÷ 2, + 4
e × 2, × 3 **f** + 1, ÷ 2
g square, − 7 **h** × b, + c

5 × 3, + 5

6 **a** ÷ 2, × 2 **b** square, × 2
c square, × 4 **d** × 2, × 3
e + 1, × 3 **f** + 1, $-\frac{1}{4}$
g ÷ 4, + $\frac{3}{4}$

What do you think?

1 **a** square, × 4 **b** × 5, square
c + 3, square **d** × 5, square

2 **a** + 2, × 2 or × 2, + 4
b ÷ 2, + 6 or + 12, ÷ 2
c × 2, + 12 or + 6, × 2
d square, × 4 or × 2, square

Consolidate

1 **a** × 10 **b** × 6 **c** × 0 **d** ÷ 10
e × 25 **f** × a **g** ÷ a

2 **a** × v **b** − 4 **c** × 9
d ÷ t **e** square

3 **a** × 5 **b** × 4
c square root **d** − 5
e − 9

4 **a** − 2, − 5 **b** × 2, − 1
c square, + 6 **d** + 6, ÷ 2
e square root, × 3
f square, × 25

Stretch

1 **a** × 2, subtract from 10
b square, ÷ 4 **c** + d, − c

2 **a** **i** × 0, − input
ii ÷ input **iii** × 1, + 0
b **i** (any), × 0 **ii** ÷ input, × 1
iii × 10, ÷ 10

Check my understanding

1 **a** 14 **b** 12 **c** 4 **d** 2

2 **a** 350 **b** 360 **c** 400 **d** 750

3 $4 - a$, $\frac{1}{2}a$, $\frac{12}{a}$, $a + 4$, a^2, $4a$

4 **a** 5, 7, 9, 11 **b** 4, 7, 12, 19
c 1, 4, 7, 10 **d** 3.5, 4, 4.5, 5
a, **c** and **d** are linear

5 **a** **i** ÷ 10, − 90 and 'square root'
ii ÷ 2, $-\frac{b}{2}$
b **i** Various answers are possible. Compare answers as a class.
ii Various answers are possible. Compare answers as a class.

Block 3 Equality and equivalence

Chapter 3.1

Are you ready?

1 6 + 3, 10 − 1, 8 + 1, 18 ÷ 2, 5 + 4
2 × 5, 6 + 4, 5 + 5
12 − 5, 2 + 5

2 $2a$, $15 - a$, $4a - 10$

3 **a** = **b** = **c** ≠ **d** ≠

4 **a** True **b** False **c** True **d** False

Practice 3.1A

1 **a** True **b** False **c** True
d True **e** True **f** True

2 **a** False **b** True **c** True
d False **e** True **f** False
g True **h** False **i** True

3 Compare answers as a class

4 One number has been reduced by 1 and the other increased by 1 so the total remains the same

5 Add 100 and subtract 1 — Add 99
Add 1 and subtract 100 — Subtract 99
Subtract 1 and subtract 10 — Subtract 11
Subtract 10 and add 1 — Subtract 9
Subtract 1 and add 10 — Add 9

6 **a** 85 **b** 16 **c** 1
d 713 **e** 103 **f** 40

What do you think?

All three claims are true.
Compare answers as a class.

Practice 3.1B

1 **a** $7 + 10 = 17$, $10 + 7 = 17$, $17 - 7 = 10$, $17 - 10 = 7$
b $23 + 59 = 82$, $59 + 23 = 82$, $82 - 23 = 59$, $82 - 59 = 23$
c $4.6 + 3.8 = 7.4$, $3.8 + 4.6 = 7.4$, $7.4 - 4.6 = 3.8$, $7.4 - 3.8 = 4.6$
d $x + 53 = 81$, $53 + x = 81$, $81 - x = 53$, $81 - 53 = x$
e $2.6 + y = 11.3$, $y + 2.6 = 11.3$, $11.3 - 2.6 = y$, $11.3 - y = 2.6$
f $9.4 + a = b$, $a + 9.4 = b$, $b - 9.4 = a$, $b - a = 9.4$
g $p + q = r$, $q + p = r$, $r - p = q$, $r - q = p$

2 **a** $96 + 72 = 168$, $168 - 72 = 96$, $168 - 96 = 72$
b $17.3 - 8.5 = 8.8$, $8.8 + 8.5 = 17.3$, $8.5 + 8.8 = 17.3$
c $104 + c = 172$, $172 - c = 104$, $172 - 104 = c$
d $130 - 51 = d$, $d + 51 = 130$, $51 + d = 130$

3 **a** $3 \times 5 = 15$, $5 \times 3 = 15$, $15 \div 5 = 3$, $15 \div 3 = 5$
b $5 \times x = 30$, $x \times 5 = 30$, $30 \div 5 = x$, $30 \div x = 5$
c $3 \times 7.1 = 21.3$, $7.1 \times 3 = 21.3$, $21.3 \div 3 = 7.1$, $21.3 \div 7.1 = 3$
d $3 \times y = 280$, $y \times 3 = 280$, $280 \div y = 3$, $280 \div 3 = y$
e $2 \times a = 18.62$, $a \times 2 = 18.62$, $18.62 \div a = 2$, $18.62 \div 2 = a$
f $4 \times p = 1.7$, $p \times 4 = 1.7$, $1.7 \div 4 = p$, $1.7 \div p = 4$

4 **a** $5 \times 72 = 360$, $360 \div 5 = 72$, $360 \div 72 = 5$
b $80 \div 20 = 4$, $4 \times 20 = 80$, $20 \times 4 = 80$
c $c \times 5 = 108$, $108 \div 5 = c$, $108 \div c = 5$
d $x \div 103.4 = 7$, $103.4 \times 7 = x$, $7 \times 103.4 = x$
e $e \times 4 = 13.2$, $13.2 \div 4 = e$, $13.2 \div e = 4$
f $12 \div 3 = f$, $f \times 3 = 12$, $3 \times f = 12$

What do you think?

1 **a** $\frac{1}{3} + \frac{1}{2} = \frac{5}{6}$, $\frac{5}{6} - \frac{1}{3} = \frac{1}{2}$, $\frac{5}{6} - \frac{1}{2} = \frac{1}{3}$
b $0.5 - 0.3 = \frac{1}{5}$, $\frac{1}{5} + 0.3 = 0.5$, $0.3 + \frac{1}{5} = 0.5$
c $12 \times \frac{3}{4} = 9$, $9 \div 12 = \frac{3}{4}$, $9 \div \frac{3}{4} = 12$
d $x \div 2 = \frac{1}{3}$, $2 \times \frac{1}{3} = x$, $\frac{1}{3} \times 2 = x$

2 Compare answers as a class

Consolidate

1 **a** True **b** True **c** True
d False **e** True **f** False

2 **a** True **b** True **c** False
d True **e** True **f** True
g True **h** True **i** False

3 $42 = a \times 6$, $42 = 6 \times a$, $42 \div a = 6$, $42 \div 6 = a$

4

86	
49	37

$86 - 49 = 37$, $37 + 49 = 86$, $49 + 37 = 86$

5 **a** $7.8 + 6.7 = 14.5$, $6.7 + 7.8 = 14.5$, $14.5 - 6.7 = 7.8$, $14.5 - 7.8 = 6.7$
b $400 - 128 = 272$, $400 - 272 = 128$, $272 + 128 = 400$, $128 + 272 = 400$
c $6 \times 9.3 = 55.8$, $9.3 \times 6 = 55.8$, $55.8 \div 6 = 9.3$, $55.8 \div 9.3 = 6$
d $836 \div 11 = 76$, $836 \div 76 = 11$, $76 \times 11 = 836$, $11 \times 76 = 836$
e $a + 9.7 = 23.4$, $9.7 + a = 23.4$, $23.4 - 9.7 = a$, $23.4 - a = 9.7$
f $305 - b = 127$, $305 - 127 = b$, $b + 127 = 305$, $127 + b = 305$
g $c \times 15 = 57$, $15 \times c = 57$, $57 \div c = 15$, $57 \div 15 = c$
h $98 \div d = 140$, $98 \div 140 = d$, $140 \times d = 98$, $d \times 140 = 98$

Stretch

1 Yes, Junaid is correct

2 **a** 10 **b** 5 **c** $\div 2$ **d** $\times 50$
Compare methods as a class

Chapter 3.2

Are you ready?

1 **a** Equal **b** Not equal
c Equal **d** Not equal

2 **a** $+ 12$ **b** $- 12$ **c** $\times 12$ **d** $\div 12$

3 **a** 60 **b** 120 **c** 2700
d 3 **e** $\frac{1}{3}$

4 **a** 659 **b** 186 **c** 473

5 **a** 5893 **b** 71 **c** 5893

Practice 3.2A

1 **a** **i** $a + 86 = 215$
ii $a + 86 = 215$, $86 + a = 215$, $215 - 86 = a$, $215 - a = 86$
iii 129
b **i** $5b = 390$
ii $5b = 390$, $b \times 5 = 390$, $390 \div b = 5$, $390 \div 5 = b$
iii 78
c **i** $c + 137 = 502$
ii $c + 137 = 502$, $137 + c = 502$, $502 - c = 137$, $502 - 137 = c$
iii 365
d **i** $4d = 131$
ii $4d = 131$, $d \times 4 = 131$, $131 \div d = 4$, $131 \div 4 = d$
iii 32.75
e **i** $65.7 + 93.8 = e$
ii $65.7 + 93.8 = e$, $93.8 + 65.7 = e$, $e - 65.7 = 93.8$, $e - 93.8 = 65.7$
iii 159.5
f **i** $3f = 293.4$
ii $3f = 293.4$, $f \times 3 = 293.4$, $293.4 \div f = 3$, $293.4 \div 3 = f$
iii 97.8
g **i** $4g = 827$
ii $4g = 827$, $g \times 4 = 827$, $827 \div g = 4$, $827 \div 4 = g$
iii 206.75
h **i** $h + 53.8 = 112$
ii $h + 53.8 = 112$, $53.8 + h = 112$, $112 - h = 53.8$, $112 - 53.8 = h$
iii 58.2
i **i** $i + 61.6 = 104$
ii $i + 61.6 = 104$, $61.6 + i = 104$, $104 - i = 61.6$, $104 - 61.6 = i$
iii 42.4
j **i** $6j = 140.4$
ii $6j = 140.4$, $j \times 6 = 140.4$, $140.4 \div j = 6$, $140.4 \div 6 = j$
iii 23.4

2 Students' answers should be supported by a suitable diagram, e.g. a bar model, balance diagram or part-whole model.
a $a = 28$ **b** $b = 36$
c $c = 17$ **d** $d = 35.6$

3 **a** $a = 22$ **b** $b = 414$
c $c = 197$ **d** $d = 1.4$
e $e = 2482$ **f** $f = 0.244$

4 **a** $a = 59$ **b** $b = 111.5$
c $c = 12.12$

5 **a** $a = 11.5$ **b** $b = 10.7$
c $c = 18.4$ **d** $d = 1.23$

6 **a** $a = 9$ **b** $b = 85$
c $c = 721$ **d** $d = 91.2$

7 **a** $a = 0.15$ **b** $b = 15.7$
c $c = 648$ **d** $d = -2.2$
e $e = 9$ **f** $f = 29.4$
g $g = 120$ **h** $h = 1710$

What do you think?

1 Kate is incorrect. $a = 0$

2 Amina is incorrect. $b = 1$

3 E.g. $a = 10$, $b = 7$

4 **a** $a = 0$ **b** $b = 0$ **c** $c = 0$
d $d = 0$ **e** $e = 0$

5 Either f or g or both must be zero

Practice 3.2B

1 **a** $a = 11.4$ **b** $b = 18.98$
c $c = 8.6$ **d** $d = 84$
e $e = 45.9$ **f** $f = 20.5$

2 **a** $a = 696$ **b** $b = 7$
c $c = 17$ **d** $d = 888$
e $e = 12$ **f** $f = 7.5$
g $g = 324$ **h** $h = 1$

3 **a** $a = 42$ **b** $b = 31$
c $c = 0.3$ **d** $d = 0.8$
e $e = 204$ **f** $f = 67$
g $g = 80$ **h** $h = 1.48$

4 **a** $a + 23 = 81$, $a = 58$
b $b - 23 = 81$, $b = 104$
c $23 - c = 12$, $c = 11$
d $\frac{d}{23} = 12$, $d = 276$
e $23e = 483$, $e = 21$
f $\frac{23}{f} = 4.6$, $f = 5$

What do you think?

1 **a** $5 + 7 = 12$; there is nothing to solve
b $\frac{b}{3}$ is not an equation
c $12c$ is not an equation
d $d + 9$ is not an equation
e $f = g$ does not give enough information. f and g could both be 1, 2, 100, 0.7, etc.

2 For $x - 103 = 24$ you need to add 103 onto 24. For $103 - x = 24$ you need to find what number has been subtracted from 103, so you need to find 103 subtract 24.

3 For $\frac{g}{12} = 8$ you need to multiply 8 by 12. For $\frac{12}{g} = 8$ you need to divide 12 by 8.

Consolidate

1 **a** **i**

a	34
51	

ii

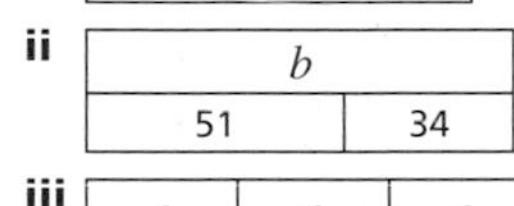

iii

c	c	c
255		

iv

d		
$\frac{d}{3}$	$\frac{d}{3}$	$\frac{d}{3}$
5.7		

b **i** $a = 17$ **ii** $b = 85$
iii $c = 85$ **iv** $d = 17.1$

2 **a** $5a = 190$, $a = 38$
b $3b = 85.8$, $b = 28.6$
c $c + 124 = 168$, $c = 44$
d $3d = 468$, $d = 156$
e $e + 47 = 513$, $e = 466$
f $f + 78 = 250$, $f = 172$
g $\frac{g}{4} = 12.8$, $g = 51.2$

3 **a** $x = 144$ **b** $x = 84$ **c** $x = 313$
d $x = 75$ **e** $x = 99$ **f** $x = 504$
g $x = 3.5$ **h** $x = 9$ **i** $x = 29$
j $x = 14.5$ **k** $x = 0.25$ **l** $x = 23.9$

4 **a** $a = 328$ **b** $b = 981$ **c** $c = 987$
d $d = 2952$ **e** $e = 981$
f $f = 987$ **g** $g = 328$

Stretch

1 **a** lw = area, $17w = 153$, $w = 9$, width = 9 cm
b l^2 = area, $l^2 = 529$, $l = 23$, length = 23 mm

2 Mario has divided 6 by 3, instead of multiplying 6 by 3

3 Examples could include equations of the form
$x - \square = \square$, $\square x = \square$,
$\square = \square x$, $\square - x = \square$,
$\frac{x}{\square} = \square$, $\frac{\square}{x} = \square$, etc.

Chapter 3.3

Are you ready?

1 They all include the letter a and the number 5, but they mean different things.
$5a$ means $5 \times a$, $\frac{5}{a}$ means $5 \div a$,
$\frac{a}{5}$ means $a \div 5$

2 They are different.
b^2 means $b \times b$, $2b$ means $2 \times b$

3 **a** 15 **b** 15 **c** 50
d $\frac{1}{2}$ **e** 2

Practice 3.3A

1 $3a$, $-5a$, $-3a$

2 **a** x^2 **b** 3 **c** $\frac{a}{b}$
d $-3t$ **e** $\frac{1}{2}$

3 **a** e.g. $-6d$, $\frac{1}{3}d$, $100d$, $\frac{d}{5}$, $-2.5d$
b e.g. $3v^2$, $-7v^2$, $\frac{3}{2}v^2$, $0.5v^2$, $\frac{7v^2}{4}$
c e.g. $5ab$, $\frac{ab}{6}$, $-2ab$, $\frac{2}{3}ab$, $3141ab$

4 4, 41, −4
$4a$, $-4a$, $\frac{1}{4}a$, $40a$
$4b$, $-14b$, $40b$, $-40b$, $\frac{1}{2}b$
$4ab$, $-4ab$, $-40ab$, $40ab$
$4a^2$, $-4a^2$
$4b^2$, $-4b^2$

5 **a** $2y$, $y + 2$, y^2, $2 + y$; $0.5y$, $\frac{y}{2}$
b $y + 2$, $2 + y$; $0.5y$, $\frac{y}{2}$
c $y + 2$, $2 + y$, $0.5y$, $\frac{y}{2}$

6 $5x$, $x + x + x + x + x$, $2x + 3x$
$x + 5$, $5 + x$
$\frac{1}{5}x$, $\frac{x}{5}$
$5 - x$, $x - 5$ and $\frac{5}{x}$ do not have matches

What do you think?

1 $3c$, $3 \times c$, $c \times 3$, $4c - c$, etc.

2 x^3 means $x \times x \times x$, but $3x$ means $3 \times x$. So they will not be equal for all values of x.

Consolidate

1 **a** e.g. $3n$, $\frac{n}{3}$, $0.4n$, $-19.4n$, $-\frac{7}{9}n$
b e.g. $7p$, $\frac{p}{5}$, $-\frac{3}{2}p$, $0.3p$, $-279p$
c e.g. $8a^3$, $\frac{a^3}{3}$, $-9a^3$, $0.5a^3$, $-5a^3$

2 $\frac{1}{2}b$, $-6b$, $-3b$, $\frac{b}{6}$

3 **a** Unlike **b** Like **c** Like
d Like **e** Unlike **f** Unlike
g Like **h** Unlike

4 **a** $3m$, $m + m + m$, m^2, $6 + m$; $3 - m$, $m - 3$
b $3m$, $m + m + m$; m^3, m^2
c $3m$, $m + m + m$

5 $2t$, $3t - t$, $t + t$
t^2, $t \times t$
$0.5t$, $\frac{t}{2}$
$t + 2$, $2 + t$
t does not have a match

Stretch

1 ab is equivalent to ba because $a \times b$ is equivalent to $b \times a$. So, they are like terms.

2 **a** If $t = 1$, then $2t - t = 2(1) - 1$, which equals 1 not 2
b $t = 2$

Chapter 3.4

Are you ready?

1 e.g. $5a$, $\frac{a}{3}$, $-0.9a$

2 **a** Like **b** Like
c Unlike **d** Like

3 **a** $3y$ **b** b^2 **c** $3bc$

4 **a** 9 **b** 7 **c** 9 **d** 3

Practice 3.4A

1 **a** $7t$ **b** $6x$ **c** $9h$ **d** $4y$
e $3n$ **f** a^2 **g** ab **h** $3g$
i fg **j** g^2

2 **a** **i** $8a$ **ii** $2a$ **iii** $2ab$
iv $2a$ **v** $4a$ **vi** $12a$
vii $8a$ **viii** $4a$
ix 0 **x** $5ab$
b **i** $15a$ **ii** $15a$ **iii** a^2
iv $2a^2$ **v** $2ab$

3 $m \times 4$, $3m + m$, $2m \times 2$, $6m - 2m$, $2 \times 2 \times m$, $8m - 6m + 2m$

4 **a** $a \times a \times a$ **b** $2b + 1$
c $c \times c$ **d** $9d - 6d$

5 **a** False. $3t - t$ means $3t$ subtract $1t$, so the answer is $2t$ not t.
b True. $4 \times y$ is the same as $y + y + y + y$.
c False. $5n$ subtract $1n$ equals $4n$ not 5.
d True. $4x$ and $3x$ are like terms so you can just add the coefficients.
e True. $2 \times 5p \equiv 10p$ and $6p + 4p \equiv 10p$, so $2 \times 5p \equiv 6p + 4p$.
f False. $7p$ add $3p$ equals $10p$ in total. There are no terms in p^2.
g False. $6ab - b$ cannot be simplified because ab and b are unlike terms.

6 **a** $10a$ **b** $4b$ **b** $4c$
b $6d^2$ **e** $4x$

What do you think?

1 $0a$ is equivalent to 0, but 0 is the simplest form. So Flo is correct.

2 e.g. $2p + 4p$, $10p - 4p$, $p + p + p + p + p + p$, $5p + p$, $7p - p$, $p \times 6$, $6 \times p$, $12p \div 2$, $6p^2 \div p$, $1.5p \times 4$

3 $2ab + 3ba \equiv 5ab$. This is because multiplication is commutative, so $3ba$ is equivalent to $3ab$.

Practice 3.4B

1 **a** $8a + 4b$ **b** $2a + 4b$ **c** $5a + 7b$
d $5a + b$ **e** $8a + 6b$ **f** $2a + 6b$
g $2a + 2b$ **h** $7a + 7b$ **i** $7a + b$
j $3a + b$ **k** $6a + 8b$ **l** $2a + 8b$
m $8b - 2a$ **n** $4a + 8b + 2$
o $4a + 2b + 2$ **p** $8b + 6$
q $2b + 6$ **r** $2b + 2$
s $5a + 3b + 6$ **t** $5a + 3b + 2$

2 **a** $8p + 6q$ **b** $2p + q$

3 $3x$ and $4x$ are like terms, but $2x^2$ is unlike them. So the correct answer is $2x^2 + 7x$.

4 **a** $3x + 6$ **b** $5x + 4$ **c** $9x$
d $7x + 2$ **e** $x^2 + 3x$ **f** $x^2 + 4x$
g $2x^2 + 4x$ **h** $3x^2 + 4x$
i $2x^2 + 4x + 2$ **j** $4x^2 + 4x$
k $2x^2 + 8x$ **l** x^2
m $4x^2 + 2x$ **n** $6x$
o $2x^2 + 8$ **p** $2x^2$

5 **a** False **b** False **c** False
d False **e** False **f** True
g False **h** True **i** False
j True

What do you think?

1 **a** m **b** n **c** $5a^2$ **d** $3ab$

2 e.g. $3 \times a + 4 \times b$, $a \times 3 + b \times 4$, $b \times 4 + 3 \times a$, $2a + a + 4b$, $5a - 2a + 4b$, $a + a + a + b + b + b + b$, $3a + 10b - 6b$, $10a - 7a + 4b$, $4a - a + 5b - b$, $6a \div 2 + 20b \div 5$

Consolidate

1 **a** $4a$ **b** $4b$ **c** $9c$ **d** d^2
e $3e$ **f** $7f$ **g** gh **h** j^3
i $6mn$

2 **a** **i** $10a$ **ii** $2a$ **iii** $4ab$
iv $5a$ **v** $7a$ **vi** $12ab$
vii $8ab$
b **i** $24a$ **ii** $24a$ **iii** ab
iv $2ab$ **v** $2ab$

3 $3n - n$, $n \times 2$, $n + n$, $4n \div 2$, $2 \times n$

4 Three x subtract two x is equivalent to a single x. So Jackson and Chloe are both correct, but Chloe's answer has been fully simplified, as $1x$ is equivalent to x.

5 **a** $4x + 2y$ **b** $2x + y$ **c** y
d $2x + y$ **e** $5x + 5y$ **f** $x + 5y$
g $x + 3y$ **h** $x + 2y$ **i** $5xy + 5y$
j $5xy + x$ **k** $5xy + x$ **l** $xy + 2y$

6 $3x^2 + 3x$, $3x + 3x^2$

7 **a** $12a$ cm **b** $(6x + 9y)$ cm
c $4n$ cm

8 **a** $4a$ **b** $9b + 7c$ **c** $5d^2 + 2d$
d $7ab + 3a + 2b$ **e** $2x^2 + 4x$

Stretch

1 $x = 2$ and $x = 0$
Generally, $x = k$ and $x = 0$

2 If $y = 5$, $2y + 10 = 2(5) + 10 = 20$ and $2(y + 5) = 2(5 + 5) = 20$
If $y = 7$, $2y + 10 = 2(7) + 10 = 24$ and $2(y + 5) = 2(7 + 5) = 24$
If $y = 4.5$, $2y + 10 = 2(4.5) + 10 = 19$ and $2(y + 5) = 2(4.5 + 5) = 19$
The expressions are equivalent, because $2(y + 5)$ means two groups of $(y + 5)$. This is equal to two groups of y added to two groups of five.
This can be written as $(2 \times y) + (2 \times 5) = 2y + 10$.
So $2(y + 5) \equiv 2y + 10$

Check my understanding

1 All of the statements are true

2 **a** $19 \times 72 = 1368$, $72 \times 19 = 1368$, $1368 \div 19 = 72$, $1368 \div 72 = 19$
b $2128 \div 38 = 56$, $2128 \div 56 = 38$, $38 \times 56 = 2128$, $56 \times 38 = 2128$
c $93 + y = x$, $y + 93 = x$, $x - y = 93$, $x - 93 = y$
d $48 - b = c$, $48 - c = b$, $c + b = 48$, $b + c = 48$

3 **a** **i** $3a = 86.1$ **ii** $4b = 4136$
iii $c + 47.7 = 84.2$
b **i** $a = 28.7$ **ii** $b = 1034$
iii $c = 36.5$

4 **a** **i** $\frac{x}{12} = 3.6$ **ii** $\frac{360}{x} = 12$
iii $36 - x = 19$
b **i** $x = 43.2$ **ii** $x = 30$
iii $x = 17$

5 **a** $5d$ **b** b^3 **c** $\frac{c}{2}$

6 $4m \div 2$, $m \times 2$, $5m - 3m$, $m + m$

7 **a** $5a + 4b$ **b** $a + 4b$
c $7a + a^2$ **d** $2a + 4a^2$
e $9ab$

Block 4 Place value and ordering

Chapter 4.1

Are you ready?

1 7, 30 638, 11 million

2 **a** One thousand and thirty-five
b 1035

3 **a** 600 **b** 6 **c** 60 **d** 6000

4 **a** 6 **b** 90, 5 **c** 16 508

5 **a** Tens **b** Hundreds
c Twos

Practice 4.1A

1 **a** 5453 **b** 14 007
c 1 200 314

2 **a**

Ones		
H	T	O
●●●●●●	●	●●

b

Thousands			Ones		
H	T	O	H	T	O
		●●			●●●●●

c

Thousands			Ones		
H	T	O	H	T	O
	●	●●●●●	●●●	●●●●	

d

Thousands			Ones		
H	T	O	H	T	O
●●●					

3 **a** 20 **b** 20 **c** 20 **d** 20

4 **a** e.g. 7560 **b** e.g. 37 560
c No

5 **a** He has not considered the place value columns.
b 20 018

6 **a** 9 **b** 3 **c** 2 **d** 0

7 **a** 563 291 **b** 280 426
c 3 107 092 **d** 3 107 092

8 **a** 20 000 + 3000 + 400 + 50 + 2
b 600 000 + 90 000 + 2000 + 7
c 4 000 000 + 100 000 + 3000 + 500 + 60
d 10 000 000 + 5 000 000 + 300 000

What do you think?

1 e.g. 206 354

Practice 4.1B

1 **a** Three hundred and fifty thousand, seven hundred and twenty-five
b Five hundred thousand, three hundred
c Twenty-six million, nine hundred and eleven thousand, one hundred and three
d One hundred and thirty-six million, two hundred and forty-eight thousand, three hundred and thirty-three

2 **a** Seven hundred and fifty-one
b Seven hundred and fifty-one thousand
c Seven hundred and fifty-one million

3 **a** Two hundred and three
b Four thousand, two hundred and three
c Fourteen thousand, two hundred and three
d Three hundred and fourteen thousand, two hundred and three

4 **a** Three thousand, five hundred and two
b Seventy-three thousand, one hundred and twelve
c Nine hundred and three thousand, seven hundred
d Seventeen million, four hundred and twenty thousand and thirty

5 **a** 36 000 **b** 157 000 000
c 384 400
d 1 000 386 000 and 9 600 000

What do you think?

1 e.g. 72 136

Practice 4.1C

1 **a** 540 and 595 **b** 4300 and 4650
c 65 000 and 67 250

2 **a** Tens **b** Fifties
c Hundreds **d** Tens

3 **a** 2520, 2550 and 2569
b 2569 is an estimate; can't be that certain

4

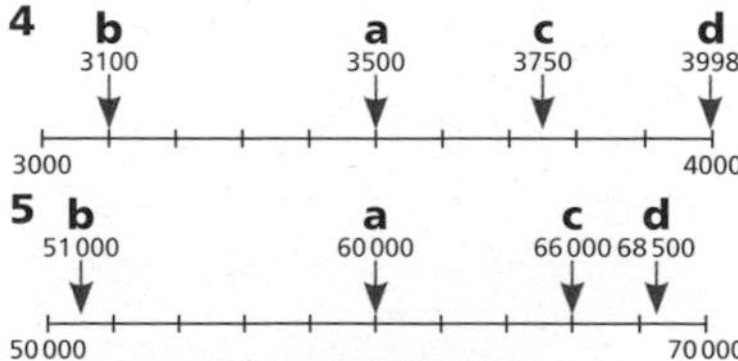

5

6 **a** Approximately 4180
b Approximately 228 000
c Approximately 3 255 000

What do you think?

1 **a** It is more than halfway, and 3350 is less than 3500
b A and B

c Divide the increment between A and B into four equal parts and select the third mark that is added

Consolidate

1 a 5 **b** 50 **c** 500
d 5 **e** 5000
2 a Thirty-two
b Six hundred and thirty-five
c Two thousand, four hundred
d Ten thousand, seven hundred and forty
e Twelve thousand, five hundred and three
3 a 57 **b** 220 **c** 332

Stretch

1 50 □□ and the two missing digits must add up to 7 so that the digit total is 12
5070, 5061, 5052, 5043, 5034, 5025, 5016, 5007
2 Harry could put the number in a place value grid.
The number is fifteen million, seven hundred and thirty-two thousand, nine hundred and eighteen
3 a

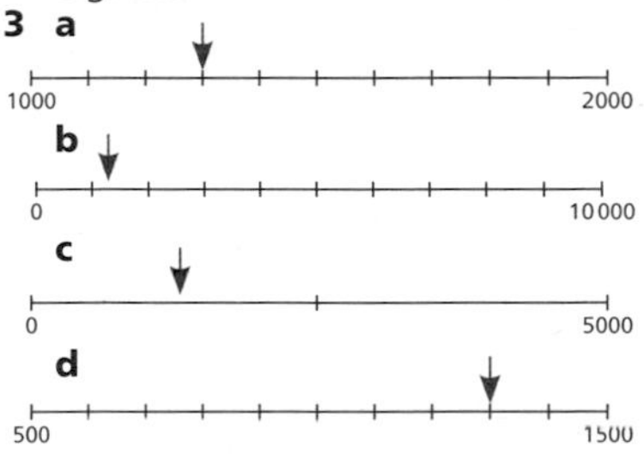

Chapter 4.2

Are you ready?

1 a 30 **b** 220 **c** 103 **d** 4000
2 a greater than **b** less than
c greater than **d** equal to
3 a 7, 17, 72 **b** 95, 100, 105
c 19, 91, 109
d 150, 1005, 1500
e Ascending
4 a 97, 96, 79 **b** 1000, 200, 30
c 190, 117, 109 **d** 1111, 111, 11
e Descending

Practice 4.2A

1 a 56 is greater than 46
b 90 is equal to 9 tens
c 104 is less than 105
d 90 is not equal to 9 ones
2 a > **b** < **c** > **d** <
3 a > **b** > **c** <
d > **e** < **f** <
4 a < **b** > **c** >
d = **e** > **f** <
5 a = **b** ≠ **c** ≠
d = **e** = **f** ≠
6 e.g. 104 007, 104 105, 104 106
7 a 0 or 1 **b** 8 or 9 **c** 9
d any number from 0 to 9

What do you think?

1 Flo is correct.
2 Faith is correct. All numbers starting 14 thousand will be less than a number starting 15 thousand.

Practice 4.2B

1 a 64 302, it is the only one with '6' in TTh column.
b i 64 302, 73 452, 74 221, 74 302
ii 74 302, 74 221, 73 452, 64 302
2 a 3265, 3282, 3350, 3611
b 999, 1062, 1521, 2655
c 5007, 5699, 5700, 5701, 5770
d 9000, 9118, 10 402, 10 420, 12 350
3 a 7220, 7200, 7105, 6980
b 5497, 5487, 5477, 5467, 5427
c 338 100, 33 810, 33 801, 3381
d 45 030, 29 760, 12 832, 8950, 979
4 a February, January, April, June, March, May
b 3600
5 Mount Everest, K2, Kangchenjunga, Lhotse, Makalu, Cho Oyu
6 a A **b** C
c Because there are two houses with the same price

What do you think?

a 0 or 1
b Any digit from 0 to 9
c 4
d 4, 5 or 6, or can be 7 if **e** is 9
e Any digit from 0 to 9 if **d** is 4, 5 or 6; or 9 if **d** is 7

Consolidate

1 a less than **b** greater than
c less than **d** greater than
e equal to **f** less than
2 a < **b** > **c** <
d > **e** = **f** <
3 a 661, 835, 974, 4365, 7299
b 8050, 8200, 8500, 9010, 9500
c 67, 73, 113, 301, 550, 703
d 12 095, 12 300, 12 320, 12 341, 12 355
4 a 470, 417, 411, 407, 401
b 5222, 5112, 5012, 512, 502
c 57 900, 57 888, 57 399, 57 000, 56 999
d 140 000, 110 400, 101 400, 100 400, 100 040

Stretch

1 a China, India, United States, Indonesia, Pakistan
b India, China, United States, Indonesia, Pakistan
c India, Pakistan, United States, Indonesia, China
2 $100c + 10b + a$, $100c + 10a + b$, $100b + 10a + c$, $100a + 10b + c$
3 $100x$, $10x$, $x + 17$, x

Chapter 4.3

Are you ready?

1 a 72 **b** 112 **c** 1050
2 a 34 **b** 9 **c** 150
3 a 19 **b** 57 **c** 22 **d** 60
4 a 341, 345, 502, 712
b 5, 71, 423, 2010
5 a 56 **b** 250 **c** 18 **d** 28

Practice 4.3A

1 a 352 g **b** 352 g
2 a She described the range but did not work it out by finding the difference.
b 103 miles
3 1222
4 23°C
5 a The values are not in order, so range isn't final value in table less first value in table
b The values have different units
6 51
7 2 minutes 6 seconds
8 a 20 loaves **b** 20 loaves

What do you think?

1 a It can take any value between 138 cm and 155 cm inclusive
b Either 157 cm or 136 cm

Practice 4.3B

1 a £25 **b** 104 mm
c **d** 85 g
2 17 points
3 68 cm
4 1678 issues
5 a 8343 m **b** 27 372 feet
6 1800 g
7 a Any number less than 12
b 10

What do you think?

1 Basketball players are taller.
2 a e.g. 1, 7, 7, 7, 7
b e.g. 3, 6, 8, 9

Consolidate

1 a 5 **b** 70
c £100 **d** 75 g
2 14
3 75
4 a 8 **b** 29
c 19 kg **d** 195 m

Stretch

1 a 33 and 46 **b** 34.5

Chapter 4.4

Are you ready?

1 b 70 and 80 **c** 90 and 100
d 270 and 280
2 b 700 and 800
c 1700 and 1800
d 3900 and 4000
3 a 150 **b** 100

Practice 4.4A

1 a 380 **b** 400
2 a 6000 **b** 5800 **c** 5780
3 He should have looked at the tens column
4 a 170, 218, 150
b 722, 715
c 3106, 3095, 2785, 2961
d 2961

5 a 2000, 1700, 1730
b 26 000, 26 400, 26 380
c 173 000, 172 500, 172 540

6

Number	Rounded to nearest 1000	Rounded to nearest 10 000	Rounded to nearest 100 000
265 380	265 000	270 000	300 000
412 119	412 000	410 000	400 000
76 289	76 000	80 000	100 000

7 a 0, 1, 2, 3 or 4 b Any digit
c 5 d 5, 6, 7, 8 or 9

8 Yes, because the value of the hundreds digits affects the rounding, but the tens and units digits do not.

9 a e.g. 623 b e.g. 596
c e.g. 2693 d e.g. 2965
e e.g. 59 632

10 Various answers

What do you think?

1 a No b 64 999 c 55 000

2 a Various answers
b Depends on part **a** answers

Consolidate

1 a 70 b 80 c 90
d 100 e 380 f 210
g 760 h 900 i 7230
j 1000 k 35 250 l 254 880

2 a 800 b 800 c 100
d 1000 e 3400 f 8200
g 10 300 h 735 700

3 a 4000 b 7000 c 1000
d 10 000 e 35 000 f 71 000
g 44 000 h 231 000

4 a 20 000 b 70 000 c 20 000
d 10 000 e 530 000 f 100 000
g 1 390 000 h 1 000 000

Stretch

1 4.85 kg

2 A is less than or equal to 4; B could be any number

3 A has to be 3 B can be 0 or 1
If B is 0, C has to be 9; if B is 1, C has to be 0
D depends on the values of B and C:
- if ABC is 309 then D has to be greater than or equal to 5
- if ABC is 310 then D has to be less than or equal to 4

E and F can be any numbers

Chapter 4.5

Are you ready?

1 a 7 ones b 7 ones
c 7 hundreds d 7 tenths

2 a 0.2 b 90 + 5 + 0.3

3 a 0.3 b 0.03

4 10

5 5.9 plus one-tenth is 6.0 not 5.10 (5.10 is the same as 5.1)

Practice 4.5A

1 a 11.71 b 1.05
c 23.156 d 203.6

2 a 1 × 1 counter, 5 × 0.1 counters
b 1 × 1 counter, 5 × 0.1 counters, 2 × 0.01 counters
c 2 × 10 counters, 6 × 1 counters, 7 × 0.1 counters, 3 × 0.01 counters
d 1 × 10 counter, 3 × 1 counters, 3 × 0.01 counters, 5 × 0.001 counters

3 a 2.3 b 2.37 c 2.8

4 a 13.7, 15.72
b Any two numbers with 7 tenths

5 a 61.13, 0.137
b Any two numbers with 3 hundredths

6 No, Flo is not correct. 213.65 has 5 hundredths, the second digit after the decimal is the hundredths.

7 a 6 tenths b 6 tenths
c 6 tenths d 6 hundreds
e 6 tens f 6 ones
g 6 tenths h 3 hundredths
i 4 thousandths

8 a 269.3 b 7205.8
c 920.14 d 1720.9
e 60.06 f 0.6 + 0.02
g 0.2 h 2

9 Huda is correct, as 10 × 0.1 counters is the same as 1 × 1 counter

10 a i 50 ii 70 iii 72 iv 195
b i 300 ii 800 iii 810 iv 815
c 70

What do you think?

1 a 5 is 5 tenths, 0 is 0 hundredths, 8 is 8 one-thousandths, 1 is 1 ten-thousandth, 9 is 9 hundred-thousandths, 2 is 2 one-millionths, 3 is 3 ten-millionths
b They all follow the pattern of one-thousandths, ten-thousandths, hundred-thousandths, one-millionth, ten-millionths, hundred-millionths.

2 As 0.1 is one tenth of £1, or 10p, the 6 is six tenths of £1, or 60p

Practice 4.5B

1 a 2.4, 2.7, 2.9
b 3.62, 3.63, 3.66, 3.68
c 1.853, 1.854, 1.855, 1.857, 1.859

2 a 12.5 b 3.75 c 0.245

3 A is 62.1, B is 62.6, C is 62.85

4 A is 7.63, B is 7.67, C is 7.695

5 Emily is correct

6

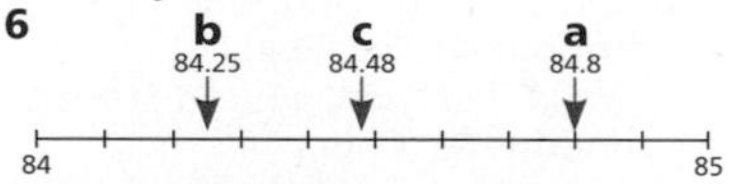

7 a e.g. 0.365, 0.37, 0.375 (less than 0.385 but greater than 0.36)
b e.g. 0.361 11, 0.369 25 (less than 0.385 but greater than 0.36 and with five digits after the decimal point)

8 a 0.78 kg b 2.75 kg

9 a 21.9°C b 21.4°C

What do you think?

1 a 7.855, 7.8575
b 7.856 is 0.001 more than 7.855 but 0.0015 less than 7.8575
c Yes. It does not matter how many digits you add on after the '6', as 7.856*** will always be less than 7.857, as we compare the thousandths digit.

2 a $\frac{3}{10}$ b Just over $\frac{8}{10}$

Consolidate

1 a 13.7
b 33.7, 15.7, 13.9

2 a 3 is 3 ones, 7 is 7 tenths
b 4 is 4 ones, 5 is 5 tenths, 8 is 8 hundredths, 1 is 1 one-thousandth
c 2 is 2 hundreds, 8 is 8 tens, 7 is 7 ones, 3 is 3 tenths, 5 is 5 hundredths

3 a 3 hundredths
b 9 ones
c 4 hundreds d 9 tenths
e 7 thousands
f 9 one-thousandths

4 There are ten hundredths in every one-tenth, so 8 tenths is 80 hundredths.

5 a 2.4, 2.7, 2.95
b 2.25, 2.5, 2.875

Stretch

1 A total of 8 numbers: 10.54, 10.63, 10.72, 10.81, 12.07, 12.34, 12.43, 12.70

2 a

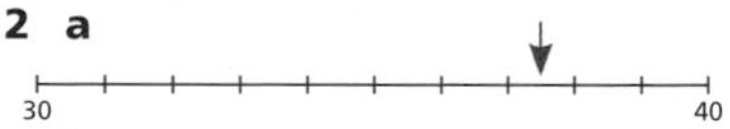

b

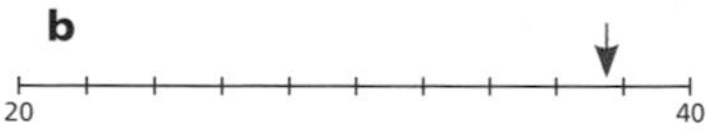

c

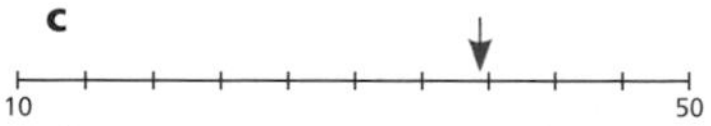

d

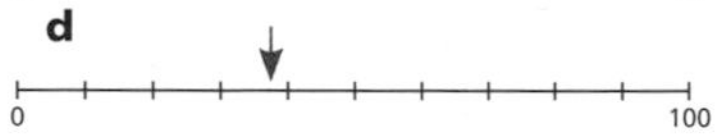

Chapter 4.6

Are you ready?

1 a 740 b 9120 c 140 d 1000

2 a 700 b 48 800 c 100
d 2000

3 Missing digit could be 0, 1, 2, 3 or 4, giving the number as 72 065, 72 165, 72 265, 72 365 or 72 465

4 a 25 b 30 c 25.20
d 25.195 18

Practice 4.6A

1 a 9 (hundreds) b 1 (ones)
c 2 (one-thousandths)

2 a 2 (tens) b 8 (tenths)
c 5 (ten-thousandths)

3 a 40 b 40 000 c 0.4

4 Emily has forgotten that the answer must be the same order of magnitude as the original number. The correct answer is 300.

5 a 40 b 2
c 3000 d 300 000
e 0.4 f 200
g 2

6 The first significant figure will not always be in the thousands column

7 a 50 000 b 5000 c 500
d 50 e 5 f 0.5
g All the numbers start with 5. This is because the second significant figure is 7, so you round the 4 to 5.

8 a 600 b 2

9 a $200 \times 4 = 800$
b $\frac{2000 \times 1000}{2} = 1\,000\,000$

10 553, 563, 573, 583 or 593. The 2nd digit must be 5 or greater.

What do you think?

1 Rounding a large number to the nearest 10, say, would still include many non-zero digits. The answer will always be the same order of magnitude as the original number. This is the same for numbers less than 1 which might round to 0 if very small.

2 25 000 and 34 999

3 $7.5 \times 1.5 = 11.25\,\text{cm}^2$

Consolidate

1 a 500 b 3000
c 6 d 0.0001

2 a 10 000 b 200 c 10 000

3 $90\,\text{m}^2$

Stretch

1 a Compare answers as a class
b You get the same answer both times, 20 000. To be of the right order of magnitude, both numbers need all the zeros.
c 0.2 and 0.20. This time you need the extra zero to show that the answer is not to 1 significant figure.

2 It could be any number between 99 950 and 100 049.

Chapter 4.7

Are you ready?

1 a 100 b 49 c 32 d 125

2 a 1 000 000 b 200 000
c 90 000 d 30 000 000

3 a > b > c < d <

4 a 90 000 b 80 000 c 900 000

Practice 4.7A

1 a 10^2 b 10^3 c 10^4 d 10^5
e The power is the same as the number of zeros

2 Agree. Because there are 8 zeros the power will be 8.

3 No; 10^7 is equal to 10 000 000 or 10 million not 1 billion

4 a 100 000 000 b 1000
c 100 000 000 000

5 a 10^8 b 10^{10}

6 10^{100}

7 a 1×10^5 b 2×10^5
c 3×10^5 d 7×10^5

8 a 2×10^3 b 3×10^2
c 9×10^4 d 5×10^9
e 3×10^6 f 8×10
g 7×10^8 h 6×10^{22}

9 a 30 000 b 300 000
c 3 000 000 d 30 000 000
e The digit 3 stays the same but the number of zeros and the size of the number is different

10 a 10 000 b 6000
c 90 000 000 d 80 000
e 900 000 000 000

11 Faith is wrong. She has compared the integers and ignored the powers.

12 a > b < c <
d < e >

13 4×10^7

What do you think?

1 a $10^5 + 10^3$ b $10^6 + 10^2$
c $10^3 + 10^2 + 10^1$

2 10^{12}

3 Yes, because they are both 10 000

4 a It must be less than 7
b It must be less than 5
c It could take any value less than 5000

5 a 6×10^4 b 7×10^6 c 8×10^7

Consolidate

1 a 3×10^4 b 1×10^4 c 9×10^3
d 7×10^6 e 5×10^6 f 6×10^{12}
g 7×10^{16} h 9×10^8

2 a 70 000 b 800 000
c 3 000 000 d 50
e 70 000 000 000
f 9 000 000 000 000

Stretch

1 10^{19}

2 Any number less than 4

3 a $6 \times 10^{24}\,\text{kg}$ b 3×10^9 hours

4 a 400 000
b 3 000 000 000 000
c 30 000

Chapter 4.8

Are you ready?

1 a 100 b 1000 c 100 000

2 a 10^4 b 10^6 c 10^8

3 a 0.1 b 0.01 c 0.001

4 10^6, 10^4, 10^3, 10

5 a Because 3 is being multiplied by 10 000 which is the same as 10^4
b 400 000 c 6

Practice 4.8A

1
10^2 — 0.01
10^1 — 100
10^0 — 0.1
10^{-1} — 10
10^{-2} — 1

2

	10^{-1}	10^{-2}	10^{-3}	10^{-4}	10^{-5}
Fraction	$\frac{1}{10}$	$\frac{1}{100}$	$\frac{1}{1000}$	$\frac{1}{10\,000}$	$\frac{1}{100\,000}$
Decimal	0.1	0.01	0.001	0.0001	0.000 01

The power matches the number of zeros

3 The answer should be 0.001. He has forgotten that a negative power gives a decimal number.

4 a 0.001 b 0.0001
c 0.000 01 d 0.000 001
e 0.000 000 1
f They are all decimals with digits zero and 1. The number of zeros after the decimal point is one less than the power.

5 a 10^4 b 10^5 c 10^8

6 a 10^{-3} b 10^{-4} c 10^{-5} d 10^{-6}

7 a 10^{-3} b 10^{-5} c 10^{-8} d 10^{-6}

8

Numbers greater than 1	Numbers between 0 and 1	Numbers less than 0
10^3	One hundredth	
10^1	10^{-5}	
	A thousandth	
	10^{-1}	

9 10^{-2}, one tenth, 10^2, 10^5

10 a 0.002 b 0.003 c 0.004
d 0.005

11 a 0.07 b 0.007 c 0.0007
d 0.000 07 e 0.000 000 07

12 3×10^2, 6×10^{-2}, 5×10^{-4}, 3×10^{-4}

What do you think?

1 As the positive powers decrease the number of zeros in the number decrease and therefore the number gets smaller. As soon as the power becomes negative the ordinary number is a decimal.

2 Emily has got the correct number of zeros but then mistakenly wrote a 5 instead of a 1. Abdullah has an extra zero. He has also made the same mistake as Emily.

3 a 10^3 b 10^{-3}
c The powers are the same apart from one is positive and one is negative.

Consolidate

1 a 10^{-3} b 10^{-6} c 10^{-7}

2 a $\frac{1}{100}$ and 0.01
b $\frac{1}{10\,000}$ and 0.0001
c $\frac{1}{100\,000}$ and 0.000 01
d $\frac{1}{10\,000\,000}$ and 0.000 000 1
e $\frac{1}{10\,000\,000\,000}$ and 0.000 000 000 1

3 Because 1 ten-thousandth has 4 zeros.

Stretch

1 a 8×10^{-3} b 4×10^{-5}
c 7×10^{-1} d 9×10^{-4}

e 7×10^{-9} f 8×10^{-5}

2 a 10^6 b 10^{-6}

c The digits are the same but one power is negative and one is positive

3 7×10^{-9}

4 It follows the pattern of powers of 10, so if $10^1 = 10$, then 10^0 must be 1

Check my understanding

1 a Seven million
b Seven hundred
c Seven thousand
d Seven million
e Seven-hundred million

2 a < b = c > d <

3 a 14 b 28 c 86

4 a 17 b 29 c 508.5

5 a 3 b 0.3 c 0.003
d 300 e 0.03

6 a 8730, 8700, 9000
b 96 410, 96 400, 96 000
c 123 690, 123 700, 124 000
d 10 000, 10 000, 10 000
e 76 600, 76 600, 77 000

7 a 9000 b 100 000
c 100 000 d 10 000
e 80 000

8 a 5×10^5 b 3×10^7
c 5×10^{-3} d 3×10^{-5}

Block 5 Fractions, decimals and percentages

Chapter 5.1

Are you ready?

1 a 0.1 b 0.01 c 0.8 d 0.03

2 a $\frac{3}{4}$ b $\frac{7}{10}$ c $\frac{5}{6}$

3 a 6 tens b 6 thousands
c 6 tenths d 6 tenths

4 Any number with 7 tenths, e.g. 3.75

5 Any number with 1 tenth and 3 hundredths, e.g. 23.134

Practice 5.1A

1 a 3 tenths, $\frac{3}{10}$ b 7 tenths, $\frac{7}{10}$
c 5 tenths, $\frac{5}{10}$ d 4 tenths, $\frac{4}{10}$
e 3 tenths, $\frac{3}{10}$ f 1 tenth, $\frac{1}{10}$

2 a 7 hundredths, $\frac{7}{100}$
b 8 hundredths, $\frac{8}{100}$
c 36 hundredths, $\frac{36}{100}$
d 7 hundredths, $\frac{7}{100}$
e 9 hundredths, $\frac{9}{100}$
f 4 hundredths, $\frac{4}{100}$

3 a A: $\frac{8}{100}$ and B: $\frac{6}{10}$
b A: $\frac{92}{100}$ and B: $\frac{4}{10}$
c The shaded and non-shaded parts add up to $\frac{100}{100}$ and $\frac{10}{10}$

4 a 20 squares shaded
b 2 squares shaded
c 7 squares shaded
d 40 squares shaded
e 10 squares shaded
f 100 squares shaded

5 b 7 tenths and 4 hundredths, 74 hundredths
c 3 tenths, 30 hundredths

6 a
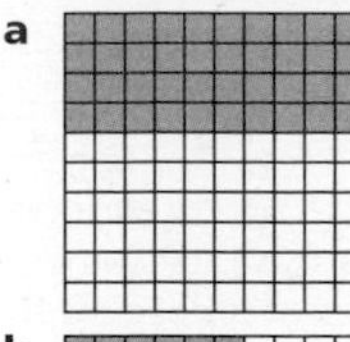

b
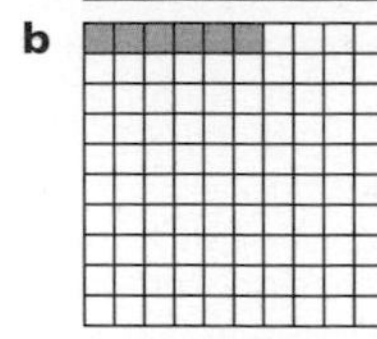

c
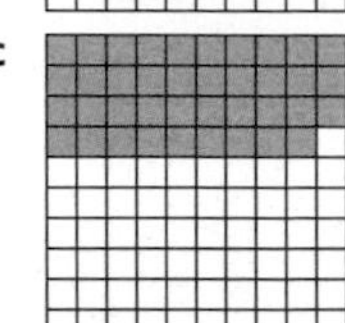

What do you think?

1 Both are correct. Sven has represented it using the shaded squares, Rhys has represented it using non-shaded squares.

2 No. One of the squares overlaps the row and column so only 1 tenth and 9 hundredths are shaded.

3 a She has shaded 7 tenths (7 whole columns) and 9 hundredths (9 single squares).
b 41 squares shaded

Practice 5.1B

1 a
0.1 0.2 0.3 0.4 0.5 0.6 0.7 0.8 0.9
0 $\frac{1}{10}$ $\frac{2}{10}$ $\frac{3}{10}$ $\frac{4}{10}$ $\frac{5}{10}$ $\frac{6}{10}$ $\frac{7}{10}$ $\frac{8}{10}$ $\frac{9}{10}$ 1

b
0.01 0.02 0.03 0.04 0.05 0.06 0.07 0.08 0.09
0 $\frac{1}{100}$ $\frac{2}{100}$ $\frac{3}{100}$ $\frac{4}{100}$ $\frac{5}{100}$ $\frac{6}{100}$ $\frac{7}{100}$ $\frac{8}{100}$ $\frac{9}{100}$ $\frac{1}{10}$

2 a 9 b $\frac{9}{10}$ c $\frac{9}{100}$

3
$\frac{1}{10}$ $\frac{2}{10}$ $\frac{3}{10}$ $\frac{4}{10}$ $\frac{5}{10}$ $\frac{6}{10}$ $\frac{7}{10}$ $\frac{8}{10}$ $\frac{9}{10}$
0 $\frac{10}{100}$ $\frac{20}{100}$ $\frac{30}{100}$ $\frac{40}{100}$ $\frac{50}{100}$ $\frac{60}{100}$ $\frac{70}{100}$ $\frac{80}{100}$ $\frac{90}{100}$ 1

4 a $\frac{53}{100}$, 0.53 b $\frac{37}{100}$, 0.37
c $\frac{46}{100}$, 0.46 d $\frac{641}{100}$, 6.41

5 a 30 b 3 c 100

6 a $0.7 = \frac{7}{10}$ b $0.07 = \frac{7}{100}$
c $0.9 = \frac{9}{10}$ d $0.17 = \frac{17}{100}$
e $\frac{8}{10} = 0.8$ or equivalent
f $0.04 = \frac{4}{100}$ or equivalent

What do you think?

1 They are all correct.

2 The scale increases by 0.2 not 0.1, so the arrow is pointing to 3.2

Consolidate

1 a 0.4 b 0.9 c 0.5

2 a $\frac{8}{10}$ or $\frac{4}{5}$ b $\frac{1}{10}$ c $\frac{2}{10}$ or $\frac{1}{5}$

3 a 0.03 b 0.06 c 0.07

4 a $\frac{1}{100}$ b $\frac{9}{100}$
c $\frac{6}{100}$ or $\frac{3}{50}$

5 b $\frac{60}{100}$ c $\frac{30}{100}$ d $\frac{70}{100}$

6 a ii $\frac{4}{10} + \frac{7}{100}$ iii $\frac{5}{10} + \frac{1}{100}$
iv $\frac{9}{10} + \frac{9}{100}$
v $\frac{2}{10} + \frac{8}{100}$ vi $\frac{7}{10} + \frac{4}{100}$
b i 0.23 ii 0.47 iii 0.51
iv 0.99 v 0.28 vi 0.74

Stretch

1 a 1.29 b 1.47 c 2.8 d 2.5

2 a $\frac{29}{100}, \frac{33}{100}, \frac{37}{100}$ or 0.29, 0.33, 0.37
b $\frac{83}{100}, \frac{75}{100}, \frac{67}{100}$ or 0.83, 0.75, 0.67
c $\frac{30}{100}, \frac{33}{100}, \frac{36}{100}$ or 0.3, 0.33, 0.36
d $\frac{22}{100}, \frac{11}{100}$, 0 or 0.22, 0.11, 0

3 a $x = \frac{1}{10}$, $x = 0.1$
b $x = \frac{31}{100}$, $x = 0.31$
c $x = \frac{4}{100}$, $x = 0.04$
d $x = \frac{62}{100}$, $x = 0.62$

4 $x = \frac{23}{100}$, $y = \frac{20}{100}$ which simplifies to $\frac{2}{10}$
or simplifies fully to $\frac{1}{5}$

Chapter 5.2

Are you ready?

1 One out of four equal parts is shaded.

2 a $\frac{7}{8}$ b $\frac{3}{5}$ c $\frac{1}{4}$
d $\frac{3}{4}$ e $\frac{5}{8}$ f $\frac{1}{8}$

3 a 0.3 b 0.03 c 0.7
d 0.43 e 0.8

4 a $\frac{29}{100}$ b $\frac{9}{10}$ c $\frac{7}{100}$
d $\frac{99}{100}$ e $\frac{1}{10}$

Practice 5.2A

1 a A $\frac{4}{5}$ B $\frac{3}{5}$ C $\frac{2}{5}$ D $\frac{1}{5}$
b A $\frac{8}{10}$ B $\frac{6}{10}$ C $\frac{4}{10}$ D $\frac{2}{10}$
c A $\frac{80}{100}$ B $\frac{60}{100}$ C $\frac{40}{100}$ D $\frac{20}{100}$
d A 0.8 B 0.6 C 0.4 D 0.2

2 a A $\frac{3}{4}$ B $\frac{2}{4}$ C $\frac{1}{4}$ D $\frac{4}{4}$
b A $\frac{75}{100}$ B $\frac{50}{100}$ C $\frac{25}{100}$ D $\frac{100}{100}$
c A 0.75 B 0.5 C 0.25 D 1

3 a A $\frac{2}{5}$ B $\frac{1}{4}$
b A $\frac{3}{5}$ B $\frac{3}{4}$
c For each diagram, the numerators in **a** and **b** add up to the denominator.

4 a $\frac{1}{5}$, 0.2 **b** $\frac{1}{4}$, 0.25 **c** $\frac{2}{4}$, 0.5
d $\frac{2}{5}$, 0.4 **e** $\frac{3}{4}$, 0.75 **f** $\frac{5}{5}$ or $\frac{4}{4}$, 1
5 a $\frac{1}{5} = \frac{2}{10} = 0.2$ **b** $\frac{2}{4} = \frac{5}{10} = 0.5$
c $\frac{8}{10} = \frac{4}{5} = 0.8$
6 a 20 squares **b** 25 squares
c One quarter. 25 hundredths is greater than 20 hundredths.
7 a $<$ **b** $>$ **c** $<$
d $=$ **e** $<$ **f** $=$

What do you think?

1 Both are correct. 20 hundredths are shaded in each.
2 Amina is correct. The whole number parts are the same, so we can just compare the fractions. $\frac{3}{4} > \frac{2}{5}$, so $175\frac{3}{4} > 175\frac{2}{5}$

Practice 5.2B

1 a $\frac{3}{1000}$, 0.003 **b** $\frac{997}{1000}$, 0.997
c $\frac{3}{8}$, 0.375 **d** $\frac{5}{8}$, 0.625
2 a 70 thousandths
b 700 thousandths
3 a 0.389 **b** 0.273 **c** 0.961
d 0.511 **e** 0.8 **f** 0.08
g 0.008 **h** 0.019
4 a $\frac{176}{1000}$ **b** $\frac{999}{1000}$ **c** $\frac{4}{1000}$ **d** $\frac{76}{1000}$
e $\frac{345}{1000}$ **f** $\frac{763}{1000}$ **g** $\frac{4}{5}$ **h** $\frac{3}{4}$
5 a $\frac{1}{4} = \frac{2}{8} = 0.25$ **b** $\frac{3}{4} = \frac{6}{8} = 0.75$
c $\frac{4}{8} = \frac{1}{2} = 0.5$ **d** $\frac{1}{8} = 0.125$
e $\frac{3}{8} = 0.375$ **f** $\frac{5}{8} = 0.625$
g $\frac{7}{8} = 0.875$ **h** $\frac{8}{8} = 1$

What do you think?

1 No. He has written '72 thousands' not '72 thousandths'. It should be 0.072.
2 a Top: 0.25, 0.375, 0.5, 0.625, 0.75, 0.875, 1.125, 1.25
Bottom: $\frac{0}{8}$, $\frac{2}{8}$, $\frac{3}{8}$, $\frac{4}{8}$, $\frac{5}{8}$, $\frac{6}{8}$, $\frac{7}{8}$, $\frac{9}{8}$, $\frac{10}{8}$
b Top: 0.002, 0.003, 0.004, 0.005, 0.006, 0.007, 0.008, 0.009, 0.01
Bottom: $\frac{0}{1000}$, $\frac{2}{1000}$, $\frac{3}{1000}$, $\frac{4}{1000}$, $\frac{5}{1000}$, $\frac{7}{1000}$, $\frac{8}{1000}$, $\frac{9}{1000}$, $\frac{1}{100}$
3 a **A** 0.1 **B** 0.01 **C** 0.001
D 0.25 **E** 0.2 **F** 0.125
G 0.5
b $\frac{1}{1000}$, $\frac{1}{100}$, $\frac{1}{10}$, $\frac{1}{8}$, $\frac{1}{5}$, $\frac{1}{4}$, $\frac{1}{2}$
c The denominators are in descending order.

Consolidate

1 a 0.2 **b** 0.4 **c** 0.6 **d** 0.8
2 a 0.25 **b** 0.5 **c** 0.75
3 a The bars representing the wholes are the same size and 2 out of 8 equal parts is equal to 1 out of 4 equal parts.
b 0.25
4 a 0.125 **b** 0.375
c 0.75 **d** 0.875
5 a $\frac{145}{1000}$ **b** $\frac{807}{1000}$ **c** $\frac{35}{1000}$ **d** $\frac{1}{1000}$

Stretch

1 a A = $\frac{1}{4}$ = 0.25, B = $\frac{7}{8}$ = 0.875, difference = $\frac{5}{8}$ = 0.625
b A = $\frac{5}{4}$ = 1.25, B = $\frac{15}{8}$ = 1.875, difference = $\frac{5}{8}$ = 0.625
c A = $\frac{1}{4}$ = 0.25, B = $\frac{15}{8}$ = 1.875, difference = $\frac{13}{8}$ = 1.625
d A = $\frac{1}{4}$ = 0.25, B = $\frac{87}{8}$ = 10.875, difference = $\frac{85}{8}$ = 10.625
2 $\frac{3}{10}y$, $\frac{3y}{10}$, $0.3y$
3 Various answers
4 a $\frac{37}{10}$ **b** $\frac{1209}{100}$ **c** $\frac{1871}{1000}$
d $\frac{43}{8}$ **e** $\frac{25}{4}$ **f** $\frac{19009}{1000}$
g $\frac{2108}{100}$ **h** $\frac{184397}{1000}$

Chapter 5.3

Are you ready?

1 a 20 **b** 9 **c** 63 **d** 97
2 a $\frac{2}{10}$ **b** $\frac{7}{10}$ **c** $\frac{9}{10}$ **d** $\frac{1}{10}$
3 a $\frac{31}{100}$ **b** $\frac{73}{100}$ **c** $\frac{97}{100}$ **d** $\frac{49}{100}$
4 a 0.3 **b** 0.19 **c** 0.99 **d** 0.5
5 a 0.25 **b** 0.2 **c** 0.75 **d** 0.6
6 a $\frac{4}{5}$ **b** $\frac{2}{5}$ **c** $\frac{1}{4}$ **d** $\frac{3}{4}$

Practice 5.3A

1 a **A** 8 **B** 7 **C** 4 **D** 9
b **A** $\frac{8}{100}$ **B** $\frac{7}{100}$ **C** $\frac{4}{100}$ **D** $\frac{9}{100}$
c **A** 8% **B** 7% **C** 4% **D** 9%
2 a **A** 20 **B** 50 **C** 30 **D** 90
b **A** $\frac{20}{100}$ **B** $\frac{50}{100}$ **C** $\frac{30}{100}$ **D** $\frac{90}{100}$
c **A** 20% **B** 50% **C** 30% **D** 90%
3 a **A** 28 **B** 57 **C** 34 **D** 99
b **A** $\frac{28}{100}$ **B** $\frac{57}{100}$ **C** $\frac{34}{100}$ **D** $\frac{99}{100}$
c **A** 28% **B** 57% **C** 34% **D** 99%
4 a 35 squares shaded
b 71 squares shaded
c 8 squares shaded
d 19 squares shaded
e 25 squares shaded
5 a 49% **b** 51% **c** 100%
d Total of **a** and **b** includes all the squares; 100% is all the squares
6 a 29% **b** 89% **c** 47%
d 99% **e** 0%

What do you think?

1 a 12% **b** 12% **c** 12% **d** 64%
e Add up the percentages and check that they make a total of 100%
2 Rob is correct, it doesn't matter which 10 squares are shaded
3 a 10% **b** 60% **c** 40% **d** 35%

Practice 5.3B

1 a $\frac{7}{100}$, 0.07, 7% **b** $\frac{9}{100}$, 0.09, 9%
c $\frac{4}{100}$, 0.04, 4% **d** $\frac{7}{100}$, 0.07, 7%
e $\frac{8}{100}$, 0.08, 8%
f $\frac{36}{100}$, 0.36, 36%
2 a $\frac{3}{10}$, 0.3, 30% **b** $\frac{7}{10}$, 0.7, 70%
c $\frac{5}{10}$, 0.5, 50% **d** $\frac{4}{10}$, 0.4, 40%
e $\frac{3}{10}$, 0.3, 30% **f** $\frac{1}{10}$, 0.1, 10%
3 a $\frac{3}{4}$, 0.75, 75%
b $\frac{2}{4}$ or $\frac{1}{2}$, 0.5, 50%
c $\frac{1}{4}$, 0.25, 25% **d** $\frac{100}{100}$, 1, 100%
4 a $\frac{4}{5}$, 0.8, 80% **b** $\frac{3}{5}$, 0.6, 60%
c $\frac{2}{5}$, 0.4, 40% **d** $\frac{1}{5}$, 0.2, 20%
5 a 0.23, 23% **b** 0.7, 70%
c 0.8, 80% **d** 0.75, 75%
e 0.5, 50%
6 a $\frac{71}{100}$, 71% **b** $\frac{3}{100}$, 3%
c $\frac{1}{4}$, 25% **d** $\frac{3}{5}$, 60%
e $\frac{9}{10}$, 90%
7 a $\frac{99}{100}$, 0.99 **b** $\frac{12}{100}$, 0.12
c $\frac{7}{100}$, 0.07 **d** $\frac{1}{5}$, 0.2
e $\frac{3}{4}$, 0.75

What do you think?

1 a False; 0.5 is equivalent to $\frac{1}{2}$, or $\frac{1}{5}$ is equivalent to 20%
b False; $\frac{3}{4}$ is equivalent to 75%, or 34% is equivalent to $\frac{34}{100}$ (or $\frac{17}{50}$)
c True
2 $\frac{2}{5}$, 0.4, 40%

Consolidate

1 a $\frac{31}{100}$, 0.31
b $\frac{72}{100}$ or $\frac{36}{50}$ or $\frac{18}{25}$, 0.72
c $\frac{80}{100}$ or $\frac{8}{10}$ or $\frac{4}{5}$, 0.8
d $\frac{90}{100}$ or $\frac{9}{10}$, 0.9
e $\frac{1}{100}$, 0.01
2 a 11%, 0.11 **b** 27%, 0.27
c 93%, 0.93 **d** 30%, 0.3
e 25%, 0.25
3 a $\frac{57}{100}$, 57% **b** $\frac{61}{100}$, 61%
c $\frac{9}{100}$, 9% **d** $\frac{1}{10}$, 10%
e $\frac{8}{10}$ or $\frac{4}{5}$, 80%

Stretch

1 a $\frac{20}{50}$ or $\frac{2}{5}$, 0.4, 40%
b 9
c $\frac{42}{100}$ or $\frac{21}{50}$, 0.42, 42%
2 a Linear: $\frac{25}{100}$ or $\frac{1}{4}$, 0.25, 25%; $\frac{3}{10}$, 0.3, 30%; $\frac{35}{100}$ or $\frac{7}{20}$, 0.35, 35%
b Not linear

c Linear: $\frac{72}{100}$ or $\frac{36}{50}$ or $\frac{18}{25}$, 0.72, 72%;

$\frac{63}{100}$, 0.63, 63%;

$\frac{54}{100}$ or $\frac{27}{50}$, 0.54, 54%

d Linear: $\frac{40}{100}$ or $\frac{4}{10}$ or $\frac{2}{5}$, 0.4, 40%;

$\frac{20}{100}$ or $\frac{2}{10}$ or $\frac{1}{5}$, 0.2, 20%;

$\frac{0}{5}$, 0, 0%

3 a 135% **b** 272%
c 609% **d** 840%

4 a $p + 0.71 = 94.5\%$,
$0.71 + p = 94.5\%$,
$94.5\% - p = 0.71$,
$94.5\% - 0.71 = p$

b $p = \frac{235}{1000}$ (or $\frac{47}{200}$), 0.235, 23.5%

5 Beth is correct. There are an infinite number of values between 0.8 and 0.82.

Chapter 5.4

Are you ready?

1 a $\frac{1}{3}$ **b** $\frac{1}{4}$ **c** $\frac{5}{7}$
d $\frac{5}{6}$ **e** $\frac{3}{8}$ **f** $\frac{1}{2}$

2 a 20% **b** 75% **c** 50%
d 100% **e** 30% **f** 50%

3 a 0.2 **b** 0.75 **c** 0.5
d 1.0 **e** 0.3 **f** 0.5

Practice 5.4A

1 a $\frac{1}{10}$, 0.1, 10%

b $\frac{4}{10}$ or $\frac{2}{5}$, 0.4, 40%

c $\frac{3}{10}$, 0.3, 30%

d $\frac{2}{10}$ or $\frac{1}{5}$, 0.2, 20%

e No, because the sector represents '3 or more' not just 3

2 a Bus: $\frac{1}{2}$, 0.5, 50%

b 30% travel by car. You can't tell how many students this represents.

c The same number of students walk as go by bike, but we can't tell how many students this number is.

3 a $\frac{1}{5}$ **b** 0.4 **c** 40% **d** 0

4 a Approximately 60%

b Approximately 40%

c Because there are only clear markings at each 25%

d The size of the sector for those who do have a driving licence is more than half of the pie chart so it must represent more than 50% of the people surveyed. 43% and 47% do not add up to 100%. More people have a driving licence than don't have one.

What do you think?

1 a Office A: $\frac{6}{10}$ or $\frac{3}{5}$, 0.6, 60%;

Office B: $\frac{3}{10}$, 0.3, 30%

b Office A: $\frac{4}{10}$ or $\frac{2}{5}$, 0.4, 40%;

Office B: $\frac{7}{10}$, 0.7, 70%

c **i** True **ii** Can't tell
iii True **iv** False

2 a

Number of clubs	Year 7	Year 8
0	30%	10%
1	50%	40%
2	10%	30%
3+	10%	20%

b

Number of clubs	Year 7	Year 8
0	$\frac{3}{10}$	$\frac{1}{10}$
1	$\frac{5}{10}$ or $\frac{1}{2}$	$\frac{4}{10}$ or $\frac{2}{5}$
2	$\frac{1}{10}$	$\frac{3}{10}$
3+	$\frac{1}{10}$	$\frac{1}{5}$

c Various answers, e.g. In both Year 7 and Year 8, a greater proportion of students go to 1 club than any other number of clubs. The proportion of Year 8 students going to 2 or more clubs is greater than the proportion of Year 7 students who attend 2 or more clubs.

d No. The proportions are the same but we don't know how many students each pie chart represents.

Consolidate

1 a Piechart Athletic. Their sector for 'Win' is the biggest compared with the other teams.

b Hamilton United. They have no sector for 'Lose'.

c Abacus FC: $\frac{3}{10}$, 0.3, 30%

Piechart Athletic: $\frac{7}{10}$, 0.7, 70%

Hamilton United: $\frac{6}{10}$ or $\frac{3}{5}$, 0.6, 60%

White Rose Rangers: $\frac{6}{10}$ or $\frac{3}{5}$, 0.6, 60%

d Abacus FC: $\frac{5}{10}$ or $\frac{1}{2}$, 0.5, 50%

Piechart Athletic: $\frac{2}{10}$ or $\frac{1}{5}$, 0.2, 20%

Hamilton United: $\frac{4}{10}$ or $\frac{2}{5}$, 0.4, 40%

White Rose Rangers: $\frac{2}{10}$ or $\frac{1}{5}$, 0.2, 20%

e Abacus FC: $\frac{2}{10}$ or $\frac{1}{5}$, 0.2, 20%

Piechart Athletic: $\frac{1}{10}$, 0.1, 10%

Hamilton United: $\frac{0}{10}$, 0, 0%

White Rose Rangers: $\frac{2}{10}$ or $\frac{1}{5}$, 0.2, 20%

Stretch

1 Rugby: $\frac{1}{10}$, 0.1, 10%

Netball: $\frac{2}{10}$ or $\frac{1}{5}$, 0.2, 20%

Football: $\frac{4}{10}$ or $\frac{2}{5}$, 0.4, 40%

Other: $\frac{3}{10}$, 0.3, 30%

2 a **10am** Under 18: $\frac{1}{10}$, 10%;

18–30: $\frac{2}{10}$ or $\frac{1}{5}$, 20%;

31–59: $\frac{3}{10}$, 30%;

60+: $\frac{4}{10}$ or $\frac{2}{5}$, 40%

6pm Under 18: $\frac{2}{10}$ or $\frac{1}{5}$, 20%;

18–30: $\frac{4}{10}$ or $\frac{2}{5}$, 40%;

31–59: $\frac{3}{10}$, 30%;

60+: $\frac{1}{10}$, 10%

b Various answers, e.g. There are more people aged 18–59 in the gym at 10am than at 6pm; many people aged 18–59 are at school or at work at 10am.

Chapter 5.5

Are you ready?

1 a Yes. The shape is split into two equal parts and one is shaded.

b Yes. The shape is split into two equal parts and one is shaded.

c No. The smaller circle is much smaller than half the larger circle.

d No. The shape is split into three equal parts and only one is shaded.

e No. The shape is split into two parts but they are not equal in size.

f Yes. The shape is split into two equal parts and one is shaded.

2 a 1s **b** 2s **c** 10s
d tenths **e** 2 tenths or fifths

3 a 15 **b** 35 **c** 126
d 147 **e** 60 **f** 112

Practice 5.5A

1 a Unequal **b** Equal
c Equal **d** Unequal
e Equal **f** Unequal

2 a $\frac{5}{12}$ **b** $\frac{4}{9}$ **c** $\frac{5}{6}$
d $\frac{2}{7}$ **e** $\frac{1}{5}$ **f** $\frac{7}{18}$

3 a $\frac{12}{12}$ **b** $\frac{5}{5}$ **c** $\frac{6}{6}$ **d** $\frac{7}{7}$

4 a $\frac{5}{11}$ **b** $\frac{7}{8}$ **c** $\frac{9}{9}$ **d** $\frac{1}{14}$

5 a Various answers, e.g. bar model split into 5 equal parts with 1 part shaded

b Various answers, e.g. bar model split into 11 equal parts with 3 parts shaded

c Various answers, e.g. bar model split into 13 equal parts with 9 parts shaded

d Various answers, e.g. bar model split into 7 equal parts with 7 parts shaded

e Various answers, e.g. bar model split into 20 equal parts with 17 parts shaded

What do you think?

1 Chloe. If you split each square into two equal pieces diagonally, there will be 14 equal parts and 1 part shaded.

So $\frac{1}{14}$ of the diagram is shaded.

2 True. Although the parts don't look equal they are. You can think of it as a half of a half; or if you split each triangle into half again, there will be 8 equal parts and 2 are shaded.

Practice 5.5B

1 a $\frac{1}{7}, \frac{2}{7}, \frac{3}{7}, \frac{4}{7}, \frac{5}{7}, \frac{6}{7}$

b $\frac{1}{5}, \frac{2}{5}, \frac{3}{5}, \frac{4}{5}$

c $\frac{1}{8}, \frac{2}{8}, \frac{3}{8}, \frac{4}{8}, \frac{5}{8}, \frac{6}{8}, \frac{7}{8}$

d $\frac{1}{11}, \frac{2}{11}, \frac{3}{11}, \frac{4}{11}, \frac{5}{11}, \frac{6}{11}, \frac{7}{11}, \frac{8}{11}, \frac{9}{11}, \frac{10}{11}$

e $\frac{1}{3}, \frac{2}{3}$

f $\frac{1}{6}, \frac{2}{6}, \frac{3}{6}, \frac{4}{6}, \frac{5}{6}$

g $\frac{1}{13}, \frac{2}{13}, \frac{3}{13}, \frac{4}{13}, \frac{5}{13}, \frac{6}{13}, \frac{7}{13}, \frac{8}{13}, \frac{9}{13}, \frac{10}{13}, \frac{11}{13}, \frac{12}{13}$

2 a $\frac{5}{7}$ **b** $\frac{3}{5}$ **c** $\frac{7}{8}$

d $\frac{3}{11}$ **e** $\frac{1}{3}$ **f** $\frac{5}{6}$

3 a Number line with 12 intervals

b Arrow pointing to the marker at the end of the fifth interval

4 a Arrow halfway between 0 and $\frac{1}{2}$km

b Arrow pointing just past half way

What do you think

1 No. There are 6 intervals, so the number line is going up in sixths.

2 A, B and C: yes. D: no because the end point is different compared to lines A, B and C.

Practice 5.5C

1 a $\frac{5}{6} = \frac{10}{12}$ **b** $\frac{2}{5} = \frac{6}{15}$

c $\frac{3}{4} = \frac{6}{8}$ **d** $\frac{1}{2} = \frac{5}{10}$

2 $\frac{1}{3} = \frac{3}{9}, \frac{2}{3} = \frac{6}{9}, \frac{3}{3} = \frac{9}{9}$

3 a Various answers, e.g. two bar models the same length, one split into 5 equal parts with 4 parts shaded, the other split into 10 equal parts with 8 parts shaded. The shaded sections should be the same length, showing equivalence.

b Various answers, e.g. two bar models the same length, one split into 3 equal parts with 1 part shaded, the other split into 15 equal parts with 5 parts shaded. The shaded sections should be the same length, showing equivalence.

c Various answers, e.g. two bar models the same length, one split into 7 equal parts with 2 parts shaded, the other split into 21 equal parts with 6 parts shaded. The shaded sections should be the same length, showing equivalence.

d Various answers, e.g. two bar models the same length, one split into 6 equal parts with 5 parts shaded, the other split into 12 equal parts with 10 parts shaded. The shaded sections should be the same length, showing equivalence.

e Various answers, e.g. two bar models the same length, one split into 9 equal parts with 5 parts shaded, the other split into 36 equal parts with 20 parts shaded. The shaded sections should be the same length, showing equivalence.

4 a 9 **b** 14 **c** 1 **d** 25

e 60 **f** 28 **g** 30 **h** 38

i 210 **j** 99

5 Various answers for each part.

What do you think?

1 $\frac{1}{8}$ and any two equivalent fractions, e.g. $\frac{2}{16}, \frac{3}{24}$

2 No. The bars are not the same length.

3 Various answers, e.g. two bar models the same length, one split into 13 equal parts with 11 parts shaded, the other split into 14 equal parts with 12 parts shaded. The shaded parts will not be the same length so the fractions are not equivalent.

Consolidate

1 a Various answers, e.g. bar model split into 5 equal parts with 1 part shaded

b Various answers, e.g. bar model split into 3 equal parts with 2 parts shaded

c Various answers, e.g. bar model split into 7 equal parts with 5 parts shaded

d Various answers, e.g. bar model split into 10 equal parts with 9 parts shaded

e Various answers, e.g. bar model split into 6 equal parts with 5 parts shaded

2 a Number line from 0 to 1 going up in fifths with arrow pointing to $\frac{1}{5}$

b Number line from 0 to 1 going up in thirds with arrow pointing to $\frac{2}{3}$

c Number line from 0 to 1 going up in sevenths with arrow pointing to $\frac{5}{7}$

d Number line from 0 to 1 going up in tenths with arrow pointing to $\frac{9}{10}$

e Number line from 0 to 1 going up in sixths with arrow pointing to $\frac{5}{6}$

3 a Various answers, e.g. $\frac{2}{10}, \frac{3}{15}, \frac{10}{50}$

b Various answers, e.g. $\frac{4}{6}, \frac{6}{9}, \frac{20}{30}$

c Various answers, e.g. $\frac{10}{14}, \frac{15}{21}, \frac{50}{70}$

d Various answers, e.g. $\frac{18}{20}, \frac{27}{30}, \frac{90}{100}$

e Various answers, e.g. $\frac{10}{12}, \frac{15}{18}, \frac{50}{60}$

Stretch

1 Arrow just to the left of halfway. Answer is an estimate as there are no marks on the number line.

2 The diagram can be split into 64 equal parts (8 × 8) and 32 are shaded. $\frac{32}{64} = \frac{1}{2}$

3 True. The denominator is always twice the numerator. The numerator has to be an integer and double any integer is always even.

4 No. $\frac{3}{4}$ is equivalent to $\frac{21}{28}$ so she did equally well on both tests.

Chapter 5.6

Are you ready?

1 a 0.2 **b** 0.1 **c** 0.01

d 0.001 **e** 0.07 **f** 0.25

2 a 0.2 **b** 0.1 **c** 0.01

d 0.001 **e** 0.07 **f** 0.25

3 a 0.27, 27% **b** 0.49, 49%

c 0.3, 30% **d** 0.75, 75%

e 0.8, 80%

4 a $\frac{79}{100}$, 79% **b** $\frac{3}{100}$, 3%

c $\frac{6}{10}$ or $\frac{3}{5}$, 60% **d** $\frac{9}{10}$, 90%

e $\frac{25}{100}$ or $\frac{1}{4}$, 25%

5 a $\frac{81}{100}$, 0.81 **b** $\frac{7}{100}$, 0.07

c $\frac{70}{100}$ or $\frac{7}{10}$, 0.7 **d** $\frac{99}{100}$, 0.99

e $\frac{371}{1000}$, 0.371

f $\frac{80}{100}$ or $\frac{8}{10}$ or $\frac{4}{5}$, 0.8

Practice 5.6A

1 a 0.03 **b** 0.3 **c** 0.17

d 1.7 **e** 0.017

2 a 0.03 **b** 0.3 **c** 0.17

d 1.7 **e** 0.017

3 9 ÷ 5 and $\frac{9}{5}$, 7 ÷ 4 and $\frac{7}{4}$, 17 ÷ 20 and $\frac{17}{20}$, 3 ÷ 8 and $\frac{3}{8}$, 19 ÷ 25 and $\frac{19}{25}$

4 a $\frac{5}{9}$ **b** $\frac{7}{11}$ **c** $\frac{6}{7}$

d $\frac{19}{20}$ **e** $\frac{20}{19}$

5 a 5 ÷ 13 **b** 1 ÷ 12 **c** 27 ÷ 5

d 19 ÷ 4 **e** 4 ÷ 19

6 a 9 ÷ 13, 0.692 307 69

b 7 ÷ 11, 0.636 363 64
c 15 ÷ 27, 0.555 555 56
d 134 ÷ 25, 5.36
e 25 ÷ 134, 0.186 567 16
f 8 ÷ 9, 0.888 888 89
g 11 ÷ 7, 1.571 428 57
h 1 ÷ 3, 0.333 333 33
i 19 ÷ 20, 0.95
j 39 ÷ 50, 0.78
k **b**, **c**, **f** and **h**

7 a $\frac{5}{8}$ b $\frac{4}{5}$ c $\frac{49}{50}$

8 a $\frac{1}{20}, \frac{14}{28}, \frac{4}{5}, \frac{17}{20}, \frac{11}{20}$
b $\frac{1}{6}, \frac{13}{20}, \frac{7}{8}, \frac{23}{25}, \frac{11}{8}$

9 $\frac{5}{7}$

What do you think?

1 a $\frac{4}{25}$ is less than 1
b She has divided 25 by 4 rather than 4 by 25
c 0.16

2 $\frac{19}{57}, \frac{126}{378}, \frac{99}{297}$

3 Zach. 15 ÷ 105 = 0.142 857 14 and 19 ÷ 133 = 0.142 857 14.
Both fractions simplify to $\frac{1}{7}$.

Practice 5.6B

1 a 0.55 b 0.84 c 0.78
d 0.875 e 0.375
2 a 55% b 84% c 78%
d 87.5% e 37.5%
3 a 55% b 84% c 78%
d 87.5% e 37.5%
4 a 15% b 55% c 70%
d 95% e 105%
5 a 0.45, 45% b 0.28, 28%
c 0.7, 70% d 0.63, 63%
e 0.817, 81.7% f 0.75, 75%
g 0.8, 80% h 0.125, 12.5%
i 0.52, 52% j 0.079, 7.9%

6 a $\frac{1}{10}$, 10% b $\frac{6}{10}$ or $\frac{3}{5}$, 60%
c $\frac{25}{100}$ or $\frac{1}{4}$, 25% d $\frac{71}{100}$, 71%
e $\frac{22}{100}$ or $\frac{11}{50}$, 22%
f $\frac{379}{1000}$, 37.9% g $\frac{99}{100}$, 99%
h $\frac{85}{100}$ or $\frac{17}{20}$, 85%
i $\frac{625}{1000}$ or $\frac{5}{8}$, 62.5%
j $\frac{7}{1000}$, 0.7%

7 a $\frac{37}{100}$, 0.37 b $\frac{41}{100}$, 0.41
c $\frac{83}{100}$, 0.83
d $\frac{95}{100}$ or $\frac{19}{20}$, 0.95
e $\frac{15}{100}$ or $\frac{3}{20}$, 0.15
f $\frac{13}{100}$, 0.13
g $\frac{20}{100}$ or $\frac{2}{10}$ or $\frac{1}{5}$, 0.2
h $\frac{75}{100}$ or $\frac{1}{4}$, 0.75
i $\frac{375}{1000}$ or $\frac{3}{8}$, 0.375
j $\frac{931}{1000}$, 0.931

8 $\frac{28}{100}$ or $\frac{7}{25}$, 0.28, 28%

What do you think?

1 a $\frac{1}{7}$ = 0.142 857 14
$\frac{2}{7}$ = 0.285 714 29
$\frac{3}{7}$ = 0.428 571 43
$\frac{4}{7}$ = 0.571 428 57
$\frac{5}{7}$ = 0.714 285 71
$\frac{6}{7}$ = 0.857 142 85
b $\frac{1}{3}$ = 0.333 333 33
$\frac{2}{3}$ = 0.666 666 67

2 a True. $\frac{3}{5} = \frac{60}{100}$ = 60%
b False. $\frac{4}{7}$ = 0.571 428 57 = 57.142 857%
c False. $\frac{7}{8}$ = 0.875 = 87.5% (It would be 88% to the nearest integer)
d True. 38% = $\frac{38}{100} = \frac{19}{50}$

3 a Faith. One value is less than a half, one is equivalent to a half and one is greater than a half.
b 72%, 0.5, $\frac{1}{5}$

Consolidate

1 a 0.35 b 0.76 c 0.77
d 0.9 e 0.4 f 0.25
2 a 0.35 b 0.76 c 0.77
d 0.9 e 0.4 f 0.25
3 a 75 b 30 c 45
d 68 e 80 f 74
4 a 75% b 30% c 45%
d 68% e 80% f 74%
5 a 0.25, 25% b 0.85, 85%
c 0.84, 84% d 0.62, 62%
e 0.625, 62.5% f 1.3, 130%

Stretch

1 a $\frac{3}{5}$, 0.6, 60%
b $\frac{54}{100}$ or $\frac{27}{50}$, 0.54, 54%
c $\frac{84}{100}$ or $\frac{21}{25}$, 0.84, 84%
d $\frac{40}{100}$ or $\frac{4}{10}$ or $\frac{2}{5}$, 0.4, 40%
e $\frac{12}{100}$ or $\frac{3}{25}$, 0.12, 12%
f $\frac{922}{1000}$ or $\frac{461}{500}$, 0.922, 92.2%

2 a $+\frac{1}{10}$, +0.1, +10%
b $-\frac{1}{20}$, –0.05, –5%
c $+\frac{3}{10}$, +0.3, +30%
d $-\frac{1}{8}$, –0.125, –12.5%

3 a $\frac{2}{5}$, 0.4, 40% then $\frac{1}{2}$, 0.5, 50% then $\frac{3}{5}$, 0.6, 60%
b $\frac{1}{2}$, 0.5, 50% then $\frac{9}{20}$, 0.45, 45% then $\frac{2}{5}$, 0.4, 40%
c $\frac{19}{20}$, 0.95, 95% then $\frac{5}{4}$, 1.25, 125% then $\frac{31}{20}$, 1.55, 155%
d $\frac{5}{8}$, 0.625, 62.5% then $\frac{1}{2}$, 0.5, 50% then $\frac{3}{8}$, 0.375, 37.5%

4 a Yes. In % terms, sequence is 0%, 10%, 10%, 20% …
Next terms: $\frac{3}{10}$, 0.3, 30% then $\frac{1}{2}$, 0.5, 50% then $\frac{4}{5}$, 0.8, 80%
b Yes. In % terms, sequence is 2%, 2%, 4% …
Next terms: $\frac{3}{50}$, 0.06, 6% then $\frac{1}{10}$, 0.1, 10% then $\frac{4}{25}$, 0.16, 16%
c No. In % terms, sequence is 5%, 5%, 20%, 50%. This is not Fibonacci.

5 a 1.15, 115% b 0.92, 92%
c 0.46, 46%
d 0.869 565 22, 86.956 522%
e 0.8, 80% f 0.4, 40%
g 1.086 956 52, 108.695 652%
h 1.25, 125% i 0.5, 50%
j 2.173 913 04, 217.391 304%
k 2.5, 250% l 2, 200%

Chapter 5.7

Are you ready?

1 $\frac{3}{5}, \frac{7}{9}, \frac{15}{17}, \frac{7}{11}, \frac{1}{8}$ The numerator is smaller than the denominator.
2 $\frac{15}{15}, \frac{3}{3}, \frac{37}{37}, \frac{4}{4}$ The numerator is equal to the denominator.
3 $\frac{21}{20}, \frac{4}{3}, \frac{11}{7}, \frac{15}{14}, \frac{9}{7}$ The numerator is greater than the denominator.
4 $\frac{7}{6}$, 2.56, 123%, 101%, 1.001, $\frac{15}{8}$, 17.001, 3076% Those given as fractions have the numerator greater than the denominator, percentages are over 100% and decimals have a positive value to the left of the decimal point.

Practice 5.7A

1 a 3 b 15 c 36 d 60
e Number of wholes × 3
2 a $2\frac{2}{5}$ b $1\frac{5}{7}$ c $4\frac{2}{3}$ d $3\frac{5}{6}$
3 a $4\frac{4}{5}$ b $4\frac{1}{4}$ c $4\frac{2}{3}$
d $2\frac{1}{9}$ e $3\frac{1}{10}$ f $9\frac{5}{8}$
4 a $\frac{16}{3}$ b $\frac{20}{7}$ c $\frac{39}{4}$
d $\frac{40}{9}$ e $\frac{162}{15}$ f $\frac{47}{3}$
5 a $\frac{3}{5}, \frac{4}{5}, \frac{5}{5}$ or 1, $\frac{6}{5}, \frac{7}{5}, \frac{8}{5}, \frac{9}{5}, \frac{10}{5}$ or 2
b 60%, 80%, 100%, 120%, 140%, 160%, 180%, 200%
c 0.6, 0.8, 1, 1.2, 1.4, 1.6, 1.8, 2
6 a 20%, 100%, 120%
b 75%, 200%, 275%
c 70%, 800%, 870%
d 71%, 300%, 371%
7 a 0.2, 1, 1.2 b 0.75, 2, 2.75
c 0.7, 8, 8.7 d 0.71, 3, 3.71
8 a $\frac{9}{10}, \frac{12}{10}, \frac{15}{10}, \frac{18}{10}, \frac{21}{10}, \frac{24}{10}, \frac{27}{10}, \frac{30}{10}$ or 3
b 75%, 100%, 125%, 150%, 175%, 200%, 225%, 250%

c $\frac{6}{7}, \frac{8}{7}, \frac{10}{7}, \frac{12}{7}, \frac{14}{7}$ or 2, $\frac{16}{7}, \frac{18}{7}, \frac{20}{7}$
d 2.25, 3, 3.75, 4.5, 5.25, 6, 6.75, 7.5
e $1\frac{6}{9}, 2\frac{2}{9}, 2\frac{7}{9}, 3\frac{3}{9}, 3\frac{8}{9}, 4\frac{4}{9}, 5, 5\frac{5}{9}$

What do you think?

1 **a** $4\frac{1}{4}, 4\frac{2}{4}$ or $4\frac{1}{2}, 4\frac{3}{4}, 5, 5\frac{1}{4}$
b $3\frac{2}{5}, 3\frac{3}{5}, 3\frac{4}{5}, 4, 4\frac{1}{5}$
c $5\frac{2}{3}, 6, 6\frac{1}{3}, 6\frac{2}{3}, 7$
2 **a** It would take a long time to draw out 22 bar models and split each into 10 equal parts
b $\frac{217}{10}$, 21.7, 2170%

Consolidate

1 **a** $\frac{40}{9}$ **b** $\frac{11}{4}$ **c** $\frac{11}{3}$ **d** $\frac{9}{8}$
2 **a** $4\frac{4}{9}$ **b** $2\frac{3}{4}$ **c** $3\frac{2}{3}$ **d** $1\frac{1}{8}$
3 **a** $2\frac{2}{5}$ **b** $3\frac{1}{4}$ **c** $3\frac{1}{6}$
d $2\frac{5}{8}$ **e** $3\frac{4}{7}$ **f** $3\frac{1}{8}$
4 **a** $\frac{15}{7}$ **b** $\frac{17}{9}$ **c** $\frac{19}{5}$
d $\frac{17}{6}$ **e** $\frac{37}{7}$ **f** $\frac{49}{10}$
5 $\frac{7}{5}$, 1.4, 140%

Stretch

1 $\frac{22}{4}, 5\frac{2}{4}$ or $5\frac{1}{2}$
2 $\frac{23}{4}, 5\frac{3}{4}$
3 $\frac{64}{8}, \frac{100}{25}, \frac{49}{7}, \frac{63}{9}$
The numerator is a multiple of the denominator.
4 **a** $1\frac{5}{7}\text{m}^2, \frac{12}{7}\text{m}^2$
b Various answers
5 **a** **i** $\frac{396}{23}$ **ii** $\frac{258}{11}$ **iii** $\frac{57}{11}$
iv $\frac{39}{17}$ **v** $\frac{270}{23}$ **vi** $\frac{63}{2}$
b **i** $3\frac{2}{5}$ **ii** $4\frac{3}{5}$ **iii** $2\frac{1}{5}$
iv $2\frac{1}{2}$ **v** $11\frac{1}{2}$ **vi** $8\frac{1}{2}$
6 **a** $\frac{101}{7}$ **b** $\frac{98 + x}{7}$ **c** $\frac{14y + 3}{y}$
d $\frac{14y + x}{y}$ **e** $\frac{zy + x}{y}$

Check my understanding

1 **a** 0.1, 10% **b** 0.71, 71%
c 0.81, 81% **d** 0.75, 75%
e 0.4, 40%
2 **a** $\frac{3}{10}$, 30% **b** $\frac{1}{4}$, 25%
c $\frac{79}{100}$, 79% **d** $\frac{3}{5}$, 60%
e $\frac{1}{2}$, 50%
3 **a** $\frac{27}{100}$, 0.27 **b** $\frac{2}{5}$, 0.4
c $\frac{9}{10}$, 0.9 **d** $\frac{99}{100}$, 0.99
e $\frac{3}{4}$, 0.75
4 **a** $\frac{3}{10}$ **b** 40% **c** 0.1
5 **a** 6 **b** 45 **c** 1 **d** 11
6 **a** 0.125 **b** 0.003 **c** 0.625
d 0.731 **e** 0.875
7 **a** $1\frac{91}{100}$, 1.91, 191%
b $3\frac{3}{10}$, 3.3, 330%
c $4\frac{3}{4}$, 4.75, 475%
d $2\frac{1}{5}$, 2.2, 220%
e $2\frac{1}{8}$, 2.125, 212.5%

Block 6 Addition and subtraction

Chapter 6.1

Are you ready?

1

+	5	7	8
4	9	11	12
6	11	13	14
7	12	14	15

2

+	2	4	7
7	9	11	14
8	10	12	15
9	11	13	16

3

−	8	3	6
9	1	6	3
7	1	4	1
4	4	1	2

4 **a** 110 **b** 1100 **c** 4000
d 1.1 **e** 0.11 **f** 0.7

Practice 6.1A

1 **a** 48 + 46 = 94, 94 − 46 = 48, 94 − 48 = 46
b **i**

48	46
94	

ii

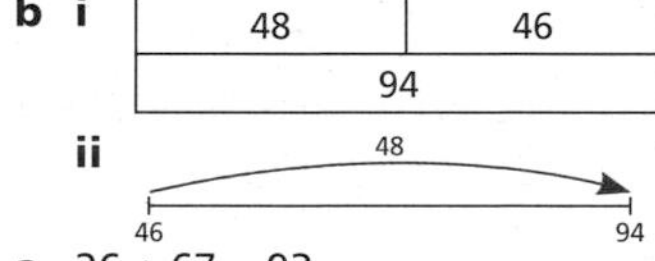

2 **a** 26 + 67 = 93
b

26	67
93	

c 93 − 67 = 26
3 30 + 40 + 50 = 120, 120 − 50 = 30 + 40, 120 − 30 = 40 + 50, 120 − 30 − 40 = 50, 120 − 40 = 30 + 50
4 **a** 80; 160 → 240
b 80 + 160 = 240, 240 − 80 = 160, 240 − 160 = 80
5 **a**

80	*x*
40	*y*

b

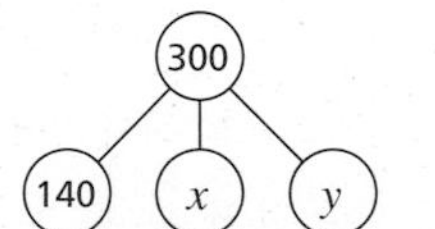

c **i** True **ii** True **iii** False
iv True **v** False **vi** True
d $y = 300 - 140 - x$
6 **a** The diagram shows that you get the same total whichever order you add the numbers
b No. The order matters in a subtraction.
7 One shows $x = y + z$ but the other shows $y = x + z$

What do you think?

1 **a** $a + b + c = x, x - a = b + c, x - (a + b) = c, x - a - b = c$
b **i** $x - b = a + c$
ii $a + b = x - c$
iii $x - (a + b + c) = 0$
2 **a** All three statements are true
b **i** $c + d - b = a$
ii $c + d - a = b$
iii $a - d + b = c$
3 **a** 63 − 49 + 29 is different because the signs of the numbers are different.
b 49 − 29 + 63 is easiest because it simplifies to 20 + 63

Practice 6.1B

1 **a** 62 + 50, 63 + 50 − 1
b 94 − 30, 93 − 20 − 9, 93 − 30 + 1
2 Compare answers as a class
3 **a** 367 **b** 217 **c** 357 **d** 500
e 360 **f** 172 **g** 117 **h** 510
i 152 **j** 197 **k** 503
4 **a** 4.8 + 10 − 0.1
b **i** 27.7 **ii** 36.7 **iii** 46.7
iv 25.9 **v** 16.9 **vi** 6.9
Compare the choice of methods as a class
5 **a** 355 **b** 521 **c** 123
d 185 **e** 76.2 **f** 18.8

What do you think?

1 **a** 934 **b** 935 **c** 935 **d** 936
e 388 **f** 545 **g** 543 **h** 3.86
2 **a** $a = 743$ **b** $b = 941$
c $c = 46.9$ **d** $d = 26.7$
e $e = 67.2$ **f** $f = 7.7$
g $g = 285$ **h** $h = 355$
i $i = 381.1$ **j** $j = 483$
k $k = 413$ **l** $l = 386.9$

Consolidate

1 **a** 182 − 73 = 109
b

182	
73	109

c 182 − 109 = 73
2 **a** 48 + 49 = 97, 97 − 48 = 49, 97 = 49 + 48
b **i** 50 **ii** 98 **iii** 96
3 **a** 236 − 51 = 87 + 98, 236 − 98 − 51 = 87, 236 − (98 + 51) = 87, 236 − 98 = 87 + 51
b **i**

236		
87	98	51

ii

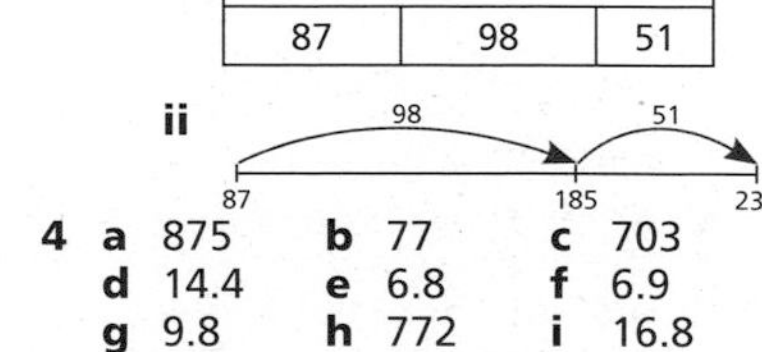

4 **a** 875 **b** 77 **c** 703
d 14.4 **e** 6.8 **f** 6.9
g 9.8 **h** 772 **i** 16.8
j 190.1 **k** 189.9
Compare the choice of methods as a class

Stretch

1 6 – (3 – 1) = 6 – 2 = 4 but (6 – 3) – 1 = 3 – 1 = 2, so subtraction is not associative

2 Yes. 8 + 6 = 14 and no other digits affect the ones column.

3 **a** 20 **b** 18 **c** 37
Compare the choice of methods as a class

Chapter 6.2

Are you ready?

1 **a** 1300 **b** 1.3 **c** 0.13
d 1.3 **e** 130

2 **a** 404 **b** 440 **c** 4040
d 4004 **e** 4.04

3 **a** $a = 19$ **b** $b = 32$ **c** $c = 368$

4 $\frac{3}{4} = 0.75$, $\frac{34}{100} = 0.34$, $\frac{1}{2} = 0.5$, $\frac{7}{100} = 0.07$, $\frac{70}{100} = 0.7$

Practice 6.2A

1 **a** **i** 579 **ii** 609
iii 580 **iv** 600
b **i** 211 **ii** 239
iii 275 **iv** 435

2 **a** 9.1 **b** 28.1 **c** 8.05 **d** 1.02

3 **a** 82.56 **b** 902 **c** 1.6

4 **a** 6.8 **b** 7.52
c 14.72 **d** 7.47

5 **a** 11 **b** 3.08 **c** 4.88
d 2.36 **e** 8.48 **f** 10.28

6 **a** 660 000 **b** 16 968
c 24 850

What do you think?

1 **a** $a = 739$ **b** $b = 1501$
c $c = 18.88$ **d** $d = 1.414$

2 Compare answers as a class

3 **a** 9.06 **b** 7.53
c 7.377 **d** 7.3617

4 The 7th term (16)

Practice 6.2B

1 **a** 203 mm **b** 22.9 cm
c 286 mm

2 **a** 42.6 cm **b** 358 mm

3 A = £805.28, B = £887.01, C = £1037.01, D = £1111, E = £1210

4 **a** £662 **b** £232
c 799 **d** 164.8 cm

5 **a** 7.6 billion **b** £157.55
c £119.91 **d** 5.66
e 33 568 **f** 3057

What do you think?

1 **a** Did not 'carry' the 1 after calculating 6 + 7 but instead put it in the answer
b Did not add the 'carry' in the hundreds column
c Not lined up properly
d Not lined up properly
e Did not 'carry' the 1 after calculating 1 + 9 but instead put it in the answer
f Did not add the 'carry' in the tens column

2 **a**

$$\begin{array}{r} 2\ 5\ 3 \\ +\ \boxed{3}\ \boxed{5}\ 1 \\ \hline 6\ 0\ 4 \end{array}$$

b

$$\begin{array}{r} 5\ \boxed{6}\ 7 \\ \boxed{9}\ 4\ 8 \\ +\ 4\ 9\ \boxed{6} \\ \hline 2\ \boxed{0}\ 1\ 1 \end{array}$$

c

$$\begin{array}{r} 4\ \boxed{0}\ 6 \\ \boxed{7}\ 1\ 6 \\ +\ 9\ 4\ \boxed{9} \\ \hline 2\ \boxed{0}\ 7\ 1 \end{array}$$

Other possible answers include 2171 and 2271

3 999 is the closest possible total to 1000, e.g. 152 + 378 + 469

Consolidate

1 **a** **i** 729 **ii** 789 **iii** 428
iv 400 **v** 11.2 **vi** 51.2
vii 50.5 **viii** 2.39
b **i** 121 **ii** 181 **iii** 983
iv 1181 **v** 8.8 **vi** 11.8
vii 21.8 **viii** 33.3

2 **a** 1632 **b** 110.1 **c** 29.56

3 **a** 890 **b** 526.4 **c** 490.04
d 89 **e** 52.64 **f** 5.264
g 0.4526

4 **a** 6922 **b** 7192
c 9892 **d** 41 887

5 **a** 131.4 cm **b** 36 cm

6 A = £657.31, B = £2157.31, C = £2295.20, D = £2576.64

Stretch

1 **a** Every 10 terms
b Every 10 terms
c Every 20 terms
Compare answers as a class.

2 Compare answers as a class.

Chapter 6.3

Are you ready?

1 **a** 50 **b** 0.5
c 0.05 **d** 50 000

2 **a** 6 **b** 6.9 **c** 6.99
d 6.999 **e** 6.9999

3 **a** $a = 12$ **b** $b = 0.3$ **c** $c = 400$

4 **a** 500 **b** 590 **c** 599
d 599.9 **e** 400 **f** 350
g 349 **h** 348.9

Practice 6.3A

1 **a** 42 **b** 36 **c** 402 **d** 393
e 396 **f** 623 **g** 603 **h** 573
i 413 **j** 283

2 **a** 5.3 **b** 4.9 **c** 4.77
d 4.67 **e** 5.66

3 **a** 262 **b** 58.4
c 47.8 **d** 274

4 **a** 5.92 **b** 6.64
c 6.12 **d** 5.97

5 **a** 5.4 **b** 2.52 **c** 0.72
d 1.8 **e** 7.92 **f** 6.12

6 **a** 1736 **b** 2186 **c** 2231

What do you think?

1 The difference is found by subtracting the smaller number from the greater, so you find 921 – 672

2 Compare answers as a class

3 **a** $a = 103$ **b** $b = 40$
c $c = 373$ **d** $d = 406.3$

4 **a** $w = 18.2$ **b** $x = 1916$
c $y = 206$ **d** $z = 5.63$

Practice 6.3B

1 **a** £45 profit **b** £40 profit
c £63 loss **d** £28 profit

2 **a** $a = 14$ **b** $b = 8.6$
c $c = 7$ **d** $d = 13.8$
e $e = 14.7$

3 A = £1146.23, B = £1127.58, C = £88.72, D = £998.87, E = £45

4 **a** 28 **b** 56 cm
c 14.4 **d** 3.36

5 **a** 133.8 cm **b** £346.72
c 146 **d** 194

6 **a** 2.4 billion **b** £117.57
c £29.45 **d** 1.56
e 6432 **f** 1743

What do you think?

1 **a** Not lined up properly
b Hundredths column incorrect. Need to write 9.4 as 9.40. Also they've not subtracted the tenths column correctly.
c Decimal columns incorrect. Need to write 6 as 6.00

2 **a**

$$\begin{array}{r} 4\ \boxed{1}\ 5 \\ -\ \boxed{1}\ 2\ \boxed{9} \\ \hline 2\ 8\ 6 \end{array}$$

b

$$\begin{array}{r} \boxed{6}\ 3\ \boxed{1} \\ -\ 3\ \boxed{9}\ 6 \\ \hline 2\ 3\ 5 \end{array}$$

c

$$\begin{array}{r} \boxed{9}\ \boxed{1}\ 3 \\ -\ 8\ 2\ \boxed{6} \\ \hline 8\ 7 \end{array}$$

3 **a** 28.5 **b** 28.5 **c** 106.5
d 57.5 **e** 106.5 **f** 57.5
g 28.5
Compare answers as a class

Consolidate

1 **a** **i** 182 **ii** 268 **iii** 95
iv 434 **v** 412 **vi** 172
b **i** 22 **ii** 17 **iii** 217
iv 352 **v** 222 **vi** 142

2 **a** 258 **b** 19.5 **c** 14.23

3 **a** 6752 **b** 6032 **c** 1829
d 1168 **e** 1213 **f** 1663

4 **a** 31 mm **b** 4.6 cm

5 A = £250, B = £376.76, C = £301.76, D = £303.23, E = £246.35, F = £765.45

Stretch

1 **a** Both numbers have been reduced by 0.01, so the difference between them is the same
b 2.57 **c** 4.28 **d** £15.17

2 There are 96 possible answers! Compare answers as a class.

Chapter 6.4

Are you ready?

1 **a** 125 **b** 8.1 **c** 523 **d** 64.1
2 **a** 54 **b** 2.8 **c** 255 **d** 17.8
3 **a** 282 **b** 84
c £54.99 **d** £35.01
4 **a** 12 **b** 3

Practice 6.4A

1 **a** 7 **b** 35 **c** 6
2 **a** 16 **b** 7 **c** 71
3 **a** 8 **b** 12 degrees
4

30
Boys 16
Girls 14

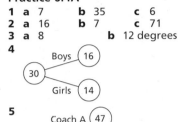

5

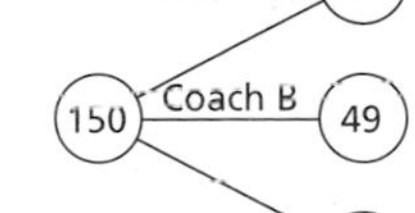

6 **a**

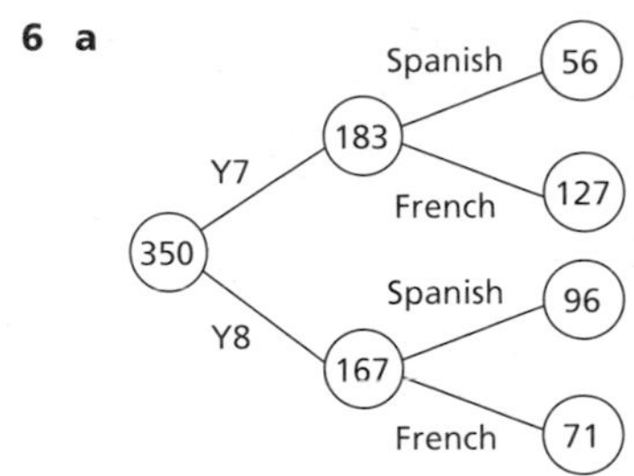

b 16 **c** 152

7 e.g.

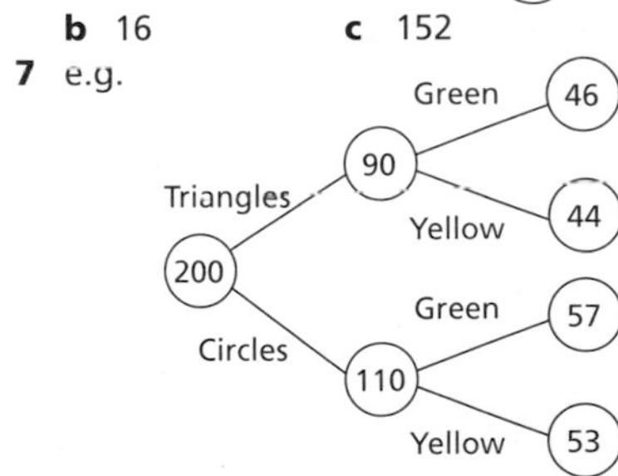

8 e.g.

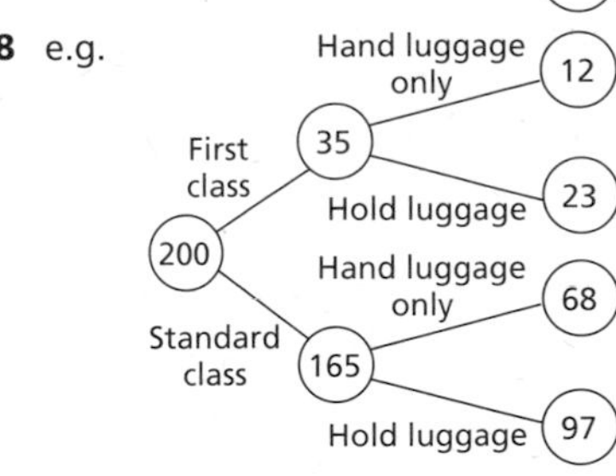

What do you think?

1 22

2

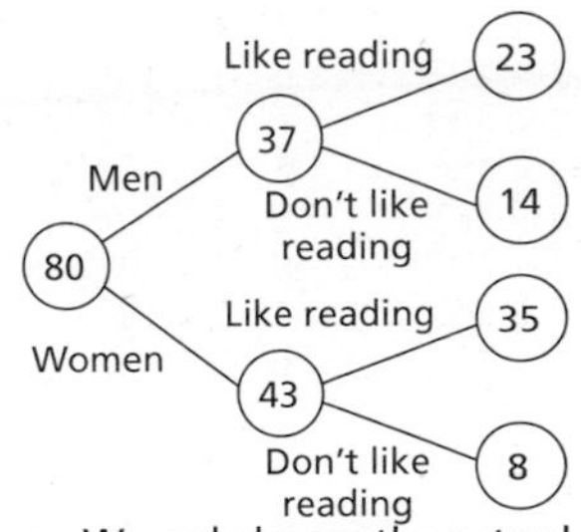

3 **a** We only know the actual temperatures at 9:00 a.m., 10:00 a.m. etc.
b So people know this is showing the pattern rather than representing values that can be read off.

4 Seb is correct in principle, but the equivalent frequency tree for a large two-way table can be very unwieldy. Compare answers as a class.

Practice 6.4B

1 **a** London and Oxford
b 362 miles **c** 154 miles
d 671 miles
2 **a** 2 hours
b 2 hours 5 minutes
c 2 hours 10 minutes
d 2 hours 50 minutes
e 3 hours 20 minutes
f 5 hours 10 minutes
g 5 hours 52 minutes
3 **a** 13:35 **b** 13:55
c 14:05 **d** 5:05 p.m.
e 5:15 p.m. **f** 18:27
4 07:46
5 **a** 09:00 **b** 26 minutes
c 42 minutes **d** 29 minutes
e 06:00, 10:30, 12:45, 19:30
6 **a** 3
b 1 hour 16 minutes
c 36 minutes **d** 3 minutes
e D
7 **a** Irbid and Tafila
b 115 miles
c 281 miles

What do you think?

1 **a**

A				
17	B			
29	**12**	C		
57	**40**	28	D	
71	54	42	**14**	E

b

A	08:40	09:46	10:51
B	08:57	10:03	11:08
C	09:09	10:15	11:20
D	09:37	10:43	11:48
E	09:51	10:57	12:02

2

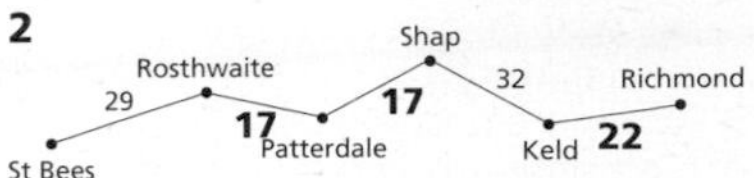

	Keld	Shap	Patterdale	Rosthwaite	St Bees
Rosthwaite					**29**
Patterdale				**17**	**46**
Shap			17	34	**63**
Keld		**32**	**49**	**66**	**95**
Richmond	**22**	54	**71**	**88**	**117**

Consolidate

1 **a**

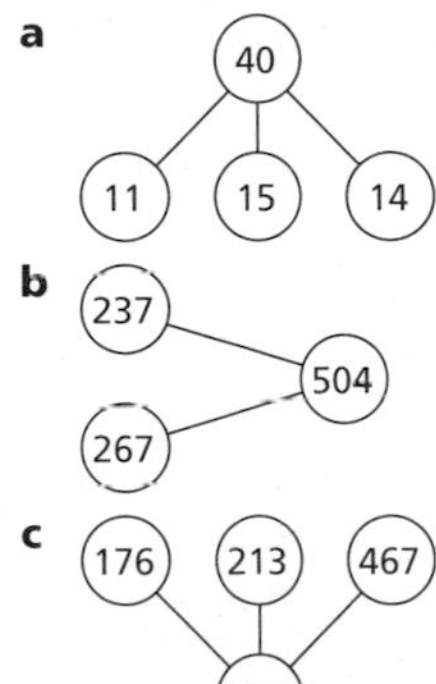

2 **a**

	Wear glasses	Don't wear glasses	Total
Boys	16	76	92
Girls	12	76	88
Total	28	152	180

b

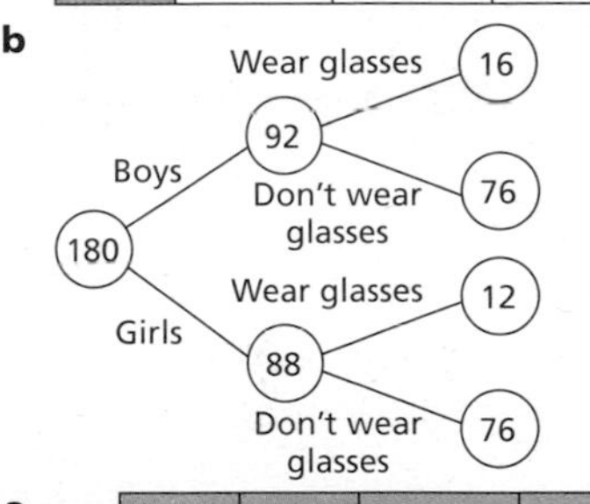

3 **a**

	Stalls	Circle	Balcony	Total
Adults	57	46	37	140
Children	31	19	42	92
Total	88	65	79	232

b 178
4 **a** 34 **b** 214
5 53 (accept any answer from 50 to 55)
6 **a** 16:47 **b** 12:08
c 18:11 **d** 13:12
7 **a** 2 hours 25 minutes
b 3 hours 55 minutes

Stretch

1 **a** You need at least four pieces of information – with at least one from each row and one from each column.
b Compare answers as a class.
2 **a** Subtract 1 hour and add 5 minutes

- **b** Add 2 hours and go back 1 minute

3 Compare answers as a class
4 Compare answers as a class

Chapter 6.5

Are you ready?

1 **a** 1×10^4 **b** 2×10^4 **c** 2×10^5 **d** 7×10^7
2 **a** 1000 **b** 100 000 000 **c** 50 000 **d** 90 000 000
3 **a** 0.1 **b** 0.5 **c** 0.000 05 **d** 0.0003
4 **a** 1×10^{-3} **b** 6×10^{-5} **c** 6×10^{-3} **d** 7×10^{-6}

Practice 6.5A

1 **a** 8×10^4 **b** 8×10^{16} **c** 8×10^{-4} **d** 8×10^{-14} **e** 2×10^4 **f** 2×10^{16} **g** 2×10^{-4} **h** 2×10^{-14}
2 **a** 360 000 **b** 570 000 **c** 0.0597 **d** 6000.003 **e** 5999.999 **f** 9×10^{-3}
3 **a** 7 080 000 **b** 6 920 000 **c** 7000.8 **d** 6999.2 **e** 0.708 **f** 0.692
4 **a** 1×10^7 **b** 1×10^5 **c** 1×10^{11} **d** 1×10^{-1} **e** 1×10^{-5} **f** 1×10^8 **g** 1×10^{-3}
5 **a** 500 000 **b** 590 000 **c** 599 000 **d** 599 900 **e** 599 990 **f** 599 999 **g** 599 999.9 **h** 599 999.99 **i** 599 999.999 **j** 599 999.9999
6 **a** 700 000 **b** 610 000 **c** 601 000 **d** 600 100 **e** 600 010 **f** 600 001 **g** 600 000.1 **h** 600 000.01 **i** 600 000.001 **j** 600 000.0001
7 **a** 5×10^{100} **b** 9×10^{100} **c** 1×10^{101} **d** 2×10^{101} **e** 4×10^{101} **f** 1×10^{102}

What do you think?

1 Benji has added the powers. The answer should be 2×10^3.
2 **a** $a = 9, b = 1$ or $a = 8, b = 2$, etc. to $a = 1, b = 9$
b $a + b = 10$ **c** 8
3 **a** 1×10^{10} **b** 1×10^{11} **c** 1×10^{19} **d** 1×10^{-7} **e** 1×10^{-8} **f** 4×10^{11} **g** 6×10^{19} **h** 6×10^{-4}
4 **a** Jackson is correct as the first sequence is geometric.
b Yes. If you keep counting in thousandths you will eventually get to all the numbers in the first sequence.

Consolidate

1 **a** 9×10^5 **b** 5×10^6 **c** 5×10^{-2} **d** 9×10^{-10} **e** 5×10^8 **f** 9×10^{47} **g** 9×10^{-9} **h** 5×10^{-9}
2 **a** 56 700 **b** 0.0765 **c** 56 000.07 **d** 6000.0705 **e** 700.0065
3 **a** 7×10^5 **b** 9×10^7 **c** 9×10^{-2} **d** 9×10^{15}
e 1×10^{16} **f** 1×10^{-1}
4 **a** 2×10^5 **b** 1×10^5 **c** 3×10^{-1} **d** 1×10^{-6} **e** 6×10^4 **f** 7×10^{-6}
5 **a** 8×10^5 **b** 6×10^6 **c** 6×10^{-2} **d** 4×10^5 **e** 4×10^5 **f** 2×10^5 **g** 1×10^6 **h** 0 **i** 1×10^{-1}

Stretch

1 **a** 10^{10} **b** 2×10^{10} **c** 4×10^{10} **d** 9 **e** 99 **f** 999
2 **a** 1×10^{12} **b** 1×10^3
3 **a** −6 **b** 9×10^{-6} **c** 4 **d** 94
4 **a** 10^9 **b** $p + q = 10^9$ **c** $f + g = 10^6$ **d** 10^{12}, one trillion

Check my understanding

1 **a** 24 390 **b** 9294
2 67.9
3 £457
4 23.1
5 The first two are equal as both numbers are reduced by 1 so the difference will be the same
6 8 − 5.23 = 2.77
7 e.g.

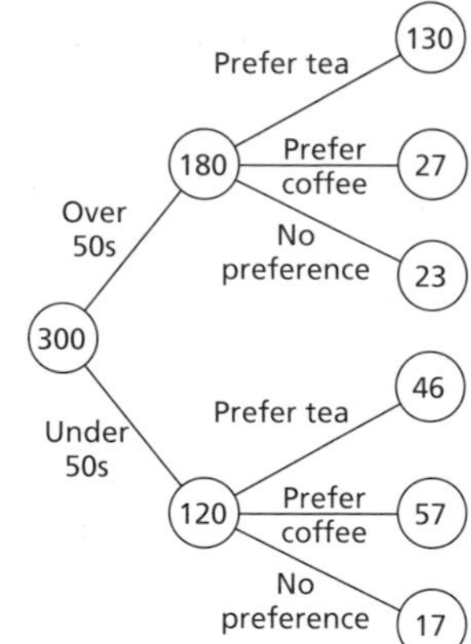

Block 7 Multiplication and division

Chapter 7.1

Are you ready?

1

×	5	7	8
2	10	14	16
4	20	28	32
6	30	42	48

2

×	4	6	8
3	12	18	24
7	28	42	56
9	36	54	72

3 **a** The last digit is a 0
b The last digit is a 0 or 5
c The last digit is a 0, 2, 4, 6 or 8
4 **a** 20, 30, 40
b 15, 20, 30, 40, 225
c 18, 20, 24, 30, 40

Practice 7.1A

1 **a** 1, 3, 5, 15 **b** 1, 2, 4, 8
c 1, 2, 3, 4, 6, 8, 12, 24
d 1, 7
e 1, 2, 3, 5, 6, 10, 15, 30
f 1, 5, 25 **g** 1, 13
h 1, 2, 4, 5, 8, 10, 20, 40
2 **a** Yes **b** Yes **c** No **d** No **e** Yes **f** Yes **g** No **h** No **i** Yes
3 1, 2, 3, 4, 5, 6, 10, 12, 15, 20
4 50, 80, 100, 200, 400
5 **a**

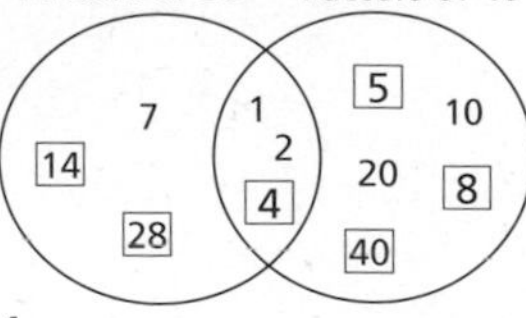

b 4
6 **a**

Factors of 20 Factors of 35
2
4
10
20
1
5
7
35

b 5
7 **a** 1, 3, 5, 15 **b** 1, 5, 25 **c** 1, 5 **d** 5
8 **a** 5 **b** 15 **c** 5 **d** 1 **e** 30 **f** 1 **g** 15 **h** 1

What do you think?

1 All positive integers have at least one factor, but numbers like 6.7 do not.
2 Any odd multiple of 3, e.g. 3 itself, 9, 33, etc.
3 **a** True **b** True **c** True **d** False (e.g. 10)
4 **a** There are many possible answers, e.g. 10 and 20, 30 and 40, 90 and 160, etc.
b There are many possible answers, e.g. 10, 20 and 30
5 **a** 48
b 60, 72, 84, 90 and 96
6 A and C

Practice 7.1B

1 **a** 7, 14, 21, 28, 35
b 8, 16, 24, 32, 40
c 10, 20, 30, 40, 50
d 9, 18, 27, 36, 45
e 20, 40, 60, 80, 100
2 **a** Yes **b** Yes **c** No **d** No **e** Yes **f** Yes **g** No **h** No **i** Yes
3 8, 16, 24, 32, 40
4 56, 63, 70, 77, 84
5 She hasn't written enough multiples. The LCM is 56.
6 **a** 10, 20, 30, 40, 50
b 15, 30, 45, 60, 75 **c** 30
7 **a** 42 **b** 24 **c** 18 **d** 12
8 **a** 40 **b** 24 **c** 60 **d** 120 **e** 45 **f** 90 **g** 90

What do you think?

1 **a** Sometimes, e.g. for 6 and 9, but not for 6 and 12 b Sometimes, e.g. for 6 and 7, but not for 6 and 9

2 a 5, 10, 15, 20, 25, 30, 35, 40. Last digits alternate between 0 and 5.
b 6, 12, 18, 24, 30, 36, 42, 48. Last digit follows a pattern 6, 2, 8, 4, 0, 6 …
c Last digit follows a pattern 8, 6, 4, 2, 0, 8 …

3 Yes, e.g. all integers are factors and multiples of themselves

4 a 30, 35, 40 b 30, 33, 36, 39
c 30, 40 d 30
e 30, 40

Consolidate

1 a 20, 40 b 20, 30, 40
c 12, 20, 24, 36, 40
d 15, 30 e 18, 36
f 12, 24

2 a 1, 5, 25
b 1, 2, 3, 4, 6, 9, 12, 18, 36
c 1, 2, 3, 6, 7, 14, 21, 42
d 1, 2, 4, 7, 14, 28
e 1, 2, 5, 10, 25, 50

3 a 10 b 8 c 12 d 20
Each smaller number is the HCF of the two numbers

4 a 20 b 16 c 24 d 40
Each larger number is the LCM of the two numbers

5 a 3 b 6 c 15 d 30

6 a 18 b 48 c 30 d 150

7 a HCF = 1, LCM = 30
b HCF = 2, LCM = 60
c HCF = 2, LCM = 60

Stretch

1 a Always b Always
c Always d Always
e Sometimes – compare examples as a class (e.g. 4000 and 1111)
f Sometimes – compare examples as a class (e.g. 4000 and 1111)

2 a Marta is correct
b 1, 2, 3, 4 and 6
c All the factors of x

3 Only if b is an integer

4 a Only one, which is 1
b All prime numbers, e.g. 2, 3, 5, 7, 11
c The squares of primes, e.g. 4, 9, 25, 49
d Numbers with exactly two prime factors, e.g. 6, 15, 34, etc.

5 The claim is true

6 Both are true for integer values of a

Chapter 7.2

Are you ready?

1 a 40 b 400 c $\frac{4}{10}$
d 4 e 40 000

2 a 0.07 b 0.7 c 1.7 d 0.17

3 a 40 b 70 c 80 d 70

4 a 100 b 1000

Practice 7.2A

1 a 370 b 3.7 c 3070
d 307 e 30.7 f 0.307
g 3700

2 a Multiplying the original number by 100
b No change
c Multiplying the original number by 1000

3 a 4600 b 406 c 4060
d 46 000 e 4600 f 4060
g 40.06 h 4600 i 460

4 a 3.7 b 0.037 c 30.7
d 3.07 e 0.307 f 0.003 07

5 a Dividing the original number by 100
b Dividing the original number by 1000
c Dividing the original number by 100

6 a 0.46 b 0.0406
c 40.6 d 4.6
e 0.0046 f 0.004 06
g 0.4006 h 0.046
i 0.46

7 a 100 b 1000 c 0.58
d 10 e 5.08 f 0.0508

8 a 100 b 100 c 920
d 1000 e 9200 f 9020

9 a × 10 b ÷ 100 c ÷ 100
d ÷ 100 e × 10 f × 1000

What do you think?

1 a $a = 78$ b $b = 0.78$
c $c = 7800$ d $d = 0.078$
e $e = 100$ f $f = 1000$

2 a 1000 b 10 c 10
d 10 000 e 100 f 530 000

3 10 and 100, or 1 and 10, or 2 and 20, etc.

4 a 720 b 420
c 11 800 d 1240

5 a 1.8 b 1.5
c 0.07 d 0.002

6 B is 1000 times smaller than C

Practice 7.2B

1 a Less than
b 36 cm = 36 ÷ 100 m = 0.36 m

2 a 7 m b 700 000 m
c 0.7 m d 70 m
e 7 000 000 m f 7 m
g 0.7 m h 70 000 m
i 0.07 m

3 a 8000 g b 80 000 g
c 80 g d 8 g
e 800 g f 0.8 g
g 8 g h 0.08 g
i 0.008 g

4 a 200 cl b 2000 cl c 20 000 cl
d 20 cl e 2 cl f 0.2 cl

5 a 0.7 m (= 70 cm) is greater than 70 mm (= 7 cm)
b i 3000 cm ii 80 cm
iii 2 km

6 Answers are opinions – compare with the class, but these are suggestions …
a cl b m c mm d g
e kg f l g cm h g
i mg j km

7 252 mm or 25.2 cm

8 3.33 kg or 3330 g

9 17.5 litres

What do you think?

1 a 250 g b 350 g c 6500 g

2 a 1 000 000 b 100 000

3 a 10 b 10 000
c i 10 ii 10 000 iii 450
iv 6 v 0.6
d i 10 ii 1
iii 8.4 iv 10 000

4 1 000 000 000

Practice 7.2C

1 × 0.1 and ÷ 10, × 0.01 and ÷ 100, × 0.001 and ÷ 1000

2 a 200 b 20.4 c 2.4
d 0.204 e 0.024 f 0.002 04

3 a 0.7 b 70 c 0.78
d 7.08 e 0.078 f 0.007 08
g 0.0078

4 18 × 0.1 and 180 ÷ 100, 1.8 × 0.01 and 18 ÷ 1000, 18 000 × 0.01 and 1800 ÷ 10

5 a 0.1 b 260
c 0.1 d 10 000

6 6.05 × 0.01, 650 ÷ 10 000, 65 ÷ 100, then the other three cards are all equal

What do you think?

1 a 86 b 3.6 c 0.93
d 8.8 e 0.52 f 4.8
g 0.63 h 0.048

2 a 60 b 2 c 62
d 0.7 e 8.3

3 a 8×10^{-3} b 8×10^{-4}
c 8×10^{-5} d 8×10^{-6}
e 1×10^{5} f 1×10^{10}
g 4×10^{6} h 4×10^{-7}

Consolidate

1 12.6 × 100 is greater

2 a 100 × 3.07 b 460 ÷ 10
c 40 × 100

3 Chloe has just added zeros and not changed the size of the number

4 a 10 b 100 c 1000
d 100 e 1000 f 1000

5 a 1.14 km b 15 cl
c 280 mg d 1 m
e 148 g f 1 m
g 30 cm h 22 cl

6 a $a = 650$ b $b = 6.5$
c $c = 650$ d $d = 0.065$
e $e = 605$ f $f = 6.05$
g $g = 6.05$ h $h = 6050$

Stretch

1 a 6×10^{24} b 6×10^{25}
c 6×10^{22} d 6×10^{20}

2 a 100 b $\frac{1}{100}$
c i 100 000 ii 10 000 000

3 Compare answers as a class

Chapter 7.3

Are you ready?

1 a 170 b 17
c 10.7 d 100.7

2 a 1.7 b 0.17
c 0.107 d 1.007

3

×	6	8	9
3	18	24	27
4	24	32	36
5	30	40	45

4 **a** 70 **b** 60 **c** 0.7 **d** 600

Practice 7.3A

1 **a** 560 **b** 560 **c** 5600
d 5600 **e** 56 000 **f** 560
g 56 000 **h** 5.6 **i** 0.56
j 0.0056
2 Both numbers have been multiplied by 10, so the answer is 4 × 5 × 100 = 2000
3 **a** 12 000 **b** 1800 **c** 2.8
d 2.4 **e** 6.3 **f** 420
g 7.2 **h** 0.21 **i** 14 000
j 0.042
4 630, 6300, 630
5 **a** 12 **b** 24 **c** 280
d 32 **e** 6300 **f** 28
g 5.6 **h** 1800 **i** 240
j 30
6 Compare answers as a class
7 **a** 187.6 **b** 18.76 **c** 1.876
d 1.876 **e** 0.1876

What do you think?

1 Compare answers as a class, e.g. 40 × 30, 80 × 15
2 Jackson is correct as one number has been multiplied by 100 and the other has been divided by 100
3 Compare answers as a class, e.g. 12 000 × 0.1, 800 × 1.5
4 Compare answers as a class, e.g. 0.38 × 46

Practice 7.3B

1 **a** 78 **b** 112 **c** 310
d 292 **e** 196 **f** 234
g 336 **h** 1038 **i** 1603
j 4104
2 **a** 7.8 **b** 11.2 **c** 3.1
d 29.2 **e** 1.96 **f** 23.4
g 3.36 **h** 1.038 **i** 1.603
j 41.04
3 **a** 22.8 **b** 20.4 **c** 33
d 57.6 **e** 1.548
4 **a** 1504 **b** 1638 **c** 1653
d 2952 **e** 2590 **f** 3060
g 1566 **h** 1431 **i** 6724
j 28 954
5 **a** **i** 3810 g **ii** 22 860 g
b **i** £44.94 **ii** £269.64
6 **a** 59.2 **b** 22.12 **c** 24.65
d 54.94 **e** 15.21
7 **a** $a = 578$ **b** $b = 147$
c $c = 1254$ **d** $d = 7.22$

What do you think?

1 **a** 80 × 60 = 4800
b On the second row he has done 75 × 6 instead of 75 × 60
2 **a** Their last digits were 1 and 7 or 3 and 9
b The last digits could also be 5 and 2, 5 and 4, 5 and 6, or 5 and 8

3 **a**
$$\begin{array}{r} 2\ \boxed{6} \\ \times\ \boxed{3}\ 6 \\ \hline 1\ 5\ 6 \\ \boxed{7}\ 8\ 0 \\ \hline 9\ \boxed{3}\ 6 \end{array}$$

b
$$\begin{array}{r} \boxed{5}\ 8 \\ \times\ 3\ \boxed{4} \\ \hline 2\ 3\ 2 \\ \boxed{1}\ 7\ 4\ \boxed{0} \\ \hline \boxed{1}\ \boxed{9}\ 7\ 2 \end{array}$$

c
$$\begin{array}{r} 7\ \boxed{1} \\ \times\ \boxed{6}\ \boxed{1} \\ \hline \boxed{7}\ 1 \\ 4\ \boxed{2}\ 6\ \boxed{0} \\ \hline \boxed{4}\ 3\ \boxed{3}\ \boxed{1} \end{array}$$

4 Compare answers as a class
5 **a** 35 442 **b** 2732.4 **c** 108.3

Practice 7.3C

1 **a** 40 cm² **b** 47 cm²
c 96 cm² **d** 187 mm²
e 315 cm² **f** 1628 cm²
2 **a** 60 cm² **b** 63 cm²
c 54 cm² **d** 76.8 cm²
e 1200 mm² **f** 21.93 cm²
3 **a** 192 cm² **b** 117 cm²
c 180 cm² **d** 337 cm²
e 96 cm² **f** 401 cm²
4 **a** 192 cm² **b** 600 cm²
c 161.4 cm²
5 **a** 57 600 ft² **b** 702 ft²
c 660 ft², 61.366 m²

What do you think?

1 **a** 1 × 72, 2 × 36, 3 × 24, 4 × 18, 6 × 12, 8 × 9
b All the numbers are the factors of 72
c 1 × 144, 2 × 72, 4 × 36, 6 × 24, 3 × 48, 8 × 18, 12 × 12, 16 × 9
2 **a** The length is 3 m (300 cm) not 3 cm. The 1 × 19 rectangle is a parallelogram, albeit a special case.
b 8400 cm² **c** 0.84 m²
d 840 000 mm²
3 Marta is correct as the only factors of 19 are 1 and 19. Benji is incorrect as you could draw a parallelogram with sides 19 cm and 2 cm with perpendicular height 1 cm.

Consolidate

1 24: 80 × 0.3, 8 × 3, 0.3 × 80, 3000 × 0.008
240: 80 × 3, 800 × 0.3, 30 × 8, 0.8 × 300
2.4: 0.8 × 3, 80 × 0.03
2 **a** 336 **b** 424 **c** 1215
d 3792 **e** 867 **f** 1482
g 2976 **h** 23 644
3 **a** 33.6 **b** 42.4 **c** 12.15
d 3792 **e** 0.867 **f** 14 820
g 2.976 **h** 2364.4
4 **a** 766 **b** 1214 **c** 535.6
5 **a** 53.66 **b** 104.69 **c** 27.416

6 **a** 266 cm² **b** 131.1 cm²
c 3634 mm² **d** 144 cm²
e 5208 mm² **f** 41.08 cm²
7 53.57 m²

Stretch

1 Compare answers as a class
2 **a** 13 × 4 = 52
b Compare answers as a class
3 **a** 370 **b** 3900 **c** 87 000
4 **a** 2100 **b** 2000 **c** 5600
5 Both numbers will be less than 100, so the product will be less than 100 × 100 (which is 10 000)
6 81, 891, 8991
Pattern is 8(9, 99, 999, etc.)1
89 991, 899 991, 8 999 991

Chapter 7.4

Are you ready?

1 **a** 12 **b** 8 **c** 6
d 4 **e** 3 **f** 2
2 **a** 10 **b** 7 **c** 4
d 2 **e** 1 **f** 0.5
3 **a** 800 ÷ 8 **b** 240 ÷ 8
4 **a** 8 **b** 7 **c** 8 **d** 83
e 68 **f** 0 **g** 60

Practice 7.4A

1 **a** 176 **b** 17.6
2 **a** 3.4 **b** 0.34
c 340 **d** 68
3 **a** 28 **b** 54 **c** 81
d 56 **e** 63.8 **f** 23.5
g 48.75 **h** 27 **i** 58
j 126
4 **a** **i** > 1 **ii** < 1 **iii** > 1
iv < 1 **v** > 1
b **i** 1.25 **ii** 0.8 **iii** 1.8
iv 0.375 **v** $3.\dot{3}$
c Use place value instead
5 All are correct
6 **a** £624 **b** 1.8 cm **c** 1.4 m

What do you think?

1 **b** 3 and 8, 8 and 3, 6 and 4
c **i** 487 **ii** 473 **iii** 47
d **i** e.g. 6 and 12, 8 and 9
ii e.g. 12 and 12, 9 and 16
iii e.g. 7 and 12, 6 and 14
2 The quotient will be less than 1
3 **a** You need to round $76.\dot{6}$ down to 76
b You round $36.\dot{1}$ up to 37

Practice 7.4B

1 **a** 13.5 cm **b** 12 cm
c 27 cm
2 **a** 42 cm² **b** 35.2 cm²
c 256.5 mm²
3 14 mm
4 **a** 246 cm² **b** 163.2 cm²
5 5.25 cm
6 **a** 180 cm² **b** 60 cm

What do you think?

1 **a** 365.4 cm² **b** 84 cm
2 Underestimate
3 **a** Compare answers as a class
b 6.8 m **c** 4.8 m

Consolidate

1 **a** 84 **b** 759 **c** 7509
d 8.4 **e** 4.2 **f** 1.25
g 1.09375

2 **a** 76 **b** 56 **c** 203 **d** 49
e 36 **f** 24 **g** 134

3 **a** $48\,cm^2$ **b** $50.4\,cm^2$
c $53.76\,cm^2$

4 **a** 6.7 cm **b** 11.4 mm
c 6.4 cm

Stretch

1 **a** 1 cm and 48 cm, 2 cm and 24 cm, 3 cm and 16 cm, 4 cm and 12 cm, 6 cm and 8 cm
b Checking you have all the factors of 48
c Compare answers as a class

2 **a** 0.5, $0.\dot{3}$, 0.25, 0.2, $0.1\dot{6}$, $0.\dot{1}4285\dot{7}$, 0.125, $0.\dot{1}$, 0.1, $0.\dot{0}\dot{9}$
b $\frac{1}{2}$, $\frac{1}{4}$, $\frac{1}{5}$, $\frac{1}{8}$ and $\frac{1}{10}$ have terminating decimals as their denominators are factors of 10, 100 or 1000

3 Compare answers as a class

4 **a** > **b** >
c Compare answers as a class, e.g. same as 270 ÷ 3

Chapter 7.5

Are you ready?

1 **a** 27 **b** 104
c 612 **d** 40.4

2 **a** 39 **b** 26 **c** 19.5
d 15.6 **e** 13

3 **a** 21.4 **b** 10.7 **c** 13.375
d 28.7 **e** 16.4

Practice 7.5A

1 **a** 5.5 **b** 6 **c** 6.5
d 7 **e** 7.5

2 **a** 23.8
b It decreases to 23
c It increases to 35

3 **a** 9.4 **b** 6.3

4 **a** 37.75 cm **b** 68.8 mm
c 8.48 cm **d** 96.4 cm

5 **a** Total is 178, so mean = $\frac{178}{5}$ = 35.6
b Larger **c** 33.8

6 9

7 180

8 Compare answers as a class – any five numbers with a total of 40

What do you think?

1 **a** Decrease **b** Stay the same
c Increase

2 **a** Increase **b** Decrease

3 **a** 3.25 **b** 13.25 **c** 103.25
d 32.5 **e** 65
Whatever has happened to the numbers in the set, this also happened to the mean.

4 **a** 35.94; add 30
b 594; × 100
c 598; add 4 to **b**
d 4.94, subtract 1

5 9

Consolidate

1 **a** 4 **b** 3 **c** 5 **d** 5

2 **a** 74%
b Mean is greater (median is 72%)
c 32%

3 **a** 60 **b** 60 **c** 690
d 69 **e** 690

4 30 196.5 cm

5 The mean is 198, so Seb is correct

6 **a** 4.2 **b** 1.9 **c** 2.3

7 63.35%

Stretch

1 **a** 13
b Compare answers as a class. In both cases the totals must be 28.

2 The third card must be 11 and the fourth card must be 13. You need one more piece of information to work out the remaining two cards (whose total must be 16).

3 Compare answers as a class, e.g. 1, 8, 8, 11 and 1, 7, 9, 11 etc.

4 1, 2, 5, 10, 25 or 50 (the factors of 50)

5 **a** $6a$ **b** $a + 7$

6 Chloe is right. You can show this algebraically (call the numbers a, $a + 1$ and $a + 2$) or by looking at bars to represent the numbers.

Chapter 7.6

Are you ready?

1 **a** 24 **b** 21 **c** 18
d 36 **e** 45 **f** 36

2 **a** 7 **b** 10 **c** 3
d 8 **e** 7 **f** 6

3 **a** 7 **b** 7 **c** 14 **d** 14
Both pairs of answers are the same

4 **a** 10 **b** 10 **c** 12 **d** 12
Both pairs of answers are the same

Practice 7.6A

1 **a** 18 **b** 18 **c** 18
d 6 **e** 18 **f** 6

2 **a** 36 **b** 36 **c** 36
d 64 **e** 64 **f** 4

3 **a** 10 **b** 28 **c** 70
d 70 **e** 16 **f** 7

4 **a** She has worked left to right instead of doing the multiplications first
b 80

5 5 + 2 × 20

6 **a** 14 **b** 24 **c** 26 **d** 54
e 104 **f** 196 **g** 27 **h** 19
i 5 **j** 13 **k** 19 **l** 49

7 **a** (7 + 3) × 4 + 2 = 42
b 7 + 3 × (4 + 2) = 25
c 7 + 3 × 4 + 2 = 21, no brackets needed
d (7 + 3) × (4 + 2) = 60

What do you think?

1 **a** Addition and subtraction have equal priority
b 11

2 Yes. Multiplication takes priority over addition so 4 × 5 would be done first anyway.

3, 4 Compare answers as a class – there are multiple solutions for each

5 150

Consolidate

1 **a** 14 **b** 20 **c** 14 **d** 8
e 20 **f** 12.5 **g** 3.5 **h** 6
i 24 **j** 50 **k** 72 **l** 2
m 5 **n** 10 **o** 2 **p** 2
q 1 **r** 2

2 (7 + 5) × 6

3 **a** 21 **b** 29 **c** 81 **d** 41

4 (15 – 7) ÷ 2; 4

5 Compare answers as a class

Stretch

All the numbers from 1 to 50 can be found, many in multiple ways. Compare answers as a class.

Chapter 7.7

Are you ready?

1 **a** $144\,cm^2$ **b** $42.4\,cm^2$
c $1104\,mm^2$

2 **a** $24\,cm^2$ **b** $3.6\,m^2$ **c** $20\,cm^2$

3 **a** $112\,cm^2$ **b** $1200\,cm^2$
c $2.75\,m^2$

4 **a** 26 **b** 13 **c** 156 **d** 78

5 **a** $2a$ **b** $2a$ **c** ab **d** ab
e a^2 **f** b^3 **g** $4ab$

Practice 7.7A

1 A, B, D, E and F

2 **a** $96\,cm^2$ **b** $30\,m^2$
c $1060\,cm^2$

3 $41\,m^2$

4 **a** $137\,cm^2$ **b** $105\,cm^2$

5 $1280\,mm^2$

What do you think?

1 **a** 8 cm **b** 6 cm

2 Compare answers as a class. Values of a, b and h need to satisfy $(a + b) \times h = 48$

Practice 7.7B

1 **a** 125 **b** 14 **c** 64
d 12.8 **e** 256 **f** 60

2 **a** $10m$ **b** $9m^2$ **c** 9
d $12m$ **e** $27m^2$ **f** 3

3 **a** $25p$ **b** $5p^2$ **c** $25p^2$
d $5pq$ **e** $25pq$ **f** $25p^2q$
g $25p^3q$ **h** $25p^2q^2$

4 **a** $12x^2y^3$ **b** $6x^2y^3$ **c** $12xy^3$
d $6x^2y^2$ **e** $4xy^2$ **f** $8y^3$
g $8x^2y$ **h** $3y$

5 **a** $8a^2b^2$ **b** 2 **c** $8a^3b^3$
d $6a^2b$ **e** $4a^2$ **f** $144a^4b^2$

6 **a** $25x^2$
b **i** $15y^2$ **ii** $14y^2$ **iii** $12y^2$
iv $12y^2$ **v** 0

What do you think?

1 **a** $32p^2q\,cm^2$ **b** $32pq^2\,cm^2$
c $16p^2q^2\,cm^2$

2 Compare answers as a class

3 You can simplify any expressions you multiply but you can only add like terms, and $3m$ and $4n$ are unlike

Consolidate

1 a 120 cm^2 b 38.5 cm^2
2 a $2xy\text{ cm}^2$ b $7xy\text{ cm}^2$
3 a $7c$ b $7bc$ c $49bc$
d $7bc^2$ e $49b^2c$ f $10fg$
g $10g$ h $10f$ i 10
j 5
4 H, F, A, C, B, D, G, E
5 a $18pq$ b $2pq$ c $72pq^2$
d $72p^2q$ e $72p^2q^2$ f $2q$
g 2 h $72p^3q^2$ i $2q$
j $2p^2q$

Stretch

1 a Perimeter b Perimeter
c Area d Area
e Area f Perimeter
g Area
2 $(9xz + 12yz)\text{ cm}^2$
3 Compare answers as a class
4 Only sometimes true as they are equal when $x = 0$

Check my understanding

1 a 6 is a factor of 6 as it divides exactly into 6. 6 is a multiple of 6 as it can be found by multiplying 6 by 1.
b No
2 a 270 b 432 c 4.32
d 0.348 e 5.8
3 a 760 b 7060
c 489.6 d 3.4
4 a 33.3 cm^2 b 7.4 cm
5 Any two numbers that together add up to 10
6 $19 - 4 \times 3 = 7$, $(19 - 4) \times 3 = 45$, $19 \times (4 - 3) = 19$, $19 \times 4 - 3 = 73$, $19 + 4^3 = 83$, so $19 + 4^3$ is largest
7 108.5 cm^2
8 $48 \div 8 = 6$ so the answer should be $6c$

Block 8 Fractions and percentages of amounts

Chapter 8.1

Are you ready?

1 a $\frac{1}{3}$ b $\frac{2}{3}$ c $\frac{1}{5}$ d $\frac{4}{5}$
2 a 15 b 10 c 7.5
d 6 e 3 f 2.5
3 a 5 b 7.5 c 12.5
d 17.5 e 27.5
4 a 15 b 25 c 35
d 10 e 280

Practice 8.1A

1 a 300 b 200 c 150
d 120 e 100 f 60
g 50 h 40 i 30
j 24 k 20 l 15
m 12 n 10 o 6
p 5 q 4 r 3
s 2
2 Compare answers as a class
3 a 20
b i 60 ii 100 iii 120
iv 40 v 80 vi 180
4 a 12 b 15 c 24
d 12 e 36 f 112
5 You can, but the answer is not an integer
6 a 7.5 b 37.5 c 6.4
d 37.5 e 12.5 f 33.6
7 a = b < c <
d = e >
8 Compare answers as a class

What do you think?

1 a 120 b 800
c 1024 d 21.6
Compare methods as a class
e 135 f 210 g 580
2 a e.g. Divide by 5 and then divide by 3
b e.g. Divide by 10 and then divide by 4
c e.g. Divide by 5 and then divide by 5
d e.g. Divide by 100 and then divide by 5

Practice 8.1B

1 a 160 b 240 c 400
d 560 e 800 f 720
2 a 20 b 80 c 50
3 a 21 b 64 c 63
d 125 e 30 f 77
4 a £30 b £20
5 24
6 a 50 b 150 c 75
d 40 e 2.25

What do you think?

1 a $a = 110$ b $b = 42$
c $c = 200$ d $d = 63$
2 a $40ab$ b $100ab$ c $50ab$
d $25ab$ e $14ab$
3 In the first question you are finding a fraction of the whole and in the second you are finding the whole.

Consolidate

1 a 45 b 20 c 6
d 20 e 7 f 90
g 3 h 2 i 4
j 4 k 2 l 200
2 a 135 b 40 c 18
d 180 e 35 f 810
g 21 h 46 i 44
j 76 k 82 l 3400
3 a 660 b 1000 c 1800
4 a 240 b 144 c 160
d 140 e 132 f 800
5 a 64 b 100

Stretch

1 a $\frac{2}{5}$ of 60 is 24 but $\frac{1}{5}$ of 30 is 6
b i 120 ii 56 iii 360
2 a $2x$ b $2x$ c $\frac{1}{2}x$
3 $\frac{7}{15}$ is less than $\frac{1}{2}$ but $\frac{11}{20}$ is greater than $\frac{1}{2}$
4 45

Chapter 8.2

Are you ready?

1 a $\frac{47}{100}$ b $\frac{81}{100}$
c $\frac{70}{100} = \frac{7}{10}$ d $\frac{7}{100}$
2 a 50% b 25% c 75%
3 a 10% b 30%
c 70% d 20%
4 a 0.64 b 0.6 c 0.06

Practice 8.2A

1 a 4 b 3 c 18
d 15 e 7.5 f 22.5
2 a 4 b 8 c 9
d 20 e 30 f 120
3 a 4 b 4.2 c 12
d 12.6 e 6 f 3
g 9 h 12 i 48
j 6.2 k 12.5 l 240
m 9 n 99 o 13.8
p 53.3
4 a £50 000 b £900
5 Compare answers as a class
6 a 20 b 20 c 3 d 30

What do you think?

1 The answer is 15, not 15%
2 Both are 24
3 Compare answers as a class
4 a i 15 ii 15 iii 36 iv 36
b 50% of 30 = 30% of 50 and 90% of 40 = 40% of 90
a% of b = b% of a
c i 34 ii 21 iii 30

Practice 8.2B

1 a 0.62 b 0.91 c 0.4
d 0.04 e 0.34 f 0.342
2 a 279 b 6370 c 52.8
d 3.52 e 2108 f 5814
3 a 50.4 cm b 81 cm
c 3.655 cm
4 30.38 kg
5 a £240 b £253.80
c £23 d £11.50
e 9 kg f 19.2 g
g 63.84 mg
6 Compare methods as a class

What do you think?

1 a 58 b 5800
c i Multiply 58 by 46
ii Multiply 58 by 81
2 e.g. Divide 60 by 4, multiply the result by 7 and then multiply this result by 0.37
3 a 1020 b 102 c 10.2
4 a 11.47 kg b 0.93 kg
c 0.248 kg

Consolidate

1 a 12 b 36 c 10
d 10 e 32 f 150
2 a 9 g b 45 g c 54 g
d 81 g e £24 f £12
g £36 h £204 i 25.2
j 36 k 42 l 49.5
m 4 n 0.4 o 2.8
p 126
Compare answers as a class
3 a £140 b £52.50
c £38.50 d £11.20

4 a 303.4 **b** 232 **c** 26
d 32.64 **e** 1314 **f** 13.02
5 a 30 **b** 19

Stretch

1 a 12.5%
b 37.5%, 62.5%, 87.5%
c Compare answers as a class
d 37.5% = $\frac{3}{8}$, $\frac{3}{8} \times 48 = 18$
62.5% = $\frac{5}{8}$, $\frac{5}{8} \times 32 = 20$, which is greater
2 a Compare methods as a class
b 8 **c** 40
3 a 180 **b** 80 **c** 500
d 480 **e** 75
4 30
5 a Filipo is correct, as 10% is the same as $\frac{1}{10}$ and 0.1
b Multiplying a number by 0.01 gives 1% of the number

Chapter 8.3

Are you ready?

1 a 18 **b** 54 **c** 126
2 150
3 a 5 **b** 15 **c** 45 **d** 32.5
4 a £88.80 **b** 6.72 kg

Practice 8.3A

1 a
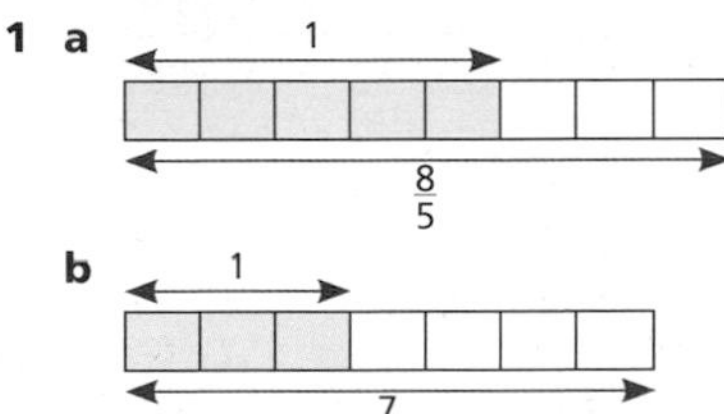

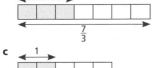

b

c
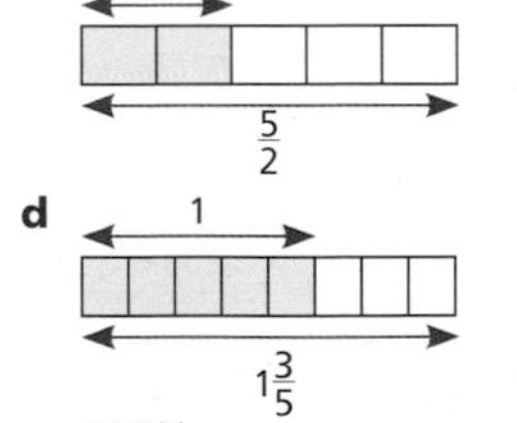

d

2 a 125%
b
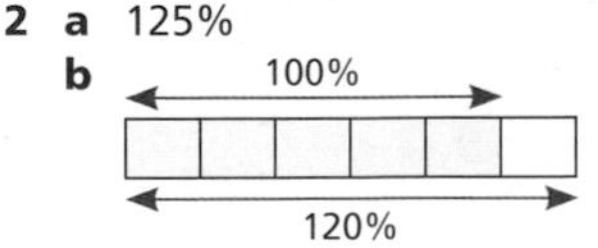

c 100% / 200%
d
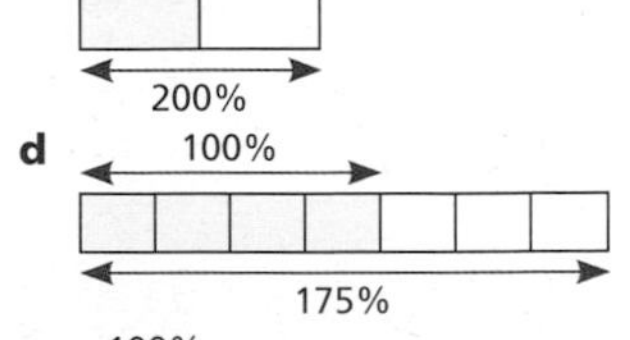

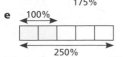

e 100% / 250%
3 Missing numbers are 30, 30 and 150
4 a 75 **b** 21 **c** 140 **d** 45
5 a 75 **b** 60 **c** 108

d 200 **e** 42
6 A and D do not make sense. B and C are possible.
7 151.8 cm

What do you think?

1 a 100% = 1 whole, 100% + 100% = 200% = 2 wholes
b i 200% **ii** 400% **iii** 150%
2 a e.g. 0.101 < 1, or multiplying by 0.101 would find 10.1%
b 1.01
3 Ed is incorrect. For example, if you increase 100 by $\frac{1}{5}$ you get 120;
decrease 120 by $\frac{1}{5}$ and you get 96 not 100.
4 a 125 **b** $166\frac{2}{3}$ **c** 14

Consolidate

1 200% = $\frac{12}{6}$, 180% = $\frac{9}{5}$, 150% = $\frac{3}{2}$,
110% = $\frac{11}{10}$, 240% = $\frac{12}{5}$
2 a 50 **b** 350 **c** 375
d 165 **e** 195 **f** 270
g 213
3 a 135 **b** 315 **c** 360
d 450 **e** 396 **f** 240
g 480
4 $133\frac{1}{3}$%
5 a 180% **b** 20%
6 a $a = 20$ **b** $b = 6$ **c** $c = 21$

Stretch

1 Compare answers as a class
2 a e.g. $0.75x = 1.2y$
b 96
3 £125 000

Check my understanding

1 a 16 **b** 100
2 a Compare answers as a class
b i 31.5 **ii** 33.3 **iii** 31.5
iv 26.95 **v** 0.18 kg
3 39
4 800
5 A is possible but B is not
6 60

Block 9 Directed number

Chapter 9.1

Are you ready?

1 a 13 **b** 9
2 2 – 8, 3 – 5, 140 – 165, 11 – 63
Second number is larger than first number
3 a 10, 9, 8, 7, 6, 5, 4, 3, 2, 1, 0, –1, –2, –3, –4, –5, –6, –7, –8, –9, –10
b 10, 8, 6, 4, 2, 0, –2, –4, –6, –8, –10
c 10, 5, 0, –5, –10, –15, –20, –25, –30
4 A = 6, B = 12, C = 17

Practice 9.1A

1 a 5 **b** –5 **c** 3
2 a –1 –1 –1 –1
b +1 +1 +1 +1 +1 +1
c –1 –1
3 a A = 4, B = 9, C = –2, D = –9
b –9, –2, 4, 9
4 a 12°C **b** –7°C
5 a A = 14, B = –5, C = 6
b D = –15, E = –5, F = 17
6 a Yasmin has read the scale incorrectly.
b
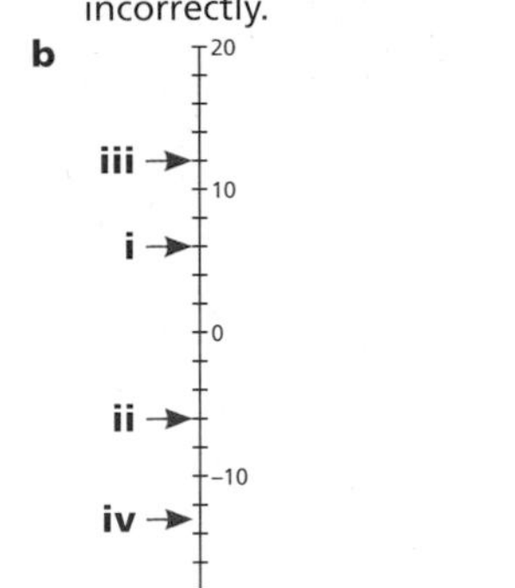

7 a 11, 0, –6, –8
b 197.5, 88.9, 14.5, –109.4, –129.6
8 a Helsinki **b** Sydney
c Any temperature between –7 and –1
9 a < **b** > **c** > **d** >
10 a –10 –9 –8 –7 –6 –5 –4 –3 –2 –1 0 1 2 3 4 5 6 7 8 9 10
b Various answers
11 a 5 **b** –2 **c** 6
12 a –5 **b** 5 **c** –6
d 3 **e** 13 **f** –9
g –7 **h** –12 **i** –6.5
j –11
13 a 9 **b** 8 **c** 4
d 13 **e** 11 **f** 14
14 –2 is wrong (the next number in the sequence should be –8). He has just followed the pattern in the digits.
15 a 8°C **b** 9°C
c It rises until 3 p.m. then begins to fall again
d –3°C

What do you think?

1 a –17.5
b The distances between 0 (zero) and A and 0 (zero) and B are the same
2 They are the same
3 a –24 e.g. only compared 23 and 47
b i –4 **ii** –1.2
4 a i –4 **ii** $-4a$ **iii** $-4x$
b –4 is in all of the answers
c i $-2p$ **ii** $3m$ **iii** $-14q$
iv $-6a$ **v** $-10b$

Consolidate

1 a –5 **b** 5 **c** –5
d –12 **e** –5 **f** –93

g −13 h −9 i −87
2 2, −1, −3, −6
3 a A = −12, B = −4, C = 6, D = 15
b 8 c 19

Stretch
1 a −3 b −2
2 −13.1
3 129.55 m
4 a −87 b −76 c −11.9
d −624 e −0.72 f −3.815
5 a 13 b 11 c 12 d 15

Chapter 9.2

Are you ready?
1 4 − 10, 2 − 8, 3.71 − 5.95
The number being subtracted is greater than the number it is subtracted from
2 a −5 b −8 c −13
d 3 e 8 f −8
3 9 °C
4 a 16 b 8 c 8 d 16

Practice 9.2A
1 Two zero pairs cancel, so 5 + −2 = 3
2 a 4 b −4 c 3
d 3 e −6 f −7
3 Jakub can add in three zero pairs and then subtract six positive counters, giving 3 − +6 = −3
4 a 3 b −2 c −5
d 9 e 4
5 9 − +3 → 3 − 9; 9 − −3 → 9 + 3; 9 + −3 → 9 + −3; −9 + 3 → −9 − 3; −9 − +3 → 3 − 9; −9 − −3 → −9 − 3 (matching lines)
6 a 3 b −5 c 7
d 15 e −9 f −15
g −4 h 4 i 29
7 Example answers are given
a 9 + 5 = 14 b 9 − 6 = 3
c −10 + −4 = −14
d 6 − 9 = −3 e 6 − −4 = 10
f 5 − 6 = −1
8 a i −3 ii 3 iii −13 iv 13
b i −8 ii 8 iii 8 iv −8
9 a 9 b −11 c 1.2 d −11
10 a x b $-10y$
11 Compare answers as a class

What do you think?
1 a 3 b 8 c 2 d −5
2 a 5 b 5 c −7
d −7 e 4 f 4
3 Jackson is incorrect. As 12 is greater than 7, the number being subtracted must be negative, so it is −5.

Consolidate
1 a 9 b −3 c −3 d −3
2 a −3 b −2 c 6 d −20
e −12 f 15 g 5 h −11
i −10 j −75 k −75 l 0
3 a 6 b −6 c 8 d −6

Stretch
1 When you add three positive counters and six negative counters there are three zero pairs, which leaves three negative counters, so the answer is −3.
To subtract six positive counters from three positive counters you need to add in three zero pairs; then you can take away six positive counters to leave three negative counters. So the answer is −3.
2 −4.3 °C − −12.9 °C = 8.6 °C
3 a −2 − −6 = 4 or +7 − +3 = 4
b −6 − −2 + +3 = −1
c +7 + +3 − −6 = 16
d −6 − +7 + −2 = −15

Chapter 9.3

Are you ready?
1 a 35 b 32 c 171
d 130 e 900 f 3.57
2 Various answers
3 a 4 b −9 c 16
d −15 e −1 f 10
4 a 5 b 12 c 24

Practice 9.3A
1 Compare observations as a class

	−3	−2	−1	0	1	2	3
3	−9	−6	−3	0	3	6	9
2	−6	−4	−2	0	2	4	6
1	−3	−2	−1	0	1	2	3
0	0	0	0	0	0	0	0
−1	3	2	1	0	−1	−2	−3
−2	6	4	2	0	−2	−4	−6
−3	9	6	3	0	−3	−6	−9

2 a Positive: −5 × −4, 11 × 17, −176 × −32, −132 ÷ −10
Negative: −3 × 8, 10 × −5, −8 ÷ 2, 64 ÷ −2
b Various answers
3 a −18 b 36 c −20
d 15 e −35 f −22
4 a 5 b −5 c 2
d −9 e 8 f −5
5 Examples are:
a −6 × −2 b 12 × −1
c −24 ÷ 4 d −6 ÷ −1
6 No. The product of a positive number and a negative number gives a negative answer.
This means that *exactly* one of the numbers is negative, not *at least* one of the numbers.
7 a i −4 ii 4 iii −4
b i −28 ii 28 iii −28
8 a $-6a$ b $6a$ c $-3a^2$

What do you think?
1 a Either both of the numbers are positive or both of the numbers are negative
b The value of the cards, and whether they are both positive or both negative
2 e.g. −5 × −2 = 10 and 10 × 3 = 30 so −5 × 3 × −2 = 30
3 a There are three negative numbers and one positive: two negative numbers will multiply to give a positive number, and then the third negative number will make the answer negative. Multiplying this by a positive number means that the answer should be negative.
b Positive. The numerator will be negative and the denominator is negative so dividing a negative number by a negative number gives a positive answer. The answer is 1649.2.

Consolidate
1 a −36 b −24 c 22
d −45 e 34 f −36
2 a i −8 ii −6
iii 12 iv −8
b i −32 ii −32
iii 32 iv 8
3 a −2 b 5 c −3
d 8 e −6 f −9
4 a Yellow and green
b Red and grey
c

	−8	−5	−3	2	7
−8	64	40	24	−16	−56
−5	40	25	15	−10	−35
−3	24	15	9	−6	−21
2	−16	−10	−6	4	14
7	−56	−35	−21	14	49

Stretch
1 −7
2 a 10 and −2 b −10 and 2
3 Both numbers must be negative.
4 a Either two numbers are positive and one is negative, or all three are negative.
b e.g. $a = 1$, $b = -2$ and $c = 10$
5

×	−13	−7	−9	8
11	−143	−77	−99	88
4	−52	−28	−36	32
−3	39	21	27	−24
−2	26	14	18	−16

Chapter 9.4

Are you ready?
1 a 6 b 1 c 11
d 36 e 2 f −5
2 a 18 b 16 c 196
d 7 e 40 f 120
3 a It multiplies the number of small packets by 3 (because there are 3 biscuits in each) and the number of large packets by 15 (because there are 15 biscuits in each) and then adds these to give the total number of biscuits.
b 81
4 a 13 b −32 c −12
d 15 e −10 f −4
g 10 h 36 i −3

Practice 9.4A

1 a Yes
  b No, because she should have worked out 4 – –9 = 13
2 a –6 + 12 = 6 and 12 + –6 = 6
  b –6 – 12 = –18 and 12 – –6 = 18
3 a 5 b –5 c 3
  d 16 e 24 f –24
  g –8 h 40 i 8
  j –80 k 400 l 16
4 10 + –2 = 8, 3 – –5 = 8
5 a 3 b 0 c 19 d 17
  e 30 f –90 g 270 h 10
  i 7 j 181
6 –53
7 a $x = 5$, $y = -2$ and $z = 3$
  b Addition is commutative so the sum of the three values does not depend on the order of addition

Practice 9.4B

1 a – 27.14 b –24.375
  c –4.6 d –77.36
2 a 41.9 b –4.965
  c –4.168 125 d 0.032 25
  e 0.251 655 6291
  f 137.933 125
3 a 59.9 °F b 12.92 °F

What do you think?

1 a $b + d$ b $b - d$ or $d - c$
  c $c \times d$
  d $d \div a$, $b \div d$, $c \div b$
2 There are several possible answers, e.g. $T = -1$, $S = -2$, $U = 3$, $V = 6$

Consolidate

1 a –2 b –7 c 17
  d –17 e –24 f –60
  g –120 h –2.4 i 184
  j –9 k 25 l 144
2 a –4 b 10 c 1
  d –17 e 30

Stretch

1 a $\frac{1}{2} - \frac{1}{4} = \frac{1}{4}$ b $\frac{1}{2} - \frac{1}{4} - 1 = -\frac{3}{4}$
2 a The sum of the negative numbers is obviously greater than the sum of the positive numbers. So the total of the numbers is negative. Dividing by +6 will give a negative answer.
  b $-4.1\dot{3}$ c 48.3
3 5.6
4 Any correct justification

Chapter 9.5

Are you ready?

1 a $x + 26 = 40$ b $x = 14$
2 a $x = 6$ b $m = 4$
  c $g = 12.5$ d $y = 35$
  e $k = 22.2$ f $f = \frac{1}{2}$ or 0.5
3 a $u = 4$ b $m = 9$ c $g = 14$
  d $n = 420$ e $y = 6$ f $f = -19$

Practice 9.5A

1 $2k + 7 = 17$
  – [7]   – [7]
  $2k$ = [10]
  ÷ [2]   ÷ [2]
  k = [5]
2 a $5 + 4y = 49$ b $y = 11$
3 a $m = 6$ b $t = 4$
  c $g = 2.5$ d $m = 0.5$
  e $m = 9$ f $k = -3$
  g $t = \frac{8}{3}$ h $k = -5$
4 a The difference between $3x$ and 4 is 11
  b $x = 5$
5 a Substitute the value of x back into the original equation and check that it gives the correct answer.
  b 5(2) – 1 = 10 – 1 = 9 (not 11)
  c He should have added 1 in his first step not subtracted it.
6 a $p = 3$ b $g = 8.5$
  c $d = 1.9$ d $k = 13.4$
  e $y = -5$
7 Various answers. Common mistakes are subtracting 10 and then writing $3m = -3$, and ignoring the sign on the $3m$.
8 a $x = 12$ b $x = 14$
  c They both have the same operations applied to x but in a different order.
9 a $u = 7.5$ b $y = 90$
  c $n = 154$ d $m = -1.8$
10 Various answers, e.g. collect like terms first to give $5a + 12 = 2$ and then solve to give $a = -2$

What do you think?

1 a The cups represent x and the counters represent ones
  b The two yellow counters on the left-hand side would be red counters
2 a A x → [× 2] → [+ 5] →
  B x → [× 5] → [+ 2] →
  C x → [÷ 2] → [+ 5] →
  D x → [+ 2] → [÷ 5] →
  E $5x + 2$
  F $\frac{x+2}{5}$
  G $2x + 5$
  H $\frac{x}{2} + 5$
  b Work backwards through the function machine with output $5x + 2$
3 a i $x = 7$ ii $x = 7$
  b i $x = 7$ ii $x = -7$
  c Check answers with a partner
4 Her equation should be $3(x + 2) = 72$ or $3x + 6 = 72$
5 Any four two-step equations that give a solution of $w = 5$. Check answers with a partner.

Consolidate

1 a $x = 4$ b $m = 4$
2 e.g. 2 + 5(9) = 2 + 45 = 47
3 a $x = 2$ b $y = 5.5$
  c $u = 16$ d $m = 1.4$
  e $t = 11$ f $p = -19$
  g $t = -2$ h $y = 1.2$
  i $h = -1.5$ j $h = -2.5$
  k $x = -1.25$ l $x = -3$

Stretch

1 Because they are the same / they are just written in a different order. Addition is commutative.
2 a $b = 2.2$ b $f = -2$
  c $g = 4.125$ d $x = 0.5$
3 You could solve them both in the same way, as they are equivalent. The fractional equation looks harder but it is the same.

Chapter 9.6

Are you ready?

1 a 2 × 3 = 6 and 6 + 5 = 11
  b 3 × 5 = 15 and 2 + 15 = 17
  c They might do 2 + 3 = 5 first and then 5 × 5 = 25
2 a $\underline{7 \times 4} + 3$ b $6 + \underline{3 \times 6}$
  c $9 - \underline{4 \div 2}$ d $2 \times (\underline{3 + 5}) + 4$
  e $20 - 3 \times 4 + 6$
f $4 \times \underline{5^2}$
3 a 31 b 24 c 7
  d 20 e 14 f 100
4 No – addition and subtraction are of equal priority, so you work from left to right.

Practice 9.6A

1 4 × –2 = –8 and 3 + – 8 gives three zero pairs so the answer is $\underline{-}5$.
2 a $4 + \underline{3 \times -6}$ b $(\underline{4 + 3}) \times 6$
  c $\underline{3 \times -6} + 4$ d $4 + (\underline{3 \times -6})$
  e **a**, **c** and **d** will give the same answer because the multiplications are the same and addition is commutative so $4 + \underline{3 \times -6} = \underline{3 \times -6} + 4 = 4 + (\underline{3 \times -6})$.
3 a i $5 + \underline{10 \times -2} = -15$
    ii $5 - \underline{10 \times -2} = 25$
  b i $-6 - \underline{3 \times 5} = -21$
    ii $6 - \underline{3 \times 5} = -9$
  c i $7 \times (\underline{3 - 5}) = -14$
    ii $-7 \times (\underline{3 - 5}) = 14$
  d i $1 + \underline{8 \times 3} - 4 = 21$
    ii $1 - \underline{8 \times 3} - 4 = -27$
  e i $11 + \underline{10 \div -2} = 6$
    ii $11 - \underline{10 \div -2} = 16$
  f i $4 - \underline{10 \div 5} = 2$
    ii $4 - \underline{10 \div -5} = 6$
  g i $(\underline{2 + 5}) \times 6 + 2 = 44$
    ii $(\underline{-2 + 5}) \times -6 + 3 = -15$
  h i $11 \times (\underline{2 - 10}) = -88$
    ii $-11 \times (\underline{2 - 10}) = 88$
4 2 × 5 = 10 (10 yellow counters) and 3 × –4 = –12 (12 red counters). There are 10 zero pairs and 2 red counters so the answer is $\underline{-}2$.
5 a Subtraction then addition from left to right (7 – 2 + –15)
  b –10
6 a i $\underline{4 \times -2} + \underline{3 \times 5}$
    ii $\underline{4 \times -2} + \underline{3 \times -5}$
    iii $\underline{4 \times -2} - \underline{3 \times 5}$
  b i $\underline{(2 - 10)} \div \underline{(9 - 7)}$
    ii $\underline{(2 - 10)} \div \underline{(7 - 9)}$
    iii $\underline{(2 - 10)} - \underline{(9 - 7)}$
  c i $3 + \underline{4 \times -6 \div 2}$
    ii $3 - \underline{4 \times -6 \div 2}$

iii $3 + \underline{4 \times -6 \div -2}$

7 a i 6 ii −2
b i 2 ii −2 iii −6
c i 7 ii 2 iii −8

8 a i $4 \times 2 - (9 + 1) = -2$
ii $4 \times (2 - 9) + 1 = -27$
b i $3 \times (5 - 3) - 10 = -4$
ii $3 \times 5 - (3 - 10) = 22$

9 a 5 × 4 is 20, add 5 to give 25, then take the square root to give 5
b Take the square root of 16 first (= 4), then 2 × 4 is 8, then subtract 3; the answer is 5

10 a −27 b 3 c 25 d −4

What do you think?

1 a Both methods give an answer of 6
b This calculation highlights the need to work from left to right. If you do the multiplication first you get a different answer.

2 a When you do the multiplications there is a positive 2 and a negative 18 in each so they both give the same answer.
b 3 × −4 = −12, adding a negative is the same as subtracting so the calculations are equivalent.

3 a Various answers
b e.g. 4 × −2 + 8
c 41 d −76

Consolidate

1 −2 × 4 gives 8 red counters; then add 6 yellow counters. There are 6 zero pairs and 2 red counters so the answer is −2

2 a 0 b −13 c −13 d 27
e −38 f 68 g 7 h 17

3 3 multiplied by 10 is 30 and 5 multiplied by −1 is −5. 30 plus −5 is 25; then take the square root to give 5

4 a $(5 + 3) \times -2 = -16$
b $5 + (3 \times -2) = -1$
c $10 \div (-2 + 12) \times 1 = 1$

Stretch

1 Multiplication is commutative so 0.765 × 3.96 = 3.96 × 0.765; then 5.61 × −7.134 will give a negative answer; adding a negative number is the same as subtracting so the calculations are equivalent.

2 9, −9 The first expression means −3 × −3, which has a positive answer, but the second expression means −(3 × 3) and its answer is negative

3 a −7 b −24 c 28 d 89

4 a −36 b 6 c −2
d −5 e 5 f 11

Chapter 9.7

Are you ready?

1 a 25 cm^2 b 64 cm^2
c 3600 mm^2 d 625 mm^2

2 a 8 b 27 c 64

3 a 9 b 49 c 81 d 144
e 9 f 49 g 81 h 144

Practice 9.7A

1 a 7 b 10 c 12
d 1 e 50

2 a $x = \pm 7$ b $y = \pm 10$
c $b = \pm 12$ d $a = \pm 1$
e $p = \pm 50$

3 a $\sqrt{49}$ missing on top, 6 and 8 missing on bottom
b i 7 and 8 ii 5 and 6
iii 6 and 7 iv 7 and 8

4 a 7.1 b 10.2 c 2.8
d 4.1 e 8.9

What do you think?

1 No, as a length can't be negative. The side length of the square is 15 cm.

2 $-4^2 = -16$ and $(-4)^2 = 16$ because -4^2 is the negative of 4^2 whereas $(-4)^2$ is −4 squared.

3 There are no numbers that give a negative answer when squared, so −25 doesn't have a square root, **or** the square of −5 is 25 not −25

Practice 9.7B

1 b 10 × 10 × 10 × 10 = 10 000
c 1 × 1 × 1 × 1 × 1 × 1 × 1 × 1 = 1
d 3 × 3 × 3 × 3 × 3 = 243

2 a 1, 1, 1, 1, 1
b 2, 4, 8, 16, 32
c 3, 9, 27, 81, 243
d 4, 16, 64, 256, 1024
e 5, 25, 125, 625, 3125
All the numbers are positive.
If the number being raised to a power is even, all the numbers are even; if the number being raised to a power is odd, all the numbers are odd.

3 a 3 b 2 c 10
d 5 e 7

4 a 5 b 1296 c 4

5 a −1, 1, −1, 1, −1
b −2, 4, −8, 16, −32
c −3, 9, −27, 81, −243
d −4, 16, −64, 256, −1024
e −5, 25, −125, 625, −3125
They are the same numbers as in question 2 but the odd powers are negative and the even powers are positive

What do you think?

1 a Even powers give a positive answer and odd powers give a negative answer. The product of two negative numbers is positive. When the power is even, the numbers can be grouped in pairs, so the result will always be positive.
b i A negative number raised to an odd power gives a negative answer.
ii A negative number raised to an even power gives a positive answer.
c i Positive ii Positive
iii Negative iv Positive

2 a Ellie is correct, as a negative number raised to an even power gives a positive answer.
b 531 441 c 531 441
−27 raised to an even power is the same as 27 raised to the same power

3 −248 832

Consolidate

1 a 25 b 49 c 81 d 100
e 64 f 25 g 49 h 81
i 100 j 64

2 b 7 × 7 × 7 × 7
c 2 × 2 × 2 × 2 × 2 × 2 × 2 × 2
d 10 × 10 × 10
f −7 × −7 × −7 × −7
g −2 × −2 × −2 × −2 × −2 × −2 × −2 × −2
h −10 × −10 × −10

3 a 243 b 2401 c 256
d 1000 e −243 f 2401
g 256 h −1000

4 a 5 b 7 c 11 d 50

5 a $x = \pm 5$ b $q = \pm 7$
c $s = \pm 11$ d $t = \pm 50$

6 a 4 b −8 c 17 d 3

Stretch

1 564 $-x$ raised to an even power is the same as x raised to the same power

2 −43 −43 raised to an odd power is the negative of 43 raised to the same power

3 $8^2 = 64$ and $9^2 = 81$ so x could be between 8 and 9, but also $(-8)^2 = 64$ and $(-9)^2 = 81$ so x could be between −8 and −9

4 a $10 < b < 11$ or $-11 < b < -10$
b $11 < y < 12$ or $-12 < y < -11$
c $2 < z < 3$ or $-3 < z < -2$
d $19 < x < 20$ or $-20 < x < -19$

5 a There is a positive solution and a negative solution
b 12.5 The calculator works outs $\sqrt{y}$, which is the positive square root, so it only gives the positive solution
c −12.5

6 $a > 0$

7 a Always true, as $2n$ is even and any number raised to an even power gives a positive answer
b Never true, as $2n$ is even and any number raised to an even power gives a positive answer
c Sometimes true, as $2n + 1$ is odd, and a positive number raised to an odd power gives a positive answer but a negative number raised to an odd power gives a negative answer

Check my understanding

1 **a** 6 **b** −6 **c** 0 **d** −4

2 **a** −24, −1, 0, 4, 15, 17
b −500, −265, −99, 127, 162, 10 600

3 **a** = −14 **b** = −19 **c** = 35
d = 23 **e** = −18 **f** = −68
g = 48 **h** = −5 **i** = 100
j = −40

4 **a** 3 **b** −14 **c** −12
d −29 **e** 40

5 **a** $x = 7$ **b** $x = 9$ **c** $x = -10$
d $x = -1$ **e** $x = -10$

6 **a** = −7 **b** = −16 **c** = −60
d = −23 **e** = −5

7 **a** $x = \pm 12$ **b** $y = \pm 15$
c $z = \pm 6$ **d** $p = \pm 10$
e $t = \pm 8$

Block 10 Fractional thinking

Chapter 10.1

Are you ready?

1 **a** $\frac{2}{5}$ **b** $\frac{2}{5}$ **c** $\frac{3}{8}$
d $\frac{15}{16}$ **e** $\frac{5}{5}$

2 **a** $\frac{1}{2}$ **b** $\frac{1}{3}$ **c** $\frac{1}{4}$ **d** $\frac{1}{5}$

3 Any diagram split into 3 equal parts with 1 part shaded

4 Any diagram split into 4 equal parts with 3 parts shaded

5 The bar is not split into equal parts.

Practice 10.1A

1 $\frac{7}{5}$

2 $1\frac{1}{7}$

3 **a** $\frac{4}{4}$ **b** $\frac{7}{7}$ **c** $\frac{3}{3}$

4 **a** 6 **b** 12 **c** 30

5 $\frac{13}{4}$

6 **a**

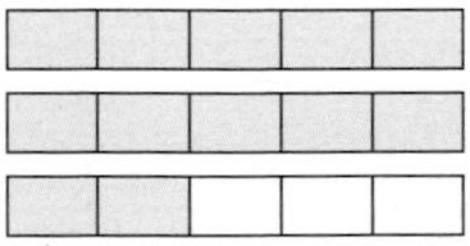

b $\frac{12}{5}$

7 **a**

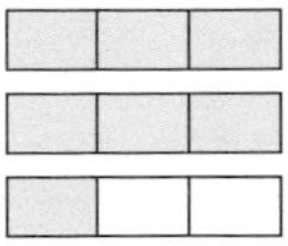

b $2\frac{1}{3}$

8 **a** $\frac{7}{4}$ **b** $\frac{22}{7}$ **c** $\frac{17}{6}$ **d** $\frac{21}{8}$

9 **a** $3\frac{1}{2}$ **b** $2\frac{2}{5}$ **c** $2\frac{3}{7}$ **d** $4\frac{2}{3}$

10 **a** > **b** = **c** < **d** <

11 **a** The 8 wholes are each worth $\frac{9}{9}$ so 8 lots of $\frac{9}{9} = \frac{72}{9}$. Then there are an extra 4 ninths.
b $\frac{87}{7}$

What do you think?

1 **a** 5, 4, 3, 2
b $5\frac{1}{4}, 4\frac{1}{4}, 3\frac{1}{4}, 2\frac{1}{4}$
c $2\frac{5}{8}, 2\frac{6}{8}, 2\frac{7}{8}, 3$
Each set is a linear sequence

2 $\frac{15}{4}$

Consolidate

1 **a** $\frac{5}{3}$ **b** $\frac{9}{4}$

2 **a** $1\frac{4}{5}$ **b** $3\frac{2}{3}$

3 **a** $2\frac{1}{5} = \frac{11}{5}$ **b** $1\frac{3}{4} = \frac{7}{4}$
c $4\frac{1}{2} = \frac{9}{2}$ **d** $\frac{24}{7} = 3\frac{3}{7}$

4 **a** $4\frac{1}{3}$ **b** $5\frac{1}{2}$ **c** $5\frac{2}{5}$ **d** $2\frac{1}{10}$

5 **a** $\frac{7}{4}$ **b** $\frac{16}{3}$ **c** $\frac{31}{10}$ **d** $\frac{17}{7}$

Stretch

1 $\frac{54}{4}, 12\frac{3}{5}, \frac{73}{6}, \frac{109}{11}$

2 **a** $5\frac{4}{7}$ **b** $4\frac{11}{7}$ **c** $3\frac{18}{7}$
d $2\frac{25}{7}$ **e** $1\frac{32}{7}$

3 12

Chapter 10.2

Are you ready?

1 $\frac{1}{8}$

2 **a** $\frac{6}{7}$ **b** $\frac{2}{3}$
c $\frac{1}{10}$ **d** $1\frac{1}{6}$ **or** $\frac{7}{6}$

3

4 $\frac{1}{12}, \frac{5}{12}, \frac{12}{12}$

5 $\frac{3}{3}, \frac{4}{4}, \frac{11}{11}$

6 **a** 0, $\frac{1}{4}$, $\frac{2}{4}$, $\frac{3}{4}$, 1
b 0, $\frac{1}{7}$, $\frac{2}{7}$, $\frac{3}{7}$, $\frac{4}{7}$, $\frac{5}{7}$, $\frac{6}{7}$, 1
c 0, $\frac{1}{5}$, $\frac{2}{5}$, $\frac{3}{5}$, $\frac{4}{5}$, 1

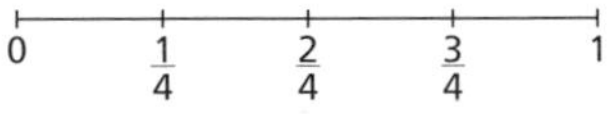

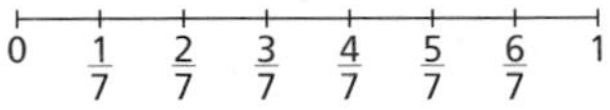

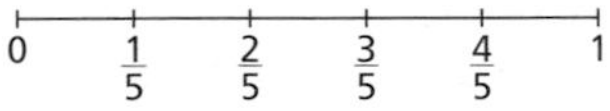

Practice 10.2A

1 **a** 
b 
c

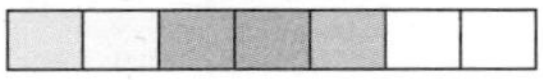

2 **a** $\frac{3}{8}$ **b** $\frac{2}{12}$ **c** $\frac{3}{74}$ **d** $\frac{1}{9}$
e $\frac{2}{13}$ **f** 0 **g** $\frac{3}{3}$

3 When the denominators are the same you are adding/subtracting parts of the same size so you can just add/subtract the numerators.

4 **a** $\frac{1}{5} + \frac{1}{5} + \frac{1}{5} + \frac{1}{5}$ **b** $\frac{1}{9} + \frac{1}{9}$
c $\frac{1}{31} + \frac{1}{31} + \frac{1}{31} + \frac{1}{31} + \frac{1}{31}$

5 **a** $\frac{3}{6} + \frac{2}{6} = \frac{5}{6}$ **b** $\frac{4}{11} + \frac{4}{11} = \frac{8}{11}$
c $\frac{3}{4} - \frac{2}{4} = \frac{1}{4}$ **d** $\frac{4}{7} + \frac{2}{7} = \frac{6}{7}$
e $\frac{4}{11} + \frac{1}{11} + \frac{3}{11} = \frac{8}{11}$
f $\frac{7}{9} - \frac{3}{9} = \frac{4}{9}$

6 Any diagram showing that $\frac{3}{8} + \frac{1}{8} = \frac{4}{8}$ **not** $\frac{4}{16}$

7 **a** **i** $\frac{6}{7}$ **ii** $\frac{11}{15}$ **iii** $\frac{1}{9}$
iv $\frac{10}{6}$ **v** 0 **vi** $\frac{6}{11}$
b **i** **iii** gave a unit fraction answer
ii **iv** gave an improper fraction answer

8 Any 4 from $\frac{4}{17} + \frac{11}{17} = \frac{15}{17}$, $\frac{11}{17} + \frac{4}{17} = \frac{15}{17}$, $\frac{15}{17} - \frac{4}{17} = \frac{11}{17}$, $\frac{15}{17} - \frac{11}{17} = \frac{4}{17}$, $\frac{15}{17} = \frac{4}{17} + \frac{11}{17}$, $\frac{15}{17} = \frac{11}{17} + \frac{4}{17}$, $\frac{11}{17} = \frac{15}{17} - \frac{4}{17}$, $\frac{4}{17} = \frac{15}{17} - \frac{11}{17}$

9 $\frac{1}{9}$

10 **a** $\frac{12}{14}$m **b** $\frac{9}{11}$cm **c** $\frac{18}{5}$m **d** $\frac{6}{7}$m

11 **a** $1\frac{1}{3}$ **b** $1\frac{3}{8}$ **c** $1\frac{4}{35}$

12 **a** −2 **b** −£2
c −2 sixths **d** $-\frac{2}{7}$

What do you think?

1 $-\frac{11}{17}$

2 **a** $a = 7$
b No you can't work out w; it doesn't matter for working out a because the denominator is the same.
c $1 < w < 9$

3 **a** $-\frac{1}{3}$ **b** $-\frac{1}{4}$ **c** $-\frac{1}{5}$ **d** $-\frac{1}{217}$
They are all negative unit fractions

Practice 10.2B

1 **a** $\frac{2}{2} = 1$ **b** $\frac{6}{6} = 1$
c $\frac{10}{10} = 1$ **d** $\frac{91}{91} = 1$
The answers are all equivalent to 1, but have different denominators

2 **a** 
b

3 **a** $\frac{1}{4}$ **b** $\frac{7}{10}$ **c** $\frac{13}{18}$
d 0 **e** 0 **f** $\frac{37}{100}$

4 **a** 1 **b** 2 **c** 3

5 **a** $1\frac{1}{6}$ **b** $2\frac{5}{6}$ **c** $1\frac{1}{3}$ **d** $1\frac{1}{6}$

6 **a** 2 cm **b** 5 cm **c** 2 cm **d** 6 cm

What do you think?

1 Flo can see that $\frac{2}{7} + \frac{5}{7} = 1$, and $\frac{3}{5} + \frac{2}{5} = 1$ so the total is 2.

2 **a** 2 **b** 1 **c** 3

Consolidate

1 **a** $\frac{1}{3} + \frac{1}{3} = \frac{2}{3}$ **b** $\frac{4}{5} + \frac{3}{5} = \frac{7}{5}$
c $\frac{8}{9} - \frac{7}{9} = \frac{1}{9}$

2 **a** 1 **b** 1 **c** 1 **d** 1
e 1 **f** 2 **g** 2 **h** 3

3 **a** $\frac{3}{5}$ **b** $\frac{5}{7}$ **c** $\frac{3}{10}$

d $\frac{6}{11}$ **e** $\frac{8}{8} = 1$ **f** $\frac{2}{3}$
g 0 **h** 0 **i** $1\frac{5}{6}$

4 $\frac{5}{11}$m
5 $\frac{23}{27}$
6 $\frac{4}{9}$

Stretch

1 6th
2 5th
3 a i $\frac{1}{12}$ **ii** $-\frac{1}{12}$
iii $-\frac{1}{12}$ **iv** $\frac{23}{12}$
b i $\frac{3}{5}$ **ii** $-\frac{3}{5}$
iii $-\frac{17}{5}$ **iv** $-\frac{3}{5}$
c i $\frac{28}{9}$ **ii** $-\frac{28}{9}$
iii $-\frac{62}{9}$ **iv** $-\frac{62}{9}$
4 a $x = \frac{18}{11}$ **b** $y = \frac{21}{15}$
c $z = -\frac{9}{23}$ **d** $a = -\frac{12}{39}$
e $b = -\frac{7}{13}$ **f** $c = -\frac{24}{51}$

Chapter 10.3

Are you ready?

1 a 5, 10, 15, 20, 25, 30
b 3, 6, 9, 12, 15, 18
c 7, 14, 21, 28, 35, 42
2 40
3 Factors are what we can multiply to get the number. Multiples are what we get after multiplying the number by an integer.
4 a i 4, 8, 12, 16, 20 and 5, 10, 15, 20, 25
ii 4, 8, 12, 16, 24 and 6, 12, 18, 24, 30
iii 6, 12, 18, 24, 30 and 12, 24, 36, 48, 60
b i 20 **ii** 12 **iii** 12

Practice 10.3A

1 a $\frac{1}{2} = \frac{2}{4}$ **b** $\frac{2}{3} = \frac{4}{6}$
c $\frac{1}{2} = \frac{5}{10}$ **d** $\frac{6}{9} = \frac{2}{3}$
e e.g. $\frac{6}{6} = \frac{4}{4}$ or $\frac{6}{8} = \frac{3}{4}$
f $1 = \frac{7}{7}$
2 a $\frac{1}{4} = \frac{2}{8}$ **b** $\frac{2}{3} = \frac{8}{12}$ **c** $\frac{4}{6} = \frac{6}{9}$
3
4 e.g. $\frac{4}{10}, \frac{6}{15}, \frac{8}{20}, \frac{10}{25}, \frac{200}{500}$
5 a = **b** = **c** ≠
d = **e** ≠ **f** ≠
6 Marta is correct: $\frac{6}{24}$ and $\frac{7}{28}$ are both equivalent to $\frac{1}{4}$; therefore $\frac{6}{24}$ is equivalent to $\frac{7}{28}$
7 True: the denominator is 4 times the numerator in each so both fractions are equivalent to $\frac{1}{4}$

8 a $\frac{1}{3} = \frac{5}{15}$ **b** $\frac{5}{6} = \frac{25}{30}$
c $\frac{3}{4} = \frac{15}{20}$ **d** $\frac{1}{7} = \frac{5}{35}$
e $\frac{3}{20} = \frac{30}{200}$ **f** $\frac{4}{12} = \frac{5}{15}$

What do you think?

1 e.g. $\frac{10}{20} = \frac{10}{20}, \frac{5}{20} = \frac{10}{40}, \frac{1}{20} = \frac{10}{200}, \frac{4}{20} = \frac{10}{50}, \frac{20}{20} = \frac{10}{10}$
2 e.g. $\frac{14}{49}, \frac{56}{196}, \frac{126}{441}, \ldots$

Practice 10.3B

1 a $\frac{5}{6}$ **b** $\frac{5}{10}$ **c** $\frac{3}{8}$ **d** $\frac{3}{12}$
2 a 8 **b** 14 **c** 9
The LCM is the greater of the two numbers in each pair
3 a $\frac{5}{8}$ **b** $\frac{13}{14}$ **c** $\frac{2}{9}$
4 a $\frac{11}{15}$ **b** $\frac{7}{15}$ **c** $1\frac{1}{3}$
d $\frac{17}{20}$ **e** $\frac{1}{8}$ **f** $\frac{1}{3}$
5 No; $\frac{10}{10} - \frac{1}{10} - \frac{6}{10} - \frac{3}{10} = 0$
6 $\frac{3}{8}$

What do you think?

1 a 15 **b** 7 **c** 4
2 a $y = \frac{3}{4}$ **b** $w = \frac{1}{2}$ **c** $x = \frac{3}{5}$

Consolidate

1 $\frac{3}{8}$
2 a $\frac{5}{6}$ **b** $\frac{1}{6}$
3 a $\frac{5}{8}$ **b** $\frac{10}{15}$ **c** $\frac{3}{10}$
d $\frac{3}{6}$ **e** $\frac{19}{30}$ **f** $\frac{7}{12}$

Stretch

1 a $1\frac{1}{2}$ **b** $1\frac{3}{14}$ **c** $2\frac{13}{48}$
2 $3\frac{7}{8}$km
3 $8\frac{1}{3}$ litres
4 $6\frac{9}{10}$

Chapter 10.4

Are you ready?

1 a $\frac{1}{2}$ **b** $\frac{3}{4}$ **c** $\frac{3}{4}$ **d** $\frac{2}{5}$
2 a $1\frac{4}{7}$ **b** $1\frac{1}{3}$ **c** $7\frac{1}{2}$ **d** $5\frac{2}{5}$
3 a $\frac{5}{3}$ **b** $\frac{15}{8}$ **c** $\frac{19}{9}$ **d** $\frac{23}{5}$

Practice 10.4A

1 a 24 **b** $\frac{19}{24}$
2 a Her answer isn't in its simplest form.
b Use 24 as the denominator (LCM of 6 and 8)
3

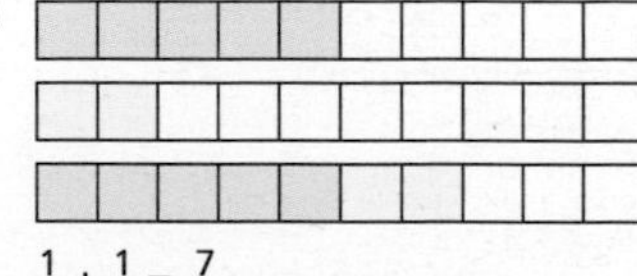

$\frac{1}{2} + \frac{1}{5} = \frac{7}{10}$
4 a $\frac{1}{3}$ **b** $\frac{17}{30}$ **c** $\frac{13}{24}$
d $\frac{1}{21}$ **e** $\frac{1}{18}$ **f** $\frac{7}{12}$
5 a $1\frac{17}{24}$ **b** $\frac{1}{24}$
6 $\frac{7}{12}$
7 3 An efficient method is to notice that $\frac{3}{7} + \frac{4}{7} = 1$, and $\frac{4}{13} + \frac{9}{13} = 1$, and $\frac{5}{8} + \frac{3}{8} = 1$ so the total is 3
8 a $\frac{7}{12}$kg
b None of them exceed 1; $\frac{11}{12}$ is closest to 1 and $\frac{1}{3}$ is the smallest

What do you think?

1 a e.g. $\frac{4}{5} + \frac{4}{5}$ **b** e.g. $\frac{29}{40} + \frac{7}{8}$
2 $\frac{19}{50}$m^2

Practice 10.4B

1 a $\frac{7}{6} + \frac{15}{4} = \frac{14}{12} + \frac{45}{12} = \frac{59}{12} = 4\frac{11}{12}$
b $1 + 3 = 4$; $\frac{1}{6} + \frac{3}{4} = \frac{11}{12}$; $1\frac{1}{6} + 3\frac{3}{4} = 4\frac{11}{12}$
c Students' own answers
d i $4\frac{7}{12}$ **ii** $4\frac{24}{35}$ **iii** $8\frac{4}{9}$
2 a $3 - 1 = 2$, but $\frac{7}{8} > \frac{2}{5}$ so subtracting the fraction parts gives a negative answer.
b $1\frac{21}{40}$
3 a $1\frac{1}{2}$ **b** $2\frac{5}{24}$ **c** $\frac{1}{45}$ **d** $3\frac{5}{16}$
4 a $7\frac{17}{28}$km **b** $2\frac{3}{28}$km
5 a $230\frac{7}{10}$m **b** $3084\frac{1}{20}$m
c $52\frac{9}{10}$m **d** $283\frac{3}{5}$m

What do you think?

1 a Both perimeters are $16\frac{4}{5}$m
b $4\frac{1}{5}$m
2 Sometimes: e.g. $1\frac{1}{5} + 3\frac{1}{10} = 4\frac{3}{10}$ (mixed number) but $1\frac{3}{5} + 3\frac{2}{5} = 5$ (integer)

Consolidate

1 a $\frac{1}{2}$ **b** $\frac{2}{3}$ **c** $\frac{17}{30}$
d $\frac{23}{30}$ **e** $1\frac{7}{12}$ **f** $1\frac{4}{15}$
g $1\frac{4}{30}$ **h** $1\frac{3}{20}$ **i** $1\frac{1}{14}$
2 a $6\frac{1}{2}$ **b** $4\frac{2}{3}$ **c** $11\frac{17}{30}$
d $35\frac{23}{30}$ **e** $201\frac{7}{12}$ **f** $7\frac{4}{15}$
g $5\frac{4}{30}$ **h** $12\frac{3}{20}$ **i** $36\frac{1}{14}$
3 a $\frac{1}{10}$ **b** $\frac{4}{15}$ **c** $\frac{11}{30}$
d $\frac{1}{6}$ **e** $\frac{1}{12}$ **f** $\frac{4}{15}$
g $\frac{4}{5}$ **h** $\frac{7}{20}$ **i** $\frac{1}{14}$
4 a $2\frac{1}{10}$ **b** $5\frac{4}{15}$ **c** $1\frac{11}{30}$
d $2\frac{1}{6}$ **e** $12\frac{1}{12}$ **f** $\frac{4}{15}$
g $18\frac{4}{5}$ **h** $\frac{7}{20}$ **i** $36\frac{1}{14}$

j $2\frac{1}{24}$

5 Compare answers as a class

6 a $\frac{19}{20}$ **b** $\frac{4}{5}$ **c** $6\frac{23}{24}$ **d** $3\frac{25}{36}$

Stretch

1 a $9\frac{8}{15}$ **b** $11\frac{1}{30}$ **c** $7\frac{9}{10}$
d $14\frac{7}{30}$ **e** $3\frac{4}{30}$ **f** $1\frac{19}{30}$
g $-3\frac{4}{30}$ **h** $-1\frac{1}{2}$

2 a $7\frac{9}{16}$ **b** $\frac{112 + 3x}{16}$
c $\frac{14x + 9}{2x}$

3 Various answers, e.g. $x = 5\frac{1}{5}$, $y = 13\frac{4}{5}$

4 $7 - 3 = 4$; $\frac{1}{5} - \frac{11}{18} = \frac{18}{90} - \frac{55}{90} = -\frac{37}{90}$, $4 - \frac{37}{90} = 3\frac{53}{90}$

Chapter 10.5

Are you ready?

1 a $x = 49.3$ **b** $x = 4.2$
c $x = 8$ **d** $x = 2312$
e $x = -3.3$ **f** $x = 4.81$
g $x = -13$ **h** $x = -31.48$

2 a $x = 24.65$ **b** $x = 0.84$
c $x = 5$ **d** $x = 1445$
e $x = -1.1$ **f** $x = 1.2025$
g $x = -14$ **h** $x = -62.96$

3 a i $\frac{4}{5}$ **ii** $\frac{1}{3}$ **iii** $\frac{43}{47}$ **iv** $\frac{466}{999}$
b i $\frac{2}{3}$ **ii** $\frac{1}{2}$ **iii** $\frac{33}{94}$ **iv** $\frac{460}{999}$
c i $1\frac{3}{20}$ **ii** $\frac{11}{24}$ **iii** $\frac{43}{63}$ **iv** $\frac{17}{48}$

4 a $2x$ **b** y
c $3z$ **d** $-w$
e $3x + 3y$ **f** $7y - z$
g $39z - 37y$ **h** $-6w + 40$

Practice 10.5A

1 a $\frac{2}{3}$ **b** $\frac{2}{15}$ **c** $\frac{6}{15}$ **d** $\frac{4}{15}$
e $\frac{4}{9}$ **f** $\frac{22}{45}$ **g** 14 **h** $5\frac{1}{5}$

2 a $\frac{1}{5}, \frac{2}{5}, \frac{3}{5}, \frac{4}{5}, \frac{5}{5}$ **b** $+\frac{1}{5}$
c Yes; it is increasing by the same amount each time.
d 40 **e** 125th

3 a Inputs are $\frac{2}{7}, \frac{3}{14}$
Outputs are $\frac{6}{7}, a + \frac{5}{7}$
b Inputs are $1\frac{2}{3}, 1\frac{7}{15}$
Outputs are $\frac{1}{6}, b - \frac{2}{3}$

4 a $x = \frac{3}{5}$ **b** $z = \frac{2}{9}$ **c** $w = \frac{3}{11}$
d $t = \frac{14}{37}$ **e** $s = \frac{1}{2}$ **f** $q = \frac{1}{2}$
g $y = \frac{41}{44}$ **h** $p = \frac{1}{2}$

5 a $a = 1\frac{1}{2}$ **b** $b = \frac{3}{4}$
c $c = 1\frac{1}{3}$ **d** $d = 1\frac{5}{9}$
e $e = 2\frac{3}{4}$ **f** $f = 34\frac{7}{10}$
g $g = 1\frac{7}{12}$ **h** $h = 12\frac{3}{20}$

What do you think?

1 a $\frac{5}{7}, \frac{10}{7}, \frac{15}{7}, \frac{20}{7}, \frac{25}{7}$
b 65th
c 7th
d Every 7th term, when the numerator is a multiple of 7
e 14

2 a $\frac{a}{b} = \frac{1}{2}, \frac{a}{b^2} = \frac{1}{4}, \frac{1}{2} > \frac{1}{4}$
b Compare answers as a class
c Always. The numerators are the same. Since b is a positive integer,
$b^2 > b$ so $\frac{a}{b} > \frac{a}{b^2}$

Practice 10.5B

1 a i $\frac{14}{17}$ **ii** $\frac{14}{23}$ **iii** $\frac{14}{x}$ **iv** $\frac{14}{11z}$
b i $\frac{4}{17}$ **ii** $\frac{4}{23}$ **iii** $\frac{4}{x}$ **iv** $\frac{4}{11z}$

2 a $\frac{32}{x}$ **b** $\frac{2}{y}$ **c** $\frac{60}{w}$ **d** $\frac{1}{z}$
e 0 **f** $\frac{20}{b}$ **g** $\frac{89}{c}$ **h** $\frac{108}{t}$

3 a i $\frac{10}{199}$ **ii** $\frac{100}{199}$ **iii** $\frac{10x}{199}$ **iv** $\frac{100x}{199}$
b i $\frac{4}{199}$ **ii** $\frac{40}{199}$ **iii** $\frac{4x}{199}$ **iv** $\frac{40x}{199}$

4 a $\frac{7y}{3}$ **b** $\frac{6d}{7}$ **c** $\frac{11a}{31}$ **d** $\frac{13b}{19}$
e $\frac{k}{21}$ **f** $\frac{17g^2}{47}$ **g** 0 **h** $\frac{8p}{9}$

5 a i $\frac{7}{10}$ **ii** $\frac{1}{8}$ **iii** $\frac{7}{2z}$ **iv** $\frac{5}{2x}$
b i $\frac{1}{2}$ **ii** $\frac{1}{40}$ **iii** $\frac{5}{2z}$ **iv** $\frac{1}{2x}$

6 a $\frac{9}{2a}$ **b** $\frac{13}{2b}$ **c** $\frac{20}{3c}$ **d** $\frac{11}{3d}$
e $\frac{89}{7e}$ **f** $\frac{3}{100x}$ **g** $\frac{103}{4y}$ **h** $\frac{93}{6k}$

What do you think?

1 Both of them are correct. The two answers are equivalent.

2 a Always true; the numerator and denominator have both been multiplied by 2
b Sometimes true; only true if $m = 1$
c Sometimes true; only true if n is positive

3 a e.g. $n = 2$ **b** e.g. $n = 17$
c $n = 10$

Consolidate

1 a $\frac{4}{11}$ **b** $\frac{4}{17}$ **c** $\frac{4}{119}$ **d** $\frac{4}{1001}$
e $\frac{4}{y}$ **f** $\frac{4}{z}$ **g** $\frac{4}{2x}$ **h** $\frac{4}{k^2}$

2 a $\frac{2}{11}$ **b** $\frac{2}{17}$ **c** $\frac{2}{119}$ **d** $\frac{2}{1001}$
e $\frac{2}{y}$ **f** $\frac{2}{z}$ **g** $\frac{2}{2x}$ **h** $\frac{2}{k^2}$

3 Compare answers as a class

4 a $\frac{3}{2z}$ **b** $\frac{5}{3z}$ **c** $\frac{13}{4z}$ **d** $\frac{39}{5z}$
e $\frac{10}{3a}$ **f** $\frac{28}{11c}$ **g** $\frac{39}{2g}$ **h** $\frac{7}{4x}$

Stretch

1 a $x = 30$ **b** $y = 63$
c $t = 11$ **d** $x = 2$

2 a $x = 9$ **b** $y = 9$
c $z = 8$ **d** $z = 4$

3 a $x = 6.5$ **b** $y = \frac{20}{3}$
c $z = \frac{13}{2}$ **d** $z = \frac{39}{11}$

4 a $\frac{1}{x}, \frac{9}{2x}, \frac{16}{2x}, \frac{23}{2x}, \frac{30}{2x}$
b $x = \frac{37}{22}$

5 a $\frac{1}{x}, \frac{-5}{2x}, \frac{-12}{2x}, \frac{-19}{2x}, \frac{-26}{2x}$
b $\frac{-33}{20}$

6 $\frac{13n}{63}$

Check my understanding

1 a $5\frac{2}{3}$ **b** $4\frac{1}{5}$ **c** $1\frac{2}{17}$
d $6\frac{6}{7}$ **e** $58\frac{1}{2}$

2 a $\frac{11}{2}$ **b** $\frac{11}{4}$ **c** $\frac{55}{6}$
d $\frac{31}{8}$ **e** $\frac{32}{19}$

3 a $4\frac{2}{3}, \frac{14}{3}$ **b** $5\frac{3}{5}, \frac{28}{5}$
c $7\frac{4}{9}, \frac{67}{9}$ **d** $9\frac{1}{4}, \frac{37}{4}$
e $11\frac{1}{10}, \frac{111}{10}$

4 a $\frac{1}{2}$ **b** $\frac{1}{9}$ **c** $\frac{7}{8}$
d $-\frac{1}{20}$ **e** $\frac{17}{21}$

5 a $7\frac{7}{10}$ **b** $\frac{13}{20}$ **c** $6\frac{17}{30}$ **d** $15\frac{1}{18}$

6 a $\frac{2}{5}$ **b** $\frac{1}{6}$ **c** $\frac{1}{2}$ **d** $\frac{3}{5}$

7 a $\frac{3x}{4}$ **b** $\frac{y}{10}$ **c** $\frac{3}{2x}$
d $\frac{29b}{35}$ **e** $\frac{122}{11a}$

Block 11 Construction and measuring

Chapter 11.1

Are you ready?

1 a 90 **b** 180

2 a 5 cm **b** 6 cm **c** 8 cm

3 Check that the length is 7.4 cm

4 °

Practice 11.1A

1 a 5.2 cm **b** 4.7 cm **c** 6.1 cm

2 a 4.2 cm **b** 6.2 cm **c** 5.7 cm

3 Both are 7.2 cm long

4 The lengths are not equal

5 Any six triangles from ABC, ABD, ABX, ACD, ADX, BCD, BCX, CDX, CXY

What do you think?

1 Ed is correct; ABC, BDE, BCE, CEF, ADF

2 a AC, BD, EH, FG
b Compare answers as a class

Practice 11.1B

1

2 Yes, as F and G are both on EG

3 Here is one possibility: a is PQA, b is PRC, c is SQB, d is QRD, e is DRS

4 a 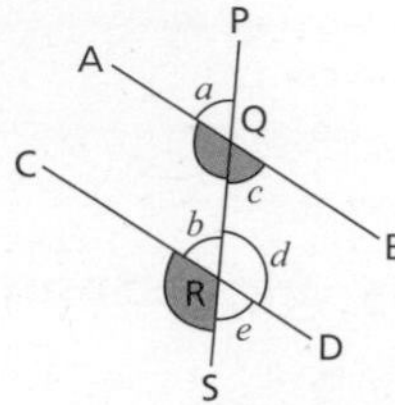

b It does exist. AQB is a straight line so the angle is 180°.

5 a 270° b 45°
c 135° d 315°

6 a 42° b 56° c 98°
d 145° e 103° f 47°

7 a Students' sketches that show ∠ABC = 104° and ∠CBD = 55°
b 159° or 49°

What do you think?

1 It is the same amount of turn but in the opposite direction

2 a Yes, you will be in the same place
b Both pairs are equivalent

3 a $\frac{1}{6}$
b i 300° anticlockwise
ii 120° clockwise
iii $\frac{4}{5}$ turn anticlockwise
iv 60° anticlockwise
v 313° clockwise

4 a i 150° ii 210°
b i 240° ii 120°
c i $2x°$ ii $360 - 2x°$

5 Both are correct: acute angle PQR = 75° and reflex angle PQR = 285°

Consolidate

1 Compare answers with a partner
2 Compare answers with a partner
3 a Compare answers with a partner
b The line segments ZO and OW are not drawn on the diagram
4 Compare answers with a partner

Stretch

1 a 360° b 180° c 240°
d 270° e 300°

2 30°

3 a 30° b 15° c 20°
d 22.5° e 25°

4 1 and 7, 2 and 8, 3 and 9, 4 and 10, 5 and 11. The numbers in each pair have a difference of 6.

5 a The hour hand is halfway between the 6 and 7 but the minute hand points to 6. The angle is 15°.
b Compare answers as a class. Possible answers include:
i 12 am ii 6 pm iii 3 pm
c Compare answers as a class

Chapter 11.2

Are you ready?

1 360°

2 It is the amount of turn between the lines, not the distance between them

3 a is BPC (or RPQ, CPQ or BPR), b is PCD, c is ABQ

4 a 90° b 180° c 270° d 135°

Practice 11.2A

1 a Acute, 62° b Obtuse, 131°
c Acute, 48° d Acute, 66°
e Obtuse, 115° f Obtuse, 140°

2 a Obtuse b Acute
c Acute d Obtuse
e Obtuse
Check accuracy with a partner

3 Check accuracy with a partner

4 Check accuracy with a partner

5 a Check answers with a partner
b 180°

6 a Check answers with a partner
b 360°

What do you think?

1 a Impossible b Impossible
c Possible d Possible
Check answers with a partner.

2 a Possible b Possible
c Possible d Possible
e Impossible
Check answers with a partner

3 a Possible b Possible
c Possible d Impossible
e Possible
Check answers with a partner

Practice 11.2B

1 Measure the amount of turn past a straight line and add 180°, or measure the non-reflex angle made and subtract from 360°

2 a 220° b 305° c 195°

3 Check accuracy with a partner

4 Check accuracy with a partner

5 a 284° + 82° = 366°, but the total should be 360°
b a is 78° and b is 282°

What do you think?

1 a All the angles are obtuse
b It is a full turn
c There are four right angles
d Right angles, acute angles and reflex angles on the arrow, a full turn for the sign as a whole

2 Check answers with a partner

Consolidate

1 a Acute, 47° b Reflex, 203°
c Reflex, 340° d Obtuse, 112°
e Acute, 35° f Reflex, 260°

2 a Acute b Obtuse
c Reflex d Obtuse
e Reflex
Check accuracy with a partner

3 Check accuracy with a partner

Stretch

1 a Sometimes true (e.g. 60° and 60°, but not 60° and 10°)
b Always true
c Sometimes true (e.g. two angles of 150° and one of 60° or two angles of 160° and one of 40°, but not two angles of 160° and one of 60°)

2 Total of three acute angles must be more than 0° and less than 270°
Total of three obtuse angles must be more than 270° and less than 540°
Total of three reflex angles must be more than 540° and less than 1080°

Chapter 11.3

Are you ready?

1 5.6 cm and 4.2 cm

2 30°, 10° and 140°

3 a 2 acute and 1 obtuse
b 3 acute, 2 reflex and 2 right angles
c 1 acute, 2 obtuse and 1 right angle

Practice 11.3A

1 a True b False
c True d True
e 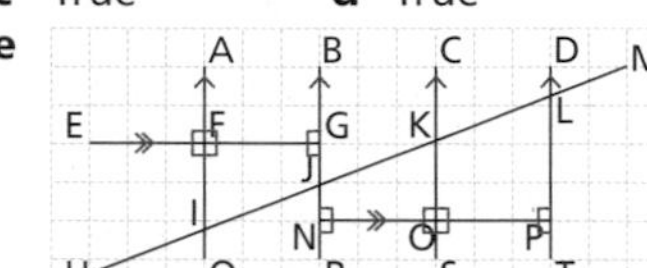

2 Huda is wrong: AB and PQ will never meet, so they are parallel

3 a AE b Equilateral
c Scalene

4 All types except equilateral are possible. Compare answers as a class.

5 There are many possibilities. Compare answers as a class.

6 AB, CD and FE are one set of parallel lines. AF, BC and DE are another set.
Perpendicular: AB and BC, BC and CD, CD and DE, DE and EF, EF and FA, FA and AB

What do you think?

1 All four statements are true

2 Compare answers as a class

3 All the triangles A to C are possible

4 Triangles A and B are possible; C is not possible

Practice 11.3B

1 a Parallelogram
b Parallelogram
c Rhombus d Trapezium
e Kite f Trapezium
g Rectangle

2 Compare answers as a class

3 a Parallelogram and square
b Parallelogram and rectangle
c Kite and rhombus

4 Compare answers as a class

5 a Parallelogram, rectangle and kite
b Trapezium and kite
c Kite
d Parallelogram, rhombus, rectangle and square

6 Possibilities include

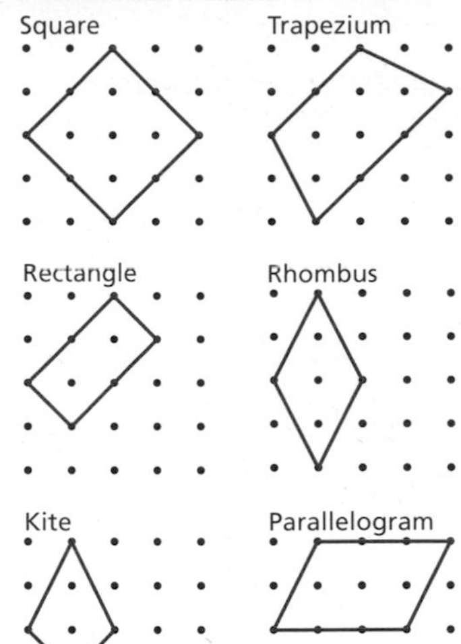

Compare answers as a class

7 The difference is that the sides of a rhombus are all equal

What do you think?

1 It is a kite with a reflex angle

2 **a** True **b** False
c True **d** True

3 Compare answers as a class

Practice 11.3C

1 All except D are hexagons

2 e.g.

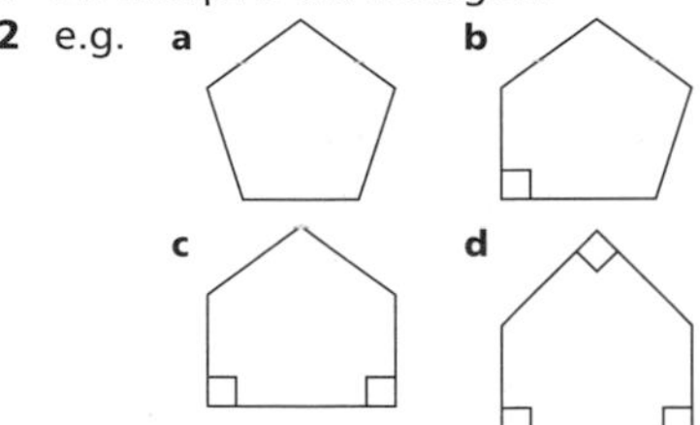

3 Flo's statement is only sometimes true.

Not true:

True:

4 **a** **i**

ii

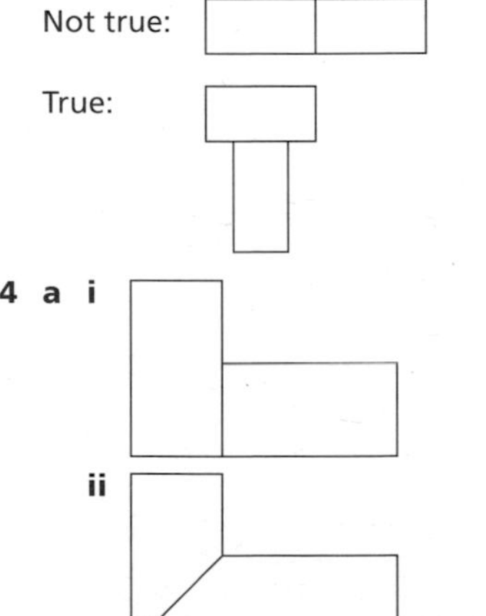

b Compare answers as a class
c Compare answers as a class

5 Equilateral triangle and square

What do you think?

1 **a** Always true
b Sometimes true

2 **a** Isosceles **b** Pentagon
c Compare answers as a class

Consolidate

1 All four statements are true

2

	One pair of equal angles	Two pairs of equal angles	All four angles equal
One pair of parallel sides	Trapezium	Isosceles trapezium	
Two pairs of parallel sides		Parallelogram Rhombus	Rectangle Square
Opposite sides equal in length		Parallelogram Rhombus	Rectangle Square
Adjacent sides equal in length	Kite	Rhombus	Square

3 **a** Isosceles **b** Isosceles
c Scalene and right-angled
d Equilateral
e Isosceles and right-angled

4 Compare answers as a class

Stretch

1 Yes, as one side will not be enough to join the two parallel sides into a closed shape

2 True

3 Compare answers as a class

4 **a** Four in a quadrilateral, three for shapes with more sides
b Compare answers as a class. For $n = 6$ to 13 you get

Sides	6	7	8	9	10	11	12	13
Maximum number of right angles	5	5	6	7	7	8	9	9

5 Discuss answers as a class

Chapter 11.4

Are you ready?

1 **a** Scalene **b** Equilateral
c Isosceles

2 **a** Rhombus
b Parallelogram
c Trapezium

3 Check accuracy with a partner

4 **a** (Irregular) pentagon
b Discuss findings as a class.

Practice 11.4A

1 **a** Check accuracy with a partner
b Check accuracy with a partner
c The lines won't meet

2 to 5 Check accuracy with a partner

What do you think?

1 A and C are impossible

2 **a** Discussion
b **i** 1 **ii** 1 **iii** 2
iv Infinite **v** 0

Practice 11.4B

1 **a** Check accuracy with a partner
b ABC = 84°, ADC = 82°

2 to 5 Check accuracy with a partner

What do you think?

1 Check accuracy with a partner

2 The lengths of both pairs of sides and the size of at least one angle

3 **a** Rhombus
b Isosceles trapezium
c Equilateral triangle, parallelogram and hexagon

Consolidate

1 **a** 75 mm
b Check accuracy with a partner

2 to 4 Check accuracy with a partner

Stretch

1 Scalene right-angled (3/4/5) and scalene non-right-angled (2/4/5 and 2/3/4). (2/3/5 is impossible.)

2 Compare answers as a class

Chapter 11.5

Are you ready?

1 Check accuracy with a partner

2 **a** 360° **b** 180° **c** 90°

3 **a** 72 **b** 120 **c** 80

4 **a** $\frac{1}{6}$ **b** $\frac{5}{12}$ **c** $\frac{3}{4}$ **d** $\frac{1}{8}$

Practice 11.5A

1 **a** 11 **b** 33

2 **a** This would mean that 10.5 people chose 'orange' and 'other', and 31.5 people chose 'banana'.
b **i** 8 **ii** 24 **iii** 24
c 96

3 **a** 32 **b** 5% **c** 16

4 **a** **i** $\frac{1}{5}$ **ii** $\frac{4}{15}$ **iii** $\frac{1}{3}$ **iv** $\frac{7}{15}$
b 120

5 **a** 10%
b **i** True **ii** Cannot tell
iii False **iv** False
c 116

What do you think?

a Two sections would be 'win' but you cannot share the remaining six sections equally so that one number is three times the other

b 16

c Compare answers as a class

Practice 11.5B

1 The angles should be Brown = 132°, Blue = 150°, Green = 48°, Other = 30°

2 **a** The angles should be Rhys = 56°, Beth = 108°, Ali = 90°, Lydia = 106°
b Ali
c No, because two sectors are very close in size

3 **a** **i** $\frac{1}{4}$ **ii** $\frac{1}{3}$ **iii** $\frac{1}{12}$ **iv** $\frac{1}{8}$
b **i** 150 **ii** 200 **iii** 50 **iv** 75

4 **a** Sprouts **b** 80°
c 4 **d** 144

5 **a** $\frac{1}{15}$

- **b** Because 150 is half of the people surveyed, so the angle is 180°, which is a straight line
- **c** The angles should be Very satisfied = 108°, Quite satisfied = 180°, Disappointed = 36°, Don't know = 12°, Never go = 24°

What do you think?

1. **a** After the fifth angle, the sixth is the remaining sector
   b $n - 1$
2. **a** 25% **b** 120° **c** $\frac{5}{12}$ **d** $\frac{4}{5}$
   e Compare answers as a class

Consolidate

1. **a** $\frac{1}{12}$
   b **i** 28 **ii** 49 **iii** 84
2. **a** Rent **b** $\frac{1}{12}$ **c** £1200
   d **i** £800 **ii** £100
3. The angles should be Guitar = 48°, Piano = 36°, Drums = 28°, Other = 92°, None = 156°
4. The angles should be Cheese = 144°, Tomato = 108°, Egg = 54°, Tuna = 18°, Other = 36°

Stretch

1. There may be a lot more students in 7X altogether. The pie charts only show the proportions of the whole class, not the actual numbers of people.
2. **a** 50
   b The 90° sector would not represent an integer number of pieces of data
   c 36
3. **a** **i** 72° **ii** 126°
   b **i** 30% **ii** 75% **iii** 85%
   c Compare answers as a class
4. Pie chart: This is useful when you are looking at the parts compared with the whole.
   Bar chart: This is useful when you want to know the actual number of values in each category.

Check my understanding

1. **a** **i** ∠ABF **ii** ∠ACD
   b ∠ADE Note: other answers are possible for **a** and **b**. Compare answers as a class.
   c AF and FD, because of the hatch marks
   d They may appear perpendicular but there is no right-angle symbol
   e 75°
2. Check accuracy with a partner
3. **a** Parallelogram
   b Isosceles triangle
   c Kite
   d (Irregular) Octagon
   e Equilateral triangle
4. Check accuracy with a partner

5. **a** The angles should be Red = 135°, Green = 108°, Blue = 18°, Black = 54°, Yellow = 45°
   b 135

Block 12 Geometric reasoning

Chapter 12.1

Are you ready?

1. **a** 220 **b** 220 **c** 211 **d** 171
2. **a** 80 **b** 50 **c** 43 **d** 116
3. **a** Obtuse **b** Acute
   c Reflex **d** Right
   e Reflex
4. **a** $x = 36$ **b** $x = 30$ **c** $x = 66$

Practice 12.1A

1. **a** $a = 210°$ **b** $b = 134°$
   c $c = 90°$ **d** $d = 60°$
   e $e = 138°$ **f** $f = 180°$
   g $g = 138°$
2. **a** $a = 25°$ **b** $b = 78°$
   c $c = 45°$ **d** $d = 78°$
   e $e = 29°$ **f** $f = 36°$
3. **a** $a = 255°$ **b** $b = 75°$
   c $c = 233°$ **d** $d = 53°$
   e $e = 124°$ **f** $f = 39°$
   g $g = 125°$
4. **a** $a = 34°$ **b** $b = 36°$
   c $c = 19°$ **d** $d = 42.5°$
   e $e = 49°$ **f** $f = 21°$

What do you think?

1. Marta is wrong; the angles are not adjacent to each other
2. **a** **i** $180° - x$ **ii** $180° - 2y$
   b No; the angles are not necessarily equal
3. The total is 364° and it should be 360°
4. **a** Straight line
   b Not a straight line
   c Not a straight line
   d Straight line

Practice 12.1B

1. Compare answers as a class
2. A and C
3. **a** $a = 57°$, $b = 123°$
   b $c = 21°$, $d = 159°$, $e = 159°$
   c $f = 94°$, $g = 86°$, $h = 86°$

What do you think?

1. e.g. both angles when added to d give 180°, so they must be the same size as each other. Discuss as a class.
2. **a** **i** x **ii** $180° - x$
   iii $180° - x$ **iv** $180° - y$
   v y
   b ∠AFB + 90° + y = 180° (angles on a straight line add up to 180°)
   So ∠AFB = 90° − y

Consolidate

1. **a** $a = 270°$ **b** $b = 200°$
   c $c = 33°$ **d** $d = 277°$
   e $e = 40°$ **f** $f = 131°$
   g $g = 143°$
   h $h = 87°$, $i = 55°$
   i $j = 80°$, $k = 20°$
2. Blue is 40°, green is 120°, red is 200°
3. **a** $a = 12.5°$ **b** $b = 28°$
   c $c = 20°$, $d = 18°$

Stretch

1. $f = 32.5°$ and $g = 25°$ so f is bigger
2. **a** $a = 120°$, $b = 90°$, $c = 72°$ and $d = 60°$
   b 2, 8, 9, 10, 12, 15, 18, 20, 24, 30, 36, 40, 45, 60, 72, 90, 120, 180 and 360
   c Because 360 has a lot of factors
3. **a** $a = 85°$, $b = 27°$, $c = 68°$
   b 180°; the same total as for the angles on a straight line

Chapter 12.2

Are you ready?

1. Scalene: all sides and angles are different.
   Isosceles: two sides and two angles are equal.
   Equilateral: all three sides and angles are equal.
2. x = ADB or BDA, y = DBC or CBD
3. $a = 42°$, $b = 112°$
4. **a** $6a$ **b** $b + 100$
   c $c + 135$ **d** $3d + 75$

Practice 12.2A

1. **a** 99° **b** 39° **c** 36°
   d Triangle A. This looks like it has two equal sides and two equal angles.
2. 2°
3. Yes; because 180° ÷ 3 = 60°
4. **a** Hatch marks show that two sides are equal
   b $a = 46°$, $b = 88°$, $c = 67°$, $d = 67°$
   c c and d are the base angles of an isosceles triangle but a and b are not
5. **a** 52.5° **b** 16°
6. **a** Both are 10°
   b 50°and 50° or 80° and 20°
   c You can't have two 160° angles, so there is only one possibility

What do you think?

- **a** He has not measured the angles accurately
- **b** The 110° angle should be 70°; perhaps she measured the adjacent angle on the other side of the straight line. The total will then be 180°.

Practice 12.2B

1. **a** **i** 156° **ii** 72°
   b Yes
2. **a** **i** 56° **ii** 40°
   b No
3. $a = 42°$ (vertically opposite angles are equal)

$b = 54°$ (adjacent angles on a straight line add up to 180°)
$c = 84°$ (angles in a triangle add up to 180°)

4 BAD = 65° (angles around a point add up to 360°)
ABD = 45° (angles in a triangle add up to 180°)
BDC = 110° (adjacent angles on a straight line add up to 180°)
CBD = 39° (angles in a triangle add up to 180°)
ABC = 84° (angles in a triangle add up to 180°)

What do you think?

You need two pieces of information in both cases

Consolidate

1 **a** 26° **b** 149° **c** 54°
2 **a** **i** a and b **ii** 77°
b **i** a and 26°
ii $a = 26°$ and $b = 128°$
3 **a** 90°
b **i** 42° **ii** 39°
iii 132° **iv** 9°

Stretch

1 **a** **i** 20° **ii** 160° **iii** 160°
b **i** 42° **ii** 138° **iii** 138°
c $d = a + b$
2 **a** **i** 40° **ii** 105° **iii** 115°
iv 140° **v** 360°
b **i** 31.5° **ii** 118.7°
iii 92.8° **iv** 148.5°
v 360°
c $x + y + z = 360°$

Chapter 12.3

Are you ready?

1 **a** $a = 286°$ **b** $b = 39°$
c $c = 22°$, $d = 158°$
2 **a** $a = 115°$
b $b = 61°$, $c = 58°$
3 **a** Parallelogram
b Rhombus
c Trapezium
4 **a** $x = 20$ **b** $x = 30$ **c** $x = 10$

Practice 12.3A

1 Compare answers as a class
2 **a** $a = 130°$ **b** $b = 79°$
c $c = 35°$
3 **a** $a = 80°$, $b = 105°$, $c = 128°$
b $d = 117°$, $e = 67°$, $f = 63°$, $g = 50°$
c $h = 76°$, $i = 76°$, $j = 81°$
4 **a** The opposite angles are equal in size
b Yes, as a rhombus is a special parallelogram
5 **a** $a = 74°$, $b = 106°$
b $c = 105°$, $d = 105°$, $e = 75°$, $f = 75°$
c $g = 140°$, $h = 40°$, $i = 40°$, $j = 140°$

What do you think?

1 **a** If one angle is a then the opposite angle will be equal to a. The other two angles will each be $(360 - 2a)/2$ (or $180 - a$). Darius is correct.
b **i** 105° **ii** 105°
iii 75° **iv** 75°
2 **a** $a = 63°$, $b = 126°$
b **i** 25° **ii** 130°
3 **a** They are equal
b 135°

Practice 12.3B

1 **a** $a = 43°$, $b = 111°$
b $c = 150°$, $d = 120°$, $e = 150°$
c $f = 45°$, $g = 135°$
d $h = 50°$
e $x = 36°$, $i = 72°$
f $j = 24°$
g $k = 60°$, $l = 110°$
2 75°
3 **a** 50° **b** 110° **c** 70°
d 60° **e** 120°

What do you think?

1 **a** 126°
b Compare answers as a class – all the angles can be found
2 **a** $90 - x$ **b** $180 - 2x$
c $2x$

Consolidate

1 **a** $a = 66°$ **b** $b = 42°$
c $c = 35°$
2 **a** $a = 60°$, $b = 10°$, $c = 10°$, $d = 160°$
b $e = 41°$, $f = 41°$, $g = 41°$, $h = 139°$
c $x = 60°$, $2x = 120°$, $y = 60°$
d $p = 35°$, $q = 50°$, $r = 243°$
e $s = 150°$

Stretch

1

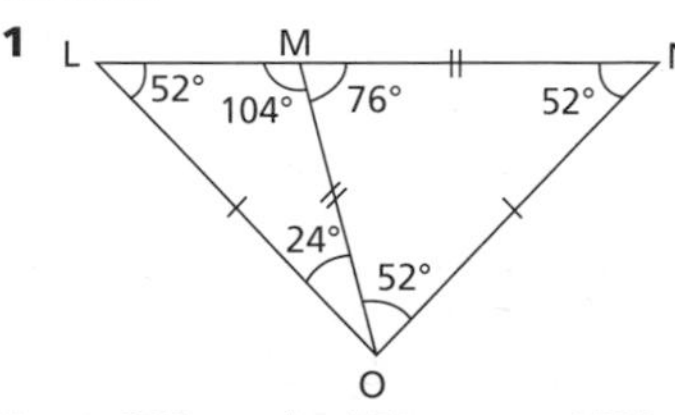

2 **a** 30° **b** 75° **c** 150°
3 **a** $a = 58°$, $b = 32°$, $c = 32°$, $d = 58°$, $e = 122°$
b b, c and e increase, but a and d decrease

Chapter 12.4

Are you ready?

1 **a** 180° **b** 360°
2 **a** 180° **b** 360°
c equal **d** equal
3 **a** $a = 5°$
b $b = 66°$, $c = 66°$, $d = 48°$

Practice 12.4A

1 **a** Compare answers as a class

b

Shape	No. of sides	No. of triangles	Sum of interior angles
Hexagon	6	4	4 × 180° = 720°
Heptagon	7	5	5 × 180° = 900°
Octagon	8	6	6 × 180° = 1080°
Nonagon	9	7	7 × 180° = 1260°
Decagon	10	8	8 × 180° = 1440°
n-sided polygon	n	$n - 2$	$(n - 2) \times 180°$

2 **a** $a = 136°$ **b** $b = 110°$
c $c = 120°$ **d** $d = 125°$
e $e = 100°$, $f = 114°$, $g = 124°$
3 **a** 120° **b** 140°
c 144° **d** 150°
4 She has made extra angles inside the shape and included them in her total

What do you think?

1 $(n - 2) \times 180/n$
2 **a** **i** $a = 108°$, $b = 36°$, $c = 36°$, $d = 36°$, $e = 72°$
ii $p = 120°$, $q = 30°$, $r = 90°$, $s = 60°$, $t = 30°$
b Compare answers as a class
c The angles are not integer values

Consolidate

1 **a** $a = 127°$ **b** $b = 95°$
c $c = 100°$ **d** $d = 140°$
e $e = 65°$, $f = 230°$, $g = 131°$, $h = 116°$
2 **a** 108° **b** 135°
c 156° **d** 162°
3 Although the sides are equal in length the angles aren't equal. The hexagon is not regular.

Stretch

1 **a** $a = 150°$, $b = 132°$
b Compare answers as a class
2 **a** 108 is not a factor of 360
b Triangles and hexagons

Chapter 12.5

Are you ready?

1 Parallel lines never meet, but perpendicular lines meet at right angles
2 **a** $a = 153°$
b $b = 76°$, $c = 76°$, $d = 104°$
c $e = 138°$
d $f = 30°$, $3f = 90°$, $4f = 120°$
e $g = 75°$, $h = 75°$, $i = 105°$, $j = 75°$

Practice 12.5A

1 Compare findings as a class. Angles a and c are equal, and b and d are equal.
2 Compare findings as a class. Angles e and h are equal, and f and g are equal.

3 Compare findings as a class. Angles p, r, u and w are equal, and q, s, t and v are equal.

What do you think?

1 ADC = 180 – a, DCB = a, CBA = 180 – a

2 a Trapezium
 b Both pairs add up to 180°
 c Yes

Practice 12.5B

1 a m b z c q
 d m e q f n

2 a = 131°, b = 131°, c = 49°, d = 49°, e = 49°, f = 49°, g = 131° with reasons
 Compare answers and reasoning with a partner

3 a a = 128° b b = 82°
 c c = 98°
 d d = 76°, e = 104°
 e f = 31°, g = 149°
 Compare answers and reasoning with a partner

4 a No b Yes

5 a a = 34°, b = c = 73°
 b d = 65°, e = 51°, f = 115°, g = 116°, h = 64° with reasons
 Compare answers and reasoning with a partner

What do you think?

1 All of the angles PQR, QPR, PRQ, PRS and RST are 60°. Compare the order and reasons you used to find them in your class

2 a x = 32.5°, y = 35°
 b z = 30°

Consolidate

1 a q b r c v
 d p e v f m

2 a = 53°, b = 53°, c = 53°, d = 127°, e = 127°, f = 127°, g = 127° with reasons
 Compare answers and reasoning with a partner

3 a a = 48°
 b b = 136°, c = 44°, d = 136°
 c e = 62°, f = 62°, g = 56°
 d h = 112°, i = 68°, j = 57°, k = 57°
 e l = 109°, m = 109°
 Compare answers and reasoning with a partner

Stretch

1 a 65° b 65°

2 They are equal. The angles adjacent to them above the parallel lines are also exterior alternate angles.

3 d is the odd one out. Compare your own sets with those of others in the class.

Chapter 12.6

Are you ready?

1 a 110° b 140°
 c 120° d 80°

2 a a = 118°, b = 118°
 b c = 68°, d = 97°

3 a a = 30°
 b b = 78°, c = 78°
 c d = 40°, e = 40°

Practice 12.6A

There are many alternative solutions to these questions. One possibility is provided for each question but you may wish to compare your methods with other people.

1 ∠ABC is marked as a right angle so is 90°
 ∠BAC + ∠BCA = 180° – 90° = 90° (angles in a triangle add up to 180°)
 ∠BAC = ∠BCA (angles at the base of an isosceles triangle are equal)
 So a = ∠BAC = 90° ÷ 2 = 45°

2 a ∠EDC = 90° – 30° = 60° (angles in a rectangle are right angles)
 ∠ECD = 90° – 40° = 50° (angles in a rectangle are right angles)
 ∠DEC = 180° – (60° + 50°) = 70° (angles in a triangle add up to 180°)
 b ∠EDC = 90° – x (angles in a rectangle are right angles)
 ∠ECD = 90° – y (angles in a rectangle are right angles)
 So ∠EDC + ∠ECD = 180° – (x + y)
 So ∠DEC = x + y (angles in a triangle add up to 180°)

3 a ∠CQX = ∠APX = a (corresponding angles are equal)
 ∠CQY = 180° – a (adjacent angles on a straight line add up to 180°)
 b ∠QPB = a (vertically opposite angles are equal)
 So ∠DQY = a (corresponding angles are equal)

4 ∠ABC = 180° – (x + y) (angles in a triangle add up to 180°)
 ∠ADC = 180° – (x + y) (angles in a triangle add up to 180°)
 So ∠ABC = ∠ADC as both are equal to 180° – (x + y)

5 a ∠ACB = 180° – (60° + 45°) = 75° (angles in a triangle add up to 180°)
 So a = 180° – 75° = 105° (adjacent angles on a straight line add up to 180°)
 b x + y + ∠ACB = 180° (angles in a triangle add up to 180°)
 So x + y = 180° – ∠ACB
 a + ∠ACB = 180° (adjacent angles on a straight line add up to 180°)
 So a = 180° – ∠ACB
 So a = x + y as both are equal to 180° – ∠ACB

What do you think?

1 a ∠XWY = ∠WYX = (180 – 2x)/2 = 90 – x
 ∠WYZ = 180 – (90 – x) = 90 + x
 ∠YWZ = ∠WZY = (90 – x)/2
 b Compare results as a class

2 ∠CBD = y (angles at the base of an isosceles triangle are equal)
 ∠BCD = 180 – 2y (angles in a triangle add up to 180°)
 ∠BCA = 2y (adjacent angles on a straight line add up to 180°)
 ∠ABC = 2y (angles at the base of an isosceles triangle are equal)
 In triangle ABC, ∠BCA + ∠ABC + ∠CAB = 180° (angles in a triangle add up to 180°)
 So 2y + 2y + x = 180°, hence x + 4y = 180°

Consolidate

1 a ∠AEB = 60° (angles in an equilateral triangle are equal)
 ∠BED = 90° (angles in a square are right angles)
 So ∠AED = 60° + 90° = 150°
 b ∠EBD = 45° (angles at the base of an isosceles triangle are equal)
 So ∠ABD = 60° + 45° = 105°

2 The angle sum of the pentagon is:
 $b + c + e + f + i + h + g + d + a$
 $= a + b + c + d + e + f + g + h + i$
 $= (a + b + c) + (d + e + f) + (g + h + i)$
 = 180° + 180° + 180° = 540°

3 a ∠ACB = x (angles at the base of an isosceles triangle are equal)
 ∠ACB + ∠ACD = 180° (adjacent angles on a straight line add up to 180°)
 So x + y = 180°, so y = 180° – x
 b ∠BAC + ∠ABC + ∠ACB = 180° (angles in a triangle add up to 180°)
 So ∠BAC + x + x = 180°, so ∠BAC = 180° – 2x
 c ∠LNM = (180 – 2y)/2 (angles at the base of an isosceles triangle are equal)
 So ∠LNM = 90° – y
 ∠LNM + ∠LNO = 180° (adjacent angles on a straight line add up to 180°)
 So 90° – y + ∠LNO = 180°, so ∠LNO = 90° + y

4 ∠BAC = ∠DAE (same angle)
 ∠ABC = ∠ADE (corresponding angles are equal)
 ∠ACB = ∠AED (corresponding angles are equal)
 So all three angles are the same.

Stretch

1 a ∠EBA = y (alternate angles are equal)
 ∠FBC = z (alternate angles are equal)
 ∠EBA + ∠EBF + ∠FBC = 180° (adjacent angles on a straight line add up to 180°)
 So x + y + z = 180°, which means the angles in triangle BEF add up to 180°
 b Yes

2 a $a = 50°$, $b = 65°$, $c = 65°$, $d = 115°$, $e = 65°$
b Compare answers as a class. Generalising, if $a = x$, then $b = c = e = (180 - x)/2$ and $d = 90 + x/2$

3 Compare answers as a class

Check my understanding

1 a $a = 33°$ (angles on a straight line add up to 180°)
$b = 147°$ (vertically opposite angles are equal)
b $c = 144°$ (angles at a point add up to 360°)

2 Other angles are 42° and 96° or 69° and 69°

3 a Compare answers as a class
b 68°

4 a $a = 50°$ (angles on a straight line add up to 180°)
$b = c = 65°$ (base angles in an isosceles triangle are equal and angles in a triangle add up to 180°)
$d = 115°$ (angles on a straight line add up to 180°)
$e = 41°$ (angles in a triangle add up to 180°)
$f = 91°$ (angles on a straight line add up to 180°)
b $g = 100°$ (angles in a quadrilateral add up to 360°)

5 ∠SPQ + ∠PQR = 180°, so PS is parallel to QR as co-interior angles add up to 180°
∠SRQ = 110° (angles on a straight line add up to 180°) so ∠PQR + ∠QRS = 180°, so PQ is parallel to SR
Both pairs of sides are parallel, so PQRS is a parallelogram

6 a 156°
b ∠CAB = x (base angles in an isosceles triangle are equal)
∠ACB = 180 − $2x$ (angles in a triangle add up to 180°)
∠ACD = $2x$ (angles on a straight line add up to 180°)
∠ADC = $2x$ (base angles in an isosceles triangle are equal)
∠DAC = 180° − $4x$ (angles in a triangle add up to 180°)

Block 13 Developing number Sense

Chapter 13.1

Are you ready?

1 a 100 **b** 884 **c** 6666 **d** 10991
2 a 6 **b** 646 **c** 4830 **d** 9623
3 a 432 **b** 1152 **c** 9152 **d** 93236
4 a 12 **b** 112 **c** 2112 **d** 352

Practice 13.1A

1 a 207 **b** 672 **c** 748 **d** 1975 **e** 13924 **f** 681 **g** 911 **h** 1575
2 a 9 **b** 474 **c** 550 **d** 1777 **e** 13726 **f** 283 **g** 313 **h** 777
3 a 78 **b** 81 **c** 171 **d** 265 **e** 36 **f** 3 **g** 85 **h** 113
4 a 142 m **b** 52 cm **c** 246 ml **d** 318 g **e** 191 km **f** 430 g **g** 2501 m **h** £1763 **i** 2761 cm

What do you think?

1 a **i**: 700, **ii**: 4800, **iii**: 7000
b **i**: 671, **ii**: 4563, **iii**: 6820
c The estimated answers are approximately equal to the actual answers

2 a Sven has used a formal written method.
Mario has added 1 to each number to keep the difference the same but make the calculation easier.
Junaid has partitioned 99999 and subtracted each part separately.
Kate has used the fact that 99999 is 1 less than 100000 to help her calculate.
b Various answers, but Mario's and Kate's methods are the most obvious mental calculations.
c Various answers

3 a 14971 + 98 = 14971 + 100 − 2 = 15069
b 14971 + 97 = 14971 + 100 − 3 = 15068
c 14971 + 199 = 14971 + 200 − 1 = 15170
d 14971 + 396 = 14971 + 400 − 4 = 15367

4 a $\frac{613}{4}$ **b** $\frac{415}{4}$ **c** $\frac{810}{7}$ **d** $\frac{1022}{9}$

Practice 13.1B

1 a **i**: You can partition 24 into 20 and 4 to break up the calculation
ii: 5 is half of 10 so 5 lots of 24 is half of 10 lots of 24
iii: If you double one value in the calculation and halve the other then the answer is the same
iv: 24 is equal to 6 × 4 so it is just rewriting the calculation
b **i** 180 **ii** 360 **iii** 95 **iv** 2420 **v** 5480 **vi** 43810

2 a 153 **b** 792 **c** 882 **d** 380 **e** 990 **f** 510 **g** 247 **h** 5994 **i** 20130 **j** 6864

3 **i**: If 75 divided by 5 is 15, then 750 divided by 5 is ten times 15
ii: You can partition 750 into 500 and 250 to make it easier to calculate
iii: There are twice as many 5s in a number as there are 10s
iv: If you double both numbers the answer remains unchanged

4 a 80 **b** 50 **c** 108 **d** 216 **e** 216 **f** 40 **g** 320 **h** 70 **i** 500 **j** 1380

5 a £50 **b** 240 g **c** 85 kg **d** 625 ml **e** £711 **f** 40 kg **g** £4950 **h** 6860 mm **i** 241 g **j** 29970 m

What do you think?

1 a 20000 cm^2 **b** 19800 cm^2
c They're approximately equal

2 A pack of 24 is better value (£4 per tile compared to £4.50)

3 a 1500 **b** 1500 **c** 15000 **d** 500 **e** 30

Consolidate

1 a 682 **b** 941 **c** £1260 **d** 2505 ml
2 a 643 **b** 601 **c** 2308 g **d** 90001 kg
3 a 2910 **b** 3960 **c** £7038 **d** 960 cm
4 a 90 **b** 8 **c** 196 ml **d** 150 g

Stretch

1 a $x = 615$ **b** $y = 480$ **c** $z = 693$ **d** $f = 1366$
2 a 92 **b** 245 **c** 400 **d** 148
3 a 5800 × 5 = 29000
b 5800 × 6 > 5801 × 5
5800 × 6 is 29000 + 5800 while 5801 × 5 is 29000 + 5

Chapter 13.2

Are you ready?

1 a 47 **b** 52 **c** 122 **d** 10605 **e** 73 **f** 69 **g** 9 **h** 8730
2 a 25 **b** 32 **c** 350 **d** 357 **e** 5 **f** 16 **g** 70 **h** 7
3 a 270 **b** 27 **c** 270 **d** 27 **e** 570 **f** 57 **g** 5.7 **h** 0.57
4 a 25 **b** 32 **c** 350 **d** 357

Practice 13.2A

1 a 8 **b** 8 trees **c** 8 tenths **d** 0.8 **e** 0.08
2 a 0.7 **b** 0.9 **c** 1.1 **d** 0.11 **e** 1.18
3 a 1 **b** 1 egg **c** 1 tenth **d** 0.1 **e** 0.01
4 a 0.5 **b** 0.05 **c** 0.54 **d** 0.87 **e** 0.79
5 a 2.4 **b** 1.4 **c** 4.8 **d** 8.1 **e** 4 **f** 0.24 **g** 0.14 **h** 0.48 **i** 0.81 **j** 0.4
6 a 8 **b** 0.8 **c** 0.08 **d** 0.6 **e** 1.9 **f** 0.06 **g** 0.3 **h** 0.07 **i** 1.07 **j** 101.07

7 a £5.96 b £15.96
c £9.51 d £2.50

What do you think?

1 a 0.3 b 0.4 c 0.8
d 0.5 e 0.2 f 4.3
g 11.4 h 99.8 i 6.5
j 98.2
2 a 9.6 b 9.6
c 0.96 d 2.4
3 a e.g. two bar models each split into 10 equal parts with 8 parts shaded in one colour to show 0.8, then the remaining two parts of that bar and 7 of the next bar another colour to show that 8 tenths plus 9 tenths is greater than 1 so can't be 0.17
b e.g. get him to think about the bar model or to think about it as tenths.
c 1.7

Practice 13.2B

1 a $\frac{3}{4}$ b $\frac{5}{8}$ c $\frac{1}{8}$ d $\frac{1}{2}$
e $\frac{5}{8}$ f $\frac{1}{8}$ g 0 h $\frac{1}{4}$
2 a 10 b 12 c 30 d 9
e 8 f 7 g 11 h 4
3 a 50 b 36 c 60 d 27
e 64 f 49 g 55 h 84
4 a £16 b £28
c £36 d £5.20

What do you think?

1 a 18 b 180
2 a £100 b £50
c £25 d £12.50
3 a $\frac{3}{8}$ b $5\frac{5}{8}$ c $16\frac{1}{8}$ d $34\frac{7}{8}$

Consolidate

1 a 5.8 b 15.8 c 10
d 10.1 e 65.6 f 1.2
g 0.4 h 5 i 2.9
j 22.2
2 a 2 b 1.8 c 2.8
d 0.3 e 0.18 f 0.2
g 0.2 h 0.9 i 0.07
j 0.002
3 a $\frac{11}{12}$ b $\frac{7}{8}$ c $\frac{13}{20}$ d $\frac{41}{100}$
e $\frac{1}{12}$ f $\frac{5}{8}$ g $\frac{1}{20}$ h $\frac{21}{100}$

Stretch

1 B = 250, C = 175
2 Range = $\frac{12}{20} = \frac{3}{5}$
3 a $x = 1.3$ b $x = 0.7$
c $x = 0.8$ d $x = 26.3$
e $x = 2.4$ f $x = 0.56$
g $x = 0.6$ h $x = 1.2$

Chapter 13.3

Are you ready?

1 a 1, 2, 3, 6
b 1, 3, 9
c 1, 2, 3, 4, 6, 12
d 1, 2, 3, 4, 6, 8, 12, 24
e 1, 2, 3, 4, 6, 9, 12, 18, 36
f 1, 3, 17, 51
g 1, 2, 4, 8, 16
h 1, 2, 3, 4, 6, 8, 12, 16, 24, 48
2 a 300 b 500 c 20
d 7 e 2000 f 0.6
g 10 h 1 000 000
3 a 800 b 160 c 20 000
d 15 e 999 700
4 a 7 b 36 c 11
d 144 e 27 f 2

Practice 13.3A

1 a 30 b 30 c 30 d 30
2 a 378 b 378 c 378
d 378 e 378
3 Any correct diagram
4 a i 80 ii 16
b 16
5 a 108 b 1080 c 1080
d 980 e 735
6 a 27 b 5 c 15
d 18 e 16

What do you think?

1 a True b True
2 The only factors of 7 are 1 and 7 so it won't simplify the calculation
3 Various answers

Practice 13.3B

1 a 800 b 1000 c 1800
d Less e 1786
2 a 17 000, less, 16 858
b 6300, greater, 6381
c 50, less, 49
d 13 000, less, 12 574
3 a 479 < 500 and 977 < 1000 therefore 479 + 977 < 1500
b 47 < 50 and 50 × 9 = 450 therefore 47 × 9 < 450
c $\sqrt{100} = 10$ therefore $\sqrt{107} > 10$
d 21 572 > 20 000 and 34 916 > 30 000 therefore 21 572 + 34 916 > 50 000
4 a 600 + 200 = 800 therefore 596 + 193 < 800
b 80 × 8 = 640 therefore 79 × 8 < 640
Zach may have rounded down to the nearest 100 rather than up. Emily may have added 8 to 640 rather than subtracting it.
5 a i 15 cm^2 ii 6400 mm^2
iii 200 m^2
b i Less ii Less iii More
6 a 20 b 64 c £60 d 22 kg

What do you think?

1 a £10
b No, she does not have enough money
The actual total is too close to the estimated total for the estimate to give a clear decision
2 It is more helpful to round the values to the nearest square number than to one significant figure
a 7, less b 8, more
c 5, more d 90, more

Consolidate

1 a 3296 b 3684
c 600 d 700
2 a 52 b 25
c 39 d 60
3 a 6000 b 3000
c 3500 d 100
4 a 5779 b 2648
c 3560 d 103
5 They are approximately equal

Stretch

1 $\frac{78}{18} = \frac{13}{3} = 4\frac{1}{3}$
2 a Various answers
b 11 000
c $c = 11\,000$, $d = 13\,000$
3 a $x \approx 100$ b $x \approx 1.25$
c $y \approx 6300$ d $b \approx 10\,000$
e $c \approx \pm 80$ f $h \approx 400$
g $p \approx \pm 70$ h $f \approx 3000$
4 a x must be prime
b y or z must be equal to 1

Chapter 13.4

Are you ready?

1 a 5 + 7 = 12, 7 + 5 = 12, 12 − 7 = 5, 12 − 5 = 7
b 300 + 700 = 1000, 700 + 300 = 1000, 1000 − 700 = 300, 1000 − 300 = 700
c 68 + 28 = 96, 28 + 68 = 96, 96 − 28 = 68, 96 − 68 = 28
2 a $x + 7 = 12$, $7 + x = 12$, $12 - 7 = x$, $12 - x = 7$
b $300 + 5x = 1000$, $5x + 300 = 1000$, $1000 - 5x = 300$, $1000 - 300 = 5x$
c $3x + y = 96$, $y + 3x = 96$, $96 - y = 3x$, $96 - 3x = y$
3 a 35 b 21 c 133
d 34 e 2 f 0

Practice 13.4A

1 a 170 b 170 c 140 d 30
2 a 91 b 91 c 13 d 7
3 a i 84 ii 84
iii 21 iv 63
b i 840 ii 840
iii 210 iv 630
c i 8400 ii 8400
iii 2100 iv 6300
d i 8.4 ii 8.4
iii 2.1 iv 6.3
e i $84x$ ii $84x$
iii $21x$ iv $63x$
4 a 3600 b 3600 c 36 000
d 36 e 36
f 360 000 000
g $360x$ h 1500
5 a 2418 b 2416
c 2407 d 1733
6 a 36 b 54 c 90
d y e 9
7 a Any correct bar model showing $210 = 5x + 25$
b i 420 ii 42 iii 185
iv 217 v 21 000
8 a 34 b 32 c 8.5

What do you think?

1 a Both numbers in the question are multiplied by 10 so the answer needs to be multiplied by 100
  b Both numbers in the question are multiplied by 100 so the answer needs to be multiplied by 10 000
  c Multiplying by 100 is the same as multiplying by 1000 then dividing by 10
  d When dividing by a smaller number the answer will be greater
2 Various answers
3 −17

Practice 13.4B

1 a Add 400 then add 5 mentally
  b 984
2 a 1769   b 91 228
  c 15 552   d 3995
3 a £34.93   b £47.70
  c £19.65
4 19 × £10 = £190, so he has enough money, as the estimate for the cost is greater than the actual cost
5 e.g. Subtract £4 then add 3p

What do you think?

1 a Various answers
  b They'll both get the same answer
2 Various answers
3 Various answers
4 She should use a formal written method or a calculator to check her answer

Consolidate

1 a 200 b 200 c 19 d 181
2 a 86 b 86 c x d 44
3 a 350 b 350 c 3500
  d 700 e 0.35
4 $2p + 2q = 16$, $3p + 3q = 24$, $4p + 4q = 32$, $8p + 8q = 64$, $10p + 10q = 80$, $2(p + q) = 16$, $\frac{1}{2}p + \frac{1}{2}q = 4$, $7(p + q) = 56$

Stretch

1 a e.g. $2(r + s)$ b e.g. $\frac{1}{2}(r + s)$
  c e.g. $5(r + s)$ d e.g. $10(r + s)$
  e e.g. $\frac{3}{2}(r + s)$ f e.g. $\frac{1}{100}(r + s)$
  g e.g. $\frac{r+s}{r+s}$
  h e.g. $3000(r + s)$
2 a 40 b 351 c 88
  d 1053 e 19.5
3 a False   b True
  c False   d False
4 a $2f - 2g$ b $\frac{1}{2}f - \frac{1}{2}g$
  c $\frac{(f-g)}{5}$ d $(f - g)^2$

Check my understanding

1 a 671   b 311
  c 4264   d 57 443

2 a 140 b 315 c 760
  d 792 e 1810
3 a 90 b 60 c 80
  d 9000 e 90
4 a 6.67   b 2.96
  c 28.07   d 10.84
  e $\frac{3}{4}$ f $\frac{1}{2}$ g $\frac{3}{8}$ h $\frac{67}{100}$
5 a 3000   b 100
  c 600   d 8000
6 a 152 b 83 c 23.5
  d $235x$ e 23 500 f 1520
  g −152 h 2.35
7 a 14   b 2940
  c 294 000   d 210

Block 14 Sets and probability

Chapter 14.1

Are you ready?

1 a 1, 3, 5, 7, 9
  b 2, 4, 6, 8, 10
  c 1, 4, 9, 16, 25
  d 3, 6, 9, 12, 15
2 a 1, 2, 3, 4, 6, 12
  b 1, 2, 4, 5, 10, 20
  c 1, 3, 7, 21
3 a 3, 5, 7, 11, 13
  b 17   c 15

Practice 14.1A

1 a Yes, same letters
  b No, B is missing a 4
  c No, A is the set of all odd numbers, B has just six odd numbers
  d No, one is sports and the other is teams
  e Yes, same values
2 Repeats are not necessary
3 a X = {21, 28, 35}
  b Y = {spring, summer, autumn, winter}
  c Z = {c, e, i, n, s]
4 a Odd numbers between 0 and 12
  b The first six powers of 10
  c Letters that make up the word CARE (or any word with just the letters CARE in)
  d The factors of 12
5 a 1, 4, 9, 16, 25, 36, 49
  b 4, 16, 36
  c All multiples of 10 are even
  d 1, 4, 9, 10, 16, 20, 25, 30, 36, 40, 49, 50

What do you think?

1 a 2.7 is not between 3 and 5, or 2.7 is not a member of the universal set
  b 5 is not between 3 and 5 (as it does not say 'inclusive')
  c 11 is not part of the universal set
  d 1, 2, 3, 5, 6, 7, 8, 9, 10
2 a X = {b, e, h, i, v}, Y = {g, a, t, e, k, p, r}
  b e
  c a, b, e, g, h, i, k, p, r, t, v
3 a The set that has no elements
  b There are no numbers that are both odd and even

Practice 14.1B

1 a

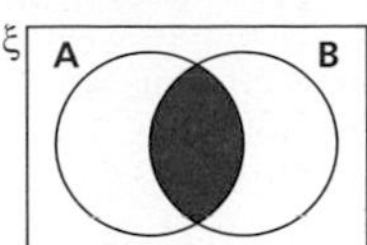

  b ξ A B

2 a A = {2, 4, 6, 8, 10}
  b B = {1, 2, 3, 4, 6}
  c 2, 4 and 6
  d

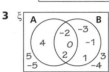

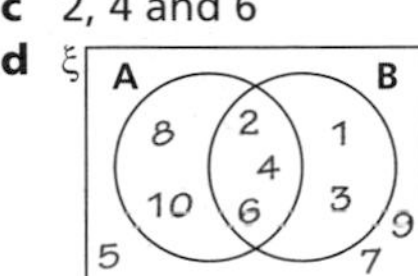

3 ξ A B 4 −2 0 2 −3 −1 1 5 −5 3 −4

4 a X = {1, 2, 5, 10}
  b Y = {1, 3, 5, 7, 9}
  c 1, 5
  d 1, 2, 3, 5, 7, 9, 10
  e 3, 4, 6, 7, 8, 9
  f 2, 4, 6, 8, 10
  g X is the factors of 10 and Y is odd numbers. Universal set is numbers between 1 and 10 inclusive.
5 a Yes. More study just geography compared with studying just history. There is no need to add in the number that study both when making the comparison.
  b 45 c 57 d 19
  e 83 f 129 g 34.9%
  h The 19 will change to 20 as she studies both subjects
6 a P = {11, 15}
  b No, he is wrong. 11 is inside set Q.
  c Q = {11, 12, 13, 14, 15, 16, 17, 18}
  d Just 1, which is 19
  e 7 elements
  f Because P is a subset (wholly inside) of Q

What do you think?

1 Diagram 2 as all the elements in B are also in A
2 a Because the number 24 is in both sets
  b Because there are numbers not in P or Q that are in the universal set
  c Yes; it is just the same as two overlapping circles
3 a There is no overlap

b Various answers, e.g. odd numbers and even numbers or odd numbers and multiples of 4

Consolidate

1 **a** A = {Jan, Feb, Mar, Apr, May, Jun, Jul, Aug, Sep, Oct, Nov, Dec}
b B = {p, a, r, l, e, o, g, m}
c C = {0, 2, 4, 6, 8, 10}
d D = {1, 2, 4, 5, 10, 20}

2 **a** A = {1, 2, 3, 4, 6, 12}
b B = {1, 2, 3, 5, 6, 10, 15, 30}
c 1, 2, 3, 6

3 **a** A = {1, 3, 5, 7, 9, 11, 13, 15, 17, 19}
b B = {2, 4, 6, 8, 10, 12, 14, 16, 18, 20}
c ξ

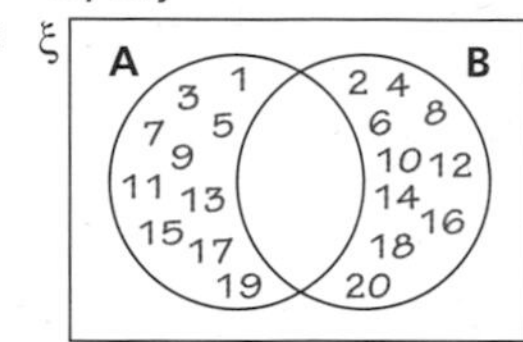

d There are no numbers that are both odd and even

Stretch

1 **a** 11 elements
b 111 elements

2 P = {multiples of 13}
Q = {even square numbers}
R = {one more than multiples of 9 above 10} or {numbers whose digits sum to 10}
S = {prime numbers between 20 and 40}
T = {the answers to 4 × 11} or {the LCM of 4 and 11}

3 Yes, because between $\frac{1}{2}$ and $\frac{3}{4}$ on the number line there are infinitely many points

4 ξ = {7, 10, 14, 15, 20, 21, 25, 28}
A = {10, 14, 15, 20, 21, 25, 28}
No elements just in B

5 **a** ξ

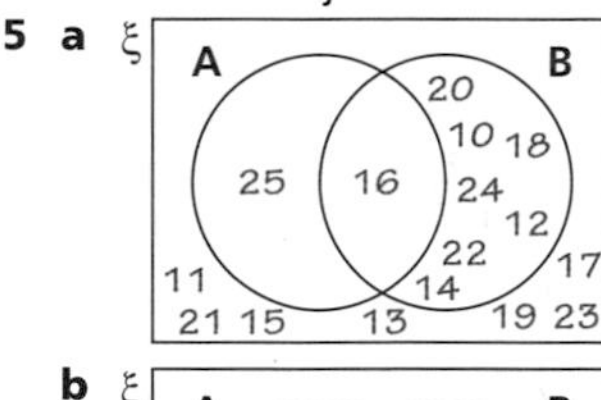

b ξ

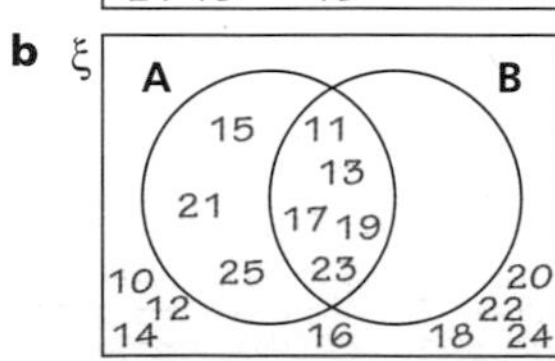

c ξ

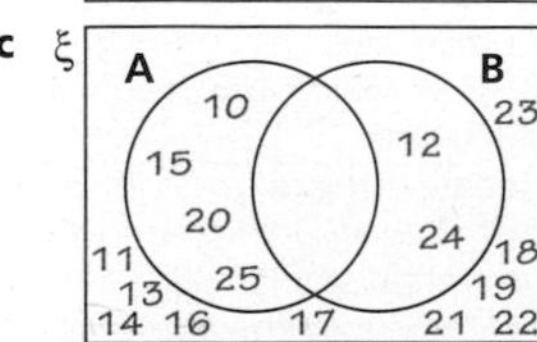

d All have 2 circles
All have elements outside the circles, which are in the universal set but not in set A or in set B
Differences: two have overlapping circles and one does not; number of elements

6 8 students

Chapter 14.2

Are you ready?

1 **a** A = {Monday, Tuesday, Wednesday, Thursday, Friday, Saturday, Sunday}
b B = {Jan, Feb, Mar, Apr, May, Jun, Jul, Aug, Sep, Oct, Nov, Dec}
c C = {1, 2, 3, 6, 9, 18}
d D = {m, a, t, h, e, i, c, s}

2 **a** A = {1, 3, 5, 7, 9, 11, 13, 15, 17, 19}
b B = {2, 4, 6, 8, 10, 12, 14, 16, 18, 20}

Practice 14.2A

1 A ∩ B = {12, 13, 14}
2 A ∩ B = {red}
3 ξ

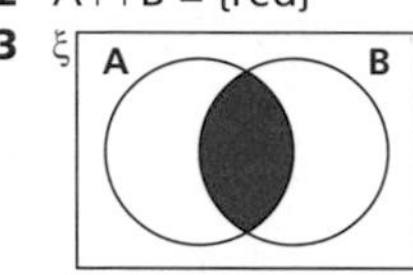

4 X ∩ Y = {20, 30}
5 **a** A ∩ B = {1, 2, 4}
b A = factors of 12, B = factors of 20.
c A ∩ B = common factors of 12 and 20.

6 Any three sets of seven integers between 1 and 100 that contain the numbers 9 and 17 but do not contain 5, 11, 29 or 33

What do you think?

1 **a** 7, 8, 9
b A ∩ B = {7, 8, 9}
The elements are the same as the intersection of those elements in A and B

2 **a** No numbers or elements that are in A and in B as there is no overlap region
b Set A could be even numbers greater than 6 and set B could be prime numbers. There are no even prime numbers greater than 6.

3 You don't write 'maths' or 'science' twice. They would just appear once in the intersection.

4 **a** Because A is a subset (wholly inside) of B
b A could be multiples of 4 and B could be even numbers

5 True, they show the same overlap region

Practice 14.2B

1 **a** M ∪ H = {tab, tan, tam, the, tea, ten, tap, tax}
b These are all the three-letter words that Mateusz and Hannah have said. Repeats are not put in twice.

2 You don't repeat the values if they are in both sets.

3 A ∪ B = {1, 2, 3, 4, 5, 6, 10, 12, 20}

4 **a** A ∪ B = {10, 15, 20, 30, 40, 45}
b A ∩ B = {30}

5 **a** A ∪ B = {2, 4, 5, 6, 8, 10}
b A ∪ C = {2, 3, 4, 5, 6, 7, 8, 10}
c B ∪ C = {2, 3, 5, 7, 10}
d (A ∪ B) ∪ C = {2, 3, 4, 5, 6, 7, 8, 10}
e Yes

6 **a** 3 **b** 23 **c** 2

What do you think?

1 **a** ξ

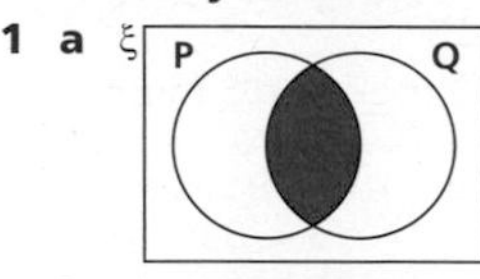

b ξ

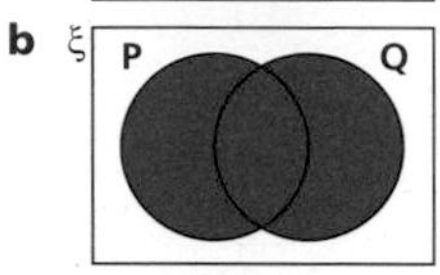

2 **a** ξ

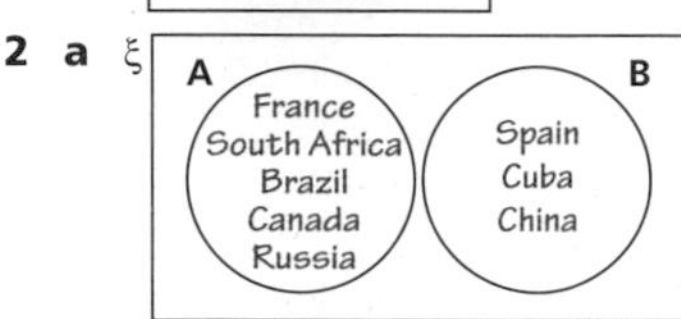

b Because the circles do not intersect

c ξ

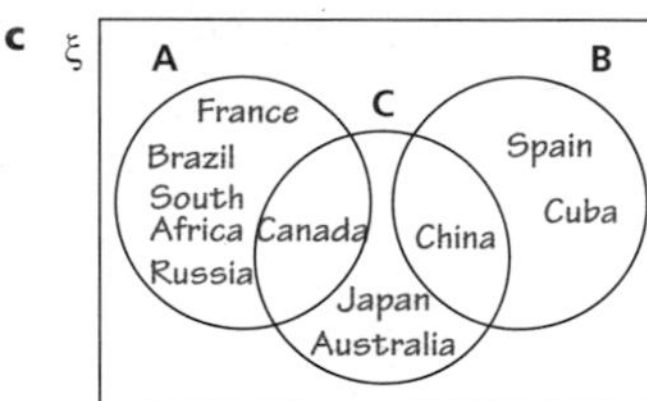

3 **a** Various answers, e.g. A = {2, 5, 6, 8} and B = {6, 8, 9, 10}
b Various answers, e.g. A = {2, 5, 6, 8} and B = {9, 10}
c Various answers, e.g. A = {2, 5, 6, 8} and B = {2, 5, 6, 8, 9, 10}

Practice 14.2C

1 A′ = {6, 10, 14}

2 **a** ξ

b ξ

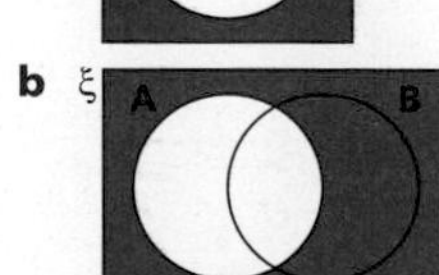

c 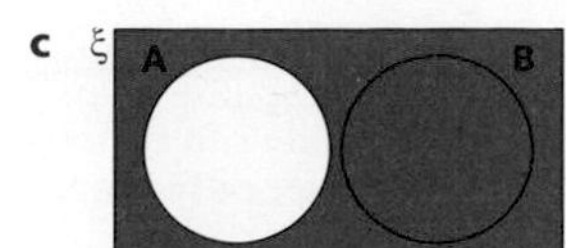

d 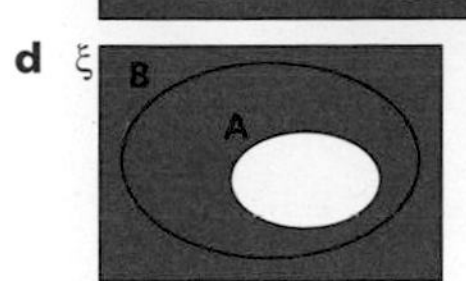

3 a A′ = {14, 28}
b B′ = {3, 5, 9, 11}

4 a A′ = {e, g, m, n}
b B′ = {n, s}
c (A ∩ B)′ = {e, g, m, n, s}
d (A ∪ B)′ = {n}

What do you think?

1 a 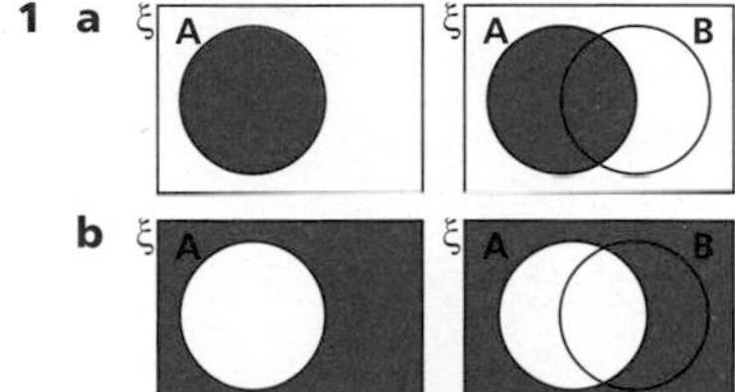

b ξ A ξ A B

c A and everything that is not in A (i.e. A and its complement) complete the whole

2 a A = {multiples of 3}
b B = {square numbers}

Consolidate

1 a A; P ∩ Q = {12, 24}
b D; P ∪ Q = {3, 4, 6, 8, 9, 12, 15, 16, 18, 20, 21, 24}
c B; P′ = {4, 8, 16, 20}
d C; Q′ = {3, 6, 9, 15, 18, 21}

2 a A ∩ B = {2, 4, 8, 12}
b A ∪ B = {1, 2, 3, 4, 6, 8, 10, 12}
c A′ = {5, 6, 7, 9, 10, 11}
d B′ = {1, 3, 5, 7, 9, 11}

Stretch

1 23 children both sing and dance

2 a X ∩ Y = {D, G}
b Y ∩ Z = {E, G}
c (X ∩ Y) ∩ Z = {G}
d Y′ = {A, B, F}
e X′ ∪ Y′ = {A, B, C, E, F}
f X ∪ Y′ = {A, B, D, F, G}
g (X ∪ Y)′ = {B}
h (X ∪ Y)′ ∩ Z′ = {A, C, D}

3 14

Chapter 14.3

Are you ready?

1 1, 2, 3, 4, 5, 6

2 Head or tail

3 A fair coin is a standard coin that hasn't been altered. A biased coin has been altered in some way, for example made heavier on one side.

Practice 14.3A

1 a You will blink today — Certain
b You will get a head when you a flip a fair coin — Even chance
c You will go to bed by midnight — Likely
d It will snow tomorrow — Unlikely
e You will get a 7 if you roll a fair dice. — Impossible

2 a Even chance **b** Very unlikely
c Certain **d** Impossible
e Very likely **f** Likely

3 a 5 **b** 1 **c** 3 and 7
d Greater than 4

4 a There are more sections on spinner B so the probability is less.
b Spinner B because $\frac{3}{8}$ is greater than $\frac{1}{4}$

5 No; he has a 1 in 500 chance of winning not a $\frac{50}{50}$ chance of winning

What do you think?

1 a e.g. the sun rising
b e.g. it snowing if it is summer time
c e.g. a newborn baby will be a boy

2 Flo might be better at the game than Faith so she might have a greater than even chance of winning, or vice versa

3 Class discussion

Practice 14.3B

1 S = {1, 2, 3, 4, 5, 6}

2 There should be no repeats in a sample space, so it should be S = {1, 2, 3}

3 a S = {H, T}
b S = {A, B, C}

4 S = {M, A, T, H, E, I, C, L}

5 a A spinner with 12 sections with 3 labelled red, 3 green, 3 yellow and 3 blue
b A spinner with 12 sections labelled red, green, yellow and blue, but with different numbers of each colour

6 a 10, 11, 12, 13, 14
b The numbers will be the same but there will need to be more of some of them. e.g. 10, 10, 11, 11, 12, 12, 13, 13, 14.
c You can't tell from the sample space how many of each number there are

What do you think?

1 S = {apple, orange, banana}

2 a S = {pink, orange, yellow}
b S = {pink, orange, yellow}

c The sample space is the same but the probabilities of picking a certain colour of ball would be different

3 Class discussion

Consolidate

1 a Likely **b** Certain
c Unlikely **d** Impossible
e Even chance

2 a Likely (unless it is Friday)
b Unlikely (unless it is winter and has been very cold)
c Even chance
d Very unlikely

3 a S = {R, B, G}
b S = {R, B, G}
They are the same because the possible outcomes are the same

Stretch

1 e.g. 4, 4, 5, 7, 8

2 a Impossible
b Certain
c We don't know what proportion of the counters have 8 or 10 on

3 S = {red, blue, yellow}

4 e.g. 1, 1, 1, 1, 3, 3, 6, 6, 12, 12, 24, 24, 48, 48, 48, 48

Chapter 14.4

Are you ready?

1 a $\frac{1}{2}$ **b** $\frac{1}{3}$ **c** $\frac{1}{4}$
d $\frac{5}{6}$ **e** $\frac{3}{10}$

2 S = {blue, green, red, yellow}

3 a

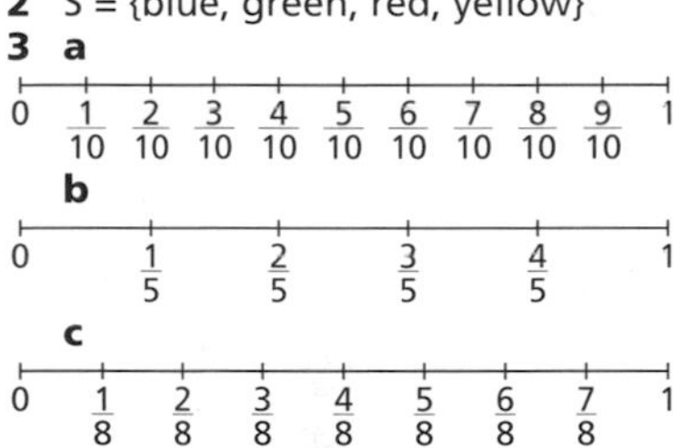

Practice 14.4A

1 a $\frac{1}{6}$ **b** 0 **c** $\frac{3}{6}$ (or $\frac{1}{2}$)
d $\frac{5}{6}$ **e** $\frac{2}{6}$ (or $\frac{1}{3}$) **f** $\frac{2}{6}$ (or $\frac{1}{3}$)
g $\frac{5}{6}$ **h** 1

2 a $\frac{4}{10}$ (or $\frac{2}{5}$) **b** $\frac{6}{10}$ (or $\frac{3}{5}$)
c 0

3 a $\frac{3}{8}$ **b** $\frac{5}{8}$ **c** $\frac{5}{8}$
d The same, because the counters that are not yellow are either red or blue

4 a $\frac{5}{9}$ **b** $\frac{3}{9}$ (or $\frac{1}{3}$)
c 0 **d** $\frac{8}{9}$

5 $\frac{25}{120}$ (or $\frac{5}{24}$)

6 a $\frac{18}{100}$ (or $\frac{9}{50}$) **b** $\frac{47}{100}$
c $\frac{52}{100}$ (or $\frac{13}{25}$) **d** $\frac{82}{100}$ (or $\frac{41}{50}$)

7 a 5 orange, 3 yellow and 4 white

b No; the probabilities add up to 1

c Because the probability of yellow or white is greater than the probability of orange

d 8; because there are 7 non-orange counters

8 a $\frac{4}{8}$ (or $\frac{1}{2}$) **b** $\frac{3}{8}$

c $\frac{2}{8}$ (or $\frac{1}{4}$)

9 The spinner will have 5 green sections, 3 red sections and 2 blue sections

What do you think?

1 No; the probability is the same

2 $\frac{1}{2}$, 0.5, 50%

3 a Benji

b Yes; Faith would now be more likely to win

c It would be an even chance

4 The dice could be eight sided and have two 3s. The dice could be six sided and be biased.

5 Yes. Jackson has a 1 in 200 chance and Tamsin has a 2 in 200 chance. $\frac{2}{200}$ is double $\frac{1}{200}$.

Practice 14.4B

1 a B **b** A **c** C

2

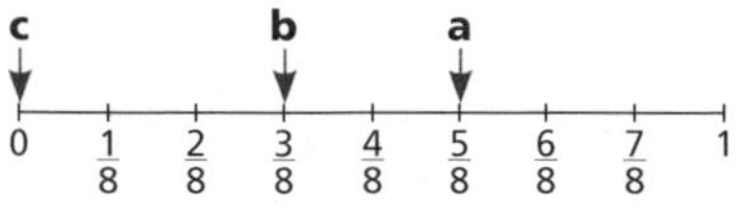

3

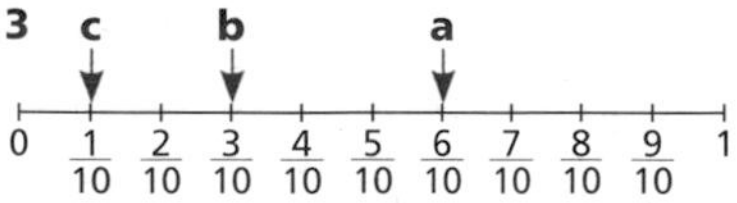

4 Jackson is incorrect; it should be at $\frac{1}{3}$

5

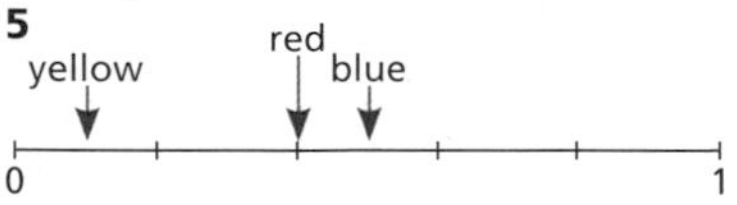

What do you think?

1

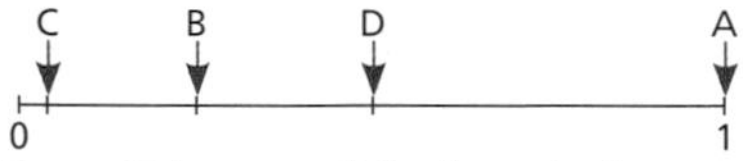

2 a Y, because it is closer to 1

b Estimated close to $\frac{3}{4}$

c Estimated close to $\frac{9}{10}$

Consolidate

1 a i 4 out of 7 counters are red

ii 3 out of 7 counters are blue

b i $\frac{4}{7}$ **ii** $\frac{3}{7}$

c

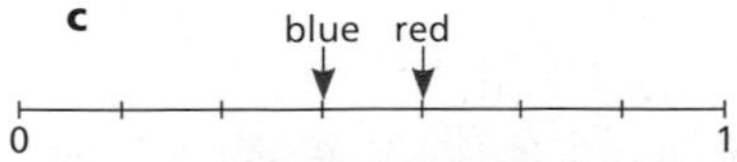

2 a i 3 out of 8 sections are red

ii 2 out of 8 sections are green

iii 3 out of 8 sections are yellow

b i $\frac{3}{8}$ **ii** $\frac{2}{8}$ **iii** $\frac{3}{8}$

c

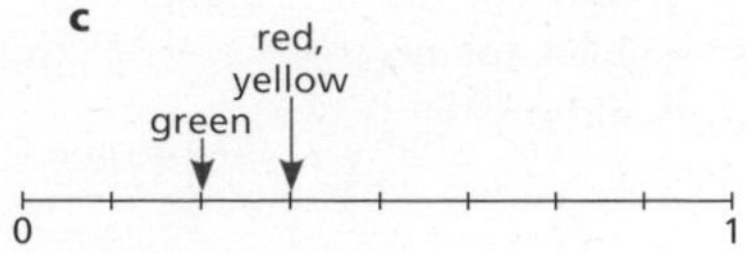

3 a $\frac{47}{100}$, 0.47, 47%

b

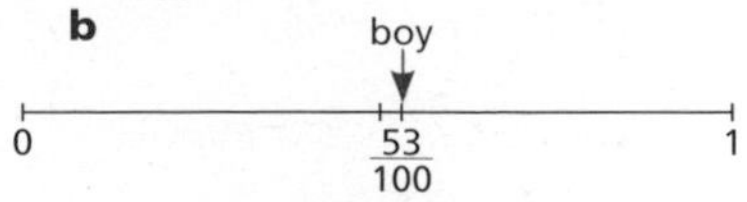

4 a P(red) = $\frac{3}{10}$, P(blue) = $\frac{1}{10}$, P(green) = $\frac{6}{10}$

b

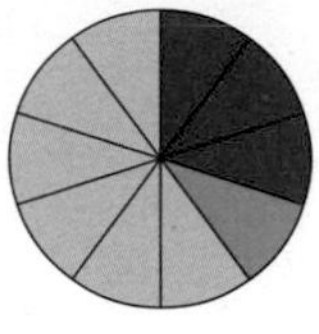

Stretch

1 Yes; 1% would be 200 winners

2 Any suitable 10-section spinner, e.g. labelled 1, 2, 3, 4, 5, 5, 7, 7, 8, 9

3 4 more cards

Chapter 14.5

Are you ready?

1 a 0.7 **b** 0.25 **c** 0.76

2 a 0.4 **b** 0.55 **c** 0.5 **d** 0.43

3 a $x = 0.8$ **b** $x = 0.2$

Practice 14.5A

1 0.6

2 $\frac{3}{10}$

3 a $\frac{7}{15}$ **b** $\frac{5}{15}$ (or $\frac{1}{3}$)

4 0.5

5 a Team A

b Chloe has incorrectly worked out 0.3 + 0.23 to be 0.26. The probability is 0.47.

6 a $\frac{3}{10}$ **b** $\frac{3}{10}$ **c** 8

7 a 0.4 **b** 0.45 **c** 0.9

8 0.47

What do you think?

1 $1 - x$

2 $x + y + z = 1$

3 Arrow for B drawn at $\frac{5}{8}$

4 Any spinner which has P(red) = 0.1, P(blue) = 0.5, P(green) = 0.2, P(yellow) = 0.2

Consolidate

1 0.9

2 $\frac{3}{5}$

3 a 0.27 **b** 0.42 **c** 0.31

Stretch

1 0.46

2 $\frac{1}{6}$

3 The probabilities are impossible. Though the probabilities sum to 1, no probability can be lower than 0 (impossible) or higher than 1 (certain).

4 15

Reflect

1 1

2 They sum to 1; e.g. 0.2, 0.5 and 0.3

Check my understanding

1 a A = {p, r, o, b, a, i, l, t, y}

b B = {1, 3, 5, 7, 9}

c C = {1, 2, 3, 4, 6, 8, 12, 16, 24, 48}

d D = {42, 45, 48, 51, 54, 57, 60, 63, 66, 69}

2 a A = {0, 2, 4, 6, 8, 10, 12, 14, 16, 18, 20}

b B = {1, 4, 9, 16}

c 4, 16

d

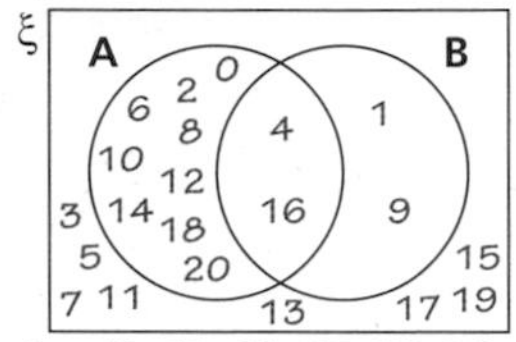

3 a A = {5, 10, 15, 20, 25, 35, 45}

b B = {1, 2, 4, 5, 10, 20}

c $A \cap B$ = {5, 10, 20}

d $A \cup B$ = {1, 2, 4, 5, 10, 15, 20, 25, 35, 45}

e A' = {1, 2, 4}

4 a $\frac{2}{5}$ **b** $\frac{3}{5}$

5 a $\frac{4}{14}$ (or $\frac{2}{7}$) **b** $\frac{3}{14}$

c $\frac{7}{14}$ (or $\frac{1}{2}$) **d** $\frac{10}{14}$ (or $\frac{5}{7}$)

e $\frac{7}{14}$ (or $\frac{1}{2}$) **f** 0

6 0.14

Block 15 Primes and proof

Chapter 15.1

Are you ready?

1 e.g. 2 × 7 = 14, 7 × 2 = 14, 14 ÷ 7 = 2, 14 ÷ 2 = 7, 7 + 7 = 14

2 a 72 **b** 72 **c** 72

d 72 **e** 72 **f** 72

3 a 6 **b** 7 **c** 7

d 9 **e** 40 **f** 111

4 b 7 r3 **c** 7 r7 **d** 9 r11

e 39 r2 **f** 110 r1

Practice 15.1A

1 a 4, 8, 12, 16, 20

b 7, 14, 21, 28, 35

c 15, 30, 45, 60, 75

d 21, 42, 63, 84, 105

2 a 8, 16, 24, 32, 40

b 14, 28, 42, 56, 70

c 30, 60, 90, 120, 150

d 42, 84, 126, 168, 210

3 a He should know that 3000 is a multiple of 6. He can start from 3000 and doesn't need to start from 6.
b e.g. 3006, 3012, 3018
4 a e.g. 105 **b** e.g. 1500
c e.g. 10 555
5 40
6 a 2, 4, 6, 8, 10, 12, 14, 16, 18, 20
b 4, 8, 12, 16, 20
c e.g. The multiples of 4 are all double the multiples of 2. They're all even.
d No; 47 is not even
7 a i 12, 20, 22, 30, 32, 40
ii 12, 15, 27, 30
iii 12, 20, 32, 40
iv 35
v 20, 30, 40
vi 15, 20, 25, 30, 35, 40
b 17 was **not** used because it is only a multiple of 1 and 17
c All multiples of 10 are also multiples of 5 because 10 is a multiple of 5

What do you think?

1 a 9, 9, 24
b They are multiples of 3
c e.g. 5001
2 48
3 a i e.g. $x = 1$ **ii** e.g. $x = 3$
b Yes, e.g. $x = 8$ or $x = 15$ for multiples of 7, $x = 33$ or $x = 18$ for multiples of both 3 and 5
4 a There is one lot of $x + 2$
b $2x + 4$
c $x + 2$, $2x + 4$, $3x + 6$, $4x + 8$, $5x + 10$

Practice 15.1B

1 a 1, 2, 3, 6 **b** 1, 2, 4, 8
c 1, 2, 4, 5, 10, 20
d 1, 2, 11, 22
2 No; e.g. 5 > 4 but 4 has more factors than 5
3 a i 1, 7 **ii** 1, 11
iii 1, 13 **iv** 1, 17
b They all have **exactly** two factors
c e.g. 15 has four factors: 1, 3, 5, 15
4 a i Array with 6 rows of 3 or 3 rows of 6
ii Array with 3 rows of 4 and 1 row of 3
b Various answers
5 a i 3, 4, 12 **ii** 12, 24, 36
b They both include 12; the first list has 12 as the largest number, the second list has 12 as the smallest number
c Flo has looked at multiples not factors
6 a 1, 2, 3, 4, 6, 8, 12, 24
b 1, 2, 3, 4, 6, 8, 12, 16, 24, 48
c False; 48 has only two more factors than 24 does

What do you think?

1 Sometimes true. Students can use various answers to support this.

2 a e.g. 2, 3, 5 **b** e.g. 4, 12, 20
3 1
4 1, 2, 4, $x + 2$, $2x + 4$, $4x + 8$

Consolidate

1 a 3, 6, 9, 12, 15
b 7, 14, 21, 28, 35
c 11, 22, 33, 44, 55
d 15, 30, 45, 60, 75
e 20, 40, 60, 80, 100
f 100, 200, 300, 400, 500
g 200, 400, 600, 800, 1000
2 a 1, 2, 4
b 1, 2, 3, 4, 6, 12
c 1, 2, 4, 8, 16
d 1, 2, 4, 7, 14, 28
e 1, 2, 4, 8, 16, 32
f 1, 2, 3, 4, 5, 6, 10, 12, 15, 20, 30, 60
g 1, 2, 4, 5, 10, 20, 25, 50, 100
3 a 35 **b** 72 **c** 900
4 a e.g. 2005 **b** e.g. 2500
c e.g. 2400 **d** e.g. 2825
e e.g. 2800 **f** e.g. 2950
g e.g. 2994

Stretch

1 a x, $2x$, $3x$, $4x$, $5x$
b $2x$, $4x$, $6x$, $8x$, $10x$
c $3y$, $6y$, $9y$, $12y$, $15y$
d $6ab$, $12ab$, $18ab$, $24ab$, $30ab$
e $7x^2$, $14x^2$, $21x^2$, $28x^2$, $35x^2$
2 a $x + 5$, $2x + 10$, $3x + 15$, $4x + 20$, $5x + 25$
b $2x - 1$, $4x - 2$, $6x - 3$, $8x - 4$, $10x - 5$
c $4 + 3y$, $8 + 6y$, $12 + 9y$, $16 + 12y$, $20 + 15y$
d $6ab + 11$, $12ab + 22$, $18ab + 33$, $24ab + 44$, $30ab + 55$
e $7x^2 - 9$, $14x^2 - 18$, $21x^2 - 27$, $28x^2 - 36$, $35x^2 - 45$
3 The area of each rectangle is $4x$ and the dimensions are 1 by $4x$, 2 by $2x$, 4 by x
4 a 1, x, 3, $3x$
b 1, $8x$, 2, $4x$, 4, $2x$, 8, x
c 1, $12y$, 2, $6y$, 3, $4y$, 4, $3y$, 6, $2y$, 12, y
d 1, $18z$, 2, $9z$, 3, $6z$, 6, $3z$, 9, $2z$, 18, z
e 1, $4ab$, 2, $2ab$, 4, ab, a, $4b$, $2a$, $2b$, $4a$, b
5 a 1, $3x + 1$
b 1, $8x + 4$, 2, $4x + 2$, 4, $2x + 1$
c 1, $12y - 6$, 2, $6y - 3$, 3, $4y - 2$, 6, $2y - 1$
d 1, $18z + 27$, 3, $6z + 9$, 9, $2z + 3$
e 1, $4ab + 2$, 2, $2ab + 1$

Chapter 15.2

Are you ready?

1 a 1, 2 **b** 1, 2, 4
c 1, 2, 3, 6, 9, 18
d 1, 5, 25 **e** 1, 29
2 a 1 **b** 4 **c** 25
d 81 **e** 225
3 a 3 **b** 6 **c** 10 **d** 15

Practice 15.2A

1 a

~~1~~	2	3	~~4~~	5	~~6~~	7	~~8~~	~~9~~	~~10~~
11	~~12~~	13	~~14~~	~~15~~	~~16~~	17	~~18~~	19	~~20~~
~~21~~	~~22~~	23	~~24~~	~~25~~	~~26~~	~~27~~	~~28~~	29	~~30~~
31	~~32~~	~~33~~	~~34~~	~~35~~	~~36~~	37	~~38~~	~~39~~	~~40~~
41	~~42~~	43	~~44~~	~~45~~	~~46~~	47	~~48~~	~~49~~	~~50~~

b 2, 3, 5, 7, 11, 13, 17, 19, 23, 29, 31, 37, 41, 43, 47
c Because all multiples of 4, 6 and 8 are also multiples of 2
2 a i 1 **ii** 1, 2
iii 1, 2, 4, 8, 16
iv 1, 19 **v** 1, 2, 11, 22
vi 1, 3 **vii** 1, 31
b 2, 19, 3, 31; they only have two factors
3 a 3 by 7 array
b 2 has only two factors (1 and 2) therefore it is prime. However, it is the only prime number that is an even number.
4 11, 37, 53
5 a False; 1 isn't prime as it has one factor not two
b False; 2 is an even prime number (but it is the only one)
c True; the only factors of 43 are 1 and 43
6 a

ξ
A: 2
A ∩ B: 3, 5, 7, 11, 13, 17, 19
B: 1, 9, 15
Outside: 4, 6, 8, 10, 12, 14, 16, 18, 20

b A = {2, 3, 5, 7, 11, 13, 17, 19}
B = {1, 3, 5, 7, 9, 11, 13, 15, 17, 19}

What do you think?

1 a It is even, so 2 is a factor
b It ends in a 5, so 5 is a factor
c It is even, so 2 is a factor
2 a $\frac{1}{2}$ **b** $\frac{15}{36}$ (or $\frac{5}{12}$)
c $\frac{30}{36}$ ($\frac{5}{6}$)
3 a $a = 71$, $b = 2$
b As the total is odd, one of the numbers must be even, so b must be 2, therefore a must be 71

Practice 15.2B

1

2

3 1, 4, 9, 16, 25, 36, 49, 64, 81, 100
4 1, 3, 6, 10, 15, 21, 28, 36, 45, 55
5 a i She has calculated 17 × 2 not 17 × 17
ii 289

b No; he hasn't made a completed square

6 a 25 **b** 121 **c** 225 **d** 400

7 a 78 **b** 91 **c** 210

What do you think?

1 The sum of two consecutive triangular numbers always gives a square number

2 No, e.g. 10 is a triangular number but half of 10 is 5 which isn't a square number

3 a False. Prime numbers have exactly two factors. Square numbers have an odd number of factors so a prime number can't be square.

b True, 36 is an example

c True, 3 is an example

Consolidate

1 a 3: 1, 3
15: 1, 3, 5, 15
100: 1, 2, 4, 5, 10, 20, 25, 50, 100
25: 1, 5, 25
11: 1, 11
16: 1, 2, 4, 8, 16
2: 1, 2
21: 1, 3, 7, 21

b 3, 11, 2 **c** 100, 25, 16

d

Prime	Square	Neither
3, 11, 2	100, 25, 16	15, 21

2 a

b 36

c 1, 3, 6, 10, 15, 21, 28, 36, 45, 55

Stretch

1 Negative 11 has four factors: 1, −11, −1, 11. A prime number is a positive integer with exactly two factors. Negative 11 isn't positive nor does it have exactly two factors.

2 a Always true **b** Always true

3 e.g. $a = 3, b = 4, c = 5$ or $a = 6, b = 8, c = 10$ or $a = 5, b = 12, c = 13$

4 a e.g. $5^2 = 25$ and $25 > 5$

b e.g. $0.5^2 = 0.25$ and $0.25 < 0.5$

c When $x > 1$ (or when $x < -1$) then $x^2 > x$

5 a $2n$ is even and greater than 2 so it can't be prime

b e.g. $n = 1$ ($2n + 1 = 3$)

c e.g. $n = 10$ ($2n + 1 = 21$)

d a must be even

Chapter 15.3

Are you ready?

1 a 16 **b** 12.7 **c** 1706

2 a 1 **b** 0.9 **c** 17.06

3 a 1, 3, 9, 27
b 1, 2, 4, 8, 16
c 1, 2, 3, 6, 9, 18
d 1, 2, 4, 7, 14, 28
e 1, 7, 49

4 a 7, 14, 21, 28, 35
b 10, 20, 30, 40, 50
c 12, 24, 36, 48, 60
d 15, 30, 45, 60, 75
e 40, 80, 120, 160, 200

Practice 15.3A

1 a 1, 2, 3, 6, 9, 18
b 1, 3, 9, 27
c 1, 3, 9
d 9

2 a 1, 2, 4, 8, 16
b 1, 2, 4, 8, 16, 32
c 1, 2, 4, 8, 16
d 16
e The HCF is one of the numbers being factorised because 32 is a multiple of 16

3 a 1, 7 **b** 1, 3, 9 **c** 1
d HCF = 1, because one of the numbers is prime.

4 a 3 **b** 5 **c** 6 **d** 1
e 8 **f** 6 **g** 12 **h** 1

5 a 1, 3, 5, 15
b 1, 2, 4, 5, 10, 20
c 1, 5, 25
d 5

6 a 3 **b** 2 **c** 1
d 15 **e** 2 **f** 4

What do you think?

1 a i 14 **ii** 5 **iii** 36 **iv** 4
b i $\frac{1}{2}$ **ii** $\frac{5}{6}$ **iii** $\frac{2}{3}$ **iv** $\frac{24}{25}$
c Knowing the HCF means you can simplify the fraction in one step

2 a 4
b i 40 **ii** 8 **iii** 2

3 a 1, 2, 4, 8, 16
b When $x = 1, y = 80, z = 128$
When $x = 2, y = 40, z = 64$
When $x = 4, y = 20, z = 32$
When $x = 8, y = 10, z = 16$
When $x = 16, y = 5, z = 8$

4 a Various answers
b Various answers
c 1, 1; from **b**, c is prime

Practice 15.3B

1 a 8, 16, 24, 32, 40, 48, 56, 64, 72, 80
b 10, 20, 30, 40, 50, 60, 70, 80, 90, 100
c 40, 80 **d** 40
e No, writing the first 5 multiples of each number revealed the LCM

2 a 12, 24, 36, 48, 60, 72, 84, 96, 108, 120
b 6, 12, 18, 24, 30, 36, 42, 48, 54, 60
c 12, 24, 36, 48, 60
d 12
e No, writing the first 2 multiples of each number revealed the LCM

3 a 40 **b** 42 **c** 12
d 22 **e** 52 **f** 126
The LCM is the product of the two numbers

4 a 12 **b** 27 **c** 15
d 30 **e** 24 **f** 51
The LCM is the greater of the two numbers

5 a 36 **b** 36 **c** 144
d 42 **e** 96 **f** 90
The LCM is a multiple of the greater number but is smaller than the product

6 a 60 **b** 60 **c** 36

What do you think?

1 Sometimes true

2 a 10 am **b** 10

3 a 13th February
b 2
c Because 4 is a multiple of 2

Consolidate

1 a 1, 3, 5, 15
b 1, 2, 3, 6, 9, 18
c 1, 5, 25
d 1, 29
e 1, 3, 5, 9, 15, 45
f 1, 2, 3, 4, 6, 12

2 a 3 **b** 5 **c** 1 **d** 15
e 3 **f** 1 **g** 1 **h** 9
i 6 **j** 1 **k** 5 **l** 1
m 1 **n** 1 **o** 3

3 a 3, 6, 9, 12, 15, 18, 21, 24, 27, 30
b 6, 12, 18, 24, 30, 36, 42, 48, 54, 60
c 8, 16, 24, 32, 40, 48, 56, 64, 72, 80
d 12, 24, 36, 48, 60, 72, 84, 96, 108, 120
e 15, 30, 45, 60, 75, 90, 105, 120, 135, 150
f 7, 14, 21, 28, 35, 42, 49, 56, 63, 70

4 a 6 **b** 24 **c** 12
d 15 **e** 21 **f** 24
g 12 **h** 30 **i** 42
j 24 **k** 120 **l** 56
m 60 **n** 84 **o** 105

Stretch

1 At least one of the numbers is prime

2 e.g. 25 and 30 or 5 and 150

3 ab

4 a $2x$ **b** 2 **c** $2x$ **d** $2x^2y$

5 a $3x$
b You can find a common denominator
c i $\frac{2}{3x}$ **ii** $\frac{7}{3x}$ **iii** $\frac{11}{6x}$

Chapter 15.4

Are you ready?

1 a True **b** False **c** True

2 2, 3, 5, 7

3 ξ Even numbers; Multiples of 5

Even numbers only: 12; Intersection: 20, 30, 40; Multiples of 5 only: 5, 25; Outside: 23, 37

4 a HCF = 1, LCM = 15
b HCF = 3, LCM = 18
c HCF = 12, LCM = 24
d HCF = 10, LCM = 60

Practice 15.4A

1 a 36 = 2 × 2 × 3 × 3
b 21 = 3 × 7
c 54 = 2 × 3 × 3 × 3
d 135 = 3 × 3 × 3 × 5
2 a 72 = 2 × 2 × 2 × 3 × 3
b 42 = 2 × 3 × 7
c 108 = 2 × 2 × 3 × 3 × 3
d 270 = 2 × 3 × 3 × 3 × 5
3 a 48 = 2 × 2 × 2 × 2 × 3
b 95 = 5 × 19
c 63 = 3 × 3 × 7
d 72 = 2 × 2 × 2 × 3 × 3
e 242 = 2 × 11 × 11
f 250 = 2 × 5 × 5 × 5
g 300 = 2 × 2 × 3 × 5 × 5
h 460 = 2 × 2 × 5 × 23
i 207 = 3 × 3 × 23
j 513 = 3 × 3 × 3 × 19
4 a 2000 = 2 × 2 × 2 × 2 × 5 × 5 × 5
b 3000 = 2 × 2 × 2 × 3 × 5 × 5 × 5
c 27 000 = 2 × 2 × 2 × 3 × 3 × 3 × 5 × 5 × 5
d 63 000 = 2 × 2 × 2 × 3 × 3 × 5 × 5 × 5 × 7
5 Both are correct; 2^3 (2 cubed) is equal to 2 × 2 × 2
6 a 9 is not a prime number
b 126 = 2 × 3 × 3 × 7

What do you think?

1 a Yes b No c Yes
d Yes e Yes
2 Sometimes true
3 5

Practice 15.4B

1 a 70 b 56 c 14 d 280
2 a 12 b 1320
3 a 48 = 2 × 2 × 2 × 2 × 3
b 64 = 2 × 2 × 2 × 2 × 2 × 2
c

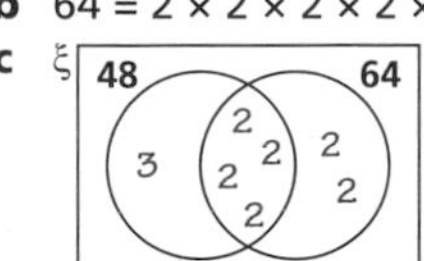

d 16 e 192
4 a 48 = 2 × 2 × 2 × 2 × 3, 28 = 2 × 2 × 7, HCF = 4, LCM = 336
b 110 = 2 × 5 × 11, 125 = 5 × 5 × 5, HCF = 5, LCM = 2750
c 315 = 3 × 3 × 5 × 7, 45 = 3 × 3 × 5, HCF = 45, LCM = 315
d 560 = 2 × 2 × 2 × 2 × 7 × 5, 140 = 2 × 2 × 5 × 7, HCF = 140, LCM = 560
e 1080 = 2 × 2 × 2 × 3 × 3 × 3 × 5, 180 = 2 × 2 × 3 × 3 × 5, HCF = 180, LCM = 1080
f 63 = 3 × 3 × 7, 721 = 7 × 103, HCF = 7, LCM = 6489

What do you think?

1 No. The HCF is 1.
2 There are two 3s in both circles that should be in the intersection.
3 a True b False
c True d True

Consolidate

1 a 24 = 2 × 2 × 2 × 3
b 72 = 2 × 2 × 2 × 3 × 3
c 100 = 2 × 2 × 5 × 5
d 125 = 5 × 5 × 5
e 840 = 2 × 2 × 2 × 3 × 5 × 7
f 48 = 2 × 2 × 2 × 2 × 3
g 216 = 2 × 2 × 2 × 3 × 3 × 3
h 700 = 2 × 2 × 5 × 5 × 7
i 750 = 2 × 3 × 5 × 5 × 5
j 84 000 = 2 × 2 × 2 × 2 × 2 × 3 × 5 × 5 × 5 × 7
2 a 49 b 50 c 10
d 1 e 40 f 10
3 a 210 b 546 c 1470
d 570 e 2520 f 903 210
4 a HCF = 24, LCM = 72
b HCF = 25, LCM = 500
c HCF = 24, LCM = 1680
d HCF = 7, LCM = 37 800
e HCF = 750, LCM = 84 000

Stretch

1 Various answers, e.g. 12 and 180 or 36 and 60
2 HCF = $9pq$, LCM = $108p^2q$
3 Various answers, e.g. 1008 and 99 792 or 9072 and 11 088
4 a $a = 2$, $b = 131$
b $c = 2$, $d = 7$, $e = 13$
5 a 1024 = 2 × 2 × 2 × 2 × 2 × 2 × 2 × 2 × 2 × 2
b (2 × 2 × 2 × 2 × 2) × (2 × 2 × 2 × 2 × 2)
c $p = \pm 32$
6 a 5832 = (2 × 3 × 3) × (2 × 3 × 3) × (2 × 3 × 3)
b $18xy^2$

Chapter 15.5

Are you ready?

1 a 5, 1, 51, 47, 11, 9, 49, 17, 21
b 16, 2, 24, 36, 100, 28, 18
c 5, 2, 47, 11, 17
d 16, 1, 36, 100, 9, 49
2 a 64 b 138 c 102
d 200 e 92 f 58
g 206 h 528
3 a 15 b 153 c 77
d 399 e 24 f 220
g 72 h 240

Practice 15.5A

1 a e.g. 5 + 7 = 12
b e.g. 2 + 3 = 5
2 a e.g. LCM of 4 and 3 is 12
b e.g. LCM of 6 and 3 is 6
3 a Any investigation
b They are both true
c No, as they are true
4 a Always true
b Sometimes true (if $x = y$)
c Always true
d Sometimes true (if $x = y$)
5 a e.g. $\frac{1}{2} + \frac{1}{3} = \frac{5}{6}$
b e.g. $\frac{1}{2} + \frac{3}{4} = \frac{5}{4}$
6 a e.g. 3 has two factors, 4 has three factors and 5 has two factors
b 1 only has one factor
7 a e.g. 3 × 4 = 12, 5 × 7 = 35, 10 × 2 = 20
b e.g. 16 × 0.5 = 8
8 a Abdullah is thinking of 2, 4, 6, 8 where the term-to-term rule is add 2 each time
b e.g. 2, 4, 8, 16 … or 2, 4, 6, 10, 16 …

What do you think?

1 a Any relevant investigations
b This conjecture is true
2 a Any relevant investigations
b This conjecture is true
3 a Various answers
b Various answers
c i 25, 36 ii 5, 8
iii 1, 0.5 iv 21, 25
More accurate as they are based on more information

Consolidate

1 a

+	1	2	3	4	5	6
1	2	3	4	5	6	7
2	3	4	5	6	7	8
3	4	5	6	7	8	9
4	5	6	7	8	9	10
5	6	7	8	9	10	11
6	7	8	9	10	11	12

b Various answers
2 a i $x + y = 10$, $xy = 21$
ii $x + y = 4$, $xy = -21$
iii $x + y = 20.5$, $xy = 10$
iv $x + y = -14$, $xy = 45$
b i **a i** and **a iv**
ii **a ii** and **a iii**
3 a i 2 ii 6 iii 1 iv 9
b i **a i**, **a ii** and **a iii**
ii **a iv**

Stretch

1 a Multiples of 2 are even; 1 more than any multiple of 2 (which is even) is odd
b $2n + (2n + 1) = 4n + 1$ which is 1 more than a multiple of 4, therefore 1 more than a multiple of 2, so it must be odd.
c If n wasn't an integer then we can't know for sure that $2n$ is even; e.g. if n was 0.5, $2n$ would be odd.
2 a e.g. 5 – 7 = –2
b $x > y$

3 **a** e.g. $f = 2$, $g = 3$, $h = 5$, $i = 17$, $j = 7$
b e.g. $f = 2$, $g = 3$, $h = 5$, $i = 7$, $j = 37$
4 **a** e.g. $x = 100$ **b** e.g. $x = 1$
5 This conjecture is true

Check my understanding

1 **a** 8, 16, 24, 32, 40
b 20, 40, 60, 80, 100
c 15, 30, 45, 60, 75
d 12, 24, 36, 48, 60
e 16, 32, 48, 64, 80
2 **a** 1, 2, 4, 8
b 1, 2, 4, 5, 10, 20
c 1, 3, 5, 15
d 1, 2, 3, 4, 6, 12
e 1, 2, 4, 8, 16
3 e.g. 13, because it only has two factors, 1 and 13
4 Array with 5 rows of 5 dots
5 225
6 **a** HCF = 1, LCM = 40
b HCF = 6, LCM = 12
c HCF = 4, LCM = 48
d HCF = 10, LCM = 100
e HCF = 15, LCM = 45
7 **a** $27 = 3 \times 3 \times 3$
b $50 = 2 \times 5 \times 5$
c $120 = 2 \times 2 \times 2 \times 3 \times 5$
d $144 = 2 \times 2 \times 2 \times 2 \times 3 \times 3$
e $960 = 2 \times 2 \times 2 \times 2 \times 2 \times 2 \times 3 \times 5$
8 **a** $720 = 2 \times 2 \times 2 \times 2 \times 3 \times 3 \times 5$
b $450 = 2 \times 3 \times 3 \times 5 \times 5$
c

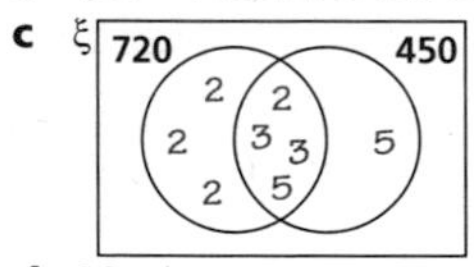

d 90
e 3600